Functional Calculus

Edited by Kamal Shah and Baver Okutmuştur

Published in London, United Kingdom

IntechOpen

Supporting open minds since 2005

Functional Calculus
http://dx.doi.org/10.5772/intechopen.77599
Edited by Kamal Shah and Baver Okutmuştur

Contributors
Mohammed Abu Saleem, Rudy Calif, Thomas Plocoste, Jie Shi, Kamal Shah, Marianna Shubov, Laszlo Kindrat, Ivan Kyrchei, Ioan R. Ciric, Fethi Ahmed Soltani, Baver Okutmuştur, Zoubir Dahmani, Meriem Mansouria Belhamiti, Elimhan Mahmudov, Thabet Abdeljawad, Hammad Khalil, Rahmat Ali Khan

First published in London, United Kingdom, 2020 by IntechOpen
IntechOpen is the global imprint of INTECHOPEN LIMITED, registered in England and Wales, registration number: 11086078, 7th floor, 10 Lower Thames Street, London, EC3R 6AF, United Kingdom
Printed in Croatia

British Library Cataloguing-in-Publication Data
A catalogue record for this book is available from the British Library

Additional hard and PDF copies can be obtained from orders@intechopen.com

Functional Calculus
Edited by Kamal Shah and Baver Okutmuştur
p. cm.
Print ISBN 978-1-83880-007-9
Online ISBN 978-1-83880-008-6
eBook (PDF) ISBN 978-1-83880-904-1

Meet the editors

Dr. Kamal Shah is an associate professor in applied mathematics at the Department of Mathematics, University of Malakand, Khyber Pakhtankhwa, Pakistan. He was born on March 1, 1983 in Deolai, a large town in the beautiful valley of Swat in Khyber Pakhtankhwa, Pakistan. He has served at the University of Malakand since 2007. He received his PhD degree in Applied Mathematics from the University of Malakand in 2016. Before joining the University of Malakand he was a lecturer in mathematics at the Government Degree College, Mingora, Swat. In November 2006, he joined Government Degree College, Kabal, Swat, and taught there until 2007. Dr. Kamal Shah has published more than 100 articles related to the qualitative theory of fractional differential equations, numerical solutions, and stability analysis in world-reputed journals. He is also involved in fixed point theory and mathematical modeling of infectious diseases. Recently, he has been working on fuzzy integral, differential, and integro-differential equations and their existence theory, and numerical/approximate solutions of fractional as well as integer orders.

Dr. Baver Okutmuştur is an assistant professor at the Department of Mathematics, Middle East Technical University, Ankara, Turkey. He received his PhD degree in Mathematics from Université Pierre et Marie Curie-UPMC (Paris 6). He has published several articles based on finite volume techniques for hyperbolic conservation laws on manifolds. Moreover, he has studied various relativistic models and their numerical analysis on different space/time geometries.

Contents

Preface

The aim of this book is to present a broad overview of the theory and applications related to functional calculus. The book is based on two main subject areas: matrix calculus and applications of Hilbert spaces.

Functional analysis is the most important branch of mathematics, whose foundation was laid by the great Persian polymath Muhammad ibn Mūsā al-Khwārizmī, also known as Algorithmi, during 973–1048. He named this branch the "Theory of Functions." Later, Newton and Leibnitz enriched this branch by introducing the concept of derivatives and integrals during 1665–1742 and thus gave birth to another name: calculus. This branch of mathematics has been recently divided into several subbranches, including differential calculus, integral calculus, stochastic calculus, etc. In mathematics, a functional calculus is a theory that permits someone to apply mathematical functions to mathematical operators. Now, functional calculus is a branch that connects operator theory, classical calculus, algebra, and functional analysis. In daily life, functionals are increasingly used to model real-world situations, for example if $f: R \rightarrow R$ is real valued functional from real to real number system. If we apply f on some function $x \in R$, then $f(x)$ makes no sense but if we write it in equation form, then it makes sense, e.g. $f(x) = x$, which represents a physical process between two quantities such that there is direct proportionality. Similar problems occur daily in our surroundings. Therefore, it is necessary to understand what criteria should be satisfied by concerned functionals and operators used in modeling or in the description of daily life problems. It is functional calculus that guides and provides us with the path to how, when, and where particular functionals and operators may be used. Mostly, integral and differential equations are used when we wish to solve a technique or procedure that converts the mentioned equations into algebraic equations of known and unknown functions and functionals. Keeping these needs in mind, the editor of this book has been motivated to welcome international mathematicians and researchers to contribute various topics that address the areas of functional calculus and its applications in both pure and applied analysis. The editor has incorporated contributions from a diverse group of leading researchers in the field of functional calculus. This book aims to provide an overview of the present knowledge that addresses applications and results related to functional calculus. The main topics covered in this book are determinantal representations of the core inverse and its generalizations, which provides a foundation to solve matrix equations. Furthermore, new series formulae for matrix exponential series have been developed, which are used in solving algebraic equations. Also covered are results on fixed point theory, which is used for mapping the satisfying condition (D_A) in Banach space. Results that address folding on chaotic graph operations and their fundamental groups are also introduced. Such algebraic structures are largely used in biology and chemistry. Elsewhere in the book, a brief review is considered of Hilbert space with its fundamental features and features of reproducing kernels in corresponding spaces. Spectral theory is an important area that is most applicable in quantum mechanics. Therefore, a number of fundamental concepts have been investigated regarding analytical applications and observations of PM10 fluctuations. Optimal control is a very important procedure, which is increasingly used in the study of mathematical models of real-world problems. It is helpful in developing future predictions and control strategies of infectious diseases. Analytic and numerical results of the Euler–Bernoulli beam model with a two-parameter family of boundary conditions are also presented, where Chebyshev polynomial approximation has been used to approximate the solution. In recent times, fractional calculus has attracted great attention. Results on fractional integral inequalities are investigated. By using the principle of functional calculus, numerical analysis for boundary value problems of fractional differential equations are studied in the final chapter.

The theory of Hilbert spaces is the center around which functional analysis has developed. Hilbert spaces have a rich geometric nature as they are endowed with an inner product that permits the concept of orthogonality of vectors. Hilbert space methods are applied to several science and engineering areas such as optimization, variational and control problems, and to problems in approximation theory, nonlinear stability, and bifurcation as well as spectral theory and quantum mechanics. That is why a part of the book is devoted to a brief presentation and applications of Hilbert spaces. For the reader who has no previous experience in the theory of normed spaces with enough background for comprehending the theory of Hilbert spaces, there two chapters based on these topics in the book. An important application of the theory of Hilbert spaces to the reproducing kernels is also analyzed in this part. Spectral theory is an important area which is most applicable in quantum mechanics. In this content, a real-life application of Hilbert space where an investigation of the pollution and air quality in Caribbean region by the help of theoretical Hilbert frame aspect is also provided. Here some observations of PM10 fluctuations are analyzed by scaling and time-frequency properties of PM10 data in Hilbert frame and compared the functioning obtained in Hilbert space. Optimal control is also very important procedure which is increasingly used in study of mathematical models of real world problems. It is helpful in developing future predictions and control strategies of infectious disease. In this issue, analytic and numerical results of the Euler-Bernoulli beam model with a two-parameter family of boundary conditions have been presented where Chebyshev polynomial approximation has been used to approximate the solution.

We hope that this book will be of benefit to mathematicians, computational mathematicians, applied mathematicians, and researchers in the field of pure mathematics as well as in analysis. The book is written basically for those who have some knowledge of classical calculus and mathematical analysis. The authors of each section convey a strong emphasis on theoretical foundations.

Kamal Shah
Associate Professor,
Department of Mathematics,
University of Malakand,
Khyber Pakhtankhawa,
Pakistan

Baver Okutmuştur
Assistant Professor,
Department of Mathematics,
Middle East Technical University (METU),
Ankara, Turkey

Chapter 1

Determinantal Representations of the Core Inverse and Its Generalizations

Ivan I. Kyrchei

Abstract

Generalized inverse matrices are important objects in matrix theory. In particular, they are useful tools in solving matrix equations. The most famous generalized inverses are the Moore-Penrose inverse and the Drazin inverse. Recently, it was introduced new generalized inverse matrix, namely the core inverse, which was late extended to the core-EP inverse, the BT, DMP, and CMP inverses. In contrast to the inverse matrix that has a definitely determinantal representation in terms of cofactors, even for basic generalized inverses, there exist different determinantal representations as a result of the search of their more applicable explicit expressions. In this chapter, we give new and exclusive determinantal representations of the core inverse and its generalizations by using determinantal representations of the Moore-Penrose and Drazin inverses previously obtained by the author.

Keywords: Moore-Penrose inverse, Drazin inverse, core inverse, core-EP inverse,
2000 AMS subject classifications: 15A15, 16W10

1. Introduction

In the whole chapter, the notations $\mathbb{R}$ and $\mathbb{C}$ are reserved for fields of the real and complex numbers, respectively. $\mathbb{C}^{m\times n}$ stands for the set of all $m \times n$ matrices over $\mathbb{C}$. $\mathbb{C}_r^{m\times n}$ determines its subset of matrices with a rank r. For $\mathbf{A} \in \mathbb{C}^{m\times n}$, the symbols $\mathbf{A}^*$ and $\mathrm{rk}(\mathbf{A})$ specify the conjugate transpose and the rank of $\mathbf{A}$, respectively, $|\mathbf{A}|$ or $det\mathbf{A}$ stands for its determinant. A matrix $\mathbf{A} \in \mathbb{C}^{n\times n}$ is Hermitian if $\mathbf{A}^* = \mathbf{A}$.

$\mathbf{A}^\dagger$ means *the Moore-Penrose inverse* of $\mathbf{A} \in \mathbb{C}^{n\times m}$, i.e., the exclusive matrix $\mathbf{X}$ satisfying the following four equations:

$$\mathbf{AXA} = \mathbf{A} \tag{1}$$

$$\mathbf{XAX} = \mathbf{X} \tag{2}$$

$$(\mathbf{AX})^* = \mathbf{AX} \tag{3}$$

$$(\mathbf{XA})^* = \mathbf{XA} \tag{4}$$

For $\mathbf{A} \in \mathbb{C}^{n\times n}$ with index $\mathrm{Ind}\,\mathbf{A} = k$, i.e., the smallest positive number such that $\mathrm{rk}\left(\mathbf{A}^{k+1}\right) = \mathrm{rk}\left(\mathbf{A}^k\right)$, *the Drazin inverse* of $\mathbf{A}$, denoted by $\mathbf{A}^d$, is called the unique matrix $\mathbf{X}$ that satisfies Eq. (2) and the following equations,

$$\mathbf{AX} = \mathbf{XA}; \tag{5}$$

$$\mathbf{XA}^{k+1} = \mathbf{A}^k \tag{6}$$

$$\mathbf{A}^{k+1}\mathbf{X} = \mathbf{A}^k. \tag{7}$$

In particular, if $\operatorname{Ind}\mathbf{A} = 1$, then the matrix $\mathbf{X}$ is called *the group inverse*, and it is denoted by $\mathbf{X} = \mathbf{A}^{\#}$. If $\operatorname{Ind}\mathbf{A} = 0$, then $\mathbf{A}$ is nonsingular and $\mathbf{A}^d = \mathbf{A}^{\dagger} = \mathbf{A}^{-1}$.

It is evident that if the condition (5) is fulfilled, then (6) and (7) are equivalent. We put both these conditions because they will be used below independently of each other and without the obligatory fulfillment of (5).

A matrix $\mathbf{A}$ satisfying the conditions $(i), (j), \ldots$ is called an $\{i, j, \ldots\}$-inverse of $\mathbf{A}$, and is denoted by $\mathbf{A}^{(i,j,\ldots)}$. The set of matrices $\mathbf{A}^{(i,j,\ldots)}$ is denoted $\mathbf{A}\{i, j, \ldots\}$. In particular, $\mathbf{A}^{(1)}$ is called the inner inverse, $\mathbf{A}^{(2)}$ is called the outer inverse, $\mathbf{A}^{(1,2)}$ is called the reflexive inverse, $\mathbf{A}^{(1,2,3,4)}$ is the Moore-Penrose inverse, etc.

For an arbitrary matrix $\mathbf{A} \in \mathbb{C}^{m\times n}$, we denote by

- $\mathcal{N}(\mathbf{A}) = \{\mathbf{x} \in \mathbb{H}^{n\times 1} : \mathbf{Ax} = 0\}$, the kernel (or the null space) of $\mathbf{A}$;

- $\mathcal{C}(\mathbf{A}) = \{\mathbf{y} \in \mathbb{H}^{m\times 1} : \mathbf{y} = \mathbf{Ax}, \mathbf{x} \in \mathbb{H}^{n\times 1}\}$, the column space (or the range space) of $\mathbf{A}$; and

- $\mathcal{R}(\mathbf{A}) = \{\mathbf{y} \in \mathbb{H}^{1\times n} : \mathbf{y} = \mathbf{xA}, \mathbf{x} \in \mathbb{H}^{1\times m}\}$, the row space of $\mathbf{A}$.

$\mathbf{P}_A := \mathbf{AA}^{\dagger}$ and $\mathbf{Q}_A := \mathbf{A}^{\dagger}\mathbf{A}$ are the orthogonal projectors onto the range of $\mathbf{A}$ and the range of $\mathbf{A}^*$, respectively.

The core inverse was introduced by Baksalary and Trenkler in [1]. Later, it was investigated by S. Malik in [2] and S.Z. Xu et al. in [3], among others.

Definition 1.1. [1] A matrix $\mathbf{X} \in \mathbb{C}^{n\times n}$ is called the core inverse of $\mathbf{A} \in \mathbb{C}^{n\times n}$ if it satisfies the conditions

$$\mathbf{AX} = \mathbf{P}_A, \textit{and } \mathcal{C}(\mathbf{X}) = \mathcal{C}(\mathbf{A}).$$

When such matrix $\mathbf{X}$ exists, it is denoted as $A^{\oplus}$.

In 2014, the core inverse was extended to the core-EP inverse defined by K. Manjunatha Prasad and K.S. Mohana [4]. Other generalizations of the core inverse were recently introduced for $n \times n$ complex matrices, namely BT inverses [5], DMP inverses [2], CMP inverses [6], etc. The characterizations, computing methods, and some applications of the core inverse and its generalizations were recently investigated in complex matrices and rings (see, e.g., [7–18]).

In contrast to the inverse matrix that has a definitely determinantal representation in terms of cofactors, for generalized inverse matrices, there exist different determinantal representations as a result of the search of their more applicable explicit expressions (see, e.g. [19–25]). In this chapter, we get new determinantal representations of the core inverse and its generalizations using recently obtained by the author determinantal representations of the Moore-Penrose inverse and the Drazin inverse over the quaternion skew field, and over the field of complex numbers as a special case [26–34]. Note that a determinantal representation of the core-EP generalized inverse in complex matrices has been derived in [4], based on the determinantal representation of an reflexive inverse obtained in [19, 20].

The chapter is organized as follows: in Section 2, we start with preliminary introduction of determinantal representations of the Moore-Penrose inverse and the

Drazin inverse. In Section 3, we give determinantal representations of the core inverse and its generalizations, namely the right and left core inverses are established in Section 3.1, the core-EP inverses in Section 3.2, the core DMP inverse and its dual in Section 3.3, and finally the CMP inverse in Section 3.4. A numerical example to illustrate the main results is considered in Section 4. Finally, in Section 5, the conclusions are drawn.

2. Preliminaries

Let $\alpha := \{\alpha_1, ..., \alpha_k\} \subseteq \{1, ..., m\}$ and $\beta := \{\beta_1, ..., \beta_k\} \subseteq \{1, ..., n\}$ be subsets with $1 \le k \le \min\{m, n\}$. By $\mathbf{A}^{\alpha}_{\beta}$, we denote a submatrix of $\mathbf{A} \in \mathbb{H}^{m \times n}$ with rows and columns indexed by α and β, respectively. Then, $\mathbf{A}^{\alpha}_{\alpha}$ is a principal submatrix of $\mathbf{A}$ with rows and columns indexed by α, and $|\mathbf{A}|^{\alpha}_{\alpha}$ is the corresponding principal minor of the determinant $|\mathbf{A}|$. Suppose that

$$L_{k,n} := \{\alpha : \alpha = (\alpha_1, ..., \alpha_k), 1 \le \alpha_1 < \cdots < \alpha_k \le n\}$$

stands for the collection of strictly increasing sequences of $1 \le k \le n$ integers chosen from $\{1, ..., n\}$. For fixed $i \in \alpha$ and $j \in \beta$, put $I_{r,m}\{i\} := \{\alpha : \alpha \in L_{r,m}, i \in \alpha\}$ and $J_{r,n}\{j\} := \{\beta : \beta \in L_{r,n}, j \in \beta\}$.

The jth columns and the ith rows of $\mathbf{A}$ and $\mathbf{A}^*$ denote $\mathbf{a}_{.j}$ and $\mathbf{a}^*_{.j}$ and $\mathbf{a}_{i.}$ and $\mathbf{a}^*_{i.}$, respectively. By $\mathbf{A}_{i.}(\mathbf{b})$ and $\mathbf{A}_{.j}(\mathbf{c})$, we denote the matrices obtained from $\mathbf{A}$ by replacing its ith row with the row $\mathbf{b}$, and its jth column with the column $\mathbf{c}$.

Theorem 2.1. [28] *If* $\mathbf{A} \in \mathbb{H}^{m \times n}_r$, *then the Moore-Penrose inverse* $\mathbf{A}^{\dagger} = \left(a^{\dagger}_{ij}\right) \in \mathbb{C}^{n \times m}$ *possesses the determinantal representations*

$$a^{\dagger}_{ij} = \frac{\sum_{\beta \in J_{r,n}\{i\}} \left|(\mathbf{A}^*\mathbf{A})_{.i}\left(\mathbf{a}^*_{.j}\right)\right|^{\beta}_{\beta}}{\sum_{\beta \in J_{r,n}} |\mathbf{A}^*\mathbf{A}|^{\beta}_{\beta}} = \tag{8}$$

$$= \frac{\sum_{\alpha \in I_{r,m}\{j\}} \left|(\mathbf{A}\mathbf{A}^*)_{j.}\left(\mathbf{a}^*_{i.}\right)\right|^{\alpha}_{\alpha}}{\sum_{\alpha \in I_{r,m}} |\mathbf{A}\mathbf{A}^*|^{\alpha}_{\alpha}}. \tag{9}$$

Remark 2.2. For an arbitrary full-rank matrix $\mathbf{A} \in \mathbb{C}^{m \times n}_r$, a row vector $\mathbf{b} \in \mathbb{H}^{1 \times m}$, and a column-vector $\mathbf{c} \in \mathbb{H}^{n \times 1}$, we put, respectively,

$$\left|(\mathbf{A}\mathbf{A}^*)_{i.}(\mathbf{b})\right| = \sum_{\alpha \in I_{m,m}\{i\}} \left|(\mathbf{A}\mathbf{A}^*)_{i.}(\mathbf{b})\right|^{\alpha}_{\alpha}, \quad i = 1, ..., m,$$

$$|\mathbf{A}\mathbf{A}^*| = \sum_{\alpha \in I_{m,m}} |\mathbf{A}\mathbf{A}^*|^{\alpha}_{\alpha}, \quad \text{when } r = m;$$

$$\left|(\mathbf{A}^*\mathbf{A})_{.j}(\mathbf{c})\right| = \sum_{\beta \in J_{n,n}\{j\}} \left|(\mathbf{A}^*\mathbf{A})_{.j}(\mathbf{c})\right|^{\beta}_{\beta}, \quad j = 1, ..., n,$$

$$|\mathbf{A}^*\mathbf{A}| = \sum_{\beta \in J_{n,n}} |\mathbf{A}^*\mathbf{A}|^{\beta}_{\beta}, \quad \text{when } r = n.$$

Corollary 2.3. [21] *Let* $\mathbf{A} \in \mathbb{C}^{m \times n}_r$. *Then, the following determinantal representations can be obtained*

i. for the projector $\mathbf{Q}_A = \left(q_{ij}\right)_{n\times n}$,

$$q_{ij} = \frac{\sum_{\beta\in J_{r,n}\{i\}} \left|(\mathbf{A}^*\mathbf{A})_{.i}(\dot{\mathbf{a}}_{.j})\right|_\beta^\beta}{\sum_{\beta\in J_{r,n}} |\mathbf{A}^*\mathbf{A}|_\beta^\beta} = \frac{\sum_{\alpha\in I_{r,n}\{j\}} \left|(\mathbf{A}^*\mathbf{A})_{j.}(\dot{\mathbf{a}}_{i.})\right|_\alpha^\alpha}{\sum_{\alpha\in I_{r,n}} |\mathbf{A}^*\mathbf{A}|_\alpha^\alpha}, \tag{10}$$

where $\dot{\mathbf{a}}_{.j}$ is the jth column and $\dot{\mathbf{a}}_{i.}$ is the ith row of $\mathbf{A}^*\mathbf{A}$; and

ii. for the projector $\mathbf{P}_A = \left(p_{ij}\right)_{m\times m}$,

$$p_{ij} = \frac{\sum_{\alpha\in I_{r,m}\{j\}} \left|(\mathbf{A}\mathbf{A}^*)_{j.}(\ddot{\mathbf{a}}_{i.})\right|_\alpha^\alpha}{\sum_{\alpha\in I_{r,m}} |\mathbf{A}\mathbf{A}^*|_\alpha^\alpha} = \frac{\sum_{\beta\in J_{r,m}\{i\}} \left|(\mathbf{A}\mathbf{A}^*)_{.i}(\ddot{\mathbf{a}}_{.j})\right|_\beta^\beta}{\sum_{\beta\in J_{r,m}} |\mathbf{A}\mathbf{A}^*|_\beta^\beta}, \tag{11}$$

where $\ddot{\mathbf{a}}_{i.}$ is the ith row and $\ddot{\mathbf{a}}_{.j}$ is the jth column of $\mathbf{A}\mathbf{A}^*$.

The following lemma gives determinantal representations of the Drazin inverse in complex matrices.

Lemma 2.4. [21] *Let* $\mathbf{A}\in\mathbb{C}^{n\times n}$ *with* $\operatorname{Ind}\mathbf{A} = k$ *and* $\operatorname{rk}\mathbf{A}^{k+1} = \operatorname{rk}\mathbf{A}^k = r$. *Then, the determinantal representations of the Drazin inverse* $\mathbf{A}^d = \left(a_{ij}^d\right)\in\mathbb{C}^{n\times n}$ *are*

$$a_{ij}^d = \frac{\sum_{\beta\in J_{r,n}\{i\}} \left|\left(\mathbf{A}^{k+1}\right)_{.i}\left(\mathbf{a}_{.j}^{(k)}\right)\right|_\beta^\beta}{\sum_{\beta\in J_{r,n}} \left|\mathbf{A}^{k+1}\right|_\beta^\beta} = \tag{12}$$

$$= \frac{\sum_{\alpha\in I_{r,n}\{j\}} \left|\left(\mathbf{A}^{k+1}\right)_{j.}\left(\mathbf{a}_{i.}^{(k)}\right)\right|_\alpha^\alpha}{\sum_{\alpha\in I_{r,n}} \left|\mathbf{A}^{k+1}\right|_\alpha^\alpha}, \tag{13}$$

where $\mathbf{a}_{i.}^{(k)}$ is the ith row and $\mathbf{a}_{.j}^{(k)}$ is the jth column of $\mathbf{A}^k$.

Corollary 2.5. [21] *Let* $\mathbf{A}\in\mathbb{C}^{n\times n}$ *with* $\operatorname{Ind}\mathbf{A} = 1$ *and* $\operatorname{rk}\mathbf{A}^2 = \operatorname{rk}\mathbf{A} = r$. *Then, the determinantal representations of the group inverse* $\mathbf{A}^\# = \left(a_{ij}^\#\right)\in\mathbb{C}^{n\times n}$ *are*

$$a_{ij}^\# = \frac{\sum_{\beta\in J_{r,n}\{i\}} \left|\left(\mathbf{A}^2\right)_{.i}(\mathbf{a}_{.j})\right|_\beta^\beta}{\sum_{\beta\in J_{r,n}} \left|\mathbf{A}^2\right|_\beta^\beta} = \frac{\sum_{\alpha\in I_{r,n}\{j\}} \left|\left(\mathbf{A}^2\right)_{j.}(\mathbf{a}_{i.})\right|_\alpha^\alpha}{\sum_{\alpha\in I_{r,n}} \left|\mathbf{A}^2\right|_\alpha^\alpha}. \tag{14}$$

3. Determinantal representations of the core inverse and its generalizations

3.1 Determinantal representations of the core inverses

Together with the core inverse in [35], the dual core inverse was to be introduced. Since the both these core inverses are equipollent and they are different only in the position relative to the inducting matrix **A**, we propose called them as the right and left core inverses regarding to their positions. So, from [1], we have the following definition that is equivalent to Definition 1.1.

Definition 3.1. A matrix $\mathbf{X} \in \mathbb{C}^{n\times n}$ is said to be the right core inverse of $\mathbf{A} \in \mathbb{C}^{n\times n}$ if it satisfies the conditions

$$\mathbf{AX} = \mathbf{P}_A, and\ \mathcal{C}(\mathbf{X}) = \mathcal{C}(\mathbf{A}).$$

When such matrix $\mathbf{X}$ exists, it is denoted as $A^{\oplus}$.

The following definition of the left core inverse can be given that is equivalent to the introduced dual core inverse [35].

Definition 3.2 A matrix $\mathbf{X} \in \mathbb{C}^{n\times n}$ is said to be the left core inverse of $\mathbf{A} \in \mathbb{C}^{n\times n}$ if it satisfies the conditions

$$\mathbf{XA} = \mathbf{Q}_A, and\ \mathcal{R}(\mathbf{X}) = \mathcal{R}(\mathbf{A}). \tag{15}$$

When such matrix $\mathbf{X}$ exists, it is denoted as $A_{\oplus}$.

Remark 3.3. In [35], the conditions of the dual core inverse are given as follows:

$$A_{\oplus}\mathbf{A} = \mathbf{P}_{A^*}, and\ \mathcal{C}(A_{\oplus}) \subseteq \mathcal{C}(\mathbf{A}^*).$$

Since $\mathbf{P}_{A^*} = \mathbf{A}^*(\mathbf{A}^*)^{\dagger} = \left(\mathbf{A}^{\dagger}\mathbf{A}\right)^* = \mathbf{A}^{\dagger}\mathbf{A} = \mathbf{Q}_A$, and $\mathcal{R}(\mathbf{A}) = \mathcal{C}(\mathbf{A}^*)$, then these conditions and (15) are analogous.

Due to [1], we introduce the following sets of quaternion matrices

$$\mathbb{C}_n^{\mathrm{CM}} = \left\{\mathbf{A} \in \mathbb{C}^{n\times n} : \mathrm{rk}\,\mathbf{A}^2 = \mathrm{rk}\,\mathbf{A}\right\},$$
$$\mathbb{C}_n^{\mathrm{EP}} = \left\{\mathbf{A} \in \mathbb{C}^{n\times n} : \mathbf{A}^{\dagger}\mathbf{A} = \mathbf{A}\mathbf{A}^{\dagger}\right\} = \{\mathcal{C}(\mathbf{A}) = \mathcal{C}(\mathbf{A}^*)\}.$$

The matrices from $\mathbb{C}_n^{\mathrm{CM}}$ are called group matrices or core matrices. If $\mathbf{A} \in \mathbb{C}_n^{\mathrm{EP}}$, then clearly $\mathbf{A}^{\dagger} = \mathbf{A}^{\#}$. It is known that the core inverses of $\mathbf{A} \in \mathbb{C}^{n\times n}$ exist if and only if $\mathbf{A} \in \mathbb{C}_n^{\mathrm{CM}}$ or $\mathrm{Ind}\,\mathbf{A} = 1$. Moreover, if $\mathbf{A}$ is nonsingular, $\mathrm{Ind}\,\mathbf{A} = 0$, then its core inverses are the usual inverse. Due to [1], we have the following representations of the right and left core inverses.

Lemma 3.4. [1] *Let* $\mathbf{A} \in \mathbb{C}_n^{\mathrm{CM}}$. *Then,*

$$\mathbf{A}^{\oplus} = \mathbf{A}^{\#}\mathbf{A}\mathbf{A}^{\dagger}, \tag{16}$$

$$\mathbf{A}_{\oplus} = \mathbf{A}^{\dagger}\mathbf{A}\mathbf{A}^{\#} \tag{17}$$

Remark 3.5. In Theorems 3.6 and 3.7, we will suppose that $\mathbf{A} \in \mathbb{C}_n^{\mathrm{CM}}$ but $\mathbf{A} \notin \mathbb{C}_n^{\mathrm{EP}}$. Because, if $\mathbf{A} \in \mathbb{C}_n^{\mathrm{CM}}$ and $\mathbf{A} \in \mathbb{C}_n^{\mathrm{EP}}$ (in particular, $\mathbf{A}$ is Hermitian), then from Lemma 3.4 and the definitions of the Moore-Penrose and group inverses, it follows that $\mathbf{A}^{\oplus} = \mathbf{A}_{\oplus} = \mathbf{A}^{\#} = \mathbf{A}^{\dagger}$.

Theorem 3.6. *Let* $\mathbf{A} \in \mathbb{C}_n^{\mathrm{CM}}$ *and* $\mathrm{rk}\,\mathbf{A}^2 = \mathrm{rk}\,\mathbf{A} = s$. *Then, its right core inverse has the following determinantal representations*

$$a_{ij}^{\oplus,r} = \frac{\sum_{\alpha\in I_{s,n}\{j\}} \left|(\mathbf{AA}^*)_{j.}\left(\mathbf{u}_{i.}^{(1)}\right)\right|_{\alpha}^{\alpha}}{\sum_{\beta\in J_{s,n}}\left|\mathbf{A}^2\right|_{\beta}^{\beta}\sum_{\alpha\in I_{s,n}}\left|\mathbf{AA}^*\right|_{\alpha}^{\alpha}} = \tag{18}$$

$$= \frac{\sum_{\beta\in J_{s,n}\{i\}} \left|(\mathbf{A}^2)_{.i}\left(\mathbf{u}_{.j}^{(2)}\right)\right|_{\beta}^{\beta}}{\sum_{\beta\in J_{s,n}}\left|\mathbf{A}^2\right|_{\beta}^{\beta}\sum_{\alpha\in I_{s,n}}\left|\mathbf{AA}^*\right|_{\alpha}^{\alpha}}, \tag{19}$$

where

$$\mathbf{u}_{i.}^{(1)} = \left[\sum_{\beta \in J_{s,n}\{i\}} \left|(\mathbf{A}^2)_{.i}(\tilde{\mathbf{a}}_{.f})\right|_{\beta}^{\beta} \right] \in \mathbb{C}^{1\times n},\ f = 1,\ldots,n$$

$$\mathbf{u}_{.j}^{(2)} = \left[\sum_{\alpha \in I_{s,n}\{j\}} \left|(\mathbf{A}\mathbf{A}^*)_{j.}(\tilde{\mathbf{a}}_{l.})\right|_{\alpha}^{\alpha} \right] \in \mathbb{C}^{n\times 1},\ l = 1,\ldots,n.$$

are the row and column vectors, respectively. Here $\tilde{\mathbf{a}}_{.f}$ and $\tilde{\mathbf{a}}_{l.}$ are the fth column and lth row of $\tilde{\mathbf{A}} := \mathbf{A}^2\mathbf{A}^*$.

Proof. Taking into account (16), we have for ${}^{\#}A$,

$$a_{ij}^{\#,r} = \sum_{l=1}^{n}\sum_{f=1}^{n} a_{il}^{\#} a_{lf} a_{fj}^{\dagger}. \tag{20}$$

By substituting (14) and (15) in (20), we obtain

$$a_{ij}^{\#,r} = \sum_{l=1}^{n} \frac{\sum_{f=1}^{n}\sum_{\beta\in J_{s,n}\{i\}}\left|(\mathbf{A}^2)_{.i}(\mathbf{a}_{.f})\right|_{\beta}^{\beta} a_{fl}}{\sum_{\beta\in J_{s,n}}\left|\mathbf{A}^2\right|_{\beta}^{\beta}} \frac{\sum_{\alpha\in I_{s,n}\{j\}}\left|(\mathbf{A}\mathbf{A}^*)_{j.}(\mathbf{a}_{l.}^*)\right|_{\alpha}^{\alpha}}{\sum_{\alpha\in I_{s,n}}\left|\mathbf{A}\mathbf{A}^*\right|_{\alpha}^{\alpha}} =$$

$$\frac{\sum_{f=1}^{n}\sum_{l=1}^{n}\sum_{\beta\in J_{s,n}\{j\}}\left|(\mathbf{A}^2)_{.j}(\mathbf{e}_{.f})\right|_{\beta}^{\beta}\tilde{a}_{fl}\sum_{\alpha\in I_{s,n}\{j\}}\left|(\mathbf{A}\mathbf{A}^*)_{j.}(\mathbf{e}_{l.})\right|_{\alpha}^{\alpha}}{\sum_{\beta\in J_{s,n}}\left|\mathbf{A}^2\right|_{\beta}^{\beta}\sum_{\alpha\in I_{s,n}}\left|\mathbf{A}\mathbf{A}^*\right|_{\alpha}^{\alpha}},$$

where $\mathbf{e}_{.l}$ and $\mathbf{e}_{l.}$ are the unit column and row vectors, respectively, such that all their components are 0, except the lth components which are 1; $\tilde{a}_{lf}$ is the (lf)th element of the matrix $\tilde{\mathbf{A}} := \mathbf{A}^2\mathbf{A}^*$.

Let

$$u_{il}^{(1)} := \sum_{f=1}^{n}\sum_{\beta\in J_{s,n}\{i\}}\left|(\mathbf{A}^2)_{.i}(\mathbf{e}_{.f})\right|_{\beta}^{\beta}\tilde{a}_{fl} = \sum_{\beta\in J_{s,n}\{i\}}\left|(\mathbf{A}^2)_{.i}(\tilde{\mathbf{a}}_{.l})\right|_{\beta}^{\beta},\quad i,l = 1,\ldots,n.$$

Construct the matrix $\mathbf{U}_1 = \left(u_{il}^{(1)}\right) \in \mathbb{H}^{n\times n}$. It follows that

$$\sum_{l} u_{il}^{(1)} \sum_{\alpha\in I_{s,n}\{j\}}\left|(\mathbf{A}\mathbf{A}^*)_{j.}(\mathbf{e}_{l.})\right|_{\alpha}^{\alpha} = \sum_{\alpha\in I_{s,n}\{j\}}\left|(\mathbf{A}\mathbf{A}^*)_{j.}\left(\mathbf{u}_{i.}^{(1)}\right)\right|_{\alpha}^{\alpha},$$

where $\mathbf{u}_{i.}^{(1)}$ is the ith row of $\mathbf{U}_1$. So, we get (18). If we first consider

$$u_{if}^{(2)} := \sum_{l}\tilde{a}_{fl}\sum_{\alpha\in I_{s,n}\{j\}}\left|(\mathbf{A}\mathbf{A}^*)_{j.}(\mathbf{e}_{l.})\right|_{\alpha}^{\alpha} = \sum_{\alpha\in I_{s,n}\{j\}}\left|(\mathbf{A}\mathbf{A}^*)_{j.}(\tilde{\mathbf{a}}_{f.})\right|_{\alpha}^{\alpha},\quad f,j = 1,\ldots,n.$$

and construct the matrix $\mathbf{U}_2 = \left(u_{if}^{(2)}\right) \in \mathbb{H}^{n\times n}$, then from

$$\sum_{f=1}^{n}\sum_{\beta\in J_{s,n}\{i\}}\left|(\mathbf{A}^2)_{.i}(\mathbf{e}_{.f})\right|_{\beta}^{\beta}u_{if}^{(2)} = \sum_{\beta\in J_{s,n}\{i\}}\left|(\mathbf{A}^2)_{.i}\left(\mathbf{u}_{.f}^{(2)}\right)\right|_{\beta}^{\beta},$$

it follows (19). □

Taking into account (17), the following theorem on the determinantal representation of the left core inverse can be proved similarly.

Theorem 3.7. *Let* $\mathbf{A} \in \mathbb{C}_n^{\mathrm{CM}}$ *and* $\operatorname{rk}\mathbf{A}^2 = \operatorname{rk}\mathbf{A} = s$. *Then for its left core inverse* $({}_{\#}A) = \left(a_{ij}^{\#,l}\right)$, *we have*

$$a_{ij}^{\#,l} = \frac{\sum_{\alpha \in I_{s,n}\{j\}} \left|\left(\mathbf{A}^2\right)_{j.}\left(\mathbf{v}_{i.}^{(1)}\right)\right|_{\alpha}^{\alpha}}{\sum_{\beta \in J_{s,n}} |\mathbf{A}^*\mathbf{A}|_{\beta}^{\beta} \sum_{\alpha \in I_{s,n}} |\mathbf{A}^2|_{\alpha}^{\alpha}} = \frac{\sum_{\beta \in J_{s,n}\{i\}} \left|\left(\mathbf{A}^*\mathbf{A}\right)_{.i}\left(\mathbf{v}_{.j}^{(2)}\right)\right|_{\beta}^{\beta}}{\sum_{\beta \in J_{s,n}} |\mathbf{A}^*\mathbf{A}|_{\beta}^{\beta} \sum_{\alpha \in I_{s,n}} |\mathbf{A}^2|_{\alpha}^{\alpha}},$$

where

$$\mathbf{v}_{i.}^{(1)} = \left[\sum_{\beta \in J_{s,n}\{i\}} \left|(\mathbf{A}^*\mathbf{A})_{.i}(\overline{\mathbf{a}}_{.f})\right|_{\beta}^{\beta}\right] \in \mathbb{C}^{1\times n},\ f = 1, \dots, n$$

$$\mathbf{v}_{.j}^{(2)} = \left[\sum_{\alpha \in I_{s,n}\{j\}} \left|\left(\mathbf{A}^2\right)_{j.}(\overline{\mathbf{a}}_{l.})\right|_{\alpha}^{\alpha}\right] \in \mathbb{C}^{n\times 1},\ l = 1, \dots, n.$$

Here $\overline{\mathbf{a}}_{.f}$ and $\overline{\mathbf{a}}_{l.}$ are the fth column and lth row of $\overline{\mathbf{A}} := \mathbf{A}^*\mathbf{A}^2$.

3.2 Determinantal representations of the core-EP inverses

Similar as in [4], we introduce two core-EP inverses.

Definition 3.8. A matrix $\mathbf{X} \in \mathbb{C}^{n\times n}$ is said to be the right core-EP inverse of $\mathbf{A} \in \mathbb{C}^{n\times n}$ if it satisfies the conditions

$$\mathbf{XAX} = \mathbf{A}, \mathit{and}\ \mathcal{C}(\mathbf{X}) = \mathcal{C}(\mathbf{X}^*) = \mathcal{C}\left(\mathbf{A}^d\right).$$

It is denoted as $\mathbf{A}^{\textcircled{\dagger}}$.

Definition 3.9. A matrix $\mathbf{X} \in \mathbb{C}^{n\times n}$ is said to be the left core-EP inverse of $\mathbf{A} \in \mathbb{C}^{n\times n}$ if it satisfies the conditions

$$\mathbf{XAX} = \mathbf{A}, \mathit{and}\ \mathcal{R}(\mathbf{X}) = \mathcal{R}(\mathbf{X}^*) = \mathcal{R}\left(\mathbf{A}^d\right).$$

It is denoted as $\mathbf{A}_{\textcircled{\dagger}}$.

Remark 3.10. Since $\mathcal{C}\left((\mathbf{A}^*)^d\right) = \mathcal{R}\left(\mathbf{A}^d\right)$, then the left core inverse $\mathbf{A}_{\textcircled{\dagger}}$ of $\mathbf{A} \in \mathbb{C}^{n\times n}$ is similar to the $*$ core inverse introduced in [4], and the dual core-EP inverse introduced in [35].

Due to [4], we have the following representations the core-EP inverses of $\mathbf{A} \in \mathbb{C}^{n\times n}$,

$$\mathbf{A}^{\textcircled{\dagger}} = \mathbf{A}^{\{2,3,6a\}}\ \text{and}\ \mathcal{C}\left(\mathbf{A}^{\textcircled{\dagger}}\right) \subseteq \mathcal{C}\left(\mathbf{A}^k\right),$$
$$\mathbf{A}_{\textcircled{\dagger}} = \mathbf{A}^{\{2,4,6b\}}\ \text{and}\ \mathcal{R}\left(\mathbf{A}_{\textcircled{\dagger}}\right) \subseteq \mathcal{R}\left(\mathbf{A}^k\right).$$

Thanks to [35], the following representations of the core-EP inverses will be used for their determinantal representations.

Lemma 3.11. *Let* $\mathbf{A} \in \mathbb{C}^{n\times n}$ *and* $\operatorname{Ind}\mathbf{A} = k$. *Then*

$$\mathbf{A}^{\textcircled{\dagger}} = \mathbf{A}^k\left(\mathbf{A}^{k+1}\right)^{\dagger}, \tag{21}$$

$$\mathbf{A}_{\textcircled{\dagger}} = \left(\mathbf{A}^{k+1}\right)^{\dagger}\mathbf{A}^k. \tag{22}$$

Moreover, if $\operatorname{Ind}\mathbf{A} = 1$, then we have the following representations of the right and left core inverses

$$A^{\textcircled{\#}} = \mathbf{A}\left(\mathbf{A}^2\right)^{\dagger}, \tag{23}$$

$$A_{\textcircled{\#}} = \left(\mathbf{A}^2\right)^{\dagger}\mathbf{A}. \tag{24}$$

Theorem 3.12. *Suppose* $\mathbf{A} \in \mathbb{C}^{n\times n}$, $\operatorname{Ind}\mathbf{A} = k$, $\operatorname{rk}\mathbf{A}^k = s$, *and there exist* $\mathbf{A}^{\textcircled{\dagger}}$ *and* $\mathbf{A}_{\textcircled{\dagger}}$. *Then* $\mathbf{A}^{\textcircled{\dagger}} = \left(a_{ij}^{\textcircled{\dagger},r}\right)$ *and* $\mathbf{A}_{\textcircled{\dagger}} = \left(a_{ij}^{\textcircled{\dagger},l}\right)$ *possess the determinantal representations, respectively,*

$$a_{ij}^{\textcircled{\dagger},r} = \frac{\sum_{\alpha\in I_{s,n}\{j\}}\left|\left(\mathbf{A}^{k+1}\left(\mathbf{A}^{k+1}\right)^*\right)_{j.}(\hat{\mathbf{a}}_{i.})\right|_{\alpha}^{\alpha}}{\sum_{\alpha\in I_{s,n}}\left|\mathbf{A}^{k+1}\left(\mathbf{A}^{k+1}\right)^*\right|_{\alpha}^{\alpha}}, \tag{25}$$

$$a_{ij}^{\textcircled{\dagger},l} = \frac{\sum_{\beta\in J_{s,n}\{i\}}\left|\left(\left(\mathbf{A}^{k+1}\right)^*\mathbf{A}^{k+1}\right)_{.i}(\check{\mathbf{a}}_{.j})\right|_{\beta}^{\beta}}{\sum_{\beta\in J_{s,n}}\left|\left(\mathbf{A}^{k+1}\right)^*\mathbf{A}^{k+1}\right|_{\beta}^{\beta}}, \tag{26}$$

where $\hat{\mathbf{a}}_{i.}$ is the ith row of $\hat{\mathbf{A}} = \mathbf{A}^k\left(\mathbf{A}^{k+1}\right)^*$ and $\check{\mathbf{a}}_{.j}$ is the jth column of $\check{\mathbf{A}} = \left(\mathbf{A}^{k+1}\right)^*\mathbf{A}^k$.

Proof. Consider $\left(\mathbf{A}^{k+1}\right)^{\dagger} = \left(a_{ij}^{(k+1,\dagger)}\right)$ and $\mathbf{A}^k = \left(a_{ij}^{(k)}\right)$. By (21),

$$a_{ij}^{\textcircled{\dagger},r} = \sum_{t=1}^{n} a_{it}^{(k)}a_{tj}^{(k+1,\dagger)}.$$

Taking into account (9) for the determinantal representation of $\left(\mathbf{A}^{k+1}\right)^{\dagger}$, we get

$$a_{ij}^{\textcircled{\dagger},r} = \sum_{t=1}^{n} a_{it}^{(k)}\frac{\sum_{\alpha\in I_{s,n}\{j\}}\left|\left(\mathbf{A}^{k+1}\left(\mathbf{A}^{k+1}\right)^*\right)_{j.}\left(\mathbf{a}_{t.}^{(k+1,*)}\right)\right|_{\alpha}^{\alpha}}{\sum_{\alpha\in I_{r,m}}\left|\mathbf{A}^{k+1}\left(\mathbf{A}^{k+1}\right)^*\right|_{\alpha}^{\alpha}},$$

where $\mathbf{a}_{t.}^{(k+1,*)}$ is the tth row of $\left(\mathbf{A}^{k+1}\right)^*$. Since $\sum_{t=1}^{n}a_{it}^{(k)}\mathbf{a}_{t.}^{(k+1,*)} = \hat{\mathbf{a}}_{i.}$, then it follows (25).

The determinantal representation (26) can be obtained similarly by integrating (8) for the determinantal representation of $\left(\mathbf{A}^{k+1}\right)^{\dagger}$ in (22). □

Taking into account the representations (23)-(24), we obtain the determinantal representations of the right and left core inverses that have more simpler expressions than they are obtained in Theorems 3.6 and 3.7.

Corollary 3.13. *Let* $\mathbf{A} \in \mathbb{C}_s^{n\times n}$, $\operatorname{Ind}\mathbf{A} = 1$, *and there exist* $A^{\oplus}$ *and* $A_{\oplus}$. *Then* $A^{\oplus} = \left(a_{ij}^{\oplus,r}\right)$ *and* $A_{\oplus} = \left(a_{ij}^{\oplus,l}\right)$ *can be expressed as follows*

$$a_{ij}^{\oplus,r} = \frac{\sum_{\alpha\in I_{s,n}\{j\}} \left|\left(\mathbf{A}^2(\mathbf{A}^2)^*\right)_{j.}(\hat{\mathbf{a}}_{i.})\right|_{\alpha}^{\alpha}}{\sum_{\alpha\in I_{s,n}} \left|\mathbf{A}^2(\mathbf{A}^2)^*\right|_{\alpha}^{\alpha}},$$

$$a_{ij}^{\oplus,l} = \frac{\sum_{\beta\in J_{s,n}\{i\}} \left|\left((\mathbf{A}^2)^*\mathbf{A}^2\right)_{.i}(\check{\mathbf{a}}_{.j})\right|_{\beta}^{\beta}}{\sum_{\beta\in J_{s,n}} \left|(\mathbf{A}^2)^*\mathbf{A}^2\right|_{\beta}^{\beta}},$$

where $\hat{\mathbf{a}}_{i.}$ is the ith row of $\hat{\mathbf{A}} = \mathbf{A}(\mathbf{A}^2)^*$ and $\check{\mathbf{a}}_{.j}$ is the jth column of $\check{\mathbf{A}} = (\mathbf{A}^2)^*\mathbf{A}$.

3.3 Determinantal representations of the DMP and MPD inverses

The concept of the DMP inverse in complex matrices was introduced in [2] by S. Malik and N. Thome.

Definition 3.14. [2] Suppose $\mathbf{A} \in \mathbb{C}^{n\times n}$ and $\operatorname{Ind}\mathbf{A} = k$. A matrix $\mathbf{X} \in \mathbb{C}^{n\times n}$ is said to be the DMP inverse of $\mathbf{A}$ if it satisfies the conditions

$$\mathbf{XAX} = \mathbf{X}, \mathbf{XA} = \mathbf{A}^d\mathbf{A}, \textit{and } \mathbf{A}^k\mathbf{X} = \mathbf{A}^k\mathbf{A}^{\dagger}. \tag{27}$$

It is denoted as $\mathbf{A}^{d,\dagger}$.

Due to [2], if an arbitrary matrix satisfies the system of Eq. (27), then it is unique and has the following representation

$$\mathbf{A}^{d,\dagger} = \mathbf{A}^d\mathbf{A}\mathbf{A}^{\dagger}. \tag{28}$$

Theorem 3.15. *Let* $\mathbf{A} \in \mathbb{C}_s^{n\times n}$, $\operatorname{Ind}\mathbf{A} = k$, *and* $\operatorname{rk}\left(\mathbf{A}^k\right) = s_1$. *Then, its DMP inverse* $\mathbf{A}^{d,\dagger} = \left(a_{ij}^{d,\dagger}\right)$ *has the following determinantal representations.*

$$a_{ij}^{d,\dagger} = \frac{\sum_{\alpha\in I_{s,n}\{j\}} \left|(\mathbf{AA}^*)_{j.}\left(\mathbf{u}_{i.}^{(1)}\right)\right|_{\alpha}^{\alpha}}{\sum_{\beta\in J_{s_1,n}} \left|\mathbf{A}^{k+1}\right|_{\beta}^{\beta} \sum_{\alpha\in I_{s,n}} \left|\mathbf{AA}^*\right|_{\alpha}^{\alpha}} = \tag{29}$$

$$= \frac{\sum_{\beta\in J_{s_1,n}\{i\}} \left|\left(\mathbf{A}^{k+1}\right)_{.i}\left(\mathbf{u}_{.j}^{(2)}\right)\right|_{\beta}^{\beta}}{\sum_{\beta\in J_{s_1,n}} \left|\mathbf{A}^{k+1}\right|_{\beta}^{\beta} \sum_{\beta\in J_{s,n}} \left|\mathbf{AA}^*\right|_{\beta}^{\beta}}, \tag{30}$$

where

$$\mathbf{u}_{i.}^{(1)} = \left[\sum_{\beta\in J_{s_1,n}\{i\}} \left|\left(\mathbf{A}^{k+1}\right)_{.i}(\tilde{\mathbf{a}}_{f})\right|_{\beta}^{\beta}\right] \in \mathbb{C}^{1\times n},\ f = 1, \ldots, n,$$

$$\mathbf{u}_{.j}^{(2)} = \left[\sum_{\alpha\in I_{s,n}\{j\}} \left|(\mathbf{AA}^*)_{j.}(\hat{\mathbf{a}}_{l.})\right|_{\alpha}^{\alpha}\right] \in \mathbb{C}^{n\times 1},\ l = 1, \ldots, n.$$

Here, $\tilde{\mathbf{a}}_{f}$ and $\hat{\mathbf{a}}_{l.}$ are the fth column and the lth row of $\tilde{\mathbf{A}} := \mathbf{A}^{k+1}\mathbf{A}^*$.

Proof. Taking into account (28) for $\mathbf{A}^{d,\dagger}$, we get

$$a_{ij}^{d,\dagger} = \sum_{l=1}^{n}\sum_{f=1}^{n} a_{il}^{d} a_{lf} a_{fj}^{\dagger}. \tag{31}$$

By substituting (12) and (9) for the determinantal representations of $\mathbf{A}^d$ and $\mathbf{A}^\dagger$ in (31), we get

$$a_{ij}^{d,\dagger} = \sum_{l=1}^{n}\sum_{f=1}^{n} \frac{\sum_{\beta\in J_{s_1,n}\{i\}} \left|\left(\mathbf{A}^{k+1}\right)_{.i}\left(\mathbf{a}_{.l}^{(k)}\right)\right|_{\beta}^{\beta}}{\sum_{\beta\in J_{s_1,n}} \left|\mathbf{A}^{k+1}\right|_{\beta}^{\beta}} a_{lf} \frac{\sum_{\alpha\in I_{s,n}\{j\}} \left|(\mathbf{A}\mathbf{A}^*)_{j.}\left(\mathbf{a}_{f.}^*\right)\right|_{\alpha}^{\alpha}}{\sum_{\alpha\in I_{s,n}} \left|\mathbf{A}\mathbf{A}^*\right|_{\alpha}^{\alpha}} = \tag{32}$$
$$\sum_{l=1}^{n}\sum_{f=1}^{n} \frac{\sum_{\beta\in J_{s_1,n}\{i\}} \left|\left(\mathbf{A}^{k+1}\right)_{.i}(\mathbf{e}_{.l})\right|_{\beta}^{\beta}}{\sum_{\beta\in J_{s_1,n}} \left|\mathbf{A}^{k+1}\right|_{\beta}^{\beta}} \tilde{a}_{lf} \frac{\sum_{\alpha\in I_{s,n}\{j\}} \left((\mathbf{A}\mathbf{A}^*)_{j.}(\mathbf{e}_{f.})\right|_{\alpha}^{\alpha}}{\sum_{\alpha\in I_{s,n}} \left|\mathbf{A}\mathbf{A}^*\right|_{\alpha}^{\alpha}},$$

where $\mathbf{e}_{.l}$ and $\mathbf{e}_{l.}$ are the lth unit column and row vectors, and $\tilde{a}_{lf}$ is the (lf)th element of the matrix $\tilde{\mathbf{A}} = \mathbf{A}^{k+1}\mathbf{A}^*$. If we put

$$u_{if}^{(1)} := \sum_{l=1}^{n} \sum_{\beta\in J_{s_1,n}\{i\}} \left|\left(\mathbf{A}^{k+1}\right)_{.i}(\mathbf{e}_{.l})\right|_{\beta}^{\beta} \tilde{a}_{lf} = \sum_{\beta\in J_{s_1,n}\{i\}} \left|\left(\mathbf{A}^{k+1}\right)_{.i}(\tilde{\mathbf{a}}_{.f})\right|_{\beta}^{\beta},$$

as the fth component of the row vector $\mathbf{u}_{i.}^{(1)} = \left[u_{i1}^{(1)}, \ldots, u_{in}^{(1)}\right]$, then from

$$\sum_{f=1}^{n} u_{if}^{(1)} \sum_{\alpha\in I_{s,n}\{j\}} \left|(\mathbf{A}\mathbf{A}^*)_{j.}(\mathbf{e}_{f.})\right|_{\alpha}^{\alpha} = \sum_{\alpha\in I_{s,n}\{j\}} \left|(\mathbf{A}\mathbf{A}^*)_{j.}\left(\mathbf{u}_{i.}^{(1)}\right)\right|_{\alpha}^{\alpha},$$

it follows (29). If we initially obtain

$$u_{lj}^{(2)} := \sum_{f=1}^{n} \tilde{a}_{lf} \sum_{\alpha\in I_{s,n}\{j\}} \left|(\mathbf{A}\mathbf{A}^*)_{j.}(\mathbf{e}_{f.})\right|_{\alpha}^{\alpha} = \sum_{\alpha\in I_{s,n}\{j\}} \left|(\mathbf{A}\mathbf{A}^*)_{j.}(\tilde{\mathbf{a}}_{l.})\right|_{\alpha}^{\alpha},$$

as the lth component of the column vector $\mathbf{u}_{.j}^{(2)} = \left[u_{1j}^{(2)}, \ldots, u_{nj}^{(2)}\right]$, then from

$$\sum_{l=1}^{n} \sum_{\beta\in J_{s_1,n}\{i\}} \left|\left(\mathbf{A}^{k+1}\right)_{.i}(\mathbf{e}_{.l})\right|_{\beta}^{\beta} u_{lj}^{(2)} = \sum_{\beta\in J_{s_1,n}\{i\}} \left|\left(\mathbf{A}^{k+1}\right)_{.i}\left(\mathbf{u}_{.j}^{(2)}\right)\right|_{\beta}^{\beta},$$

it follows (30). □

The name of the DMP inverse is in accordance with the order of using the Drazin inverse (D) and the Moore-Penrose (MP) inverse. In that connection, it would be logical to consider the following definition.

Definition 3.16. Suppose $\mathbf{A}\in\mathbb{C}^{n\times n}$ and $\operatorname{Ind}\mathbf{A} = k$. A matrix $\mathbf{X}\in\mathbb{C}^{n\times n}$ is said to be the MPD inverse of $\mathbf{A}$ if it satisfies the conditions

$$\mathbf{XAX} = \mathbf{X}, \mathbf{AX} = \mathbf{A}\mathbf{A}^d, \textit{and } \mathbf{X}\mathbf{A}^k = \mathbf{A}^\dagger\mathbf{A}^k.$$

It is denoted as $\mathbf{A}^{\dagger,d}$.

The matrix $\mathbf{A}^{\dagger,d}$ is unique, and it can be represented as

$$\mathbf{A}^{\dagger,d} = \mathbf{A}^{\dagger}\mathbf{A}\mathbf{A}^{d}. \tag{33}$$

Theorem 3.17. *Let* $\mathbf{A}\in\mathbb{C}_s^{n\times n}$, $\operatorname{Ind}\mathbf{A}=k$, *and* $\operatorname{rk}\mathbf{A}^k=s_1$. *Then, its MPD inverse* $\mathbf{A}^{\dagger,d}=\left(a_{ij}^{\dagger,d}\right)$ *has the following determinantal representations*

$$a_{ij}^{\dagger,d} = \frac{\sum_{\beta\in J_{s,n}\{i\}}\left|(\mathbf{A}^*\mathbf{A})_{.i}\left(\mathbf{v}_{.j}^{(1)}\right)\right|_{\beta}^{\beta}}{\sum_{\beta\in J_{s,n}}|\mathbf{A}^*\mathbf{A}|_{\beta}^{\beta}\sum_{\beta\in I_{s_1,n}}\left|\mathbf{A}^{k+1}\right|_{\alpha}^{\alpha}} = \frac{\sum_{\alpha\in I_{s_1,n}\{j\}}\left|\left(\mathbf{A}^{k+1}\right)_{j.}\left(\mathbf{v}_{i.}^{(2)}\right)\right|_{\alpha}^{\alpha}}{\sum_{\alpha\in I_{s_1,n}}|\mathbf{A}^*\mathbf{A}|_{\beta}^{\beta}\sum_{\alpha\in I_{s,n}}\left|\mathbf{A}^{k+1}\right|_{\alpha}^{\alpha}},$$

where

$$\mathbf{v}_{.j}^{(1)} = \left[\sum_{\alpha\in I_{s_1,n}\{j\}}\left|\left(\mathbf{A}^{k+1}\right)_{j.}(\hat{\mathbf{a}}_{l.})\right|_{\alpha}^{\alpha}\right]\in\mathbb{C}^{n\times 1},\quad l=1,\ldots,n$$

$$\mathbf{v}_{i.}^{(2)} = \left[\sum_{\beta\in J_{s,n}\{i\}}\left|(\mathbf{A}^*\mathbf{A})_{.i}(\hat{\mathbf{a}}_{.f})\right|_{\beta}^{\beta}\right]\in\mathbb{C}^{1\times n},\quad l=1,\ldots,n.$$

Here, $\hat{\mathbf{a}}_{l.}$ and $\hat{\mathbf{a}}_{.f}$ are the lth row and the fth column of $\hat{\mathbf{A}}:=\mathbf{A}^*\mathbf{A}^{k+1}$.
Proof. The proof is similar to the proof of Theorem 3.15. □

3.4 Determinantal representations of the CMP inverse

Definition 3.18. [6] Suppose $\mathbf{A}\in\mathbb{C}^{n\times n}$ has the core-nilpotent decomposition $\mathbf{A}=\mathbf{A}_1+\mathbf{A}_2$, where $\operatorname{Ind}\mathbf{A}_1=\operatorname{Ind}\mathbf{A}$, $\mathbf{A}_2$ is nilpotent, and $\mathbf{A}_1\mathbf{A}_2=\mathbf{A}_2\mathbf{A}_1=0$. The CMP inverse of $\mathbf{A}$ is called the matrix $\mathbf{A}^{c,\dagger}:=\mathbf{A}^{\dagger}\mathbf{A}_1\mathbf{A}^{\dagger}$.

Lemma 3.19. [6] *Let* $\mathbf{A}\in\mathbb{C}^{n\times n}$. *The matrix* $\mathbf{X}=\mathbf{A}^{c,\dagger}$ *is the unique matrix that satisfies the following system of equations:*

$$\mathbf{XAX}=\mathbf{X},\ \mathbf{AXA}=\mathbf{A}_1,\ \mathbf{AX}=\mathbf{A}_1\mathbf{A}^{\dagger},\ and\ \mathbf{XA}=\mathbf{A}^{\dagger}\mathbf{A}_1.$$

Moreover,

$$\mathbf{A}^{c,\dagger}=\mathbf{Q}_A\mathbf{A}^d\mathbf{P}_A. \tag{34}$$

Taking into account (34), it follows the next theorem about determinantal representations of the quaternion CMP inverse.

Theorem 3.20. *Let* $\mathbf{A}\in\mathbb{C}_s^{n\times n}$, $\operatorname{Ind}\mathbf{A}=m$, *and* $\operatorname{rk}(\mathbf{A}^m)=s_1$. *Then, the determinantal representations of its CMP inverse* $\mathbf{A}^{c,\dagger}=\left(a_{ij}^{c,\dagger}\right)$ *can be expressed as*

$$a_{ij}^{c,\dagger} = \frac{\sum_{\beta\in J_{s,n}\{i\}}\left|(\mathbf{A}^*\mathbf{A})_{.i}\left(\mathbf{v}_{.j}^{(l)}\right)\right|_{\beta}^{\beta}}{\left(\sum_{\beta\in J_{s,n}}|\mathbf{A}^*\mathbf{A}|_{\beta}^{\beta}\right)^2\sum_{\beta\in J_{s_1,n}}\left|\mathbf{A}^{m+1}\right|_{\beta}^{\beta}} \tag{35}$$

$$a_{ij}^{c,\dagger} = \frac{\sum_{\alpha \in I_{s,n}\{j\}} \left| (\mathbf{AA}^*)_{j.}\left(\mathbf{w}_{i.}^{(l)}\right) \right|_{\alpha}^{\alpha}}{\left(\sum_{\alpha \in I_{s,n}} |\mathbf{AA}^*|_{\alpha}^{\alpha}\right)^2 \sum_{\beta \in J_{s_1,n}} |\mathbf{A}^{m+1}|_{\beta}^{\beta}} \tag{36}$$

for all $l = 1, 2$, where

$$\mathbf{v}_{.j}^{(1)} = \left[\sum_{\alpha \in I_{s,n}\{j\}} \left| (\mathbf{AA}^*)_{j.}(\hat{\mathbf{u}}_{t.}) \right|_{\alpha}^{\alpha} \right] \in \mathbb{C}^{n\times 1}, t = 1, ..., n, \tag{37}$$

$$\mathbf{w}_{i.}^{(1)} = \left[\sum_{\beta \in J_{s,n}\{i\}} \left| (\mathbf{A}^* \mathbf{A})_{.i}(\hat{\mathbf{u}}_{.k}) \right|_{\beta}^{\beta} \right] \in \mathbb{C}^{1\times n}, k = 1, ..., n, \tag{38}$$

$$\mathbf{v}_{.j}^{(2)} = \left[\sum_{\alpha \in I_{s,n}\{j\}} \left| (\mathbf{A}^* \mathbf{A})_{j.}(\tilde{\mathbf{g}}_{t.}) \right|_{\alpha}^{\alpha} \right] \in \mathbb{C}^{n\times 1}, t = 1, ..., n, \tag{39}$$

$$\mathbf{w}_{i.}^{(2)} = \left[\sum_{\beta \in J_{s,n}\{i\}} \left| (\mathbf{A}^* \mathbf{A})_{.i}(\tilde{\mathbf{g}}_{.k}) \right|_{\beta}^{\beta} \right] \in \mathbb{C}^{1\times n}, k = 1, ..., n. \tag{40}$$

Here, $\hat{\mathbf{u}}_{t.}$ is the tth row and $\hat{\mathbf{u}}_{.k}$ is the kth column of $\hat{\mathbf{U}} := \mathbf{UAA}^*$, $\tilde{\mathbf{g}}_{t.}$ is the tth row and $\tilde{\mathbf{g}}_{.k}$ is the kth column of $\tilde{\mathbf{G}} := \mathbf{A}^* \mathbf{AG}$, and the matrices $\mathbf{U} = (u_{ij}) \in \mathbb{H}^{n\times n}$ and $\mathbf{G} = \left(g_{ij}\right) \in \mathbb{H}^{n\times n}$ are such that

$$u_{ij} = \sum_{\alpha \in I_{s_1,n}\{j\}} \left| \left(\mathbf{A}^{m+1}\right)_{j.}(\hat{\mathbf{a}}_{i.}) \right|_{\alpha}^{\alpha}, \quad g_{ij} = \sum_{\beta \in J_{s_1,n}\{i\}} \left| \left(\mathbf{A}^{m+1}\right)_{.i}(\tilde{\mathbf{a}}_{.j}) \right|_{\beta}^{\beta},$$

where $\hat{\mathbf{a}}_{i.}$ is the ith row of $\hat{\mathbf{A}} := \mathbf{A}^* \mathbf{A}^{m+1}$ and $\tilde{\mathbf{a}}_{.j}$ is the jth column of $\tilde{\mathbf{A}} := \mathbf{A}^{m+1}\mathbf{A}^*$.
Proof. Taking into account (34), we get

$$a_{ij}^{c,\dagger} = \sum_{l=1}^{n} \sum_{k=1}^{n} q_{il}^{A} a_{lk}^{d} p_{kj}^{A}, \tag{41}$$

where $\mathbf{Q}_A = (q_{il}^A)$, $\mathbf{A}^d = (a_{il}^d)$, and $\mathbf{P}_A = (p_{il}^A)$.

a. Taking into account the expressions (13), (10), and (11) for the determinantal representations of $\mathbf{A}^d$, $\mathbf{Q}_A$, and $\mathbf{P}_A$, respectively, we have

$$a_{ij}^{c,\dagger} = \sum_{l}\sum_{t} \frac{\sum_{\beta \in J_{s,n}\{i\}} \left| (\mathbf{A}^* \mathbf{A})_{.i}(\dot{\mathbf{a}}_{.t}) \right|_{\beta}^{\beta}}{\sum_{\beta \in J_{s,n}} |\mathbf{A}^* \mathbf{A}|_{\beta}^{\beta}} \frac{\sum_{\alpha \in I_{s_1,n}\{l\}} \left| \left(\mathbf{A}^{m+1}\right)_{l.}\left(\mathbf{a}_{t.}^{(m)}\right) \right|_{\alpha}^{\alpha}}{\sum_{\alpha \in I_{s_1,n}} |\mathbf{A}^{m+1}|_{\alpha}^{\alpha}} \frac{\sum_{\alpha \in I_{s,n}\{j\}} \left| (\mathbf{AA}^*)_{j.}(\ddot{\mathbf{a}}_{l.}) \right|_{\alpha}^{\alpha}}{\sum_{\alpha \in I_{s,n}} |\mathbf{AA}^*|_{\alpha}^{\alpha}},$$

where $\dot{\mathbf{a}}_{.t}$ is the tth column of $\mathbf{A}^* \mathbf{A}$, $\ddot{\mathbf{a}}_{l.}$ is the lth row of $\mathbf{AA}^*$, and $\mathbf{a}_{t.}^{(m)}$ is the tth row of $\mathbf{A}^m$. So, it is clear that

$$a_{ij}^{c,\dagger} = \sum_{l}\sum_{t}\sum_{k} \frac{\sum_{\beta \in J_{s,n}\{i\}} \left| (\mathbf{A}^* \mathbf{A})_{.i}(\mathbf{e}_{.t}) \right|_{\beta}^{\beta} \hat{a}_{tk} \sum_{\alpha \in I_{s_1,n}\{l\}} \left| \left(\mathbf{A}^{m+1}\right)_{l.}(\mathbf{e}_{k.}) \right|_{\alpha}^{\alpha}}{\sum_{\beta \in J_{s,n}} |\mathbf{A}^* \mathbf{A}|_{\beta}^{\beta} \sum_{\alpha \in I_{s_1,n}} |\mathbf{A}^{m+1}|_{\alpha}^{\alpha}} \frac{\sum_{\alpha \in I_{s,n}\{j\}} \left| (\mathbf{AA}^*)_{j.}(\ddot{\mathbf{a}}_{l.}) \right|_{\alpha}^{\alpha}}{\sum_{\alpha \in I_{s,n}} |\mathbf{AA}^*|_{\alpha}^{\alpha}},$$

where $\mathbf{e}_{.t}$ is the tth unit column vector, $\mathbf{e}_{k.}$ is the kth row vector, and $\hat{a}_{tk}$ is the (tk)th element of $\hat{\mathbf{A}} = \mathbf{A}^*\mathbf{A}^{m+1}$.

Denote

$$u_{tl} := \sum_k \hat{a}_{tk} \sum_{\alpha \in I_{s_1,n}\{j\}} \left|(\mathbf{A}^{m+1})_{l.}(\mathbf{e}_{k.})\right|_\alpha^\alpha = \sum_{\alpha \in I_{s_1,n}\{j\}} \left|(\mathbf{A}^{m+1})_{l.}(\hat{\mathbf{a}}_{t.})\right|_\alpha^\alpha \tag{42}$$

as the tth component of a column vector $\mathbf{u}_{.l} = [u_{1l}, \ldots, u_{nl}]$. Then from

$$\sum_t \sum_{\beta \in J_{s,n}\{i\}} |(\mathbf{A}^*\mathbf{A})_{.i}(\mathbf{e}_{.t})|_\beta^\beta u_{tl} = \sum_{\beta \in J_{s,n}\{i\}} |(\mathbf{A}^*\mathbf{A})_{.i}(\mathbf{u}_{.l})|_\beta^\beta,$$

we have

$$a_{ij}^{c,\dagger} = \sum_l \frac{\sum_{\beta \in J_{s,n}\{i\}} |(\mathbf{A}^*\mathbf{A})_{.i}(\mathbf{u}_{.l})|_\beta^\beta \sum_{\alpha \in I_{s,n}\{j\}} \left|(\mathbf{A}\mathbf{A}^*)_{j.}(\ddot{\mathbf{a}}_{l.})\right|_\alpha^\alpha}{\sum_{\beta \in J_{r,n}} |\mathbf{A}^*\mathbf{A}|_\beta^\beta \sum_{\alpha \in I_{s_1,n}} |\mathbf{A}^{m+1}|_\alpha^\alpha \sum_{\alpha \in I_{s,n}} |\mathbf{A}\mathbf{A}^*|_\alpha^\alpha}.$$

Construct the matrix $\mathbf{U} = (u_{tl}) \in \mathbb{H}^{n\times n}$, where u_{tl} is given by (42), and denote $\hat{\mathbf{U}} := \mathbf{U}\mathbf{A}\mathbf{A}^*$. Then, taking into account that $|\mathbf{A}^*\mathbf{A}|_\beta^\beta = |\mathbf{A}\mathbf{A}^*|_\alpha^\alpha$, we have

$$a_{ij}^{c,\dagger} = \frac{\sum_t \sum_k \sum_{\beta \in J_{s,n}\{i\}} |(\mathbf{A}^*\mathbf{A})_{.i}(\mathbf{e}_{.t})|_\beta^\beta \hat{u}_{tk} \sum_{\alpha \in I_{s,n}\{j\}} \left|(\mathbf{A}\mathbf{A}^*)_{j.}(\mathbf{e}_{k.})\right|_\alpha^\alpha}{\left(\sum_{\beta \in J_{s,n}} |\mathbf{A}^*\mathbf{A}|_\beta^\beta\right)^2 \sum_{\alpha \in I_{s_1,n}} |\mathbf{A}^{m+1}|_\alpha^\alpha}.$$

If we put that

$$v_{tj}^{(1)} := \sum_k \hat{u}_{tk} \sum_{\alpha \in I_{s,n}\{j\}} \left|(\mathbf{A}\mathbf{A}^*)_{j.}(\mathbf{e}_{k.})\right|_\alpha^\alpha = \sum_{\alpha \in I_{s,n}\{j\}} \left|(\mathbf{A}\mathbf{A}^*)_{j.}(\hat{\mathbf{u}}_{t.})\right|_\alpha^\alpha$$

is the tth component of a column vector $\mathbf{v}_{.j}^{(1)} = \left[v_{1j}^{(1)}, \ldots, v_{nj}^{(1)}\right]$, then from

$$\sum_t \sum_{\beta \in J_{s,n}\{i\}} |(\mathbf{A}^*\mathbf{A})_{.i}(\mathbf{e}_{.t})|_\beta^\beta v_{tj}^{(1)} = \sum_{\beta \in J_{s,n}\{i\}} \left|(\mathbf{A}^*\mathbf{A})_{.i}\left(\mathbf{v}_{.j}^{(1)}\right)\right|_\beta^\beta,$$

it follows (35) with $\mathbf{v}_{.j}^{(1)}$ given by (37). If we initially put

$$w_{ik}^{(1)} := \sum_t \sum_{\beta \in J_{s,n}\{i\}} |(\mathbf{A}^*\mathbf{A})_{.i}(\mathbf{e}_{.t})|_\beta^\beta \hat{u}_{tk} = \sum_{\beta \in J_{s,n}\{i\}} |(\mathbf{A}^*\mathbf{A})_{.i}(\hat{\mathbf{u}}_{.k})|_\beta^\beta$$

as the kth component of the row vector $\mathbf{w}_{i.}^{(1)} = \left[w_{i1}^{(1)}, \ldots, w_{in}^{(1)}\right]$, then from

$$\sum_k w_{ik}^{(1)} \sum_{\alpha \in I_{s,n}\{j\}} \left|(\mathbf{A}^2)_{j.}(\mathbf{e}_{k.})\right|_\alpha^\alpha = \sum_{\alpha \in I_{s,n}\{j\}} \left|(\mathbf{A}^2)_{j.}\left(\mathbf{w}_{i.}^{(1)}\right)\right|_\alpha^\alpha,$$

it follows (36) with $\mathbf{w}_{i.}^{(1)}$ given by (38).

b. By using the determinantal representation (12) for $\mathbf{A}^{d}$ in (41), we have

$$a_{ij}^{c,\dagger}=\sum_{k}\sum_{t}\frac{\sum_{\beta\in J_{s,n}\{i\}}\left|(\mathbf{A}^{*}\mathbf{A})_{.i}(\dot{\mathbf{a}}_{.t})\right|_{\beta}^{\beta}}{\sum_{\beta\in J_{s,n}}|\mathbf{A}^{*}\mathbf{A}|_{\beta}^{\beta}}\frac{\sum_{\beta\in J_{s_1,n}\{t\}}\left|\left(\mathbf{A}^{m+1}\right)_{.t}\left(\mathbf{a}_{.k}^{(m)}\right)\right|_{\beta}^{\beta}}{\sum_{\beta\in J_{s,n}}\left|\mathbf{A}^{m+1}\right|_{\beta}^{\beta}}\frac{\sum_{\alpha\in I_{s,n}\{j\}}\left|(\mathbf{A}\mathbf{A}^{*})_{j.}(\ddot{\mathbf{a}}_{k.})\right|_{\alpha}^{\alpha}}{\sum_{\alpha\in I_{s,n}}|\mathbf{A}\mathbf{A}^{*}|_{\alpha}^{\alpha}}.$$

Therefore,

$$a_{ij}^{c,\dagger}=\sum_{l}\sum_{k}\sum_{t}\frac{\sum_{\beta\in J_{s,n}\{i\}}\left((\mathbf{A}^{*}\mathbf{A})_{.i}(\dot{\mathbf{a}}_{.t})\right)_{\beta}^{\beta}}{\sum_{\beta\in J_{s,n}}|\mathbf{A}^{*}\mathbf{A}|_{\beta}^{\beta}}\times$$

$$\frac{\sum_{\beta\in J_{s_1,n}\{t\}}\left|\left(\mathbf{A}^{m+1}\right)_{.t}(\mathbf{e}_{.k})\right|_{\beta}^{\beta}}{\sum_{\beta\in J_{s_1,n}}\left|\mathbf{A}^{m+1}\right|_{\beta}^{\beta}}\tilde{a}_{kl}\frac{\sum_{\alpha\in I_{s,n}\{j\}}\left|(\mathbf{A}\mathbf{A}^{*})_{j.}(\mathbf{e}_{l.})\right|_{\alpha}^{\alpha}}{\sum_{\alpha\in I_{s,n}}|\mathbf{A}\mathbf{A}^{*}|_{\alpha}^{\alpha}}.$$

where $\mathbf{e}_{.k}$ is the kth unit column vector, $\mathbf{e}_{l.}$ is the lth unit row vector, and $\tilde{a}_{kl}$ is the (kl)th element of $\tilde{\mathbf{A}}=\mathbf{A}^{m+1}\mathbf{A}^{*}$.

If we denote

$$g_{tl}:=\sum_{l}\sum_{\beta\in J_{s_1,n}\{t\}}\left|\left(\mathbf{A}^{m+1}\right)_{.t}(\mathbf{e}_{.k})\right|_{\beta}^{\beta}\tilde{a}_{kl}=\sum_{\beta\in J_{s_1,n}\{t\}}\left|\left(\mathbf{A}^{m+1}\right)_{.t}(\tilde{\mathbf{a}}_{.l})\right|_{\beta}^{\beta}\tag{43}$$

as the lth component of a row vector $\mathbf{g}_{t.}=\left[g_{t1},\ldots,g_{tn}\right]$, then

$$\sum_{l}g_{tl}\sum_{\alpha\in I_{s,n}\{j\}}\left|(\mathbf{A}\mathbf{A}^{*})_{j.}(\mathbf{e}_{l.})\right|_{\alpha}^{\alpha}=\sum_{\alpha\in I_{s,n}\{j\}}\left|(\mathbf{A}\mathbf{A}^{*})_{j.}(\mathbf{g}_{t.})\right|_{\alpha}^{\alpha}.$$

From this, it follows that

$$a_{ij}^{c,\dagger}=\sum_{t}\frac{\sum_{\beta\in J_{s,n}\{i\}}\left|(\mathbf{A}^{*}\mathbf{A})_{.i}(\dot{\mathbf{a}}_{.t})\right|_{\beta}^{\beta}\sum_{\alpha\in I_{s,n}\{j\}}\left|(\mathbf{A}\mathbf{A}^{*}\right|_{j.}(\mathbf{g}_{t.}))_{\alpha}^{\alpha}}{\sum_{\beta\in J_{r,n}}|\mathbf{A}^{*}\mathbf{A}|_{\beta}^{\beta}\sum_{\alpha\in I_{s_1,n}}\left|\mathbf{A}^{m+1}\right|_{\alpha}^{\alpha}\sum_{\alpha\in I_{s,n}}|\mathbf{A}\mathbf{A}^{*}|_{\alpha}^{\alpha}}.$$

Construct the matrix $\mathbf{G}=\left(g_{tl}\right)\in\mathbb{H}^{n\times n}$, where g_{tl} is given by (43). Denote $\tilde{\mathbf{G}}:=\mathbf{A}^{*}\mathbf{A}\mathbf{G}$. Then,

$$a_{ij}^{c,\dagger}=\frac{\sum_{t}\sum_{k}\sum_{\beta\in J_{s,n}\{i\}}\left|(\mathbf{A}^{*}\mathbf{A})_{.i}(\mathbf{e}_{.t})\right|_{\beta}^{\beta}\tilde{g}_{tk}\sum_{\alpha\in I_{s,n}\{j\}}\left|(\mathbf{A}\mathbf{A}^{*})_{j.}(\mathbf{e}_{k.})\right|_{\alpha}^{\alpha}}{\sum_{\beta\in J_{s,n}}\left(|\mathbf{A}^{*}\mathbf{A}|_{\beta}^{\beta}\right)^{2}\sum_{\alpha\in I_{s_1,n}}\left|\mathbf{A}^{m+1}\right|_{\alpha}^{\alpha}}.$$

If we denote

$$v_{tj}^{(2)}:=\sum_{k}\tilde{g}_{tk}\sum_{\alpha\in I_{s,n}\{j\}}\left|(\mathbf{A}\mathbf{A}^{*})_{j.}(\mathbf{e}_{k.})\right|_{\alpha}^{\alpha}=\sum_{\alpha\in I_{s,n}\{j\}}\left|(\mathbf{A}\mathbf{A}^{*})_{j.}(\tilde{\mathbf{g}}_{t.})\right|_{\alpha}^{\alpha}$$

as the tth component of a column vector $\mathbf{v}_{.j}^{(2)} = \left[v_{1j}^{(2)}, \dots, v_{nj}^{(2)}\right]$, then

$$\sum_t \sum_{\beta \in J_{s,n}\{i\}} \left|(\mathbf{A}^*\mathbf{A})_{.i}(\mathbf{e}_{.t})\right|_\beta^\beta v_{tj}^{(2)} = \sum_{\beta \in J_{s,n}\{i\}} \left|(\mathbf{A}^*\mathbf{A})_{.i}\left(\mathbf{v}_{.j}^{(2)}\right)\right|_\beta^\beta.$$

Thus, we have (35) with $\mathbf{v}_{.j}^{(2)}$ given by (39).
If, now, we denote

$$w_{ik}^{(2)} := \sum_t \sum_{\beta \in J_{s,n}\{i\}} \left|(\mathbf{A}^*\mathbf{A})_{.i}(\mathbf{e}_{.t})\right|_\beta^\beta \tilde{g}_{tk} = \sum_{\beta \in J_{s,n}\{i\}} \left|(\mathbf{A}^*\mathbf{A})_{.i}(\tilde{\mathbf{g}}_{.k})\right|_\beta^\beta$$

as the kth component of a row vector $\mathbf{w}_{i.}^{(2)} = \left[w_{i1}^{(2)}, \dots, w_{in}^{(2)}\right]$, then

$$\sum_k w_{ik}^{(2)} \sum_{\alpha \in I_{s,n}\{j\}} \left|(\mathbf{A}\mathbf{A}^*)_{j.}(\mathbf{e}_{k.})\right|_\alpha^\alpha = \sum_{\alpha \in I_{s,n}\{j\}} \left|(\mathbf{A}\mathbf{A}^*)_{j.}\left(\mathbf{w}_{i.}^{(2)}\right)\right|_\alpha^\alpha.$$

So, finally, we have (36) with $\mathbf{w}_{i.}^{(2)}$ given by (40).

4. An example

Given the matrix

$$\mathbf{A} = \begin{bmatrix} 2 & 0 & 0 \\ -i & i & i \\ -i & -i & -i \end{bmatrix}.$$

Since

$$\mathbf{A}\mathbf{A}^* = \begin{bmatrix} 4 & 2\mathbf{i} & 2\mathbf{i} \\ -2\mathbf{i} & 3 & -1 \\ -2\mathbf{i} & -1 & 3 \end{bmatrix}, \ \mathbf{A}^2 = \begin{bmatrix} 4 & 0 & 0 \\ 2-2i & 0 & 0 \\ -2-2i & 0 & 0 \end{bmatrix}, \ \mathbf{A}^3 = \begin{bmatrix} 8 & 0 & 0 \\ 4-4i & 0 & 0 \\ -4-4i & 0 & 0 \end{bmatrix},$$

then $\operatorname{rk}\mathbf{A} = 2$ and $\operatorname{rk}\mathbf{A}^2 = \operatorname{rk}\mathbf{A}^3 = 1$, and $k = \operatorname{Ind}\mathbf{A} = 2$ and $r_1 = 1$. So, we shall find $\mathbf{A}^{\textcircled{\dagger}}$ and $\mathbf{A}_{\textcircled{\dagger}}$ by (25) and (26), respectively.
Since

$$\hat{\mathbf{A}} = \mathbf{A}^2(\mathbf{A}^3)^* = 16\begin{bmatrix} 2 & 1+\mathbf{i} & -1+\mathbf{i} \\ 1-\mathbf{i} & 1 & \mathbf{i} \\ -1-\mathbf{i} & \mathbf{i} & 1 \end{bmatrix},$$

then by (25),

$$a_{11}^{\textcircled{\dagger},r} = \frac{\sum_{\alpha \in I_{1,3}\{1\}} \left|\left(\mathbf{A}^3(\mathbf{A}^3)^*\right)_{1.}(\hat{\mathbf{a}}_{1.})\right|_\alpha^\alpha}{\sum_{\alpha \in I_{1,3}} \left|\mathbf{A}^3(\mathbf{A}^3)^*\right|_\alpha^\alpha} = \frac{1}{4}.$$

By similarly continuing, we get

$$\mathbf{A}^{\oplus} = \frac{1}{8}\begin{bmatrix} 2 & 1+\mathbf{i} & -1+\mathbf{i} \\ 1-\mathbf{i} & 1 & \mathbf{i} \\ -1-\mathbf{i} & \mathbf{i} & 1 \end{bmatrix}.$$

By analogy, due to (26), we have

$$\mathbf{A}_{\oplus} = \frac{1}{2}\begin{bmatrix} 1 & 0 & 0 \\ 0 & 0 & 0 \\ 0 & 0 & 0 \end{bmatrix}.$$

The DMP inverse $\mathbf{A}^{d,\dagger}$ can be found by Theorem 3.15. Since

$$\tilde{\mathbf{A}} = \mathbf{A}^3\mathbf{A}^* = 4\begin{bmatrix} 4 & 2\mathbf{i} & 2\mathbf{i} \\ 2-2\mathbf{i} & 1+\mathbf{i} & 1+\mathbf{i} \\ -2-2\mathbf{i} & 1-\mathbf{i} & 1-\mathbf{i} \end{bmatrix}.$$

and $\mathrm{rk}\left(\mathbf{A}^3\right) = 1$, then

$$\mathbf{u}_1^{(1)} = \tilde{\mathbf{a}}_{1.}, \ \mathbf{u}_2^{(1)} = \tilde{\mathbf{a}}_{2.}, \ \mathbf{u}_3^{(1)} = \tilde{\mathbf{a}}_{3.}.$$

Furthermore, by (29),

$$a_{11}^{d,\dagger} = \frac{\sum_{\alpha\in I_{2,3}\{1\}}\left|(\mathbf{AA}^*)_{1.}\left(\mathbf{u}_{1.}^{(1)}\right)\right|_{\alpha}^{\alpha}}{\sum_{\beta\in J_{1,3}}\left|\mathbf{A}^3\right|_{\beta}^{\beta}\sum_{\alpha\in I_{2,3}}\left|\mathbf{AA}^*\right|_{\alpha}^{\alpha}} = \frac{1}{192}\left(det\begin{bmatrix} 16 & 8\mathbf{i} \\ -2\mathbf{i} & 3 \end{bmatrix} + det\begin{bmatrix} 16 & 8\mathbf{i} \\ -2\mathbf{i} & 3 \end{bmatrix}\right) = \frac{1}{3}.$$

By similarly continuing, we get

$$\mathbf{A}^{d,\dagger} = \frac{1}{12}\begin{bmatrix} 4 & 2\mathbf{i} & 2\mathbf{i} \\ 2-2\mathbf{i} & 1+\mathbf{i} & 1+\mathbf{i} \\ -2-2\mathbf{i} & 1-\mathbf{i} & 1-\mathbf{i} \end{bmatrix}.$$

Similarly by Theorem 3.17, we get

$$\mathbf{A}^{\dagger,d} = \frac{1}{4}\begin{bmatrix} 2 & 0 & 0 \\ -\mathbf{i} & 0 & 0 \\ -\mathbf{i} & 0 & 0 \end{bmatrix}.$$

Finally, by theorem, we find the CMP inverse $\mathbf{A}^{c,\dagger} = \left(a_{ij}^{c,\dagger}\right)$. Since $\mathrm{rk}\,\mathbf{A}^3 = 1$, then $\mathbf{G} = \tilde{\mathbf{A}}$ and

$$\tilde{\mathbf{G}} = \mathbf{A}^*\mathbf{A}\tilde{\mathbf{A}} = 16\begin{bmatrix} 6 & 3\mathbf{i} & 3\mathbf{i} \\ -2\mathbf{i} & 1 & 1 \\ -2\mathbf{i} & 1 & 1 \end{bmatrix}.$$

Furthermore, by (40),

$$w_{11}^{(2)} = \sum_{\beta \in J_{2,3}\{1\}} \left|(\mathbf{A}^*\mathbf{A})_{.1}(\tilde{\mathbf{g}}_{.1})\right|_\beta^\beta = \left(det\begin{bmatrix} 6 & 0 \\ -2\mathbf{i} & 2 \end{bmatrix} + det\begin{bmatrix} 6 & 0 \\ -2\mathbf{i} & 2 \end{bmatrix}\right) = 24.$$

By similar calculations, we get

$$\mathbf{w}_{1.}^{(2)} = [384, 96\mathbf{i}, 96\mathbf{i}], \;\; \mathbf{w}_{2.}^{(2)} = [-192\mathbf{i}, 96, 96], \;\; \mathbf{w}_{3.}^{(2)} = [-192\mathbf{i}, 96, 06].$$

So, by (36), we get

$$a_{11}^{c,\dagger} = \frac{\sum_{\alpha \in I_{2,3}\{1\}} \left|(\mathbf{AA}^*)_{1.}\left(\mathbf{w}_{1.}^{(2)}\right)\right|_\alpha^\alpha}{\left(\sum_{\alpha \in I_{2,3}} |\mathbf{AA}^*|_\alpha^\alpha\right)^2 \sum_{\beta \in J_{1,3}} |\mathbf{A}^3|_\beta^\beta}$$
$$= \frac{1}{4608}\left(det\begin{bmatrix} 384 & 192\mathbf{i} \\ -2\mathbf{i} & 3 \end{bmatrix} + det\begin{bmatrix} 384 & 192\mathbf{i} \\ -2\mathbf{i} & 3 \end{bmatrix}\right) = \frac{1}{3}.$$

By similarly continuing, we derive

$$\mathbf{A}^{c,\dagger} = \frac{1}{12}\begin{bmatrix} 4 & 2\mathbf{i} & 2\mathbf{i} \\ -2\mathbf{i} & 1 & 1 \\ -2\mathbf{i} & 1 & 1 \end{bmatrix}.$$

5. Conclusions

In this chapter, we get the direct method to find the core inverse and its generalizations that are based on their determinantal representations. New determinantal representations of the right and left core inverses, the right and left core-EP inverses, the DMP, MPD, and CMP inverses are derived.

Author details

Ivan I. Kyrchei
Pidstryhach Institute for Applied Problems of Mechanics and Mathematics, NAS of Ukraine, Lviv, Ukraine

*Address all correspondence to: kyrchei@online.ua

References

[1] Baksalary OM, Trenkler G. Core inverse of matrices. Linear and Multilinear Algebra. 2010;**58**:681-697

[2] Malik S, Thome N. On a new generalized inverse for matrices of an arbitrary index. Applied Mathematics and Computation. 2014;**226**:575-580

[3] Xu SZ, Chen JL, Zhang XX. New characterizations for core inverses in rings with involution. Frontiers in Mathematics China. 2017;**12**:231-246

[4] Prasad KM, Mohana KS. Core EP inverse. Linear and Multilinear Algebra. 2014;**62**(3):792-802

[5] Baksalary OM, Trenkler G. On a generalized core inverse. Applied Mathematics and Computation. 2014; **236**:450-457

[6] Mehdipour M, Salemi A. On a new generalized inverse of matrices. Linear and Multilinear Algebra. 2018;**66**(5): 1046-1053

[7] Chen JL, Zhu HH, Patrićio P, Zhang YL. Characterizations and representations of core and dual core inverses. Canadian Mathematical Bulletin. 2017;**60**:269-282

[8] Gao YF, Chen JL. Pseudo core inverses in rings with involution. Communications in Algebra. 2018;**46**: 38-50

[9] Guterman A, Herrero A, Thome N. New matrix partial order based on spectrally orthogonal matrix decomposition. Linear and Multilinear Algebra. 2016;**64**(3):362-374

[10] Ferreyra DE, Levis FE, Thome N. Maximal classes of matrices determining generalized inverses. Applied Mathematics and Computation. 2018;**333**:42-52

[11] Ferreyra DE, Levis FE, Thome N. Revisiting the core EP inverse and its extension to rectangular matrices. Quaestiones Mathematicae. 2018;**41**(2): 265-281

[12] Liu X, Cai N. High-order iterative methods for the DMP inverse. Journal of Mathematics. 2018;**8175935**:6

[13] Ma H, Stanimirović PS. Characterizations, approximation and perturbations of the core-EP inverse. Applied Mathematics and Computation. 2019;**359**:404-417

[14] Mielniczuk J. Note on the core matrix partial ordering. Discussiones Mathematicae Probability and Statistics 2011;**31**:71-75

[15] Mosić D, Deng C, Ma H. On a weighted core inverse in a ring with involution. Communications in Algebra. 2018;**46**(6):2332-2345

[16] Prasad KM, Raj MD. Bordering method to compute core-EP inverse. Special Matrices. 2018;**6**:193-200

[17] Rakić DS, Dinčić ČN, Djordjević DS. Group, Moore-Penrose, core and dual core inverse in rings with involution. Linear Algebra and its Applications. 2014;**463**:115-133

[18] Wang HX. Core-EP decomposition and its applications. Linear Algebra and its Applications. 2016;**508**:289-300

[19] Bapat RB, Bhaskara Rao KPS, Prasad KM. Generalized inverses over integral domains. Linear Algebra and its Applications. 1990;**140**:181-196

[20] Bhaskara Rao KPS. Generalized inverses of matrices over integral domains. Linear Algebra and its Applications. 1983;**49**:179-189

[21] Kyrchei I. Analogs of the adjoint matrix for generalized inverses and corresponding Cramer rules. Linear and Multilinear Algebra. 2008;**56**(4): 453-469

[22] Kyrchei I. Explicit formulas for determinantal representations of the Drazin inverse solutions of some matrix and differential matrix equations. Applied Mathematics and Computation. 2013;**219**:7632-7644

[23] Kyrchei I. Cramer's rule for generalized inverse solutions. In: Kyrchei I, editor. Advances in Linear Algebra Research. New York: Nova Science Publ; 2015. pp. 79-132

[24] Stanimirović PS. General determinantal representation of pseudoinverses of matrices. Matematichki Vesnik. 1996;**48**:1-9

[25] Stanimirović PS, Djordjevic DS. Full-rank and determinantal representation of the Drazin inverse. Linear Algebra and its Applications. 2000;**311**:131-151

[26] Kyrchei I. Determinantal representations of the Moore-Penrose inverse over the quaternion skew field. Journal of Mathematical Sciences. 2012; **180**(1):23-33

[27] Kyrchei I. Determinantal representations of the Moore-Penrose inverse over the quaternion skew field and corresponding Cramer's rules. Linear and Multilinear Algebra. 2011; **59**(4):413-431

[28] Kyrchei I. Determinantal representations of the Drazin inverse over the quaternion skew field with applications to some matrix equations. Applied Mathematics and Computation. 2014;**238**:193-207

[29] Kyrchei I. Determinantal representations of the W-weighted Drazin inverse over the quaternion skew field. Applied Mathematics and Computation. 2015;**264**:453-465

[30] Kyrchei I. Explicit determinantal representation formulas of W-weighted Drazin inverse solutions of some matrix equations over the quaternion skew field. Mathematical Problems in Engineering. 2016;**8673809**:13

[31] Kyrchei I. Explicit determinantal representation formulas for the solution of the two-sided restricted quaternionic matrix equation. Journal of Applied Mathematics and Computing. 2018;**58** (1-2):335-365

[32] Kyrchei I. Determinantal representations of the Drazin and W-weighted Drazin inverses over the quaternion skew field with applications. In: Griffin S, editor. Quaternions: Theory and Applications. New York: Nova Sci. Publ.; 2017. pp. 201-275

[33] Kyrchei I. Weighted singular value decomposition and determinantal representations of the quaternion weighted Moore-Penrose inverse. Applied Mathematics and Computation. 2017;**309**:1-16

[34] Kyrchei I. Determinantal representations of the quaternion weighted Moore-Penrose inverse and its applications. In: Baswell AR, editor. Advances in Mathematics Research 23. New York: Nova Science Publ; 2017. pp. 35-96

[35] Zhou M, Chen J, Li T, Wang D. Three limit representations of the core-EP inverse. Univerzitet u Nišu. 2018;**32**: 5887-5894

Chapter 2

New Matrix Series Formulae for Matrix Exponentials and for the Solution of Linear Systems of Algebraic Equations

Ioan R. Ciric

Abstract

The solution of certain differential equations is expressed using a special type of matrix series and is directly related to the solution of general systems of algebraic equations. Efficient formulae for matrix exponentials are derived in terms of rapidly convergent series of the same type. They are essential for two new solution methods, especially beneficial for large linear systems, namely an iterative method and a method based on an exact matrix product formula. The computational complexity of these two methods is analysed, and for both of them, the number of matrix exponential-vector multiplications required for an imposed accuracy can be predetermined in terms of the system condition. The total number of arithmetic operations involved is roughly proportional to n^2, where n is the matrix dimension. The common feature of all the series in the results presented is that starting with a first term that is already well-conditioned, each subsequent term is computed by multiplication with an even better conditioned matrix, tending quickly to the identity matrix. This contributes substantially to the stability of the numerical computation. A very efficient method based on the numerical integration of a special kind of differential equations, applicable to even ill-conditioned systems, is also presented.

Keywords: matrix equations, matrix exponentials, numerical solutions

1. Introduction

New matrix series expressions were recently derived by the author [1] for the solution of simple first order differential equations associated with general systems of linear algebraic equations. These differential equations describe the orthogonal trajectories of a family of hypersurfaces that represent a quadratic functional related to the linear algebraic system. The solution of the latter can be obtained by minimizing the functional along an orthogonal trajectory instead of applying various techniques based on minimization along conjugate gradient directions or based on minimized iterations [2]. Since the solutions of the differential equations considered are simply related to the solutions of the corresponding algebraic systems through matrix exponentials, there is the possibility to develop efficient solution methods if only the matrix exponentials could be used in numerical calculations accurately and with a small computational effort. A survey of various existent algorithms for computing matrix exponentials and a useful discussion of the difficulties involved are presented in [3].

In the present work, we use new formulae for arbitrary matrix exponentials that contain highly convergent infinite series which allow accurate and stable numerical computations. Employing these formulae, two new solution methods are proposed which are particularly efficient for large-scale general linear algebraic systems.

2. Differential equations associated with linear systems of algebraic equations

We start with simple vector differential equations whose solutions are related to the solution of general systems of linear algebraic equations. Later, in Section 5 we construct differential equations that allow an efficient numerical integration in order to obtain the solution of these systems.

2.1 Matrix series solution of some vector differential equations

Consider a first order vector differential equation of the form

$$\frac{dx}{dv} = f(v)(Ax - b) \tag{1}$$

where $A \in R^{n\times n}$ is a general nonsingular matrix, $b \in R^n$ is a given n-dimensional vector, $x = (x_1, x_2, ..., x_n)^T$ is the unknown n-dimensional vector, T indicates the transpose, and $f(v)$ is a continuous function over some interval of the real variable v. With the condition

$$x(v_o) = x_o \tag{2}$$

x_o being a given vector, (1) has a unique solution over the interval considered [4]. Integrating both sides of (1) from v_o to v yields

$$Ax(v) - b = e^{A[g(v)-g(v_o)]}(Ax_o - b) \tag{3}$$

where

$$g(v) \equiv \int f(v)\,dv \tag{4}$$

is a primitive for $f(v)$, i.e., $f(v) = dg/dv$. Thus, (1) can be written in the form

$$\frac{dx}{dv} = f(v)e^{A[g(v)-g(v_o)]}(Ax_o - b) \tag{5}$$

Integrating again both sides from v_o to v gives

$$x(v) = x_o + e^{-Ag(v_o)}\sum_{k=0}^{\infty}\frac{A^k}{(k+1)!}\left[(g(v))^{k+1} - (g(v_o))^{k+1}\right](Ax_o - b) \tag{6}$$

Two particular cases are presented.

If $f(v)$ is chosen to be $f(v) = -1$ then $g(v) = -v$. Choosing $v_o = 0$ gives $g(v_o) = 0$ and (6) becomes

$$x(v) = e^{-Av}x_o + v\sum_{k=0}^{\infty}(-1)^k \frac{(Av)^k}{(k+1)!}b \tag{7}$$

If $f(v) = 1/v$ then $g(v) = \ln v$. With $v_o = 1$ and $g(v_o) = 0$ (6) becomes now

$$x(v) = x_o + (\ln v)\sum_{k=0}^{\infty}\frac{(A\ln v)^k}{(k+1)!}(Ax_o - b) \tag{8}$$

The solution of (1) with $f(v) = 1/v$ can also be obtained by employing a Taylor series expansion about $x_o = 1$. Indeed,

$$x(v) = x(1) + \frac{(v-1)}{1!}x^{(1)}(1) + \frac{(v-1)^2}{2!}x^{(2)}(1) + \cdots + \frac{(v-1)^k}{k!}x^{(k)}(1) + \cdots \tag{9}$$

and evaluating the derivatives yields

$$x(v) = x_o - (1-v)\left[I + \sum_{k=1}^{\infty}\frac{(1-v)^k}{k+1}(I-A)\left(I-\frac{A}{2}\right)\cdots\left(I-\frac{A}{k}\right)\right](Ax_o - b) \tag{10}$$

where I is the identity matrix. The series in (10) is convergent for $v \in (0, 2)$ but its rate of convergence is, in general, very small for v very close to zero.

2.2 Relationship with linear systems of algebraic equations

Consider now a system of equations written in matrix form as

$$Ax = b \tag{11}$$

and assume that x is a continuous function of the real variable v over a certain interval. The solution of (11) can be obtained from (3) for a $v \equiv v_S$ for which

$$Ax(v_S) - b = 0 \tag{12}$$

To satisfy this condition $g(v)$ in (3) must be chosen such that

$$e^{A[g(v_s) - g(v_o)]} = 0 \tag{13}$$

For positive definite matrices A and a finite $g(v_o)$ this is achieved if $g(v) \to -\infty$ for $v \to v_S$. Thus, the solution of (11) cannot be computed directly from (6). On the other hand, in the particular case of $f(v) = 1/v, g(v) = \ln v$ the solution of (11) can be obtained from (10) with $v = v_S = 0$,

$$x_s = x_o - \left[I + \sum_{k=1}^{\infty}\frac{1}{k+1}(I-A)\left(I-\frac{A}{2}\right)\cdots\left(I-\frac{A}{k}\right)\right](Ax_o - b) \tag{14}$$

The expression in the brackets is just the inverse of A,

$$A^{-1} = I + \sum_{k=1}^{\infty} \frac{1}{k+1}(I - A)\left(I - \frac{A}{2}\right)\cdots\left(I - \frac{A}{k}\right) \tag{15}$$

The rate of convergence of the series in (14) and (15) is very small and they cannot be practically used in numerical computation for arbitrary matrices.

We note that the solution of (11) can be formally expressed from (6) as

$$x_s = x_o + e^{Ag(v_o)}\left(e^{Ag(v_o)} - e^{Ag(v)}\right)^{-1}\left(x(v) - x_o\right) \tag{16}$$

which is valid for any $v \neq v_o$ in the interval considered. To compute x_s with (16) would require the computation of $\left(e^{Ag(v_o)} - e^{Ag(v)}\right)^{-1}$. Equation (16) will be used in Subsection 4.1.

Note. The matrix product in the terms of the series in (10), (14) and (15) can be expressed as a matrix polynomial using the relationship with the Stirling numbers of the first kind $S_{k+1}^{(m)}$ [5]

$$(I - A)\left(I - \frac{A}{2}\right)\cdots\left(I - \frac{A}{k}\right) = (-1)^k \frac{1}{k!}\sum_{m=1}^{k+1} S_{k+1}^{(m)} A^{m-1} \tag{17}$$

While each new term in such series as those in (10), (14) and (15) is calculated through a multiplication with a matrix that becomes more and more well-conditioned as k increases, the computation with the expression in (17) would require successive multiplications with the same original matrix and, for each k, a new polynomial is to be constructed and new Stirling numbers have to be generated. The formulae derived in the next two sections contain the same type of series and, therefore, are simpler and more efficient to be used for numerical computations.

3. New formulae for computing matrix exponentials

In this section, we derive matrix exponential expressions which contain highly convergent infinite series that allow accurate and stable numerical computations in numerous applications. They shall also be used in the next section.

3.1 Series expressions for matrix exponentials

Consider the matrix function $v^A \equiv e^{A \ln v}$ where v is a positive real variable and A a general square matrix with real number entries. By integration,

$$\int_1^v v^{A-I} dv = \left.\frac{v^A}{A}\right|_{v=1}^{v} = \frac{1}{A}\left(e^{A\ln v} - I\right) \tag{18}$$

For $v \in (0, 2)$ the integrand can be expanded in a power series as

$$\begin{aligned} v^{A-I} &= \left[1-(1-v)\right]^{A-I} = I - \frac{(1-v)}{1!}(A-I) + \frac{(1-v)^2}{2!}(A-I)(A-2I) \\ &\quad - \frac{(1-v)^3}{3!}(A-I)(A-2I)(A-3I) + \cdots \end{aligned} \tag{19}$$

that can be integrated term by term. From (18) and (19)

$$e^{A\ln v} = I-(1-v)A\left[I+\sum_{k=1}^{\infty}\frac{(1-v)^k}{k+1}(I-A)\left(I-\frac{A}{2}\right)\cdots\left(I-\frac{A}{k}\right)\right] \tag{20}$$

which is valid for any A, positive definite or not, of arbitrary condition. One can see that, for a positive definite A and for $v \to 0$, the expression in the brackets of (20) gives a series expansion for the inverse A^{-1} (see (14) and (15)). Various expressions for the matrix exponential are obtained by giving particular values to v in (20). For example, for any $v = e^{-q}, q > -\ln 2$, we have

$$e^{-qA} = I-(1-e^{-q})A\left[I+\sum_{k=1}^{\infty}\frac{(1-e^{-q})^k}{k+1}(I-A)\left(I-\frac{A}{2}\right)\cdots\left(I-\frac{A}{k}\right)\right] \tag{21}$$

the series becoming less convergent as q increases above $q = 0$. On the other hand, with $v = 1/e$ in (20) and then A replaced with qA we obtain for any real number q

$$e^{-qA} = I-(1-e^{-1})qA\left[I+\sum_{k=1}^{\infty}\frac{(1-e^{-1})^k}{k+1}(I-qA)\left(I-\frac{qA}{2}\right)\cdots\left(I-\frac{qA}{k}\right)\right] \tag{22}$$

this series being more convergent than that in (21) for greater values of q.

For any $v \in (1,2)$, the terms in the series expressions derived from (20) have coefficients that are alternately positive and negative. With $v = e^{1/2} = 1.64872127$, e.g., and then replacing A with $2A$ we have

$$e^{A} = I+2(e^{1/2}-1)A\left[I+\sum_{k=1}^{\infty}\frac{(-1)^k(e^{1/2}-1)^k}{k+1}(I-2A)\left(I-\frac{2A}{2}\right)\cdots\left(I-\frac{2A}{k}\right)\right] \tag{23}$$

As well, by replacing A by $-qA$ in (23) one obtains instead of (22) an expression with alternating in sign series coefficients.

3.2 Rapidly convergent series formulae

From the basic Eq. (20) we derive now formulae which contain series that have a higher rate of convergence than those presented in the previous subsection.

Firstly, it is obvious that for values of v close to 1 the series in (20) has a high rate of convergence. For instance, with $v = 1+10^{-q}, 10^{-q} \ll 1$, and replacing $A\ln(1+10^{-q})$ by A we obtain

$$e^{A} = I+10^{-q}c_qA\left[I+\sum_{k=1}^{\infty}\frac{(-1)^k10^{-qk}}{k+1}(I-c_qA)\left(I-\frac{c_qA}{2}\right)\cdots\left(I-\frac{c_qA}{k}\right)\right] \tag{24}$$

where $c_q \equiv 1/\ln(1+10^{-q})$. This series is rapidly convergent.

Secondly, the convergence of the series in the expressions derived from (20) can further be improved by successive integrations. Indeed, integrating both sides of (20) from 1 to v, we have

$$\begin{aligned} ve^{A\ln v} = I+(1-v)(I+A)\Bigg[&-I+\frac{1-v}{2}A \\ &+A\sum_{k=1}^{\infty}\frac{(1-v)^{k+1}}{(k+1)(k+2)}(I-A)\left(I-\frac{A}{2}\right)\cdots\left(I-\frac{A}{k}\right)\Bigg] \end{aligned} \tag{25}$$

Integrating repeatedly we obtain the identity

$$
\begin{aligned}
v^p e^{A\ln v} &= I + p!\sum_{k=1}^{p}\frac{(-1)^k(1-v)^k}{k!(p-k)!}\left(I+\frac{A}{p}\right)\left(I+\frac{A}{p-1}\right)\cdots\left(I+\frac{A}{p-k+1}\right)\\
&+(-1)^{p+1}(1-v)^{p+1}A(I+A)\left(I+\frac{A}{2}\right)\cdots\left(I+\frac{A}{p}\right)\left[\frac{1}{p+1}I\right.\\
&\left.+p!\sum_{k=1}^{\infty}\frac{(1-v)^k}{(k+1)(k+2)\cdots(k+p+1)}(I-A)\left(I-\frac{A}{2}\right)\cdots\left(I-\frac{A}{k}\right)\right],\\
&\qquad p=1,2,\ldots;\quad v\in(0,2)
\end{aligned}
\tag{26}
$$

Same result is obtained by replacing A with $pI + A$ in (20). This identity contains an infinite series whose coefficients decrease rapidly as p increases. Obviously, for a given A and v, (26) generates more efficient computational formulae than those in the previous subsection. As before, for $v \in (1, 2)$ the infinite series have coefficients that alternate in sign. For example, with $v = e^{1/2}$ in (26) and then replacing A by $2A$ and p by $2p$ we have instead of (23)

$$
\begin{aligned}
e^A &= e^{-p}\left\{I+(2p)!\sum_{k=1}^{2p}\frac{(e^{1/2}-1)^k}{k!(2p-k)!}\left(I+\frac{2A}{2p}\right)\left(I+\frac{2A}{2p-1}\right)\cdots\left(I+\frac{2A}{2p-k+1}\right)\right.\\
&+2(e^{1/2}-1)^{2p+1}A(I+2A)\left(I+\frac{2A}{2}\right)\cdots\left(I+\frac{2A}{2p}\right)\left[\frac{1}{2p+1}I\right.\\
&\left.\left.+(2p)!\sum_{k=1}^{\infty}\frac{(-1)^k(e^{1/2}-1)^k}{(k+1)(k+2)\cdots(k+2p+1)}(I-2A)\left(I-\frac{2A}{2}\right)\cdots\left(I-\frac{2A}{k}\right)\right]\right\},\\
&\qquad p=1,2,\ldots
\end{aligned}
\tag{27}
$$

Taking $v = 1 + 10^{-q}$, with $q>0$ and $c_q \equiv 1/\ln(1 + 10^{-q})$, (26) gives (compare with (24))

$$
\begin{aligned}
e^A &= (1+10^{-q})^{-p}\left\{I+p!\sum_{k=1}^{p}\frac{10^{-qk}}{k!(p-k)!}\left(I+\frac{c_qA}{p}\right)\left(I+\frac{c_qA}{p-1}\right)\cdots\left(I+\frac{c_qA}{p-k+1}\right)\right.\\
&+10^{-(p+1)q}c_qA(I+c_qA)\left(I+\frac{c_qA}{2}\right)\cdots\left(I+\frac{c_qA}{p}\right)\left[\frac{1}{p+1}I\right.\\
&\left.\left.+p!\sum_{k=1}^{\infty}\frac{(-1)^k10^{-qk}}{(k+1)(k+2)\cdots(k+p+1)}(I-c_qA)\left(I-\frac{c_qA}{2}\right)\cdots\left(I-\frac{c_qA}{k}\right)\right]\right\},\\
&\qquad p=1,2,\ldots
\end{aligned}
\tag{28}
$$

Notice that, in the new formulae derived from (26) the infinite series are very rapidly convergent, with their rate of convergence increasing when the parameter p increases. Highly accurate numerical results can be generated with only a small number of terms retained in the infinite series of these formulae (see Section 4).

Note. All the formulae presented in this section remain valid if A is changed in $-A$. Obviously, in all these expressions A can be replaced by a real number and the identity matrix I by 1, yielding a few novel identities and summation formulae for series of real numbers. Also, the expression in the brackets of (26) for $v \to 0$ is just $(I + A/p)^{-1}/p$ if $I + A/p$ is positive definite and, thus, we obtain another identity, i.e.,

$$p!\sum_{k=1}^{p}\frac{(-1)^{k+1}}{k!(p-k)!}\left(I+\frac{A}{p}\right)\left(I+\frac{A}{p-1}\right)\cdots\left(I+\frac{A}{p-k+1}\right)$$
$$=I+\frac{(-1)^{p+1}}{p}A(I+A)\left(I+\frac{A}{2}\right)\cdots\left(I+\frac{A}{p-1}\right),\quad p=2,3,\ldots \tag{29}$$

which reduces for $A = 0$ to an elementary binomial sum.

4. Solution of general linear systems

In what follows, we apply the matrix exponential formulae from the previous section and present a new iteration procedure and a matrix product formula for the solution of large systems of linear algebraic equations.

4.1 An iterative method

Equation (16) can be written in the form

$$x(v)=x_o+\left(I-e^{A[g(v)-g(v_o)]}\right)\left(x_S-x_o\right) \tag{30}$$

where $x_S = x(v_S)$ is the solution of (11), $x_o = x(v_o)$, A is a positive definite matrix and $g(v)$ is a function of the real variable v such that $g(v) \to -\infty$ for $v \to v_s$ (see Subsection 2.2). To get $x(v)$ for values of v very close to v_S we choose an adequate $g(v)$ and a formula for the matrix exponential from Section 3. When $g(v_o) = 0$, applying (22) for instance gives

$$x(v)=x_o+(1-e^{-1})g(v)\left[I+\sum_{k=1}^{\infty}\frac{(1-e^{-1})^k}{k+1}(I+Ag(v))\right.$$
$$\left.\times\left(I+\frac{A}{2}g(v)\right)\cdots\left(I+\frac{A}{k}g(v)\right)\right](Ax_o-b) \tag{31}$$

If $g(v) \equiv \ln v$, with $v_o = 1$ and $v_s = 0$, we compute $x\left(e^{-N}\right)$ for $N \gg 1$ which is closer to the solution x_S,

$$x(e^{-N})\equiv x_S^{(1)}=x_S^{(0)}-N(1-e^{-1})\left[I+\sum_{k=1}^{\infty}\frac{(1-e^{-1})^k}{k+1}(I-NA)\right.$$
$$\left.\times\left(I-\frac{N}{2}A\right)\cdots\left(I-\frac{N}{k}A\right)\right](Ax_S^{(0)}-b) \tag{32}$$

where $x_S^{(0)} \equiv x_o$. This equation is applied iteratively by replacing $x_S^{(1)}$ and $x_S^{(0)}$, respectively, with $x_S^{(i)}$ and $x_S^{(i-1)}$, $i = 2, 3, \ldots$, until $x_S^{(i)}$ satisfies (11) with a desired accuracy.

To evaluate the amount of computation necessary to obtain the solution of (11) with a certain accuracy, let us take N such that $N\|A\| = 10$ when one needs about 30 terms in the infinite series, i.e., 30 matrix-vector multiplications. The number of iterations increases with the condition number of A. To see this and to determine the corresponding number of iterations, consider (11) with $b = 0$ and A replaced with a diagonal matrix whose entries are positive numbers, the greatest of these being 1, and whose condition is the same as that of A. The solution of this system is

$x_S = 0$ and the components of the solution of (1), with $f(v) = 1/v$, are $x_{ok}v^{\lambda_k}$ where $\lambda_k, k = 1, 2, ..., n$, are the entries in the diagonal matrix. In order to make the magnitudes of all these components at least 1%, e.g., of the corresponding magnitudes of the initial components, one needs no iteration if the condition number is less than 2, but 5 and, respectively, 46 iterations are needed if the condition number is 10 and 100.

What is remarkable in the iterative method based on (32) is that, for matrices with same condition number and same norm, the number of iterations required is the same, independently of the size of the matrices. Considering approximately $2n^2$ arithmetic operations for one matrix-vector multiplication, where n is the number of equations and unknowns in (11), the total number of arithmetic operations required is, thus, proportional to only n^2. In the examples given above one has to perform, respectively, $60n^2$, $300n^2$ and $2760n^2$ arithmetic operations. Assuming only $2n^3/3$ arithmetic operations for the Gaussian elimination procedure, the method presented in this subsection requires less computation for the same examples if, respectively, $n>90$, $n>450$ and $n>4140$. One can also notice that the application of Eq. (32) leads to the actual solution of (11) independently of the small error introduced in the computation at each iteration.

4.2 A matrix product formula

The original general system (11) is replaced with an equivalent system such that its solution is obtained in terms of matrix exponentials for which highly convergent and accurate series formulae have been derived in Section 3.

Namely, instead of (11) we use the system

$$\left(I - e^{-\alpha A}\right)x = b_\alpha \tag{33}$$

where α is a real scalar to be chosen, $\alpha \neq 0$, and

$$b_\alpha = \alpha\left[1 - \frac{(\alpha A)}{2!} + \frac{(\alpha A)^2}{3!} - \frac{(\alpha A)^3}{4!} + \cdots\right]b \tag{34}$$

Assuming A to be positive definite, α is taken positive. Then, since $\|e^{-\alpha A}\| < 1$ for a normal matrix, the solution can be expressed as

$$x_s = \left(I - e^{-\alpha A}\right)^{-1} b_\alpha = \left[\sum_{k=0}^{\infty} e^{-k\alpha A}\right] b_\alpha \tag{35}$$

with the norm of the matrix exponentials decreasing when k increases [6]

$$\left\|e^{-k\alpha A}\right\| < e^{-k\alpha\lambda} \tag{36}$$

where λ is the smallest eigenvalue of A. b_α can be accurately computed by using instead of (34) an equivalent expression, for instance (see (22))

$$b_\alpha = (1 - e^{-1})\alpha\left[I + \sum_{k=1}^{\infty} \frac{(1 - e^{-1})^k}{k+1}(I - \alpha A)\left(I - \frac{\alpha A}{2}\right)\cdots\left(I - \frac{\alpha A}{k}\right)\right]b \tag{37}$$

If the infinite series in (35) is truncated to $k = N_S$ the rest of the series has a norm

$$\left\| R_{N_s} \right\| < \frac{e^{-(N_s+1)\alpha\lambda}}{1-e^{-\alpha\lambda}} \tag{38}$$

Much less numerical computation (see below) is needed if the infinite series in (35) is transformed into an infinite product using the identity [5]

$$(1-u)^{-1} = \prod_{k=0}^{\infty}(1+u^{2^k}), \quad 0<u<1 \tag{39}$$

which is also valid for matrices whose norm is less than 1. Thus, (35) becomes

$$x_s = \left[\prod_{k=0}^{\infty}(1+e^{-2^k\alpha A}) \right] b_\alpha \tag{40}$$

with the norm of the exponentials $e^{-2^k\alpha A}$ decreasing very rapidly when k increases. Truncating the infinite product to $k = N_P$, i.e., $N_P + 1$ factors, leaves a remaining factor

$$F_{N_P} = \prod_{k=N_P+1}^{\infty} (I+e^{-2^k\alpha A}) \tag{41}$$

whose norm is

$$\left\| F_{N_P} \right\| < \prod_{k=N_P+1}^{\infty} (1+e^{-2^k\alpha\lambda}) = \frac{1}{1-e^{-\alpha\lambda}} \frac{1}{\prod_{k=0}^{N_P}(1+e^{-2^k\alpha\lambda})} \tag{42}$$

Let us compare the maximum value of the norm of the truncated matrix in the brackets of (35) and (40) with that of the corresponding untruncated matrix in order to get a rough estimate of the numbers N_S and N_P of matrix exponentials involved in the numerical computation to achieve a certain accuracy. This will also allow to estimate the total number of matrix-vector multiplications necessary to obtain the solution. The ratio of the maximum norm of the truncated matrix to the maximum norm of the untruncated matrix in the brackets of (35) and (40) is, respectively, (see (38) and (42))

$$\rho = 1-e^{-(N_S+1)\alpha\lambda} \tag{43}$$

and

$$\rho = (1-e^{-\alpha\lambda})\prod_{k=0}^{N_P}(1+e^{-2^k\alpha\lambda}) \tag{44}$$

To illustrate the computation complexity when using (35) or (40), assume that $\|A\| = 1$ and $\alpha = 20$. If ρ is imposed to be ≈ 0.99, for example, one needs $N_S = 2$, i.e., three terms in (35) and $N_P = 1$, i.e., two factors in (40) if $\lambda = 10^{-1}$. If $\lambda = 10^{-2}$ these numbers increase to $N_S = 23$ and $N_P = 4$, and when $\lambda = 10^{-3}$ one gets $N_S = 230$, but N_P only increases to $N_P = 7$. It is clear that applying the formula (40) the number of exponentials needed in the numerical computation is much smaller than that if formula (35) would be applied. For all the matrix exponentials involved in the numerical computation we use the formula (28) containing a highly convergent series, such that

$$e^{-\alpha A}=(1+10^{-q})^{-p}\left\{I+p!\sum_{k=1}^{p}\frac{10^{-qk}}{k!(p-k)!}\left(I-\frac{c_q\alpha A}{p}\right)\left(I-\frac{c_q\alpha A}{p-1}\right)\cdots\left(I-\frac{c_q\alpha A}{p-k+1}\right)\right.$$
$$-10^{-(p+1)q}c_q\alpha A(I-c_q\alpha A)\left(I-\frac{c_q\alpha A}{2}\right)\cdots\left(I-\frac{c_q\alpha A}{p}\right)\left[\frac{I}{p+1}\right. \tag{45}$$
$$\left.\left.+p!\sum_{k=1}^{\infty}\frac{(-1)^k10^{-qk}k!}{(k+p+1)!}(I+c_q\alpha A)\left(I+\frac{c_q\alpha A}{2}\right)\cdots\left(I+\frac{c_q\alpha A}{k}\right)\right]\right\},$$
$$p=1,2,\ldots$$

where $q>o$ and $c_q\equiv 1/\ln(1+10^{-q})$. With $\alpha\|A\|=20$ and choosing $q=1$ and $p=10$, for instance, $e^{-\alpha A}$ is determined accurately by retaining 50 terms in the infinite series and, thus, to multiply $e^{-\alpha A}$ with a vector one needs 71 matrix-vector multiplications. To compute x_s from (40) one has to use repeatedly the multiplication of e^{-20A} with a vector. For a matrix A with $\lambda=10^{-1}$ one has to retain two factors in (40) and, thus, to multiply e^{-20A} and $e^{-2\times 20A}$ with a vector. This means to use repeatedly 3 times the multiplication of e^{-20A} with a vector which requires, therefore, $3\times 71=213$ matrix-vector multiplications. When $\lambda=10^{-2}$ the infinite product in (40) is truncated at $k=N_P=4$ and this requires the multiplication of $e^{-2^k\times 20A}$, $k=0,1,2,3,4$, with a vector, i.e., to use repeatedly 31 times the multiplication of e^{-20A} with a vector, for a total of $31\times 71=2201$ matrix-vector multiplications. We also have to add the matrix-vector multiplications required to compute b_α in (37). A very accurate result for b_α when $\alpha=20$ can be achieved by applying four times the series in the brackets of (37) for $\alpha=5$, each time retaining 30 terms. This requires a total of about 120 multiplications of a matrix $I-5A/k$ with a vector. In all the matrix-vector multiplications involved when applying (45), the matrices are in the form $I\pm c_q\alpha A/k$ and become better and better conditioned as k increases.

Adding up the number of arithmetic operations involved shows that, with respect to the classical Gaussian elimination method, the procedure presented in this subsection is advantageous for very large systems (11). Namely, assuming same accuracy and only $2n^3/3$ arithmetic operations for the Gaussian elimination, with the data given above, one has to have $n>3\times(213+120)=999$ equations and unknowns if $\lambda=10^{-1}$ and $n>3\times(2201+120)=6963$ equations and unknowns if $\lambda=10^{-2}$ for the proposed method to be more advantageous. For a given $\alpha\|A\|$, one application of $e^{-\alpha A}$ requires a determined finite number of matrix-vector multiplications, independently of the size of A. It is remarkable, as for the iterative method in the previous subsection, that for a given condition of A, one has to apply $e^{-\alpha A}$ a well-determined number of times and, thus, the total number of arithmetic operations necessary to compute the solution with an imposed accuracy is proportional to only n^2.

It should be noted that, since the infinite series in the expression (45) is truncated and thus determined with a finite accuracy, the accuracy of the solution x_S becomes compromised after a too big a number of matrix exponential-vector multiplications. This is why, the worse conditioned systems (11) should be appropriately preconditioned. Practically, the computation with (40) is continued factor by factor and the accuracy of x is checked after each step.

5. Solution of general linear systems by numerical integration of differential equations

In this section, we introduce first order differential equations whose numerical integration allows to efficiently find the solution of linear systems of algebraic equations. Differential equations of the type of those in (1), with $f(v) = -1$ or $f(v) = 1/v$, cannot be used for this purpose due to the fact that the first and higher order derivatives of $x(v)$ tend to infinite values as x tends to the solution x_S of (11) (see Section 2).

Here below, we construct ordinary differential equations for $x(v)$ which satisfy the condition that the first few derivatives are finite when $x(v)$ tends to x_S and, therefore, are particularly useful for an accurate computation of x_S. Let us consider the system (11) with a symmetric positive definite matrix. A quadratic functional

$$F(x) = \frac{1}{2} x^T A x - x^T b + \frac{1}{2} b^T A^{-1} b \tag{46}$$

is associated with (11) [6] whose minimum value is $F(x_S) = 0$. Define now a real variable $v, v \geq 0$, such that

$$F(x) = v^r \tag{47}$$

where r is a real number to be chosen, $r>0$, with $v = 0$ corresponding to the solution $x = x_S$ and $v = v_o$ to an initial point $x_o, F(x_o) = v_o^r$. Then,

$$\frac{dF}{dv} = r v^{r-1} = (Ax - b)^T \frac{dx}{dv} \tag{48}$$

and, thus,

$$\frac{dx}{dv} = r v^{r-1} \frac{Ax - b}{\|Ax - b\|^2} \tag{49}$$

This is the differential equation to be integrated from $v = v_o$ to $v = v_S = 0$. The second derivative of x is obtained in the form

$$\begin{aligned} \frac{d^2x}{dv^2} = r(r-1)v^{r-2} \frac{Ax - b}{\|Ax - b\|^2} + (r v^{r-1})^2 \frac{1}{\|Ax - b\|^4} \Bigg\{ A \\ -2 \frac{\left[(Ax - b)^T A (Ax - b) \right]}{\|Ax - b\|^2} \Bigg\} (Ax - b) \end{aligned} \tag{50}$$

Higher order derivatives can be worked out if needed.

In order to see the behaviour of the derivatives close to the solution x_S, Eqs. (49) and (50) are rewritten as

$$\frac{dx}{dv}=r\frac{(F(x))^{1-\frac{1}{r}}(Ax-b)}{\left\|Ax-b\right\|^2} \tag{51}$$

and

$$\frac{d^2x}{dv^2}=r(r-1)\frac{(F(x))^{1-\frac{2}{r}}(Ax-b)}{\left\|Ax-b\right\|^2}+r^2\frac{(F(x))^{2-\frac{2}{r}}}{\left\|Ax-b\right\|^4}\left\{A-2\frac{\left[(Ax-b)^T A(Ax-b)\right]}{\left\|Ax-b\right\|^2}\right\}(Ax-b) \tag{52}$$

with $F(x)$ in (46) put in the form

$$F(x)=\frac{1}{2}(Ax-b)^T(x-x_S) \tag{53}$$

Notice that as x tends to x_s, when $\|Ax-b\|\equiv\varepsilon$ tends to zero,

$$\left\|\frac{dx}{dv}\right\|\sim K_1(v)\varepsilon^{1-\frac{2}{r}},\quad \left\|\frac{d^2x}{dv^2}\right\|\sim K_2(v)\varepsilon^{1-\frac{4}{r}} \tag{54}$$

where $K_1(v)$ and $K_2(v)$ are finite when $v\rightarrow 0$. Therefore, as $x\rightarrow x_S$, $\|dx/dv\|\rightarrow 0$ if $r>2$ and $\|d^2x/dv^2\|\rightarrow 0$ if $r>4$.

Another differential equation we present here is

$$\frac{dx}{ds}=\frac{Ax-b}{\left\|Ax-b\right\|} \tag{55}$$

with the second derivative

$$\frac{d^2x}{ds^2}=\frac{1}{\left\|Ax-b\right\|^2}\left\{A-\frac{[(Ax-b)^T A(Ax-b)]}{\left\|Ax-b\right\|^2}\right\}(Ax-b) \tag{56}$$

In this case, always

$$\left\|\frac{dx}{ds}\right\|=1 \tag{57}$$

even for $x\rightarrow x_S$, but the second derivative tends to an infinite value when $x\rightarrow x_S$

$$\left\|\frac{d^2x}{ds^2}\right\|\sim K(s)\varepsilon^{-1} \tag{58}$$

where $K(s)$ is finite and $\varepsilon\equiv\|Ax-b\|$. The relationship between the differentials of the variables v and s in (49) and (55) is

$$ds = \frac{r(F(x))^{1-\frac{1}{r}}}{\|Ax - b\|} dv \tag{59}$$

and for $x \to x_S$ we have (see (54))

$$\frac{ds}{dv} \sim K_1(v)\varepsilon^{1-\frac{2}{r}} \tag{60}$$

The differential Eqs. (49) in v and (55) in s require practically the same amount of computation for their right-hand sides, i.e., one matrix-vector multiplication. The first derivatives dx/dv for $r = 2$ and dx/ds remain finite when x tends to x_S, while the second derivatives increase to infinite values as in (54) and (58). For $r = 4$ the second derivative in (52) remains finite when x tends to x_S (see (54)), while the first derivative and the ratio ds/dv tend to zero as in (54) and (60), respectively. If $r>4$ even $\|d^2x/dv^2\|$ tends to zero as in (54).

Equations (49) and (55) can be integrated by classical numerical methods. Since we are not looking for an accurate solution of these equations all along from x_o to x_S but for finding accurately the final value $x = x_S$, we can use a lower order method, for instance, even Euler's method [7]. This yields an approximate value of x_S which is to be used as initial point for repeating the numerical integration procedure. As we get closer to the solution x_S, we decrease the step size in order to reduce the error. In the case of Euler's method the error is determined in terms of the norm of the second derivative. Higher order numerical integration methods can also be used in order to increase the computation efficiency.

To find a starting point for the integration procedure which is reasonably close to the solution point, one can minimize $F(x)$ in (46) along the normal direction, followed by a minimization of the distance to the solution point x_S along the direction of the normal to F [8]. These two preliminary steps are repeated a few times as needed.

Numerical experiments have been performed applying Euler's method to (49) for $r = 2$, $r = 4$ and $r = 8$, and to (55). Systems (11) of various sizes have been automatically generated and the differential Eqs. (49) for $r = 2$ and $r = 4$, and (55) have produced results with the least amount of computation when imposing an accuracy of 1%.

For matrices which are not symmetric positive definite, (46) is replaced with $F(x) = 1/2\,(Ax - b)^T(Ax - b)$.

6. Conclusions

A special type of matrix series are used in Section 2 to express the relationship between some first order ordinary differential equations and systems of linear algebraic equations and, also, in Section 3 to derive efficient formulae for matrix exponentials that allow accurate and stable numerical computations in various applications. The main feature of these series consists in the fact that, starting with their first term which is already a matrix substantially better conditioned than the original problem matrix, each of the subsequent terms is obtained through a multiplication with a better and better conditioned matrix that tends to the identity matrix. The new matrix exponential formulae contain very rapidly convergent series and can be applied to general, arbitrarily conditioned, positive definite or not matrices. They are used in Section 4 for two new methods of solution for general linear algebraic systems. One is an iterative method which corresponds to the solution of the differential Eq. (1) with $f(v) = 1/v$. It is based on the exact analytical expressions (30)–(32) that always yield results converging finally to the exact solution of the system (11). In a second method, the original algebraic system (11) is replaced with an equivalent system containing a matrix exponential $e^{-\alpha A}$ such that instead of inverting the system matrix A we have now to invert $I - e^{-\alpha A}$. The exact analytical solution is obtained in the form of a series of matrix exponentials which is transformed into an infinite matrix product in order to reduce substantially the necessary amount of computation. It should be remarked that, since the number of matrix-vector multiplications required for the application of one matrix exponential-vector multiplication only depends on the norm of the matrix while the number of matrix exponential-vector multiplications depends on the condition of the system matrix, the total number of arithmetic operations needed to achieve an imposed accuracy when applying each of the two methods is practically proportional to n^2, where n is the dimension of the matrix. The two methods require a comparable total amount of computation. It is also remarkable that for both methods the necessary amount of computation can be roughly predicted beforehand in terms of the system size, the system condition and the desired accuracy.

In Section 5, a powerful method is presented based on the numerical integration of specially constructed ordinary differential equations.

Author details

Ioan R. Ciric
The University of Manitoba, Winnipeg, Canada

*Address all correspondence to: ioan.ciric@umanitoba.ca

References

[1] Ciric IR. Rapidly convergent matrix series for solving linear systems of equations. In: Proceedings of the 17th International Symposium on Antenna Technology and Applied Electromagnetics (ANTEM); 10-16 July 2016; Montréal, Canada: IEEE; 2016. DOI: 10.1109/ANTEM.2016.755022/ 978-1-4673-8478-0

[2] Lanczos C. Applied Analysis. New York: Dover Publications, Inc.; 1988. ISBN: 0-486-65656-X (Paperback). CIP: 88-3961

[3] Moler CB, Van Loan CF. Nineteen dubious ways to compute the exponential of a matrix. SIAM Review. 1978;**20**:801-836. DOI: 10.1137/ S00361445024180

[4] Hurewicz W. Lectures on Ordinary Differential Equations. New York: Dover; 1990. ISBN: 0-486-66420-I (Paperback). QA372.H93

[5] Gradshteyn IS, Ryzhik IM. Table of Integrals, Series and Products. 5th ed. New York: Academic Press; 1994. ISBN: 0-12-294755-X. QA55.G6613

[6] Golub GH, Van Loan CF. Matrix Computations. 2nd ed. Baltimore: Johns Hopkins; 1989. ISBN: 0-8018-3772-3, 0-8018-3739-1 (Paperback). 88-4504

[7] Burden RL, Faires JD. Numerical Analysis. 5th ed. Boston: PWS-KENT; 1993. ISBN: 0-534-93219-3. CIP: 92-32192

[8] Ciric IR. A geometric property of the functional associated with general systems of algebraic equations. In: Proceedings of the International Symposium on Antenna Technology and Applied Electromagnetics and Canadian Radio Sciences Conference (ANTEM/URSI); 17-19 July 2006; Montréal, Canada: IEEE; 2006. p. 311-315. ISBN: 978-0-9738425-1-7

Chapter 3

Fixed Point Theorems of a New Generalized Nonexpansive Mapping

Shi Jie

Abstract

This paper introduces a $T-(D_a)$ mapping that is weaker than the nonexpansive mapping. It introduces several iterations for the fixed point of the $T-(D_a)$ mapping. It gives fixed point theorems and convergence theorems for the $T-(D_a)$ mapping in Banach space, instead of uniformly convex Banach space. This paper gives some basic properties on the $T-(D_a)$ mapping and gives the example to show the existence of $T-(D_a)$ mapping. The results of this paper are obtained in general Banach spaces. It considers some sufficient conditions for convergence of fixed points of mappings in general Banach spaces under higher iterations.

Keywords: iteration, convergence theorems, nonexpansive mapping, fixed point

2010 MSC: 47H09, 47H10

1. Introduction

In this paper, E is a Banach space, C is a nonempty closed convex subset of E, and $Fix(T) = x \in C : Tx = x$.

Definition 1. T is contraction mapping if there is $r \in [0, 1)$

$$\|Tx - Ty\| \leq r\|x - y\| \text{ for all } x \ \ y \in C.$$

Definition 2. T is nonexpansive mapping if

$$\|Tx - Ty\| \leq \|x - y\| \text{ for all } x \ \ y \in C.$$

Definition 3. T is quasinonexpansive mapping if

$$\|Tx - Ty\| \leq \|x - y\| \text{ for all } x \in C, y \in F(T).$$

Definition 4. $T : C \to C$ is a $T-(D_a)$ mapping on a subset C, if there is $a \in \left(\frac{1}{2}, 1\right)$, $\|Tx - Ty\| \leq \|x - y\|$ for all $\alpha \in [a, 1], x \in C, y \in C(T, x, \alpha)$, where $C(T, x, \alpha) = \{y \in C | y = (1-\alpha)p + \alpha Tp, p \in C, \|Tp - p\| \leq \|Tx - x\|\}$.

In 2008 Suzuki [1] defined a mapping T in Banach space: $\frac{1}{2}\|Tx - Ty\| \leq \|x - y\|$ implies $\|Tx - Ty\| \leq \|x - y\|$. And T is said to satisfy condition (C). Suzuki [1] showed that the mapping satisfying condition (C) is weaker than nonexpansive mapping and stronger than quasinonexpansive mapping.

Suzuki [1] proved the theorem T is a mapping in Banach space, T satisfies condition (C), and $\{x_n\}$ is the sequence defined by the iteration process:

$$\begin{cases} x_1 = x \in C, \\ x_{n+1} = (1-\alpha_n)x_n + \alpha_n T x_n, \end{cases} \tag{1}$$

then $\{x_n\}$ converges to a fixed point of T.

Suzuki [1] gave this convergence theorem in an ordinary Banach space, and the mapping satisfying condition (C) is weaker than nonexpansive mapping.

In 2016, Thakur [2] proved the theorem T is a mapping in uniformly convex Banach space, T satisfies condition (C), and $\{x_n\}$ is the sequence defined by iteration process:

$$\begin{cases} x_1 = x \in C, \\ x_{n+1} = T y_n, \\ y_n = T z_n, \\ z_n = (1-\alpha_n)x_n + \alpha_n T x_n, \end{cases} \tag{2}$$

then $\{x_n\}$ converges to a fixed point of T.

Thakur [2] claimed that the rate of iteration is fastest of known iterations. However, the disadvantage is that their results must be in uniformly convex Banach space, instead of the ordinary Banach space.

The aim of this article is there exists a generalized nonexpansive mapping, which makes the sequence generated by Thakur's iteration converge to a fixed point in a general Banach space.

The following propositions are obvious:

Proposition 1. If T is nonexpansive, then T satisfies condition (D_a).

Proposition 2. If T is $T-(D_a)$ mapping, then T is quasinonexpansive.

Proposition 3. Suppose $T: C \to C$ is a $T-(D_a)$ mapping. Then, for $x, y \in C$:

(1) $\|T^2x - Tx\| \leq \|Tx - x\|$ for all $x \in C$.

(2) $\|T^2x - Ty\| \leq \|Tx - y\|$ or $\|T^2y - Tx\| \leq \|Ty - x\|$ for all $x, y \in C$.

Proof:

(1) Since $\|Tx - x\| \leq \|Tx - x\|$, $Tx \in C(T, x, 1)$, we have $\|T^2x - Tx\| \leq \|Tx - x\|$.

(2) For all $x, y \in C$, $\|Tx - x\| \leq \|Ty - y\|$ or $\|Ty - y\| \leq \|Tx - x\|$.

Then $Tx \in C(T, y, \alpha)$ or $Ty \in C(T, x, \alpha)$.

It follows that $\|T^2x - Ty\| \leq \|Tx - y\|$ or $\|T^2y - Tx\| \leq \|Ty - x\|$.

Example 1

$$Tx = \begin{cases} \begin{pmatrix} 1.1 & x_2 \\ x_3 & x_4 \end{pmatrix}, & x_1 = 3, \\ \begin{pmatrix} 0 & x_2 \\ x_3 & x_4 \end{pmatrix}, & x_1 \neq 3, \end{cases}$$

where

$$x = \begin{pmatrix} x_1 & x_2 \\ x_3 & x_4 \end{pmatrix}, x_1 \in [0,3], x_2 \in [0, 0.01], x_3 \in [0, 0.01], x_4 \in [0, 0.01].$$

$$\|x\|_1 = \max\{|x_1| + |x_3|, |x_2| + |x_4|\}$$

Set

$$x = \begin{pmatrix} 3 & 0 \\ 0 & 0 \end{pmatrix}$$

and

$$y = \begin{pmatrix} 2.5 & 0 \\ 0 & 0 \end{pmatrix}$$

We see that

$$\|Tx - Ty\|_1 = 1.1 > \|x - y\|_1.$$

Hence, T is not a nonexpansive mapping.
To verify that T is a $T - (D_a)$ mapping, consider the following cases:
Case 1:

$$\alpha \in \left[\frac{11}{19}, 1\right], x = \begin{pmatrix} x_1 & x_2 \\ x_3 & x_4 \end{pmatrix}, x_1 \neq 3.$$
$$y = \begin{pmatrix} y_1 & y_2 \\ y_3 & y_4 \end{pmatrix} \in C(T, x, \alpha),$$

then $y_1 \neq 3$. We have

$$\|Ty - Tx\| = \|\begin{pmatrix} 0 & y_2 - x_2 \\ y_3 - x_3 & y_4 - x_4 \end{pmatrix}\| \leq \|\begin{pmatrix} y_1 - x_1 & y_2 - x_2 \\ y_3 - x_3 & y_4 - x_4 \end{pmatrix}\| = \|y - x\|$$

Case 2:

$$\alpha \in \left[\frac{11}{19}, 1\right], x = \begin{pmatrix} x_1 & x_2 \\ x_3 & x_4 \end{pmatrix}, x_1 = 3.$$
$$y = \begin{pmatrix} y_1 & y_2 \\ y_3 & y_4 \end{pmatrix} \in C(T, x, \alpha),$$

then $y_1 \in [0, 1.9]$. We have

$$\|Ty - Tx\| = \|\begin{pmatrix} 1.1 & y_2 - x_2 \\ y_3 - x_3 & y_4 - x_4 \end{pmatrix}\| \leq 1.11 \leq \|y - x\|$$

Hence, T is a $T - (D_a)$ mapping, and T is not nonexpansive.

2. Fixed point

In this section, we prove convergence theorems for fixed point of the $T - (D_a)$ mapping in Banach space.

Lemma 1. Let C be bounded convex subset of a Banach space B. Assume that $T : C \to C$ is $T - (D_a)$ mapping and $\{x_n\}$, $\{y_n\}$, $\{z_n\}$ are sequences generated by iteration:

$$\begin{cases} x_1 = x \in C, \\ x_{n+1} = Ty_n, \\ y_n = Tz_n, \\ z_n = (1-\alpha_n)x_n + \alpha_n Tx_n, \end{cases} \tag{3}$$

where $\frac{1}{2} < a \le \alpha_n \le b < 1$. Then
(1) $\|Tx_{n+1} - x_{n+1}\| \le \|Ty_n - y_n\| \le \|Tz_n - z_n\| \le \|Tx_n - x_n\|$.
(2) $\lim_{n\to\infty} \|Tx_n - x_n\| = \lim_{n\to\infty} \|Ty_n - y_n\| = \lim_{n\to\infty} \|Tz_n - z_n\| = r \ge 0$.

Proof: (1) From Proposition 3 and $z_n = (1-\alpha_n)x_n + \alpha_n Tx_n$, we have

$$\begin{aligned} \|Tx_{n+1} - x_{n+1}\| &\le \|T^2 y_n - Ty_n\| \\ &\le \|Ty_n - y_n\| = \|T^2 z_n - Tz_n\| \\ &\le \|Tz_n - z_n\| \\ &= \|Tz_n - Tx_n + (1-\alpha_n)(Tx_n - x_n)\| \\ &\le \|z_n - x_n\| + (1-\alpha_n)\|Tx_n - x_n\| \\ &= \|Tx_n - x_n\|. \end{aligned}$$

(2) From (1), we have $0 \le \|Tx_{n+1} - x_{n+1}\| \le \|Tx_n - x_n\|$. So $\lim_{n\to\infty} \|Tx_n - x_n\| = r \ge 0$. Now, we have $\lim_{n\to\infty} \|Tx_n - x_n\| = \lim_{n\to\infty} \|Ty_n - y_n\| = \lim_{n\to\infty} \|Tz_n - z_n\| = r \ge 0$.

Lemma 2. Assume that $T : C \to C$ is a $T - (D_a)$ mapping and $\{x_n\}$, $\{y_n\}$, $\{z_n\}$ are sequences generated by iteration (3). $\frac{1}{2} < a \le \alpha_n \le b < 1$. Let $\{u_m\}$ satisfy $u_{3n-2} = x_n, u_{3n-1} = z_n, u_{3n} = y_n$. Then, for all $n \ge 1, p \ge 1$

$$\begin{aligned} \left(1 + \sum_{k=3n-2}^{3n+p-3} \beta_k\right) \|Tu_{3n-2} - u_{3n-2}\| &\le \|Tu_{3n-2+p} - u_{3n-2}\| \\ &+ \left(\prod_{k=n}^{n+p-1} \frac{2}{1-\alpha_k}\right)\left(\|Tu_{3n-2} - u_{3n-2}\| - \|Tu_{3n-2+3p} - u_{3n-2+3p}\|\right), \end{aligned} \tag{4}$$

where

$$\beta_k = \begin{pmatrix} \alpha_n & k = 3n-2 \\ 1 & k \ne 3n-2 \end{pmatrix}$$

Proof: From **Lemma 1**, we have

$$\begin{aligned} &\|Tx_{n+1} - x_{n+1}\| \\ &\le \|Tz_n - z_n\| \\ &= \|Tz_n - (1-\alpha_n)x_n - \alpha_n Tx_n\| \\ &\le (1-\alpha_n)\|Tz_n - x_n\| + \alpha_n\|Tz_n - Tx_n\| \\ &\le (1-\alpha_n)\|Tz_n - x_n\| + \alpha_n\|z_n - x_n\| \\ &= (1-\alpha_n)\|Tz_n - x_n\| + \alpha_n^2\|Tx_n - x_n\|. \end{aligned}$$

So, for $p = 1$ and all $n \geq 1$

$$\begin{aligned}
&(1+\beta_{3n-2})\|Tu_{3n-2}-u_{3n-2}\| \\
&= (1+\alpha_n)\|Tx_n - x_n\| \\
&\leq \|Tx_n - x_n\| + \left(\frac{1}{1-\alpha_n}\right)(\|Tx_n - x_n\| - \|Tx_{n+1} - x_{n+1}\|) \\
&= \|Tu_{3n-1} - u_{3n-2}\| + \left(\frac{1}{1-\alpha_n}\right)(\|Tu_{3n-2} - u_{3n-2}\| - \|Tu_{3n+1} - u_{3n+1}\|) \\
&\leq \|Tu_{3n-1} - u_{3n-2}\| + \left(\frac{2}{1-\alpha_n}\right)(\|Tu_{3n-2} - u_{3n-2}\| - \|Tu_{3n+1} - u_{3n+1}\|).
\end{aligned}$$

(4) holds.

We make the inductive assumption that (4) holds for a given $p > 1$ and all $n > 0$ and obtain, upon replacing n with $n+1$

$$\begin{aligned}
&\left(1+\sum_{k=3n+1}^{3n+p}\beta_k\right)\|Tu_{3n+1}-u_{3n+1}\| \leq \|Tu_{3n+1+p} - u_{3n+1}\| \\
&+\left(\prod_{k=n+1}^{n+p}\frac{2}{1-\alpha_k}\right)\left(\|Tu_{3n+1}-u_{3n+1}\| - \|Tu_{3n+1+3p} - u_{3n+1+3p}\|\right).
\end{aligned} \tag{5}$$

And obviously

$$k \geq 3n-2,\ \|Tu_{k+1} - Tu_k\| \leq \beta_k\|Tu_{3n-2} - u_{3n-2}\|, \tag{6}$$

$$k > t,\ \|Tu_k - Tu_t\| \leq \|u_k - u_t\|. \tag{7}$$

Case 1: $p = 3m, m \geq 1$. From (6) and (7)

$$\begin{aligned}
&\|Tu_{3n+1+p} - u_{3n+1}\| \\
&= \|Tx_{n+m+1} - x_{n+1}\| \\
&= \|Tx_{n+m+1} - Ty_n\| \\
&\leq \|x_{n+m+1} - y_n\| \\
&= \|Ty_{n+m} - Tz_n\| \\
&\leq \|y_{n+m} - z_n\| \\
&= \|Tz_{n+m} - (1-\alpha_n)x_n - \alpha_n Tx_n\| \\
&\leq (1-\alpha_n)\|Tz_{n+m} - x_n\| + \alpha_n\|Tz_{n+m} - Tx_n\| \\
&\leq (1-\alpha_n)\|Tz_{n+m} - x_n\| \\
&+\alpha_n\left(\|Tz_{n+m} - Tx_{n+m}\| + \|Tx_{n+m} - Ty_{n+m-1}\| + \cdots + \|Ty_n - Tz_n\| + \|Tz_n - Tx_n\|\right) \\
&= (1-\alpha_n)\|Tu_{3n-1+p} - u_{3n-2}\| + \alpha_n\sum_{k=3n-2}^{3n-2+p}\|Tu_{k+1} - Tu_k\| \\
&\leq (1-\alpha_n)\|Tu_{3n-1+p} - u_{3n-2}\| + \alpha_n\sum_{k=3n-2}^{3n-2+p}\beta_k\|Tu_{3n-2} - u_{3n-2}\|.
\end{aligned}$$

It follows that

$$\|Tu_{3n+1+p} - u_{3n+1}\| \leq (1-\alpha_n)\|Tu_{3n-1+p} - u_{3n-2}\| + \alpha_n\sum_{k=3n-2}^{3n-2+p}\beta_k\|Tu_{3n-2} - u_{3n-2}\| \tag{8}$$

Using (5) and (8), we have

$$\left(1+\sum_{k=3n+1}^{3n+p}\beta_k\right)\|Tu_{3n+1}-u_{3n+1}\|$$
$$\leq(1-\alpha_n)\|Tu_{3n-1+p}-u_{3n-2}\|+\alpha_n\sum_{k=3n-2}^{3n-2+p}\beta_k\|Tu_{3n-2}-u_{3n-2}\|$$
$$+\left(\prod_{k=n+1}^{n+p}\frac{2}{1-\alpha_k}\right)\left(\|Tu_{3n+1}-u_{3n+1}\|-\|Tu_{3n+1+3p}-u_{3n+1+3p}\|\right).$$

From $\left(1+\sum_{k=3n+1}^{3n+p}\beta_k\right)\leq\left(\prod_{k=n+1}^{n+p}\frac{1}{1-\alpha_k}\right)$ and $\|Tu_{3n+1}-u_{3n+1}\|\leq\|Tu_{3n-2}-u_{3n-2}\|$, we have

$$\left(1+\sum_{k=3n+1}^{3n+p}\beta_k\right)\|Tu_{3n-2}-u_{3n-2}\|$$
$$\leq(1-\alpha_n)\|Tu_{3n-1+p}-u_{3n-2}\|+\alpha_n\sum_{k=3n-2}^{3n-2+p}\beta_k\|Tu_{3n-2}-u_{3n-2}\|$$
$$+\left(\prod_{k=n+1}^{n+p}\frac{2}{1-\alpha_k}\right)\left(\|Tu_{3n-2}-u_{3n-2}\|-\|Tu_{3n+1+3p}-u_{3n+1+3p}\|\right).$$

Then

$$\frac{\left(1+\sum_{k=3n+1}^{3n+p}\beta_k-\alpha_n\sum_{k=3n-2}^{3n-2+p}\beta_k\right)}{1-\alpha_n}\|Tu_{3n-2}-u_{3n-2}\|$$
$$\leq\|Tu_{3n-1+p}-u_{3n-2}\|$$
$$+\left(\prod_{k=n}^{n+p}\frac{2}{1-\alpha_k}\right)\left(\|Tu_{3n-2}-u_{3n-2}\|-\|Tu_{3n+1+3p}-u_{3n+1+3p}\|\right).$$

It follows that

$$\left(1+\sum_{k=3n-2}^{3n-2+p}\beta_k\right)\|Tu_{3n-2}-u_{3n-2}\|$$
$$\leq\|Tu_{3n-1+p}-u_{3n-2}\|$$
$$+\left(\prod_{k=n}^{n+p}\frac{2}{1-\alpha_k}\right)\left(\|Tu_{3n-2}-u_{3n-2}\|-\|Tu_{3n+1+3p}-u_{3n+1+3p}\|\right).$$

Thus, for $n, p+1$, (4) holds.
Case 2: $p=3m+1, m\geq 0$. From (6) and (7), we have

$$\|Tu_{3n+1+p}-u_{3n+1}\|$$
$$=\|Tz_{n+m+1}-x_{n+1}\|$$
$$=\|Tz_{n+m+1}-Ty_n\|$$
$$\leq\|z_{n+m+1}-y_n\|$$
$$=\|(1-\alpha_{m+n+1})x_{m+n+1}+\alpha_{m+n+1}Tx_{m+n+1}-Tz_n\|$$

$$\le (1-\alpha_{m+n+1})\|x_{m+n+1}-Tz_n\|+\alpha_{m+n+1}\|Tx_{m+n+1}-Tz_n\|$$
$$\le (1-\alpha_{m+n+1})\|Ty_{m+n}-Tz_n\|+\alpha_{m+n+1}\|x_{m+n+1}-z_n\|$$
$$\le (1-\alpha_{m+n+1})(\|Ty_{m+n}-Tz_{m+n}\|+\|Tz_{m+n}-Tx_{n+m}\|+\cdots+\|Tx_{n+1}-Ty_n\|$$
$$+\|Ty_n-Tz_n\|)+\alpha_{m+n+1}\|x_{m+n+1}-z_n\|$$
$$=(1-\alpha_{m+n+1})\sum_{k=3n-1}^{3n-2+p}\|Tu_{k+1}-Tu_k\|+\alpha_{m+n+1}\|Ty_{m+n}-(1-\alpha_n)x_n-\alpha_nTx_n\|$$
$$\le (1-\alpha_{m+n+1})\sum_{k=3n-1}^{3n-2+p}\beta_k\|Tu_{3n-2}-u_{3n-2}\|$$
$$+\alpha_{m+n+1}\left((1-\alpha_n)\|Ty_{m+n}-x_n\|+\alpha_n\|Ty_{m+n}-Tx_n\|\right)$$
$$\le (1-\alpha_{m+n+1})\sum_{k=3n-1}^{3n-2+p}\beta_k\|Tu_{3n-2}-u_{3n-2}\|+\alpha_{m+n+1}(1-\alpha_n)\|Ty_{m+n}-x_n\|$$
$$+\alpha_{m+n+1}\alpha_n\left(\|Ty_{m+n}-Tz_{m+n}\|+\|Tz_{m+n}-Tx_{m+n}\|+\cdots+\|Ty_n-Tz_n\|+\|Tz_n-Tx_n\|\right)$$
$$=(1-\alpha_{m+n+1})\sum_{k=3n-1}^{3n-2+p}\beta_k\|Tu_{3n-2}-u_{3n-2}\|+\alpha_{m+n+1}(1-\alpha_n)\|Ty_{m+n}-x_n\|$$
$$+\alpha_{m+n+1}\alpha_n\sum_{k=3n-2}^{3n-2+p}\|Tu_{k+1}-Tu_k\|$$
$$\le (1-\alpha_{m+n+1})\sum_{k=3n-1}^{3n-2+p}\beta_k\|Tu_{3n-2}-u_{3n-2}\|+\alpha_{m+n+1}(1-\alpha_n)\|Ty_{m+n}-x_n\|$$
$$+\alpha_{m+n+1}\alpha_n\sum_{k=3n-2}^{3n-2+p}\beta_k\|Tu_k-u_k\|$$
$$=(1-\alpha_{m+n+1})\sum_{k=3n-1}^{3n-2+p}\beta_k\|Tu_{3n-2}-u_{3n-2}\|+\alpha_{m+n+1}(1-\alpha_n)\|Tu_{3n-1+p}-u_{3n-2}\|$$
$$+\alpha_{m+n+1}\alpha_n\sum_{k=3n-2}^{3n-2+p}\beta_k\|Tu_k-u_k\|$$
$$=(1-\alpha_{m+n+1}+\alpha_{m+n+1}\alpha_n)\sum_{k=3n-2}^{3n-2+p}\beta_k\|Tu_{3n-2}-u_{3n-2}\|$$
$$-\alpha_n(1-\alpha_{m+n+1})\|Tu_{3n-2}-u_{3n-2}\|+\alpha_{m+n+1}(1-\alpha_n)\|Tu_{3n-1+p}-u_{3n-2}\|.$$

It follows that

$$\begin{aligned}&\|Tu_{3n+1+p}-u_{3n+1}\|\\&\le (1-\alpha_{m+n+1}+\alpha_{m+n+1}\alpha_n)\sum_{k=3n-2}^{3n-2+p}\beta_k\|Tu_{3n-2}-u_{3n-2}\|\\&-\alpha_n(1-\alpha_{m+n+1})\|Tu_{3n-2}-u_{3n-2}\|+\alpha_{m+n+1}(1-\alpha_n)\|Tu_{3n-1+p}-u_{3n-2}\|\end{aligned}\tag{9}$$

Using (5) and (9), we have

$$\left(1+\sum_{k=3n+1}^{3n+p}\beta_k\right)\|Tu_{3n+1}-u_{3n+1}\|$$
$$\le (1-\alpha_{m+n+1}+\alpha_{m+n+1}\alpha_n)\sum_{k=3n-2}^{3n-2+p}\beta_k\|Tu_{3n-2}-u_{3n-2}\|$$

$$-\alpha_n(1-\alpha_{m+n+1})\|Tu_{3n-2}-u_{3n-2}\|+\alpha_{m+n+1}(1-\alpha_n)\|Tu_{3n-1+p}-u_{3n-2}\|$$
$$+\left(\prod_{k=n+1}^{n+p}\frac{2}{1-\alpha_k}\right)(\|Tu_{3n+1}-u_{3n+1}\|-\|Tu_{3n+1+3p}-u_{3n+1+3p}\|).$$

From $\left(1+\sum_{k=3n+1}^{3n+p}\beta_k\right)\leq\left(\prod_{k=n+1}^{n+p}\frac{1}{1-\alpha_k}\right)$ and $\|Tu_{3n+1}-u_{3n+1}\|\leq\|Tu_{3n-2}-u_{3n-2}\|$, we have

$$\left(1+\sum_{k=3n+1}^{3n+p}\beta_k\right)\|Tu_{3n-2}-u_{3n-2}\|$$
$$\leq(1-\alpha_{m+n+1}+\alpha_{m+n+1}\alpha_n)\sum_{k=3n-2}^{3n-2+p}\beta_k\|Tu_{3n-2}-u_{3n-2}\|$$
$$-\alpha_n(1-\alpha_{m+n+1})\|Tu_{3n-2}-u_{3n-2}\|+\alpha_{m+n+1}(1-\alpha_n)\|Tu_{3n-1+p}-u_{3n-2}\|$$
$$+\left(\prod_{k=n+1}^{n+p}\frac{2}{1-\alpha_k}\right)(\|Tu_{3n-2}-u_{3n-2}\|-\|Tu_{3n+1+3p}-u_{3n+1+3p}\|).$$

Then

$$\frac{\left(1+\sum_{k=3n+1}^{3n+p}\beta_k+\alpha_n(1-\alpha_{m+n+1})-(1-\alpha_{m+n+1}+\alpha_{m+n+1}\alpha_n)\sum_{k=3n-2}^{3n-2+p}\beta_k\right)}{\alpha_{m+n+1}(1-\alpha_n)}$$
$$\|Tu_{3n-2}-u_{3n-2}\|\leq\|Tu_{3n-1+p}-u_{3n-2}\|$$
$$+\left(\prod_{k=n}^{n+p}\frac{2}{1-\alpha_k}\right)(\|Tu_{3n-2}-u_{3n-2}\|-\|Tu_{3n+1+3p}-u_{3n+1+3p}\|).$$

It follows that

$$\left(1+\sum_{k=3n-2}^{3n-2+p}\beta_k\right)\|Tu_{3n-2}-u_{3n-2}\|$$
$$\leq\|Tu_{3n-1+p}-u_{3n-2}\|$$
$$+\left(\prod_{k=n}^{n+p}\frac{2}{1-\alpha_k}\right)(\|Tu_{3n-2}-u_{3n-2}\|-\|Tu_{3n+1+3p}-u_{3n+1+3p}\|).$$

Thus, for $n, p+1$, (4) holds.
Case 3: $p=3m+2, m\geq 0$. From (6) and (7), we have

$$\|Tu_{3n+1+p}-u_{3n+1}\|$$
$$=\|Ty_{n+m+1}-Ty_n\|$$
$$\leq\|y_{n+m+1}-y_n\|$$
$$=\|Tz_{n+m+1}-Tz_n\|$$
$$\leq\|z_{n+m+1}-z_n\|$$
$$\leq\|z_{n+m+1}-(1-\alpha_n)x_n-\alpha_nTx_n\|$$
$$\leq(1-\alpha_n)\|z_{n+m+1}-x_n\|+\alpha_n\|z_{n+m+1}-Tx_n\|$$
$$=(1-\alpha_n)\|(1-\alpha_{n+m+1})x_{n+m+1}+\alpha_{n+m+1}Tx_{n+m+1}-x_n\|$$
$$+\alpha_n\|(1-\alpha_{n+m+1})x_{n+m+1}+\alpha_{n+m+1}Tx_{n+m+1}-Tx_n\|$$

$$
\begin{aligned}
&\leq (1-\alpha_n)((1-\alpha_{m+n+1})\|x_{n+m+1}-x_n\| + \alpha_{m+n+1}\|Tx_{n+m+1}-x_n\|) \\
&+\alpha_n((1-\alpha_{m+n+1})\|x_{n+m+1}-Tx_n\| + \alpha_{m+n+1}\|Tx_{n+m+1}-Tx_n\|) \\
&\leq (1-\alpha_n)\alpha_{m+n+1}\|Tx_{n+m+1}-x_n\| \\
&+((1-\alpha_{m+n+1})(1-\alpha_n)+\alpha_{m+n+1}\alpha_n)\|x_{n+m+1}-x_n\| \\
&+\alpha_n(1-\alpha_{m+n+1})\|Ty_{n+m+1}-Tx_n\| \\
&\leq (1-\alpha_n)\alpha_{m+n+1}\|Tx_{n+m+1}-x_n\| \\
&+((1-\alpha_{m+n+1})(1-\alpha_n)+\alpha_{m+n+1}\alpha_n)(\|Ty_{n+m}-Tz_{n+m}\|+\cdots+\|Tz_n-Tx_n\|+\|Tx_n-x_n\|) \\
&+\alpha_n(1-\alpha_{m+n+1})(\|Ty_{n+m+1}-Tz_{n+m+1}\|+\cdots+\|Ty_n-Tz_n\|+\|Tz_n-Tx_n\|) \\
&\leq (1-\alpha_n)\alpha_{m+n+1}\|Tx_{n+m+1}-x_n\| \\
&+((1-\alpha_{m+n+1})(1-\alpha_n)+\alpha_{m+n+1}\alpha_n)\left(\sum_{k=3n-2}^{3n-3+p}\|Tu_{k+1}-Tu_k\|+\|Tu_{3n-2}-x_{3n-2}\|\right) \\
&+(1-\alpha_{m+n+1})\alpha_n\sum_{k=3n-2}^{3n-3+p}\|Tu_{k+1}-Tu_k\| \\
&\leq (1-\alpha_n)\alpha_{m+n+1}\|Tx_{n+m+1}-x_n\| \\
&+((1-\alpha_{m+n+1})(1-\alpha_n)+\alpha_{m+n+1}\alpha_n)\sum_{k=3n-2}^{3n-2+p}\beta_k\|Tx_n-x_n\| \\
&+(1-\alpha_{m+n+1})\alpha_n\sum_{k=3n-2}^{3n-3+p}\beta_k\|Tx_n-x_n\| \\
&\leq (1-\alpha_n)\alpha_{m+n+1}\|Tu_{3n-1+p}-u_{3n-2}\| \\
&+\left((1-\alpha_{m+n+1}+\alpha_n\alpha_{m+n+1})\sum_{k=3n-2}^{3n-2+p}\beta_k-\alpha_n(1-\alpha_{m+n+1})\right)\|Tu_{3n-2}-u_{3n-2}\|.
\end{aligned}
$$

It follows that

$$
\begin{aligned}
&\|Tu_{3n+1+p}-u_{3n+1}\| \\
&\leq (1-\alpha_n)\alpha_{m+n+1}\|Tu_{3n-1+p}-u_{3n-2}\| \\
&+\left((1-\alpha_{m+n+1}+\alpha_n\alpha_{m+n+1})\sum_{k=3n-2}^{3n-2+p}\beta_k-\alpha_n(1-\alpha_{m+n+1})\right)\|Tu_{3n-2}-u_{3n-2}\|
\end{aligned}
\tag{10}
$$

Using (5) and (10), we have

$$
\begin{aligned}
&\left(1+\sum_{k=3n+1}^{3n+p}\beta_k\right)\|Tu_{3n+1}-u_{3n+1}\| \\
&\leq (1-\alpha_n)\alpha_{m+n+1}\|Tu_{3n-1+p}-u_{3n-2}\| \\
&+\left((1-\alpha_{m+n+1}+\alpha_n\alpha_{m+n+1})\sum_{k=3n-2}^{3n-2+p}\beta_k-\alpha_n(1-\alpha_{m+n+1})\right)\|Tu_{3n-2}-u_{3n-2}\| \\
&+\left(\prod_{k=n+1}^{n+p}\frac{2}{1-\alpha_k}\right)(\|Tu_{3n+1}-u_{3n+1}\|-\|Tu_{3n+1+3p}-u_{3n+1+3p}\|).
\end{aligned}
$$

From $\left(1+\sum_{k=3n+1}^{3n+p}\beta_k\right)\leq\left(\prod_{k=n+1}^{n+p}\frac{2}{1-\alpha_k}\right)$ and $\|Tu_{3n+1}-u_{3n+1}\|\leq\|Tu_{3n-2}-u_{3n-2}\|$, we have

$$\left(1+\sum_{k=3n+1}^{3n+p}\beta_k\right)\|Tu_{3n-2}-u_{3n-2}\|$$
$$\le(1-\alpha_n)\alpha_{m+n+1}\|Tu_{3n-1+p}-u_{3n-2}\|$$
$$+\left((1-\alpha_{m+n+1}+\alpha_n\alpha_{m+n+1})\sum_{k=3n-2}^{3n-2+p}\beta_k-\alpha_n(1-\alpha_{m+n+1})\right)\|Tu_{3n-2}-u_{3n-2}\|$$
$$+\left(\prod_{k=n+1}^{n+p}\frac{2}{1-\alpha_k}\right)(\|Tu_{3n-2}-u_{3n-2}\|-\|Tu_{3n+1+3p}-u_{3n+1+3p}\|).$$

Then

$$\frac{(1+\sum_{k=3n+1}^{3n+p}\beta_k-\left((1-\alpha_{m+n+1}+\alpha_n\alpha_{m+n+1})\sum_{k=3n-2}^{3n-2+p}\beta_k-\alpha_n(1-\alpha_{m+n+1})\right)}{(1-\alpha_n)\alpha_{m+n+1}}\|Tu_{3n-2}-u_{3n-2}\|$$
$$\le\|Tu_{3n-1+p}-u_{3n-2}\|$$
$$+\left(\prod_{k=n}^{n+p}\frac{2}{1-\alpha_k}\right)(\|Tu_{3n-2}-u_{3n-2}\|-\|Tu_{3n+1+3p}-u_{3n+1+3p}\|).$$

It follows that

$$\left(1+\sum_{k=3n-2}^{3n-2+p}\beta_k\right)\|Tu_{3n-2}-u_{3n-2}\|$$
$$\le\|Tu_{3n-1+p}-u_{3n-2}\|$$
$$+\left(\prod_{k=n}^{n+p}\frac{2}{1-\alpha_k}\right)(\|Tu_{3n-2}-u_{3n-2}\|-\|Tu_{3n+1+3p}-u_{3n+1+3p}\|).$$

Thus, for $n, p+1$, (4) holds. This completes the induction.
Lemma 3. $T:C\to C$ is a $T-(D_a)$ mapping, $\|Tx-x\|\le\|Ty-y\|$. Then

$$\|x-Ty\|\le3\|Tx-x\|+\|x-y\|.$$

Proof: Since $\|Tx-x\|\le\|Ty-y\|$, we have $Tx\in C(T,y,\alpha)$. Then

$$\|T^2x-Ty\|\le\|Tx-y\|.$$

It follows that

$$\|x-Ty\|\le\|x-Tx\|+\|T^2x-Tx\|+\|T^2x-Ty\|.$$

From **Proposition 3**, we have

$$\|x-Ty\|\le2\|Tx-x\|+\|Tx-y\|\le2\|Tx-x\|+\|Tx-x\|+\|x-y\|=3\|Tx-x\|+\|x-y\|.$$

Theorem 1. Assume that $T:C\to C$ is a $T-(D_a)$ mapping and $\{x_n\}$, $\{y_n\}$, $\{z_n\}$ are sequences generated by iteration (3), $\frac{1}{2}<a\le\alpha_n\le b<1$. Then $\lim_{n\to\infty}\|Tx_n-x_n\|=0$.

Proof: Since C is bounded, there must exists $d>0$, for every $x\in C$, $\|x\|\le d$. Let $\{u_m\}$ satisfy $u_{3n-2}=x_n$, $u_{3n-1}=z_n$, $u_{3n}=y_n$. From Lemma 1, $\lim_{k\to\infty}\|Tu_k-u_k\|=r\ge0$. Assume $r>0$. Let ε satisfy

$$e^{\frac{6}{1-b}\left(\frac{d}{r}+1\right)}\varepsilon < r$$

and choose n so that for every $p > 0$

$$\|Tu_{3n-2} - u_{3n-2}\| - \|Tu_{3n-2+3p} - u_{3n-2+3p}\| < \varepsilon.$$

Now choose p so that $r\left(\sum_{k=3n-2}^{3n+p-4} \beta_k\right) \le d \le r\left(\sum_{k=3n-2}^{3n+p-3} \beta_k\right)$.

Since $\frac{1}{2} < a \le \alpha_n \le b < 1$, for every k, t, we have $1 + \alpha_k < 3\alpha_k$, $\alpha_t < 2\alpha_k$. From **Lemma 2** and $r \le \|Tu_{3n-2} - u_{3n-2}\|$, we have

$$\begin{aligned}
d + r &\le r\left(1 + \sum_{k=3n-2}^{3n+p-3} \beta_k\right) \\
&\le \left(1 + \sum_{k=3n-2}^{3n+p-3} \beta_k\right)\|Tu_{3n-2} - u_{3n-2}\| \\
&\le \|Tu_{3n-2+p} - u_{3n-2}\| \\
&+ \left(\prod_{k=n}^{n+p-1} \frac{2}{1-\alpha_k}\right)\left(\|Tu_{3n-2} - u_{3n-2}\| - \|Tu_{3n-2+3p} - u_{3n-2+3p}\|\right) \\
&< d + \left(\prod_{k=n}^{n+p-1} \frac{2}{1-\alpha_k}\right)\varepsilon \\
&= d + e^{\sum_{k=n}^{n+p-1} \ln\left(1+\frac{1+\alpha_k}{1-\alpha_k}\right)}\varepsilon \\
&\le d + e^{\sum_{k=n}^{n+p-1} \frac{1+\alpha_k}{1-\alpha_k}}\varepsilon \\
&\le d + e^{\frac{3}{1-b}\sum_{k=n}^{n+p-1} \alpha_k}\varepsilon \\
&\le d + e^{\frac{6}{1-b}\sum_{k=3n-2}^{3n+p-3} \beta_k}\varepsilon \\
&\le d + e^{\frac{6}{1-b}\left(\sum_{k=3n-2}^{3n+p-4} \beta_k + 1\right)}\varepsilon \\
&< d + e^{\frac{6}{1-b}\left(\frac{d}{r}+1\right)}\varepsilon < d + r.
\end{aligned}$$

This is a contradiction. So $\lim_{k\to\infty} \|Tu_k - u_k\| = 0$. That is to say, $\lim_{n\to\infty} \|Tx_n - x_n\| = 0$. This completes the proof.

Theorem 2. Assume that $T : C \to C$ is a $T - (D_a)$ mapping and $\{x_n\}$ is generated by iteration (3), $\frac{1}{2} < a \le \alpha_n \le b < 1$. Then the sequence $\{x_n\}$ converges to a fixed point of T.

Proof: Since C is compact, there exists a subsequence $\{x_{n_k}\} \subset \{x_n\}$ which converges to some $z \in C$. By **Lemma 3**, we have

$\|x_{n_k} - Tz\| \leq 3\|Tx_{n_k} - x_{n_k}\| + \|x_{n_k} - z\|$. Since $\lim_{n_k \to \infty} \|Tx_{n_k} - x_{n_k}\| = 0$ and $\lim_{n_k \to \infty} \|x_{n_k} - z\| = 0$, we have $\lim_{n_k \to \infty} \|x_{n_k} - Tz\| = 0$. This implies that $z = Tz$. On the other hand, from **Proposition 3**

$$\begin{aligned}\|x_{n+1} - z\| &\leq \|y_n - z\| \leq \|z_n - z\| \\ &\leq \alpha_n \|Tx_n - z\| + (1 - \alpha_n)\|x_n - z\| \\ &\leq \|x_n - z\|.\end{aligned}$$

So, $\lim_{n \to \infty} \|x_n - z\|$ exists. Therefore, $\lim_{n \to \infty} \|x_n - z\| = 0$. This completes the proof.

Author details

Shi Jie
School of Mathematics and Statistics, Hubei Engineering University, Xiaogan, Hubei, China

*Address all correspondence to: shijie@whu.edu.cn

References

[1] Suzuki T. Fixed point theorems and convergence theorems for some generalized nonexpansive mappings. Journal of Mathematical Analysis and Applications. 2008;**340**:1088-1095

[2] Thakur BS, Thakur D, Postolache M. A new iterative scheme for numerical reckoning fixed points of Suzuki's generalized nonexpansive mappings. Journal of Applied Mathematics and Computing. 2016;**275**:147-155

Chapter 4

Folding on the Chaotic Graph Operations and Their Fundamental Group

Mohammed Abu Saleem

Abstract

Our aim in the present chapter is to introduce a new type of operations on the chaotic graph, namely, chaotic connected edge graphs under the identification topology. The concept of chaotic foldings on the chaotic edge graph will be discussed from the viewpoint of algebra and geometry. The relation between the chaotic homeomorphisms and chaotic foldings on the chaotic connected edge graphs and their fundamental group is deduced. The fundamental group of the limit chaotic chain of foldings on chaotic. Many types of chaotic foldings are achieved. Theorems governing these relations are achieved. We also discuss some applications in chemistry and biology.

Keywords: chaotic graph, edge graph, chaotic folding, limit folding fundamental group
2010 Mathematics Subject Classification: 51H20, 57N10, 57M05, 14F35, 20F34

1. Introduction and definitions

During the past few decades, examinations of social, biological, and communication networks have taken on enhanced attention throughout these examinations; graphical representations of those networks and systems have been evident to be terribly helpful. Such representations are accustomed to confirm or demonstrate the interconnections or relationships between parts of those networks [1, 2].

A graph is an ordered G = (V(G), E(G)) where V(G) $\neq \varphi$, E(G) is a set disjoint from V(G), elements of V(G) are called the vertices of G, and elements of E(G) are called the edges. The foundation stone of graph theory was laid by Euler in 1736 by solving a puzzle called Königsberg seven-bridge problem as in **Figure 1** [1, 3].

There are many graphs with which one can construct a new graph from a given graph or set of graphs, such as the Cartesian product and the line graph. A graph G is a finite non-empty set V of objects called vertices (the singular is vertex) together with a set E of two-element subsets of V called edges. The number of vertices in a graph G is the order of G, and the number of edges is the size of G. To indicate that a graph G has vertex set V and edge set E, we sometimes write G = (V, E). To emphasize that V is the vertex set of a graph G, we often write V as V(G). For the same reason, we also write E as E(G). A graph H is said to be a subgraph of a graph G if V(H) $\subseteq$ V(G) and E(H) $\subseteq$ E(G). The complete graph with n-vertices will be denoted by K_n. A null graph is a graph containing no edges; the null graph with

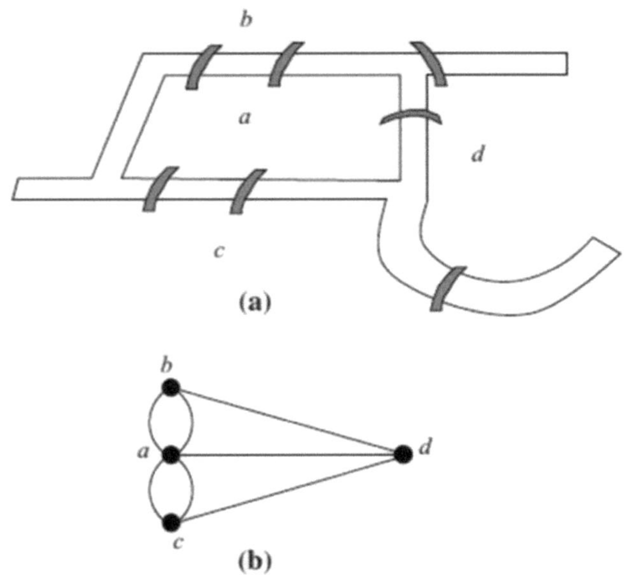

Figure 1.
Königsberg seven-bridge problem.

n-vertices is denoted by N_n. A cycle graph is a graph consisting of a single cycle, the cycle graph with n-vertices is denoted by C_n. The path graph is a graph consisting of a single path; the path graph with n-vertices is denoted by P_n [1–11]. Let G and H be two graphs. A function $\varphi : V(G) \to V(H)$ is a homomorphism from G to H if it preserves edges, that is, if for every edge $e \in E(G), f(e) \in E(H)$ [12, 13]. A core is a graph which does not retract to a proper subgraph. Any graph is homomorphically equivalent to a unique core [7].

The folding is a continuous function $f : G \to H$ such that for each $v \in V(G), f(v) \in V(H)$, and for each $e \in E(G)$, $f(e) \in E(H)$ [14]. Let X be a space, and let I be the unit interval [0,1] in R, a homotopy of paths in X is a family $g_t : I \to X$, $0 \le t \le 1$such that (i) the endpoints $g_t(0) = x_0$ and $g_t(1) = x_1$ are independent of t and (ii) the associated map $G : I \times I \to X$ defined by $G(s,t) = g_t(s)$ is continuous [15]. Given spaces X and Y with chosen points $x_0 \in X$, and $y_0 \in Y$, the wedge sum X∨Y is the quotient of the disjoint union X∪Y obtained identifying x_0 and y_0 to a single point [15]. Two spaces X and Y are of the “same homotopy type” if there exist continuous maps $f : X \to Y$ and $g : Y \to X$ such that $g \circ f \cong I_X : X \to X$ and $f \circ g \cong I_Y : Y \to Y$ [16]. The fundamental group briefly consists of equivalence classes of homotopic closed paths with the law of composition following one path to another. However, the set of homotopy classes of loops based at the point x_0 with the product operation $[f][g] = [f \cdot g]$ is called the fundamental group and denoted by $\pi_1(X, x_0)$ [4, 17–24]. Over many years, chaos has been shown to be an interesting and even common phenomenon in nature. Chaos has been shown to exist in a wide variety of settings: in fluid dynamics such as Raleigh-Bernard convection, in chemistry such as the Belousov-Zhabotinsky reaction, in nonlinear optics in certain lasers, in celestial mechanics, in electronics in the flutter of an overdriven airplane wing, some models of population dynamics, and likely in meteorology, physiological oscillations such as certain heart rhythms, as well as brain patterns [17, 24–30]. AI algorithms related to adjacency matrices on the operations of the graph are discussed in [31, 32].

DOI: http://dx.doi.org/10.5772/intechopen.88553

2. The main results

First, we will introduce the following:

Definition 1. *The chaotic edge $\bar{e}$ is a geometric edge e_1 that carries many other edges $(e_2, e_3, \ldots)$, each one of them homotopic to the original one as in* **Figure 2**. *Also the chaotic vertices of $\bar{e}$ are $\bar{v} = (v_1, v_2, \ldots)$ and $\bar{u} = (u_1, u_2, \ldots)$. For chaotic edge $\bar{e}$, we have two cases:*

Case 1 (1) *e_1, e_2, e_3, … are of the same physical properties.*

Case 2 (2) *e_1, e_2, e_3, … represent different physical properties; for example, e_1 represents density, e_2 represents hardness, e_3 represents magnetic fields, and so on.*

Definition 2. *A chaotic graph $\overline{G}$ is a collection of finite non-empty set $\overline{V}$ of objects called chaotic vertices together with a set $\overline{E}$ of two-element subsets of $\overline{V}$ called chaotic edges. The number of chaotic edges is the size of $\overline{G}$.*

Definition 3. *Given chaotic connected graphs $\overline{G}_1$ and $\overline{G}_2$ with given edges $\bar{e}_1 \in \overline{G}_1$ and $\bar{e}_2 \in \overline{G}_2$, then the chaotic connected edge graph $\overline{G}_1 \ \overline{G}_2$ is the quotient of disjoint union $\overline{G}_1 \cup \overline{G}_2$ acquired by identifying two chaotic edges $\bar{e}_1$ and $\bar{e}_2$ to a single chaotic edge (up to chaotic isomorphism) as in* **Figure 3**.

Definition 4. *A chaotic graph $\overline{H}$ is called a chaotic subgraph of a chaotic graph $\overline{G}$ if $\overline{V}(\overline{H}) \subseteq \overline{V}(\overline{G})$ and $\overline{E}(\overline{H}) \subseteq \overline{E}(\overline{G})$.*

Definition 5. *Let $\overline{G}$ and $\overline{H}$ be two chaotic graphs. A function $\overline{\varphi} : \overline{V}(\overline{G}) \to \overline{V}(\overline{H})$ is chaotic homomorphism from $\overline{G}$ to $\overline{H}$ if it preserves chaotic edges, that is, if for any chaotic edge $[\bar{u}, \bar{v}]$ of $\overline{G}$, $[\overline{\varphi}(\bar{u}), \overline{\varphi}(\bar{v})]$ is a chaotic edge of $\overline{H}$.*

Definition 6. *A chaotic folding of a graph $\overline{G}$ is a chaotic subgraph $\overline{H}$ of $\overline{G}$ such that there exists a chaotic homomorphism $\bar{f} : \overline{G} \to \overline{H}$, called chaotic folding with $\bar{f}(\bar{x}) = \bar{x}$ for every chaotic vertex $\bar{x}$ of $\overline{H}$.*

Definition 7. *A chaotic core is a chaotic graph which does not chaotic retract to chaotic proper subgraph.*

Theorem 1. *Let $\overline{G}_1$ and $\overline{G}_2$ be two chaotic connected graphs. Then $\bar{\pi}_1(\overline{G}_1 \veebar \overline{G}_2) = \bar{\pi}_1(\overline{G}_1) * \bar{\pi}_1(\overline{G}_2)$.*

Proof. Let $\overline{G}_1$ and $\overline{G}_2$ be two chaotic connected graphs. Since $\overline{G}_1 \veebar \overline{G}_2$ and $\overline{G}_1 \vee \overline{G}_2$ are of same chaotic homotopy type, it follows that $\bar{\pi}_1(\overline{G}_1 \veebar \overline{G}_2) \approx \bar{\pi}_1(\overline{G}_1) * \bar{\pi}_1(\overline{G}_2)$. Hence, $\bar{\pi}_1(\overline{G}_1 \veebar \overline{G}_2) = \bar{\pi}_1(\overline{G}_1) * \bar{\pi}_1(\overline{G}_2)$.

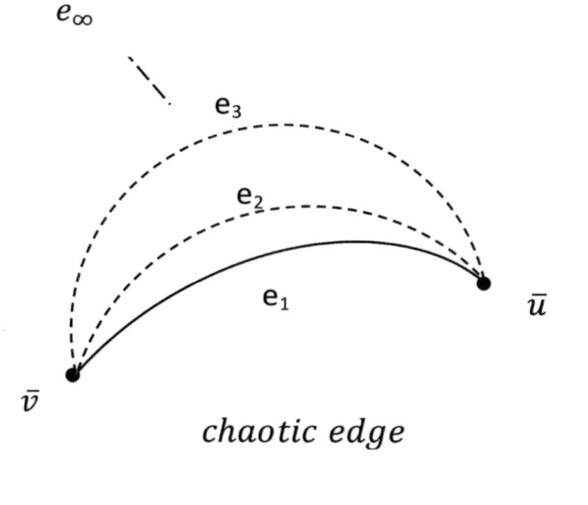

Figure 2.
Chaotic edge.

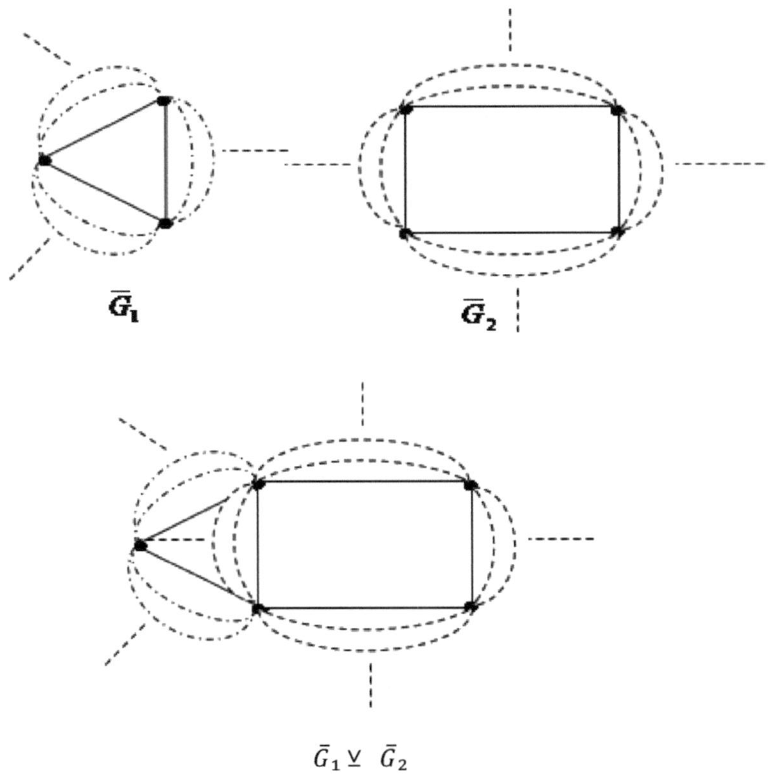

Figure 3.
Chaotic connected edge.

Theorem 2. *The chaotic graphs* $\overline{G}_1$ *and* $\overline{G}_2$*are chaotic subgraphs of* $\overline{G}_1 \veebar \overline{G}_2$*. Also, for any chaotic tree* $\overline{G}_1$ *and* $\overline{G}_2$*,* $\overline{G}_1 \veebar \overline{G}_2$ *is also chaotic tree and* $\overline{\pi}_1(\overline{G}_1 \veebar \overline{G}_2) = \overline{0}$*.*

Proof. The proof of this theorem is clear.

Theorem 3. *If* $\overline{G}_1, \overline{G}_2, \ldots, \overline{G}_n$ *are connected graphs, and* $\left\langle \overline{f}_1, \overline{f}_2, \ldots, \overline{f}_n \right\rangle$ *is a sequence of chaotic topological foldings of* $\veebar_{i=1}^{n} \overline{G}_i$ *into itself, then there is an induced sequence* $\left\langle \overline{\overline{f}}_1, \overline{\overline{f}}_2, \ldots, \overline{\overline{f}}_n \right\rangle$ *of non-trivial chaotic topological folding* $\overline{\overline{f}}_j : *_{i=1}^{n} \overline{\pi}_1(\overline{G}_{ii}) \rightarrow *_{i=1}^{n} \overline{\pi}_1(\overline{G}_{ii})$*,* $j = 1, 2, \ldots, n$ *such that* $\overline{\overline{f}}_j(*_{i=1}^{n} \overline{\pi}_1(\overline{G}_{ii}))$ *reduces the rank of* $*_{i=1}^{n} \overline{\pi}_1(\overline{G}_{ii})$*.*

Proof. Consider the following sequence of topological foldings $\left\langle \overline{f}_1, \overline{f}_2, \ldots, \overline{f}_n \right\rangle$, where $\overline{f}_1 : \veebar_{i=1}^{n} \overline{G}_i \rightarrow \veebar_{i=1}^{n} \overline{G}_i$, is a topological folding from $\vee_{i=1}^{n} \overline{G}_i$ into itself such that $\overline{f}_1(\veebar_{i=1}^{n} \overline{G}_i) = \overline{G}_1 \veebar \overline{G}_2 \veebar \ldots \veebar \overline{f}_1(\overline{G}_s) \veebar \ldots \veebar \overline{G}_n$ *for* $s = 1, 2, \ldots n$.

Since $size\left(\overline{f}_1(\overline{G}_s)\right) \leq size(\overline{G}_s)$ and $\overline{\overline{f}}_1(\overline{\pi}_1(\overline{G}_i)) = \overline{\pi}_1\left(\overline{f}_1(\overline{G}_i)\right)$, it follows that $\text{rank}\left(\overline{\overline{f}}_1(\overline{\pi}_1(\overline{G}_s))\right) = \text{rank}\left(\overline{\pi}_1\left(\overline{f}_1(\overline{G}_s)\right)\right) \leq \text{rank}(\overline{\pi}_1(\overline{G}_s))$, and so $\overline{\overline{f}}_1$ reduces the rank of $*_{i=1}^{n} \overline{\pi}_1(\overline{G}_{ii})$. Also, if $\overline{f}_2(\veebar_{i=1}^{n} \overline{G}_i) = \overline{G}_1 \veebar \overline{G}_2 \veebar \ldots \veebar \overline{f}_2(\overline{G}_s) \veebar \ldots \veebar \overline{f}_2(\overline{G}_k)$ $\veebar \ldots \veebar \overline{G}_n$ *for* $k = 1, 2, \ldots n$ and $s < k$ and $size\left(\overline{f}_2(\overline{G}_s)\right) \leq size(\overline{G}_s)$ *and* $size\left(\overline{f}_2(\overline{G}_k)\right)$ $\leq size(\overline{G}_k)$, we haverank$\left(\overline{\overline{f}}_2(\overline{\pi}_1(\overline{G}_s))\right) = \text{rank}\left(\overline{\pi}_1\left(\overline{f}_2(\overline{G}_s)\right)\right) \leq \text{rank}(\overline{\pi}_1(\overline{G}_s))$, $\text{rank}\left(\overline{\overline{f}}_2(\overline{\pi}_1(\overline{G}_k))\right) = \text{rank}\left(\overline{\pi}_1\left(\overline{f}_2(\overline{G}_k)\right)\right) \leq \text{rank}(\overline{\pi}_1(\overline{G}_k))$; thus $\overline{\overline{f}}_2$ reduces the rank of $*_{i=1}^{n} \overline{\pi}_1(\overline{G}_{ii})$. Moreover, by continuing with this procedure if $\overline{f}_n(\veebar_{i=1}^{n} \overline{G}_i) = \veebar_{i=1}^{n}\left(\overline{f}_n(\overline{G}_i)\right)$, then $\overline{\overline{f}}_n(*_{i=1}^{n} \overline{\pi}_1(\overline{G}_{ii})) = \overline{\pi}_1\left(\overline{f}_n(\veebar_{i=1}^{n} \overline{G}_i)\right) =$ $\overline{\pi}_1\left(\veebar_{i=1}^{n} \overline{f}_n(\overline{G}_i)\right) \approx *_{i=1}^{n} \overline{\pi}_1\left(\overline{f}_n(\overline{G}_{ii})\right)$. Hence, $\overline{\overline{f}}_n$ reduces the rank of $*_{i=1}^{n} \overline{\pi}_1(\overline{G}_{ii})$.

Theorem 4. *Let* $\overline{G}_1$ *and* $\overline{G}_2$ *be two chaotic graphs; then there is a chaotic homomorphism* $\overline{f} : \overline{G}_1 \to \overline{G}_2$ *which induces* $\overline{\overline{f}} : \overline{\pi}_1(\overline{G}_1) \to \overline{\pi}_1(\overline{G}_2)$ *if* $\overline{\pi}_1(\overline{G}_2)$ *is a chaotic folding of* $\overline{\pi}_1(\overline{G}_1 \veebar \overline{G}_2)$.

Proof. Let $\overline{f} : \overline{G}_1 \to \overline{G}_2$ be a chaotic homomorphism. Since $\overline{G}_2$ is chaotic subgraph of $\overline{G}_1 \veebar \overline{G}_2$, then there exists a chaotic homomorphism $\overline{f} : \overline{G}_1 \veebar \overline{G}_2 \to \overline{G}_2$ with $\overline{f}(\overline{v}) = \overline{v}$ for any chaotic vertex $\overline{v}$ of $\overline{G}_2$ which induces $\overline{\overline{f}} : \overline{\pi}_1(\overline{G}_1) \to \overline{\pi}_1(\overline{G}_2)$. What follows from $\overline{G}_2$ is a chaotic folding of $\overline{G}_1 \veebar \overline{G}_2$ in that $\overline{\pi}_1(\overline{G}_2)$ is a chaotic folding of $\overline{\pi}_1(\overline{G}_1 \veebar \overline{G}_2)$. Conversely, assume that $\overline{G}_2$ is a chaotic folding of $\overline{G}_1 \veebar \overline{G}_2$; thus $\overline{f} : \overline{G}_1 \veebar \overline{G}_2 \to \overline{G}_2$ is a chaotic homomorphism with $\overline{f}(\overline{v}) = \overline{v}$ for any chaotic vertex $\overline{v}$ of $\overline{G}_2$, and so there is a chaotic homomorphism $\overline{f} : \overline{G}_1 \to \overline{G}_2$ which induce $\overline{\overline{f}} : \overline{\pi}_1(\overline{G}_1) \to \overline{\pi}_1(\overline{G}_2)$.

Theorem 5. *For any chaotic path graphs* $\overline{P}_n, \overline{P}_m, n, m \geq 2$, *there is a sequence of topological foldings with variation curvature* $\left\{\overline{f}_i : i = 1, 2, \ldots k\right\}$ *on* $\overline{P}_n \veebar \overline{P}_m$ *which induce a sequence of topological foldings* $\left\{\overline{\overline{f}}_i : i = 1, 2, \ldots k\right\}$ *such that* $\overline{\overline{f}}_k\left(\overline{\overline{f}}_{k-1}\left(\ldots\left(\overline{\overline{f}}_1\right.\right.\right.$ $\left(\overline{\pi}_1(\overline{P}_n \veebar \overline{P}_m)\ldots\right\} = \overline{Z}$ *and* $\lim_{k\to\infty}\left(\overline{\overline{f}}_k\left(\overline{\overline{f}}_{k-1}\left(\ldots\left(\overline{\overline{f}}_1\left(\overline{\pi}_1(\overline{P}_n \veebar \overline{P}_m)\ldots\right)\right) = \overline{0}\right.\right.\right.$.

Proof. Consider the following sequence of chaotic topological foldings with variation curvature, $\overline{f}_1 : \overline{P}_n \veebar \overline{P}_m \to (\overline{P}_n \veebar \overline{P}_m)_1$, where $(\overline{P}_n \veebar \overline{P}_m)_1$ is a chaotic subgraph with decreasing inner curvature between every two adjacent chaotic edges in $\overline{P}_n \veebar \overline{P}_m$ and $\overline{f}_2 : \overline{f}_1(\overline{P}_n \veebar \overline{P})_m \to \overline{f}_1((\overline{P}_n \veebar \overline{P}_m)_1)$ where $\overline{f}_2\left(\overline{f}_1((\overline{P}_n \veebar \overline{P}_m)_1)\right)$ is a chaotic subgraph with decreasing inner curvature between every two adjacent chaotic edges in $\overline{f}_1((\overline{P}_n \veebar \overline{P}_m)_1)$, and so on, such that $\overline{f}_k\left(\overline{f}_{k-1}\left(\overline{f}_{k-2}\left(\cdots\left(\overline{f}_1(\overline{P}_n \veebar \overline{P}_m)\cdots\right) = \right.\right.\right.$ $\overline{C}_{n+m-2}$ and $\lim_{k\to\infty}\left(\overline{f}_k\left(\overline{f}_{k-1}\left(\overline{f}_{k-2}\left(\ldots\left(\overline{f}_1(\overline{P}_n \veebar \overline{P}_m)\ldots\right) = \overline{N}_1\right.\right.\right.\right.$, thus $\overline{\overline{f}}_k\left(\overline{\overline{f}}_{k-1}\left(\overline{\overline{f}}_{k-2}(\ldots\right.\right.$ $\left(\overline{\overline{f}}_1(\overline{\pi}_1(\overline{P}_n \veebar \overline{P}_m)\ldots\right) = \overline{\pi}_{11}(\overline{C}_{n+m-2}) = \overline{Z}$. Also, $\lim_{k\to\infty}\left(\overline{\overline{f}}_k\left(\overline{\overline{f}}_{k-1}\left(\overline{\overline{f}}_{k-2}\left(\ldots\left(\overline{\overline{f}}_1(\overline{\pi}_1\right.\right.\right.\right.\right.$ $(\overline{P}_n \veebar \overline{P}_m)\ldots) = \overline{\pi}_1(\overline{N}_1) = \overline{0}$.

Theorem 6. *For every two chaotic connected graphs* $\overline{G}_1$ *and* $\overline{G}_2$, *the fundamental group of the limit of chaotic topological folding of* $\overline{G}_1 \veebar \overline{G}_2 = \overline{0}$.

Proof. Let $\overline{G}_1$ and $\overline{G}_2$ be two chaotic connected graphs; then we have two cases:

Case (1): If $\overline{f}_1 : \overline{G}_1 \veebar \overline{G}_2 \to \overline{G}_1 \veebar \overline{G}_2$ is a chaotic topological folding such that $\overline{f}_1(\overline{G}_1 \veebar \overline{G}_2)$ consists of chaotic cycles, so we can define a sequence of chaotic topological folding $\overline{f}_2 : \overline{f}_1(\overline{G}_1 \veebar \overline{G}_2) \to \overline{f}_1(\overline{G}_1 \veebar \overline{G}_2)$ where $\overline{f}_2\left(\overline{f}_1(\overline{G}_1 \veebar \overline{G}_2)\right)$ is a chaotic tree with $\overline{n}$chaotic edges, $\overline{f}_3 : \overline{f}_2\left(\overline{f}_1(\overline{G}_1 \veebar \overline{G}_2)\right) \to \overline{f}_2\left(\overline{f}_1(\overline{G}_1 \veebar \overline{G}_2)\right)$, such that $\overline{f}_3\left(\overline{f}_2\left(\overline{f}_1(\overline{G}_1 \veebar \overline{G}_2)\ldots\right)\right.$ is a chaotic tree with $\overline{k} < \overline{n}$ chaotic edges, chaotic edges by continuing this process we get $\overline{f}_k : \overline{f}_{k-1}\left(\overline{f}_{k-2}\left(\ldots\left(\overline{f}_1(\overline{G}_1 \veebar \overline{G}_2)\cdots\right) \to\right.\right.$ $\overline{f}_{k-1}\left(\overline{f}_{k-2}\left(\ldots\left(\overline{f}_1(\overline{G}_1 \veebar \overline{G}_2)\cdots\right)\right.\right.$ such that $\lim_{k\to\infty}\left(\overline{f}_k\left(\overline{f}_{k-1}\left(\overline{f}_{k-2}\left(\ldots\left(\overline{f}_1(\overline{G}_1 \veebar \overline{G}_2)\ldots\right)\right.\right.\right.\right.$ is a chaotic edge, and so $\overline{\pi}_1\left(\lim_{k\to\infty}\left(\overline{f}_k\left(\overline{f}_{k-1}\left(\overline{f}_{k-2}\left(\ldots\left(\overline{f}_1(\overline{G}_1 \veebar \overline{G}_2)\ldots\right)\right)\right.\right.\right.\right. = \overline{0}$.

Case (2): If $\overline{g}_1 : \overline{G}_1 \veebar \overline{G}_2 \to \overline{G}_1 \veebar \overline{G}_2$ is a chaotic topological folding such that. $\overline{g}_1(\overline{G}_1 \veebar \overline{G}_2)$ has no chaotic cycles, then clearly $\lim_{k\to\infty}(\overline{g}_k(\overline{g}_{k-1}(\overline{g}_{k-2}(\cdots(\overline{g}_1$ $(\overline{G}_1 \veebar \overline{G}_2)\ldots)$ is a chaotic edge and $\overline{\pi}_1(\lim_{k\to\infty}(\overline{g}_k(\overline{g}_{k-1}(\overline{g}_{k-2}(\ldots(\overline{g}_1(\overline{G}_1 \veebar \overline{G}_2)\ldots)) = \overline{0}$.

Theorem 7. *If* $\overline{G}_1$ *and* $\overline{G}_2$ *are chaotic connected and not chaotic cores graphs, then* $\overline{\pi}_1\left(\lim_{n\to\infty}\overline{f}_n(\overline{G}_1 \veebar \overline{G}_2)\right) = \overline{\pi}_1\left(\lim_{n\to\infty}\overline{f}_n(\overline{G}_1)\right) * \overline{\pi}_1\left(\lim_{n\to\infty}\overline{f}_n(\overline{G}_2)\right)$.

Proof. If $\overline{G}_1$ and $\overline{G}_2$ are chaotic connected and not chaotic cores graphs, then we get the following chaotic induced graphs $\lim_{n\to\infty}\overline{f}_n(\overline{G}_1 \veebar \overline{G}_2)$, $\lim_{n\to\infty}\overline{f}_n(\overline{G}_1)$, $\lim_{n\to\infty}\overline{f}_n(\overline{G}_2)$, and each of them are isomorphic to $\overline{k}_2$. Since $\overline{k}_2 \approx \overline{k}_2 \veebar \overline{k}_2$ it follows that $\lim_{n\to\infty}\overline{f}_n(\overline{G}_1 \veebar \overline{G}_2) = \lim_{n\to\infty}\overline{f}_n(\overline{G}_1) \veebar \lim_{n\to\infty}\overline{f}_n(\overline{G}_2)$ and $\overline{\pi}_1\left(\lim_{n\to\infty}\overline{f}_n(\overline{G}_1 \veebar \overline{G}_2)\right) = \overline{\pi}_1\left(\lim_{n\to\infty}\overline{f}_n(\overline{G}_1)\right) * \overline{\pi}_1\left(\lim_{n\to\infty}\overline{f}_n(\overline{G}_2)\right)$.

3. Some applications

i. A polymer is composed of many repeating units called monomers. Starch, cellulose, and proteins are natural polymers. Nylon and polyethylene are synthetic polymers. Polymerization is the process of joining monomers. Polymers may be formed by addition polymerization; furthermore, one essential advance likewise polymerization is mix as in **Figure 4**, which happens when the polymer's development is halted by free electrons from two developing chains that join and frame a solitary chain. The accompanying chart portrays mix, with the image (R) speaking to whatever remains of the chain.

ii. Chemical nature of enzymes, all known catalysts are proteins. They are high atomic weight mixes made up primarily of chains of amino acids connected together by peptide bonds as in **Figure 5**.

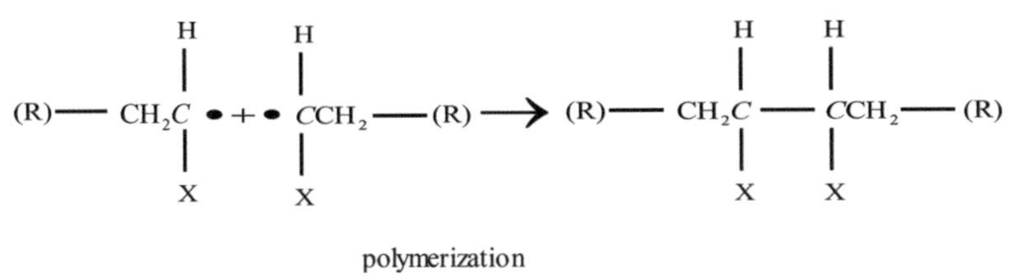

Figure 4.
Polymerization.

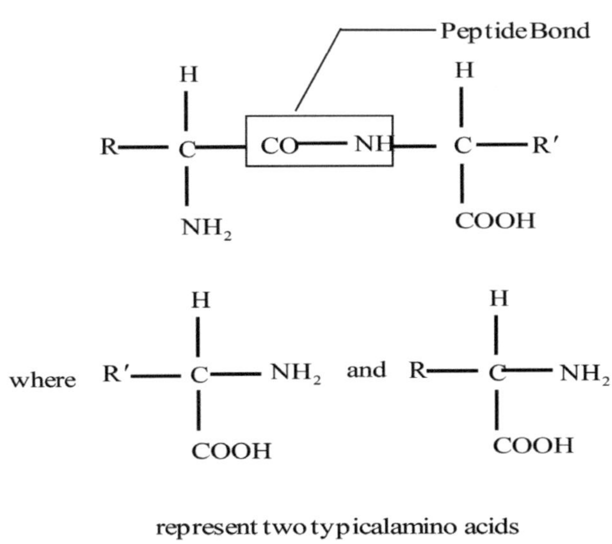

Figure 5.
Typical amino acids.

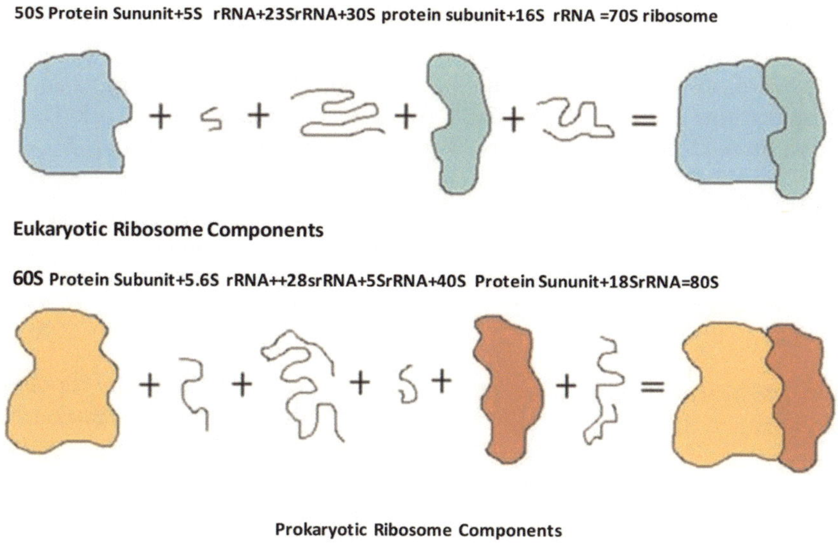

Figure 6.
Prokaryotic ribosome components.

iii. There are two types of the subunit structure of ribosomes as in **Figure 6** which is represented by the different connected types of protein subunit and rRNA to form a new type of ribosomes.

4. Conclusion

In this chapter, the fundamental group of the limit chaotic foldings on chaotic connected edge graphs is deduced. Also, we can deduce some algorithms from a new operation of a graph by using the adjacency matrices.

Author details

Mohammed Abu Saleem
Department of Mathematics, Faculty of Science, Al-Balqa Applied University, Salt, Jordan

*Address all correspondence to: m_abusaleem@bau.edu.jo

References

[1] Bailey S. Dot Product Graphs and Their Applications to Ecology. Logan, Utah: Utah State University; 2013

[2] Pirzada S, Dharwadker A. Applications of graph theory. The Journal of the Korean Society for Industrial and Applied Mathematics. 2007;**11**(4):19-38

[3] Bondy JA, Murty USR. Graph Theory. Springer; 2008

[4] Abu-Saleem M. Folding on the wedge sum of graphs and their fundamental group. Applied Sciences. 2010;**12**:14-19

[5] Abu-Saleem M. Retractions and homomorphisms on some operations of graphs. Journal of Mathematics. 2018: 1-4

[6] Biggs NL, Lloyd EK, Wilson RJ. Graph Theory. Oxford: Oxford University Press; 1976. pp. 1736-1936

[7] Chartrand G, Zhang P. Chromatic Graph Theory. First Edition. Taylor & Francis Group; 2009

[8] Giblin J. Graphs, Surfaces, and Homology: An Introduction to Algebraic Topology. Second Edition. London: Chapman and Hall Ltd.; 1977

[9] Rahman MS. Basic Graph Theory. Springer International Publishing AG; 2017

[10] White AT. Graphs, Groups and Surfaces. Amsterdam: North-Holland Publishing Company; 1973

[11] Wilson RJ, Watkins JJ. Graphs, An Introductory Approach, A First Course in Discrete Mathematics. Canada John Wiley & sons, Inc.; 1990

[12] Hammack R, Imrich W, Klavžar S. Handbook of Product Graphs. Boca Raton: Taylor & Francis Group; 2011

[13] Hell P, Nesetfil J. Graphs and Homomorphisms. New York: Oxford University Press; 2004

[14] El-kholy E, El-Esawy A. Graph folding of some special graphs. Journal of Mathematics and Statistics. 2005; **1**(1):66-70

[15] Massey WS. Algebraic Topology, an Introduction. New York: Harcourt Brace and World; 1967

[16] Jones D. Chaos Math Online. Available from: http://www.mcasco.c om [Accessed: 11 November 2001]

[17] Abu-Saleem M. Folding on the chaotic Cartesian product of manifolds and their fundamental group. Tamkang Journal of Mathematics. 2008;**39**: 353-361

[18] Abu-Saleem M. Dynamical manifold and their fundamental group. Advanced Studies in Contemporary Mathematics. 2010;**20**(1):125-131

[19] Abu-Saleem M. Conditional fractional folding of a manifold and their fundamental group. Advanced Studies in Contemporary Mathematics. 2010;**20**(2):271-277

[20] Abu-Saleem M. Folding on manifolds and their fundamental group. Journal of Taibah University for Science. United Kingdom; 2018:1-4

[21] El-Ghoul M, El-Ahmady AE, Abu-Saleem M. Folding on the Cartesian product of manifolds and their fundamental group. Applied Sciences. Romania; 2007;**9**:86-91

[22] El-Ghoul M, El-Ahmady AE, Rafat H, Abu-Saleem M. The fundamental group of the connected sum of manifolds and their foldings. International Journal of Pure and

Applied Mathematics. 005;**22**(2): 557-567

[23] Frigerio R. Hyperbolic manifold with geodesic boundary which are determine by their fundamental group. Topology and its Applications. 2004;**45**: 69-81

[24] Neto O, Silva PC. The fundamental group of an algebraic link. Comptes Rendus de l'Académie des Sciences, Paris, Série I. 2005;**340**:141-146

[25] Abu-Saleem M. On dynamical chaotic de sitter spaces and their deformation retracts. Proceedings of the Jangjeon Mathematical Society. 2011;**14**: 231-238

[26] Abu-Saleem M. Dynamical chaotic homotopy group and its applications. International Journal of Nonlinear Science. 2011;**11**(12):206-212

[27] Alligood KT, Sauer TD, Yorke JA. Chaos: An Introduction to Dynamical Systems. New York, Heidelberg, Berlin: Springer-Verlag; 1990

[28] Devaney RL. An Introduction to Chaotic Dynamical Systems. 2nd ed. New York: Addison-Wesley; 1989

[29] Peitgen H-O, Jürgen H, Saupe D. Chaos and Fractals: New Frontiers of Science. New York: Springer-Verlag; 1992

[30] Rafat H. Conditional foldings of chaotic manifolds and their retractions [thesis]. Tanta, Egypt: University of Tanta; 2004

[31] Kutschenreiter-Praszkiewicz I. Graph-based decision making in industry. In: Sirmacek B, editor. Graph Theory—Advanced Algorithms and Applications. IntechOpen; 2017. DOI: 10.5772/intechopen.72145. Available from: https://www.intechopen.com/books/graph-theory-advanced-algorithms-and-applications/graph-based-decision-making-in-industry

[32] Mayer DC. Modeling rooted in-trees by finite p-groups. In: Sirmacek B, editor. Graph Theory—Advanced Algorithms and Applications. United Kingdom: IntechOpen; 2017. DOI: 10.5772/intechopen.68703. Available from: https://www.intechopen.com/books/graph-theory-advanced-algorithms-and-applications/modeling-rooted-in-trees-by-finite-p-groups

Chapter 5

A Survey on Hilbert Spaces and Reproducing Kernels

Baver Okutmuştur

Abstract

The main purpose of this chapter is to provide a brief review of Hilbert space with its fundamental features and introduce reproducing kernels of the corresponding spaces. We separate our analysis into two parts. In the first part, the basic facts on the inner product spaces including the notion of norms, pre-Hilbert spaces, and finally Hilbert spaces are presented. The second part is devoted to the reproducing kernels and the related Hilbert spaces which is called the reproducing kernel Hilbert spaces (RKHS) in the complex plane. The operations on reproducing kernels with some important theorems on the Bergman kernel for different domains are analyzed in this part.

Keywords: Hilbert spaces, norm spaces, reproducing kernels, reproducing kernel Hilbert spaces (RKHS), operations on reproducing kernels, sesqui-analytic kernels, analytic functions, Bergman kernel

1. Framework

This chapter consists of introductory concept on the Hilbert space theory and reproducing kernels. We start by presenting basic definitions, propositions, and theorems from functional analysis related to Hilbert spaces. The notion of linear space, norm, inner product, and pre-Hilbert spaces are in the first part. The second part is devoted to the fundamental properties of the reproducing kernels and the related Hilbert spaces. The operations with reproducing kernels, inclusion property, Bergman kernel, and further properties with examples of the reproducing kernels are analyzed in the latter section.

2. Introduction to Hilbert spaces

We start by the definition of a vector space and related topics. Let $\mathbb{C}$ be the complex field. The following preliminaries can be considered as fundamental concepts of the Hilbert spaces.

2.1 Vector spaces and inner product spaces

Vector space. A vector space is a linear space that is closed under vector addition and scalar multiplication. More precisely, if we denote our linear space by $\mathcal{H}$ over the field $\mathbb{C}$, then it follows that

i. if $x, y, z \in \mathcal{H}$, then

$$x + y = y + x \in \mathcal{H}, \quad x + (y + z) = (x + y) + z \in \mathcal{H};$$

ii. if k is scalar, then $kx \in \mathcal{H}$.

Inner product. Let $\mathcal{H}$ be a linear space over the complex field $\mathbb{C}$. An *inner product* on $\mathcal{H}$ is a two variable function

$$\langle \cdot, \cdot \rangle : \mathcal{H} \times \mathcal{H} \to \mathbb{C}, \text{ satisfying}$$

i. $\langle f, g \rangle = \overline{\langle g, f \rangle}$ for $f, g \in \mathcal{H}$.

ii. $\langle \alpha f + \beta g, h \rangle = \alpha \langle f, h \rangle + \beta \langle g, h \rangle$ and $\langle f, \alpha g + \beta h \rangle = \overline{\alpha} \langle f, g \rangle + \overline{\beta} \langle f, h \rangle$ for $\alpha, \beta \in \mathbb{C}$ and $f, g, h \in \mathcal{H}$.

iii. $\langle f, f \rangle \geq 0$ for $f \in \mathcal{H}$ and $\langle f, f \rangle = 0 \Leftrightarrow f = 0$.

Pre-Hilbert space. A *pre-Hilbert space* $\mathcal{H}$ is a linear space over the complex field $\mathbb{C}$ with an inner product defined on it.

Norm space or inner product space. A norm on an inner product space $\mathcal{H}$ denoted by $\| \cdot \|$ is defined by

$$\|f\| = \langle f, f \rangle^{1/2} \text{ or } \|f\|_{\mathcal{H}} = \langle f, f \rangle_{\mathcal{H}}^{1/2}$$

where $f \in \mathcal{H}$ and $\langle \cdot, \cdot \rangle = \langle \cdot, \cdot \rangle_{\mathcal{H}}$ denote the inner product on $\mathcal{H}$. The corresponding space is called as the inner product space or the norm space.

Properties of norm. For all $f, g \in \mathcal{H}$, and $\lambda \in \mathbb{C}$, we have

- $\|f\| \geq 0$. (Observe that the equality occurs only if $f = 0$).
- $\|\lambda f\| = |\lambda| \|f\|$.

Schwarz inequality. For all $f, g \in \mathcal{H}$, it follows that

$$|\langle f, g \rangle| \leq \|f\| \|g\|. \tag{1}$$

In case if f and g are linearly dependent, then the inequality becomes equality.

Triangle inequality. For all $f, g \in \mathcal{H}$, it follows that

$$\|f + g\| \leq \|f\| + \|g\|. \tag{2}$$

In case if f and g are linearly dependent, then the inequality becomes equality.

Polarization identity. For all $f, g \in \mathcal{H}$, it follows that

$$\langle f, g \rangle = \frac{1}{4} \left(\|f + g\|^2 - \|f - g\|^2 + i\|f + ig\|^2 - \|f - ig\|^2 \right) \text{ for } f, g \in \mathcal{H}. \tag{3}$$

Parallelogram identity. For all $f, g \in \mathcal{H}$, it follows that

$$\|f + g\|^2 + \|f - g\|^2 = 2\|f\|^2 + 2\|g\|^2. \tag{4}$$

Metric. A metric on a set X is a function d: $X \times X \to \mathbb{R}$ satisfying the properties.

- $d(x,y) \geq 0$ and $d(x,y) = 0$ only if $x = y$;
- $d(x,y) = d(y,x)$;
- $d(x,y) \leq d(x,z) + d(z,y)$;

for all $x, y, z \in X$. Moreover the space (X, d) is the associated metric space. If we rearrange the metric with its properties for the inner product space $\mathcal{H}$, then it follows that for all $f, g, h \in \mathcal{H}$ and for all $\lambda \in \mathbb{C}$, where d satisfies all requirements to be a metric, we have

- $d(f,g) \geq 0$ and equality occurs only if $f = g$.
- $d(f,g) = d(g,f)$.
- $d(f,g) \leq d(f,h) + d(h,g)$.
- $d(f-h, g-h) = d(f,g)$.
- $d(\lambda f, \lambda g) = |\lambda| \cdot d(f,g)$.

Note. The binary function d given in the metric definition above represents the metric topology in $\mathcal{H}$ which is called *strong topology* or *norm topology*. As a result, a sequence $(f)_{n \geq 0}$ in the pre-Hilbert space $\mathcal{H}$ converges strongly to f if the condition

$$\|f_n - f\| \to 0 \text{ whenever } n \to \infty$$

is satisfied.

2.2 Introduction to linear operators

Linear operator. A map L from a linear space to another linear space is called *linear operator* if

$$L(\alpha f + \beta g) = \alpha L f + \beta L g$$

is satisfied for all α, $\beta \in \mathbb{C}$ and for all $f, g \in \mathcal{H}$.

Continuous operator. An operator L is said to be continuous if it is continuous at each point of its domain. Notice that the domain and range spaces must be convenient for appropriate topologies.

Lipschitz constant of a linear operator. If L is a linear operator from $\mathcal{H}$ to $\mathcal{G}$ where $\mathcal{H}$ and $\mathcal{G}$ are pre-Hilbert spaces, then the Lipschitz constant for L is its norm $\|L\|$ and it is defined by

$$\|L\| = \sup\{\|Lf\|_{\mathcal{G}} / \|f\|_{\mathcal{H}} : 0 \neq f \in \mathcal{H}\}. \tag{5}$$

Theorem 1. Let L be a linear operator from the pre-Hilbert spaces $\mathcal{H}$ to $\mathcal{G}$. Then the followings are mutually equivalent:

i. L is continuous.

ii. L is bounded, that is,

$$\sup\{\|Lf\|_G : \|f\|_H \leq k\} < \infty$$

for $0 \leq k < \infty$.

iii. L is Lipschitz continuous, that is,

$$\|Lf - Lg\|_G \leq \lambda \|f - g\|_{\mathcal{H}},$$

where $0 \leq \lambda < \infty$ and $f, g \in \mathcal{H}$.

Some properties of linear operators. Let $B(\mathcal{H}, \mathcal{G})$ be the collection of all continuous linear operators from the pre-Hilbert spaces $\mathcal{H}$ to $\mathcal{G}$. Then

- $B(\mathcal{H}, \mathcal{G})$ is a linear space with respect to the natural addition and scalar multiplication satisfying

$$(\alpha L + \beta M)f = \alpha Lf + \beta Mf,$$

where L and M are linear operators, $f \in \mathcal{H}$ and $\alpha, \beta \in \mathbb{C}$.

- Whenever $\mathcal{H} = \mathcal{G}$, then $B(\mathcal{H}, \mathcal{G})$ is denoted by $B(\mathcal{H})$.

- If $\mathcal{K}$ is another pre-Hilbert space, $L \in B(\mathcal{H}, \mathcal{G})$ and $K \in B(\mathcal{G}, \mathcal{K})$. Then the product

$$(KL)f = K(Lf) \text{ for } f \in \mathcal{H} \in B(\mathcal{H}, \mathcal{K}).$$

In addition,

i. $K(\xi L + \zeta M) = \xi KL + \zeta KM$

ii. $\|\xi L\| = |\xi| \cdot \|L\|$

iii. $\|L + M\| \leq \|L\| + \|M\|$ and

iv. $\|KL\| \leq \|K\| \|L\|$.

are also satisfied.

2.3 Hilbert spaces and linear operators

Linear form (or linear functional). A linear operator from the pre-Hilbert space $\mathcal{H}$ to the scalar field $\mathbb{C}$ is called a *linear form* (or *linear functional*).

Hilbert spaces. A pre-Hilbert space $\mathcal{H}$ is said to be a *Hilbert space* if it is complete in metric. In other words if f_n is a Cauchy sequence in $\mathcal{H}$, that is, if

$$\|f_n - f_m\| \to 0 \text{ whenever } n, m \to \infty,$$

then there is $f \in \mathcal{H}$ such that

$$\|f_n - f\| \to 0 \text{ whenever } n \to \infty.$$

Note. Every subspace of a pre-Hilbert space is also a pre-Hilbert space with respect to the induced inner product. However, the reverse is not always true. For a subspace of a Hilbert space to be also a Hilbert space, it must be closed.

Completion. The canonical method for which a pre-Hilbert space $\mathcal{H}$ is embedded as a dense subspace of a Hilbert space $\tilde{\mathcal{H}}$ so that

$$\langle f, g \rangle_{\tilde{\mathcal{H}}} = \langle f, g \rangle_{\mathcal{H}} \text{ for } f, g \in \mathcal{H}$$

is called *completion*.

Note. If L is a continuous linear operator from a dense subspace $\mathcal{M}$ of a Hilbert space $\mathcal{H}$ to a Hilbert space $\mathcal{G}$, then it can be extended uniquely to a continuous linear operator from $\mathcal{H}$ to $\mathcal{G}$ with preserving norm.

Theorem 2. Let $\mathcal{M}$ and $\mathcal{N}$ be dense subspaces of the Hilbert spaces $\mathcal{H}$ and $\mathcal{G}$, respectively. For $f \in \mathcal{H}, g \in \mathcal{M}$ and $0 \leq \lambda < \infty$, if a linear operator L from $\mathcal{M}$ to $\mathcal{G}$ satisfies

$$|\langle Lf, g\rangle_{\mathcal{G}}| \leq |\lambda| \|f\|_{\mathcal{H}} \|g\|_{\mathcal{G}}, \tag{6}$$

then L is uniquely extended to a continuous linear operator from $\mathcal{M}$ to $\mathcal{G}$ with norm $\leq \lambda$ where the norm coincides with the minimum of such λ.

Theorem 3. Let (Ω, μ) denotes a measure space so that Ω is the union of subsets of finite positive measure and $L^2(\Omega, \mu)$ consists of all measurable functions $f(\omega)$ on Ω such that

$$\int_{\Omega} |f(\omega)|^2 d\mu(\omega) < \infty. \tag{7}$$

Then $L^2(\Omega, \mu)$ is a Hilbert space with respect to the inner product

$$\langle f, g\rangle := \int_{\Omega} f(\omega)\overline{g(\omega)}\, d\mu(\omega). \tag{8}$$

Theorem 4 (F. Riesz). For each continuous linear functional φ on a Hilbert space $\mathcal{H}$, there exists uniquely $g \in \mathcal{H}$ such that

$$\varphi(f) = \langle f, g\rangle \ for\, f \in \mathcal{H}. \tag{9}$$

Theorem 5. Let $\mathcal{M}$ be a closed subspace of a Hilbert space $\mathcal{H}$. Then the algebraic direct sum relation

$$\mathcal{H} = \mathcal{M} \oplus \mathcal{M}^{\perp}$$

is satisfied. In other words, $\forall f \in \mathcal{H}$ can be uniquely written by

$$f = f_{\mathcal{M}} + f_{\mathcal{M}^{\perp}} \text{ with } f_{\mathcal{M}} \in \mathcal{M},\ f_{\mathcal{M}^{\perp}} \in \mathcal{M}^{\perp}. \tag{10}$$

In addition, $\|f_{\mathcal{M}}\|$ coincides with the distance from f to $\mathcal{M}^{\perp}$

$$\|f_{\mathcal{M}}\| = \min\left\{\|f - g\| : g \in \mathcal{M}^{\perp}\right\}. \tag{11}$$

Remark. In a Hilbert space, the closed linear span of any subset A of a Hilbert space $\mathcal{H}$ coincides with $\left(A^{\perp}\right)^{\perp}$.

Total subset of a Hilbert space. A subset $\mathcal{A}$ of a Hilbert space $\mathcal{H}$ is called *total* in $\mathcal{H}$ if 0 is the only element that is orthogonal to all elements of $\mathcal{A}$. In other words,

$$\mathcal{A}^{\perp} = \{0\}.$$

As a result, $\mathcal{A}$ is total if and only if every element of $\mathcal{H}$ can be approximated by linear combinations of elements of $\mathcal{A}$.

Orthogonal projection. If $\mathcal{M}$ is a closed subspace of $\mathcal{H}$, the map $f \mapsto f_{\mathcal{M}}$ gives a linear operator from $\mathcal{H}$ to $\mathcal{M}$ with norm ≤ 1. We call this operator as the *orthogonal projection* to $\mathcal{M}$ and denote it by $P_{\mathcal{M}}$.

Note. If I is the identity operator on $\mathcal{H}$, then $I - P_{\mathcal{M}}$ denotes the orthogonal projection to $\mathcal{M}^{\perp}$, and the relation

$$\|f\|^2 = \|P_{\mathcal{M}}f\|^2 + \|(I - P_{\mathcal{M}})f\|^2 \tag{12}$$

is satisfied for all $f \in \mathcal{H}$.

Weak topology. The weakest topology that makes continuous all linear functionals of the form $f \mapsto \langle f, g\rangle$ is called *the weak topology* of a Hilbert space $\mathcal{H}$.

Note. If $f \in \mathcal{H}$, then with respect to the weak topology, a fundamental system of neighborhoods of f is composed of subsets of the form

$$U(f; A, \epsilon) = \{h : |\langle f, g\rangle - \langle h, g\rangle| < \epsilon \text{ for } g \in \mathcal{A}\},$$

where $\mathcal{A}$ is a finite subset of $\mathcal{H}$ and $\epsilon > 0$. Then a directed net $\{f_\lambda\}$ converges weakly to f if and only if

$$\langle f_\lambda, g\rangle \xrightarrow{\lambda} \langle f, g\rangle \text{ for all } g \in \mathcal{H}.$$

Operator weak topology. The weakest topology that makes continuous all linear functionals of the form

$$L \mapsto \langle Lf, g\rangle \text{ for } f \in \mathcal{H}, g \in \mathcal{G}$$

is called *the operator weak topology* in the space $B(\mathcal{H}, \mathcal{G})$ of continuous linear operators from $\mathcal{H}$ to $\mathcal{G}$. In addition, a directed net $\{L_\lambda\}$ converges weakly to L if

$$\langle L_\lambda f, g\rangle \xrightarrow{\lambda} \langle Lf, g\rangle.$$

Operator strong topology. The weakest topology that makes continuous all linear operators of the form

$$L \mapsto Lf \text{ for } f \in \mathcal{H}$$

is called *the operator strong topology*. Moreover a directed net $\{L_\lambda\}$ converges strongly to L if

$$\|L_\lambda f - Lf\| \xrightarrow{\lambda} 0 \text{ for all } f \in \mathcal{H}.$$

Theorem 6. Let $\mathcal{H}$ and $\mathcal{G}$ be Hilbert spaces and $B(\mathcal{H}, \mathcal{G})$ be a continuous linear operator from $\mathcal{H}$ to $\mathcal{G}$. Then

- the closed unit ball $U := \{f : \|f\| \le 1\}$ of $\mathcal{H}$ is weakly compact;
- the closed unit ball $\{L : \|L\| \le 1\}$ of $B(\mathcal{H}, \mathcal{G})$ is weakly compact.

Theorem 7. Let $\mathcal{H}$ be a Hilbert space and $A \subseteq \mathcal{H}$. Then if A is weakly bounded in the sense

$$\sup_{f \in A} |\langle f, g\rangle| < \infty \text{ for } g \in \mathcal{H}, \tag{13}$$

then it is strongly bounded, that is, $\sup_{f \in A} \|f\| < \infty$.

Theorem 8. If $\mathcal{H}$ and $\mathcal{G}$ are Hilbert spaces and L is a linear operator from $\mathcal{H}$ to $\mathcal{G}$, then the strong continuity and weak continuity for L are equivalent.

Theorem 9. Let $\mathcal{H}$ and $\mathcal{G}$ be Hilbert spaces. Then the following statements for $\mathbf{L} \subseteq B(\mathcal{H}, \mathcal{G})$ are mutually equivalent:

(i) $\mathbf{L}$ is weakly bounded; that is, for $f \in \mathcal{H},\ g \in \mathcal{G}$, we have

$$\sup_{L \in \mathbf{L}} |\langle Lf, g \rangle| < \infty$$

(ii) $\mathbf{L}$ is strongly bounded; that is, for $f \in \mathcal{H}$, we have

$$\sup_{L \in \mathbf{L}} \|Lf\| < \infty.$$

(iii) $\mathbf{L}$ is norm bounded (or uniformly bounded); that is,

$$\sup_{L \in \mathbf{L}} \|L\| < \infty.$$

Theorem 10. A linear operator L from the Hilbert spaces $\mathcal{H}$ to $\mathcal{G}$ is said to be *closed* if its *graph*

$$G_L := \{ f \oplus Lf : f \in \mathcal{H} \} \tag{14}$$

is a closed subspace of the direct sum space $\mathcal{H} \oplus \mathcal{G}$, that is, whenever $n \to \infty$,

$$\|f_n - f\| \to 0 \ in\ \mathcal{H} \ \ and \ \ \|Lf_n - g\| \to 0 \ \ in \ \ \mathcal{G} \Rightarrow g = Lf.$$

Theorem 11. If L is a closed linear operator with a domain of a Hilbert space $\mathcal{H}$ to another Hilbert space $\mathcal{G}$, then it is continuous.

Sesqui-linear form. A function $\Phi : \mathcal{H} \times \mathcal{G} \to \mathbb{C}$ is a *sesqui-linear form* (or *sesqui-linear function*) if for $f, h \in \mathcal{H}, g, k \in \mathcal{G}$ and $\alpha, \beta \in \mathbb{C}$,

$$\text{(i) } \Phi(\alpha f + \beta h, g) = \alpha \Phi(f, g) + \beta \Phi(h, g) \tag{15}$$

$$\text{(ii) } \Phi(f, \alpha g + \beta k) = \overline{\alpha} \Phi(f, g) + \overline{\beta} \Phi(f, k) \tag{16}$$

are satisfied where $\mathcal{H}$ and $\mathcal{G}$ are Hilbert spaces.

Remark. If $L \in B(\mathcal{H}, \mathcal{G})$, then the sesqui-linear form Φ defined by

$$\Phi(f, g) = \langle Lf, g \rangle_G \tag{17}$$

is bounded in the sense that

$$|\Phi(f, g)| \leq \lambda \|f\|_{\mathcal{H}} \|g\|_{\mathcal{G}} \text{ for } f \in \mathcal{H},\ g \in \mathcal{G}, \tag{18}$$

where $\lambda \geq \|L\|$.

Remark. If a sesqui-linear form Φ satisfies the condition (18), then for $f \in \mathcal{H}$, the linear functional

$$g \mapsto \overline{\Phi(f, g)}$$

is continuous on $\mathcal{G}$. If we apply the Riesz theorem, then there exists uniquely $f' \in \mathcal{G}$ satisfying

$$\|f'\|_{\mathcal{G}} \le \lambda \|f\|_{\mathcal{H}} \text{ and } \Phi(f,g) = \langle f',g\rangle_{\mathcal{G}} \text{ for } g \in \mathcal{G}.$$

Hence $f \mapsto f'$ becomes linear, and as a result we obtain

$$\Phi(f,g) = \langle f',g\rangle_{\mathcal{G}} = \langle Lf,g\rangle_{\mathcal{G}}.$$

Adjoint operator. If $L \in B(\mathcal{H},\mathcal{G})$, then the unique operator $L^* \in B(\mathcal{G},\mathcal{H})$ satisfying

$$\Phi(f,g) = \langle f, L^* g\rangle_{\mathcal{H}} \text{ for } f \in \mathcal{H}, g \in \mathcal{G} \tag{19}$$

is called the *adjoint* of L.

Remark. By the definitions of L and L^*, it follows that

$$\langle Lf,g\rangle_{\mathcal{G}} = \langle f, L^* g\rangle_{\mathcal{H}} \text{ for } f \in \mathcal{H},\ g \in \mathcal{G}. \tag{20}$$

Isometric property. The adjoint operation is isometric if

$$\|L\| = \|L^*\| \text{ is satisfied.} \tag{21}$$

Remark. Let $\mathcal{H}, \mathcal{G}$, and $\mathcal{K}$ be Hilbert spaces and $K \in B(\mathcal{G},\mathcal{K})$ and $L \in B(\mathcal{H},\mathcal{G})$ be given. Then

$$KL \in B(\mathcal{H},\mathcal{K}) \text{ and } (KL)^* = L^* K^* \tag{22}$$

$$\mathrm{Ker}(L) = (\mathrm{Ran}(L^*))^{\perp} \text{ and } (\mathrm{Ker}(L))^{\perp} = \mathrm{Clos}\{\mathrm{Ran}(L)^*\} \tag{23}$$

where $\mathrm{Ker}(L)$ is the kernel of L and $\mathrm{Ran}(L)$ is the range of L.

Theorem 12. If $L, M \in B(\mathcal{H},\mathcal{G})$, then the following statements are mutually equivalent.

i. $\mathrm{Ran}(M) \subseteq \mathrm{Ran}(L)$.

ii. There exists $K \in B(\mathcal{H})$ such that $M = LK$.

iii. There exists $0 \le \lambda < \infty$ such that

$$\|M^* g\| \le \lambda \|L^* g\| \text{ for } g \in \mathcal{G}.$$

Quadric form. Let $\mathcal{H}$ be a Hilbert space. A function

$$\varphi : \mathcal{H} \to \mathbb{C}$$

is a *quadratic form* if for all $f \in \mathcal{H}$ and $\zeta \in \mathbb{C}$,

$$\varphi(\zeta f) = |\zeta|^2 \varphi(f) \tag{24}$$

and

$$\varphi(f+g) + \varphi(f-g) = 2\{\varphi(f) + \varphi(g)\} \tag{25}$$

are satisfied.

Note. If $L \in B(\mathcal{H})$, the quadratic form φ on $\mathcal{H}$ is defined by

$$\varphi(f) = \langle Lf, f \rangle \text{ for } f \in \mathcal{H}, \tag{26}$$

and it is bounded

$$|\varphi(f)| \leq \lambda \|f\|^2 \text{ for } f \in \mathcal{H}, \tag{27}$$

where $\lambda \geq \|L\|$.

Remark. The sesqui-linear form Φ associated with L can be recovered from the quadratic form φ by the equation

$$\Phi(f, g) = \frac{1}{4}\{\varphi(f+g) - \varphi(f-g)\} + \{\varphi(f+ig) - \varphi(f-ig)\} \tag{28}$$

for all $f, g \in \mathcal{H}$.

Self-adjoint operator. A continuous linear operator L on a Hilbert space $\mathcal{H}$ is said to be *self-adjoint* if $L = L^*$.

Remark. L is self-adjoint if and only if the associated sesqui-linear form Φ is Hermitian.

Remark. If L is self-adjoint, then the norm of L coincides with the minimum of λ given in (27) for the related quadratic form

Theorem 13. If L is a continuous self-adjoint operator, then

$$\|L\| = \sup\{|\langle Lf, f \rangle| : \|f\| \leq 1\}. \tag{29}$$

Positive definite operator. A self-adjoint operator $L \in B(\mathcal{H})$ is said to be *positive* (or *positive definite*) if

$$\langle Lf, f \rangle \geq 0 \quad \text{for all } f \in \mathcal{H}.$$

If $\langle Lf, f \rangle = 0$ only when $f = 0$, then L is said to be *strictly positive* (or, *strictly positive definite*).

Note. For any positive operator $L \in B(\mathcal{H})$, the Schwarz inequality holds in the following sense

$$|\langle Lf, g \rangle|^2 \leq \langle Lf, f \rangle \cdot \langle Lg, g \rangle. \tag{30}$$

Theorem 14. Let L and M be continuous positive operators on $\mathcal{H}$ and $\mathcal{G}$, respectively. Then a continuous linear operator K from $\mathcal{H}$ to $\mathcal{G}$ satisfies the inequality

$$\left|\langle Kf, g \rangle_{\mathcal{G}}\right|^2 \leq \langle Lf, f \rangle_{\mathcal{H}} \langle Mg, g \rangle_{\mathcal{G}} \text{ for } f \in \mathcal{H}, \in \mathcal{G} \tag{31}$$

if and only if the continuous linear operator

$$\begin{bmatrix} L & K^* \\ K & M \end{bmatrix}$$

on the direct sum Hilbert space $\mathcal{H} \oplus \mathcal{G}$ with

$$f \oplus g \mapsto (Lf + K^* g) \oplus (Kf + Mg)$$

is positive definite.

Theorem 15. Let L be a continuous positive definite operator. Then there exists a unique positive definite operator called the *square root* of L, denoted by $L^{1/2}$, such that $(L^{1/2})^2 = L$.

Modulus operator. The square root of the positive definite operator L^*L is called the *modulus* (*operator*) of L if L is a continuous linear operator.

Isometry. A linear operator U between Hilbert spaces $\mathcal{H}$ and $\mathcal{G}$ is called *isometric* or an *isometry* if

$$\|Uf\|_{\mathcal{G}} = \|f\|_{\mathcal{H}} \text{ for } f \in \mathcal{H} \tag{32}$$

is satisfied, that is, it preserves the norm.

Note. Eq. (32) implies that a continuous linear operator U is isometric if and only if $U^*U = I_{\mathcal{H}}$; in other words,

$$\langle Uf, Ug\rangle_{\mathcal{G}} = \langle f, g\rangle_{\mathcal{H}} \text{ for } f, g \in \mathcal{H}, \tag{33}$$

that is, U preserves the inner product.

Unitary operator. A surjective isometry linear operator $U : \mathcal{H} \to \mathcal{H}$ is called a *unitary* (*operator*).

Note. Observe that if $U \in B(H)$ is a unitary operator, then $U^* = U^{-1}$.

Partial isometry. A continuous linear operator U between Hilbert spaces $\mathcal{H}$ and $\mathcal{G}$ is called a *partial isometry* if

$$f \in (\text{Ker}U)^{\perp} = \text{Ran}(U^*) \Rightarrow \|Uf\| = \|f\|.$$

The spaces $(\text{Ker}U)^{\perp}$ and $Ran(U)$ are called the *initial space* of U and the *final space* of U, respectively.

Note. If U is a partial isometry, then its adjoint U^* is also a partial isometry.

Theorem 17. Every continuous linear operator L on $\mathcal{H}$ admits a unique decomposition

$$L = U\tilde{L}, \tag{34}$$

where $\tilde{L}$ is a positive definite operator and U is a partial isometry with initial space the closure of $\text{Ran}(\tilde{L})$.

3. Reproducing kernels and RKHS

We continue our analysis on the abstract theory of reproducing kernels.

3.1 Definition and fundamental properties

Reproducing kernels. Let $\mathcal{H}$ be a Hilbert space of functions on a nonempty set X with the inner product $\langle f, g\rangle$ and norm $\|f\| = \langle f, f\rangle^{1/2}$ for f and $g \in \mathcal{H}$. Then the complex valued function $K(y, x)$ of y and x in X is called a *reproducing kernel of* $\mathcal{H}$ if

i. For all $x \in X$, it follows that $K_x(\cdot) = K(\cdot, x) \in \mathcal{H}$,

ii. For all $x \in X$ and all $f \in \mathcal{H}$,

$$f(x) = \langle f, K_x\rangle, \tag{35}$$

are satisfied.

Note. Let K be a reproducing kernel. Applying (35) to the function K_x at y, we get

$$K_x(y) = K(y,x) = \langle K_y, K_x \rangle, \text{ for } x, y \in X. \tag{36}$$

Then, for any $x \in X$, we obtain

$$\|K_x\| = \langle K_x, K_x \rangle^{1/2} = K(x,x)^{1/2}. \tag{37}$$

Note. Observe that the subset $\{K_x\}_{x\in X}$ is total in $\mathcal{H}$, that is, its closed linear span coincides with $\mathcal{H}$. This follows from the fact that, if $f \in \mathcal{H}$ and $f \perp K_x$ for all $x \in X$, then

$$f(x) = \langle f, K_x \rangle = 0 \text{ for all } x \in X,$$

and hence f is the 0 element in $\mathcal{H}$. As a result, $\{0\}^{\perp} = \mathcal{H}$.

RKHS. A Hilbert space $\mathcal{H}$ of functions on a set X is called a RKHS if there exists a reproducing kernel K of $\mathcal{H}$.

Theorem 18. If a Hilbert space $\mathcal{H}$ of functions on a set X admits a reproducing kernel K, then this reproducing kernel K is unique.

Theorem 19. There exists a reproducing kernel K for $\mathcal{H}$ for a Hilbert space $\mathcal{H}$ of functions on X, if and only if for all $x \in X$, the linear functional $\mathcal{H} \ni f \mapsto f(x)$ of evaluation at x is bounded on $\mathcal{H}$.

Hermitian and positive definite kernel. Let X be an arbitrary set and K be a kernel on X, that is, $K : X \times X \to \mathbb{C}$. The kernel K is called *Hermitian* if for any finite set of points $\{y_1, \dots, y_n\} \subseteq X$, we have

$$\sum_{i,j=1}^{n} \overline{\epsilon}_j \epsilon_i K\left(y_j, y_i\right) \in \mathbb{R}.$$

It is called *positive definite*, if for any complex numbers $\epsilon_1, \dots, \epsilon_n$, we have

$$\sum_{i,j=1}^{n} \overline{\epsilon}_j \epsilon_i K\left(y_j, y_i\right) \geq 0.$$

Note. From the previous inequality, it follows that for any finitely supported family of complex numbers $\{\epsilon_x\}_{x\in X}$, we have

$$\sum_{x,y\in X} \overline{\epsilon}_y \epsilon_x K(y,x) \geq 0. \tag{38}$$

Theorem 20. The reproducing kernel K of a reproducing kernel Hilbert space $\mathcal{H}$ is a positive definite matrix in the sense of E.H. Moore.

Properties of RKHS. Given a reproducing kernel Hilbert space $\mathcal{H}$ and its kernel $K(y,x)$ on X, then for all $x, y \in X$, we have

i. $K(y,y) \geq 0$.

ii. $K(y,x) = \overline{K(x,y)}$.

iii. $|K(y,x)|^2 \leq K(y,y)K(x,x)$ (Schwarz inequality).

iv. Let $x_0 \in X$. Then the following statements are equivalent:

a. $K(x_0, x_0) = 0$.

b. $K(y, x_0) = 0$ for all $y \in X$.

c. $f(x_0) = 0$ for all $f \in \mathcal{H}$.

Theorem 21. For any positive definite kernel K on X, there exists a unique Hilbert space $\mathcal{H}_K$ of functions on X with reproducing kernel K.

Theorem 22. Every sequence of functions $(f_n)_{n\geq 1}$ that converges strongly to a function f in $\mathcal{H}_K(X)$ converges also in the pointwise sense, i.e., for any point $x \in X$,

$$\lim_{n\to\infty} f_n(x) = f(x).$$

In addition, this convergence is uniform on every subset of X on which $x \mapsto K(x,x)$ is bounded.

Theorem 23. A complex valued function g on X belongs to the reproducing kernel Hilbert space $\mathcal{H}_K(X)$ if and only if there exists $0 \leq \lambda < \infty$ such that,

$$\left[g(y)\overline{g(x)}\right] \leq \lambda^2 [K(y,x)] \text{ on } X. \tag{39}$$

$\|g\|$ coincides with the minimum of all such λ.

Theorem 24. If $K^{(1)}(y,x)$ and $K^{(2)}(y,x)$ are two positive definite kernels on X, then the following statements are mutually equivalent:

i. $\mathcal{H}_{K^{(1)}}(X) \subseteq \mathcal{H}_{K^{(2)}}(X)$.

ii. There exists $0 \leq \lambda < \infty$ such that

$$\left[K^{(1)}(y,x)\right] \leq \lambda^2 \left[K^{(2)}(y,x)\right].$$

Note. For any map φ from a set X to a Hilbert space $\mathcal{H}$, with the notation $x \mapsto \varphi_x$, a kernel K can be defined by

$$K(y,x) = \left\langle \varphi_x, \varphi_y \right\rangle \text{ for } x, y \in X. \tag{40}$$

Theorem 25. Let $\varphi : X \mapsto \mathcal{H}$ be an arbitrary map and for $x, y \in X$ let K be defined as

$$K(y,x) = \left\langle \varphi_x, \varphi_y \right\rangle.$$

Then K is a positive definite kernel.

Theorem 26. Let T be the linear operator from $\mathcal{H}$ to the space of functions on X, defined by

$$(Tf)(x) = \langle f, \varphi_x \rangle \text{ for } x \in X, f \in \mathcal{H}.$$

Then $\mathrm{Ran}(T)$ coincides with $\mathcal{H}_K(X)$ and

$$\|Tf\|_K = \|P_{\mathcal{M}} f\| \text{ for } f \in \mathcal{H},$$

where $\mathcal{M}$ is the orthogonal complement of $\text{Ker}(T)$, $P_{\mathcal{M}}$ is the orthogonal projection onto $\mathcal{M}$, and $\|\cdot\|_K$ denotes the norm in $\mathcal{H}_K(X)$.

Kolmogorov decomposition. Let $K(y,x)$ be a positive definite kernel on an abstract set X. Then there exists a Hilbert space $\mathcal{H}$ and a function $\varphi : X \to \mathcal{H}$ such that

$$K(y,x) = \left\langle \varphi_x, \varphi_y \right\rangle \text{ for } x,y \in X.$$

3.2 Operations with RKHSs

Theorem 27. Let $K^{(0)}$ be the restriction of the positive definite kernel K to a nonempty subset X_0 of X and let $\mathcal{H}_{K^{(0)}}(X)$ and $\mathcal{H}_K(X)$ be the RKHS corresponding to $K^{(0)}$ and K, respectively. Then

$$\mathcal{H}_{K^{(0)}}(X_0) = \{ f|_{X_0} : f \in \mathcal{H}_K(X) \} \tag{41}$$

and

$$\|h\|_{K^{(0)}} = \min \left\{ \|f\|_K : f|_{X_0} = h \right\} \text{ for all } h \in \mathcal{H}_{K^{(0)}}(X_0). \tag{42}$$

Remark. If $K^{(1)}(y,x)$ and $K^{(2)}(y,x)$ are two positive definite kernels, then

$$K(y,x) = K^{(1)}(y,x) + K^{(2)}(y,x)$$

is also a positive definite kernel.

Remark. Let $\mathcal{H}_{K^{(1)}}, \mathcal{H}_{K^{(2)}}$, and $\mathcal{H}_K$ be RKHSs with reproducing kernels $K^{(1)}(y,x)$, $K^{(2)}(y,x)$, and $K(y,x)$, respectively, and let $K = K^{(1)} + K^{(2)}$. Then

$$\mathcal{H}_K(X) = \mathcal{H}_{K^{(1)}}(X) + \mathcal{H}_{K^{(2)}}(X),$$

and for $f \in \mathcal{H}_{K^{(1)}}(X)$ and $g \in \mathcal{H}_{K^{(2)}}(X)$, it follows that

$$\|f+g\|_K^2 = \min \left\{ \|f+h\|_{K^{(1)}}^2 + \|g-h\|_{K^{(2)}}^2 : h \in \mathcal{H}_{K^{(1)}}(X) \cap \mathcal{H}_{K^{(2)}}(X) \right\}. \tag{43}$$

Theorem 28. The intersection $\mathcal{H}_{K^{(1)}}(X) \cap \mathcal{H}_{K^{(2)}}(X)$ of Hilbert spaces $\mathcal{H}_{K^{(1)}}(X)$ and $\mathcal{H}_{K^{(2)}}(X)$ is again a Hilbert space of functions on X with respect to the norm

$$\|f\|^2 := \|f\|_{K^{(1)}}^2 + \|f\|_{K^{(2)}}^2.$$

In addition the intersection Hilbert space is a RKHS.

Theorem 29. The reproducing kernel of the space

$$\mathcal{H}_K(X) = \mathcal{H}_{K^{(1)}}(X) \cap \mathcal{H}_{K^{(2)}}(X)$$

is determined, as a quadratic form, by

$$\begin{aligned} \sum_{x,y} \overline{\varepsilon_y} \varepsilon_x K(y,x) = \inf \{ &\sum_{x,y} \overline{\eta_y} \eta_x K^{(1)}(y,x) + \sum_{x,y} \overline{\zeta_y} \zeta_x K^{(2)}(y,x) : [\varepsilon_x] \\ &= [\eta_x] + [\zeta_x] \}, \end{aligned}$$

where $[\epsilon_x]$, $[\eta_x]$, $[\zeta_x]$ are an arbitrary complex valued function on X with finite support.

Theorem 30. The tensor product Hilbert space

$$\mathcal{H}_{K^{(1)}}(X) \otimes \mathcal{H}_{K^{(2)}}(X)$$

is a RKHS on $X \times X$.

Theorem 31. The RKHS $\mathcal{H}_K(X)$ of the kernel $K(y,x) = K^{(1)}(y,x) \cdot K^{(2)}(y,x)$ consists of all functions f on X for which there are sequences $(g_n)_{n \geq 0}$ of functions in $\mathcal{H}_{K^{(1)}}(X)$ and $(h_n)_{n \geq 0}$ of functions in $\mathcal{H}_{K^{(2)}}(X)$ so that

$$\sum_1^{\infty} \|g_n\|_{K^{(1)}}^2 \|h_n\|_{K^{(2)}}^2 < \infty, \quad \sum_1^{\infty} g_n(x)h_n(x) = f(x), \ x \in X, \tag{44}$$

and the norm is given by

$$\|f\|_K^2 = \min\left\{\sum_1^{\infty} \|g_n\|_{K^{(1)}}^2 \|h_n\|_{K^{(2)}}^2\right\},$$

where the minimum is taken over the set of all sequences $(g_n)_{n \geq 0}$ and $(h_n)_{\geq 0}$ satisfying (44).

3.3 Examples of RKHS. Bergman and Hardy spaces

Bergman space. The space of all analytic functions f on Ω for which

$$\int\int_{\Omega} |f(z)|^2 dxdy < \infty, \qquad (z = x + iy)$$

is satisfied is called the *Bergman space* on Ω and denoted by $A^2(\Omega)$.

Remark. $A^2(\Omega)$ is a *RKHS* with respect to the inner product

$$\langle f, g\rangle \equiv \langle f, g\rangle_{\Omega} := \int\int_{\Omega} f(z)\overline{g(z)} dxdy,$$

and its kernel is called the *Bergman kernel* on Ω and denoted by $B^{(\Omega)}(w,z)$.

Bergman kernel for the unit disc. The Bergman kernel for the open unit disc $\mathbb{D}$ is given by

$$B^{(\mathbb{D})}(w,z) = \frac{1}{\pi} \frac{1}{(1 - w\overline{z})^2} \qquad \text{for } w, z \in \mathbb{D}. \tag{45}$$

Bergman kernel of a simply connected domain. The Bergman kernel of a simply connected domain $\Omega(\neq \mathbb{C})$ is given by

$$B^{(\Omega)}(w,z) = \frac{1}{\pi} \frac{\varphi'(w)\overline{\varphi'(z)}}{\left(1 - \varphi(w)\overline{\varphi(z)}\right)^2} \text{ for } w, z \in \Omega, \tag{46}$$

where φ is any conformal mapping function from Ω onto $\mathbb{D}$.

Theorem 32. A conformal mapping from Ω to $\mathbb{D}$ can be recovered from the Bergman kernel of Ω.

Jordan curve. A *Jordan curve* is a continuous $1-1$ image of $\{|\xi| = 1\}$ in $\mathbb{C}$.

Green function. A *Green function* $G(w,z)$ of Ω is a function harmonic in Ω except at z, where it has logarithmic singularity, and continuous in the closure $\overline{\Omega}$, with boundary values $G(w,z) = 0$ for all $w \in \partial\Omega$, where Ω is a finitely connected domain of the complex plane.

Theorem 33. Let Ω be a finitely connected domain bounded by analytic Jordan curves, and let $G(w,z)$ be the Green's function of Ω. Then the Bergman kernel function is

$$B^{(\Omega)}(w,z) = -\frac{2}{\pi}\frac{\partial^2 G}{\partial w \partial \overline{z}}(w,z), \; w \neq z. \tag{47}$$

Hardy space. The closed linear span of $\{\varphi_n : n = 0, 1, \ldots\}$ in $L^2((T)$ is called the (Hilbert type) *Hardy space* on $\mathbb{T}$ and is denoted by $H^2(\mathbb{T})$. Here $\varphi_n(\xi) = \xi^n$.

Remark. $f \in L^2(\mathbb{T})$ belongs to the Hardy space $H^2(\mathbb{T})$ if and only if it is orthonormal to all φ_n $(n < 0)$, that is, all Fourier coefficients of f with negative indices vanish. Then we have

$$\langle f, g \rangle_{L^2} = \sum_{n=0}^{\infty} a_n \overline{b}_n \text{ for } f, g \in H^2(\mathbb{T}), \tag{48}$$

where

$$a_n = \langle f, \varphi_n \rangle_{L^2} \text{ and } b_n = \langle g, \varphi_n \rangle_{L^2} \quad (n = 0, 1, \ldots).$$

Szegö kernel. The kernel $S(\xi, z) := \frac{1}{1-\xi\overline{z}}$ for $\xi \in \mathbb{T}, z \in \mathbb{D}$, or its analytic extension $\tilde{S}(w,z) := \frac{1}{1-w\overline{z}}$ for $w, z \in \mathbb{D}$ is called the *Szegö kernel*.

Notes

This chapter intends to offer a sample survey for the fundamental concepts of Hilbert spaces and provide an introductory theory of reproducing kernels. We present the basic properties with important theorems and sometimes with punctual notes and remarks to support the subject. However, due to the limit of content and pages, we skipped the proofs of the theorems. The proofs of the first part can be found in [1, 2] and in most of the basic functional analysis books. Besides, the proofs of the second part (related with the reproducing kernels) can easily be found in [3]. The Hilbert space and functional analysis parts of this chapter are based on the books by J.B. Conway [1] and R.G. Douglas [2]. On the other hand, the reproducing kernel part is based on the lecture notes of T. Ando [4] and N. Aronszajn [5], the book of S. Saitoh and Y. Sawano [6], and the book of B. Okutmustur and A. Gheondea [3]. Moreover, the details of Bergman and Hardy spaces are widely explained in the books [7–9].

Author details

Baver Okutmuştur
Department of Mathematics, Middle East Technical University (METU), Ankara, Turkey

*Address all correspondence to: baver@metu.edu.tr

References

[1] Conway JB. A Course in Functional Analysis. Berlin-Heidelberg, New York: Springer-Verlag; 1989

[2] Douglas RG. Banach Algebra Techniques in Operator Theory. New York: Springer-Verlag, Academic Press; 1972

[3] Okutmustur B, Gheondea A. Reproducing Kernel Hilbert Spaces: The Basics, Bergman Spaces, and Interpolation Problems of Reproducing Kernels and its Applications. Riga-Latvia: Lap Lambert Academic Publishing; 2010

[4] Ando T. Reproducing Kernel Spaces and Quadratic Inequalities, Lecture Notes. Sapporo, Japan: Hokkaido University, Research Institute of Applied Electricity, Division of Applied Mathematics; 1987

[5] Aronszajn N. Theory of reproducing kernels. Transactions of the American Mathematical Society. 1950;**68**:337-404

[6] Saitoh S, Sawano Y. Theory of Reproducing Kernels and Applications. Singapore: Springer; 2016

[7] Duren PL. Theory of H_p Spaces. New York: Academic Press, Inc.; 1970

[8] Duren PL, Schuster A. Bergman Spaces. Providence, R.I.: American Mathematical Society; 2004

[9] Koosis P. Introduction to H_p Spaces. Cambridge: Cambridge Mathematical Press; 1970

Chapter 6

Analytical Applications on Some Hilbert Spaces

Fethi Soltani

Abstract

In this paper, we establish an uncertainty inequality for a Hilbert space H. The minimizer function associated with a bounded linear operator from H into a Hilbert space K is provided. We come up with some results regarding Hardy and Dirichlet spaces on the unit disk $\mathbb{D}$.

Keywords: Hilbert space, Hardy space, Dirichlet space, uncertainty inequality, minimizer function

1. Introduction

Hilbert spaces are the most important tools in the theories of partial differential equations, quantum mechanics, Fourier analysis, and ergodicity. Apart from the classical Euclidean spaces, examples of Hilbert spaces include spaces of square-integrable functions, spaces of sequences, Sobolev spaces consisting of generalized functions, and Hardy spaces of holomorphic functions. Saitoh et al. applied the theory of Hilbert spaces to the Tikhonov regularization problems [1, 2]. Matsuura et al. obtained the approximate solutions for bounded linear operator equations with the viewpoint of numerical solutions by computers [3, 4]. During the last years, the theory of Hilbert spaces has gained considerable interest in various fields of mathematical sciences [5–9]. We expect that the results of this paper will be useful when discussing (in Section 2) uncertainty inequality for Hilbert space H and minimizer function associated with a bounded linear operator T from H into a Hilbert space K. As applications, we consider Hardy and Dirichlet spaces as follows.

Let $\mathbb{C}$ be the complex plane and $\mathbb{D} = \{z \in \mathbb{C} : |z| < 1\}$ the open unit disk. The Hardy space $H(\mathbb{D})$ is the set of all analytic functions f in the unit disk $\mathbb{D}$ with the finite integral:

$$\int_0^{2\pi} |f(e^{i\theta})|^2 \mathrm{d}\theta. \tag{1}$$

It is a Hilbert space when equipped with the inner product:

$$\langle f, g \rangle_{H(\mathbb{D})} = \frac{1}{2\pi} \int_0^{2\pi} f(e^{i\theta}) \overline{g(e^{i\theta})} \mathrm{d}\theta. \tag{2}$$

Over the years, the applications of Hardy space $H(\mathbb{D})$ play an important role in various fields of mathematics [5, 10] and in certain parts of quantum mechanics

[11, 12]. And this space is the background of some applications. For example, in Section 3, we study on $H(\mathbb{D})$ the following two operators:

$$\nabla f(z) = f'(z), \quad Lf(z) = z^2 f'(z) + zf(z), \tag{3}$$

and we deduce uncertainty inequality for this space. Next, we establish the minimizer function associated with the difference operator:

$$T_1 f(z) = \frac{1}{z}(f(z) - f(0)). \tag{4}$$

In Section 4, we consider the Dirichlet space $\mathcal{D}(\mathbb{D})$, which is the set of all analytic functions f in the unit disk $\mathbb{D}$ with the finite Dirichlet integral:

$$\int_{\mathbb{D}} |f'(z)|^2 \frac{dx\,dy}{\pi}, \quad z = x + iy. \tag{5}$$

It is also a Hilbert space when equipped with the inner product:

$$\langle f, g \rangle_{\mathcal{D}(\mathbb{D})} = f(0)\overline{g(0)} + \int_{\mathbb{D}} f'(z)\overline{g'(z)} \frac{dx\,dy}{\pi}, \quad z = x + iy. \tag{6}$$

This space is the objective of many applicable works [5, 13–17] and plays a background to our contribution. For example, we study on $\mathcal{D}(\mathbb{D})$ the following two operators:

$$\Lambda f(z) = f'(z) - f'(0), \quad Xf(z) = z^2 f'(z), \tag{7}$$

and we deduce the uncertainty inequality for this space $\mathcal{D}(\mathbb{D})$. And we establish the minimizer function associated with the difference operator:

$$T_2 f(z) = \frac{1}{z}(f(z) - zf'(0) - f(0)). \tag{8}$$

2. Generalized results

Let H be a Hilbert space equipped with the inner product $\langle ., . \rangle_H$. And let A and B be the two operators defined on H. We define the commutator $[A, B]$ by

$$[A, B] := AB - BA. \tag{9}$$

The adjoint of A denoted by A^* is defined by

$$\langle Af, g \rangle_H = \langle f, A^* g \rangle_H, \tag{10}$$

for $f \in \mathrm{Dom}(A)$ and $g \in \mathrm{Dom}(A^*)$.

Theorem 2.1. For $f \in \mathrm{Dom}(AA^*) \cap \mathrm{Dom}(A^*A)$, one has

$$\|A^* f\|_H^2 = \|Af\|_H^2 + \langle [A, A^*] f, f \rangle_H. \tag{11}$$

Proof. Let $f \in \mathrm{Dom}(AA^*) \cap \mathrm{Dom}(A^*A)$. Then AA^*f and A^*Af belong to H. Therefore $[A, A^*] f \in H$. Hence one has

$$\|A^*f\|_H^2 = \langle AA^*f, f\rangle_H = \langle A^*Af, f\rangle_H + \langle [A, A^*]f, f\rangle_H \tag{12}$$
$$= \|Af\|_H^2 + \langle [A, A^*]f, f\rangle_H. \quad \square \tag{13}$$

The following result is proved in [18, 19].

Theorem 2.2. Let A and B be the self-adjoint operators on a Hilbert space H. Then

$$\|(A-a)f\|_H \|(B-b)f\|_H \geq \frac{1}{2} |\langle [A, B] f, f\rangle_H|, \tag{14}$$

for all $f \in \text{Dom}(AB) \cap \text{Dom}(BA)$, and all $a, b \in \mathbb{R}$.

Theorem 2.3. Let $f \in \text{Dom}(AA^*) \cap \text{Dom}(A^*A)$. For all $a, b \in \mathbb{R}$, one has

$$\|(A + A^* - a)f\|_H \|(A - A^* + ib)f\|_H \geq |\|Af\|_H^2 - \|A^*f\|_H^2|, \tag{15}$$

where i is the imaginary unit.

Proof. Let us consider the following two operators on $\text{Dom}(AA^*) \cap \text{Dom}(A^*A)$ by

$$P = A + A^*, \quad Q = i(A - A^*). \tag{16}$$

It follows that, for $f \in \text{Dom}(AA^*) \cap \text{Dom}(A^*A)$, we have $Pf, Qf \in H$. The operators P and Q are self-adjoint and $[P, Q] = -2i[A, A^*]$. Thus the inequality (15) follows from Theorems 2.1 and 2.2. □

Theorem 2.4. Let $f \in \text{Dom}(AA^*) \cap \text{Dom}(A^*A)$. Then

$$\Delta_H^+(f)\Delta_H^-(f) \geq \|f\|_H^4 \left(\|Af\|_H^2 - \|A^*f\|_H^2\right)^2, \tag{17}$$

where

$$\Delta_H^\pm(f) = \|f\|_H^2 \|(A \pm A^*)f\|_H^2 - |\langle (A \pm A^*)f, f\rangle_H|^2. \tag{18}$$

Proof. Let $f \in \text{Dom}(AA^*) \cap \text{Dom}(A^*A)$. The operator P given by (16) is self-adjoint; then for any real a, we have

$$\|(P-a)f\|_H^2 = \|Pf\|_H^2 + a^2\|f\|_H^2 - 2a\langle Pf, f\rangle_H. \tag{19}$$

This shows that

$$\min_{a \in \mathbb{R}} \|(P-a)f\|_H^2 = \|Pf\|_H^2 - \frac{|\langle Pf, f\rangle_H|^2}{\|f\|_H^2}, \tag{20}$$

and the minimum is attained when $a = \frac{\langle Pf, f\rangle_H}{\|f\|_H^2}$. In other words, we have

$$\min_{a \in \mathbb{R}} \|(A + A^* - a)f\|_H^2 = \|(A + A^*)f\|_H^2 - \frac{|\langle (A + A^*)f, f\rangle_H|^2}{\|f\|_H^2}. \tag{21}$$

Similarly

$$\min_{b \in \mathbb{R}} \|(A - A^* + ib)f\|_H^2 = \|(A - A^*)f\|_H^2 - \frac{|\langle (A - A^*)f, f\rangle_H|^2}{\|f\|_H^2}. \tag{22}$$

Then by (15), (21), and (22), we deduce the inequality (17). □

Let $\lambda > 0$ and let $T : H \to K$ be a bounded linear operator from H into a Hilbert space K. Building on the ideas of Saitoh [2], we examine the minimizer function associated with the operator T.

Theorem 2.5. For any $k \in K$ and for any $\lambda > 0$, the problem

$$\inf_{f \in H} \{\lambda \|f\|_H^2 + \|Tf - k\|_K^2\} \tag{23}$$

has a unique minimizer given by

$$f_{\lambda,k}^* = (\lambda I + T^* T)^{-1} T^* k. \tag{24}$$

Proof. The problem (23) is solved elementarily by finding the roots of the first derivative $D\Phi$ of the quadratic and strictly convex function $\Phi(f) = \lambda \|f\|_H^2 + \|Tf - k\|_K^2$. Note that for convex functions, the equation $D\Phi(f) = 0$ is a necessary and sufficient condition for the minimum at f. The calculation provides

$$D\Phi(f) = 2\lambda f + 2T^* (Tf - k), \tag{25}$$

and the assertion of the theorem follows at once. □

Theorem 2.6. If $T : H \to K$ is an isometric isomorphism; then for any $k \in K$ and for any $\lambda > 0$, the problem

$$\inf_{f \in H} \{\lambda \|f\|_H^2 + \|Tf - k\|_K^2\} \tag{26}$$

has a unique minimizer given by

$$f_{\lambda,h}^* = \frac{1}{\lambda + 1} T^{-1} k. \tag{27}$$

Proof. We have $T^* = T^{-1}$ and $T^* T = I$. Thus, by (24), we deduce the result. □

3. The Hardy space $H(\mathbb{D})$

Let $\mathbb{C}$ be the complex plane and $\mathbb{D} = \{z \in \mathbb{C} : |z| < 1\}$ the open unit disk. The Hardy space $H(\mathbb{D})$ is the set of all analytic functions f in the unit disk $\mathbb{D}$ with the finite integral:

$$\int_0^{2\pi} |f(e^{i\theta})|^2 \, d\theta. \tag{28}$$

It is a Hilbert space when equipped with the inner product:

$$\langle f, g \rangle_{H(\mathbb{D})} = \frac{1}{2\pi} \int_0^{2\pi} f(e^{i\theta}) \overline{g(e^{i\theta})} \, d\theta. \tag{29}$$

If $f, g \in H(\mathbb{D})$ with $f(z) = \sum_{n=0}^{\infty} a_n z^n$ and $g(z) = \sum_{n=0}^{\infty} b_n z^n$, then

$$\langle f, g \rangle_{H(\mathbb{D})} = \sum_{n=0}^{\infty} a_n \overline{b_n}. \tag{30}$$

The set $\{z^n\}_{n=0}^{\infty}$ forms an Hilbert's basis for the space $H(\mathbb{D})$.

The Szegő kernel S_z given for $z \in \mathbb{D}$, by

$$S_z(w) = \sum_{n=0}^{\infty} \bar{z}^n w^n = \frac{1}{1 - \bar{z}w}, \quad w \in \mathbb{D}, \tag{31}$$

is a reproducing kernel for the Hardy space $H(\mathbb{D})$, meaning that $S_z \in H(\mathbb{D})$, and for all $f \in H(\mathbb{D})$, we have $\langle f, S_z \rangle_{H(\mathbb{D})} = f(z)$.

For $z \in \mathbb{D}$, the function $u(z) = S_{\bar{z}}(w)$ is the unique analytic solution on $\mathbb{D}$ of the initial problem:

$$u'(z) = w(zu'(z) + u(z)), \quad w \in \mathbb{D}, \quad u(0) = 1. \tag{32}$$

In the next of this section, we define the operators ∇, $\mathfrak{R}$, and L on $H(\mathbb{D})$ by

$$\nabla f(z) = f'(z), \quad \mathfrak{R}f(z) = zf'(z), \quad Lf(z) = z^2 f'(z) + zf(z). \tag{33}$$

These operators satisfy the commutation rule:

$$[\nabla, L] = \nabla L - L\nabla = 2\mathfrak{R} + I, \tag{34}$$

where I is the identity operator.

We define the Hilbert space $U(\mathbb{D})$ as the space of all analytic functions f in the unit disk $\mathbb{D}$ such that

$$\|f\|_{U(\mathbb{D})}^2 = \frac{1}{2\pi} \int_0^{2\pi} |f'(e^{i\theta})|^2 \mathrm{d}\theta < \infty. \tag{35}$$

If $f \in U(\mathbb{D})$ with $f(z) = \sum_{n=0}^{\infty} a_n z^n$, then

$$\|f\|_{U(\mathbb{D})}^2 = \sum_{n=1}^{\infty} n^2 |a_n|^2. \tag{36}$$

Thus, the space $U(\mathbb{D})$ is a subspace of the Hardy space $H(\mathbb{D})$.

Theorem 3.1.

i. For $f \in U(\mathbb{D})$, then ∇f, $\mathfrak{R}f$ and Lf belong to $H(\mathbb{D})$.

ii. $\nabla^* = L$.

iii. For $f \in U(\mathbb{D})$, one has

$$\|Lf\|_{H(\mathbb{D})}^2 = \|\nabla f\|_{H(\mathbb{D})}^2 + \|f\|_{H(\mathbb{D})}^2 + 2\langle \mathfrak{R}f, f \rangle_{H(\mathbb{D})}. \tag{37}$$

Proof.

i. Let $f \in U(\mathbb{D})$ with $f(z) = \sum_{n=0}^{\infty} a_n z^n$. Then

$$\nabla f(z) = \sum_{n=0}^{\infty} (n+1) a_{n+1} z^n, \quad \mathfrak{R}f(z) = \sum_{n=1}^{\infty} n a_n z^n, \tag{38}$$

and

$$Lf(z) = \sum_{n=1}^{\infty} n a_{n-1} z^n. \tag{39}$$

Therefore

$$\|\nabla f\|_{H(\mathbb{D})}^2 = \sum_{n=0}^{\infty} (n+1)^2 |a_{n+1}|^2 = \|f\|_{U(\mathbb{D})}^2, \tag{40}$$

$$\|\mathfrak{R}f\|_{H(\mathbb{D})}^2 = \sum_{n=1}^{\infty} n^2 |a_n|^2 = \|f\|_{U(\mathbb{D})}^2, \tag{41}$$

and

$$\|Lf\|_{H(\mathbb{D})}^2 = \sum_{n=0}^{\infty} (n+1)^2 |a_n|^2 \le |f(0)|^2 + 4\|f\|_{U(\mathbb{D})}^2. \tag{42}$$

Consequently ∇f, $\mathfrak{R}f$, and Lf belong to $H(\mathbb{D})$.

ii. For $f, g \in U(\mathbb{D})$ with $f(z) = \sum_{n=0}^{\infty} a_n z^n$ and $g(z) = \sum_{n=0}^{\infty} b_n z^n$, one has

$$\langle \nabla f, g \rangle_{H(\mathbb{D})} = \sum_{n=0}^{\infty} (n+1) a_{n+1} \overline{b_n} = \sum_{n=1}^{\infty} n a_n \overline{b_{n-1}} = \langle f, Lg \rangle_{H(\mathbb{D})}. \tag{43}$$

Thus $\nabla^* = L$.

iii. Let $f \in U(\mathbb{D})$. By (ii) and (34), we deduce that

$$\|Lf\|_{H(\mathbb{D})}^2 = \langle \nabla L f, f \rangle_{H(\mathbb{D})} \tag{44}$$

$$= \langle L \nabla f, f \rangle_{H(\mathbb{D})} + \langle [\nabla, L] f, f \rangle_{H(\mathbb{D})} \tag{45}$$

$$= \|\nabla f\|_{H(\mathbb{D})}^2 + \|f\|_{H(\mathbb{D})}^2 + 2\langle \mathfrak{R}f, f \rangle_{H(\mathbb{D})}. \quad \square \tag{46}$$

Theorem 3.2. Let $f \in U(\mathbb{D})$. For all $a, b \in \mathbb{R}$, one has

$$\|(\nabla + L - a) f\|_{H(\mathbb{D})} \|(\nabla - L + ib) f\|_{H(\mathbb{D})} \ge \|f\|_{H(\mathbb{D})}^2 + 2\langle \mathfrak{R}f, f \rangle_{H(\mathbb{D})}. \tag{47}$$

Theorem 3.3. Let T_1 be the difference operator defined on $H(\mathbb{D})$ by

$$T_1 f(z) = \frac{1}{z}(f(z) - f(0)). \tag{48}$$

i. The operator T_1 maps continuously from $H(\mathbb{D})$ to $H(\mathbb{D})$, and

$$\|T_1 f\|_{H(\mathbb{D})} \le \|f\|_{H(\mathbb{D})}. \tag{49}$$

ii. For $f \in H(\mathbb{D})$ and $z \in \mathbb{D}$, we have

$$T_1^* f(z) = z f(z), \quad T_1^* T_1 f(z) = f(z) - f(0). \tag{50}$$

iii. For any $h \in H(\mathbb{D})$ and for any $\lambda > 0$, the problem

$$\inf_{f \in H(\mathbb{D})} \left\{ \lambda \|f\|_{H(\mathbb{D})}^2 + \|T_1 f - h\|_{H(\mathbb{D})}^2 \right\} \tag{51}$$

has a unique minimizer given by

$$f_{\lambda,h}^*(z) = \frac{1}{\lambda + 1} z h(z), \quad z \in \mathbb{D}. \tag{52}$$

Proof.

i. If $f \in H(\mathbb{D})$ with $f(z) = \sum_{n=0}^{\infty} a_n z^n$, then $T_1 f(z) = \sum_{n=0}^{\infty} a_{n+1} z^n$ and

$$\|T_1 f\|_{H(\mathbb{D})}^2 = \sum_{n=1}^{\infty} |a_n|^2 \leq \|f\|_{H(\mathbb{D})}^2. \tag{53}$$

ii. If $f, g \in H(\mathbb{D})$ with $f(z) = \sum_{n=0}^{\infty} a_n z^n$ and $g(z) = \sum_{n=0}^{\infty} b_n z^n$, then

$$\langle T_1 f, g \rangle_{H(\mathbb{D})} = \sum_{n=0}^{\infty} a_{n+1} \overline{b_n} = \sum_{n=1}^{\infty} a_n \overline{b_{n-1}} = \langle f, T_1^* g \rangle_{H(\mathbb{D})}, \tag{54}$$

where $T_1^* g(z) = zg(z)$, for $z \in \mathbb{D}$. And therefore

$$T_1^* T_1 f(z) = z T_1 f(z) = f(z) - f(0). \tag{55}$$

iii. From Theorem 2.5 we have

$$(\lambda I + T_1^* T_1) f_{\lambda,h}^*(z) = T_1^* h(z). \tag{56}$$

By (ii) we deduce that

$$(\lambda + 1) f_{\lambda,h}^*(z) - f_{\lambda,h}^*(0) = zh(z). \tag{57}$$

And from this equation, $f_{\lambda,h}^*(0) = 0$. Hence

$$f_{\lambda,h}^*(z) = \frac{1}{\lambda + 1} zh(z). \quad \square \tag{58}$$

4. The Dirichlet space $\mathcal{D}(\mathbb{D})$

The Dirichlet space $\mathcal{D}(\mathbb{D})$ is the set of all analytic functions f in the unit disk $\mathbb{D}$ with the finite Dirichlet integral:

$$\int_{\mathbb{D}} |f'(z)|^2 \frac{\mathrm{d}x\,\mathrm{d}y}{\pi}, \quad z = x + iy. \tag{59}$$

It is a Hilbert space when equipped with the inner product:

$$\langle f, g \rangle_{\mathcal{D}(\mathbb{D})} = f(0)\overline{g(0)} + \int_{\mathbb{D}} f'(z)\overline{g'(z)} \frac{\mathrm{d}x\,\mathrm{d}y}{\pi}, \quad z = x + iy. \tag{60}$$

If $f, g \in \mathcal{D}(\mathbb{D})$ with $f(z) = \sum_{n=0}^{\infty} a_n z^n$ and $g(z) = \sum_{n=0}^{\infty} b_n z^n$, then

$$\langle f, g \rangle_{\mathcal{D}(\mathbb{D})} = a_0 \overline{b_0} + \sum_{n=1}^{\infty} n a_n \overline{b_n}. \tag{61}$$

The set $\left\{1, \frac{z^n}{\sqrt{n}}\right\}_{n=1}^{\infty}$ forms an Hilbert's basis for the space $\mathcal{D}(\mathbb{D})$.
The function K_z given for $z \in \mathbb{D}$, by

$$K_z(w) = 1 + \log\left(\frac{1}{1 - \bar{z}w}\right), \quad w \in \mathbb{D}, \tag{62}$$

is a reproducing kernel for the Dirichlet space $\mathcal{D}(\mathbb{D})$, meaning that $K_z \in \mathcal{D}(\mathbb{D})$, and for all $f \in \mathcal{D}(\mathbb{D})$, we have $\langle f, K_z \rangle_{\mathcal{D}(\mathbb{D})} = f(z)$.

For $z \in \mathbb{D}$, the function $u(z) = K_{\bar{z}}(w)$ is the unique analytic solution on $\mathbb{D}$ of the initial problem:

$$\frac{u'(z) - u'(0)}{z} = wu'(z), \quad w \in \mathbb{D}, \quad u(0) = 1. \tag{63}$$

In the next of this section, we define the operators Λ, $\mathfrak{R}$, and X on $\mathcal{D}(\mathbb{D})$ by

$$\Lambda f(z) = f'(z) - f'(0), \quad \mathfrak{R} f(z) = zf'(z), \quad Xf(z) = z^2 f'(z). \tag{64}$$

These operators satisfy the following commutation relation:

$$[\Lambda, X] = \Lambda X - X\Lambda = 2\mathfrak{R}. \tag{65}$$

We define the Hilbert space $V(\mathbb{D})$ as the space of all analytic functions f in the unit disk $\mathbb{D}$ such that

$$\|f\|^2_{V(\mathbb{D})} = \int_{\mathbb{D}} |f'(z)|^2 |z|^2 \frac{dx\,dy}{\pi} < \infty, \quad z = x + iy. \tag{66}$$

If $f \in V(\mathbb{D})$ with $f(z) = \sum_{n=0}^{\infty} a_n z^n$, then

$$\|f\|^2_{V(\mathbb{D})} = \sum_{n=1}^{\infty} n^3 |a_n|^2. \tag{67}$$

Thus, the space $V(\mathbb{D})$ is a subspace of the Dirichlet space $\mathcal{D}(\mathbb{D})$.

Theorem 4.1.

i. For $f \in V(\mathbb{D})$, then Λf, $\mathfrak{R} f$, and Xf belong to $\mathcal{D}(\mathbb{D})$.

ii. $\Lambda^* = X$.

iii. For $f \in V(\mathbb{D})$, one has

$$\|Xf\|^2_{\mathcal{D}(\mathbb{D})} = \|\Lambda f\|^2_{\mathcal{D}(\mathbb{D})} + 2\langle \mathfrak{R} f, f\rangle_{\mathcal{D}(\mathbb{D})}. \tag{68}$$

Proof.

i. Let $f \in V(\mathbb{D})$ with $f(z) = \sum_{n=0}^{\infty} a_n z^n$. Then

$$\Lambda f(z) = \sum_{n=1}^{\infty} (n+1) a_{n+1} z^n, \quad \mathfrak{R} f(z) = \sum_{n=1}^{\infty} n a_n z^n, \tag{69}$$

and

$$Xf(z) = \sum_{n=2}^{\infty} (n-1) a_{n-1} z^n. \tag{70}$$

Therefore

$$\|\Lambda f\|^2_{\mathcal{D}(\mathbb{D})} = \sum_{n=1}^{\infty} n(n+1)^2 |a_{n+1}|^2 \leq \sum_{n=2}^{\infty} n^3 |a_n|^2 \leq \|f\|^2_{V(\mathbb{D})}, \tag{71}$$

$$\|\Re f\|^2_{\mathcal{D}(\mathbb{D})} = \sum_{n=1}^{\infty} n^3 |a_n|^2 = \|f\|^2_{V(\mathbb{D})}, \tag{72}$$

and

$$\|Xf\|^2_{\mathcal{D}(\mathbb{D})} = \sum_{n=1}^{\infty} (n+1)n^2 |a_n|^2 \leq 2\|f\|^2_{V(\mathbb{D})}. \tag{73}$$

Consequently Λf, $\Re f$, and Xf belong to $\mathcal{D}(\mathbb{D})$.

ii. For $f, g \in V(\mathbb{D})$ with $f(z) = \sum_{n=0}^{\infty} a_n z^n$ and $g(z) = \sum_{n=0}^{\infty} b_n z^n$, one has

$$\langle \Lambda f, g \rangle_{\mathcal{D}(\mathbb{D})} = \sum_{n=1}^{\infty} n(n+1) a_{n+1} \overline{b_n} = \sum_{n=2}^{\infty} n(n-1) a_n \overline{b_{n-1}} = \langle f, Xg \rangle_{\mathcal{D}(\mathbb{D})}. \tag{74}$$

iii. Let $f \in V(\mathbb{D})$. By (ii) and (65), we deduce that

$$\|Xf\|^2_{\mathcal{D}(\mathbb{D})} = \langle \Lambda X f, f \rangle_{\mathcal{D}(\mathbb{D})} \tag{75}$$

$$= \langle X \Lambda f, f \rangle_{\mathcal{D}(\mathbb{D})} + \langle [\Lambda, X] f, f \rangle_{\mathcal{D}(\mathbb{D})} \tag{76}$$

$$= \|\Lambda f\|^2_{\mathcal{D}(\mathbb{D})} + 2\langle \Re f, f \rangle_{\mathcal{D}(\mathbb{D})}. \quad \square \tag{77}$$

Theorem 4.2. Let $f \in V(\mathbb{D})$. For all $a, b \in \mathbb{R}$, one has

$$\|(\Lambda + X - a) f\|_{\mathcal{D}(\mathbb{D})} \|(\Lambda - X + ib) f\|_{\mathcal{D}(\mathbb{D})} \geq 2\langle \Re f, f \rangle_{\mathcal{D}(\mathbb{D})}. \tag{78}$$

Theorem 4.3. Let T_2 be the difference operator defined on $\mathcal{D}(\mathbb{D})$ by

$$T_2 f(z) = \frac{1}{z}\left(f(z) - z f'(0) - f(0) \right). \tag{79}$$

i. The operator T_2 maps continuously from $\mathcal{D}(\mathbb{D})$ to $\mathcal{D}(\mathbb{D})$, and

$$\|T_2 f\|_{\mathcal{D}(\mathbb{D})} \leq \|f\|_{\mathcal{D}(\mathbb{D})}. \tag{80}$$

ii. For $f \in \mathcal{D}(\mathbb{D})$ with $f(z) = \sum_{n=0}^{\infty} a_n z^n$, we have

$$T_2^* f(z) = \sum_{n=2}^{\infty} \frac{n-1}{n} a_{n-1} z^n, \quad T_2^* T_2 f(z) = \sum_{n=2}^{\infty} \frac{n-1}{n} a_n z^n. \tag{81}$$

iii. For any $d \in \mathcal{D}(\mathbb{D})$ and for any $\lambda > 0$, the problem

$$\inf_{f \in \mathcal{D}(\mathbb{D})} \left\{ \lambda \|f\|^2_{\mathcal{D}(\mathbb{D})} + \|T_2 f - d\|^2_{\mathcal{D}(\mathbb{D})} \right\} \tag{82}$$

has a unique minimizer given by

$$f^*_{\lambda,d}(z) = \langle d, \Psi_z \rangle_{\mathcal{D}(\mathbb{D})}, \quad z \in \mathbb{D}, \tag{83}$$

$$\Psi_z(w) = \sum_{n=1}^{\infty} \frac{\overline{z}^{n+1}}{\lambda(n+1) + n} w^n, \quad w \in \mathbb{D}. \tag{84}$$

Proof.

i. If $f \in \mathcal{D}(\mathbb{D})$ with $f(z) = \sum_{n=0}^{\infty} a_n z^n$, then $T_2 f(z) = \sum_{n=1}^{\infty} a_{n+1} z^n$ and

$$\|T_2 f\|_{\mathcal{D}(\mathbb{D})}^2 = \sum_{n=2}^{\infty} (n-1)|a_n|^2 \leq \sum_{n=2}^{\infty} n|a_n|^2 \leq \|f\|_{\mathcal{D}(\mathbb{D})}^2. \tag{85}$$

ii. If $f, g \in \mathcal{D}(\mathbb{D})$ with $f(z) = \sum_{n=0}^{\infty} a_n z^n$ and $g(z) = \sum_{n=0}^{\infty} b_n z^n$, then

$$\langle T_2 f, g \rangle_{\mathcal{D}(\mathbb{D})} = \sum_{n=1}^{\infty} n a_{n+1} \overline{b_n} = \sum_{n=2}^{\infty} (n-1) a_n \overline{b_{n-1}} = \left\langle f, T_2^* g \right\rangle_{\mathcal{D}(\mathbb{D})}, \tag{86}$$

where

$$T_2^* g(z) = \sum_{n=2}^{\infty} \frac{n-1}{n} b_{n-1} z^n, \quad z \in \mathbb{D}. \tag{87}$$

And therefore

$$T_2^* T_2 f(z) = \sum_{n=2}^{\infty} \frac{n-1}{n} a_n z^n. \tag{88}$$

iii. We put $d(z) = \sum_{n=0}^{\infty} d_n z^n$ and

$$f_{\lambda,d}^*(z) = \sum_{n=0}^{\infty} c_n z^n. \tag{89}$$

From (ii) and the equation

$$(\lambda I + T_2^* T_2) f_{\lambda,d}^*(z) = T_2^* d(z), \tag{90}$$

we deduce that

$$c_1 = c_0 = 0, \quad c_n = \frac{n-1}{\lambda n + n - 1} d_{n-1}, \quad n \geq 2. \tag{91}$$

Thus

$$f_{\lambda,d}^*(z) = \sum_{n=1}^{\infty} \frac{n d_n}{\lambda(n+1) + n} z^{n+1} = \langle d, \Psi_z \rangle_{\mathcal{D}(\mathbb{D})}, \quad z \in \mathbb{D}. \quad \square \tag{92}$$

Author details

Fethi Soltani[1,2]

1 Université de Tunis El Manar, Faculté des Sciences de Tunis, Laboratoire d'Analyse Mathématique et Applications LR11ES11, Tunis, Tunisia

2 Université de Carthage, Ecole Nationale d'Ingénieurs de Carthage, Tunis, Tunisia

*Address all correspondence to: fethi.soltani@fst.utm.tn

References

[1] Saitoh S. The Weierstrass transform and an isometry in the heat equation. Applicable Analysis. 1983;**16**:1-6

[2] Saitoh S. Best approximation, Tikhonov regularization and reproducing kernels. Kodai Mathematical Journal. 2005;**28**:359-367

[3] Matsuura T, Saitoh S, Trong DD. Inversion formulas in heat conduction multidimensional spaces. Journal of Inverse and Ill-Posed Problems. 2005;**13**: 479-493

[4] Matsuura T, Saitoh S. Analytical and numerical inversion formulas in the Gaussian convolution by using the Paley-Wiener spaces. Applicable Analysis. 2006;**85**:901-915

[5] Paulsen VI. An Introduction to the Theory of Reproducing Kernel Hilbert Spaces. Cambridge: Cambridge University Press; 2016

[6] Saitoh S. Approximate real inversion formulas of the Gaussian convolution. Applicable Analysis. 2004; **83**:727-733

[7] Soltani F. Operators and Tikhonov regularization on the Fock space. Integral Transforms and Special Functions. 2014;**25**(4):283-294

[8] Soltani F. Inversion formulas for the Dunkl-type Segal-Bargmann transform. Integral Transforms and Special Functions. 2015;**26**(5):325-339

[9] Soltani F. Dunkl multiplier operators on a class of reproducing kernel Hilbert spaces. Journal of Mathematical Research with Applications. 2016;**36**(6): 689-702

[10] Tuan VK, Hong NT. Interpolation in the Hardy space. Integral Transforms and Special Functions. 2013;**24**(8): 664-671

[11] Bohm A, Bui HV. The marvelous consequences of Hardy spaces in quantum physics. Geometric Methods in Physics. 2013;**30**(1):211-228

[12] Mouayn Z. Resolution of the identity of the classical Hardy space by means of Barut-Girardello coherent states. Mathematical Physics. 2012;**2012**: Article ID 530473, 12 p

[13] Arcozzi N, Rochberg R, Sawyer ET, Wick BD. The Dirichlet space: A survey. New York Journal of Mathematics. 2011; **17a**:45-86

[14] Chartrand R. Toeplitz operators on Dirichlet-type spaces, J. Operator Theory. 2002;**48**(1):3-13

[15] Geng L, Tong C, Zeng H. Some linear isometric operators on the Dirichlet space. Applied Mathematics & Information Sciences. 2012;**6**(1):265-270

[16] Geng LG, Zhou ZH, Dong XT. Isometric composition operators on weighted Dirichlet-type spaces. Journal of Inequalities and Applications. 2012; **2012**:23

[17] Martin MJ, Vukotic D. Isometries of the Dirichlet space among the composition operators. Proceedings of the American Mathematical Society. 2005;**134**:1701-1705

[18] Folland G. Harmonic analysis on phase space. In: Annals of Mathematics Studies. Vol. 122. Princeton, New Jersey: Princeton University Press; 1989

[19] Gröchenig K. Foundations of Time-frequency Analysis. Boston: Birkhäuser; 2001

Chapter 7

Spectral Observations of PM10 Fluctuations in the Hilbert Space

Thomas Plocoste and Rudy Calif

Abstract

During the last 20 years, many megacities have experienced air pollution leading to negative impacts on human health. In the Caribbean region, air quality is widely affected by African dust which causes several diseases, particularly, respiratory diseases. This is why it is crucial to improve the understanding of PM10 fluctuations in order to elaborate strategies and construct tools to predict dust events. A first step consists to characterize the dynamical properties of PM10 fluctuations, for instance, to highlight possible scaling in PM10 density power spectrum. For that, the scale-invariant properties of PM10 daily time series during 6 years are investigated through the theoretical Hilbert frame. Thereafter, the Hilbert spectrum in time-frequency domain is considered. The choice of theoretical frame must be relevant. A comparative analysis is also provided between the results achieved in the Hilbert and Fourier spaces.

Keywords: PM10 data, empirical mode decomposition, Hilbert spectral analysis, time-frequency representation, Fourier space

1. Introduction

Generally, the concentration of air pollutants varies and is impacted by the local pollutant emission levels and meteorological and topographical conditions [1, 2]. Particulate matter (PM) is a complex mixture of elemental and organic carbon, ammonium, nitrates, sulfates, mineral dust, trace elements, and water [3]. PM with an aerodynamic diameter of <10 μm, i.e., PM10, are well known for their impact on human health [4]. Many studies have highlighted that exposure to PM increases the number of hospital admissions for cardiovascular disease, acute bronchitis, asthma attacks, respiratory disease, and congestive heart failure [5–8]. In the Caribbean area, one of the main emitters of PM10 is from large-scale sources, i.e., African dust [9]. Knowledge of the dynamics of PM10 process is crucial to elaborate strategies and construct tools to predict dust events. The time-frequency distribution of a signal provides information about how the spectral content of a signal evolves with time, thus providing an ideal tool to dissect, analyze, and interpret nonstationary signals [10]. Contrary to classical methods, the need of a time-frequency representation (TFR) is stemmed from the inadequacy of either time domain or frequency domain analysis to fully describe the nature of nonstationary signals [10]. In literature, there are numerous methods to obtain energy density as a function of time and frequency simultaneously as the short-time Fourier transform (STFT), Hilbert-Huang transform (HHT), and wavelet transform (WT) [10–12].

In this study, the scaling properties of PM10 data are firstly analyzed, and then the TFR is investigated. In order to highlight the performance of the Hilbert space, an analysis of PM10 data was also performed in the Fourier space.

This chapter is organized as follows. Section 2 presents PM10 data analyzed in this study. Section 3 describes the methods applied in order to investigate PM10 dynamics. Section 4 comments on the results obtained and then discusses them.

2. Experimental data

Guadeloupe archipelago is a French West Indies island located in the middle of the Caribbean basin, i.e., 16.25°N latitude and 61.58°W longitude, which experiences a tropical and humid climate [13, 14]. The time series analyzed here belong to Guadeloupe air quality network which is managed by the Gwad'Air agency (http://www.gwadair.fr/). PM10 concentrations are measured at Pointe-à-Pitre (16.2422°N 61.5414°W) using the Thermo Scientific tapered element oscillating microbalance (TEOM) models 1400ab and 1400-FDMS. Hourly PM10 concentrations were sampled during the period from 1 January 2005 to 31 December 2010. We processed these data into daily average concentrations. In total, there are 2150 daily averaged data points available continuously for 6 years. **Figure 1** displays PM10 daily signal illustrating huge fluctuations and thus indicating a strong variability. These strong oscillations observed in the middle of each year are attributed to PM10 related to dust outbreaks coming from the African coast from May to September [9]. For the rest of the year, PM10 is mainly generated by anthropogenic pollution [15].

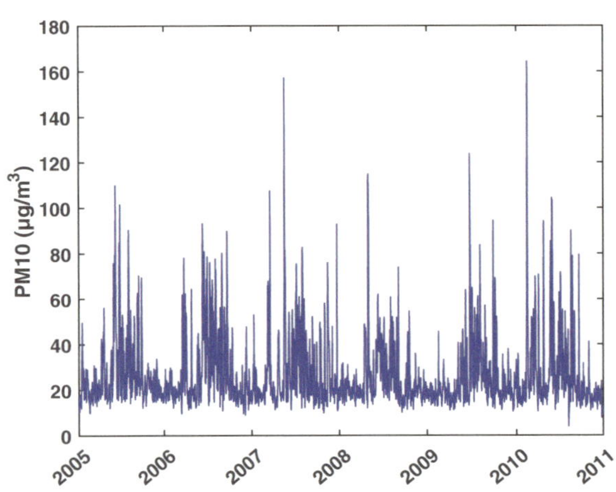

Figure 1.
Illustration of PM10 daily average concentrations between 2005 and 2010, highlighting intermittent burst events with huge fluctuations.

3. Methods

3.1 Scaling analysis (1D representation)

The description of natural phenomena by the study of statistical scale laws is not recent [16]. Self-similarity of complex systems has been widely observed in nature and is the simplest form of scale invariance. A scale invariance can be detected by computing of power spectral density (PSD). The PSD separates and measures the amount of variability occurring in different frequency bands. In this study, PSD are estimated through the Fourier and Hilbert spaces.

3.1.1 Fourier analysis

In order to investigate the scaling properties of PM10 data, classically the discrete Fourier transform of the times series considered is computed. The expression of Fourier transform $X(f)$ for a process $x(t)$ is recalled here. An N point-long interval is used to construct the value at frequency domain point f, X_f [17]:

$$X(f) = \int_{-T}^{+T} x(t)e^{-2\pi ift}\mathrm{dt} \tag{1}$$

Thus, the analytical expression of $X(f)$ is [18]

$$|X(f)| = \sqrt{\mathcal{R}e^2(X(f)) + \mathcal{I}m^2(X(f))} \tag{2}$$

Consequently the power spectral density $E(f)$ is estimated by computing the following expression:

$$E(f) = |X(f)|^2 \tag{3}$$

3.1.2 Hilbert analysis

To determine the scale invariance of a given time series in a joint amplitude-frequency space, the Hilbert-Huang transform [19, 20] is performed. HHT can be summarized in two steps: (i) empirical mode decomposition (EMD) and (ii) Hilbert spectral analysis (HSA). Empirical mode decomposition is a powerful tool to separate a nonlinear and nonstationary time series into a sum of intrinsic mode functions (IMF) without a priori basis as required by traditional Fourier-based method [19–21]. An IMF must satisfy the following two conditions: (i) the difference between the number of local extrema and the number of zero-crossings must be zero or one, and (ii) the local maxima and the envelope defined by the local minima are close to zero. Therefore, the original signal $x(\underline{t})$ is decomposed into a sum of $n1$ IMF modes with the residual $r_n(t)$:

$$x(t) = \sum_{m=1}^{n-1} C_m(t) + r_n(t) \tag{4}$$

To obtain a physically significant IMF, this selection process must be stopped by a certain criterion. For more details, EMD decomposition is widely described in the literature [19–23].

To characterize the time-frequency energy distribution from the original signal $x(t)$, HSA is applied on each obtained IMF component $C_m(t)$ to extract the instantaneous amplitude and frequency [19, 24]. The Hilbert transform is defined by:

$$\tilde{C}_m(t) = \frac{1}{\pi} P \int_{-\infty}^{+\infty} \frac{C_m(t')}{t-t'}\mathrm{dt}' \tag{5}$$

with P the Cauchy principal value [24, 25]. We can specify an analytical signal z for each IMF mode $C_m(t)$ with

$$Z_m(t) = C_m(t) + j\tilde{C}_m(t) = A_m(t)e^{j\varphi_m(t)} \tag{6}$$

where $A_m(t) = |Z_m(t)| = \sqrt{C_m(t)^2 + \tilde{C}_m(t)^2}$ describes an amplitude and $\varphi_m(t) = \arg(z) = \arctan\left[\frac{\tilde{C}_m(t)}{C_m(t)}\right]$ represents the phase function of IMF modes. Consequently, the instantaneous frequency $\omega_m(t)$ is defined from the phase $\varphi_m(t)$ by

$$\omega_m(t) = \frac{1}{2\pi}\frac{d\varphi_m(t)}{\mathrm{dt}} \tag{7}$$

Thus, the original signal $x(t)$ can be expressed as

$$x(t) = \mathcal{R}\mathrm{e}\sum_{m=1}^{N} A_m(t)e^{j\varphi_m(t)} = \mathcal{R}\mathrm{e}\sum_{m=1}^{N} A_m(t)e^{j\int_{-\infty}^{t}\omega_m(t)\mathrm{dt}} \tag{8}$$

where $\mathcal{R}e$ is a part real [19, 20, 26].

Due to the simultaneous representation of frequency modulation and amplitude modulation, the HHT can be considered as a generalization of the Fourier transform [19, 20]. The energy in a time-frequency space is designated as the Hilbert spectrum with $H(\omega, t) = A^2(\omega, t)$. The Hilbert marginal spectrum $h(\omega)$ is defined by

$$h(\omega) = \frac{1}{T}\int_0^T H(\omega, t)\mathrm{dt} \tag{9}$$

where T is the total data length. The Hilbert spectrum $H(\omega,t)$ gives a measure of amplitude from each frequency and time, while the marginal spectrum $h(\omega)$ gives a measure of the total amplitude from each frequency [27]. As a result, the marginal spectrum can be compared to the Fourier spectrum [19, 20].

In conclusion, for a scale-invariant process, the Fourier $E(f)$ and the Hilbert $h(\omega)$ spectral densities obtained follow a power law over a range of frequencies:

$$E(f) \sim f^{-\beta_f} \tag{10}$$

$$h(\omega) \sim \omega^{-\beta_h} \tag{11}$$

where f and ω are the frequencies and β_f and β_h are the spectral exponents, respectively, in the Fourier and Hilbert spaces. It reveals the scale-free memory effect as a power law dependence of the frequency distribution. Consequently, β_f and β_h contain information about the degree of stationarity of the studied parameter [16, 28, 29]:

- If β_f or β_h <1, the process is stationary.
- If β_f or β_h >1, the process is nonstationary.
- If 1< β_f or β_h <3, the process is nonstationary with increments stationary.
- Spectral analysis has been widely applied in various research fields [30–34].

3.2 Time-frequency representation (2D representation)

3.2.1 Spectrogram

The spectrogram (SPEC) of a signal $x(t)$ is defined as the squared magnitudes of the STFT as shown in Eq. (12) [12]:

$$\mathrm{SPEC}_x(t,f) = |S_x(t,f)|^2 \tag{12}$$

where $S_x(t,f) = \int_{-\infty}^{+\infty} x(\tau)w(\tau - t)e^{-jf\tau}d\tau$ is the STFT of $x(t)$, $w(\tau)$ is a window (e.g., Hanning, rectangular, Hamming), t is time, and f is frequency.

As depicted in Eq. (13), SPEC roughly describes the energy density of the signal at point (t,f) [12]:

$$\int_{-\infty}^{+\infty}\int_{-\infty}^{+\infty} \mathrm{SPEC}_x(t,f)\mathrm{dtdf} = \int_{-\infty}^{+\infty} |x(t)|^2\mathrm{dt} \tag{13}$$

The SPEC has been applied successfully in various research fields [12, 35–37]. The main advantages of SPEC are an easily understanding interpretation, and it allows a fast computation. However, the main drawback of SPEC is the same as that of the STFT [12]. Indeed, there is a trade-off between time and frequency resolution.

3.2.2 Hilbert spectrum

The Hilbert spectrum (HS) is a joint time-frequency representation introduced by [19]. It is important to notice that the two important tools (i.e., EMD and HS) for exploratory analysis of the data are provided by HSA method. This approach was applied successfully in various research fields as fault diagnosis for rolling bearing [11], turbulence [38], environment [34, 39], and geophysics [40], to cite a few.

4. Results

4.1 Scaling properties

In order to identify the presence of scaling in PM10 time series, the PSD is estimated in the Hilbert and Fourier spaces. **Figure 2** depicts the power spectral density provided by the Hilbert transform and the Fourier transform. On this figure, we try to detect a power law behavior of the form $h(\omega) \sim \omega^{-\beta_h}$ and $E(f) \sim f^{-\beta_f}$ where β_h and β_f are, respectively, the spectral exponents in the Hilbert and Fourier spaces. On the frequency range $2.09 \times 10^{-7} \leqslant f \leqslant 4.57 \times 10^{-5}$ Hz which corresponds to time scales 6.1 hours$\leqslant T \leqslant$55.4 days, a power law behavior is clearly

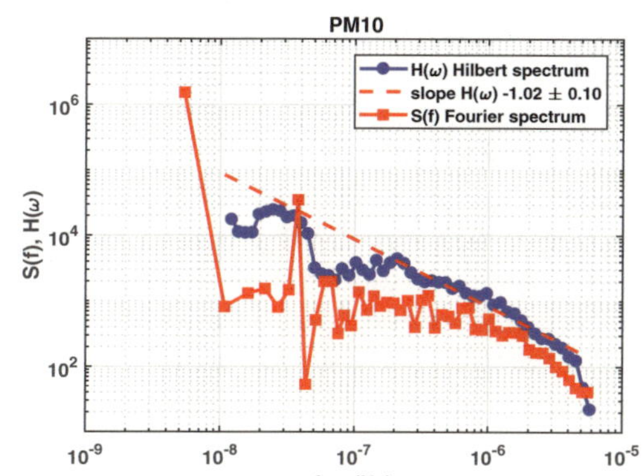

Figure 2.
The spectrum of PM10 time series in the Hilbert space and the Fourier space. A power law behavior is significant only in the Hilbert space.

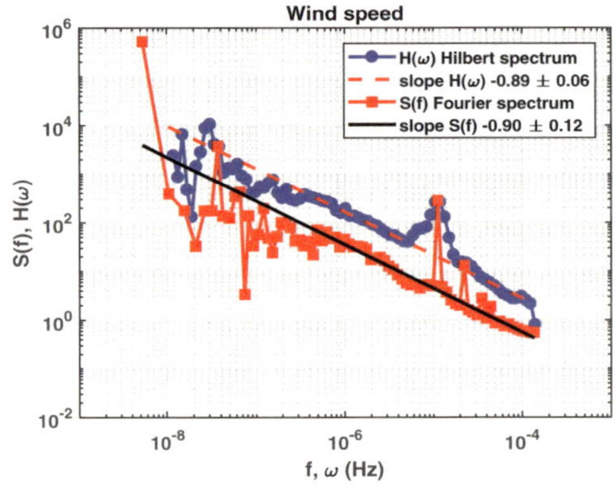

Figure 3.
The spectrum of wind speed in the Hilbert space and the Fourier space. A power law behavior is observed in both spaces.

noticed in the Hilbert space with an estimated spectral exponent $\beta_h = 1.02 \pm 0.10$. β_h is equal to 1 power law scaling observed in the mesoscale range [41]. In the Fourier space, this power law is not significant. This is due to the existence of intermittent dust events with huge fluctuations in PM10 data (see **Figure 1**). Indeed, the Fourier transform is a linear asymptotic approach which requires high-order harmonic components to mimic nonlinear and nonstationary process [42]. Thus, the high-order harmonics may lead an artificial energy transfer flux from a large scale (low frequency) to a small scale (high frequency) in the Fourier space. Consequently, the Fourier-based spectrum may be contaminated by this artificial energy flux [42]. The artificial energy transfer may give a less steep power spectrum as we observed in **Figure 2**. By contrast, combined with the EMD method, HSA has very local abilities both in physical and spectral domains and does not require any higher-order harmonic components to simulate the nonlinear and nonstationary events. As a consequence, HSA method may provide a more accurate scaling exponent and singularity spectrum [42].

According to [43], wind speed dominates the amount of pollutant dispersion in the atmospheric boundary layer. In addition, this meteorological parameter could also transport PM10 from large-scale sources, i.e., African dust [9]. To complete our results, we used hourly wind speed measurements provided by the French weather office (Météo France Guadeloupe) located at Abymes (16.2630°N 61.5147°W). PM10 and wind speed measurements are very close, i.e., ≈8.1 km of distance, and performed at the center of the island under the same atmospheric conditions [2]. **Figure 3** illustrates the PSD provided by the Hilbert transform and the Fourier transform for wind speed data. This time, a power law behavior is observed in both spaces on the same frequency range $3.54 \times 10^{-7} \leqslant f \leqslant 1.36 \times 10^{-4}$ Hz which corresponds to time scales 2.1 hours$\leqslant T \leqslant$32.7 days. Contrary to PM10 which is a passive scalar, wind speed is a vector quantity. The estimated spectral exponents are identical with, respectively, 0.89 ± 0.06 and 0.90 ± 0.12 in the Hilbert and Fourier spaces. As PM10, spectral exponent values are also close to -1. For wind speed, at low frequencies, a spectrum close to the 1 power law is likely occurs close to a rough surface, due to a strong interaction between the mean flow vorticity and the fluctuating vorticity [44, 45].

4.2 Time-frequency domain

The TFR in the Fourier and Hilbert spaces are, respectively, illustrated in **Figures 4** and **5**. Both figures show a color gradient from strong energy (in red) to weak energy (in blue). This highlights the energy activity related to PM10 concentrations during the study period. Such an approach gives the possibility of tracking the evolution of PM10 data spectral content in time, which is typically represented by variations of the amplitudes and frequencies of the components from which the signal is composed [46].

On **Figure 4**, strong energies are observed throughout the years with slight fluctuations on the frequency range $0 \leqslant f \leqslant 1 \times 10^{-6}$ Hz. For $f > 4 \times 10^{-6}$ Hz, strong energies are also noticed in the middle of each year and at the beginning of 2010 with more fluctuations. In **Figure 5**, energy distributions are more localized. On the frequency range $0 \leqslant f \leqslant 1 \times 10^{-6}$ Hz, we can observe the influence of small-scale event on energy behavior. As noticed, this energy may be weak or null. As an example, the impact of a general strike in early 2009 that paralyzed Guadeloupean archipelago at least 2 months is highlighted by zero energy due to the lack of PM10 sources, i.e., industrial activity and road traffic. For $f > 1 \times 10^{-6}$ Hz, one can see more precisely energy variation related to dust events from mesoscale to large scale. Contrary to SPEC, HS clearly illustrates localized energy fluctuations due to small-scale event. In fact, the STFT makes an assumption that any signal as piecewise stationary and uses suitable window function to produce the short-time spectral characteristics of the signal. However, in

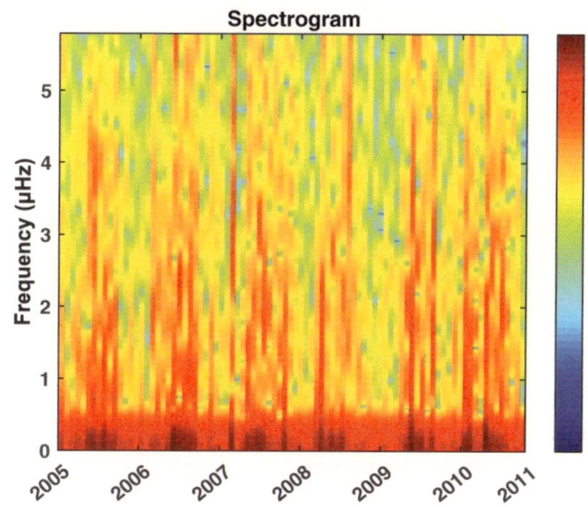

Figure 4.
Spectrogram of PM10 time series with a color gradient from strong energy (in red) to weak energy (in blue).

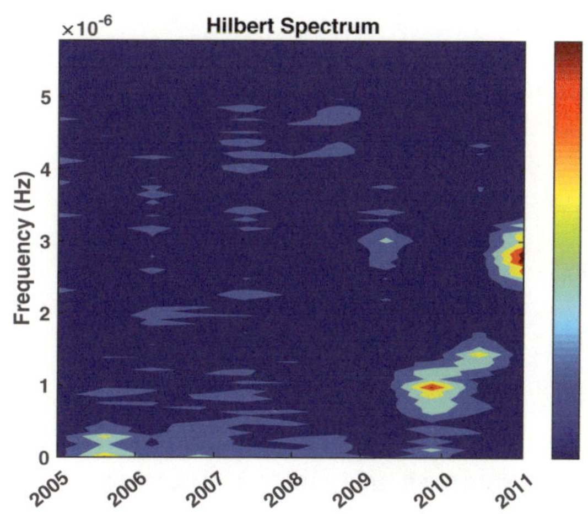

Figure 5.
Hilbert spectrum of PM10 time series with a color gradient from strong energy (in red) to weak energy (in blue).

reality, most of air pollution signals are usually nonstationary [9, 14, 47]. The Fourier transform-based technique treats the signal as a sum of predefined basis functions. If the analyzing signal is well matched with the bases, it performs better; otherwise the performance is degraded [10]. Here, the SPEC highlight energy fluctuations linked to PM10 coming from African dust between May and September (large-scale sources) [9] and from the eruption of Soufrière on Montserrat in February 2010 (mesoscale sources) [48]. However, the SPEC does not detect energy fluctuation related to anthropogenic pollution, i.e., local sources. This shows HS is a robust method in time-frequency domain. Indeed, based on the EMD method, this TFR is fully data adaptive, and the signal decomposition is performed without any predefined basis functions. These results confirm the superiority of HS over STFT in TFR.

5. Conclusion

In this paper, we investigated scaling and time-frequency properties of PM10 data in Hilbert frame. The performances obtained in the Hilbert space are compared with those achieved in the Fourier space. Firstly, with the Hilbert spectral analysis (HSA), a power law behavior is clearly observed on the frequency range $2.09 \times 10^{-7} \leqslant f \leqslant 4.57 \times 10^{-5}$ Hz which corresponds to time scales 6.1 hours$\leqslant T \leqslant$55.4 days with an estimated spectral exponent $\beta_h = 1.02 \pm 0.10$. As HSA methodology has a very local ability in both physical and spectral spaces, the influence of intermittent dust events with huge fluctuations is included in the amplitude-frequency space which is not the case in Fourier spectrum. Thereafter, PM10 data are illustrated in time-frequency representations with the Hilbert spectrum and spectrogram. The results provide the evidence that HS-based TFR performs better than SPEC. The higher resolution in TFR offers better fluctuations of PM10 energy for $f <$ 1μ Hz. This is due to the fact that it is impossible to increase the TF resolution at the desired level in SPEC. The major asset of HS is that the time resolution can be as precise as the sampling period and the frequency resolution depends on the choice up to the Nyquist limit. In addition, contrary to SPEC which introduces a noticeable amount of cross-spectral energy terms during the use of window function with overlapping, HS is fully adaptive to datasets due to the decomposition of the signals. These first results suggest a substantial possibility to perform a profound dynamical analysis of PM10 concentrations for the Caribbean area in order to quantify the origin and the threshold pollution.

Acknowledgements

The authors would like to thank Guadeloupe air quality network (Gwad'Air) and the French Met Office (Météo France Guadeloupe) for providing air quality and meteorological data.

Conflict of interest

The authors declare no conflict of interest.

Abbreviations

PM10	particulate matter with an aerodynamic diameter 10 μm or less
PSD	power spectral density
SPEC	spectrogram

EMD	empirical mode decomposition
IMF	intrinsic mode function
HSA	Hilbert spectral analysis
HHT	Hilbert-Huang transform
TFR	time-frequency representation
HS	Hilbert spectrum
STFT	short-time Fourier transform
WT	wavelet transform

Nomenclature

$E(f)$	Fourier spectral density
f	frequency (Hz)
β	spectral exponent
$\mathcal{A}$	instantaneous amplitude
$C(t)$	intrinsic mode function component
$h(\omega)$	Hilbert spectrum
ω	instantaneous frequency (Hz)
j	scale index
N	total length of a sequence
$x(t)$	particulate matter signal ($\mu g/m^3$)
$r(t)$	residual of the intrinsic mode function
φ	phase function of the intrinsic mode function

Author details

Thomas Plocoste[1*†] and Rudy Calif[2†]

1 Department of Research in Geoscience, KaruSphère SASU, Abymes, Guadeloupe, France

2 EA 4539—LaRGE (Laboratoire de Recherche en Géosciences et Energies), Département de Physique, Université des Antilles, Pointe-à-Pitre, Guadeloupe, France

*Address all correspondence to: karusphere@gmail.com

† These authors contributed equally.

References

[1] Xi W, Chen RJ, Chen BH, Kan HD. Application of statistical distribution of PM10 concentration in air quality management in 5 representative cities of China. Biomedical and Environmental Sciences. 2013;**26**(8):638-646

[2] Plocoste T, Dorville JF, Monjoly S, Jacoby-Koaly S, André M. Assessment of nitrogen oxides and ground-level ozone behavior in a dense air quality station network: Case study in the Lesser Antilles arc. Journal of the Air & Waste Management Association. 2018;**68**(12): 1278-1300

[3] Hueglin C, Gehrig R, Baltensperger U, Gysel M, Monn C, Vonmont H. Chemical characterisation of PM2.5, PM10 and coarse particles at urban, near-city and rural sites in Switzerland. Atmospheric Environment. 2005;**39**(4): 637-651

[4] Momtazan M, Geravandi S, Rastegarimehr B, Valipour A, Ranjbarzadeh A, Yari AR, et al. An investigation of particulate matter and relevant cardiovascular risks in Abadan and Khorramshahr in 2014–2016. Toxin Reviews. 2018:1-8

[5] Cairncross EK, John J, Zunckel M. A novel air pollution index based on the relative risk of daily mortality associated with short-term exposure to common air pollutants. Atmospheric Environment. 2007;**41**(38):8442-8454

[6] Kassomenos P, Papaloukas C, Petrakis M, Karakitsios S. Assessment and prediction of short term hospital admissions: The case of Athens, Greece. Atmospheric Environment. 2008; **42**(30):7078-7086

[7] Geravandi S, Sicard P, Khaniabadi YO, De Marco A, Ghomeishi A, Goudarzi G, et al. A comparative study of hospital admissions for respiratory diseases during normal and dusty days in Iran. Environmental Science and Pollution Research. 2017;**24**(22): 18152-18159

[8] Khaniabadi YO, Fanelli R, De Marco A, Daryanoosh SM, Kloog I, Hopke PK, et al. Hospital admissions in Iran for cardiovascular and respiratory diseases attributed to the middle eastern dust storms. Environmental Science and Pollution Research. 2017;**24**(20): 16860-16868

[9] Plocoste T, Calif R, Jacoby-Koaly S. Temporal multiscaling characteristics of particulate matter PM10 and ground-level ozone O_3 concentrations in Caribbean region. Atmospheric Environment. 2017;**169**:22-35

[10] Molla MKI, Hirose K. Hilbert spectrum in time-frequency representation of audio signals considering disjoint orthogonality. Advances in Adaptive Data Analysis. 2010;**2**(03):313-336

[11] Peng Z, Peter WT, Chu F. A comparison study of improved Hilbert-Huang transform and wavelet transform: Application to fault diagnosis for rolling bearing. Mechanical Systems and Signal Processing. 2005;**19**(5): 974-988

[12] Andrade AO, Kyberd P, Nasuto SJ. The application of the Hilbert spectrum to the analysis of electromyographic signals. Information Sciences. 2008; **178**(9):2176-2193

[13] Plocoste T, Jacoby-Koaly S, Molinié J, Petit R. Evidence of the effect of an urban heat island on air quality near a landfill. Urban Climate. 2014;**10**:745-757

[14] Plocoste T, Calif R, Jacoby-Koaly S. Multi-scale time dependent correlation between synchronous measurements of ground-level ozone and meteorological parameters in the Caribbean Basin.

Atmospheric Environment. 2019;**211**: 234-246

[15] Euphrasie-Clotilde L, Molinié J, Feuillard T, Brute F. The relationship between coastal West African dust level and Caribbean island dust. WIT Transactions on Ecology and the Environment. 2017;**211**:121-127

[16] Mandelbrot B. The Fractal Geometry of Nature. New York: Freeman and Company; 1982

[17] Bracewell R. The Fourier Transform and Its Applications. 3rd ed. New York: McGraw-Hill Science; 1999

[18] Sengupta R. Interpretation of magnetic anomalies of a two-dimensional fault by Fourier integral. Canadian Journal of Exploration Geophysicists. 1975;**11**:65-71

[19] Huang NE, Shen Z, Long SR, Wu MC, Shih HH, Zheng Q, et al. The empirical mode decomposition and the Hilbert spectrum for nonlinear and non-stationary time series analysis. Proceedings of the Royal Society of London, Series A: Mathematical, Physical and Engineering Sciences. 1998;**454**(1971):903-995

[20] Huang NE, Shen Z, Long SR. A new view of nonlinear water waves: The Hilbert spectrum. Annual Review of Fluid Mechanics. 1999;**31**(1): 417-457

[21] Flandrin P, Goncalves P. Empirical mode decompositions as data-driven wavelet-like expansions. International Journal of Wavelets, Multiresolution and Information Processing. 2004; **2**(04):477-496

[22] Huang NE, Wu MLC, Long SR, Shen SS, Qu W, Gloersen P, et al. A confidence limit for the empirical mode decomposition and Hilbert spectral analysis. Proceedings of the Royal Society of London, Series A: Mathematical, Physical and Engineering Sciences. 2003;**459**(2037):2317-2345

[23] Rilling G, Flandrin P. One or two frequencies? The empirical mode decomposition answers. IEEE Transactions on Signal Processing. 2008;**56**(1):85-95

[24] Long SR, Huang N, Tung C, Wu M, Lin R, Mollo-Christensen E, et al. The Hilbert techniques: An alternate approach for non-steady time series analysis. GRSL Geoscience and Remote Sensing Letters. 1995;**3**:6-11

[25] Cohen L. Time-Frequency Analysis. Vol. 778. New Jersey: Prentice Hall; 1995

[26] Calif R, Schmitt FG, Huang Y. Multifractal description of wind power fluctuations using arbitrary order Hilbert spectral analysis. Physica A: Statistical Mechanics and its Applications. 2013;**392**(18):4106-4120

[27] Calif R, Schmitt FG, Huang Y. The scaling properties of the turbulent wind using empirical mode decomposition and arbitrary order Hilbert spectral analysis. In: Wind Energy-Impact of Turbulence. Berlin: Springer; 2014. pp. 43-49

[28] Schertzer D, Lovejoy S. Physical modeling and analysis of rain and clouds by anisotropic scaling multiplicative processes. Journal of Geophysical Research-Atmospheres. 1987;**92**: 9693-9714

[29] Marshak A, Davis A, Cahalan R, Wiscombe W. Bounded cascade models as nonstationary multifractals. Physical Review E. 1994;**49**:55-69

[30] Windsor H, Toumi R. Scaling and persistence of UK pollution. Atmospheric Environment. 2001;**35**: 4545-4556

[31] Marr LC, Harley RA. Spectral analysis of weekday–weekend

differences in ambient ozone, nitrogen oxide, and non-methane hydrocarbon time series in California. Atmospheric Environment. 2002;**36**:2327-2335

[32] Messina AR, Vittal V. Nonlinear, non-stationary analysis of interarea oscillations via Hilbert spectral analysis. IEEE Transactions on Power Systems. 2006;**21**(3):1234-1241

[33] Choi YS, Ho CH, Chen D, Noh YH, Song CK. Spectral analysis of weekly variation in PM10 mass concentration and meteorological conditions over China. Atmospheric Environment. 2008;**42**:655-666

[34] Huang Y, Schmitt FG. Time dependent intrinsic correlation analysis of temperature and dissolved oxygen time series using empirical mode decomposition. Journal of Marine Systems. 2014;**130**:90-100

[35] Kingsbury BE, Morgan N, Greenberg S. Robust speech recognition using the modulation spectrogram. Speech Communication. 1998;**25**(1–3): 117-132

[36] Sussillo D, Kundaje A, Anastassiou D. Spectrogram analysis of genomes. EURASIP Journal on Advances in Signal Processing. 2004;**2004**(1):29-42

[37] Fulop SA, Fitz K. Algorithms for computing the time-corrected instantaneous frequency (reassigned) spectrogram, with applications. The Journal of the Acoustical Society of America. 2006;**119**(1):360-371

[38] Huang Y, Schmitt FG, Lu Z, Liu Y. An amplitude-frequency study of turbulent scaling intermittency using empirical mode decomposition and Hilbert spectral analysis. EPL (Europhysics Letters). 2008;**84**(4): 40010

[39] Ismail DKB, Lazure P, Puillat I. Advanced spectral analysis and cross correlation based on the empirical mode decomposition: Application to the environmental time series. IEEE Geoscience and Remote Sensing Letters. 2015;**12**(9):1968-1972

[40] Li X, Li Z, Wang E, Feng J, Kong X, Chen L, et al. Analysis of natural mineral earthquake and blast based on Hilbert–Huang transform (HHT). Journal of Applied Geophysics. 2016; **128**:79-86

[41] Kader B, Yaglom A. Spectra and correlation functions of surface layer atmospheric turbulence in unstable thermal stratification. In: Turbulence and Coherent Structures. Dordrecht: Springer; 1991. pp. 387-412

[42] Huang Y, Schmitt FG, Hermand JP, Gagne Y, Lu Z, Liu Y, et al. Arbitrary-order Hilbert spectral analysis for time series possessing scaling statistics: Comparison study with detrended fluctuation analysis and wavelet leaders. Physical Review E. 2011;**84**(1): 016208

[43] Zhang C, Ni Z, Ni L. Multifractal detrended cross-correlation analysis between PM2.5 and meteorological factors. Physica A: Statistical Mechanics and its Applications. 2015;**438**:114-123

[44] Tchen C. On the spectrum of energy in turbulent shear flow. Journal of Research of the National Bureau of Standards. 1953;**50**:51-62

[45] Katul GG, Porporato A, Nikora V. Existence of k^{-1} power-law scaling in the equilibrium regions of wall-bounded turbulence explained by Heisenberg's eddy viscosity. Physical Review E. 2012; **86**(6):066311

[46] Iatsenko D, McClintock PV, Stefanovska A. Linear and synchrosqueezed time–frequency representations revisited: Overview, standards of use, resolution, reconstruction, concentration, and

algorithms. Digital Signal Processing. 2015;**42**:1-26

[47] Varotsos C, Ondov J, Efstathiou M. Scaling properties of air pollution in Athens, Greece and Baltimore, Maryland. Atmospheric Environment. 2005;**39**(22):4041-4047

[48] Prospero JM, Collard FX, Molinié J, Jeannot A. Characterizing the annual cycle of African dust transport to the Caribbean Basin and South America and its impact on the environment and air quality. Global Biogeochemical Cycles. 2014;**28**(7):757-773

Chapter 8

Optimal Control of Evolution Differential Inclusions with Polynomial Linear Differential Operators

Elimhan N. Mahmudov

Abstract

In this chapter, we studied a new class of problems in the theory of optimal control defined by polynomial linear differential operators. As a result, an interesting Mayer problem arises with higher order differential inclusions. Thus, in terms of the Euler-Lagrange and Hamiltonian type inclusions, sufficient optimality conditions are formulated. In addition, the construction of transversality conditions at the endpoints of the considered time interval plays an important role in future studies. To this end, the apparatus of locally adjoint mappings is used, which plays a key role in the main results of this chapter. The presented method is demonstrated by the example of the linear optimal control problem, for which the Weierstrass-Pontryagin maximum principle is derived.

Keywords: Euler-Lagrange, differential inclusion, set-valued mapping, polynomial differential operators, linear problem, transversality, Weierstrass-Pontryagin maximum principle

1. Introduction

This chapter concerns with the special kind of optimal control problem with differential inclusions, where the left-hand side of the evolution inclusion is polynomial linear differential operators with variable coefficients; in fact, the main difficulty in the considered problems is to construct the Euler-Lagrange type higher order adjoint inclusions and the transversality conditions. That is why in the whole literature, only the qualitative properties of second-order differential inclusions are investigated (see [1–3] and references therein).

The paper [1] gives necessary and sufficient conditions ensuring the existence of a solution to the second-order differential inclusion with Cauchy initial value problem. Furthermore, second-order interior tangent sets are introduced and studied to obtain such conditions. The paper [2] studies, in the context of Banach spaces, the problem of three boundary conditions for both second-order differential inclusions and second-order ordinary differential equations. The results are obtained in several new settings of Sobolev type spaces involving Bochner and Pettis integrals. In the paper [3], the existence of viable solutions to the Cauchy problem $x'' \in F(x, x'), x(0) = x_0, x'(0) = y_0$ is proved, where F is a set-valued map defined on a locally compact set $M \subset \mathbb{R}^{2n}$, contained in the Frechet subdifferential of a φ-convex function of order two.

Some qualitative properties and optimization of first-order discrete and continuous time processes with lumped and distributed parameters have been expanding in all directions at an astonishing rate during the last few decades (see [4–13] and their references).

The optimization of higher order differential inclusions was first developed by Mahmudov in [14–21]. Since then this problem has attracted many author's attentions (see [22] and their references). The paper [14] studies a new class of problems of optimal control theory with Sturm-Liouville type differential inclusions involving second-order linear self-adjoint differential operators. By using the discretization method guaranteeing transition to continuous problem, the discrete and discrete-approximate inclusions are investigated. Necessary and sufficient conditions, containing both the Euler-Lagrange and Hamiltonian type inclusions, and "transversality" conditions are derived. The paper [15] deals with the optimization of the Bolza problem with third-order differential inclusions and arbitrary higher order discrete inclusions. The work [16] is devoted to the Bolza problem of optimal control theory given by second-order convex differential inclusions with second-order state variable inequality constraints. According to the proposed discretization method, problems with discrete-approximate inclusions and inequalities are investigated. Necessary and sufficient conditions of optimality including distinctive "transversality" condition are proved in the form of Euler-Lagrange inclusions. The paper [17] is concerned with the necessary and sufficient conditions of optimality for second-order polyhedral optimization described by polyhedral discrete and differential inclusions. The paper [18] is devoted to the study of optimal control theory with higher order differential inclusions and a varying time interval. Essentially, under a more general setting of problems and endpoint constraints, the main goal is to establish sufficient conditions of optimality for higher order differential inclusions. Thus with the use of Euler-Lagrange and Hamiltonian type of inclusions and transversal conditions on the "initial" sets, the sufficient conditions are formulated. The paper [21] studies a new class of problems of optimal control theory with state constraints and second-order delay discrete and delay differential inclusions. Under the "regularity" condition by using discrete approximations as a vehicle, in the forms of Euler-Lagrange and Hamiltonian type inclusions, the sufficient conditions of optimality for delay DFIs, including the peculiar transversality ones, are proved.

The present chapter is ordered in the following manner.

In Section 2 the necessary facts and supplementary results from the book of Mahmudov are given [23]; Hamiltonian function and locally adjoint mapping are introduced, and the problems with initial point constraints for polynomial linear differential operators governed by time-dependent set-valued mapping are formulated. In Section 3, we present the main results; on the basis of "transversality" conditions at the endpoints of the considered time interval, the sufficient conditions of optimality for differential inclusions with polynomial linear differential operators and with initial point constraints are proved. In particular, it is shown that our problems involve optimization of the so-called Sturm-Liouville type differential inclusions. To the best of our knowledge, there is no paper which considers optimality conditions for these problems in the literature, and we aim to fill this gap. Therefore, the novelty of our formulation of the problem is justified. To establish the Euler-Lagrange and Hamiltonian inclusions and the transversality conditions, we use the construction of a suitable rewriting of the primal polynomial linear differential operator and the rearrangement of its integration. The case of variable coefficients of polynomial linear differential operators turns out to be more complicated, unless transversality assumptions at the endpoints of the considered time interval are applied. It should be noted that the main proof can be easily generalized to the nonconvex case. Then, using the new approach given in Section 4 of this chapter, we construct the Weierstrass-Pontryagin maximum condition [24] for the linear optimal control problem. Consequently, in the particular case, the maximum principle follows from the Euler-Lagrange inclusion.

In Section 5 the optimality conditions are given for convex problem with second-order differential inclusions and endpoint constraints. By using second-order suitable Euler-Lagrange type adjoint inclusions and transversality conditions, Theorem 5.1 is proved.

The main results in this section can be extended to the case of Hilbert spaces ℓ_2, L_2^n. We remind that a Hilbert space H is a real or complex inner product space that is also a complete metric space with respect to the distance function induced by the inner product [2]. By definition, every Hilbert space is also a Banach space. Furthermore, in every Hilbert space, the following parallelogram identity $\|x+y\|^2+\|x-y\|^2=2\left(\|x\|^2+\|y\|^2\right)$ holds. Conversely, every Banach space in which the parallelogram identity holds is a Hilbert space. Remember that ℓ_2 is a space of numerical sequences, such that if $x=\{x_i\}$, then $\sum_{i=1}^{\infty}x_i^2<\infty$. In fact ℓ_2 is an infinity dimensional coordinate-wise Hilbert space with the corresponding inner product $\langle x,y\rangle=\sum_{i=1}^{\infty}x_iy_i$. Endowing a relevant norm, we have a Banach space. Obviously, optimization of problem with PLDOs can be reduced to problem with geometric constraints in such finite-dimensional Hilbert space. As is known with all the pairs of elements of this space, a certain finite number is associated, i.e., inner product, existence of which is guaranteed by applying the familiar Cauchy Schwarz-Bunyakovskii [25] inequality. We remark that in our case for $x=\{x_0,x_1,x_2,\ldots\}\in\ell_2$ and $x^*=\{x_0^*,x_1^*,x_2^*,\ldots\}\in\ell_2^*$, the inner product $\langle x,x^*\rangle=\sum_{i=1}^{\infty}x_ix_i^*$ is finite numbers since this series is convergent. Besides it is known [25] that ℓ_2 is a self-adjoint space, i.e., $\ell_p=\ell_q^*$, and $1/p+1/q=1$, and so $\ell_2=\ell_2^*$ for $p=2$. Thus a dual cone constructed can be defined. The set of square integrable functions $L_2^n([0,1])$ is a Hilbert space with inner product $\langle x(t),y(t)\rangle=\int_0^1x(t)y(t)dt$.

2. Preliminaries and problem statements

The basic concepts given in this section can be found in the book [23]; let $\mathbb{R}^n$ be a n-dimensional Euclidean space, $\langle x,v\rangle$ be an inner product of elements $x,v\in\mathbb{R}^n$, and (x,v) be a pair of x,v. Let $F:\mathbb{R}^n\rightrightarrows\mathbb{R}^n$ be a set-valued mapping from $\mathbb{R}^n$ into the set of subsets of $\mathbb{R}^n$. Therefore F is a convex set-valued mapping, if its graph $gph\ F=\{(x,v):v\in F(x)\}$ is a convex subset of $\mathbb{R}^{2n}$. A set-valued mapping F is called closed if its $gph\ F$ is a closed subset in $\mathbb{R}^{2n}$. The domain of a set-valued mapping F is denoted by $dom\,F$ and is defined as $dom\,F=\{x:F(x)\neq\varnothing\}$. A set-valued mapping F is convex-valued if $F(x)$ is a convex set for each $x\in dom\,F$.

The Hamiltonian function and argmaximum set corresponding to a set-valued mapping F are defined by the relations correspondingly:

$$H_F(x,v^*)=\sup_v\{\langle v,v^*\rangle:v\in F(x)\},\quad v^*\in\mathbb{R}^n,$$

$$F_A(x,v^*)=\{v\in F(x):\langle v,v^*\rangle=H_F(x,v^*)\}$$

We set $H_F(x,v^*)=-\infty$ if $F(x)=\varnothing$. The interior and relative interior of a set $M\subset\mathbb{R}^{2n}$ are denoted by int$M$ and ri$M$, respectively.

A convex cone $K_M(z_0)$, $z_0 \in M$ is a cone of tangent directions if from $\bar{z} = (\bar{x}, \bar{v}) \in K_M(z_0)$ it follows that $\bar{z}$ is a tangent vector to the set M at a point $z_0 \in M$, i.e., there exists such function $q : \mathbb{R}^1 \to \mathbb{R}^{2n}$ that $z_0 + \alpha\bar{z} + q(\alpha) \in M$ for sufficiently small $\alpha > 0$ and $\alpha^{-1}q(\alpha) \to 0$, as $\alpha \downarrow 0$.

For a set-valued mapping F, the set-valued mapping $F^* : \mathbb{R}^n \rightrightarrows \mathbb{R}^n$ is defined by

$$F^*(v^*; (x, v)) := \{x^* : (x^*, -v^*) \in K^*_{gph\,F}(x, v)\}$$

$$K_{gphF}(x, v) = \text{cone}\,[\,gph\,F - (x, v)], \forall (x^1, v^1) \in gph\,F.$$

It is called the LAM to F at a point $(x, v) \in gph\,F$, where $K^* = \{z^* : \langle \bar{z}, z^* \rangle \geq 0, \forall \bar{z} \in K\}$ denotes the dual cone to the cone K, as usual. Below by using the Hamiltonian function, associated to a set-valued mapping F, we will define another LAM. Thus, the LAM to "nonconvex" mapping F is defined as follows:

$$F^*(v^*; (x, v)) := \{x^* : H_F(x^1, v^*) - H_F(x, v^*) \leq \langle x^*, x^1 - x \rangle, \forall x^1 \in \mathbb{R}^n\},$$
$$(x, v) \in gph\,F, v \in F_A(x, v^*).$$

Clearly, for the convex mapping F, the Hamiltonian function $H_F(\cdot, \cdot, v^*)$ is concave, and the latter definition of LAM coincides with the previous definition of LAM ([23], p. 62). Note that prior to the LAM, the notion of coderivative has been introduced for set-valued mappings in terms of the basic normal cone to their graphs by Mordukhovich [26] and for the smooth convex maps, the two notions are equivalent.

The aim of Section 3 is to obtain the Euler-Lagrange type adjoint inclusion and sufficient optimality conditions for a problem with polynomial linear differential operators:

$$\text{Minimize } \varphi\left(x(1), x'(1), \ldots, x^{(s-1)}(1)\right), \tag{1}$$

$$(P_V)\ Lx(t) \in F(x(t), t), \ \text{a.e. } t \in [0, 1], \tag{2}$$

$$x(0) \in Q_0,\ x'(0) \in Q_1 x''(0) \in Q_2, \ldots, x^{(s-1)}(0) \in Q_{s-1} \tag{3}$$

where $Lx = \sum_{k=1}^{s} p_k(t) D^k x$ is a PLDO of degree s with variable coefficients $p_k : [0, 1] \to \mathbb{R}^1$ and $D^k, k = 1, \ldots, s$ is the operator of kth-order derivatives. In what follows for each k, a scalar function p_k is kth-order continuously differentiable function, $p_s(t) \neq 0$ on $[0, 1]$ identically, $F(\cdot, t) : \mathbb{R}^n \rightrightarrows \mathbb{R}^n$ is time-dependent set-valued mapping, $\varphi : (\mathbb{R}^n)^s \to \mathbb{R}^1$ is continuous function, $Q_j \subseteq \mathbb{R}^n, j = 0, 1, \ldots, s-1$ are nonempty subsets of $\mathbb{R}^n$, and s $(s \geq 2)$ is an arbitrary fixed natural number. It is required to find an arc $\tilde{x}(t)$ of the problem Eqs. (1)–(3) for the sth-order differential inclusions satisfying Eq. (2) almost everywhere (a.e.) on a considered time interval and minimizing the functional $\varphi(x(1), \ldots, x^{(s-1)}(1))$. An arc $x(\cdot)$ is absolutely continuous and $s-1$ order differentiable function, where $x^{(s)}(\cdot) \equiv \frac{d^s x(\cdot)}{dt^s} \in L_1^n([0, 1])$. Obviously, such class of functions is a Banach space, endowed with the different equivalent norms.

Remark 2.1. Notice that to get sufficient condition of optimality for the Mayer problem (P_V) described by ordinary evolution differential inclusions with PLDOs and with initial point constraints, using the discretized method, we consider the following sth-order discrete-approximate problem instead of the problem (P_V):

$$\text{minimize } \varphi\left(x(1-(s-1)h), \Delta x(1-(s-1)h), \dots, \Delta^{s-1}x(1-(s-1)h)\right),$$

$$\sum_{k=1}^{s} p_k(t)\Delta^k x(t) \in F(x(t),t), t = 0, h, 2h, \dots, 1-sh;$$

$$\Delta^k x(0) \in Q_k, k = 0, \dots, s-1.$$

Here kth-order difference operator is defined as follows:

$$\Delta^k x(t) = \frac{1}{h^k}\sum_{s=0}^{k}(-1)^s C_k^s x(t+(k-s)h), \ C_k^s = \frac{k!}{s!(k-s)!}, \ t = 0, h, \dots, 1-h.$$

Thus by using the method of approximation [23, 26, 27], we can establish necessary and sufficient conditions for the rather complicated sth-order discrete-approximate problem. Then by passing to the limit in necessary and sufficient conditions of this problem as $h \to 0$, we can construct the optimality condition for the Mayer problem (P_V) described by higher order differential inclusions with PLDOs and with initial point constraints. But in this chapter to avoid long calculations, derivations of these conditions are omitted.

3. Optimization of evolution differential inclusions with PLDOs

In the present section, we study sufficient optimality conditions for the problem (P_V). Before all, we formulate the so-called sth-order Euler-Lagrange type differential inclusion and the transversality conditions:

i. $L^* x^*(t) \in F^*(x^*(t); (\tilde{x}(t), L\tilde{x}(t)), t)$, a.e. $t \in [0,1]$,

where $L^* x^*(t) = \sum_{k=1}^{s}(-1)^k D^k[p_k(t)x^*(t)]$ is the adjoint PLDO of the primal operator L.

ii. $\sum_{k=0}^{s-1}(-1)^{s-k}D^{s-k-1}[p_{s-k}(0)x^*(0)] \in K^*_{Q_0}(\tilde{x}(0))$;

$\sum_{k=0}^{s-2}(-1)^{s-k-1}D^{s-k-2}[p_{s-k}(0)x^*(0)] \in K^*_{Q_1}(\tilde{x}'(0))$;
$D[p_s(0)x^*(0)] - p_{s-1}(0)x^*(0) \in K^*_{Q_{s-2}}\left(\tilde{x}^{(s-2)}(0)\right)$;
$-p_s(0)x^*(0) \in K^*_{Q_{s-1}}\left(\tilde{x}^{(s-1)}(0)\right)$.

iii. $\left(\sum_{k=0}^{s-1}(-1)^{s-k}D^{s-k-1}[p_{s-k}(1)x^*(1)], \sum_{k=0}^{s-2}(-1)^{s-k-1}D^{s-k-2}[p_{s-k}(1)x^*(1)], \dots,\right.$
$\left. D[p_s(1)x^*(1)] - p_{s-1}(1)x^*(1), p_s(1)x^*(1)\right) \in \partial\varphi\left(\tilde{x}(1), \tilde{x}'(1), \dots, \tilde{x}^{(s-1)}(1)\right)$.

Later on we assume that $x^*(t), t \in [0,1]$ is absolutely a continuous function with the higher order derivatives until $s-1$ and $x^{*(s)}(\cdot) \in L_1^n([0,1])$. The following condition ensures that the LAM F^* is nonempty:

iv. $L\tilde{x}(t) \in F_A(\tilde{x}(t), x^*(t), t)$, a.e. $t \in [0,1]$ or, equivalently,
$\langle L\tilde{x}(t), x^*(t)\rangle = H_F(\tilde{x}(t), x^*(t)), L\tilde{x}(t) \in F(\tilde{x}(t), t)$.

The following are sufficient optimality conditions for evolution differential inclusions with PLDOs.

Theorem 3.1. Let φ be a continuous and convex function and $F(\cdot,t)$ a convex set-valued mapping. Moreover, let

$$Q_j, j = 0, \ldots, s-1$$

be convex sets. Then for optimality of the trajectory $\tilde{x}(\cdot)$ in the problem (P_V) with evolution differential inclusions and PLDOs, it is sufficient that there exists an absolutely continuous function $x^*(\cdot)$ with the higher order derivatives until $s-1$, satisfying a.e. the Euler-Lagrange type differential inclusion with PLDOs Eqs. (i) and (iv) and transversality conditions Eqs. (ii) and (iii) at the endpoints $t=0$ and $t=1$.

Proof. Using Theorem 2.1 ([23], p. 62), the definition of the Hamiltonian function and condition Eq. (i), we obtain

$$H_F(x(t), x^*(t), t) - H_F(\tilde{x}(t), x^*(t), t) \le \langle L^* x^*(t), x(t) - \tilde{x}(t)\rangle$$

which can be rewritten as follows

$$H_F(x(t), x^*(t), t) - H_F(\tilde{x}(t), x^*(t), t) \le \left\langle \sum_{k=1}^{s} (-1)^k \left(p_k(t) x^*(t)\right)^{(k)}, x(t) - \tilde{x}(t) \right\rangle. \tag{4}$$

Further using the definition of the Hamiltonian function, Eq. (4) can be converted to the inequality

$$0 \ge \langle Lx(t) - L\tilde{x}(t), x^*(t)\rangle - \langle L^* x^*(t), x(t) - \tilde{x}(t)\rangle$$

or

$$0 \ge \left\langle \sum_{k=1}^{s} p_k(t)(x(t) - \tilde{x}(t))^{(k)}, x^*(t) \right\rangle - \left\langle \sum_{k=1}^{s} (-1)^k \left(p_k(t) x^*(t)\right)^{(k)}, x(t) - \tilde{x}(t) \right\rangle. \tag{5}$$

Integrating Eq. (5) over the interval $[0,1]$, we have

$$0 \ge \int_0^1 \left[\left\langle \sum_{k=1}^{s} p_k(t)(x(t) - \tilde{x}(t))^{(k)}, x^*(t) \right\rangle - \left\langle \sum_{k=1}^{s} (-1)^k \left(p_k(t) x^*(t)\right)^{(k)}, x(t) - \tilde{x}(t) \right\rangle \right] dt \tag{6}$$

Let us denote

$$B = \sum_{k=1}^{s} \left\langle (x(t) - \tilde{x}(t))^{(k)}, p_k(t) x^*(t) \right\rangle - \sum_{k=1}^{s} \left\langle (-1)^k \left(p_k(t) x^*(t)\right)^{(k)}, x(t) - \tilde{x}(t) \right\rangle$$

In what follows our approach lies in reducing B in a relationship consisting of s sums from k ($k = 1, \dots, s$) to s of suitable derivatives of scalar products; thus, after some transformations we can deduce an important representation for a first term of B as follows:

$$
\begin{aligned}
\sum_{k=1}^{s}\left\langle (x(t)-\tilde{x}(t))^{(k)}, p_k(t)x^*(t)\right\rangle &= \sum_{k=1}^{s}\left[\frac{d}{dt}\left\langle x^{(k-1)}(t)-\tilde{x}^{(k-1)}(t), p_k(t)x^*(t)\right\rangle\right] \\
&\quad -\sum_{k=2}^{s}\left[\frac{d}{dt}\left\langle x^{(k-2)}(t)-\tilde{x}^{(k-2)}(t), \left(p_k(t)x^*(t)\right)'\right\rangle\right] \\
&\quad +\sum_{k=3}^{s}\left[\frac{d}{dt}\left\langle x^{(k-3)}(t)-\tilde{x}^{(k-3)}(t), \left(p_k(t)x^*(t)\right)''\right\rangle\right] \\
&\quad -\cdots+\sum_{k=s-2}^{s}\left[\frac{d}{dt}\left\langle x^{(k-s+2)}(t)-\tilde{x}^{(k-s+2)}(t), (-1)^{m-3}\left(p_k(t)x^*(t)\right)^{(s-3)}\right\rangle\right] \\
&\quad +\sum_{k=s-1}^{s}\left[\frac{d}{dt}\left\langle x^{(k-s+1)}(t)-\tilde{x}^{(k-s+1)}(t), (-1)^{s-2}\left(p_k(t)x^*(t)\right)^{(s-2)}\right\rangle\right] \\
&\quad +\frac{d}{dt}\left\langle x(t)-\tilde{x}(t), (-1)^{m-1}\left(p_s(t)x^*(t)\right)^{(s-1)}\right\rangle \\
&\quad +\sum_{k=1}^{s}\left[\left\langle x(t)-\tilde{x}(t), (-1)^{k}\left(p_k(t)x^*(t)\right)^{(k)}\right\rangle\right].
\end{aligned}
\tag{7}
$$

Then in view of Eq. (7) in the definition of B, we have an efficient formula:

$$
\begin{aligned}
B &= \sum_{k=1}^{s}\left[\frac{d}{dt}\left\langle x^{(k-1)}(t)-\tilde{x}^{(k-1)}(t), p_k(t)x^*(t)\right\rangle\right] \\
&\quad -\sum_{k=2}^{s}\left[\frac{d}{dt}\left\langle x^{(k-2)}(t)-\tilde{x}^{(k-2)}(t), \left(p_k(t)x^*(t)\right)'\right\rangle\right] \\
&\quad +\sum_{k=3}^{s}\left[\frac{d}{dt}\left\langle x^{(k-3)}(t)-\tilde{x}^{(k-3)}(t), \left(p_k(t)x^*(t)\right)''\right\rangle\right] \\
&\quad -\cdots+\sum_{k=s-2}^{s}\left[\frac{d}{dt}\left\langle x^{(k-s+2)}(t)-\tilde{x}^{(k-s+2)}(t), (-1)^{s-3}\left(p_k(t)x^*(t)\right)^{(s-3)}\right\rangle\right] \\
&\quad +\sum_{k=s-1}^{s}\left[\frac{d}{dt}\left\langle x^{(k-s+1)}(t)-\tilde{x}^{(k-s+1)}(t), (-1)^{s-2}\left(p_k(t)x^*(t)\right)^{(s-2)}\right\rangle\right] \\
&\quad +\frac{d}{dt}\left\langle x(t)-\tilde{x}(t), (-1)^{s-1}\left(p_s(t)x^*(t)\right)^{(s-1)}\right\rangle.
\end{aligned}
\tag{8}
$$

Then taking into account the structure of B in Eq. (8), we can compute the integral on the right-hand side of Eq. (6) as follows:

$$
\int_0^1 B\,dt = \sum_{k=1}^{s}\left[\int_0^1 d\left\langle x^{(k-1)}(t)-\tilde{x}^{(k-1)}(t), p_k(t)x^*(t)\right\rangle\right]
$$

$$-\sum_{k=2}^{s}\left[\int_0^1 d\left\langle x^{(k-2)}(t)-\tilde{x}^{(k-2)}(t),\left(p_k(t)x^*(t)\right)'\right\rangle\right]$$

$$+\sum_{k=3}^{s}\left[\int_0^1 d\left\langle x^{(k-3)}(t)-\tilde{x}^{(k-3)}(t),\left(p_k(t)x^*(t)\right)''\right\rangle\right]$$

$$-\cdot\cdots+\sum_{k=s-2}^{s}\left[\int_0^1 d\left\langle x^{(k-s+2)}(t)-\tilde{x}^{(k-s+2)}(t),(-1)^{s-3}\left(p_k(t)x^*(t)\right)^{(s-3)}\right\rangle\right]$$

$$+\sum_{k=s-1}^{s}\left[\int_0^1 d\left\langle x^{(k-s+1)}(t)-\tilde{x}^{(k-s+1)}(t),(-1)^{s-2}\left(p_k(t)x^*(t)\right)^{(s-2)}\right\rangle\right]$$

$$+\int_0^1 d\left\langle x(t)-\tilde{x}(t),(-1)^{s-1}\left(p_s(t)x^*(t)\right)^{(s-1)}\right\rangle.$$

Thus, integrating B, we can obtain

$$\int_0^1 Bdt=\sum_{k=1}^{s}\left[\left\langle x^{(k-1)}(1)-\tilde{x}^{(k-1)}(1),p_k(1)x^*(1)\right\rangle\right.$$

$$\left.-\left\langle x^{(k-1)}(0)-\tilde{x}^{(k-1)}(0),p_k(0)x^*(0)\right\rangle\right]$$

$$-\sum_{k=2}^{s}\left[\left\langle x^{(k-2)}(1)-\tilde{x}^{(k-2)}(1),\left(p_k(1)x^*(1)\right)'\right\rangle-\left\langle x^{(k-2)}(0)-\tilde{x}^{(k-2)}(0),\left(p_k(0)x^*(0)\right)'\right\rangle\right]$$

$$+\sum_{k=3}^{s}\left[\left\langle x^{(k-3)}(1)-\tilde{x}^{(k-3)}(1),\left(p_k(1)x^*(1)\right)''\right\rangle\right.$$

$$\left.-\left\langle x^{(k-3)}(0)-\tilde{x}^{(k-3)}(0),\left(p_k(0)x^*(0)\right)''\right\rangle\right]$$

$$-\cdots+\sum_{k=s-2}^{s}\left[\left\langle x^{(k-s+2)}(1)-\tilde{x}^{(k-s+2)}(1),(-1)^{s-3}\left(p_k(1)x^*(1)\right)^{(s-3)}\right\rangle\right.$$

$$\left.-\left\langle x^{(k-s+2)}(0)-\tilde{x}^{(k-s+2)}(0),(-1)^{s-3}\left(p_k(0)x^*(0)\right)^{(s-3)}\right\rangle\right]$$

$$+\sum_{k=s-1}^{s}\left[\left\langle x^{(k-s+1)}(1)-\tilde{x}^{(k-s+1)}(1),(-1)^{s-2}\left(p_k(1)x^*(1)\right)^{(s-2)}\right\rangle\right.$$

$$\left.-\left\langle x^{(k-s+1)}(0)-\tilde{x}^{(k-s+1)}(0),(-1)^{s-2}\left(p_k(0)x^*(0)\right)^{(s-2)}\right\rangle\right]$$

$$+\left\langle x(1)-\tilde{x}(1),(-1)^{s-1}\left(p_s(1)x^*(1)\right)^{(s-1)}\right\rangle$$

$$-\left\langle x(0)-\tilde{x}(0),(-1)^{s-1}\left(p_s(0)x^*(0)\right)^{(s-1)}\right\rangle.$$

Here by suitable rearrangement and necessary simplification, we have

$$\begin{aligned}
\int_0^1 B dt &= \sum_{k=1}^{s}\left[\left\langle x^{(k-1)}(1)-\tilde{x}^{(k-1)}(1), p_k(1)x^*(1)\right\rangle\right] \\
&-\sum_{k=2}^{s}\left[\left\langle x^{(k-2)}(1)-\tilde{x}^{(k-2)}(1), \left(p_k(1)x^*(1)\right)'\right\rangle\right] \\
&+\sum_{k=3}^{s}\left[\left\langle x^{(k-3)}(1)-\tilde{x}^{(k-3)}(1), \left(p_k(1)x^*(1)\right)''\right\rangle\right] \\
&-\dots+\sum_{k=s-2}^{s}\left[\left\langle x^{(k-s+2)}(1)-\tilde{x}^{(k-s+2)}(1), (-1)^{s-3}\left(p_k(1)x^*(1)\right)^{(s-3)}\right\rangle\right] \\
&+\sum_{k=s-1}^{s}\left[\left\langle x^{(k-s+1)}(1)-\tilde{x}^{(k-s+1)}(1), (-1)^{s-2}\left(p_k(1)x^*(1)\right)^{(s-2)}\right\rangle\right] \qquad (9)\\
&+\left\langle x(1)-\tilde{x}(1), (-1)^{s-1}\left(p_s(1)x^*(1)\right)^{(s-1)}\right\rangle \\
&+\left\langle x(0)-\tilde{x}(0), \sum_{k=0}^{s-1}(-1)^{s-k}D^{s-k-1}\left[p_{s-k}(0)x^*(0)\right]\right\rangle \\
&+\left\langle x'(0)-\tilde{x}'(0), \sum_{k=0}^{s-2}(-1)^{s-k-1}D^{s-k-2}\left[p_{s-k}(0)x^*(0)\right]\right\rangle \\
&+\left\langle x''(0)-\tilde{x}''(0), \sum_{k=0}^{s-3}(-1)^{s-k-2}D^{s-k-3}\left[p_{s-k}(0)x^*(0)\right]\right\rangle \\
&+\cdot\dots-\left\langle x^{(s-2)}(0)-\tilde{x}^{(s-2)}(0), -D\left[p_s(0)x^*(0)\right]+p_{s-1}(0)x^*(0)\right\rangle \\
&-\left\langle x^{(s-1)}(0)-\tilde{x}^{(s-1)}(0), p_s(0)x^*(0)\right\rangle.
\end{aligned}$$

In order to make use of the transversality condition Eq. (ii), we rewrite it in a more relevant form:

$$\begin{aligned}
&\left\langle x(0)-\tilde{x}(0), \sum_{k=0}^{s-1}(-1)^{s-k}D^{s-k-1}\left[p_{s-k}(0)x^*(0)\right]\right\rangle \\
&+\left\langle x'(0)-\tilde{x}'(0), \sum_{k=0}^{s-2}(-1)^{s-k-1}D^{s-k-2}\left[p_{s-k}(0)x^*(0)\right]\right\rangle \\
&+\left\langle x''(0)-\tilde{x}''(0), \sum_{k=0}^{s-3}(-1)^{s-k-2}D^{s-k-3}\left[p_{s-k}(0)x^*(0)\right]\right\rangle \\
&+\cdot\dots-\left\langle x^{(s-2)}(0)-\tilde{x}^{(s-2)}(0), -D\left[p_s(0)x^*(0)\right]+p_{s-1}(0)x^*(0)\right\rangle \\
&-\left\langle x^{(s-1)}(0)-\tilde{x}^{(s-1)}(0), p_s(0)x^*(0)\right\rangle \geq 0; \forall x^{(k)}(0)\in K_{Q_k}\left(\tilde{x}^{(k)}(0)\right),
\end{aligned}$$

$k = 0, \dots, s-1$. Thus, from Eqs. (6) and (9), we have

$$
\begin{aligned}
0 \geq & \sum_{k=1}^{s}\left[\left\langle x^{(k-1)}(1)-\tilde{x}^{(k-1)}(1), p_k(1)x^*(1)\right\rangle\right] \\
& -\sum_{k=2}^{s}\left\langle x^{(k-2)}(1)-\tilde{x}^{(k-2)}(1), \frac{d\left(p_k(1)x^*(1)\right)}{dt}\right\rangle \\
& +\sum_{k=3}^{s}\left[\left\langle x^{(k-3)}(1)-\tilde{x}^{(k-3)}(1), \frac{d^2\left(p_k(1)x^*(1)\right)}{dt^2}\right\rangle\right]-\cdots \\
& +\sum_{k=s-2}^{s}\left[\left\langle x^{(k-s+2)}(1)-\tilde{x}^{(k-s+2)}(1), (-1)^{s-3}\frac{d^{s-3}\left(p_k(1)x^*(1)\right)}{dt^{s-3}}\right\rangle\right] \\
& +\sum_{k=s-1}^{s}\left[\left\langle x^{(k-s+1)}(1)-\tilde{x}^{(k-s+1)}(1), (-1)^{s-2}\frac{d^{s-2}\left(p_k(1)x^*(1)\right)}{dt^{s-2}}\right\rangle\right] \\
& +\left\langle x(1)-\tilde{x}(1), (-1)^{s-1}\frac{d^{s-1}\left(p_s(1)x^*(1)\right)}{dt^{s-1}}\right\rangle.
\end{aligned}
$$

Using the derivative operator D, it is not hard to see that the relation described above can be expressed in a more compact form:

$$
\begin{aligned}
0 \geq & \left\langle \sum_{k=0}^{s-1}(-1)^{s-k-1}D^{s-k-1}\left[p_{s-k}(1)x^*(1)\right], x(1)-\tilde{x}(1)\right\rangle \\
& +\left\langle \sum_{k=0}^{s-2}(-1)^{s-k-2}D^{s-k-2}\left[p_{s-k}(1)x^*(1)\right], x'(1)-\tilde{x}'(1)\right\rangle+\cdots \\
& -\left\langle D\left[p_s(1)x^*(1)\right]-p_{s-1}(1)x^*(1), x^{(s-2)}(1)-\tilde{x}^{(s-2)}(1)\right\rangle \\
& +\left\langle p_s(1)x^*(1), x^{(s-1)}(1)-\tilde{x}^{(s-1)}(1)\right\rangle.
\end{aligned}
\tag{10}
$$

Furthermore, applying the definition of the transversality condition Eq. (iii) for all feasible arc $x(\cdot)$, we have

$$
\begin{aligned}
& \varphi\left(x(1), x'(1), \ldots, x^{(s-1)}(1)\right)-\varphi\left(\tilde{x}(1), \tilde{x}'(1), \ldots, \tilde{x}^{(s-1)}(1)\right) \\
& \geq \left\langle \sum_{k=0}^{s-1}(-1)^{s-k}D^{s-k-1}\left[p_{s-k}(1)x^*(1)\right], x(1)-\tilde{x}(1)\right\rangle \\
& +\left\langle \sum_{k=0}^{s-2}(-1)^{s-k-1}D^{s-k-2}\left[p_{s-k}(1)x^*(1)\right], x'(1)-\tilde{x}'(1)\right\rangle+\cdots \\
& +\left\langle D\left[p_s(1)x^*(1)\right]-p_{s-1}(1)x^*(1), x^{(s-2)}(1)-\tilde{x}^{(s-2)}(1)\right\rangle \\
& -\left\langle p_s(1)x^*(1), x^{(s-1)}(1)-\tilde{x}^{(s-1)}(1)\right\rangle
\end{aligned}
\tag{11}
$$

Then from the last two inequalities Eqs. (10) and (11) for all feasible arc $x(\cdot)$, we have immediately $\varphi\left(x(1), x'(1), \ldots, x^{(s-1)}(1)\right) \geq \varphi\left(\tilde{x}(1), \tilde{x}'(1), \ldots, \tilde{x}^{(s-1)}(1)\right)$, that is, $\tilde{x}(\cdot)$ is an optimal trajectory. □

Remark 3.1. It can be noted that in the particular case, if $p_2(t) = p(t), p_1(t) = p'(t)$, where $p(\cdot) : [0,1] \to (0, \infty)$, the second-order linear differential operator $Lx = p_2(t)x'' + p_1(t)x'$ is a well-known self-adjoint Sturm-Liouville operator $Lx \equiv (px')'$.

Corollary 3.1. Let $F(\cdot, t)$ be a closed set-valued mapping. Then under the assumptions of Theorem 3.1, the conditions Eqs. (i) and (iii) can be rewritten in terms of Hamiltonian function as follows:

$$L^* x^*(t) \in \partial_x H_F(\tilde{x}(t), x^*(t), t);\ .L\tilde{x}(t) \in \partial_{v^*} H(\tilde{x}(t), x^*(t), t), \text{a.e.} t \in [0,1]$$

Proof. Indeed, by Theorem 2.1 ([23], p. 62) and Lemma 5.1 [14], we can write.

$$F^*(v^*; (x, v), t) = \partial_x H_F(x, v^*, t), \text{and } F_A(x, v^*, t) = \partial_{v^*} H_F(x, v^*, t)$$

respectively. Then it is easy to see that the result of corollary are equivalent with the conditions Eqs. (i) and (iv) of Theorem 3.1. □

Below nonconvexity of a set-valued mapping $F(\cdot, t)$ means that its Hamilton function in general is a nonconcave function satisfying the condition Eq. (a).

Theorem 3.2. Suppose that we have the "nonconvex" problem (P_V), that is, $\varphi : (\mathbb{R}^n)^s \to \mathbb{R}^1$ and $F(\cdot, t) : \mathbb{R}^n \rightrightarrows \mathbb{R}^n$ in general are nonconvex function and set-valued mapping, respectively. Moreover, suppose that $K_{Q_j}\left(\tilde{x}^{(j)}(0)\right), \tilde{x}^{(j)}(0) \in Q_j$ is the cones of tangent directions to $Q_j, j = 0, \dots, s-1$.

Then for optimality of the trajectory $\tilde{x}(\cdot)$, it is sufficient that there exists an absolutely continuous function $x^*(\cdot)$, satisfying the following conditions:

a. $L^* x^*(t) \in F^*(x^*(t); (\tilde{x}(t), L\tilde{x}(t)), t)$, a.e. $t \in [0,1]$,

b. $\sum_{k=0}^{s-1-j} (-1)^{s-k-j} D^{s-k-1-j} [p_{s-k}(0) x^*(0)] \in K^*_{Q_j}\left(\tilde{x}^{(j)}(0)\right), j = 0, \dots, s-1,$

c. $\varphi(v_0, v_1, \dots, v_{s-1}) - \varphi\left(\tilde{x}(1), \tilde{x}'(1), \dots, \tilde{x}^{(s-1)}(1)\right)$

$$\geq \sum_{j=0}^{s-1} \left\langle \sum_{k=0}^{s-1-j} (-1)^{s-k-j} D^{s-k-1-j} [p_{s-k}(1) x^*(1)], v_j - \tilde{x}^{(j)}(1) \right\rangle,$$

$\forall v_j \in \mathbb{R}^n, j = 0, \dots, s-1,$

d. $\langle L\tilde{x}(t), x^*(t) \rangle = H_F(\tilde{x}(t), x^*(t), t)$, a.e. $t \in [0,1]$.

Proof. In the proof of Theorem 3.1, we have used the following inequality:

$$H_F(x(t), x^*(t), t) - H_F(\tilde{x}(t), x^*(t), t) \leq \left\langle \sum_{k=1}^{s} (-1)^k (p_k(t) x^*(t))^{(k)}, x(t) - \tilde{x}(t) \right\rangle. \tag{12}$$

Hence, from the inequality Eq. (12), immediately we have the inequality Eq. (10). Moreover, setting $v_j = \tilde{x}^{(j)}(1)\ (j = 0, \dots, s-1)$ for all feasible trajectories $x(\cdot)$, it is not hard to see that for nonconvex φ the following inequality holds:

$$\varphi\left(x(1),x'(1),\ \dots,x^{(s-1)}(1)\right)-\varphi\left(\tilde{x}(1),\tilde{x}'(1),\ \dots,\tilde{x}^{(s-1)}(1)\right)$$
$$\geq\sum_{j=0}^{s-1}\left\langle\sum_{k=0}^{s-1-j}(-1)^{s-k-j}D^{s-k-1-j}\left[p_{s-k}(1)x^{*}(1)\right],x^{(j)}(1)-\tilde{x}^{(j)}(1)\right\rangle,j=0,1,\ \dots,s-1,$$

Then for the furthest proof, we proceed by analogy with the preceding derivation of Theorem 3.1. □

4. Some applications to optimal control problems with PLDOs

In this section we give two applications of our results. The first one is the particular Mayer problem for differential inclusions involving PLDOs with constant coefficients, and the second one concerns optimization of "linear" differential inclusions with PLDOs and constant coefficients. Thus, suppose now we have the following optimization problem (for simplicity we consider a convex problem) with sth-order PLDO with constant coefficients:

$$\text{Minimize } \varphi_0(x(1)),$$
$$(P_C)\ Lx(t)\in F(x(t),t),\ \text{ a.e. } t\in[0,1],Lx=D^{s}x+p_1D^{s-1}x+\ \dots\ +p_{s-1}Dx$$
$$x(0)=\alpha_0,x'(0)=\alpha_1,x''(0)=\alpha_2,\ \dots,x^{(s-1)}(0)=\alpha_{s-1},. \tag{13}$$

where L is the sth-order polynomial operator, $p_k, k=1,\ \dots,s-1$ are some real constants, $F(\cdot,t):\mathbb{R}^n\rightrightarrows\mathbb{R}^n$ is a convex set-valued mapping, $\varphi_0:\mathbb{R}^n\to\mathbb{R}^1$ is a continuous convex function, and $\alpha_j\in\mathbb{R}^n,\ j=0,\ \dots,s-1$ are fixed n-dimensional vectors. It is known that the multiplication operation is commutative for polynomial linear differential operators with constant coefficients. On the other hand, the sth-order adjoint operator is defined as follows:

$$L^{*}x^{*}=(-1)^{s}D^{s}x^{*}+(-1)^{s-1}p_1D^{s-1}x^{*}+\cdots-p_{s-1}Dx^{*}.$$

Corollary 4.1. Let φ_0 and $F(\cdot,t)$ be a convex function and a set-valued mapping, respectively. Then, for the trajectory $\tilde{x}(\cdot)$ to be optimal in the problem (P_C), it is sufficient that there exists an absolutely continuous function $x^{*}(\cdot)$ satisfying the Euler-Lagrange type differential inclusion.

$$L^{*}x^{*}(t)\in F^{*}(x^{*}(t);(\tilde{x}(t),L\tilde{x}(t)),t);\langle L\tilde{x}(t),x^{*}(t)\rangle=H_F(\tilde{x}(t),x^{*}(t)),$$
$$\text{a.e. } t\in[0,1],L^{*}x^{*}=(-1)^{s}x^{*(s)}+(-1)^{s-1}p_1x^{*(s-1)}+\cdots-p_{s-1}x^{*\prime}$$

and transversality condition at the endpoint $t=1$.

$$(-1)^{s}x^{*(s-1)}(1)\in\partial\varphi_0(\tilde{x}(1)),x^{*(j)}(1)=0,j=0,\ \dots,m-2.$$

Proof. We conclude this proof by returning to the conditions Eqs. (i)–(iii) of Theorem 3.1. Clearly, a problem (P_C) can be reduced to the problem of form (P_V), where

$$\varphi\left(x(1),x'(1),\ \dots,x^{(s-1)}(1)\right)\equiv\varphi_0(x(1)).$$

It follows that $\partial\varphi\left(x(1), x'(1), \dots, x^{(s-1)}(1)\right) = \partial_x\varphi_0(x(1)) \times (\underbrace{0, \dots, 0}_{s-1})$. On the other hand, since $p_s(t) \equiv 1, p_j(t) = p_{s-j}$, and $j = 1, \dots, s-1$ are constants, by sequentially substitution in the transversality condition Eq. (iii), we derive that

$$\sum_{k=0}^{s-1-j} (-1)^{s-k-j} D^{s-k-1-j}\left[p_{s-k}(1)x^*(1)\right] = x^{*(s-1-j)}(1) = 0, j = 1, \dots, s-1,$$

and therefore for $j = 0$.

$$\sum_{k=0}^{s-1} (-1)^{s-k} D^{s-k-1}\left[p_{s-k}(1)x^*(1)\right] = (-1)^s x^{*(s-1)}(1) \in \partial\varphi_0(\tilde{x}(1)). \square$$

Suppose now that we have the so-called linear Mayer problem with PLDOs:

$$\text{Minimize } \varphi_0(x(1)), \tag{14}$$

$$Lx(t) \in F(x(t), t), \text{ a.e. } t \in [0, 1],$$

$$x^{(j)}(0) = \alpha_j, j = 0, \dots, s-1, F(x, t) = A(t)x + B(t)U \tag{15}$$

where φ_0 differentiable convex function; $A(t)$ and $B(t)$ are $n \times n$ and $n \times r$ continuous matrices, respectively; U is a convex compact of $\mathbb{R}^r$; $\alpha_j, j = 0, \dots, s-1$ are constant vectors. It is required to find a control function $\tilde{u}(\cdot)$ such that the corresponding trajectory $\tilde{x}(\cdot)$ minimizes the Mayer functional $\varphi_0(x(1))$.

In fact, this is optimization of Cauchy problem for "linear" differential inclusions with PLDO. The controlling parameter $u(\cdot)$ is called admissible if it only takes values in the given control set U which is a nonempty, convex compact.

Theorem 4.1. The arc $\tilde{x}(t)$ corresponding to the controlling parameter $\tilde{u}(t)$ is a solution to Eqs. (14) and (15) if there exists an absolutely continuous function $x^*(\cdot)$, satisfying the Euler-Lagrange type differential equation, the transversality condition, and Weierstrass-Pontryagin maximum principle:

$$L^*x^*(t) = A^*(t)x^*(t), \text{a.e. } t \in [0, 1],$$

$$(-1)^s x^{*(s-1)}(1) = \varphi_0'(\tilde{x}(1)), x^{*(j)}(1) = 0, j = 0, \dots, s-2$$

$$\langle B(t)\tilde{u}(t), x^*(t)\rangle = \max_{u \in U} \langle B(t)u, x^*(t)\rangle.$$

Proof. Obviously, the Hamiltonian is

$$H_F(x, v^*, t) = \langle A(t)x, v^*\rangle + \max_{u \in U} \langle B(t)u, v^*\rangle.$$

Hence,

$$F^*(v^*; (x, \tilde{v}), t) = \partial_x H_F(x, v^*, t) = A^*(t)v^*, \tilde{v} \in F_A(x, v^*, t)\tilde{v} = A(t)x + B(t)\tilde{u}$$

where the argmaximum inclusion $\tilde{v} \in F_A(x, v^*, t)$ implies that $\langle B(t)\tilde{u}, v^*\rangle = \max_{u \in U}\langle B(t)u, v^*\rangle$ and $F^*(v^*; (x, \tilde{v}), t) \neq \varnothing$. Thus, applying Theorem 3.1, we obtain

$$L^*x^*(t) = A^*(t)x^*(t), \; L\tilde{x}(t) \in F_A(\tilde{x}(t), x^*(t), t),$$

$$\langle B(t)\tilde{u}(t), x^*(t)\rangle = \max_{u\in U} \langle B(t)u, x^*(t)\rangle.$$

Consequently, the transversality condition Eq. (ii) of Theorem 3.1 is unnecessary and by Corollary 4.1 $(-1)^s D^{s-1}x^*(1) = \varphi_0'(\tilde{x}(1))$, $D^j x^*(1) = 0, j = 0, \ldots, s-2$. □

Remark 4.1. Suppose that in the definition of sth-order PLDO (see Eq. (13)) $s = 1, p_i = 0, i = 1, .., s-1$ and U is a convex closed polyhedron. Then we have linear equations with variable coefficients $x' = A(t)x + B(t)u, u \in U$ in the finite time interval $t \in [0, 1]$. Obviously, for such problems an adjoint Euler-Lagrange type differential equation and transversality condition at a point $t = 1$ consist of the following: $x^{*\prime}(t) = -A^*(t)x^*(t), x^*(1) = -\varphi_0'(\tilde{x}(1))$. We remind that along with Pontryagin's maximum principle (see, e.g., [24]) under the condition for generality of position for time-optimal problem, the existence results of optimal control are proved.

Example 4.1. Let us consider the following Mayer problem with second-order PLDO $Lx = D^2x = x''$:

$$\text{Infimum } \varphi(x(1), x'(1)) \text{ is subject to } x'' = u, u \in [-1, 1], x(0) \in Q_0, x'(0) \in Q_1. \tag{16}$$

Here $\varphi(x(1), x'(1)) = x'^2(1) - x(1)$ and $Q_0 = \{0\}, Q_1 = \{1\}$.

It should be noted that substituting $F(t) = u(t), x''(t) = a(t), m = 1$ into Newton's second law $F(t) = ma(t)$, we have $x'' = u$.

Obviously, in this problem $F(x, t) \equiv F(x) = \{u : |u| \leq 1\}, s = 2$.

Then Eq. (16) has the form:

$$\text{Infimum } \varphi(x(1), x'(1)) \text{ is subject to } Lx \in F(x), t \in [0, 1], x(0) = 0, x'(0) = 1. \tag{17}$$

It can be easily seen that in the adjoint inclusion Eq. (i).
$-D(p_1(t)x^*(t)) + D^2(p_2(t)x^*(t)) \in F^*(x^*(t); (\tilde{x}(t), L\tilde{x}(t)))$
of Corollary 3.1 $p_2(t) \equiv 0$ and $p_2(t) \equiv 1$, and so we have

$$x^{*\prime\prime}(t) \in F^*(x^*(t); (\tilde{x}(t), \tilde{x}''(t))).$$

Now, it is not hard to see that

$$H_F(x, v^*) = \max_u \{uv^* : |u| \leq 1\} = |v^*| \tag{18}$$

and

$$F^*(v^*; (x, v)) = \partial_x H_F(x, v^*) \equiv 0, \ v \in F_A(x, v^*) = \{-1, +1\}. \tag{19}$$

Then taking into account $Lx^* = d^2x^*/dt^2$, as a result of Theorem 3.1 (see also Corollary 3.1) from Eq. (19), we deduce that

$$x^{*\prime\prime} = 0, \ t \in [0, 1],$$

for which the solution is a linear function of the form $x^*(t) = C_1 t + C_2$, where C_1, C_2 are arbitrary constants. Then Eq. (18) implies that $\tilde{u}(t)x^*(t) = |x^*(t)|$ or

$$\tilde{u}(t) = \begin{cases} \operatorname{sgn} x^*(t), & \text{if } x^*(t) \neq 0, \\ \forall u_0 \in [-1, 1], & \text{if } x^*(t) = 0. \end{cases} \tag{20}$$

Further, from the linearity of $x^*(\cdot)$ and from Eq. (20), we insure that each optimal control function is a piecewise constant function.

In addition, by the transversality condition Eq. (iii) of Corollary 3.1, we can write.

$(x^{*\prime}(1), -x^*(1)) \in \partial\varphi(\tilde{x}(1), \tilde{x}'(1))$. On the other hand, it is not hard to see that $\varphi(x, y) = y^2 - x$ is a convex function; in fact, the 2×2 Hessian matrix

$$\varphi''(x, y) = \begin{bmatrix} \varphi''_{xx}(x, y) & \varphi''_{xy}(x, y) \\ \varphi''_{yx}(x, y) & \varphi''_{yy}(x, y) \end{bmatrix} = \begin{bmatrix} 0 & 0 \\ 0 & 2 \end{bmatrix}$$

is a positive semidefinite, that is, all eigenvalues of $\varphi''(x, y)$ are nonnegative. Indeed, denoting this matrix by A, we see that the characteristic equation $|A - \lambda E| = \lambda^2 - 2\lambda = 0$ (E is a 2×2 unique square matrix) has two real nonnegative eigenvalues $\lambda_1 = 0, \lambda_2 = 2$. Consequently, $\varphi(x, y)$ is convex and $\partial\varphi(x, y) = (-1, 2y)$. It follows that $\partial\varphi(\tilde{x}(1), \tilde{x}'(1)) = (-1, 2\tilde{x}'(1))$. Comparing this relation with $(x^{*\prime}(1), -x^*(1)) \in \partial\varphi(\tilde{x}(1), \tilde{x}'(1))$, we immediately have $x^{*\prime}(1) = -1, x^*(1) = -2\tilde{x}'(1)$. Then from a general solution of the adjoint Euler-Lagrange type inclusion (equation) $x^*(t) = C_1 t + C_2$, we have $-2\tilde{x}'(1) = x^*(1) = C_1 + C_2$, $-1 = x^{*\prime}(1) = C_1$ (C_1, C_2 are arbitrary constants), and so $x^*(t) = 1 - t - 2\tilde{x}'(1)$, whence $x^*(t) \neq 0$, if $t \neq \tau = 1 - 2\tilde{x}'(1)$. Therefore, Eq. (20) implies that for optimal control $\tilde{u}(\cdot)$, there are four possibilities:

$$\tilde{u}(t) = 1, \;\; x^*(t) > 0, \;\; t \in [0, 1]. \tag{21}$$

$$\tilde{u}(t) = -1, \;\; x^*(t) < 0, t \in [0, 1]. \tag{22}$$

$$\tilde{u}(t) = \begin{cases} 1, & \text{if } \; 0 \le t < \tau, \\ -1, & \text{if } \; \tau < t \le 1. \end{cases} \tag{23}$$

$$\tilde{u}(t) = \begin{cases} -1, & \text{if } \; 0 \le t < \tau, \\ 1, & \text{if } \; \tau < t \le 1. \end{cases} \tag{24}$$

(observe that τ is a point of discontinuity of $\tilde{u}(\cdot)$ and the values of the control functions $\tilde{u}(\cdot)$ at a point of discontinuity τ are unessential). As a consequence, it follows that either the sign of the linear function $x^*(t)$ does not change for the whole interval $[0, 1]$ or $x^*(t) > 0, 0 \le t < \tau; x^*(t) < 0, \tau < t \le 1$ for a some τ in the interval $0 < \tau < 1$ (the case Eq. (24) is excluded). Therefore, since $\tilde{u}(t)$ is a piecewise constant function, having not more than two intervals of constancy, we have either the cases Eqs. (21) and (22) or the case Eq. (23). In general, using Eqs. (21)–(23), by solving the Cauchy problem

$$x''(t) = u(t), \;\; x(0) = 0, x'(0) = 1 \tag{25}$$

we have a unique solution of the initial value problem Eq. (25). Thus for the time interval on which $u = 1$, we have $x'(t) = t + c_1$; $x(t) = t^2/2 + c_1 t + c_2$ (c_1, c_2 are constants). From Eq. (25) we obtain.

$$x'(t) = t + 1; x(t) = t^2/2 + t. \tag{26}$$

By analogy, for $u = -1$ we have.

$$x'(t) = 1 - t;\ x(t) = -t^2/2 + t. \qquad (27)$$

Now, let $x_1(t)$ and $x_2(t)$ be parabolas of Eqs. (26) and (27), respectively. Here, in the case Eq. (21), $\tilde{u}(t) = 1, t \in [0,1]$, and so from Eq. (26), we have $\tilde{x}_1(1) = 0.5 + 1 = 1.5;\ \tilde{x}_1'(1) = 2$. Consequently, the value of problem Eq. (16) is $\varphi(\tilde{x}_1(1), \tilde{x}_1'(1)) = \tilde{x}_1'^2(1) - \tilde{x}_1(1) = 2^2 - (1.5) = 2.5$, if $\tilde{u}(t) = 1, t \in [0,1]$. By a similar way, for a control function $\tilde{u}(t) = -1, t \in [0,1]$ from Eq. (27), we obtain that $\tilde{x}_2(1) = -1^2/2 + 1 = 0.5, \tilde{x}_2'(1) = 0$ and $\varphi(\tilde{x}_2(1), \tilde{x}_2'(1)) = 0^2 - (0.5) = -0.5$.

On the other hand, in the case Eq. (23), the control function $\tilde{u}(t)$ first is equal to $+1$ and then equal to -1, and the trajectory $\tilde{x}(t)$ consists of two pieces of parabolas $\tilde{x}_1(t)$ and $\tilde{x}_2(t)$ ($\tilde{x}(t)$ is continuous and piecewise smooth on the interval $0 \le t \le 1$). Then the solution of the equation Eq. (25) on the interval $0 \le t \le \tau$ is given by Eq. (26); at a point τ are satisfied $x_1(\tau) = \tau^2/2 + \tau$, $x_1'(\tau) = 1 + \tau$. Consider now the initial value problem:

$$x_2''(t) = -1,\ x_2(\tau) = \tau^2/2 + \tau, x_2'(\tau) = 1 + \tau, t \in [\tau, 1]. \qquad (28)$$

It is clear that $(\tau^2/2) + \tau = x_2(\tau) = -(\tau^2/2) + c_1\tau + c_2$ and $1 + \tau = x_2'(\tau) = -\tau + c_1$ from which we obtain that the solution of the initial value problem Eq. (28) is $\tilde{x}_2(t) = -(t^2/2) + (1 + 2\tau)t - \tau^2$. Substituting the value $\tau = 1 - 2\tilde{x}'(1)$ into equation $\tilde{x}_2'(t) = 1 - t + 2\tau$, we have $5\tilde{x}_2'(1) = 2$, $\tilde{x}_2'(1) = \tilde{x}'(1) = 0.4$ (it follows that $\tau = 0.2$). Moreover, $\tilde{x}_2(1) = 2\tau - \tau^2 + (0.5)$ and $\tilde{x}_2(1) = \tilde{x}(1) = (3/2) - 4\tilde{x}'^2(1) = 0.86$. Thus, the value of our Mayer problem is $\varphi(\tilde{x}(1), \tilde{x}'(1)) = \tilde{x}'^2(1) - \tilde{x}(1) = (2/5)^2 - (43/50) = -0.7$, where $\tilde{u}(t)$ is defined as in Eq. (23). Comparing the values 2.5, $-$ 0.5, $-$ 0.7, we believe that the value of Mayer problem is -0.7.

5. Sufficient conditions of optimality for second-order evolution differential inclusions with endpoint constraints

Note that in this section the optimality conditions are given for second-order convex differential inclusions (P_M) with convex endpoint constraints. These conditions are more precise than any previously published ones since they involve useful forms of the Weierstrass-Pontryagin condition and second-order Euler-Lagrange type adjoint inclusions. In the reviewed results, this effort culminates in Theorem 5.1:

$$\text{Minimize}\, g(x(1), x'(1)),$$

$$(P_M)\ x''(t) \in F(x(t), x'(t), t), \text{a.e. } t \in [0,1],$$

$$x(0) = x_0, x'(0) = x_1; x(1) \in M_0,\ x'(1) \in M_1,$$

where g is a convex continuous function, $F(\cdot, t) : \mathbb{R}^{2n} \rightrightarrows \mathbb{R}^n$ is convex set-valued mapping, and $M_0, M_1 \subseteq \mathbb{R}^n$ are convex sets.

The following adjoint inclusion is the second-order Euler-Lagrange type inclusion for the problem (P_M):

a$_1$. $(x^{*\prime\prime}(t) + v^{*\prime}(t), v^*(t)) \in F^*(x^*(t); (\tilde{x}(t), \tilde{x}'(t), \tilde{x}''(t)), t)$, a.e. $t \in [0,1]$,

where

b$_1$. $\tilde{x}''(t) \in F_A(\tilde{x}(t), \tilde{x}'(t); x^*(t), t)$, a.e. $t \in [0,1]$.

In what follows we assume that $x^*(t), t \in [0,1]$ is absolutely continuous function together with the first-order derivatives for which $x^{*\prime\prime}(\cdot) \in L_1^n([0,1])$. Besides the auxiliary function $v^*(t), t \in [0,1]$ is absolutely continuous and $v^{*\prime}(\cdot) \in L_1^n([0,1])$.

The transversality conditions at the endpoint $t = 1$ consist of the following:

c$_1$. $(v^*(1) + x^{*\prime}(1), -x^*(1)) \in \partial_{(x,u)} g(\tilde{x}(1), \tilde{x}'(1)) - K_{M_0}^*(\tilde{x}(1)) \times K_{M_1}^*(\tilde{x}'(1))$.

Now we are ready to formulate the following theorem of optimality.

Theorem 5.1. Suppose that g is a continuous and convex function, $F(\cdot, t)$ is a convex set-valued mapping, and M_0, M_1 are convex sets. Then for optimality of the feasible trajectory $\tilde{x}(t)$ in the problem (P_M), it is sufficient that there exists a pair of absolutely continuous functions:

$$\{x^*(t), v^*(t)\}, t \in [0,1]$$

satisfying a.e. the second-order Euler-Lagrange type inclusions Eqs. (a_1) and (b_1) and the transversality condition Eq. (c_1) at the endpoint $t = 1$.

Proof. By the proof idea of Theorem 3.1 from Eqs. (a_1) and (b_1), we obtain the adjoint differential inclusion of second order:

$$(x^{*\prime\prime}(t) + v^{*\prime}(t), v^*(t)) \in \partial_{(x,u)} H_F(\tilde{x}(t), \tilde{x}'(t), x^*(t), t), t \in [0,1].$$

On the definition of subdifferential set of the Hamiltonian function $H_F(\cdot, t)$ for all feasible trajectory $x(t), t \in [0,1]$, we rewrite the last relation in the equivalent form:

$$H_F(x(t), x'(t), x^*(t), t) - H_F(\tilde{x}(t), \tilde{x}'(t), x^*(t), t)$$
$$\leq \langle x^{*\prime\prime}(t) + v^{*\prime}(t), x(t) - \tilde{x}(t) \rangle + \langle v^*(t), x'(t) - \tilde{x}'(t) \rangle. \quad (29)$$

Now by using definition of the Hamiltonian function, the inequality Eq. (29) can be reduced to the inequality

$$0 \geq \langle (x(t) - \tilde{x}(t))'', x^*(t) \rangle - \langle x^{*\prime\prime}(t), x(t) - \tilde{x}(t) \rangle - \frac{d}{dt} \langle v^*(t), x(t) - \tilde{x}(t) \rangle. \quad (30)$$

Integrating of the inequality Eq. (30) over the interval $[0,1]$, we derive that

$$0 \geq \int_0^1 [\langle (x(t) - \tilde{x}(t))'', x^*(t) \rangle - \langle x^{*\prime\prime}(t), x(t) - \tilde{x}(t) \rangle] dt$$
$$+ \langle v^*(0), x(0) - \tilde{x}(0) \rangle - \langle v^*(1), x(1) - \tilde{x}(1) \rangle. \quad (31)$$

For convenience we transform the expression in the square parentheses on the right-hand side of Eq. (31) as follows

$$\langle (x(t) - \tilde{x}(t))'', x^*(t) \rangle - \langle x^{*\prime\prime}(t), x(t) - \tilde{x}(t) \rangle$$

$$= \frac{d}{dt}\langle (x(t)-\tilde{x}(t))', x^*(t)\rangle - \frac{d}{dt}\langle x^{*\prime}(t), x(t)-\tilde{x}(t)\rangle.$$

Thus by elementary property of the definite integrals, we can compute the integral on the right-hand side of Eq. (31):

$$\int_0^1 \left[\langle (x(t)-\tilde{x}(t))'', x^*(t)\rangle - \langle x^{*\prime\prime}(t), x(t)-\tilde{x}(t)\rangle\right] dt$$
$$= \langle x'(1)-\tilde{x}'(1), x^*(1)\rangle - \langle x'(0)-\tilde{x}'(0), x^*(0)\rangle$$
$$-\langle x^{*\prime}(1), x(1)-\tilde{x}(1)\rangle + \langle x^{*\prime}(0), x(0)-\tilde{x}(0)\rangle. \tag{32}$$

Then substituting Eq. (32) into Eq. (31), we have

$$0 \geq \langle x'(1)-\tilde{x}'(1), x^*(1)\rangle - \langle x'(0)-\tilde{x}'(0), x^*(0)\rangle$$
$$-\langle v^*(1)+x^{*\prime}(1), x(1)-\tilde{x}(1)\rangle + \langle v^*(0)+x^{*\prime}(0), x(0)-\tilde{x}(0)\rangle. \tag{33}$$

Now, remember that $x(\cdot), \tilde{x}(\cdot)$ are feasible trajectories and $x(0)=\tilde{x}(0)=x_0$ and $x'(0)=\tilde{x'}(0)$.
$=x_1$. Then it follows from Eq. (33) that

$$0 \geq \langle x'(1)-\tilde{x}'(1), x^*(1)\rangle - \langle v^*(1)+x^{*\prime}(1), x(1)-\tilde{x}(1)\rangle. \tag{34}$$

Now, thanking to the transversality conditions Eq. (c_1) at the endpoint $t=1$, we can rewrite

$$g(x(1),x'(1)) - g(\tilde{x}(1),\tilde{x}'(1)) \geq \langle v^*(1)+x^{*\prime}(1)+x^*(1), x(1)-\tilde{x}(1)\rangle$$
$$+\langle x^{*\prime}(1)-x^*(1), x'(1)-\tilde{x}'(1)\rangle,\ x^*(1) \in K^*_{M_0}(\tilde{x}(1)),\ x^{*\prime}(1) \in K^*_{M_1}(\tilde{x}'(1))$$

or, in other words

$$g(x(1),x'(1)) - g(\tilde{x}(1),\tilde{x}'(1))$$
$$\geq \langle v^*(1)+x^{*\prime}(1), x(1)-\tilde{x}(1)\rangle - \langle x^*(1), x'(1)-\tilde{x}'(1)\rangle \tag{35}$$

Thus, summing the inequalities Eqs. (34) and (35) for all feasible trajectories $x(\cdot)$, satisfying the initial conditions $x(0)=x_0$ and $x'(0)=x_1$ and endpoint constraints $x(1) \in M_0, x'(1) \in M_1$, we have the needed inequality:
$g(x(1),x'(1)) - g(\tilde{x}(1),\tilde{x}'(1)) \geq 0$ or $g(x(1),x'(1)) \geq g(\tilde{x}(1),\tilde{x}'(1))$. □

6. Conclusion

According to proposed method, the problem with the differential inclusions described by polynomial linear differential operators is investigated. Obviously, this problem is an important generalization of problems with first-order differential inclusions. Thus, sufficient conditions of optimality for such problems are deduced. Here the existence of nonfunctional initial point or endpoint constraints generates different kinds of transversality conditions. Besides, there can be no doubt that investigations of optimality conditions of problems with second- and fourth-order Sturm-Liouville type differential inclusions can play an important role in the development of modern optimization and there is every reason to believe that this role will be even more significant in the future. Thus, the suggested problem with linear differential operators and variable coefficients can be used in various forms in applied problems.

Author details

Elimhan N. Mahmudov[1,2]

1 Department of Mathematics, Istanbul Technical University, Istanbul, Turkey

2 Azerbaijan National Academy of Sciences, Institute of Control Systems, Baku, Azerbaijan

*Address all correspondence to: elimhan22@yahoo.com

References

[1] Auslender A, Mechler J. Second order viability problems for differential inclusions. Journal of Mathematical Analysis and Applications. 1994;**181**: 205-218

[2] Azzam-Laouir D, Castaing C, Thibault L. Three boundary value problems for second order differential inclusion in Banach spaces. Control and Cybernetics. 2002;**31**:659-693

[3] Cernea A. On the existence of viable solutions for a class of second order differential inclusions. Discussiones Mathematicae. Differential Inclusions, Control and Optimization. 2002;**22**: 67-78

[4] Barbu V, Precupanu T. Convex Control Problems in Banach Spaces. 4th ed. Netherlands: Springer; 2012. 368 p

[5] Blagodatskikh VI, Filippov AF. Differential inclusions and optimal control. Trudy Matematicheskogo Instituta Imeni VA Steklova. 1985;**169**: 194-252

[6] Buttazzo G, Drakhlin ME, Freddi L, Stepanov E. Homogenization of optimal control problems for functional differential equations. Journal of Optimization Theory and Applications. 1997;**93**:103-119

[7] Cannarsa P, Sinestrari C. Semiconcave Functions, Hamilton-Jacobi Equations and Optimal Control. Boston: Birkhäuser; 2004. 304p

[8] Clarke FH. Functional analysis, calculus of variations and optimal control. Graduate Texts in Mathematics. London: Springer; 2013. 264p

[9] Kolmogorov AN, Fomin SV. Elements of the Theory of Functions and Functional Analysis. (Dover Books on Mathematics) Paperback. New York: Dover Publications; 1999. 128p

[10] Lasiecka I, Triggiani R. Control Theory for Partial Differential Equations: Vol. 1, Abstract Parabolic Systems: Continuous and Approximation Theories. Cambridge: Cambridge University Press; 2000. 672p

[11] Papageorgiou NS, Rădulescu VD. Periodic solutions for time-dependent subdifferential evolution inclusions. Evolution Equations & Control Theory. 2017;**6**:277-297

[12] Long DQ, Dang Quang A, Luan VT. Iterative method for solving a fourth order differential equation with nonlinear boundary condition. Applied Mathematical Sciences. 2010;**4**: 3467-3481

[13] Zhou Y, Vijayakumar V, Murugesu R. Controllability for fractional evolution inclusions without compactness. Evolution Equations and Control Theory. 2015;**4**:507-524

[14] Mahmudov EN. Optimization of Mayer problem with Sturm–Liouville-type differential inclusions. Journal of Optimization Theory and Applications. 2018;**177**:345-375

[15] Mahmudov EN. Approximation and optimization of higher order discrete and differential inclusions. Nonlinear Differential Equations and Applications NoDEA. 2014;**21**:1-26

[16] Mahmudov EN. Convex optimization of second order discrete and differential inclusions with inequality constraints. Journal of Convex Analysis. 2018;**25**:1-26

[17] Mahmudov EN. Mathematical programming and polyhedral optimization of second order discrete

and differential inclusions. Pacific Journal of Optimization. 2015;**11**:495-525

[18] Mahmudov EN. Free time optimization of higher order differential inclusions with endpoint constraints. Applicable Analysis. 2018;**97**:2071-2084

[19] Mahmudov EN. Optimization of boundary value problems for certain higher-order differential inclusions. Journal of Dynamical and Control Systems. 2019;**25**:17-27

[20] Mahmudov EN. Optimization of second order differential inclusions with boundary value conditions. Journal of Nonlinear and Convex Analysis. 2017; **18**:1653-1664

[21] Mahmudov EN. Optimal control of second order delay-discrete and delay differential inclusions with state constraints. Evolution Equations & Control Theory. 2018;**7**:501-529. DOI: 10.3934/eect.2018024

[22] Bors D, Majewski M. On Mayer problem for systems governed by second-order ODE. Optimization. 2014; **63**:239-254

[23] Mahmudov EN. Approximation and Optimization of Discrete and Differential Inclusions. Waltham, USA: Elsevier; 2011. 396p

[24] Pontryagin LS, Boltyanskii VG, Gamkrelidze RV, Mishchenko EF. The Mathematical Theory of Optimal Processes. New York/London/Sydney: John Wiley & Sons, Inc.; 1965. 360p

[25] Mahmudov EN. Optimal control of Cauchy problem for first order discrete and partial differential inclusions. Journal of Dynamical and Control Systems. 2009;**15**:587-610

[26] Mordukhovich BS. Variational Analysis and Generalized Differentiation, I: Basic Theory. II: Applications, Grundlehren Series (Fundamental Principles of Mathematical Sciences). Vol. 330 and 331. New York: Springer; 2006. 579p

[27] Mahmudov EN. Necessary and sufficient conditions for discrete and differential inclusions of elliptic type. Journal of Mathematical Analysis and Applications. 2006;**323**:768-789

Chapter 9

Spectral Analysis and Numerical Investigation of a Flexible Structure with Nonconservative Boundary Data

Marianna A. Shubov and Laszlo P. Kindrat

Abstract

Analytic and numerical results of the Euler-Bernoulli beam model with a two-parameter family of boundary conditions have been presented. The co-diagonal matrix depending on two control parameters (k_1 and k_2) relates a two-dimensional input vector (the shear and the moment at the right end) and the observation vector (the time derivatives of displacement and the slope at the right end). The following results are contained in the paper. First, high accuracy numerical approximations for the eigenvalues of the discretized differential operator (the dynamics generator of the model) have been obtained. Second, the formula for the number of the deadbeat modes has been derived for the case when one control parameter, k_1, is positive and another one, k_2, is zero. It has been shown that the number of the deadbeat modes tends to infinity, as $k_1 \to 1^+$ and $k_2 = 0$. Third, the existence of *double* deadbeat modes and the asymptotic formula for such modes have been proven. Fourth, numerical results corroborating all analytic findings have been produced by using Chebyshev polynomial approximations for the continuous problem.

Keywords: matrix differential operator, eigenvalues, Chebyshev polynomials, numerical scheme, boundary control

1. Introduction

The present paper is concerned with the spectral analysis and numerical investigation of the eigenvalues of the Euler-Bernoulli beam model. The beam is clamped at the left end and subject to linear feedback-type conditions with a non-dissipative feedback matrix [1, 2]. Depending on the boundary parameters k_1 and k_2, the model can be either conservative, dissipative, or completely non-dissipative. We focus on the non-dissipative case, i.e., when the energy of a vibrating system is not a decreasing (or nonincreasing) function of time. In our approach, the initial-boundary value problem describing the beam dynamics is reduced to the first order in time evolution equation in the state Hilbert space $\mathcal{H}$. The evolution of the system is completely determined by the dynamics generator $\mathcal{L}_{k_1,k_2}$, *which is an unbounded non-self-adjoint matrix differential operator* (see Eqs. (2), (3), and (8)).

IntechOpen

The eigenmodes and the mode shapes of the flexible structure are defined as the eigenvalues (up to a multiple i) and the generalized eigenvectors of $\mathcal{L}_{k_1,k_2}$.

Based on the results of [1, 2], the dynamics generator has a purely discrete spectrum, whose location on the complex plane is determined by the controls k_1 and k_2. Having in mind the practical applications of the asymptotic formulas [3–5], we discuss the case of $k_1 \geq 0$ and $k_2 \geq 0$, such that $|k_1| + |k_2| > 0$ (see Proposition 2). As shown in [2], even though the operator $\mathcal{L}_{k_1,k_2}$ is non-dissipative, for the case $k_1 > 0$ and $k_2 = 0$ (or $k_1 = 0$ and $k_2 > 0$), the entire set of eigenvalues is located in the closed upper half of the complex plane $\mathbb{C}$, which means that all eigenmodes are stable or neutrally stable. (We recall that to obtain an elastic mode from an eigenvalues of $\mathcal{L}_{k_1,k_2}$, one should multiply the eigenvalue by a factor i).

In the paper we address the question of accuracy of the asymptotic formulas for the eigenvalues. *Namely, under what conditions the leading asymptotic terms in formulas (20) and (21) can be used for practical estimation of the actual frequencies of the flexible beam?* Numerical simulations show that the accuracy of the asymptotic formulas is really high; the leading asymptotic terms can be used by practitioners almost immediately, i.e., almost from the first vibrational mode. The second question is concerned with the role of the *deadbeat modes*. A deadbeat mode is a purely negative elastic mode that generates a solution of the evolution equation exponentially decaying in time. The deadbeat modes are important in engineering applications. As we prove in the paper, when the boundary parameter k_1 is close to 1 (while $k_2 = 0$), the number of the deadbeat modes is so large that the corresponding mode shapes become important for the description of the beam dynamics. More precisely, the number of deadbeat modes tends to infinity as $k_1 \to 1^+$.

We have also shown that there exists a sequence of values of the parameter k_1, i.e., $\left\{k_1^{(n)}\right\}_{n=1}^{\infty}$, such that for each $k_1 = k_1^{(n)}$ there exist a finite number of deadbeat modes and each corresponds to a *double eigenvalue* of the dynamics generator $\mathcal{L}_{k_1,k_2}$. For each value $k_1^{(n)}$, the operator $\mathcal{L}_{k_1,k_2}$ has a two-dimensional root subspace spanned by an eigenvector and an associate vector. This result means that for a double deadbeat mode (corresponding to $k_1^{(n)}$), there exists a mode shape and an associate mode shape. This fact indicates that for some values of k_1 and k_2, there exists a significant number of associate vectors of $\mathcal{L}_{k_1,k_2}$. Therefore, if one can prove that the set of the generalized eigenvectors (eigenvectors and associate vectors together) forms an unconditional basis for the state space, then construction of the bi-orthogonal basis [6] would be a more complicated problem than for the case when no associate vectors exist.

Finally, we mention that the feedback control of beams is a well-studied area [6], with multiple applications to the control of robotic manipulators, long and slender aircraft wings, propeller blades, large space structure [7, 8], and the dynamics of carbon nanotubes [9]. The analysis of a classical beam model with nonstandard feedback control law that originated in engineering literature [4, 10–12] may be of interest for both analysts and practitioners.

This paper is organized as follows. In Section 1 we formulate the initial-boundary value problem for the Euler-Bernoulli beam model. In Section 2, we reformulate the problem as an evolution equation in the Hilbert space of Cauchy data (the energy space). The dynamics generator $\mathcal{L}_{k_1,k_2}$, which is *a non-self-adjoint matrix differential operator depending on two parameters, k_1 and k_2, is the main object of interest*. The eigenvalues and the generalized eigenvectors of $\mathcal{L}_{k_1,k_2}$ correspond to the modes and the mode shapes of the beam. We also give numerical approximations and graphical representations of the eigenvalues of a discrete approximation of the main operator (see **Tables 1** and **2** and **Figures 1** and **2**). In Section 3,

$k_1 = 0, k_2 = 0.5, EI = 1, \rho = 0.1, N = 64, \varepsilon_f = 10^{-20}$					
No.	**Numerical**	**Analytic**	**No.**	**Numerical**	**Analytic**
1.	$-7988.1 + 237.00i$	$-7988.1 + 237.00i$	18.	$28.860 + 13.515i$	$29.453 + 14.812i$
2.	$-7020.6 + 222.19i$	$-7020.6 + 222.19i$	19.	$123.16 + 29.723i$	$123.08 + 29.625i$
3.	$-6115.5 + 207.37i$	$-6115.5 + 207.37i$	20.	$279.13 + 44.431i$	$279.14 + 44.437i$
4.	$-5272.8 + 192.56i$	$-5272.8 + 192.56i$	21.	$497.61 + 59.250i$	$497.61 + 59.250i$
5.	$-4492.5 + 177.75i$	$-4492.5 + 177.75i$	22.	$778.50 + 74.062i$	$778.50 + 74.062i$
6.	$-3774.7 + 162.94i$	$-3774.7 + 162.94i$	23.	$1121.8 + 88.875i$	$1121.8 + 88.875i$
7.	$-3119.3 + 148.12i$	$-3119.3 + 148.12i$	24.	$1527.6 + 103.69i$	$1527.6 + 103.69i$
8.	$-2526.3 + 133.31i$	$-2526.3 + 133.31i$	25.	$1995.7 + 118.50i$	$1995.7 + 118.50i$
9.	$-1995.7 + 118.50i$	$-1995.7 + 118.50i$	26.	$2526.3 + 133.31i$	$2526.3 + 133.31i$
10.	$-1527.6 + 103.69i$	$-1527.6 + 103.69i$	27.	$3119.3 + 148.12i$	$3119.3 + 148.12i$
11.	$-1121.8 + 88.875i$	$-1121.8 + 88.875i$	28.	$3774.7 + 162.94i$	$3774.7 + 162.94i$
12.	$-778.5 + 74.062i$	$-778.5 + 74.062i$	29.	$4492.5 + 177.75i$	$4492.5 + 177.75i$
13.	$-497.61 + 59.250i$	$-497.61 + 59.250i$	30.	$5272.8 + 192.56i$	$5272.8 + 192.56i$
14.	$-279.13 + 44.431i$	$-279.14 + 44.437i$	31.	$6115.5 + 207.37i$	$6115.5 + 207.37i$
15.	$-123.16 + 29.723i$	$-123.08 + 29.625i$	32.	$7020.6 + 222.19i$	$7020.6 + 222.19i$
16.	$-28.860 + 13.515i$	$-29.453 + 14.812i$	33.	$7988.1 + 237.00i$	$7988.1 + 237.00i$
17.	$-2.2 \cdot 10^{-17} + 4.6007i$				

Table 1.
Approximations of the eigenvalues for the discrete and "continuous" operators ($\mathcal{K} > 1$).

we study the deadbeat modes and derive the estimates for the number of the deadbeat modes from below and above for different values of the boundary parameters (see **Figure 5**). Section 4 is concerned with the asymptotic approximation for the set of double deadbeat modes (see **Tables 3** and **4** and **Figures 6** and 7). In Section 5, we outline the numerical scheme used for the spectral analysis of the finite-dimensional approximation of the dynamics generator.

1.1 The initial-boundary value problem for the Euler-Bernoulli beam model of a unit length

The Lagrangian of the system is defined by [10, 11]

$$\frac{1}{2}\int_0^1 \left[\varrho(x)A(x)h_t^2(x,t) - E(x)I(x)h_{xx}^2(x,t)\right]dx, \tag{1}$$

where $h(x,t)$ is the transverse deflection, $E(x)$ is the modulus of elasticity, $I(x)$ is the area moment of inertia, $\varrho(x)$ is the linear density, and $A(x)$ is the cross-sectional area of the beam.

Assuming that the beam is clamped at the left end ($x = 0$) and free at the right end ($x = 1$), and applying Hamilton's variational principle to the action functional defined by (1), we obtain the equation of motion

$$\varrho(x)A(x)h_{tt}(x,t) + (E(x)I(x)h_{xx}(x,t))_{xx} = 0, \qquad 0 \le x \le 1, \quad t > 0, \tag{2}$$

$k_1 = 1.3, k_2 = 1.2, EI = 10, \rho = 0.1, N = 64, \epsilon_f = 10^{-20}$					
No.	**Numerical**	**Analytic**	**No.**	**Numerical**	**Analytic**
1.	$-25266 - 229.07i$	$-25266 - 229.07i$	18.	$99.467 - 19.816i$	$98.177 - 14.317i$
2.	$-22206 - 214.76i$	$-22206 - 214.76i$	19.	$394.17 - 28.149i$	$394.26 - 28.634i$
3.	$-19344 - 200.44i$	$-19344 - 200.44i$	20.	$887.75 - 42.983i$	$887.75 - 42.951i$
4.	$-16679 - 186.12i$	$-16679 - 186.12i$	21.	$1578.6 - 57.267i$	$1578.6 - 57.268i$
5.	$-14212 - 171.81i$	$-14212 - 171.81i$	22.	$2466.9 - 71.586i$	$2466.9 - 71.585i$
6.	$-11942 - 157.49i$	$-11942 - 157.49i$	23.	$3552.5 - 85.902i$	$3552.5 - 85.903i$
7.	$-9869.1 - 143.17i$	$-9869.1 - 143.17i$	24.	$4835.6 - 100.22i$	$4835.6 - 100.22i$
8.	$-7993.9 - 128.85i$	$-7993.9 - 128.85i$	25.	$6316.0 - 114.54i$	$6316.0 - 114.54i$
9.	$-6316.0 - 114.54i$	$-6316.0 - 114.54i$	26.	$7993.9 - 128.85i$	$7993.9 - 128.85i$
10.	$-4835.6 - 100.22i$	$-4835.6 - 100.22i$	27.	$9869.1 - 143.17i$	$9869.1 - 143.17i$
11.	$-3552.5 - 85.902i$	$-3552.5 - 85.903i$	28.	$11942 - 157.49i$	$11942 - 157.49i$
12.	$-2466.9 - 71.586i$	$-2466.9 - 71.585i$	29.	$14212 - 171.81i$	$14212 - 171.81i$
13.	$-1578.6 - 57.267i$	$-1578.6 - 57.268i$	30.	$16679 - 186.12i$	$16679 - 186.12i$
14.	$-887.75 - 42.983i$	$-887.75 - 42.951i$	31.	$19344 - 200.44i$	$19344 - 200.44i$
15.	$-394.17 - 28.149i$	$-394.26 - 28.634i$	32.	$22206 - 214.76i$	$22206 - 214.76i$
16.	$-99.467 - 19.816i$	$-98.177 - 14.317i$	33.	$25266 - 229.07i$	$25266 - 229.07i$
17.	$1.5 \cdot 10^{-17} + 7.7256i$				

Table 2.
Approximations of the eigenvalues for the discrete and "continuous" operators ($\mathcal{K} < -1$).

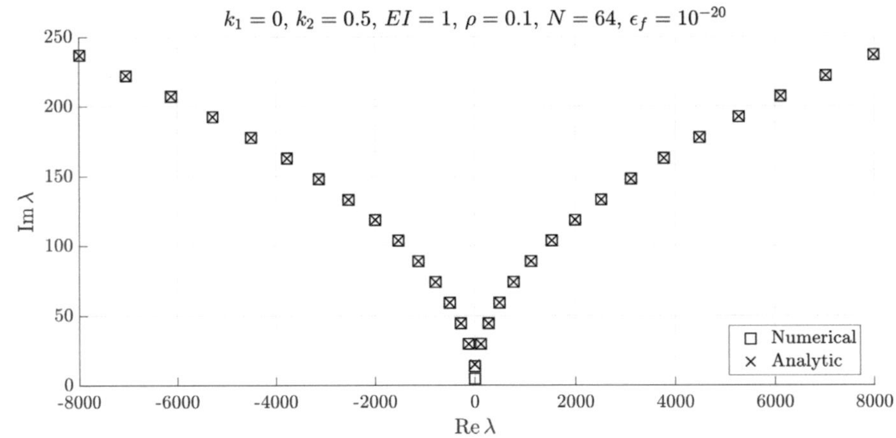

Figure 1.
Graphical representation of the eigenvalues of the discrete and "continuous" operators ($\mathcal{K} > 1$).

and the boundary conditions

$$h(0,t) = h_x(0,t) = 0 \qquad \text{and} \qquad M(1,t) = Q(1,t) = 0, \tag{3}$$

where $M(x,t)$ and $Q(x,t)$ are the moment and the shear, respectively [10]:

$$M(x,t) = E(x)I(x)h_{xx}(x,t) \qquad \text{and} \qquad Q(x,t) = M_x(x,t). \tag{4}$$

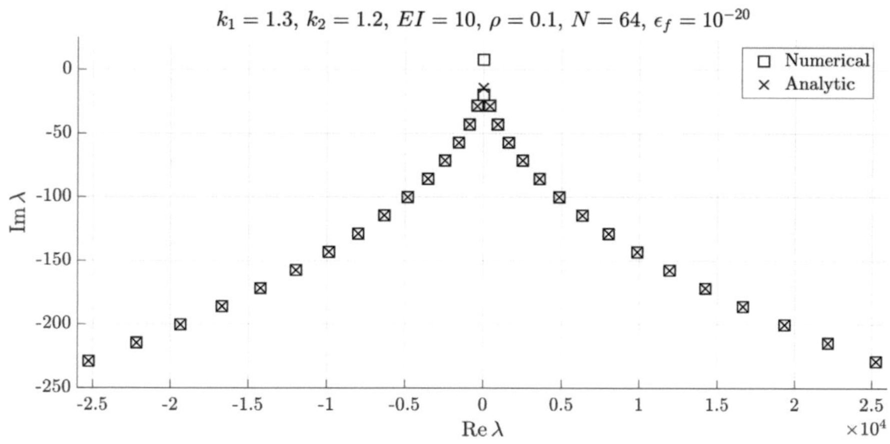

Figure 2.
Graphical representation of the eigenvalues of the discrete and "continuous" operators ($\mathcal{K} < -1$).

$k_2 = 0, EI = 1, \rho = 1, N = 64, \varepsilon_f = 10^{-30}$			
No.	$k_1 = 1 + 10^{-4}$	$k_1 = 1 + 10^{-7}$	$k_1 = 1 + 10^{-10}$
1.	$-222.22 + 155.56i$	$-176.06 + 264.07i$	$-106.90 + 372.41i$
2.	$-133.38 + 124.45i$	$-87.723 + 211.27i$	$-2.4378 \cdot 10^{-16} + 254.44i$
3.	$-64.540 + 93.396i$	$-3.9609 \cdot 10^{-18} + 162.37i$	$-3.0717 \cdot 10^{-17} + 123.66i$
4.	$-9.8081 + 58.559i$	$-5.7977 \cdot 10^{-19} + 116.23i$	$-1.3012 \cdot 10^{-17} + 4.9349i$
5.	$-7.7725 \cdot 10^{-21} + 5.0488i$	$-6.3661 \cdot 10^{-20} + 44.182i$	$-8.9232 \cdot 10^{-18} + 44.421i$
6.	$4.9365 \cdot 10^{-21} + 38.994i$	$-6.0066 \cdot 10^{-20} + 4.9383i$	$8.9143 \cdot 10^{-18} + 44.406i$
7.	$7.8007 \cdot 10^{-21} + 4.8257i$	$6.0114 \cdot 10^{-20} + 4.9313i$	$1.3012 \cdot 10^{-17} + 4.9347i$
8.	$9.8081 + 58.559i$	$6.6801 \cdot 10^{-20} + 44.651i$	$2.9907 \cdot 10^{-17} + 123.09i$
9.	$64.540 + 93.396i$	$2.9381 \cdot 10^{-18} + 139.20i$	$1.1286 \cdot 10^{-16} + 234.06i$
10.	$133.38 + 124.45i$	$87.723 + 211.27i$	$3.2280 \cdot 10^{-16} + 315.13i$
11.	$222.22 + 155.56i$	$176.06 + 264.07i$	$106.90 + 372.41i$

Table 3.
Eigenvalues closest to the imaginary axis as $k_1 \to 1^+$.

Now we replace the free right-end conditions from Eq. (3) with the following boundary feedback control law [2, 4]. Define the input and the output as

$$U(t) = [-Q(1,t),\ M(1,t)]^T \qquad \text{and} \qquad Y(t) = [h_t(1,t),\ h_{xt}(1,t)]^T, \tag{5}$$

where T stands for transposition. The feedback control law is given by

$$U(t) = KY(t), \tag{6}$$

where K is the 2×2 feedback matrix. We select

$$K = \text{codiag}(-k_2, -k_1), \qquad k_1, k_2 \geq 0, \tag{7}$$

$k_2 = 0, EI = 1, \rho = 1, N = 64, \varepsilon_f = 10^{-30}$			
#	$k_1 = 1 - 10^{-4}$	$k_1 = 1 - 10^{-7}$	$k_1 = 1 - 10^{-10}$
1.	$-175.34 + 140.01i$	$-129.20 + 237.56i$	$-60.143 + 338.80i$
2.	$-96.394 + 108.84i$	$-50.206 + 186.90i$	$-8.6602 + 240.01i$
3.	$-36.896 + 78.778i$	$-8.0431 + 121.15i$	$-0.28608 + 123.37i$
4.	$-6.0769 + 42.171i$	$-0.23455 + 44.410i$	$-0.0074192 + 44.413i$
5.	$-0.11141 + 4.9324i$	$-0.0035253 + 4.9348i$	$-0.00011148 + 4.9348i$
6.	$0.11141 + 4.9324i$	$0.0035253 + 4.9348i$	$0.00011148 + 4.9348i$
7.	$6.0769 + 42.171i$	$0.23455 + 44.410i$	$0.0074192 + 44.413i$
8.	$36.896 + 78.778i$	$8.0431 + 121.15i$	$0.28608 + 123.37i$
9.	$96.394 + 108.84i$	$50.206 + 186.90i$	$8.6602 + 240.01i$
10.	$175.34 + 140.01i$	$129.20 + 237.56i$	$60.143 + 338.80i$

Table 4.
Eigenvalues closest to the imaginary axis as $k_1 \to 1^-$.

with k_1, k_2 being the control parameters. The feedback (6) can be written as

$$E(1)I(1)h_{xx}(1,t) = -k_1 h_t(1,t) \quad \text{and} \quad (E(x)I(x)h_{xx}(x,t))_x|_{x=1} = k_2 h_{xt}(1,t). \tag{8}$$

Finally, we arrive to the following initial-boundary value problem: the equation of motion (2), the boundary conditions (3), and the standard initial conditions $h(x,0) = h_0(x),\ \ h_t(x,0) = h_1(x)$.

Notice that the choice of a feedback matrix K defines whether the system is dissipative or not. Indeed, let $\mathcal{E}(t)$ be the energy of the system, defined by representation (1). Evaluating $\mathcal{E}_t(t)$ on the solutions of Eq. (2) satisfying the left-end conditions from Eqs. (3), we obtain

$$\begin{aligned}\mathcal{E}_t(t) &= \int_0^1 [\varrho(x)A(x)h_t(x,t)h_{tt}(x,t) + E(x)I(x)h_{xx}(x,t)h_{xxt}(x,t)]x \\ &= -(E(x)I(x)h_{xx}(xt))_x h_t(xt)|_{x=1} + E(1)I(1)h_{xx}(1,t)h_{xt}(1,t).\end{aligned} \tag{9}$$

Taking into account Eqs. (4) and (6), we represent the right-hand side of Eq. (9) as the dot product in $\mathbb{R}^2$:

$$\mathcal{E}_t(t) = -Q(1,t)h_t(1,t) + M(1,t)h_{xt}(1,t) = U(t) \cdot Y(t) = KY(t) \cdot Y(t). \tag{10}$$

With the choice of K as in Eq. (7), we have

$$\mathcal{E}_t(t) = \begin{bmatrix} 0 & -k_2 \\ -k_1 & 0 \end{bmatrix} \begin{bmatrix} 2h_t(1,t) \\ h_{xt}(1,t) \end{bmatrix} \cdot \begin{bmatrix} 2h_t(1,t) \\ h_{xt}(1,t) \end{bmatrix} = -(k_1 + k_2)h_t(1,t)h_{xt}(1,t). \tag{11}$$

Thus the system is not dissipative for all nonnegative values of k_1 and k_2.

2. Operator form of the problem

In what follows, we incorporate the cross-sectional area $A(x)$ into the density, write $\rho(x)$ instead of $\varrho(x)A(x)$, and also abbreviate $EI(x) \equiv E(x)I(x)$. Let $\mathcal{H}$ be the

Hilbert space of two-component vector functions $U(x) = [u_0(x), u_1(x)]^T$ equipped with the following norm:

$$\|U\|^2_{\mathcal{H}} = \frac{1}{2}\int_0^1 \left[EI(x)|u_0''(x)|^2 + \rho(x)\,|u_1(x)|^2\right]x. \tag{12}$$

Assuming that $EI, \rho \in C^2(0,1)$ are positive functions, we obtain that the closure of smooth functions $U(x) = [u_0(x), u_1(x)]^T$ satisfying $u_0(0) = u_0'(0) = 0$ will produce the energy space $\mathcal{H} = H_0^2(0,1) \times L^2(0,1)$. Here $H_0^2(0,1) = \{u \in H^2(0,1) : u(0) = u'(0) = 0\}$, and the equality of function spaces is understood in the sense of a Hilbert-space isomorphism.

Problem (2) with conditions (3) can be represented as the time evolution problem:

$$U_t(x,t) = i(\mathcal{L}_{k_1,k_2}U)(x,t) \qquad \text{and} \qquad U(x,0) = [u_0(x), u_1(x)]^T, \tag{13}$$

where $0 \geq x \geq 1,\ \ t \geq 0$. The dynamics generator $\mathcal{L}_{k_1,k_2}$ is given by the following matrix differential expression:

$$\mathcal{L}_{k_1,k_2} = -i\begin{bmatrix} 0 & 1 \\ -\dfrac{1}{\rho(x)}\dfrac{\partial^2}{\partial x^2}\left(EI(x)\dfrac{\partial^2}{\partial x^2}\right) & 0 \end{bmatrix}, \tag{14}$$

defined on the domain

$$\mathcal{D}(\mathcal{L}_{k_1,k_2}) = \{U = (u_0, u_1)^T \in \mathcal{H} : u_0 \in H^4(0,1),\ \ u_1 \in H_0^2(0,1);\ \ u_1(0) = u_1'(0) = 0;$$
$$EI(1)u_0''(1) = -k_1u_1(1),\ \ (EI(x)u_0''(x))'\big|_{x=1} = k_2u_1'(1)\}. \tag{15}$$

For any $(k_1, k_2) \in \mathbb{R}^2$, the adjoint operator $\mathcal{L}^*_{k_1,k_2}$ [13] is given by

$$\mathcal{L}^*_{k_1,k_2} = \mathcal{L}_{-k_2,-k_1}, \tag{16}$$

i.e., $\mathcal{L}^*_{k_1,k_2}$ is defined by the same differential expression (14) on the domain described in Eq. (15), where k_1 and k_2 are replaced by $(-k_2)$ and $(-k_1)$, respectively. It follows from Eq. (16) that $\mathcal{L}_{0,0}$ is self-adjoint in $\mathcal{H}$ and thus $\mathcal{L}_{0,0}$ is the dynamics generator of the clamped-free beam model. For the reader's convenience, we summarize the properties of $\mathcal{L}_{k_1,k_2}$ from [1, 2] needed for the present work.

Proposition 1:

1. $\mathcal{L}_{k_1,k_2}$ is an unbounded operator with compact resolvent, whose spectrum consists of a countable set of normal eigenvalues (i.e., isolated eigenvalues, each of finite algebraic multiplicity [6, 13]).

2. For each $(k_1, k_2) \in \mathbb{R}^2,\ \ |k_1| + |k_2| > 0$, the operator $\mathcal{L}_{k_1,k_2}$ is a rank-two perturbation of the self-adjoint operator $\mathcal{L}_{0,0}$ in the sense that the operators $\mathcal{L}^{-1}_{k_1,k_2}$ and $\mathcal{L}^{-1}_{0,0}$ exist and are related by the rule

$$\mathcal{L}^{-1}_{k_1,k_2} = \mathcal{L}^{-1}_{0,0} + \mathcal{T}_{k_1,k_2}, \tag{17}$$

where $\mathcal{T}_{k_1,k_2}$ is a rank-two operator. A similar decomposition is valid for the adjoint operator, i.e.,

$$\left(\mathcal{L}_{k_1,k_2}^{-1}\right)^* = \mathcal{L}_{0,0}^{-1} + \mathcal{T}_{k_1,k_2}^*, \qquad \mathcal{T}_{k_1,k_2}^* = \mathcal{T}_{-k_2,-k_1}. \tag{18}$$

From now on, we assume that the structural parameters are constant. In the case of variable parameters, the spectral asymptotics will have the same leading terms and remainder terms depending on parameter smoothness.

Proposition 2: Assume that $k_1, k_2 > 0$ and $k_1k_2 \neq EI\rho$. Let

$$\mathcal{K} = \frac{k_1 + k_2}{A - k_1k_2/A}, \qquad A = \sqrt{EI\rho}, \qquad \text{and} \qquad |\mathcal{K}| \neq 1. \tag{19}$$

The following asymptotic approximations for the eigenvalues λ_n (as $|n| \to \infty$) of the operator $\mathcal{L}_{k_1,k_2}$ hold:

1. If $1 < |\mathcal{K}| < \infty$, then for $|n| \to \infty$ one has

$$\lambda_n = \text{sign}(\mathcal{K}n)\sqrt{\frac{EI}{\rho}}\left[(\pi n)^2 - \frac{1}{4}\ln^2\left(\frac{\mathcal{K}+1}{\mathcal{K}-1}\right) + i\pi n \ln\left(\frac{\mathcal{K}+1}{\mathcal{K}-1}\right)\right] + O\left(ne^{-\pi|n|}\right). \tag{20}$$

2. If $0 < \mathcal{K} < 1$, then for $n \to \infty$ one has

$$\lambda_n = \text{sign}(\mathcal{K}n)\sqrt{\frac{EI}{\rho}}\left[\left(\tfrac{2n+1}{2}\pi\right)^2 - \frac{1}{4}\ln^2\left(\frac{\mathcal{K}+1}{\mathcal{K}-1}\right) + i\pi\left(\frac{2n+1}{2}\right)\ln\left(\frac{\mathcal{K}+1}{\mathcal{K}-1}\right)\right] + O\left(ne^{-\pi|n|}\right). \tag{21}$$

First of all, we address the question of accuracy of the asymptotic formulas (20) and (21). By its nature, formula (20) (as well as formula (21)) means that for any small $\varepsilon > 0$, one can find a positive integer N, such that all eigenvalues λ_n with $|n| \geq N + 1$ satisfy the estimate

$$\left|\lambda_n - \text{sign}(\mathcal{K}n)\sqrt{\frac{EI}{\rho}}\left[(\pi n)^2 - \frac{1}{4}\ln^2\left(\frac{\mathcal{K}+1}{\mathcal{K}-1}\right) + i\pi n \ln\left(\frac{\mathcal{K}+1}{\mathcal{K}-1}\right)\right]\right| \leq \varepsilon \tag{22}$$

for the case when $1 < |\mathcal{K}| < \infty$ and

$$\left|\lambda_n - \text{sign}(\mathcal{K}n)\sqrt{\frac{EI}{\rho}}\left[\left(\frac{2n+1}{2}\pi\right)^2 - \frac{1}{4}\ln^2\left(\frac{\mathcal{K}+1}{\mathcal{K}-1}\right) + i\pi\left(\frac{2n+1}{2}\right)\ln\left(\frac{\mathcal{K}+1}{\mathcal{K}-1}\right)\right]\right| \leq \varepsilon \tag{23}$$

for the case when $0 < |\mathcal{K}| < 1$. The following important question holds: *From which index N can the eigenvalues be approximated by the leading asymptotic terms with acceptable accuracy?* In other words, can one claim that the asymptotic formulas (20) and (21) are valuable to practitioners, or are they just important mathematical results of the spectral analysis?

The results of numerical simulations (see **Tables 1** and 2 and **Figures 1** and 2) show that the asymptotic formulas are indeed quite accurate. That is, if one places on the complex plane the numerically produced sets of the eigenvalues, then the theoretically predicted distribution of eigenvalues can be seen almost immediately. To obtain these results, we used the numerical procedure based on Chebyshev polynomial approximations [14–16], as outlined in Section 5.

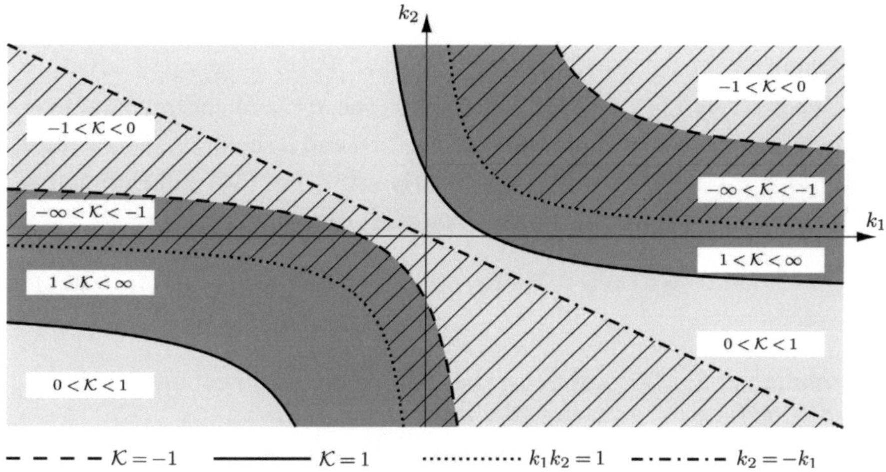

Figure 3.
Regions of $\mathcal{K}$ on the k_1, k_2-plane, $A = \sqrt{EI\rho} = 1$.

In **Figures 1** and **2**, we represent the graphical distribution of the eigenvalues corresponding to the discretized operator ("numerical" eigenvalues) and the leading asymptotic terms from Eqs. (20) and (21) ("analytic" eigenvalues). In **Tables 1** and **2**, the numerical values of the corresponding graphical points on **Figures 1** and **2** are listed. We have used the following notations: $N = 64$ is the number of grid points on $[0, 1]$, and ε_f is the filtering parameter as described in Eq. (69). It can be easily seen from the graphs and tables that the two sets of data coincide almost immediately, i.e., the leading asymptotic terms in the approximations are very close to the numerically approximated eigenvalues.

Figure 3 shows the sub-domains of the k_1, k_2-plane, which correspond to different intervals for the values of $\mathcal{K}$ defined by Eq. (19). On the sub-domain with dark gray color $\mathcal{K}$ such that $|\mathcal{K}| > 1$, i.e., to evaluate the asymptotic approximation for the eigenvalues, one needs formula (20), while on the complementary sub-domain, one needs formula (21).

3. The deadbeat modes

An eigenvalue λ_n of the dynamics generator $\mathcal{L}_{k_1,k_2}$ is called a deadbeat mode if $\lambda_n = i\beta_n, \ \beta_n > 0$. If the corresponding eigenfunction is $\Phi_n(x)$, then the evolution problem (13) has a solution given in the form $e^{i\lambda_n t}\Phi_n(x) = e^{-\beta_n t}\Phi_n(x)$, which tends to zero without any oscillation.

As shown in paper [2], for the case when one of the control parameters is zero and the other one is positive, the entire set of the eigenvalues is located in the closed upper half plane. This result is not obvious since the operator is not dissipative; in fact, it requires a fairly nontrivial proof. However, due to this fact, we assume that any deadbeat mode can be given in the form $i\beta$, with $\beta > 0$. To deal with the deadbeat modes analytically, we rewrite the spectral equation $(\mathcal{L}_{k_1,k_2}\Phi)(x) = \lambda\Phi(x)$ in the form of an equivalent problem for an operator pencil [17] as

$$\begin{aligned} EI\varphi''''(x) &= \lambda^2\rho\varphi(x), \qquad & \varphi(0) &= \varphi'(0) = 0, \\ EI\varphi''(1) &= -i\lambda k_1\varphi(1), \qquad & EI\varphi'''(1) &= i\lambda k_2\varphi'(1). \end{aligned} \tag{24}$$

If λ_n and $\varphi_n(x)$ are an eigenvalue and eigenfunction of the pencil (24), then λ_n is also an eigenvalue of $\mathcal{L}_{k_1,k_2}$ with the eigenfunction $\Phi_n(x) = \left[\frac{1}{i\lambda_n}\varphi_n(x), \varphi_n(x)\right]^T$.

To solve problem (24), we first redefine the spectral and control parameters to eliminate ρ and EI from Eq. (24). We define $\tilde{\lambda}$, $\tilde{k}_1$, and $\tilde{k}_2$ by $\lambda = \sqrt{EI/\rho}\tilde{\lambda}$ and $\tilde{k}_j = \sqrt{EI\rho}k_j, j = 1, 2$. Substituting these relations into Eq. (24) and eliminating the "tilde," we obtain the following Sturm-Liouville eigenvalue problem:

$$\varphi''''(x) = \lambda^2\varphi(x), \quad \varphi(0) = \varphi'(0) = 0, \quad \varphi''(1) = -i\lambda k_1\varphi(1), \quad \varphi'''(1) = i\lambda k_2\varphi'(1). \tag{25}$$

The solution of Eq. (25) satisfying the left-end boundary conditions $\varphi(0) = \varphi'(0) = 0$ can be written in the form

$$\varphi(\lambda, x) = \mathcal{A}(\lambda)\left[\cosh\left(\sqrt{\lambda}x\right) - \cos\left(\sqrt{\lambda}x\right)\right] + \mathcal{B}(\lambda)\left[\sinh\left(\sqrt{\lambda}x\right) - \sin\left(\sqrt{\lambda}x\right)\right]. \tag{26}$$

Substituting formula (26) into the right-end boundary conditions of Eq. (25), one gets a system for the coefficients $\mathcal{A}(\lambda)$ and $\mathcal{B}(\lambda)$:

$$\begin{aligned} &\mathcal{A}(\lambda)\left[(1+ik_1)\cosh\sqrt{\lambda} - (1-ik_1)\cos\sqrt{\lambda}\right] + \\ &\quad \mathcal{B}(\lambda)\left[(1+ik_1)\sinh\sqrt{\lambda} - (1-ik_1)\sin\sqrt{\lambda}\right] = 0, \\ &\mathcal{A}(\lambda)\left[(1-ik_2)\sinh\sqrt{\lambda} - (1+ik_2)\sin\sqrt{\lambda}\right] + \\ &\quad \mathcal{B}(\lambda)\left[(1-ik_2)\cosh\sqrt{\lambda} + (1+ik_2)\cos\sqrt{\lambda}\right] = 0. \end{aligned} \tag{27}$$

Let $\Delta(\lambda)$ be the determinant of the matrix of coefficients for $\mathcal{A}(\lambda)$ and $\mathcal{B}(\lambda)$ in Eqs. (27). System (27) has nontrivial solutions if and only if $\Delta(\lambda) = 0$, i.e.,

$$(1 + k_1k_2) + (1 - k_1k_2)\cosh\sqrt{\lambda}\cos\sqrt{\lambda} + i(k_1 + k_2)\sinh\sqrt{\lambda}\sin\sqrt{\lambda} = 0. \tag{28}$$

Theorem 1: The following results hold in the case when $k_1 > 0$ and $k_2 = 0$. Similar results hold in the case when $k_1 = 0$ and $k_2 > 0$.

1. For $0 < k_1 < 1$, the deadbeat modes do not exist.

2. For $k_1 = 1$, there exist infinitely many deadbeat modes given explicitly by

$$\lambda_n = \mu_n^2, \qquad \mu_n = x_n(1+i), \qquad x_n = \frac{\pi}{2}(2n+1), \qquad n = 0, 1, 2, \ldots. \tag{29}$$

3. For any $k_1 > 1$, there exist a finite number $\mathcal{N}(k_1)$ of deadbeat modes. Each mode has the form $\lambda = \mu^2$, $\mu = x(1+i)$, where x is a root of the function

$$H(x; k_1) \equiv 2 + (1 + k_1)\cos 2x + (1 - k_1)\cosh 2x, \quad x > 0. \tag{30}$$

Let $X(k_1) = \frac{1}{2\pi}\cosh^{-1}\left(1 + \frac{4}{k_1 - 1}\right)$, and then $\mathcal{N}(k_1)$ satisfies the estimate

$$2[X(k_1)] + 1 \le \mathcal{N}(k_1) \le 2[X(k_1)] + 3. \tag{31}$$

Hence $\mathcal{N}(k_1) \to \infty$ as $k_1 \to 1^+$. (By $[X]$ we denote the greatest integer less than or equal to X).

Proof: Let $\mu = \sqrt{\lambda} = x(1+i), x > 0$. Taking into account the relations

$$2\cosh\mu\cos\mu = \cosh(1+i)\mu + \cosh(1-i)\mu,$$
$$2i\sinh\mu\sin\mu = \cosh(1+i)\mu - \cosh(1-i)\mu,$$

we reduce Eq. (28) to the following form:

$$2 + (1+k_1)\cos 2x + (1-k_1)\cosh 2x = 0, \quad x > 0. \tag{32}$$

It can be readily seen that if $0 < k_1 < 1$, then $2 + (1+k_1)\cos 2x > 0$ and $(1-k_1)\cosh 2x > 0$, which means that Eq. (32) has no solutions. Statement (1) is shown. Statement (2) follows immediately if one considers Eq. (32) for $k_1 = 1$.

To prove Statement (3), we rewrite Eq. (32) in the form

$$\cosh 2x = \frac{2}{k_1 - 1} + \left(1 + \frac{2}{k_1 - 1}\right)\cos 2x, \quad x > 0, \quad k_1 > 1. \tag{33}$$

The left-hand side of Eq. (33) is monotonically increasing, while the right-hand side is sinusoidal, with maximum $\left(1 + \frac{4}{k_1 - 1}\right)$ and minimum (-1), and period π. So the graphs of the left- and right-hand side have intersections only on the interval $\left[0, \frac{1}{2}\cosh^{-1}\left(1 + \frac{4}{k_1 - 1}\right)\right]$. There are two intersections for each full period of the

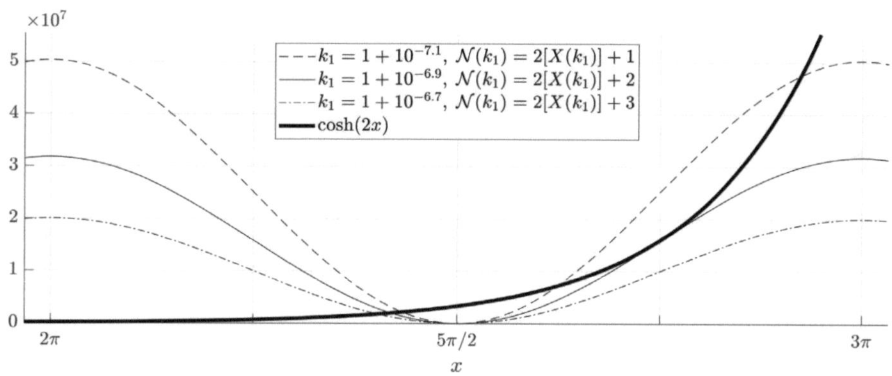

Figure 4.
Left- and right-hand side of Eq. (33) for different values of k_1.

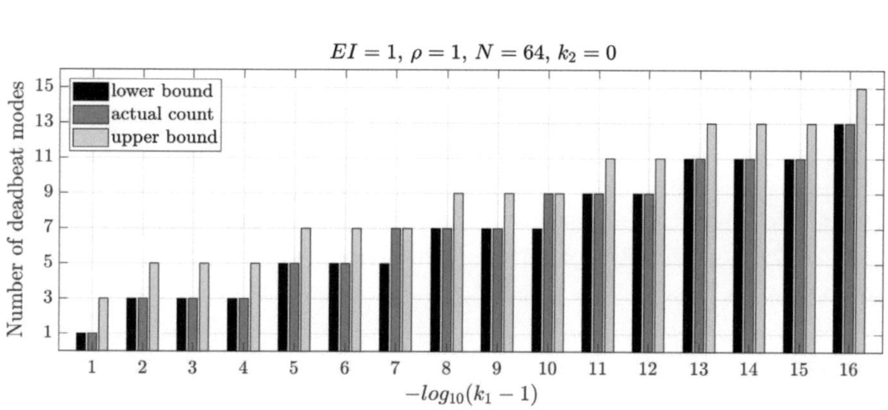

Figure 5.
Estimates and actual count of deadbeat modes based on numerical simulations.

right-hand side that fits into the above interval (**Figure 4**). As it can be seen in **Figure 4**, one should add at least one more intersection for the first half-period after the full periods. Depending on the value of k_1, the two graphs can have two intersections, one tangential intersection or no intersections on the second half-period. This leads to estimate (31). ■

A graphical illustration of the result of Theorem 1 is shown in **Figure 5**.

4. Structure of the deadbeat mode set

The main result on the existence and distribution of double roots of the function $H(x;k_1)$ is presented in the statement below.

Theorem 2: For a given $k_1 > 1$, the multiplicity of each root of $H(x;k_1)$ does not exceed 2. There exists a sequence $\left\{k_1^{(n)}; n = 0, 1, 2, \ldots\right\}$, such that the function $H(x;k_1)$ has a double root if and only if $k_1 = k_1^{(n)}$ for some n. So the original spectral problem with $k_1 = k_1^{(n)}$, $k_2 = 0$ has a double deadbeat mode $\lambda_n = \mu_n^2 = 2ix_n^2$. The following asymptotic formulas hold

$$x_n = \frac{3\pi}{4} + \pi n + P_n^{-1} + O\left(P_n^{-2}\right), \qquad \text{where} \qquad P_n = \exp\left\{\frac{3\pi}{2} + 2\pi n\right\}, \tag{34}$$

and

$$k_1^{(n)} = 1 + 4P_n^{-1} + O\left(P_n^{-2}\right). \tag{35}$$

Proof: If x is a double root of H, then $H(x;k_1) = H'(x;k_1) = 0$, i.e., separating the real and imaginary parts, we have

$$2 + (k_1 + 1)\cos 2x - (k_1 - 1)\cosh 2x = 0, \tag{36}$$

$$(k_1 + 1)\sin 2x + (k_1 - 1)\sinh 2x = 0. \tag{37}$$

Eliminating k_1 from system given by (36) and (37), we obtain that the following equation has to be satisfied:

$$G(x) \equiv (1 + \cos 2x)\sinh 2x + (1 + \cosh 2x)\sin 2x = 0, \qquad x > 0. \tag{38}$$

Rewriting Eq. (37) in the form $(k_1 + 1)\sin 2x = -(k_1 - 1)\sinh 2x$, and taking into account that $k_1 > 1$, we obtain that if x is the solution of Eq. (38), then $\sin 2x < 0$.

Now we show that when $\cos 2x < 0$ and $\sin 2x < 0$, i.e.,

$$\pi(2n + 1) < 2x < \frac{3\pi}{2} + 2\pi n, \qquad n \in \{0, 1, 2, \ldots\},$$

Eq. (38) does not have any solutions. Indeed, in the above range of x, we have $\cos 2x + \sin 2x = \sqrt{2}\sin(2x + \pi/4) < -1$ and $|\sin 2x - \cos 2x| < 1$. With such estimates we obtain that

$$G(x) = \sin 2x + \sinh 2x + \frac{1}{2}e^{2x}(\cos 2x + \sin 2x) + \frac{1}{2}e^{-2x}(\sin 2x - \cos 2x) < \sin 2x < 0, \tag{39}$$

which mean that Eq. (38) cannot be satisfied.

Now we consider the case when $\cos 2x > 0$ and $\sin 2x < 0$, i.e.,

$$\frac{3\pi}{2} + 2\pi n < 2x < 2\pi(n+1), \qquad n \in \{0, 1, 2, \ldots\}.$$

It is convenient to rewrite system given by (36) and (37) in the form $2x = 3\pi/2 + 2\pi n + s$, where $n \in \{0, 1, 2, \ldots\}$, $0 < s < \pi/2$. If $g(s) \equiv G(3\pi/4 + \pi n + s)$, then Eq. (38) generates the following equation for g:

$$g(s) \equiv (1 + \sin s)\sinh\left(\frac{3\pi}{2} + 2\pi n + s\right) - \left[1 + \cosh\left(\frac{3\pi}{2} + 2\pi n + s\right)\right]\cos s = 0. \tag{40}$$

Let us show that for each n, Eq. (40) has a unique solution. For $s = 0$ we have

$$g(0) = \sinh\left(\frac{3\pi}{2} + 2\pi n\right) - 1 - \cosh\left(\frac{3\pi}{2} + 2\pi n\right) < 0,$$

and for $s = \pi/2$ we have $g(\pi/2) = 2\sinh(2\pi(n+1)) > 0$. Evaluating g' we have

$$g'(s) = \sin s + (1 + 2\sin s)\cosh\left(\frac{3\pi}{2} + 2\pi n + s\right) > 0. \tag{41}$$

Thus $g(s)$ is a monotonically increasing function, such that $g(0) < 0 < g(\pi/2)$, which means that g has a unique root on $[0, \pi/2]$.

Finally we show that the multiplicity of a multiple root cannot exceed 2. Using a contradiction argument, assume that there exists x_0, such that in addition to Eqs. (36) and (37), one has $H''(x_0; k_1) = 0$, i.e., the multiplicity of x_0 is at least 3. The system $H'(x_0; k_1) = 0$ and $H''(x_0; k_1) = 0$ can be written as

$$\begin{aligned} (k_1 + 1)\sin 2x_0 + (k_1 - 1)\sinh 2x_0 &= 0, \\ (k_1 + 1)\cos 2x_0 + (k_1 - 1)\cosh 2x_0 &= 0. \end{aligned} \tag{42}$$

Since $k_1 > 1$, the second equation of (42) yields $\cos 2x_0 < 0$. Also, since x_0 is a multiple root, we must have $\sin 2x_0 < 0$. Then $2x_0$ is in the third quadrant, which means that $G(x_0) \neq 0$, as we have seen above. This contradicts our assumption that x_0 is a root of Eqs. (36) and (37).

To derive asymptotic distribution of the roots of Eq. (40), we check that with P_n from Eq. (34), the following approximations are valid:

$$\begin{aligned} \sin\left(2P_n^{-1}\right) &= 2P_n^{-1} + O\left(P_n^{-3}\right), \quad \cos\left(2P_n^{-1}\right) = 1 - 2P_n^{-2} + O\left(P_n^{-4}\right), \\ 2\sinh\left(\frac{3\pi}{2} + 2\pi n + 2P_n^{-1}\right) &= P_n + 2 + P_n^{-1} + O\left(P_n^{-2}\right), \\ 2\cosh\left(\frac{3\pi}{2} + 2\pi n + 2P_n^{-1}\right) &= P_n + 2 + 3P_n^{-1} + O\left(P_n^{-2}\right). \end{aligned} \tag{43}$$

Evaluating $g(s)$ from Eq. (40) for $s = 2P_n^{-1}$ and using Eq. (43), we get

$$\begin{aligned} g\left(2P_n^{-1}\right) = &\left[1 + \sin\left(2P_n^{-1}\right)\right]\sinh\left(\frac{3\pi}{2} + 2\pi n + 2P_n^{-1}\right) - \\ &\left[1 + \cosh\left(\frac{3\pi}{2} + 2\pi n + 2P_n^{-1}\right)\right]\cos\left(2P_n^{-1}\right) = 2P_n^{-1} + O\left(P_n^{-2}\right). \end{aligned} \tag{44}$$

Representation (44) implies that there exists n_0, such that for all $n \geq n_0$, we have $g(2P_n^{-1}) > 0$. Taking into account that $g(0) < 0$, we obtain that the root s_n, $n \geq n_0$, of the function $g(s)$ is located on the interval $(0, 2P_n^{-1})$. To find the location of this root more precisely [18], we use linear interpolation. Namely, substituting Eq. (43) into the expression for $g'(s)$ from Eq. (41) yields

$$g'(2P_n^{-1}) = \frac{P_n}{2} + O(1). \tag{45}$$

Replacing $g(s)$ by the linear function tangential to $g(s)$ at the point $(2P_n^{-1}, g(2P_n^{-1}))$, and finding the root of this function, we get

$$s_n = 2P_n^{-1} - \frac{g(2P_n^{-1})}{g'(2P_n^{-1})} + O(P_n^{-2}) = 2P_n^{-1} + O(P_n^{-2}). \tag{46}$$

Having this approximation for s_n, we immediately get

$$x_n = \frac{3\pi}{4} + \pi n + \frac{s_n}{2} = \frac{3\pi}{4} + \pi n + P_n^{-1} + O(P_n^{-2}). \tag{47}$$

From the equation $H\left(x_n; k_1^{(n)}\right) = 0$, we obtain the formula for $k_1^{(n)}$ as

$$k_1^{(n)} = \frac{\sinh(3\pi/2 + 2\pi n + s_n) + \cos s_n}{\sinh(3\pi/2 + 2\pi n + s_n) - \cos s_n}. \tag{48}$$

Substituting formulas (43) and (46) into formula (48), we obtain representation (35). ■

Corollary 1: Let $k_1 = k_1^{(n)}$ for some $n \in \mathbb{N}^+ \cup \{0\}$, and let x_n be the corresponding double root of the function $H\left(x; k_1^{(n)}\right)$. Then $\lambda_0 = 2ix_n^2$ is an eigenvalue of the operator $\mathcal{L}_{k_1^{(n)}, 0}$, such that the geometric multiplicity of λ_0 is 1 and its algebraic multiplicity is 2. Therefore there exists a unique eigenvector Φ and one associate vector Ψ, such that

$$\mathcal{L}_{k_1^{(n)}, 0}\Phi = \lambda_0\Phi, \qquad \mathcal{L}_{k_1^{(n)}, 0}\Psi - \lambda_0\Psi = \Phi. \tag{49}$$

Proof: It suffices to show that problem (24) does not have two linearly independent eigenvectors corresponding to λ_0 [18, 19]. Using contradiction argument we assume that for some λ_0 the boundary-value problem (25) with $k_2 = 0$ has two linearly independent solutions ψ and χ. Each function satisfies the problem

$$\varphi''''(x) = \lambda_0^2\varphi(x), \qquad \varphi(0) = \varphi'(0) = 0, \qquad \varphi'''(1) = 0, \qquad \varphi''(1) = i\lambda_0 k_1\varphi(1). \tag{50}$$

First we observe that $\psi(1)\chi(1) \neq 0$. Indeed, if $\psi(1) = 0$, then we have

$$\int_0^1 |\psi''(\sigma)|^2 d\sigma = \int_0^1 \psi''''(\sigma)\overline{\psi(\sigma)} d\sigma = \lambda_0^2 \int_0^1 |\psi(\sigma)|^2 d\sigma. \tag{51}$$

Since λ_0 is purely imaginary, Eq. (51) is not valid. We define a new function:

$$g(x) = \psi(x) - \frac{\psi(1)}{\chi(1)}\chi(x). \tag{52}$$

One can readily check that g satisfies the following boundary-value problem:

$$\varphi''''(x) = \lambda_0^2\varphi(x), \qquad \varphi(0) = \varphi'(0) = 0, \qquad \varphi(1) = \varphi''(1) = 0, \tag{53}$$

and therefore

$$\int_0^1 |g''(\sigma)|^2 d\sigma = \int_0^1 g''''(\sigma)\overline{g(\sigma)}\sigma = \lambda_0^2 \int_0^1 |g(\sigma)|^2 d\sigma. \tag{54}$$

Eq. (54) is valid if and only if $\lambda_0^2 > 0$; however, for a deadbeat mode, $\lambda_0^2 < 0$. The obtained contradiction means that for each double mode, there is one eigenfunction and one associate function. ■

4.1 Deadbeat mode behavior as $k_1 \to 1$

As k_1 approaches 1, the spectral branches are moving upward and toward the imaginary axis (**Figures 6** and 7). As a result of this motion, eigenvalues approach the imaginary axis at different rates depending on whether k_1 approaches 1 from above or below.

As follows from **Table 3**, the real parts of the eigenvalues decrease steadily as $k_1 \to 1^+$, to a point where the eigenvalue becomes a deadbeat mode. An increase in the number of deadbeat modes can be seen as $k_1 \to 1^+$, which is in agreement with Statement (3) of Theorem 1. One can see from **Table 3** that there are pairs of modes such that the distance between them tends to zero as $k_1 \to 1^+$. (Compare modes no.5 and no.7 for $|k_1 - 1| = 10^{-4}$, modes no.4 and no.7 for $|k_1 - 1| = 10^{-6}$, modes no.5 and no.8 for $|k_1 - 1| = 10^{-8}$, and modes no.4 and no.7 for $|k_1 - 1| = 10^{-10}$). Such behavior indicates convergence of the two simple deadbeat modes to a double mode, which is consistent with Theorem 2.

Analyzing **Table 4**, one can see that the eigenvalues get closer to the imaginary axis as $k_1 \to 1^-$. However the rate at which their real parts approach zero is

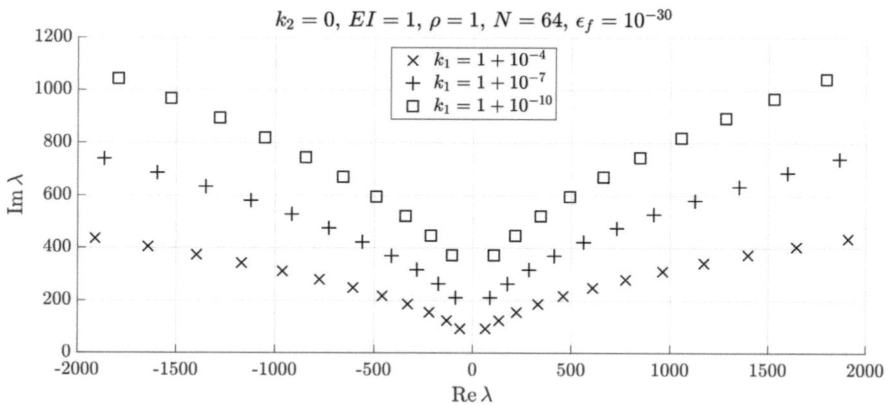

Figure 6.
Eigenvalues with $|\mathrm{Re}\lambda| > 10$ *as* $k_1 \to 1^+$.

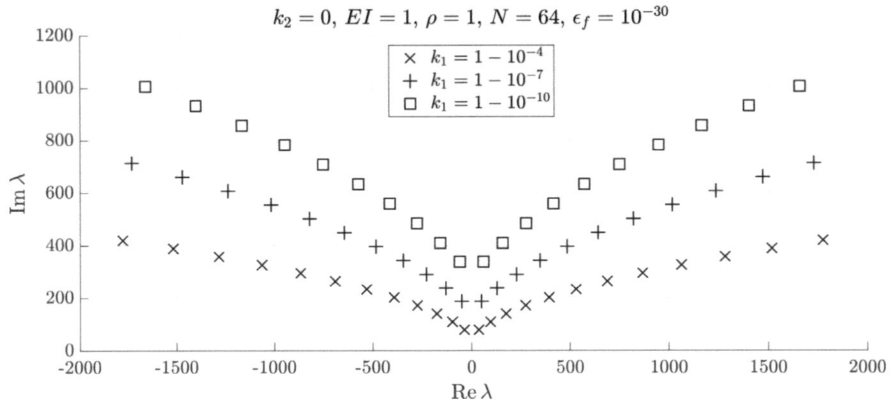

Figure 7.
Eigenvalues with $|\mathrm{Re}\lambda| > 10$ *as* $k_1 \to 1^-$.

significantly lower than in the case $k_1 \to 1^+$. Even at $k_1 = 1 - 10^{-10}$, the eigenvalue closest to the imaginary axis has a real part of about 10^{-4}, which means that it is not a deadbeat mode (see Statement (1) of Theorem 1).

The eigenvalues near the imaginary axis approach the same double deadbeat modes in both cases when $k_1 \to 1^{\pm}$ (see Statement (2) of Theorem 1). In conclusion, one can claim that the eigenvalues are indeed approaching the imaginary axis; however, the rate of this approach is different for $k_1 \to 1^-$ and $k_1 \to 1^+$. In the former case, an eigenvalue's distance from the imaginary axis decreases very slowly; in the latter case, the eigenvalues quickly "jump" on the imaginary axis and turn into deadbeat modes.

5. Outline of the numerical scheme

To carry out the numerical analysis of the differential operator $\mathcal{L}_{k_1,k_2}$, we use the Chebyshev collocation method and cardinal functions [14–16].

Recall that the Nth Chebyshev polynomial of the first kind is defined by

$$T_N(\xi) = \cos N\theta, \quad -1 \le \xi \le 1 \quad \text{where} \quad \xi = \cos\theta, \quad \theta \in [0, \pi]. \tag{55}$$

The cardinal functions, $\psi_k(\xi)$, and the Chebyshev-Gauss-Lobatto (CGL) grid points $\{\xi_k\}$ are defined as follows:

$$\psi_k(\xi) = (-1)^k \frac{(1-\xi^2)T'_{N-1}(\xi)}{c_k(N-1)^2(\xi - \xi_k)}, \qquad \xi_k = \cos\frac{(k-1)\pi}{N-1}, \quad \text{for} \quad 1 \le k \le N, \tag{56}$$

where coefficients c_k are such that $c_1 = c_N = 2$ and $c_k = 1$ for $1 < k < N$. The main property of cardinal functions is $\psi_k\left(\xi_j\right) = \delta_{kj}$ (using the Kronecker delta). The family $\{\psi_k\}_{k=1}^N$ forms a basis in the space of polynomials of degree $(N-1)$, i.e., if f is such polynomial, then f and f' can be written in the forms

$$f(\xi) = \sum_{k=1}^{N} f(\xi_k)\psi_k(\xi) \quad \text{and} \quad f'(\xi) = \sum_{k=1}^{N} f'(\xi_k)\psi_k(\xi). \tag{57}$$

If $f = [f(\xi_1),\ f(\xi_2),\ \dots,\ f(\xi_N)]^T$ and $g = [f'(\xi_1),\ f'(\xi_2),\ \dots,\ f'(\xi_N)]^T$, then $g = Df$, where D is the Chebyshev derivative matrix with the elements

$$D_{11} = -D_{NN} = \frac{1+2(N-1)^2}{6}, \qquad D_{kk} = -\frac{\xi_k}{2(1-\xi_k^2)} \quad \text{for} \quad 1<k<N,$$
$$D_{j,k} = \frac{c_j(-1)^{j+k}}{c_k\left(\xi_j-\xi_k\right)} \quad \text{for} \quad j \neq k. \tag{58}$$

5.1 Discretization of $\mathcal{L}_{k_1,k_2}$

Rescaling the independent variable x as $\xi = 2x - 1$, we rewrite the operator and its domain, representations (14) and (15), in the form

$$\mathcal{L}_{k_1,k_2} = -i\begin{bmatrix} 0 & 1 \\ -\frac{16}{\rho(\xi)}\frac{\partial^2}{\partial\xi^2}\left(EI(\xi)\frac{\partial^2}{\partial\xi^2}\right) & 0 \end{bmatrix}, \tag{59}$$

and

$$\mathcal{D}(\mathcal{L}_{k_1,k_2}) = \{(u_0,u_1)^T \in \mathcal{H}:\ u_0 \in H^4(-1,1),\ u_1 \in H_0^2(-1,1);\ u_1(-1) = u_1'(-1) = 0;$$
$$4EI(1)u_0''(1) = -k_1u_1(1),\ 4\left(EI(\xi)u_0''(\xi)\right)'\Big|_{\xi=1} = k_2u_1'(1)\}, \tag{60}$$

where $\mathcal{H} = H_0^2(-1,1) \times L^2(-1,1)$, equipped with the norm

$$\|U\|_{\mathcal{H}}^2 = \frac{1}{4}\int_{-1}^{1}\left[16EI(\xi)\left|u_0''(\xi)\right|^2 + \rho(\xi)\left|u_1(\xi)\right|^2\right]d\xi. \tag{61}$$

We approximate the action of $\mathcal{L}_{k_1,k_2}$ on the finite-dimensional subspace $\mathcal{H}_N \subset \mathcal{H}$ of polynomials of degree at most $(N-1)$. Using the CGL grid and the cardinal functions, we substitute for u_0 and u_1 their truncated expansions:

$$u_0(\xi) \approx \sum_{k=1}^{N}\Phi_k\psi_k(\xi), \quad \Phi_k = u_0(\xi_k), \qquad u_1(\xi) \approx \sum_{k=1}^{N}\Theta_k\psi_k(\xi), \quad \Theta_k = u_1(\xi_k). \tag{62}$$

Let $\mathbf{\Phi}$ and $\mathbf{\Theta}$ be N-dim vectors and $\mathbf{\Psi}$ be a $2N$-dim vector defined by

$$\mathbf{\Phi} = [\Phi_1,\ \Phi_2,\ \dots,\ \Phi_N]^T, \qquad \mathbf{\Theta} = [\Theta_1,\ \Theta_2,\ \dots,\ \Theta_N]^T, \qquad \mathbf{\Psi} = \begin{bmatrix} \mathbf{\Phi} \\ \mathbf{\Theta} \end{bmatrix}. \tag{63}$$

Let $\boldsymbol{L}$ be the finite-dimensional approximation of the differential operator $\mathcal{L}_{k_1,k_2}$. The discretized operator $\boldsymbol{L}$ induced by $\mathcal{L}_{k_1,k_2}$ can be given by

$$\boldsymbol{L} = -i\begin{bmatrix} \mathbf{0} & I_{N\times N} \\ -16\frac{EI}{\rho}D^4 & \mathbf{0} \end{bmatrix}, \tag{64}$$

where $I_{N\times N}$ is the $N \times N$ identity matrix and D is the derivative matrix (58).

5.2 Incorporating the boundary conditions

Discretization of the boundary conditions in the domain description (60) yields

$$\begin{aligned} &\Phi_N = 0, \quad [D\Phi]_N = 0, \quad \Theta_N = 0, \quad [D\Theta]_N = 0, \\ &4EI\left[D^2\Phi\right]_1 + k_1\Theta_1 = 0, \quad 4EI\left[D^3\Phi\right]_1 - k_2[D\Theta]_1 = 0. \end{aligned} \tag{65}$$

Let $r_N, z_N, l_N \in \mathbb{R}^N$ be auxiliary row-vectors

$$r_N = [0 \;\; 0 \;\; \cdots \;\; 0 \;\; 1], \quad z_N = [0 \;\; 0 \;\; \cdots \;\; 0 \;\; 0], \quad l_N = [1 \;\; 0 \;\; \cdots \;\; 0 \;\; 0] \tag{66}$$

and D_j^n designate the jth row of the nth derivative matrix D^n. Using Eqs. (66) we represent Eqs. (65) as the following matrix equation:

$$\mathbb{K}\Psi \equiv \begin{bmatrix} r_N & z_N \\ D_N^1 & z_N \\ z_N & r_N \\ z_N & D_N^1 \\ 4EID_1^2 & k_1 l_N \\ 4EID_1^3 & -k_2 D_1^1 \end{bmatrix} \begin{bmatrix} \Phi \\ \Theta \end{bmatrix} = 0. \tag{67}$$

$\mathbb{K}$ is called the boundary operator. Let $\mathcal{K}_N$ be the kernel of $\mathbb{K}$, i.e., $\mathcal{K}_N = \{v \in \mathbb{R}^{2N} : \mathbb{K}v = 0\}$. We have to identify all eigenvalues of the operator $\mathbf{L}$, when its domain is restricted to $\mathcal{K}_N$. It is clear that $\mathcal{K}_N$ is isomorphic to $\mathbb{R}^k$ with $k \equiv \dim\mathcal{K}_N = \dim\mathcal{H}_N - \operatorname{rank}\mathbb{K} = 2N - 6$. Let B be the matrix consisting of column vectors that form an orthonormal basis in $\mathcal{K}_N$. It is clear that B^TB is the identity matrix on $\mathbb{R}^k$ and BB^T is the identity matrix on $\mathbb{K}$. The following result holds: if λ is an eigenvalue of the operator $\boldsymbol{L}$, and the corresponding eigenvector Ψ satisfies Eq. (67), then the same λ is an eigenvalue of the matrix $\left(B^T\boldsymbol{L}B\right)$. However, the inverse statement is not necessarily true. Indeed, we observe that BB^T is the identity in $\mathcal{K}_N$, which is not equivalent to the identity in $\mathcal{H}_N$. Assume now that λ is an eigenvalue of $B^T\boldsymbol{L}B$ with corresponding eigenvector $v \in \mathbb{R}^k$. If $\Psi = Bv$, we have

$$BB^T\boldsymbol{L}\Psi = BB^T\boldsymbol{L}Bv = \lambda Bv = \lambda\Psi, \tag{68}$$

but $BB^T\boldsymbol{L} \neq \boldsymbol{L}$, which indicates that fake eigenvalues may exist.

5.3 Filtering of spurious eigenvalues

In order to decide which eigenvalues of $B^T\boldsymbol{L}B$ should be discarded, we impose the following condition. Let Λ be the spectrum of $B^T\boldsymbol{L}B$ and V be the set of its eigenfunctions. We construct the set of "trusted" eigenvalues [14, 15], for some $\varepsilon_f > 0$ filtering precision, as

$$\Lambda_\varepsilon = \left\{\lambda \in \Lambda : \|\boldsymbol{L}Bv_\lambda - \lambda Bv_\lambda\|_C < \varepsilon_f, \text{for corresponding eigenvector} \, v_\lambda \in V\right\}, \tag{69}$$

where $\|\cdot\|_C$ is a discrete approximation to the integral norm defined in Eq. (61). (The subscript C is short for Chebyshev). Using the CGL quadrature, we obtain the following formula for the norm of a vector Ψ defined as in Eq. (63):

$$\|\mathbf{\Psi}\|_C = \frac{\pi/4}{N-1}\sum_{k=1}^{N}\sqrt{1-\xi_k^2}\Big[16EI(\xi_k)\,\big|\big[D^2\mathbf{\Phi}\big]_k\big|^2 + \rho(\xi_k)\,|\Theta_k|^2\Big].$$

6. Conclusions

In this work we have considered the spectral properties of the Euler-Bernoulli beam model with special feedback-type boundary conditions. The dynamics generator of the model is a non-self-adjoint matrix differential operator acting in a Hilbert space of two-component Cauchy data. This operator has been approximated by a "discrete" operator using Chebyshev polynomial approximation. We have shown that the eigenvalues of the main operator can be approximated by the eigenvalues of its discrete counterpart with high accuracy. This means that the leading asymptotic terms in formulas (20) and (21) can be used by practitioners who need the elastic modes.

Further results deal with existence and formulas of the deadbeat modes. It has been shown that for the case when one control parameter, k_1, is such that $k_1 \to 1^+$ and the other one $k_2 = 0$, the number of deadbeat modes approaches infinity. The formula for the rate at which the number of the deadbeat modes tends to infinity has been derived. It has also been established that there exists a sequence $\left\{k_1^{(n)}\right\}_{n=1}^{\infty}$ of the values of parameter k_1, such that the corresponding deadbeat mode has a multiplicity 2, which yields the existence of the associate mode shapes for the operator $\mathcal{L}_{k_1,k_2}$. The formulas for the double deadbeat modes and asymptotics for the sequence $\left\{k_1^{(n)}\right\}$ as $n \to \infty$ have been derived.

Acknowledgements

Partial support of the National Science Foundation award DMS-1810826 is highly appreciated by the first author.

Author details

Marianna A. Shubov*† and Laszlo P. Kindrat†
University of New Hampshire, Durham, NH, USA

*Address all correspondence to: marianna.shubov@gmail.com

† These authors contributed equally.

References

[1] Shubov MA, Kindrat LP. Spectral analysis of the Euler–Bernoulli beam model with fully nonconservative feedback matrix. Mathematicsl Methods in the Applied Sciences. 2018;**41**(12): 4691-4713. DOI: 10.1002/mma.4922

[2] Shubov MA, Shubov VI. Stability of a flexible structure with destabilizing boundary conditions. Proceedings of the Royal Society A. 2016;**472**(2191): 20160109. DOI: 10.1098/rspa.2016.0109

[3] Shubov MA. Exact controllability of nonselfadjoint Euler-Bernoulli beam model via spectral decomposition method. IMA Journal of Mathematical Control and Information. 2008;**25**(2): 185-203. DOI: 10.1093/imamci/dnm018

[4] Guiver CH, Opmeer MR. Non-dissipative boundary feedback for Rayleigh and Timoshenko beams. Systems and Control Letters. 2010; **59**(9):578-586. DOI: 10.1016/j.sysconle.2010.07.002

[5] Shubov MA. Exact controllability of coupled Euler-Bernoulli and Timoshenko beam model. IMA Journal of Mathematical Control and Information. 2006;**23**(3):279-300. DOI: 10.1093/imamci/dnm059

[6] Curtain RF, Zwart HJ. An Introduction to Infinite Dimensional Linear Systems Theory. New York, NY: Springer; 1995

[7] Dowell EH, editor. A Modern Course on Aeroelasticity. 4th Revised ed. Boston, MA: Klumer Academic Publishers; 2004

[8] Balakrishnan AV, Shubov MA. Asymptotic behaviour of aeroelastic modes for aircraft wing model in subsonic air flow. Proceedings of the Royal Society A. 2004; **460**(2044):1057-1091. DOI: 10.1098/rspa.2003.1217

[9] Shubov MA, Rojas-Araneza M. Four-branch vibrational spectrum of double-walled carbon nanotube model. Proceedings of the Royal Society A. 2011;**467**(2125):99-126. DOI: 10.1098/rspa.2010.0100

[10] Benaroya H. Mechanical Vibration: Analysis, Uncertainties, and Control. Upper Saddle River, NJ: Prentice Hall; 1998

[11] Han SM, Benaroya H, Wei T. Dynamics of transversely vibrating beams using four engineering theories. Journal of Sound and Vibration. 1999; **225**(5):935-988. DOI: 10.1006/jsvi.1999.2257

[12] Shubov MA. Generation of Gevrey class semigroup by non-selfadjoint Euler-Bernoulli beam model. Mathematicsl Methods in the Applied Sciences. 2006;**29**(18):2181-2199. DOI: 10.1002/mma.768

[13] Gohberg IT, Krein MG. Introduction to the Theory of Nonselfadjoint Operators in Hilbert Space. Translations of Mathematical Monographs No. 18. Providence, RI: AMS; 1996

[14] Mason JC, Handscomb DC. Chebyshev Polynomials. Boca Raton, FL: Chapman & Hall/CRC; 2003

[15] Trefethen LN, Bau D III. Numerical Linear Algebra. Philadelphia, PA: SIAM; 1997

[16] Shubov MA, Wineberg S, Holt R. Numerical investigation of aeroelastic mode distribution for aircraft wing model in subsonic air flow. Mathematical Problems in Engineering. 2010;**2010**:879519. DOI: 10.1155/2010/879519

[17] Marcus AS. Introduction to the Spectral Theory of Polynomial Operator Pencils. Translations of Mathematical Monographs No. 71. Providence, RI: AMS; 1988

[18] Fedoryuk MV. Asymptotic Analysis. Berlin, Heidelberg: Springer-Verlag; 1993

[19] Naimark MA. Linear Differential Operators. New York, NY: Ungar Pub Co; 1967

Chapter 10

Integral Inequalities and Differential Equations via Fractional Calculus

Zoubir Dahmani and Meriem Mansouria Belhamiti

Abstract

In this chapter, fractional calculus is used to develop some results on integral inequalities and differential equations. We develop some results related to the Hermite-Hadamard inequality. Then, we establish other integral results related to the Minkowski inequality. We continue to present our results by establishing new classes of fractional integral inequalities using a family of positive functions; these classes of inequalities can be considered as generalizations of order n for some other classical/fractional integral results published recently. As applications on inequalities, we generate new lower bounds estimating the fractional expectations and variances for the beta random variable. Some classical covariance identities, which correspond to the classical case, are generalised for any $\alpha \geq 1, \beta \geq 1$. For the part of differential equations, we present a contribution that allow us to develop a class of fractional chaotic electrical circuit. We prove recent results for the existence and uniqueness of solutions for a class of Langevin-type equation. Then, by establishing some sufficient conditions, another result for the existence of at least one solution is also discussed.

Keywords: fractional calculus, fixed point, Riemann-Liouville integral, Caputo derivative, integral inequality

1. Introduction

During the last few decades, fractional calculus has been extensively developed due to its important applications in many field of research [1–4]. On the other hand, the integral inequalities are very important in probability theory and in applied sciences. For more details, we refer the reader to [5–12] and the references therein. Moreover, the study of integral inequalities using fractional integration theory is also of great importance; we refer to [1, 13–17] for some applications.

Also, boundary value problems of fractional differential equations have occupied an important area in the fractional calculus domain, since these problems appear in several applications of sciences and engineering, like mechanics, chemistry, electricity, chemistry, biology, finance, and control theory. For more details, we refer the reader to [3, 18–20].

In this chapter, we use the Riemann-Liouville integrals to present some results related to Minkowski and Hermite-Hadamard inequalities [21]. We continue to present our results by establishing several classes of fractional integral inequalities using a family of positive functions; these classes of inequalities can be considered as generalizations for some other fractional and classical integral results published recently [22]. Then, as applications, we generate new lower bounds estimating the fractional expectations and variances for the beta random variable. Some classical covariance identities, which correspond to $\alpha = 1$, are generalized for any $\alpha \geq 1$ and $\beta \geq 1$; see [23].

For the part of differential equations, with my coauthor, we present a contribution that allows us to develop a class of fractional differential equations generalizing the chaotic electrical circuit model. We prove recent results for the existence and uniqueness of solutions for a class of Langevin-type equations. Then, by establishing some sufficient conditions on the data of the problem, another result for the existence of at least one solution is also discussed. The considered class has some relationship with the good paper in [20].

The chapter is structured as follows: In Section 2, we recall some preliminaries on fractional calculus that will be used in the chapter. Section 3 is devoted to the main results on integral inequalities as well as to some estimates on continuous random variables. The Section 4 deals with the class of differential equations of Langevin type: we study the existence and uniqueness of solutions for the considered class by means of Banach contraction principle, and then using Schaefer fixed point theorem, an existence result is discussed. At the end, the Conclusion follows.

2. Preliminaries on fractional calculus

In this section, we present some definitions and lemmas that will be used in this chapter. For more details, we refer the reader to [2, 13, 15, 24].

Definition 1.1. The Riemann-Liouville fractional integral operator of order $\alpha \geq 0$, for a continuous function f on $[a, b]$ is defined as

$$J_a^\alpha[f(t)] = \frac{1}{\Gamma(\alpha)} \int_a^t (t-\tau)^{\alpha-1} f(\tau) d\tau, \alpha > 0,\ a < t \leq b, \tag{1}$$

$$J_a^0[f(t)] = f(t),$$

where $\Gamma(\alpha) := \int_0^\infty e^{-u} u^{\alpha-1} du$.

Note that for $\alpha > 0, \beta > 0$, we have

$$J^\alpha J^\beta f(t) = J^{\alpha+\beta} f(t), \tag{2}$$

and

$$J_a^\alpha J_a^\beta [f(t)] = J_a^\beta J_a^\alpha [f(t)]. \tag{3}$$

In the rest of this chapter, for short, we note a probability density function by $p.d.f$. So, let us consider a positive continuous function ω defined on $[a, b]$. We recall the $\omega-$concepts:

Definition 1.2. The fractional $\omega-$weighted expectation of order $\alpha > 0$, for a random variable X with a positive $p.d.f.f$ defined on $[a, b]$, is given by

$$E_{\alpha,\omega}(X) := J_a^\alpha[t\omega f](b) = \frac{1}{\Gamma(\alpha)} \int_a^b (b-\tau)^{\alpha-1} \tau\omega(\tau) f(\tau) d\tau, \alpha > 0, a < t \leq b, \tag{4}$$

Definition 1.3. The fractional $\omega-$weighted variance of order $\alpha > 0$ for a random variable X having a $p.d.f.f$ on [a, b] is given by

$$\sigma^2_{\alpha,\omega}(X) = V_{\alpha,\omega}(X) := \frac{1}{\Gamma(\alpha)} \int_a^b (b-\tau)^{\alpha-1} (\tau - E(X))^2 \omega(\tau) f(\tau) d\tau, \alpha > 0. \quad (5)$$

Definition 1.4. The fractional $\omega-$weighted moment of orders $r > 0, \alpha > 0$ for a continuous random variable X having a $p.d.f.f$ defined on [a, b] is defined by the quantity:

$$E_{\alpha,\omega}(X^r) := \frac{1}{\Gamma(\alpha)} \int_a^b (b-\tau)^{\alpha-1} \tau^r \omega(\tau) f(\tau) d\tau, \alpha > 0. \quad (6)$$

We introduce the covariance of fractional order as follows.

Definition 1.5. Let f_1 and f_2 be two continuous on $[a, b]$. We define the fractional $\omega-$weighted covariance of order $\alpha > 0$ for $(f_1(X), f_2(X))$ by

$$Cov_{\alpha,\omega}(f_1(X), f_2(X)) := \frac{1}{\Gamma(\alpha)} \int_a^b (b-\tau)^{\alpha-1} (f_1(\tau) - f_1(\mu))(f_2(\tau) - f_2(\mu)) \omega(\tau) f(\tau) d\tau, \alpha > 0, \quad (7)$$

where μ is the classical expectation of X.

It is to note that when $\omega(x) = 1, x \in [a, b]$, then we put

$$Var_{\alpha,\omega}(X) := Var_\alpha(X), Cov_{\alpha,\omega}(X) := Cov_\alpha(X), E_{\alpha,\omega}(X) := E_\alpha(X)$$

Definition 1.6. For a function $K \in C^n([a, b], \mathbb{R})$ and $n - 1 < \alpha \le n$, the Caputo fractional derivative of order α is defined by

$$D^\alpha K(t) = J^{n-\alpha} \frac{d^n}{dt^n}(K(t))$$
$$= \frac{1}{\Gamma(n-\alpha)} \int_a^t (t-s)^{n-\alpha-1} K^{(n)}(s) ds.$$

We recall also the following properties.

Lemma 1.7. Let $n \in \mathbb{N}^*$, and $n - 1 < \alpha < n$. The general solution of $D^\alpha y(t) = 0$, $t \in [a, b]$ is given by

$$y(t) = \sum_{i=0}^{n-1} c_i (t-a)^i, \quad (8)$$

where $c_i \in \mathbb{R}, i = 0, 1, 2, .., n-1$.

Lemma 1.8. Let $n \in \mathbb{N}^*$ and $n - 1 < \alpha < n$. Then

$$J^\alpha D^\alpha y(t) = y(t) + \sum_{i=0}^{n-1} c_i (t-a)^i, t \in [a, b], \quad (9)$$

for some $c_i \in \mathbb{R}, i = 0, 1, 2, .., n-1$.

3. Some integral inequalities

3.1 On Minkowski and Hermite-Hadamard fractional inequalities

In this subsection, we present some fractional integral results related to Minkowski and Hermite-Hadamard integral inequalities. For more details, we refer the reader to [21].

Theorem 1.9. Let $\alpha > 0, p \geq 1$ and let f, g be two positive functions on $[0, \infty[$, such that for all $t > 0, J^{\alpha}f^{p}(t) < \infty, J^{\alpha}g^{p}(t) < \infty$. If $0 < m \leq \frac{f(\tau)}{g(\tau)} \leq M, \tau \in [0, t]$, then we have

$$[J^{\alpha}f^{p}(t)]^{\frac{1}{p}} + [J^{\alpha}g^{p}(t)]^{\frac{1}{p}} \leq \frac{1 + M(m+2)}{(m+1)(M+1)}[J^{\alpha}(f+g)^{p}(t)]^{\frac{1}{p}}. \tag{10}$$

Proof: We use the hypothesis $\frac{f(\tau)}{g(\tau)} < M, \tau \in [0, t], t > 0$. We can write

$$\begin{aligned} &\frac{(M+1)^{p}}{\Gamma(\alpha)}\int_{0}^{t}(t-\tau)^{\alpha-1}f^{p}(\tau)d\tau \\ &\leq \frac{M^{p}}{\Gamma(\alpha)}\int_{0}^{t}(t-\tau)^{\alpha-1}(f+g)^{p}(\tau)d\tau. \end{aligned} \tag{11}$$

Hence, we have

$$J^{\alpha}f^{p}(t) \leq \frac{M^{p}}{(M+1)^{p}}J^{\alpha}(f+g)^{p}(t). \tag{12}$$

Thus, it yields that

$$[J^{\alpha}f^{p}(t)]^{\frac{1}{p}} \leq \frac{M}{M+1}[J^{\alpha}(f+g)^{p}(t)]^{\frac{1}{p}}. \tag{13}$$

In the same manner, we have

$$\left(1 + \frac{1}{m}\right)g(\tau) \leq \frac{1}{m}(f(\tau) + g(\tau)). \tag{14}$$

And then,

$$[J^{\alpha}g^{p}(t)]^{\frac{1}{p}} \leq \frac{1}{m+1}[J^{\alpha}(f+g)^{p}(t)]^{\frac{1}{p}}. \tag{15}$$

Combining (13) and (15), we achieve the proof.

Remark 1.10. Applying the above theorem for $\alpha = 1$, we obtain Theorem 1.2 of [25] on $[0, t]$.

With the same arguments as before, we present the following theorem.

Theorem 1.11. Let $\alpha > 0, p \geq 1$ and let f, g be two positive functions on $[0, \infty[$, such that for all $t > 0, J^{\alpha}f^{p}(t) < \infty, J^{\alpha}g^{p}(t) < \infty$. If $0 < m \leq \frac{f(\tau)}{g(\tau)} \leq M, \tau \in [0, t]$, then we have

$$[J^\alpha f^p(t)]^{\frac{1}{p}} + [J^\alpha g^p(t)]^{\frac{1}{p}} \le \frac{1+M(m+2)}{(m+1)(M+1)}[J^\alpha(f+g)^p(t)]^{\frac{1}{p}}. \tag{16}$$

Remark 1.12. Taking $\alpha = 1$ in this second theorem, we obtain Theorem 2.2 in [26] on $[0, t]$.

Using the notions of concave and L^p−functions, we present to the reader the following result.

Theorem 1.13. Suppose that $\alpha > 0, p > 1, q > 1$ and let f, g be two positive functions on $[0, \infty[$. If f^p, g^q are two concave functions on $[0, \infty[$, then we have

$$\begin{aligned} &2^{-p-q}(f(0)+f(t))^p(g(0)+g(t))^q(J^\alpha(t^{\alpha-1}))^2 \\ &\le J^\alpha(t^{\alpha-1}f^p(t))J^\alpha(t^{\alpha-1}g^q(t)). \end{aligned} \tag{17}$$

The proof of this theorem is based on the following auxiliary result.

Lemma 1.14. Let h be a concave function on $[a, b]$. Then for any $x \in [a, b]$, we have

$$h(a) + h(b) \le h(b+a-x) + h(x) \le 2h\left(\frac{a+b}{2}\right). \tag{18}$$

3.2 A family of fractional integral inequalities

We present to the reader some integral results for a family of functions [22]. These results generalize some integral inequalities of [27]. We have

Theorem 1.15. Suppose that $(f_i)_{i=1,\dots n}$ are n positive, continuous, and decreasing functions on $[a, b]$. Then, the following inequality

$$\frac{J^\alpha\left[\prod_{i\neq p}^n f_i^{\gamma_i} f_p^\beta(t)\right]}{J^\alpha\left[\prod_{i=1}^n f_i^{\gamma_i}(t)\right]} \ge \frac{J^\alpha\left[(t-a)^\delta \prod_{i\neq p}^n f_i^{\gamma_i} f_p^\beta(t)\right]}{J^\alpha\left[(t-a)^\delta \prod_{i=1}^n f_i^{\gamma_i}(t)\right]} \tag{19}$$

holds for any $a < t \le b, \alpha > 0, \delta > 0, \beta \ge \gamma_p > 0$, where p is a fixed integer in $\{1, 2, \dots, n\}$.

Proof: It is clear that

$$\left((\rho-a)^\delta - (\tau-a)^\delta\right)\left(f_p^{\beta-\gamma_p}(\tau) - f_p^{\beta-\gamma_p}(\rho)\right) \ge 0, \tag{20}$$

for any fixed $p \in \{1, \dots n\}$ and for any $\beta \ge \gamma_p > 0, \delta > 0, \tau, \rho \in [a, t]; a < t \le b$.

Taking

$$K_p(\tau,\rho) := \frac{(t-\tau)^{\alpha-1}}{\Gamma(\alpha)} \prod_{i=1}^n f_i^{\gamma_i}(\tau)\left((\rho-a)^\delta - (\tau-a)^\delta\right)\left(f_p^{\beta-\gamma_p}(\tau) - f_p^{\beta-\gamma_p}(\rho)\right), \tag{21}$$

we observe that

$$K_p(\tau,\rho) \ge 0. \tag{22}$$

Also, we have

$$0 \leq \int_a^t K_p(\tau,\rho)d\tau = (\rho-a)^\delta J^\alpha\left[\prod_{i\neq p}^{n} f_i^{\gamma_i} f_p^{\beta}(t)\right] + f_p^{\beta-\gamma_p}(\rho) J^\alpha\left[(t-a)^\delta \prod_{i=1}^{n} f_i^{\gamma_i}(t)\right]$$
$$-J^\alpha\left[(t-a)^\delta \prod_{i\neq p}^{n} f_i^{\gamma_i} f_p^{\beta}(t)\right] - (\rho-a)^\delta f^{\beta-\gamma_p}(\rho) J^\alpha\left[\prod_{i=1}^{n} f_i^{\gamma_i}(t)\right]. \tag{23}$$

Hence, we get

$$J^\alpha\left[(t-a)^\delta \prod_{i=1}^{n} f_i^{\gamma_i}(t)\right] J^\alpha\left[\prod_{i\neq p}^{n} f_i^{\gamma_i} f_p^{\beta}(t)\right] \geq J^\alpha\left[\prod_{i=1}^{n} f_i^{\gamma_i}(t)\right] J^\alpha\left[(t-a)^\delta \prod_{i\neq p}^{n} f_i^{\gamma_i} f_p^{\beta}(t)\right]. \tag{24}$$

The proof is thus achieved.

Remark 1.16. Applying Theorem 1.15 for $\alpha = 1, t = b, n = 1,$ we obtain Theorem 3 in [27].

Using other sufficient conditions, we prove the following generalization.

Theorem 1.17. Suppose that $(f_i)_{i=1,\dots n}$ are positive, continuous, and decreasing functions on $[a,b]$. Then for any fixed p in $\{1, 2, \dots, n\}$ and for any $a < t \leq b, \alpha > 0, \omega > 0, \delta > 0, \beta \geq \gamma_p > 0,$ we have

$$\frac{J^\alpha\left[\prod_{i\neq p}^{n} f_i^{\gamma_i} f_p^{\beta}(t)\right] J^\omega\left[(t-a)^\delta \prod_{i=1}^{n} f_i^{\gamma_i}(t)\right] + J^\omega\left[\prod_{i\neq p}^{n} f_i^{\gamma_i} f_p^{\beta}(t)\right] J^\alpha\left[(t-a)^\delta \prod_{i=1}^{n} f_i^{\gamma_i}(t)\right]}{J^\alpha\left[(t-a)^\delta \prod_{i\neq p}^{n} f_i^{\gamma_i} f_p^{\beta}(t)\right] J^\omega\left[\prod_{i=1}^{n} f_i^{\gamma_i}(t)\right] + J^\omega\left[(t-a)^\delta \prod_{i\neq p}^{n} f_i^{\gamma_i} f_p^{\beta}(t)\right] J^\alpha\left[\prod_{i=1}^{n} f_i^{\gamma_i}(t)\right]} \geq 1. \tag{25}$$

Proof: Multiplying both sides of (23) by $\frac{(t-\rho)^{\omega-1}}{\Gamma(\omega)} \prod_{i=1}^{n} f_i^{\gamma_i}(\rho), \omega > 0,$ then integrating the resulting inequality with respect to ρ over $(a,t), a < t \leq b$ and using Fubini's theorem, we obtain the desired inequality.

Remark 1.18.

i. Applying Theorem 1.17 for $\alpha = \omega,$ we obtain Theorem 1.15.

ii. Applying Theorem 1.17 for $\alpha = \omega = 1, t = b, n = 1,$ we obtain Theorem 3 of [27].

Introducing a positive increasing function g to the family $(f_i)_{i=1,\dots n}$, we establish the following theorem.

Theorem 1.19. Let $(f_i)_{i=1,\dots n}$ and g be positive continuous functions on $[a,b]$, such that g is increasing and $(f_i)_{i=1,\dots n}$ are decreasing on $[a,b]$. Then, the following inequality

$$\frac{J^\alpha\left[\prod_{i\neq p}^{n} f_i^{\gamma_i} f_p^{\beta}(t)\right] J^\alpha\left[g^\delta(t) \prod_{i=1}^{n} f_i^{\gamma_i}(t)\right]}{J^\alpha\left[g^\delta(t) \prod_{i\neq p}^{n} f_i^{\gamma_i} f_p^{\beta}(t)\right] J^\alpha\left[\prod_{i=1}^{n} f_i^{\gamma_i}(t)\right]} \geq 1 \tag{26}$$

holds for any $a < t \leq b, \alpha > 0, \delta > 0, \beta \geq \gamma_p > 0,$ where p is a fixed integer in $\{1, 2, \dots, n\}$.

Remark 1.20. Applying Theorem 1.19 for $\alpha = 1, t = b, n = 1,$ we obtain Theorem 4 of [27].

3.3 Some estimations on random variables

3.3.1 Bounds for fractional moments of beta distribution

In what follows, we present some fractional results on the beta distribution [23]. So let us prove the following $\alpha-$version.

Theorem 1.21. Let $X, Y, U,$ and V be four random variables, such that $X \sim B(p,q), Y \sim B(m,n), U \sim B(p,n),$ and $V \sim B(m,q)$. If $(p-m)(q-n) \leq 0$, then

$$\frac{E_\alpha(X^r)E_\alpha(Y^r)}{E_\alpha(U^r)E_\alpha(V^r)} \geq \frac{B(p,n)B(m,q)}{B(p,q)B(m,n)}, \alpha \geq 1.$$

For the proof of this result, we can apply a weighted version of the fractional Chebyshev inequality as is mentioned in [1].

Remark 1.22. The above theorem generalizes Theorem 3.1 of [7].

We propose also the following $(\alpha,\beta)-$version that generalizes the above result. We have

Theorem 1.23. Let $X, Y, U,$ and V be four random variables, such that $X \sim B(p,q), Y \sim B(m,n), U \sim B(p,n),$ and $V \sim B(m,q)$. If $(p-m)(q-n) \leq 0$, then

$$\frac{E_\alpha(X^r)E_\beta(Y^r) + E_\beta(X^r)E_\alpha(Y^r)}{E_\alpha(U^r)E_\beta(V^r) + E_\beta(U^r)E_\alpha(V^r)} \geq \frac{B(p,n)B(m,q)}{B(p,q)B(m,n)}, \alpha, \beta \geq 1.$$

Remark 1.24. If $\alpha = \beta = 1$, then the above theorem reduces to Theorem 3.1 of [7].

3.3.2 Identities and lower bounds

In the following theorem, the fractional covariance of X and $g(X)$ is expressed with the derivative of $g(X)$. It can be considered as a generalization of a covariance identity established by the authors of [28]. So, we prove the result:

Theorem 1.25. Let X be a random variable having a $p.d.f$ defined on $[a,b]$; $\mu = E(X)$. Then, we have

$$Cov_\alpha(X, g(X)) = \frac{1}{\Gamma(\alpha)} \int_a^b g'(x)dx \int_a^x (b-t)^{\alpha-1}(\mu - t)f(t)dt, \alpha \geq 1. \tag{27}$$

We can prove this result by the application of the covariance definition in the case where $\omega(x) = 1$.

The following theorem establishes a lower bound for $Var_\alpha(g(X))$ of any function $g \in C^1([a,b])$. We have

Theorem 1.26. Let X be a random variable having a $p.d.f$ defined on $[a,b]$, such that $\mu = E(X)$. Then, we have

$$Var_\alpha(g(X)) \geq \frac{1}{Var_{X,\alpha}} \left(\frac{1}{\Gamma(\alpha)} \int_a^b g'(x)dx \int_a^x (b-t)^{\alpha-1}(\mu - t)f(t)dt \right)^2, \tag{28}$$

for any $g \in C^1([a,b])$.

To prove this result, we use fractional Cauchy-Schwarz inequality established in [29].

Remark 1.27. Let us consider $\Omega \in C([a,b])$ that satisfies $\int_a^x (b-t)^{\alpha-1}(\mu-t) f(t)dt = (b-x)^{\alpha-1}\sigma^2\Omega(x)f(x)$. Then, we present the following result.

Theorem 1.28. Let X be a random variable having a $p.d.f.$ defined on $[a,b]$, such that $\mu = E(X), \sigma^2 = Var(X)$ and $\Omega \in C([a,b])$; $\int_a^x (b-t)^{\alpha-1}(\mu-t)f(t)dt = (b-x)^{\alpha-1}\sigma^2\Omega(x)f(x)$. Then, we have

$$Var_\alpha(g(X)) \geq \frac{\sigma^4(X)}{Var_\alpha(X)} E_\alpha^2(g'(X)\Omega(X)). \tag{29}$$

Proof: We have

$$Cov_\alpha^2(X,g(X)) = \left[\frac{1}{\Gamma(\alpha)}\int_a^b g'(x)dx(b-x)^{\alpha-1}\sigma^2\Omega(x)f(x)dx\right]^2. \tag{30}$$

On the other hand, we can see that

$$\left[\frac{1}{\Gamma(\alpha)}\int_a^b g'(x)dx(b-x)^{\alpha-1}\sigma^2\Omega(x)f(x)dx\right]^2 = \sigma^4 E_\alpha^2(g'(X)\Omega(X)) \tag{31}$$

Thanks to the fractional version of Cauchy Schwarz inequality [29], and using the fact that

$$Cov_\alpha^2(X,g(X)) \leq Var_\alpha(X)Var_\alpha(g(X)), \tag{32}$$

we obtain

$$\sigma^4 E_\alpha^2(g'(X)\Omega(X)) \leq Var_\alpha(X)Var_\alpha(g(X)). \tag{33}$$

This ends the proof.

Remark 1.29. Thanks to (30) and (31), we obtain the following fractional covariance identity

$$\sigma^2 E_\alpha(g'(X)\Omega(X)) = Cov_\alpha(X,g(X)).$$

It generalizes the good standard identity obtained in [28] that corresponds to $\alpha = 1$ and it is given by

$$\sigma^2 E(g'(X)\Omega(X)) = Cov(X,g(X)).$$

We end this section by proving the following fractional integral identity between covariance and expectation in the fractional case.

Theorem 1.30. Let X be a continuous random variable with a $p.d.f.$ having a support an interval $[a,b]$, $E(X) = \mu$. Then, for any $\alpha \geq 1$, the following general covariance identity holds

$$Cov_\alpha(h(X),g(X)) = E_\alpha(g'(X)Z(X)), \tag{34}$$

where $g \in C^1([a,b])$, with $E|Z(X)g'(X)| < \infty$, $h(x)$ is a given function and $Z(x)f(x)\frac{(b-x)^{\alpha-1}}{\Gamma(\alpha)} = \int_a^x (E(h(X)) - h(t))\frac{(b-t)^{\alpha-1}}{\Gamma(\alpha)}f(t)dt$.

Proof: We have

$$Cov_\alpha(h(X),g(X)) = \frac{1}{\Gamma(\alpha)}\int_a^b (b-x)^{\alpha-1}(h(x)-h(\mu))(g(x)-g(\mu))f(x)dx \tag{35}$$

and

$$E_\alpha(g'(X)Z(X)) = \frac{1}{\Gamma(\alpha)}\int_a^b (b-x)^{\alpha-1}g'(x)Z(x)f(x)dx. \tag{36}$$

The definition of $Z(X)$ implies that

$$\begin{aligned} E_\alpha(g'(X)Z(X)) &= \frac{1}{\Gamma(\alpha)}\int_a^\mu (g(\mu)-g(t))(b-t)^{\alpha-1}(h(\mu)-h(t))f(t)dt \\ &+ \frac{1}{\Gamma(\alpha)}\int_\mu^b (g(t)-g(\mu))(b-t)^{\alpha-1}(h(t)-h(\mu))f(t)dt. \end{aligned} \tag{37}$$

Hence, we obtain

$$E_\alpha(Z(X)g'(X)) = Cov_\alpha(g(X),h(X)). \tag{38}$$

Remark 1.31. Taking $\alpha = 1$, in the above theorem, we obtain Theorem 2.2 of [10].

4. A class of differential equations of fractional order

Inspired by the work in [4, 20], in what follows we will be concerned with a more general class of Langevin equations of fractional order. The considered class will contain a nonlinearity that depends on a fractional derivative of order δ. So, let us consider the following problem:

$$\begin{aligned} &{}^cD^\alpha(D^2+\lambda^2)u(t) = f(t,u(t),{}^cD^\delta u(t)), \\ &t \in [0,1], \quad \lambda \in \mathbb{R}_+^* \\ &0 < \alpha \le 1, \quad 0 \le \delta < \alpha, \end{aligned} \tag{39}$$

associated with the conditions

$$u(0) = 0, \quad u''(0) = 0, \quad u(1) = \beta u(\eta), \eta \in (0,1), \tag{40}$$

where ${}^cD^\alpha$ denotes the Caputo fractional derivative of fractional order α, D^2 is the two-order classical derivative, $f : [0,1] \times \mathbb{R} \times \mathbb{R} \to \mathbb{R}$ is a given function, and $\beta \in \mathbb{R}$, such that $\beta sin(\lambda\eta) \ne sin(\lambda)$.

4.1 Integral representation

We recall the following result [20]:

Lemma 1.32. Let θ be a continuous function on $[0,1]$. The unique solution of the problem

$$\begin{aligned} &{}^{c}D^{\alpha}(D^{2}+\lambda^{2})u(t)=\theta(t),\\ &t\in[0,1], \quad \lambda\in\mathbb{R}_{+}^{*}\\ &n-1<\alpha\leq n, \quad n\in\mathbb{N}^{*}, \end{aligned} \tag{41}$$

is given by

$$u(t)=\frac{1}{\lambda}\int_{0}^{t} sin\lambda(t-s)\left(\int_{0}^{s}\frac{(s-\tau)^{\alpha-1}}{\Gamma(\alpha)}\theta(\tau)d\tau+\sum_{i=1}^{n-1}c_{i}s^{i}\right)ds+c_{n}\cos(\lambda t)+c_{n+1}sin(\lambda t), \tag{42}$$

where $c_i \in \mathbb{R}, i = 1 \ldots n+1$.

Thanks to the above lemma, we can state that

The class of Langevin equations (39) and (40) has the following integral representation:

$$\begin{aligned} u(t)=&\frac{1}{\lambda}\int_{0}^{t} sin\lambda(t-s)\left(\int_{0}^{s}\frac{(s-\tau)^{\alpha-1}}{\Gamma(\alpha)}f(\tau,u(\tau),D^{\delta}(\tau))d\tau\right)ds\\ &+\frac{\sin(\lambda t)}{\Delta}\left[\beta\int_{0}^{\eta} sin\lambda(\eta-s)\left(\int_{0}^{s}\frac{(s-\tau)^{\alpha-1}}{\Gamma(\alpha)}f(\tau,u(\tau),D^{\delta}(\tau))d\tau\right)ds\right.\\ &\left.-\int_{0}^{1} sin\lambda(1-s)\left(\int_{0}^{s}\frac{(s-\tau)^{\alpha-1}}{\Gamma(\alpha)}f(\tau,u(\tau),D^{\delta}(\tau))d\tau\right)ds\right], \end{aligned} \tag{43}$$

where

$$\Delta := \lambda(sin\lambda - \beta sin\lambda\eta). \tag{44}$$

4.2 Existence and uniqueness of solutions

Using the above integral representation (43), we can prove the following existence and uniqueness theorem.

Theorem 1.33. Assume that the following hypotheses are valid:

(H1): The function $f : [0,1] \times \mathbb{R} \times \mathbb{R} \to \mathbb{R}$ is continuous, and there exist two constants $\Lambda_1, \Lambda_2 > 0$, such that for all $t \in [0,1]$ and $u_i, v_i \in \mathbb{R}, i = 1,2$,

$$|f(t,u_1,u_2) - f(t,v_1,v_1)| \leq \Lambda_1|u_1 - v_1| + \Lambda_2|u_2 - v_2|. \tag{45}$$

(H2): Suppose that $\Lambda \leq \frac{1}{(\Phi+\Upsilon)}$,

where

$$\Phi := \frac{\Delta_1+\lambda+\beta_1\lambda\eta^{\alpha+1}}{\Gamma(\alpha+2)\lambda\Delta_1}, \Psi := \frac{\Delta_1 s^{\alpha}(\alpha+1)+\lambda^2+\beta_1\lambda^2\eta^{\alpha+1}}{\Gamma(\alpha+2)\lambda\Delta_1}, \Upsilon := \frac{\Psi}{\Gamma(2-\delta)},$$

$$\Lambda := max\,(\Lambda_1, \Lambda_2), \quad \Delta_1 = |\Delta|, \quad \beta_1 = |\beta|.$$

Then problem (39) and (40) has a unique solution on $[0, 1]$.
Proof: We introduce the space

$$E = \{u; u \in C([0,1]), D^{\delta}u \in C([0,1])\},$$

endowed with the norm $\|u\|_E := \|u\|_\infty + \|D^\delta u\|_\infty$.
Then, $(E, \|.\|_E)$ is a Banach space.
Also, we consider the operator $T : E \to E$ defined by

$$\begin{aligned}(Tu)(t) := \frac{1}{\lambda}\int_0^t sin\lambda(t-s)J_0^\alpha f\left(s, u(s), D^\delta u(s)\right)ds \\ + \frac{\sin(\lambda t)}{\Delta}\left[\beta\int_0^\eta sin\lambda(\eta-s)J_0^\alpha f(s, u(s), D^\delta(s))ds\right. \\ \left. - \int_0^1 sin\lambda(1-s)J_0^\alpha f(s, u(s), D^\delta u(s))ds\right]\end{aligned} \tag{46}$$

We shall prove that the above operator is contractive over the space E.
Let $u_1, u_2 \in E$. Then, for each $t \in [0, 1]$, we have

$$\begin{aligned}|Tu_1(t) - Tu_2(t)| \le \frac{1}{\lambda}\int_0^t |sin\lambda(t-s)| \quad J_0^\alpha |f\left(s, u_1(s), D^\delta u_1(s)\right) - f\left(s, u_2(s), D^\delta u_2(s)\right)|ds \\ + \frac{|\sin(\lambda t)|}{|\Delta|}\left[|\beta|\int_0^\eta |sin\lambda(\eta-s)| \quad J_0^\alpha|f(s, u_1(s), D^\delta u_1(s)) - f(s, u_2(s), D^\delta u_2(s))|ds\right. \\ \left. + \int_0^1 |sin\lambda(1-s) \quad |J_0^\alpha f(s, u_1(s), D^\delta u_1(s)) - f(s, u_2(s), D^\delta u_2(s))|ds\right] := \mathcal{A}\end{aligned}$$

By (H1), we have

$$\mathcal{A} \le \frac{\Lambda}{\Gamma(\alpha+2)}\left(\frac{1}{\lambda} + \frac{1}{|\Delta|} + \frac{|\beta|}{|\Delta|}\eta^{\alpha+1}\right)\left(|u_1 - u_2| + |D^\delta u_1 - D^\delta u_2|\right).$$

Hence, it yields that

$$\|Tu_1 - Tu_2\|_\infty \le \Lambda\Phi\|u_1 - u_2\|_E. \tag{47}$$

With the same arguments as before, we can write.

$$\begin{aligned}|T'u_1(t) - T'u_2(t)| \le \frac{1}{\lambda}|sin\lambda(t-s)|J_0^\alpha|f\left(s, u_1(s), D^\delta u_1(s)\right) - f\left(s, u_2(s), D^\delta u_2(s)\right)|ds \\ + \frac{|\lambda cos(\lambda t)|}{|\Delta|}\left[|\beta|\int_0^\eta |sin\lambda(\eta-s)|J_0^\alpha|f\left(s, u_1(s), D^\delta u_1(s)\right) - f\left(s, u_2(s), D^\delta u_2(s)\right)|ds\right. \\ \left. + \int_0^1 |sin\lambda(1-s)|J_0^\alpha|f(s, u_1(s), D^\delta u_1(s)) - f(s, u_2(s), D^\delta u_2(s))|ds\right] := \mathcal{B}.\end{aligned}$$

Again, by (H1), we obtain

$$\mathcal{B} \le \Lambda\left(\frac{\Delta_1 s^{\alpha}(\alpha+1) + \lambda^2 + \beta_1\lambda^2\eta^{\alpha+1}}{\Gamma(\alpha+2)\lambda\Delta_1}\right)\left(|u_1 - u_2| + |D^{\delta}u_1 - D^{\delta}u_2|\right).$$

Consequently, we get

$$\|T'u_1 - T'u_2\|_{\infty} \le \Lambda\Psi\|u_1 - u_2\|_E.$$

This implies that

$$\|D^{\delta}Tu_1 - D^{\delta}Tu_2\|_{\infty} \le \Lambda\Upsilon\|u_1 - u_2\|_E. \tag{48}$$

Using (47) and (48), we can state that

$$\|Tu_1 - Tu_2\|_E \le \Lambda(\Phi + \Upsilon)\|u_1 - u_2\|_E.$$

Thanks to (H2), we can say that the operator T is contractive.

Hence, by Banach fixed point theorem, the operator has a unique fixed point which corresponds to the unique solution of our Langevin problem.

4.3 Existence of solutions

We prove the following theorem.

Theorem 1.34. Assume that the following conditions are satisfied:

(H3): The function $f : [0,1] \times \mathbb{R} \times \mathbb{R} \to \mathbb{R}$ is jointly continuous.

(H4): There exists a positive constant $\mathcal{M}$; $|f(t,u,v)| \le \mathcal{M}$ for any $t \in [0,1], u, v \in \mathbb{R}$.

Then the problem (39), (40) has at least one solution on $[0,1]$.

Proof: We use Schaefer fixed point theorem to prove this result. So we proceed into three steps.

Step 1: We prove that T is continuous and bounded.

Since the function f is continuous by (H3), then the operator is also continuous; this proof is trivial and hence it is omitted.

Let $\Omega \subset E$ be a bounded set. We need to prove that $T(\Omega)$ is a bounded set.

Let $u \in \Omega$. Then, for any $t \in [0,1]$, we have

$$|(Tu)(t)| \le \left(\frac{1}{\lambda} + \frac{1}{|\Delta|}\right)\int_0^1 J_0^{\alpha}|f(s,u(s),D^{\delta}u(s))|ds + \frac{|\beta|}{|\Delta|}\int_0^{\eta} J_0^{\alpha}|f(s,u(s),D^{\delta}(s))|ds := C$$

Using (H4), we get

$$\|Tu\|_{\infty} \le \Phi\mathcal{M}. \tag{49}$$

In the same manner, we find that

$$\|D^{\delta}Tu\|_{\infty} \le \Upsilon\mathcal{M}. \tag{50}$$

From (49) and (50), we have

$$\|Tu\|_E \le (\Phi + \Upsilon)\mathcal{M}.$$

The operator is thus bounded.

Step 2: Equicontinuity.
Let $u \in E$. Then, for each $t_1, t_2 \in [0, 1]$, we have

$$\begin{aligned}|Tu(t_2) - Tu(t_1)| \le |\frac{1}{\lambda}\left[\int_0^{t_2} sin\lambda(t_2 - s)J_0^\alpha f(s, u(s), D^\delta u(s))ds - \int_0^{t_1} sin\lambda(t_1 - s)J_0^\alpha f(s, u(s), D^\delta u(s))ds\right] \\ + \frac{\sin(\lambda t_2) - \sin(\lambda t_1)}{\Delta}\left[\beta\int_0^\eta sin\lambda(\eta - s)J_0^\alpha f(s, u(s), D^\delta u(s))ds\right. \\ \left. + \int_0^1 sin\lambda(1 - s)J_0^\alpha f(s, u(s), D^\delta u(s))ds\right]| \\ \le \frac{|\Theta|}{|\Delta|}|\sin(\lambda t_2) - \sin(\lambda t_1)| + \frac{1}{\lambda}\int_{t_1}^{t_2} sin\lambda(t_2 - s)J_0^\alpha|f(s, u(s), D^\delta u(s))|ds \\ + \frac{1}{\lambda}\int_0^{t_2}|(sin\lambda(t_2 - s) - sin\lambda(t_1 - s)|J_0^\alpha|f(s, u(s), D^\delta u(s))|ds,\end{aligned} \tag{51}$$

where

$$\Theta := \beta\int_0^\eta |sin\lambda(\eta - s)J_0^\alpha|f(s, u(s), D^\delta u(s))|ds + \int_0^1 |sin\lambda(1 - s)|J_0^\alpha|f(s, u(s), D^\delta u(s))|ds.$$

Analogously, we can obtain

$$|T'u(t_2) - T'u(t_1)| \le \lambda\frac{|\Theta|}{|\Delta|}|cos(\lambda t_2) - cos(\lambda t_1)| + \frac{1}{\lambda}|sin\lambda(t_2 - s) - sin\lambda(t_1 - s)|J_0^\alpha|f(s, u(s), D^\delta u(s))|.$$

Consequently, we can write

$$|D^\delta Tu(t_2) - D^\delta Tu(t_1)| \le J^{1-\delta}|T'u(t_2) - T'u(t_1)| \tag{52}$$

As $t_1 \to t_2$, the right-hand sides of (51) and (52) tend to zero.
Therefore,

$$\|Tu(t_2) - Tu(t_1)\|_E \to 0.$$

The operator T is thus equicontinuous.
As a consequence of Step 1 and Step 2 and thanks to Arzela-Ascoli theorem, we conclude that T is completely continuous.
Step 3: We prove that $\Sigma := \{u \in E; u = \lambda Tu, 0 < \lambda < 1\}$ is a bounded set.
Let $u \in \Sigma$. Then, for each $t \in [0, 1]$, the following two inequalities are valid:

$$|u(t)| = |\lambda Tu(t)| \le |Tu(t)| \le \mathcal{M}\Phi$$

and

$$|D^\delta u(t)| = |\lambda D^\delta Tu(t)| \le |D^\delta Tu(t)| \le \mathcal{M}\Upsilon.$$

Therefore,

$$\|u\|_E \leq \mathcal{M}(\Upsilon + \Phi).$$

Thanks to steps 1, 2, and 3 and by Schaefer fixed point theorem, the operator T has at least one fixed point. This ends the proof of the above theorem.

5. Conclusions

In this chapter, the fractional calculus has been applied for some classes of integral inequalities. In fact, using Riemann-Liouville integral, some Minkowski and Hermite-Hadamard-type inequalities have been established. Several other fractional integral results involving a family of positive functions have been also generated. The obtained results generalizes some classical integral inequalities in the literature. In this chapter, we have also presented some applications on continuous random variables; new identities have been established, and some estimates have been discussed.

The existence and the uniqueness of solutions for nonlocal boundary value problem including the Langevin equations with two fractional parameters have been studied. We have used Caputo approach together with Banach contraction principle to prove the existence and uniqueness result. Then, by application of Schaefer fixed point theorem, another existence result has been also proved. Our approach is simple to apply for a variety of real-world problems.

Author details

Zoubir Dahmani[1*] and Meriem Mansouria Belhamiti[2]

1 Laboratory LMPA, Faculty of SEI, UMAB, University of Mostaganem, Algeria

2 Department of Math-Info, Faculty of SEI, University of Mostaganem, Algeria

*Address all correspondence to: zzdahmani@yahoo.fr

References

[1] Belarbi S, Dahmani Z. On some new fractional integral inequalities. Journal of Inequalities in Pure and Applied Mathematics. 2009;**10**(3):1-12

[2] Kilbas AA, Srivastava M, Trujillo JJ. Theory and Applications of Fractional Differential Equations. Amsterdam: Elsevier Science B.V; 2006

[3] Podlubny I. Fractional Differential Equations: An Introduction to Fractional Derivatives, Fractional Differential Equations, to Methods of Their Solution and Some of Their Applications. San Diego: Academic Press; 1999

[4] Salem A, Alzahrani F, Almaghamsi L. Fractional Langevin equation with nonlocal integral boundary condition. Mathematics. 2019;7:110

[5] Barnett NS, Cerone P, Dragomir SS, Roumeliotis J. Some inequalities for the expectation and variance of a random variable whose PDF is n-time diferentiable. Journal of Inequalities in Pure and Applied Mathematics. 2000; **1**(21):1-29

[6] Barnett NS, Cerone P, Dragomir SS, Roumeliotis J. Some inequalities for the dispersion of a random variable whose PDF is defined on a finite interval. Journal of Inequalities in Pure and Applied Mathematics. 2001;**2**(1):1-18

[7] Kumar P, Singh SP, Dragomir SS. Some inequalities involving beta and gamma functions. Nonlinear Analysis Forum. 2001;**6**(1):143-150

[8] Kumar P. Moment inequalities of a random variable defined over a finite interval. Journal of Inequalities in Pure and Applied Mathematics. 2002;**3**(3): 1-24

[9] Kumar P. Inequalities involving moments of a continuous random variable defined over a finite interval. Computers and Mathematics with Applications. 2004;**48**:257-273

[10] Mohtashami Borzadaran GR, Shanbhag DN. Further results based on Chernoff-type inequalities. Statistics & Probability Letters. 1998;**39**:109-117

[11] Niezgoda M. New bounds for moments of continuous random variables. Computers & Mathematcs with Applications. 2010;**60**(12): 3130-3138

[12] Sharma R, Devi S, Kapoor G, Ram S, Barnett NS. A brief note on some bounds connecting lower order moments for random variables defined on a finite interval. International Journal of Applied Science and Research. 2009; **1**(2):83-85

[13] Dahmani Z. Fractional integral inequalities for continuous random variables. Malaya Journal of Matematik. 2014;**2**(2):172-179

[14] Dahmani Z, Tabharit L. On weighted Gruss type inequalities via fractional integrals. Journal of Advanced Research in Pure Mathematics. 2010; **2**(4):31-38

[15] Dahmani Z, Bouziane AE, Houas M, Sarikaya MZ. New w-weighted concepts for continuous random variables with applications. Note di Matematica. 2017; **37**(1):2340

[16] Sarikaya MZ, Yaldiz H. New generalization fractional inequalities of Ostrowski-Gruss type. Lobachevskii Journal of Mathematics. 2013;**34**(4): 326-331

[17] Sarikaya MZ, Yaldiz H, Basak N. New fractional inequalities of Ostrowski-Gruss type. Le Matematiche. 2014;**69**:227-235

[18] Dahmani M, Anber A, Dahmani Z. Speech movements on vocal tract:

Fractional nonlinear dynamics. Journal of Information and Optimization Sciences. 2019;**40**(6):1307-1315

[19] Mainradi F, Pironi P. The fractional Langevin equation: Brownian motion revisited. Extracta Mathematicae. 1996; **11**(1):140-154

[20] Salem A, Alzahrani F, Almaghamsi L. Langevin equation involving one fractional order with three point boundary conditions. Journal of Nonlinear Sciences and Applications. 2019;**12**:791-798

[21] Dahmani Z. On Minkowski and Hermite-Hadamard integral inequalities via fractional integration. Annals of Functional Analysis. 2010;**1**(1):51-58

[22] Dahmani Z. New classes of integral inequalities of fractional order. Le Matematiche. 2014;**LXIX**(1):237-247

[23] Dahmani Z. New identities and lower bounds for CUD and beta distributions. ROMAI Journal. 2019

[24] Gorenflo R, Mainardi F. Fractional Calculus: Integral and Differential Equations of Fractional Order. Wien: Springer Verlag; 1997. pp. 223-276

[25] Bougoffa L. On Minkowski and Hardy integral inequality. Journal of Inequalities in Pure and Applied Mathematics. 2003;7(2):60

[26] Set E, Ozdemir ME, Dragomir SS. On the Hermite-Hadamard inequality and other integral inequalities involving two functions. Journal of Inequalities and Applications. 2010:148102. DOI: 10.1155/2010/148102

[27] Liu W, Ngo QA, Huy VN. Several interesting integral inequalities. Journal of Mathematical Inequalities. 2009; **10**(2):201-212

[28] Caccoulos T, Papathanasiou V. Characterizations of distribution by variance bounds. Statistics & Probability Letters. 1989;7:351-356

[29] Dahmani Z. About some integral inequalities using Riemann-Liouville integrals. General Mathematics. 2012; **20**(4):63-69

Chapter 11

Approximate Solutions of Some Boundary Value Problems by Using Operational Matrices of Bernstein Polynomials

Kamal Shah, Thabet Abdeljawad, Hammad Khalil and Rahmat Ali Khan

Abstract

In this chapter, we develop an efficient numerical scheme for the solution of boundary value problems of fractional order differential equations as well as their coupled systems by using Bernstein polynomials. On using the mentioned polynomial, we construct operational matrices for both fractional order derivatives and integrations. Also we construct a new matrix for the boundary condition. Based on the suggested method, we convert the considered problem to algebraic equation, which can be easily solved by using Matlab. In the last section, numerical examples are provided to illustrate our main results.

Keywords: Bernstein polynomials, coupled systems, fractional order differential equations, operational matrices of integration, approximate solutions
2010 MSC: 34L05, 65L05, 65T99, 34G10

1. Introduction

Generalization of classical calculus is known as fractional calculus, which is one of the fastest growing area of research, especially the theory of fractional order differential equations because this area has wide range of applications in real-life problems. Differential equations of fractional order provide an excellent tool for the description of many physical biological phenomena. The said equations play important roles for the description of hereditary characteristics of various materials and genetical problems in biological models as compared with integer order differential equations in the form of mathematical models. Nowadays, most of its applications are found in bio-medical engineering as well as in other scientific and engineering disciplines such as mechanics, chemistry, viscoelasticity, control theory, signal and image processing phenomenon, economics, optimization theory, etc.; for details, we refer the reader to study [1–9] and the references there in. Due to these important applications of fractional order differential equations, mathematicians are taking interest in the study of these equations because their models are more realistic and practical. In the last decade, many researchers have studied the existence and uniqueness of solutions to boundary value problems and their coupled systems for fractional order differential equations (see [10–17]). Hence the area devoted to existence theory has been very well explored. However, every fractional differential equation cannot be solved for its analytical solutions easily due to the complex nature of fractional derivative; so, in such a situation, approximate solutions to such a problem is most efficient and helpful. Recently, many methods such as finite difference method, Fourier series method, Adomian decomposition method (ADM), inverse Laplace technique (ILT), variational iteration methods (VIM), fractional transform method (FTM), differential transform method (DTM), homotopy analysis method (HAM), method of radial base function (MRBM), wavelet techniques (WT), spectral methods and many more (for more details, see [9, 18–38]) have been developed for obtaining numerical solutions of such differential equations. These methods have their own merits and demerits. Some of them provide a very good approximation. However, in the last few years, some operational matrices were constructed to achieve good approximation as in [39]. After this, a variety of operational matrices were developed for different wavelet methods. This method uses operational matrices, where every operation, for example differentiation and integration, involved in these equations is performed with the help of a matrix. A large variety of operational matrices are available in the literature for different orthogonal polynomials like Legendre, Laguerre, Jacobi and Bernstein polynomials [40–48]. Motivated by the above applications and uses of fractional differential equations, in this chapter, we developed a numerical scheme based on operational matrices via Bernstein polynomials. Our proof is more generalized and there is no need to convert the Bernstein polynomial function vector to another basis like block pulse function or Legendre polynomials. To the best of our knowledge, the method we consider provides a very good approximation to the solution. By the use of these operational matrices, we apply our scheme to a single fractional order differential equation with given boundary conditions as

$$\begin{cases} D^{\alpha}y(t) + AD^{\mu}y(t) + By(t) = f(t), \quad 1 < \alpha \leq 2, \quad 0 < \mu \leq 1, \\ y(0) = a, \quad y(1) = b, \end{cases} \tag{1}$$

where $f(t)$ is the source term, A, B are any real numbers; then we extend our method to solve a boundary value problem of coupled system of fractional order differential equations of the form

$$\begin{cases} D^{\alpha}x(t) + A_1D^{\mu_1}x(t) + B_1D^{\nu_1}y(t) + C_1x(t) + D_1y(t) = f(t), \quad 1 < \alpha \leq 2, \quad 0 < \mu_1, \nu_1 \leq 1, \\ D^{\beta}y(t) + A_2D^{\mu_2}x(t) + B_2D^{\nu_2}y(t) + C_2x(t) + D_2y(t) = g(t), \quad 1 < \beta \leq 2, \quad 0 < \mu_2, \nu_2 \leq 1, \\ y(0) = a, \; y(1) = b, \; y(0) = c, \; y(1) = d, \end{cases} \tag{2}$$

where $f(t), g(t)$ are source terms of the system, $A_i, B_i, C_i, D_i (i = 1, 2)$ are any real constants. Also we compare our approximations to exact values and approximations of other methods like Jacobi polynomial approximations and Haar wavelets methods to evaluate the efficiency of the proposed method. We also provide some examples for the illustration of our main results.

This chapter is designed in five sections. In the first section of the chapter, we have cited some basic works related to the numerical and analytical solutions of differential equations of arbitrary order by various methods. The necessary definitions and results related to fractional calculus and Bernstein polynomials along with the construction of some operational matrices are given in Section 2. In Section 3, we have discussed the main theory for the numerical procedure. Section 4 contains some interesting practical examples and their images. Section 5 describes the conclusion of the chapter.

2. Basic definitions and results

In this section, we recall some fundamental definitions and results from the literature, which can be found in [10–16].

Definition 2.1. *The fractional integral of order* $\gamma \in \mathbb{R}_+$ *of a function* $y \in L^1([0,1],\mathbb{R})$ *is defined as*

$$I^{\gamma}_{0+}y(t) = \frac{1}{\Gamma(\gamma)}\int_0^t (t-\tau)^{\gamma-1}y(\tau)d\tau.$$

Definition 2.2. *The Caputo fractional order derivative of a function y on the interval* $[0,1]$ *is defined by*

$$D^{\gamma}_{0+}y(t) = \frac{1}{\Gamma(n-\gamma)}\int_0^t (t-\tau)^{n-\gamma-1}y^{(n)}(\tau)d\tau,$$

where $n = [\gamma] + 1$ and $[\gamma]$ represents the integer part of γ.

Lemma 2.1. *The fractional differential equation of order* $\gamma > 0$

$$D^{\gamma}y(t) = 0, n-1 < \gamma \leq n,$$

has a unique solution of the form $y(t) = d_0 + d_1t + d_2t^2 + \dots + d_{n-1}t^{n-1}$, where $d_k \in R$ and $k = 0, 1, 2, 3, \dots, n-1$..

Lemma 2.2. *The following result holds for fractional differential equations*

$$I^{\gamma}D^{\gamma}y(t) = y(t) + d_0 + c_1t + d_2t^2 + \dots + d_{n-1}t^{n-1},$$

for arbitrary $d_k \in R$, $k = 0, 1, 2, \dots, n-1$.

Hence it follows that

$$D^{\gamma}t^k = \frac{\Gamma(k+1)}{\Gamma(k-\gamma+1)}t^{k-\gamma}, \quad I^{\gamma}t^k = \frac{\Gamma(k+1)}{\Gamma(k+\gamma+1)}t^{k+\gamma} \text{ and } D^{\gamma}[constant] = 0.$$

2.1 The Bernstein polynomials

The Bernstein polynomials $B_{i,m}(t)$ on $[0,1]$ can be defined as

$$B_{i,m}(t) = \binom{m}{i} t^i(1-t)^{m-i}, \text{for } i = 0, 1, 2 \dots m,$$

where $\binom{m}{i} = \frac{m!}{(m-i)!i!}$, which on further simplification can be written in the most simplified form as

$$B_{i,m}(t) = \sum_{k=0}^{m-i} \Theta_{(i,k,m)} t^{k+i}, \quad i = 0, 1, 2 \dots m, \tag{3}$$

where

$$\Theta_{(i,k,m)} = (-1)^k \binom{m}{i}\binom{m-i}{k}.$$

Note that the sum of the Bernstein polynomials converges to 1.

Lemma 2.3. ***Convergence Analysis:*** *Assume that the function $g \in C^{m+1}[0,1]$ that is $m+1$ times continuously differentiable function and let $X = \langle B_{0,m}, B_{1,m}, \dots, B_{m,m}\rangle$. If $C^T\Psi(x)$ is the best approximation of g out of X, then the error bound is presented as*

$$\|g - C^T\Psi\|_2 \le \frac{\sqrt{2}MS^{\frac{2m+3}{2}}}{\Gamma(m+2)\sqrt{2m+3}},$$

where $M = \max_{x\in[0,1]}|g^{(m+1)}(x)|,\ \ S = \max\{1-x_0, x_0\}$.

Proof. In view of Taylor polynomials, we have

$$F(x) = g(x_0) + (x-x_0)g^{(1)}(x_0) + \frac{(x-x_0)^2}{\Gamma 3}g^{(2)} + \dots + \frac{(x-x_0)^m}{\Gamma(m+1)}g^{(m)},$$

from which we know that

$$|g - F(x)| = |g^{(m+1)}(\eta)|\frac{(x-x_0)^{m+1}}{\Gamma(m+2)}, \quad \text{there exist } \eta \in (0,1).$$

Due the best approximation $C^T\Psi(x)$ of g, we have

$$\begin{aligned}
\|g - C^T\Psi(x)\|_2^2 &\le \|g - F\|_2^2 \\
&= \int_0^1 (g(x) - F(x))^2 dx \\
&= \int_0^1 \left[|g^{(m+1)}(\eta)|\frac{(x-x_0)^{m+1}}{\Gamma(m+2)}\right]^2 dx \\
&\le \frac{M^2}{\Gamma^2(m+2)}\int_0^1 (x-x_0)^{2m+2}dx \\
&\le \frac{2M^2S^{2m+3}}{\Gamma^2(m+2)(2m+3)}.
\end{aligned}$$

Hence we have

$$\|g - C^T\Psi(x)\|_2 \le \frac{\sqrt{2}MS^{\frac{2m+3}{2}}}{\Gamma(m+2)\sqrt{(2m+3)}}. \quad \square$$

Let $H = L^2[0,1]$ be a Hilbert space, then the inner product can be defined as

$$<f,g> = \int_0^1 f(x).g(x)dx$$

and

$$Y = span\{B_{0,m}, B_{1,m}, \dots B_{m,m}\}$$

is a finite dimensional and closed subspace. So if $f \in H$ is an arbitrary element then its best approximation is unique in Y. This terminology can be achieved by using $y_0 \in Y$ and for all $y \in Y$, we have $\|f - y_0\| \le \|f - y\|$. Thus any function can be approximated in terms of Bernstein polynomials as

$$f(t) = \sum_{i=0}^{m} c_i B_{(i,m)}, \tag{4}$$

where coefficient c_i can easily be calculated by multiplying (4) by $B_{(j,m)}(t),\ j = 0, 1, 2, \dots m$ and integrating over $[0, 1]$ by using inner product and $d_i = \int_0^1 B_{(i,m)}(t) f(t) dt,\ \theta_{(i,j)} = \int_0^1 B_{(i,m)}(t) B_{(j,m)}(t) dt,\ i,j = 0, 1, 2. \dots m,$ we have

$$\int_0^1 f(t) B_{(j,m)}(t) dt = \int_0^1 \sum_{i=0}^{m} c_i B_{(i,m)}(t).B_{(j,m)}(t) dt, \quad j = 0, 1, 2 \dots m,$$

which implies that $\int_0^1 f(t) B_{(j,m)}(t) dt = \sum_{i=0}^{m} c_i \int_0^1 B_{(i,m)}(t).B_{(j,m)}(t) dt, \quad j = 0, 1, 2 \dots m$

which implies that $[d_0\ d_1\ \dots.\ d_m] = [c_0\ c_1 . \dots c_m] \begin{bmatrix} \theta_{(0,0)} & \theta_{(0,1)} & \cdots & \theta_{(0,r)} & \cdots & \theta_{(0,m)} \\ \theta_{(1,0)} & \theta_{(1,1)} & \cdots & \theta_{(1,r)} & \cdots & \theta_{(1,m)} \\ \vdots & \vdots & \vdots & \vdots & \vdots & \vdots \\ \theta_{(r,0)} & \theta_{(r,1)} & \cdots & \theta_{(r,r)} & \cdots & \theta_{(r,m)} \\ \vdots & \vdots & \vdots & \vdots & \vdots & \vdots \\ \theta_{(m,0)} & \theta_{(m,1)} & \cdots & \theta_{(m,r)} & \cdots & \theta_{(m,m)} \end{bmatrix}.$ (5)

Let $X_M = [d_0\ d_1\ \dots.\ d_m],\ C_M = [c_0\ c_1 . \dots c_m]$, where $M = m + 1$ where M is the scale level and $\Phi_{M\times M} = \begin{bmatrix} \theta_{(0,0)} & \theta_{(0,1)} & \cdots & \theta_{(0,r)} & \cdots & \theta_{(0,m)} \\ \theta_{(1,0)} & \theta_{(1,1)} & \cdots & \theta_{(1,r)} & \cdots & \theta_{(1,m)} \\ \vdots & \vdots & \vdots & \vdots & \vdots & \vdots \\ \theta_{(r,0)} & \theta_{(r,1)} & \cdots & \theta_{(r,r)} & \cdots & \theta_{(r,m)} \\ \vdots & \vdots & \vdots & \vdots & \vdots & \vdots \\ \theta_{(m,0)} & \theta_{(m,1)} & \cdots & \theta_{(m,r)} & \cdots & \theta_{(m,m)} \end{bmatrix}$, so

$$X_M = C_M \Phi_{M\times M} \Rightarrow C_M = X_M \Phi_{M\times M}^{-1}. \tag{6}$$

where $\Phi_{m\times M}$ is called the dual matrix of the Bernstein polynomials. After calculating $c_{i,}$ (4) can be written as

$$f(t) = C_M B_M^T(t), \quad C_M \text{ is coefficient matrix}$$

where

$$B_M(t) = \left[B_{(0,m)}, B_{(1,m)}, \dots B_{(m,m)}\right]. \tag{7}$$

Lemma 2.4. *Let $B_M^T(t)$ be the function vector defined in* (3), *then the fractional order integration of $B_M^T(t)$ is given by*

$$I^{\alpha}B_M^T(t) = P_{M\times M}^{\alpha}B_M^T(t), \tag{8}$$

where $P_{M\times M}^{\alpha}$ is the fractional integration's operational matrix defined as

$$P_{M\times M}^{\alpha} = \hat{P}_{M\times M}^{\alpha}\Phi_{M\times M}^{-1}$$

and $\Phi_{M\times M}^{-1}$ *is given in* (3) *and* $P_{M\times M}^{\alpha}$ *is given by*

$$\hat{P}_{M\times M}^{\alpha} = \begin{bmatrix} \Psi_{(0,0)} & \Psi_{(0,1)} & \cdots & \Psi_{(0,r)} & \cdots & \Psi_{(0,m)} \\ \Psi_{(1,0)} & \Psi_{(1,1)} & \cdots & \Psi_{(1,r)} & \cdots & \Psi_{(1,m)} \\ \vdots & \vdots & \vdots & \vdots & \vdots & \vdots \\ \Psi_{(r,0)} & \Psi_{(r,1)} & \cdots & \Psi_{(r,r)} & \cdots & \Psi_{(r,m)} \\ \vdots & \vdots & \vdots & \vdots & \vdots & \vdots \\ \Psi_{(m,0)} & \Psi_{(m,1)} & \cdots & \Psi_{(m,r)} & \cdots & \Psi_{(m,m)} \end{bmatrix}, \tag{9}$$

where

$$\Psi_{i,j} = \sum_{k=0}^{m-i}\sum_{l=0}^{m-j}\theta_{(i,k,m)}\theta_{(j,l,m)}\frac{\Gamma(k+i+1)}{(i+j+k+l+\alpha+1)\Gamma(k+i+\alpha+1)}. \tag{10}$$

Proof. Consider

$$B_{i,m}(t) = \sum_{k=0}^{m-i}\theta_{(i,k,m)}t^{k+i} \tag{11}$$

taking fractional integration of both sides, we have

$$I^{\alpha}B_{i,m}(t) = \sum_{k=0}^{m-i}\theta_{(i,k,m)}I^{\alpha}t^{k+i} = \sum_{k=0}^{m-i}\theta_{(i,k,m)}\frac{\Gamma(k+i+\alpha)}{\Gamma(k+i+\alpha+1)}t^{k+i+\alpha}. \tag{12}$$

Now to approximate right-hand sides of above

$$\sum_{k=0}^{m-i}\theta_{(i,k,m)}\frac{\Gamma(k+i+\alpha)}{\Gamma(k+i+\alpha+1)}t^{k+i+\alpha} = C_M^{(i)}B_M^T(t) \tag{13}$$

where $C_M^{(i)}$ can be approximated by using (3) as

$$C_M^{(i)} = X_M^{(i)}\Phi_{M\times M}^{-1}, \tag{14}$$

where entries of the vector $X_M^{(i)}$ can be calculated in generalized form as

$$X_M^{(j)} = \int_0^1\sum_{k=0}^{m-i}\theta_{(i,k,m)}\frac{\Gamma(k+i+\alpha)}{\Gamma(k+i+\alpha+1)}t^{k+i+\alpha}B_{j,m}(t)dt, \; j = 0,1,2.\dots m$$

$$\Rightarrow X_M^{(j)} = \int_0^1\sum_{k=0}^{m-i}\theta_{(i,k,m)}\frac{\Gamma(k+i+\alpha)}{\Gamma(k+i+\alpha+1)}\sum_{l=0}^{m-j}\theta_{(j,l,m)}t^{k+i+\alpha}.t^{l+j}dt, \; j = 0,1,2.\dots m$$

$$= \sum_{k=0}^{m-i}\theta_{(i,k,m)}\sum_{l=0}^{m-j}\theta_{(j,l,m)}\frac{\Gamma(k+i+\alpha)}{\Gamma(k+i+\alpha+1)}\frac{1}{(k+l+j+i+\alpha+1)}, \; j = 0,1,2,\dots.m \tag{15}$$

evaluating this result for i = 0,1,2....m, we have

$$\begin{bmatrix} I^{\alpha}B_{0,m}(t) \\ I^{\alpha}B_{1,m}(t) \\ \vdots \\ \vdots \\ \vdots \\ I^{\alpha}B_{m,m}(t) \end{bmatrix} = \begin{bmatrix} X_M^{(0)}\Phi_{M\times M}^{-1}B_M^T(t) \\ X_M^{(1)}\Phi_{M\times M}^{-1}B_M^T(t) \\ \vdots \\ \vdots \\ \vdots \\ X_M^{(m)}\Phi_{M\times M}^{-1}B_M^T(t)) \end{bmatrix} \tag{16}$$

further writing

$$\Psi_{i,j} = \sum_{k=0}^{m-i}\sum_{l=0}^{m-j}\theta_{(i,k,m)}.\theta_{(j,l,m)}\frac{\Gamma(k+i+\alpha)}{\Gamma(k+i+\alpha+1)}\frac{1}{(k+l+j+i+\alpha+1)}$$

we get

$$I^{\alpha}B_M^T(t) = \hat{P}^{\alpha}_{M\times M}.\Phi^{-1}_{M\times M}.B_M^T(t). \tag{17}$$

Let us represent

$$\hat{P}^{\alpha}_{M\times M}.\Phi^{-1}_{M\times M} = P^{\alpha}_{M\times M}$$

thus

$$I^{\alpha}B_M^T(t) = P^{\alpha}_{M\times M}.B_M^T(t). \tag{18}$$

□

Lemma 2.5. *Let $B_M^T(t)$ be the function vector as defined in* (3), *then fractional order derivative is defined as*

$$D^{\alpha}B_M^T(t) = G^{\alpha}_{M\times M}.B_M^T(t) \tag{19}$$

where $G^{\alpha}_{M\times M}$ is the operational matrix of fractional order derivative given by

$$G^{\alpha}_{M\times M} = \hat{G}^{\alpha}_{M\times M}\Phi^{-1}_{M\times M}, \tag{20}$$

where $\Phi_{M\times M}$ is the dual matrix given in (3) *and*

$$\hat{G}^{\alpha}_{M\times M} = \begin{bmatrix} \Psi_{(0,0)} & \Psi_{(0,1)} & \cdots & \Psi_{(0,r)} & \cdots & \Psi_{(0,m)} \\ \Psi_{(1,0)} & \Psi_{(1,1)} & \cdots & \Psi_{(1,r)} & \cdots & \Psi_{(1,m)} \\ \vdots & \vdots & \vdots & \vdots & \vdots & \vdots \\ \Psi_{(r,0)} & \Psi_{(r,1)} & \cdots & \Psi_{(r,r)} & \cdots & \Psi_{(r,m)} \\ \vdots & \vdots & \vdots & \vdots & \vdots & \vdots \\ \Psi_{(m,0)} & \Psi_{(m,1)} & \cdots & \Psi_{(m,r)} & \cdots & \Psi_{(m,m)} \end{bmatrix}, \tag{21}$$

where $\Psi_{(i,j)}$ is defined for two different cases as
Case I: $(i < [\alpha])$

$$\Psi_{i,j} = \sum_{k=[\alpha]}^{m-i} \sum_{l=0}^{m-j} \theta_{(i,k,m)} . \theta_{(j,l,m)} \frac{\Gamma(k+i-\alpha)}{\Gamma(k+i-\alpha+1)} \frac{1}{(k+l+j+i-\alpha+1)} \tag{22}$$

Case II: $(i \geq [\alpha])$

$$\Psi_{i,j} = \sum_{k=0}^{m-i} \sum_{l=0}^{m-j} \theta_{(i,k,m)} . \theta_{(j,l,m)} \frac{\Gamma(k+i-\alpha)}{\Gamma(k+i-\alpha+1)} \frac{1}{(k+l+j+i-\alpha+1)}. \tag{23}$$

Proof. Consider the general element as

$$D^\alpha B_{i,m}(t) = D^\alpha \left(\sum_{k=0}^{m-i} \theta_{(i,k,m)} . t^{k+i} \right) = \sum_{k=0}^{m-i} \theta_{(i,k,m)} D^\alpha t^{k+i}. \tag{24}$$

It is to be noted in the polynomial function $B_{i,m}$ the power of the variable 't' is an ascending order and the lowest power is 'i' therefore the first $[\alpha - 1]$ terms becomes zero when we take derivative of order α.

Case I: $(i < [\alpha])$By the use of definition of fractional derivative

$$D^\alpha B_{i,m}(t) = \sum_{k=[\alpha]}^{m-i} \theta_{(i,k,m)} \frac{\Gamma(k+i+1)}{\Gamma(k+i-\alpha+1)} t^{k+i-\alpha}. \tag{25}$$

Now approximating RHS of (25) as

$$\sum_{k=[\alpha]}^{m-i} \theta_{(i,k,m)} \frac{\Gamma(k+i+1)}{\Gamma(k+i-\alpha+1)} t^{k+i-\alpha} = C_M^{(i)} B_M^T(t) \tag{26}$$

further implies that

$$X_M^{(j)} = \int_0^1 \sum_{k=[\alpha]}^{m-i} \theta_{(i,k,m)} \frac{\Gamma(k+i+1)}{\Gamma(k+i-\alpha+1)} t^{k+i-\alpha} B_{j,m}(t) dt, \; j = 0, 1, 2. \dots m$$

$$\Rightarrow X_M^{(j)} = \sum_{k=[\alpha]}^{m-i} \theta_{(i,k,m)} \sum_{l=0}^{m-j} \theta_{(j,l,m)} \frac{\Gamma(k+i+1)}{\Gamma(k+i-\alpha+1)(k+i+l+j-\alpha+1)}, \; j = 0, 1, 2. \dots m \tag{27}$$

Case II: $(i \geq [\alpha])$ if $i \leq [\alpha]$ then

$$X_M^{(j)} = \sum_{k=0}^{m-i} \theta_{(i,k,m)} \sum_{l=0}^{m-j} \theta_{(j,l,m)} \frac{\Gamma(k+i+1)}{\Gamma(k+i-\alpha+1)(k+i+l+j-\alpha+1)}, \; j = 0, 1, 2. \dots m. \tag{28}$$

After careful simplification, we get

$$\begin{bmatrix} D^{\alpha}B_{0,m}(t) \\ D^{\alpha}B_{1,m}(t) \\ \vdots \\ \vdots \\ \vdots \\ D^{\alpha}B_{m,m}(t) \end{bmatrix} = \begin{bmatrix} X_M^{(0)}\Phi_{M\times M}^{-1}B_M^T(t) \\ X_M^{(1)}\Phi_{M\times M}^{-1}B_M^T(t) \\ \vdots \\ \vdots \\ \vdots \\ X_M^{(m)}\Phi_{M\times M}^{-1}B_M^T(t)) \end{bmatrix}. \tag{29}$$

On further simplification, we have

$$\Psi_{i,j} = \sum_{k=[\alpha]}^{m-i}\sum_{l=0}^{m-j}\theta_{(i,k,m)}.\theta_{(j,l,m)}\frac{\Gamma(k+i+1)}{\Gamma(k+i-\alpha+1)}\frac{1}{(k+l+j+i-\alpha+1)} \quad (i<[\alpha])$$

$$\Psi_{i,j} = \sum_{k=0}^{m-i}\sum_{l=0}^{m-j}\theta_{(i,k,m)}.\theta_{(j,l,m)}\frac{\Gamma(k+i+1)}{\Gamma(k+i-\alpha+1)}\frac{1}{(k+l+j+i-\alpha+1)}$$

we get $D^{\alpha}B_M^T(t) = \hat{G}_{M\times M}^{\alpha}\Phi_{M\times M}^{-1}.B_M^T(t)$.

$$\tag{30}$$

Let

$$\hat{G}_{M\times M}^{\alpha}\Phi_{M\times M}^{-1} = G_{M\times M}^{\alpha}$$

so

$$D^{\alpha}B_M^T(t) = G_{M\times M}^{\alpha}B_M^T(t)$$

which is the desired result. □

Lemma 2.6. *An operational matrix corresponding to the boundary condition by taking $B_M^T(t)$ is function vector and K is coefficient vector by taking the approximation*

$$u(t) = K\hat{B}(t)$$

is given by

$$Q_{M\times M}^{\alpha,\phi} = \begin{bmatrix} \Omega_{(0,0)} & \Omega_{(0,1)} & \cdots & \Omega_{(0,r)} & \cdots & \Omega_{(0,m)} \\ \Omega_{(1,0)} & \Omega_{(1,1)} & \cdots & \Omega_{(1,r)} & \cdots & \Omega_{(1,m)} \\ \vdots & \vdots & \vdots & \vdots & \vdots & \vdots \\ \Omega_{(r,0)} & \Omega_{(r,1)} & \cdots & \Omega_{(r,r)} & \cdots & \Omega_{(r,m)} \\ \vdots & \vdots & \vdots & \vdots & \vdots & \vdots \\ \Omega_{(m,0)} & \Omega_{(m,1)} & \cdots & \Omega_{(m,r)} & \cdots & \Omega_{(m,m)} \end{bmatrix}, \tag{31}$$

where

$$\Omega_{i,j} = \int_0^1 \Delta_{i,m}\phi(t)B_j(t)dt, \quad i,j = 0,1,2.\dots m.$$

Proof. Let us take $u(t) = K\hat{B}(t)$, then

$$
{}_0I_1^{\alpha}K\hat{B}(t) = K{}_0I_1^{\alpha}\hat{B}(t) = K\begin{bmatrix} {}_0I_1^{\alpha}B_0(t) \\ {}_0I_1^{\alpha}B_1(t) \\ \vdots \\ \vdots \\ \vdots \\ {}_0I_1^{\alpha}B_m(t) \end{bmatrix}.
$$

Let us evaluate the general terms

$$
\begin{aligned}
{}_0I_1^{\alpha}B_i(t)dt &= \frac{1}{\Gamma\alpha}\int_0^1 (1-s)^{\alpha-1}B_{i,m}(s)ds \\
&= \frac{1}{\Gamma\alpha}\sum_{k=0}^{m-i}\Theta_{i,k,m}\int_0^1 (1-s)^{\alpha-1}s^{k+i}ds.
\end{aligned} \tag{32}
$$

Since by

$$
L\left(\int_0^1 (1-s)^{\alpha-1}s^{k+i}ds\right) = \frac{\Gamma\alpha\Gamma(k+i+1)}{\tau^{k+\alpha+i}}
$$

taking inverse Laplace of both sides, we get

$$
\int_0^1 (1-s)^{\alpha-1}.s^{k+i}ds = L^{-1}\left[\frac{\Gamma\alpha.\Gamma(k+i+1)}{\tau^{k+\alpha+i}}\right] = \frac{\Gamma\alpha.\Gamma(k+i+1)}{\Gamma(k+i+\alpha+1)}
$$

now Eq. (32) implies that

$$
{}_0I_1^{\alpha}B_i(t)dt = \sum_{k=0}^{m-i}\Theta_{i,k,m}\frac{\Gamma(k+i+1)}{\Gamma(k+i+\alpha+1)} = \Delta_{i,m}. \tag{33}
$$

Now using the approximation $\Delta_{i,m}\phi(t) = \sum_{i=0}^{m}\hat{c}_iB_i(t) = C_M^iB_M^T$, and using Eq. (3) we get $C_M^i = K_M^i\Phi_{M\times M}^{-1}B_M^T$ and using $c_j = \int_0^1\phi(t)B_j(t)dt$,

$$
\begin{aligned}
\phi(t)KI^{\alpha}\hat{B}(t) &= K\begin{bmatrix} \Delta_{0,m}\phi(t) \\ \Delta_{1,m}\phi(t) \\ \vdots \\ \vdots \\ \vdots \\ \Delta_{m,m}\phi(t) \end{bmatrix} = K\begin{bmatrix} C_M^0\Phi_{M\times M}^{-1}B_M^T(t) \\ C_M^1\Phi_{M\times M}^{-1}B_M^T(t) \\ \vdots \\ \vdots \\ \vdots \\ C_M^m\Phi_{M\times M}^{-1}B_M^T(t) \end{bmatrix} \\
&= K\begin{bmatrix} c_0^0 & c_1^0 & \cdots & c_r^0 & \cdots & c_m^0 \\ c_0^1 & c_1^1 & \cdots & c_r^1 & \cdots & c_m^1 \\ \vdots & \vdots & \vdots & \vdots & \vdots & \vdots \\ c_0^r & c_1^r & \cdots & c_r^r & \cdots & c_m^r \\ \vdots & \vdots & \vdots & \vdots & \vdots & \vdots \\ c_0^m & c_1^m & \cdots & c_r^m & \cdots & c_m^m \end{bmatrix}\begin{bmatrix} \Phi_{M\times M}^{-1}B_M^T(t) \\ \Phi_{M\times M}^{-1}B_M^T(t) \\ \vdots \\ \vdots \\ \vdots \\ \Phi_{M\times M}^{-1}B_M^T(t) \end{bmatrix}.
\end{aligned} \tag{34}
$$

On further simplification, we get

$$\phi(t)KI^{\alpha}\hat{B}(t) = K\begin{bmatrix} \Omega_{(0,0)} & \Omega_{(0,1)} & \cdots & \Omega_{(0,r)} & \cdots & \Omega_{(0,m)} \\ \Omega_{(1,0)} & \Omega_{(1,1)} & \cdots & \Omega_{(1,r)} & \cdots & \Omega_{(1,m)} \\ \vdots & \vdots & \vdots & \vdots & \vdots & \vdots \\ \Omega_{(r,0)} & \Omega_{(r,1)} & \cdots & \Omega_{(r,r)} & \cdots & \Omega_{(r,m)} \\ \vdots & \vdots & \vdots & \vdots & \vdots & \vdots \\ \Omega_{(m,0)} & \Omega_{(m,1)} & \cdots & \Omega_{(m,r)} & \cdots & \Omega_{(m,m)} \end{bmatrix}\begin{bmatrix} B_0(t) \\ B_1(t) \\ \vdots \\ \vdots \\ \vdots \\ B_m(t) \end{bmatrix}. \tag{35}$$

So

$$\phi(t)_0I_1^{\alpha}u(t) = KQ^{\alpha,\phi}_{M\times M}B^T_M(t),$$

and

$$\Omega_{i,j} = \int_0^1 \Delta_{i,m}\phi(t)B_j(t)dt, \quad i,j = 0,1,2.\dots m. \tag{36}$$

which is the required result. □

3. Applications of operational matrices

In this section, we are going to approximate a boundary value problem of fractional order differential equation as well as a coupled system of fractional order boundary value problem. The application of obtained operational matrices is shown in the following procedure.

3.1 Fractional differential equations

Consider the following problem in generalized form of fractional order differential equation

$$\begin{aligned} &D^{\alpha}y(t) + AD^{\mu}y(t) + By(t) = f(t), \;\; 1<\alpha\leq 2, \;\; 0<\mu\leq 1, \\ &\text{subject to the boundary conditions } y(0) = a, \quad y(1) = b, \end{aligned} \tag{37}$$

where $f(t)$ is a source term; A, B are any real constants and $y(t)$ is an unknown solution which we want to determine. To obtain a numerical solution of the above problem in terms of Bernstein polynomials, we proceed as

$$\text{Let } D^{\alpha}y(t) = K_M B^T_M(t). \tag{38}$$

Applying fractional integral of order α we have

$$y(t) = K_M P^{\alpha}_{M\times M}B^T_M(t) - c_0 + c_1 t$$

using boundary conditions, we have

$$c_0 = a, \quad c_1 = b - a - K_M P^{\alpha}_{M\times M}B^T_M(t)\big|_{t=1}.$$

Using the approximation and Lemma 2.2

$$a + t(b-a) = F_M^{(1)} B_M^T(t), \quad tP_{M\times M}^{\alpha} B_M^T(t)\Big|_{t=1} = Q_{M\times M}^{\alpha,\phi} B_M^T(t).$$

Hence

$$y(t) = K_M P_{M\times M}^{\alpha} B_M^T(t) + a + t(b-a) - tK_M P_{M\times M}^{\alpha} B_M^T(t)\big|_{t=1},$$

$$\text{which gives } y(t) = K_M P_{M\times M}^{\alpha} B_M^T(t) + F_M^{(1)} B_M^T(t) - Q_{M\times M}^{\alpha,\phi} B_M^T(t) \qquad (39)$$

$$= K_M\left(P_{M\times M}^{\alpha} - Q_{M\times M}^{\alpha,\phi}\right) B_M^T(t) + F_M^{(1)} B_M^T(t).$$

Now

$$D^{\mu} y(t) = D^{\mu}\left[K_M\left(P_{M\times M}^{\alpha} - Q_{M\times M}^{\alpha,\phi}\right) B_M^T(t) + F_M^{(1)} B_M^T(t)\right]$$

$$= K_M\left(P_{M\times M}^{\alpha} - Q_{M\times M}^{\alpha,\phi}\right) G_{M\times M}^{\mu} B_M^T(t) + F_M^{(1)} G_{M\times M}^{\mu} B_M^T(t) \qquad (40)$$

and

$$f(t) = F_M^{(2)} B_M^T(t). \qquad (41)$$

Putting Eqs. (38)–(41) in Eq. (37), we get

$$K_M B_M^T(t) + AK_M\left(P_{M\times M}^{\alpha} - Q_{M\times M}^{\alpha,\phi}\right) G_{M\times M}^{\mu} B_M^T(t) + AF_M^{(1)} G_{M\times M}^{\mu} B_M^T(t)$$
$$+ BK_M\left(P_{M\times M}^{\alpha} - Q_{M\times M}^{\alpha,\phi}\right) B_M^T(t) + BF_M^{(1)} B_M^T(t) = F_M^{(2)} B_M^T(t). \qquad (42)$$

In simple form, we can write (42) as

$$K_M B_M^T(t) + AK_M\left(P_{M\times M}^{\alpha} - Q_{M\times M}^{\alpha,\phi}\right) G_{M\times M}^{\mu} B_M^T(t) + AF_M^{(1)} G_{M\times M}^{\mu} B_M^T(t)$$
$$+ BK_M\left(P_{M\times M}^{\alpha} - Q_{M\times M}^{\alpha,\phi}\right) B_M^T(t) + BF_M^{(1)} B_M^T(t) - F_M^{(2)} B_M^T(t) = 0 \qquad (43)$$
$$K_M + K_M\left(A\hat{P}_{M\times M}^{\alpha} G_{M\times M}^{\mu} + B\hat{P}_{M\times M}^{\alpha}\right) + AF_M^{(1)} G_{M\times M}^{\mu} + BF_M^{(1)} - F_M^{(2)},$$

where

$$\hat{P}_{M\times M}^{\alpha} = P_{M\times M}^{\alpha} - Q_{M\times M}^{\alpha,\phi}.$$

Eq. (43) is an algebraic equation of Lyapunov type, which can be easily solved for the unknown coefficient vector K_M. When we find K_M, then putting this in Eq. (39), we get the required approximate solution of the problem.

3.2 Coupled system of boundary value problem of fractional order differential equations

Consider a coupled system of a fractional order boundary value problem

$$D^{\alpha} x(t) + A_1 D^{\mu_1} x(t) + B_1 D^{\nu_1} y(t) + C_1 x(t) + D_1 y(t) = f(t), \quad 1 < \alpha \le 2, 0 < \mu_1, \nu_1 \le 1,$$
$$D^{\beta} y(t) + A_2 D^{\mu_2} x(t) + B_2 D^{\nu_2} y(t) + C_2 x(t) + D_2 y(t) = g(t), \quad 1 < \beta \le 2, 0 < \mu_2, \nu_2 \le 1, \qquad (44)$$

subject to the boundary conditions

$$x(0) = a, \quad x(1) = b \quad y(0) = c, \quad y(1) = d, \tag{45}$$

where $A_i, B_i, C_i, D_i (i = 1, 2)$ are any real constants, $f(t), g(t)$ are given source terms. We approximate the solution of the above system in terms of Bernstein polynomials such as

$$D^{\alpha}x(t) = K_M B_M^T(t), \quad D^{\beta}y(t) = L_M B_M^T(t)$$
$$x(t) = K_M P^{\alpha}_{M\times M} B_M^T(t) + c_0 + c_1 t, \quad y(t) = L_M \left(P^{\beta}_{M\times M} B_M^T(t) + d_0 + d_1 t \right.$$

applying boundary conditions, we have

$$x(t) = K_M (P^{\alpha}_{M\times M} B_M^T(t) + a + t(b-a) - tK_M P^{\alpha}_{M\times M} B_M^T(t)\big|_{t=1},$$
$$y(t) = K_M (P^{\beta}_{M\times M} B_M^T(t) + c + t(d-c) - tK_M P^{\beta}_{M\times M} B_M^T(t)\Big|_{t=1}.$$

let us approximate

$$a + t(b-a) = F^1_M B_M^T(t), c + t(d-c) = F^2_M B_M^T(t)$$
$$tP^{\alpha}_{M\times M} B_M^T(t)\Big|_{t=1} = Q^{\alpha,\phi}_{M\times M} B_M^T(t), tP^{\beta}_{M\times M} B_M^T(t)\Big|_{t=1} = Q^{\beta,\phi}_{M\times M} B_M^T(t)$$

then

$$x(t) = K_M P^{\alpha}_{M\times M} B_M^T(t) + F^{(1)}_M B_M^T(t) - K_M Q^{\alpha,\phi}_{M\times M} B_M^T(t)$$

$$y(t) = L_M P^{\beta}_{M\times M} B_M^T(t) + F^{(2)}_M B_M^T(t) - L_M Q^{\beta,\phi}_{M\times M} B_M^T(t)$$

$$D^{\mu_1}x(t) = \left[K_M P^{\alpha}_{M\times M} B_M^T(t) + F^{(1)}_M B_M^T(t) - K_M Q^{\alpha,\phi}_{M\times M} B_M^T(t) \right]$$

$$= K_M \left(P^{\alpha}_{M\times M} - Q^{\alpha,\phi}_{M\times M} \right) G^{\mu_1}_{M\times M} + F^{(1)}_M G^{\mu_1}_{M\times M} B_M^T(t)$$

$$D^{\nu_1}y(t) = D^{\nu_1} \left[L_M P^{\beta}_{M\times M} B_M^T(t) + F^{(2)}_M B_M^T(t) - L_M Q^{\beta,\phi}_{M\times M} B_M^T(t) \right]$$

$$= L_M \left(P^{\beta}_{M\times M} - Q^{\beta,\phi}_{M\times M} \right) G^{\nu_1}_{M\times M} + F^{(2)}_M G^{\nu_1}_{M\times M} B_M^T(t)$$

$$D^{\mu_2}x(t) = D^{\mu_2} \left[K_M P^{\alpha}_{M\times M} B_M^T(t) + F^{(1)}_M B_M^T(t) - K_M Q^{\alpha,\phi}_{M\times M} B_M^T(t) \right]$$

$$= K_M \left(P^{\alpha}_{M\times M} - Q^{\alpha,\phi}_{M\times M} \right) G^{\mu_2}_{M\times M} + F^{(1)}_M G^{\mu_2}_{M\times M} B_M^T(t)$$

and

$$D^{\nu_2}y(t) = D^{\nu_2} \left[K_M P^{\beta}_{M\times M} B_M^T(t) + F^{(2)}_M B_M^T(t) - K_M Q^{\beta,\phi}_{M\times M} B_M^T(t) \right]$$

$$= L_M \left(P^{\beta}_{M\times M} - Q^{\beta,\phi}_{M\times M} \right) G^{\nu_2}_{M\times M} + F^{(2)}_M G^{\nu_2}_{M\times M} B_M^T(t)$$

$$f(t) = F^{(3)} B_M^T(t), \quad g(t) = F^{(4)} B_M^T(t).$$

Thus system (44) implies that

$$
\begin{aligned}
&K_M B_M^T(t) + A_1 K_M \left(P^{\alpha}_{M\times M} - Q^{\alpha,\phi}_{M\times M}\right) G^{\mu_1}_{M\times M} + A_1 F_M^{(1)} G^{\mu_1}_{M\times M} B_M^T(t) \\
&\qquad + B_1 L_M \left(P^{\beta}_{M\times M} - Q^{\beta,\phi}_{M\times M}\right) G^{\nu_1}_{M\times M} + B_1 F_M^{(2)} G^{\nu_1}_{M\times M} B_M^T(t) + C_1 K_M P^{\alpha}_{M\times M} B_M^T(t) \\
&\qquad + C_1 F_M^{(1)} B_M^T(t) - C_1 K_M Q^{\alpha,\phi}_{M\times M} B_M^T(t) + D_1 L_M P^{\beta}_{M\times M} B_M^T(t) + D_1 F_M^{(2)} B_M^T(t) \\
&\qquad - D_1 L_M Q^{\beta,\phi}_{M\times M} B_M^T(t) = F^{(3)} B_M^T(t) \\
&L_M B_M^T(t) + A_2 K_M \left(P^{\alpha}_{M\times M} - Q^{\alpha,\phi}_{M\times M}\right) G^{\mu_2}_{M\times M} + A_2 F_M^{(1)} G^{\mu_2}_{M\times M} B_M^T(t) \\
&\qquad + B_2 L_M \left(P^{\beta}_{M\times M} - Q^{\beta,\phi}_{M\times M}\right) G^{\nu_2}_{M\times M} + B_2 F_M^{(2)} G^{\nu_2}_{M\times M} B_M^T(t) + C_2 K_M P^{\alpha}_{M\times M} B_M^T(t) \\
&\qquad + C_2 F_M^{(1)} B_M^T(t) - C_2 K_M Q^{\alpha,\phi}_{M\times M} B_M^T(t) + D_2 L_M P^{\beta}_{M\times M} B_M^T(t) + D_2 F_M^{(2)} B_M^T(t) \\
&\qquad - D_2 L_M Q^{\beta,\phi}_{M\times M} B_M^T(t) = F^{(4)} B_M^T(t).
\end{aligned}
\tag{46}
$$

Rearranging the terms in the above system and using the following notation for simplicity in Eq. (46)

$$
\begin{aligned}
\hat{Q}^{\alpha}_{M\times M} &= A_1\left(P^{\alpha}_{M\times M} - Q^{\alpha,\phi}_{M\times M}\right) G^{\mu_1}_{M\times M} + C_1\left(P^{\alpha}_{M\times M} - Q^{\alpha,\phi}_{M\times M}\right) \\
\hat{Q}^{\beta}_{M\times M} &= B_1\left(P^{\beta}_{M\times M} - Q^{\beta,\phi}_{M\times M}\right) G^{\nu_1}_{M\times M} + D_1\left(P^{\beta}_{M\times M} - Q^{\beta,\phi}_{M\times M}\right) \\
\hat{R}^{\alpha}_{M\times M} &= A_2\left(P^{\alpha}_{M\times M} - Q^{\alpha,\phi}_{M\times M}\right) G^{\mu_2}_{M\times M} + C_2\left(P^{\alpha}_{M\times M} - Q^{\alpha,\phi}_{M\times M}\right) \\
\hat{R}^{\beta}_{M\times M} &= B_2\left(P^{\beta}_{M\times M} - Q^{\beta,\phi}_{M\times M}\right) G^{\nu_2}_{M\times M} + D_2\left(P^{\beta}_{M\times M} - Q^{\beta,\phi}_{M\times M}\right) \\
F_M &= A_1 F_M^{(1)} G^{\mu_1}_{M\times M} + B_1 F_M^{(2)} G^{\nu_1}_{M\times M} + C_1 F_M^{(1)} + F_M^{(2)} - D_1 F_M^{(3)} \\
G_M &= A_2 F_M^{(1)} G^{\mu_2}_{M\times M} + B_2 F_M^{(2)} G^{\nu_2}_{M\times M} + C_2 F_M^{(1)} + D_2 F_M^{(2)} - F_M^{(4)},
\end{aligned}
$$

the above system (46) becomes

$$
\begin{aligned}
&K_M B_M^T(t) + K_M \hat{Q}^{\alpha}_{M\times M} B_M^T(t) + L_M \hat{Q}^{\beta}_{M\times M} B_M^T(t) + F_M B_M^T(t) = 0 \\
&L_M B_M^T(t) + K_M \hat{R}^{\alpha}_{M\times M} B_M^T(t) + L_M \hat{R}^{\beta}_{M\times M} B_M^T(t) + G_M B_M^T(t) = 0 \\
&[K_M\ L_M]\begin{bmatrix} B_M^T(t) & 0 \\ 0 & B_M^T(t)\end{bmatrix} + [K_M \quad L_M]\begin{bmatrix} \hat{Q}^{\alpha}_{M\times M} & 0 \\ 0 & \hat{R}^{\beta}_{M\times M}\end{bmatrix}\begin{bmatrix} B_M^T(t) & 0 \\ 0 & B_M^T(t)\end{bmatrix} \\
&+[K_M \quad L_M]\begin{bmatrix} 0 & \hat{R}^{\alpha}_{M\times M} \\ \hat{Q}^{\beta}_{M\times M} & 0\end{bmatrix}\begin{bmatrix} B_M^T(t) & 0 \\ 0 & B_M^T(t)\end{bmatrix} + [F_M \quad G_M]\begin{bmatrix} B_M^T(t) & 0 \\ 0 & B_M^T(t)\end{bmatrix} = 0 \\
&\qquad [K_M \quad L_M] + [K_M \quad L_M]\begin{bmatrix} \hat{Q}^{\alpha}_{M\times M} & \hat{R}^{\alpha}_{M\times M} \\ \hat{Q}^{\beta}_{M\times M} & \hat{R}^{\beta}_{M\times M}\end{bmatrix} + [F_M \quad G_M] = 0,
\end{aligned}
\tag{47}
$$

which is an algebraic equation that can be easily solved by using Matlab functional solver or Mathematica for unknown matrix $[K_M \quad L_M]$. Calculating the coefficient matrix K_M, L_M and putting it in equations

$$x(t) = K_M P^{\alpha}_{M\times M} B^T_M(t) + F^{(1)}_M B^T_M(t) - K_M Q^{\alpha,\phi}_{M\times M} B^T_M(t)$$

$$y(t) = L_M P^{\beta}_{M\times M} B^T_M(t) + F^{(2)}_M B^T_M(t) - L_M Q^{\beta,\phi}_{M\times M} B^T_M(t),$$

we get the required approximate solution.

4. Applications of our method to some examples

Example 4.1. *Consider*

$$D^{\alpha}y(t) + c_1 D^{\nu}y(t) + c_2 y(t) = f(t), \quad 1<\alpha<2 \tag{48}$$

subject to the boundary conditions

$$y(0) = 0, \quad y(1) = 0.$$

Solution: *We solve this problem under the following parameters sets defined as* $S_1 = \{\alpha = 2, \nu = 1, c_1 = 1, c_2 = 1\}$, $S_2 = \{\alpha = 1.8, \nu = 0.8, c_1 = 10, c_2 = 100\}$, $S_3 = \{\alpha = 1.5, \nu = 0.5, c_1 = 1/10, c_2 = 1/100\}$, *and select source term for* S_1 *as*

$$f_1(t) = t^6(t-1)^3 + t^6((72t - 168) + 126) - 30t^4 + 3t^5(3t-2)(t-1)^2 \tag{49}$$

$$\begin{aligned} f_2(t) = {} & \frac{111476825837237703125 t^{\frac{21}{5}}(1750t^3 - 4200t^2 + 3255t - 806)}{40654894556198941 4912} \\ & + \frac{278692064593092578125 t^{\frac{26}{5}}(5250t^3 - 14350t^2 + 12915t - 3813)}{250027601520623490 17088} \\ & + 100t^6(t-1)^3, \end{aligned} \tag{50}$$

$$\begin{aligned} f_3(t) = {} & \frac{5081767996463981 t^{\frac{9}{2}}(1344t^3 - 3360t^2 + 2730t - 715)}{264146673456906240} \\ & + \frac{5081767996463981 t^{\frac{11}{2}}(1344t^3 - 3808t^2 + 3570t - 1105)}{2245246724383703 0400} + \frac{t^6(t-1)^3}{100}. \end{aligned} \tag{51}$$

The exact solution of the above problem is

$$y(t) = t^6(t-1)^3.$$

We solve this problem with the proposed method under different sets of parameters as defined in S_1, S_2, S_3. *The observation and simulation demonstrate that the solution obtained with the proposed method is highly accurate. The comparison of exact solution with approximate solution obtained using the parameters set* S_1 *is displayed in* ***Figure 1*** *subplot (a), while in* ***Figure 1*** *subplot (b) we plot the absolute difference between the exact and approximate solutions using different scale levels. One can easily observe that the absolute error is much less than* 10^{-12}. *The order of derivatives in this set is an integer.*

By solving the problem under parameters set S_2 *and* S_3, *we observe the same phenomena. The approximate solution matches very well with the exact solution. See* ***Figures 2*** *and* ***3*** *respectively.*

Example 4.2. *Consider*

$$D^{\alpha}y(t) - 2D^{0.9}y(t) - 3y(t) = -4\cos(2t) - 7\sin(2t) \tag{52}$$

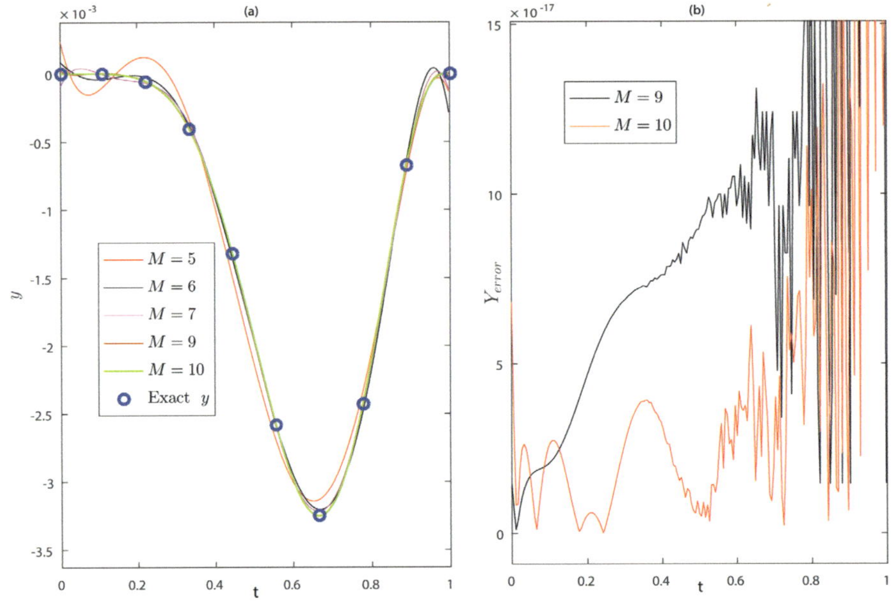

Figure 1.
(a) Comparison of exact and approximate solution of Example 4.1, under parameters set S_1*. (b) Absolute error in the approximate solution of Example 4.1, under parameters set* S_1*.*

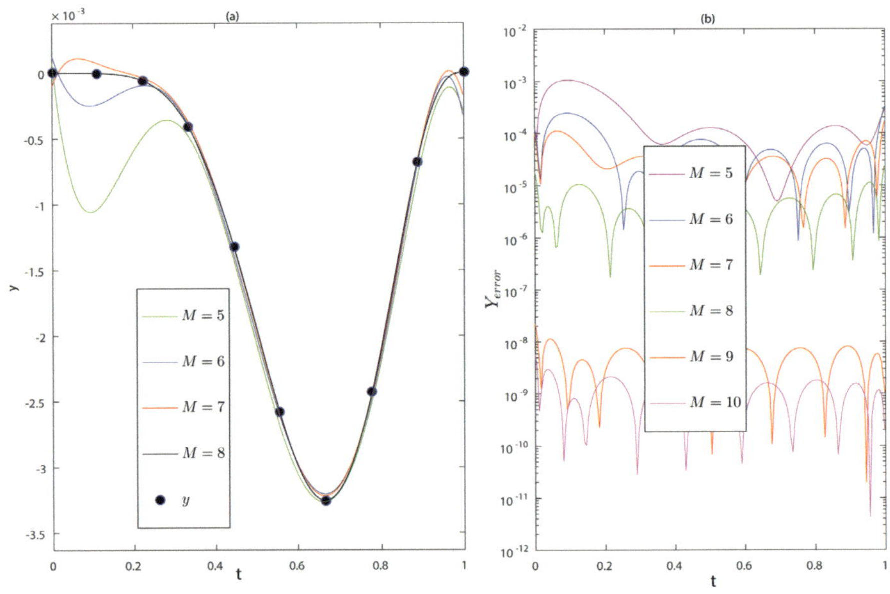

Figure 2.
(a) Comparison of exact and approximate solution of Example 4.1, under parameters set S_2*. (b) Absolute error in the approximate solution of Example 4.1, under parameters set* S_2*.*

subject to the boundary conditions

$$y(0) = 0, \qquad y(1) = \sin(2).$$

Solution: *The exact solution of the above problem is* $y(t) = \sin(2t)$*, when* $\alpha = 2$*. However the exact solution at fractional order is not known. We use the well-known property of FDEs that when* $\alpha \to 2$*, the approximate solution approaches the exact solution for the evaluations of approximate solutions and check the accuracy by using different scale levels. By increasing the scale level M, the accuracy is also increased. By the*

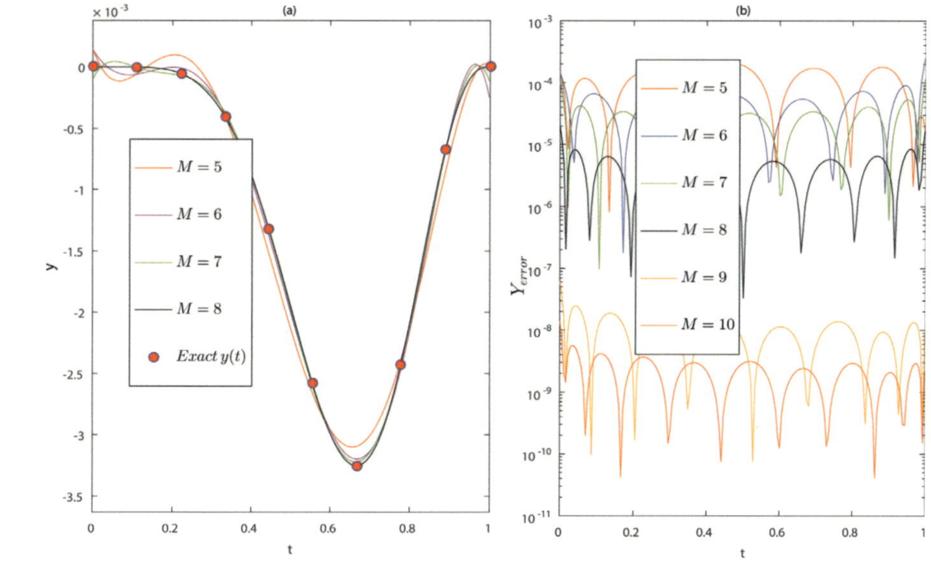

Figure 3.
(a) Comparison of exact and approximate solution of Example 4.1, under parameters set S_3*. (b) Absolute error in the approximate solution of Example 4.1, under parameters set* S_3*.*

proposed method, the graph of exact and approximate solutions for different values of M and at $\alpha = 1.7$ *is shown in* **Figure 4**. *From the plot, we observe that the approximate solution becomes equal to the exact solution at* $\alpha = 2$. *We approximate the error of the method at different scale levels and record that when scale level increases the absolute error decreases as shown in* **Figure 4** *subplot (b) and accuracy approaches* 10^{-9}, *which is a highly acceptable figure. For convergence of our proposed method, we examined the quantity* $\int_0^1 |y_{exact} - y_{approx}| dt$ *for different values of M and observed that the norm of error decreases with a high speed with the increase of scale level M as shown in* **Figure 4b**.

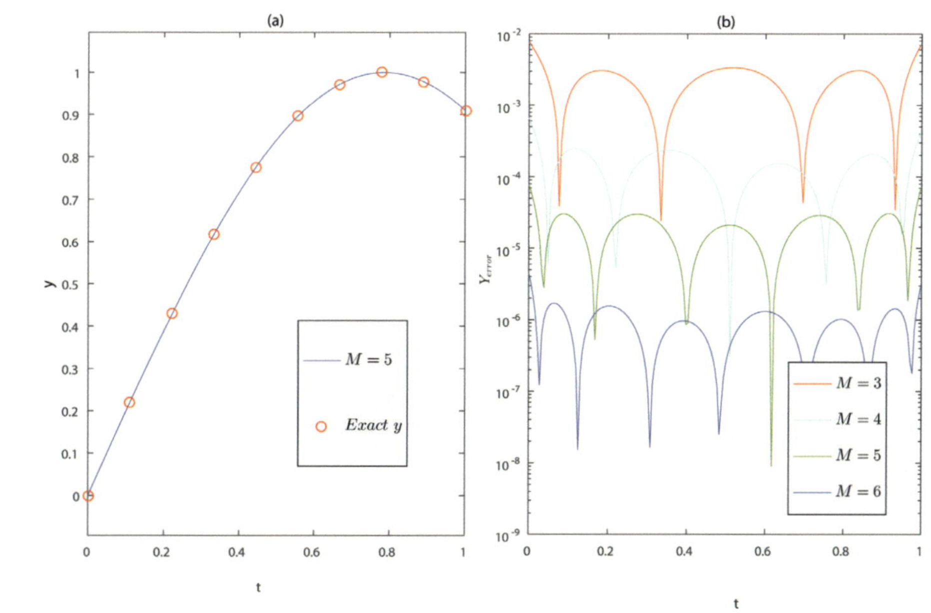

Figure 4.
(a) Comparison of exact and approximate solution of Example 4.2. (b) Absolute error for different scale level M of Example 4.2.

Example 4.3. *Consider the following coupled system of fractional differential equations*

$$\begin{aligned} D^{1.8}x(t) + Dx(t) + 9D^{0.8}y(t) + 2x(t) - y(t) &= f(t) \\ D^{1.8}y(t) - 6D^{0.8}x(t) + Dy(t) - x(t) &= g(t) \end{aligned} \tag{53}$$

subject to the boundary conditions

$$x(0) = 1, \quad x(1) = 2 \quad \textit{and} \quad y(0) = 2, \quad y(1) = 2.$$

Solution: *The exact solution is*

$$x(t) = t^5(1-t), \qquad y(t) = t^4(1-t).$$

We approximate the solution of this problem with this new method. The source terms are given by

$$f(t) = 2t^5(t-1) - t^4(t-1) + t^4(6t-5) - \frac{2229536516744740625t^{\frac{16}{5}}(10t-7)}{1008806316530991104} + \frac{1337721910046844375t^{\frac{16}{5}}(25t-21)}{1008806316530991104} \tag{54}$$

$$g(t) = t^3(5t-4) - t^5(t-1) - \frac{1114768258372370312 5t^{\frac{21}{5}}(15t-13)}{6557241057451442176} - \frac{891814606697896 25t^{\frac{11}{5}}(25t-16)}{144115188075855872}. \tag{55}$$

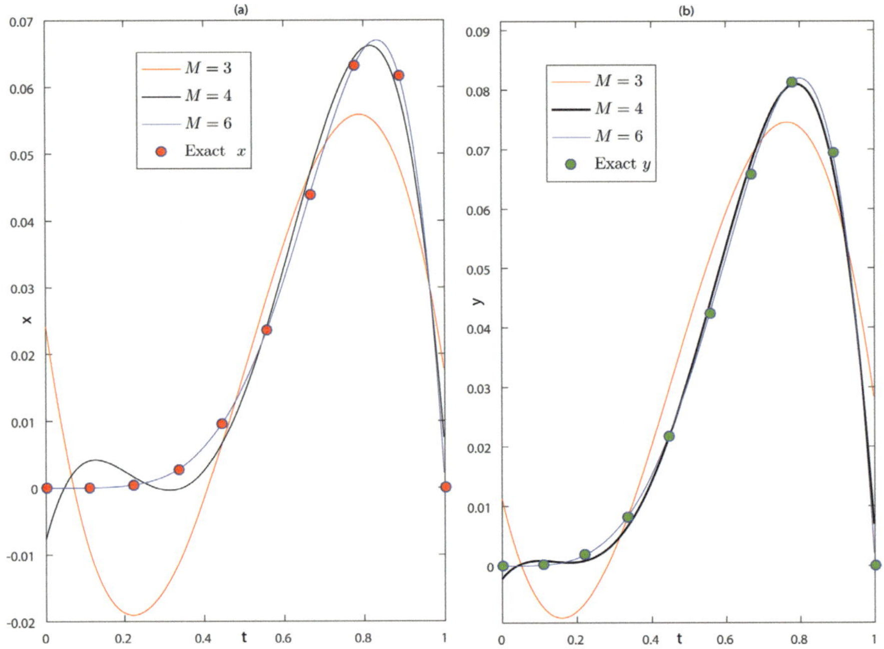

Figure 5.
Comparison of exact and approximate solution of Example 4.3 for different scale level M.

In the given **Figure 5**, *we have shown the comparison of exact* $x(t), y(t)$ *and approximate* $x(t), y(t)$ *in subplot (a) and (b) respectively.*

As expected, the method provides a very good approximation to the solution of the problem. At first, we approximate the solutions of the problem at $\alpha = 2$ *because the exact solution at* $\alpha = 2$ *is known. We observe that at very small scale levels, the method provides a very good approximation to the solution. We approximate the absolute error by the formula*

$$X_{error} = |x_{exact} - x_{approx}|.$$

and

$$Y_{error} = |y_{exact} - y_{approx}|.$$

We approximate the absolute error at different scale level of M, and observe that the absolute error is much less than 10^{-10} *at scale level* $M = 7$, *see* **Figure 6**. *We also approximate the solution at some fractional value of* α *and observe that as* $\alpha \to 2$ *the approximate solution approaches the exact solution, which guarantees the accuracy of the solution at fractional value of* α. **Figure 6** *shows this phenomenon. In* **Figure 6**, *the subplot (a) represents the absolute error of* $x(t)$ *and subplot (b) represents the absolute error of* $y(t)$.

Example 4.4. *Consider the following coupled system*

$$\begin{aligned} D^{1.8}x(t) - x(t) + 3y(t) &= f(t) \\ D^{1.8}y(t) + 4x(t) - 2y(t) &= g(t), \end{aligned} \tag{56}$$

subject to the boundary conditions

$$x(0) = -1, \quad x(1) = -1 \quad and \quad y(0) = -1, \quad y(1) = -1.$$

Solution: *The exact solution for* $\alpha = \beta = 2$ *is*

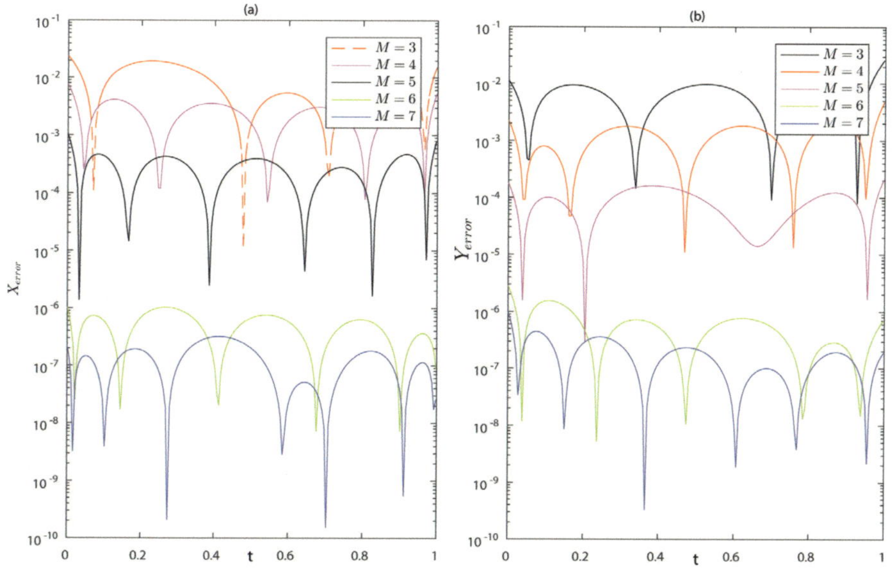

Figure 6.
Absolute error in approximate solutions at different scale level M = 3:7 *for Example 4.3.*

$$x(t) = t^5 - t^4 - 1, \quad and \quad y(t) = t^4 - t^3 - 2. \tag{57}$$

The source terms are given by

$$f(t) = \frac{445907303348948125t^{3.2}(25t - 21)}{302641894959297331 2} + t^3 - 4t^4 + 3t^5 - 2,$$

$$g(t) = \frac{89181460669789625t^{2.5}(5t - 4)}{14411518807585872} - 4t^3 + 6t^4 - 2t^5 - 2.$$

Approximating the solution with the proposed method, we observe that our scheme gives high accuracy of approximate solution. In **Figure 7**, *we plot the exact solutions together with the approximate solutions in* **Figure 7(a)** and **(b)** *for $x(t)$ and $y(t)$, respectively. We see from the subplots (a) and (b) that our approximations have close agreement to that of exact solutions. This accuracy may be made better by increasing scale level. Further, one can observe that absolute error is below 10^{-10} in* **Figure 8**, *which indicates better accuracy of our proposed method for such types of practical problems of applied sciences.*

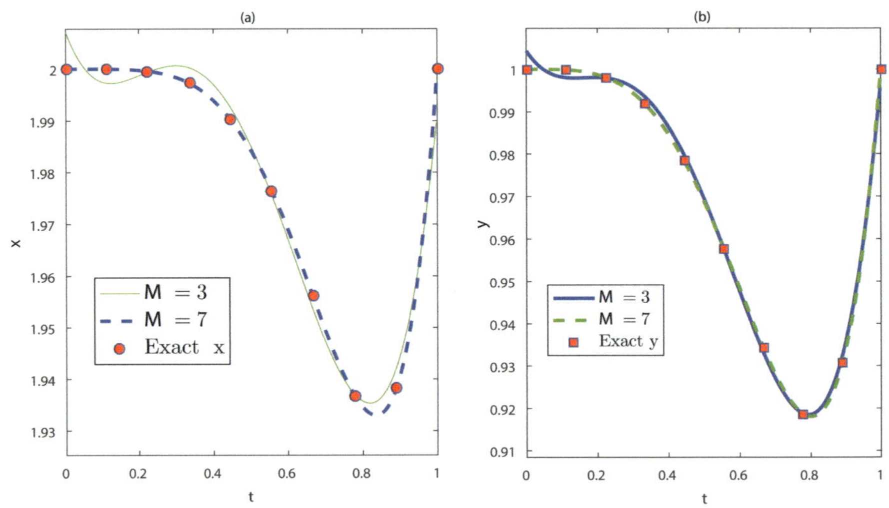

Figure 7.
Comparison of exact and approximate solution at scale level M = *3, 7 for Example 4.4.*

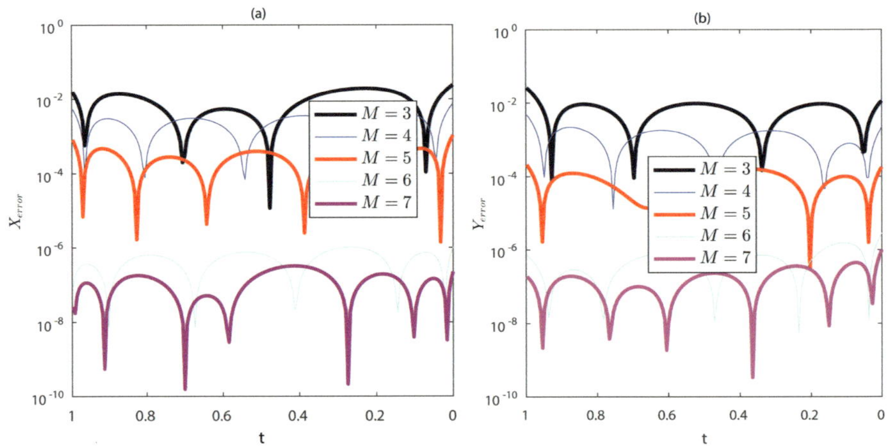

Figure 8.
Absolute error for different scale level M = *3:7 for Example 4.4.*

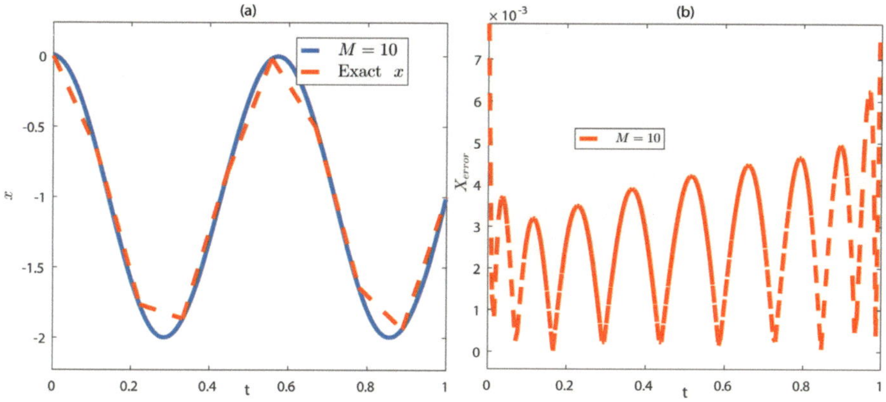

Figure 9.
(a) Comparison of exact and approximate solution at scale level $M = 10, \omega = 3.5, \alpha = 2, \nu = 1$ *for Example 4.5. (b) Absolute error at* $M = 10$.

ω	M	α	ν	$\|x_{app} - x_{ex}\|$ at BM	$\|x_{app} - x_{ex}\|$ at WM	$\|x_{app} - x_{ex}\|$ at JM
0.5	10	2	1	7.000(−3)	2.966(−1)	1.500(−2)
1.5	15	1.6	0.9	6.091(−3)	4.918(−2)	1.623(−1)
2.0	20	1.8	0.8	1.237(−3)	2.108(−2)	2.723(−2)
3.5	25	1.9	0.7	1.008(−3)	5.795(−2)	1.813(−3)

Table 1.
Comparison of solution between Legendre wavelet method (LWM) [47], Jacobi polynomial method (JM) and Bernstein polynomials method (BM) for Example 4.5.

In **Figure 8**, *the subplot (a) represents absolute error for* $x(t)$ *while subplot (b) represents the same quantity for* $y(t)$. *From the subplots, we see that maximum absolute error for our proposed method for the given problem (4.4) is below* 10^{-10}. *This is very small and justifies the efficiency of our constructed method.*

Example 4.5. *Consider the boundary value problem*

$$\begin{aligned} &D^{\alpha}x(t) + (\omega\pi)^2 D^{\nu}x(t) + x(t) = -\omega\pi(\sin(\omega\pi t) + \omega\pi) \\ &x(0) = 0, \quad x(1) = -2. \end{aligned} \tag{58}$$

Taking $\alpha = 2, \nu = 1$ *and* $\omega = 1, 3, 5, \ldots,$ *the exact solution is given by*

$$x(t) = \cos(\omega\pi t) - 1.$$

We plot the comparison between exact and approximate solutions to the given example at $M = 10$ *and corresponding to* $\omega = 3.5, \alpha = 2, \beta = 1$. *Further, we approximate the solution through Legendre wavelet method (LWM) [47], Jacobi polynomial method (JM) and Bernstein polynomials method (BM), as shown in* **Figure 9**.

From **Table 1**, we see that Bernstein polynomials also provide excellent solutions to fractional differential equations [48].

5. Conclusion and future work

The above analysis and discussion take us to the conclusion that the new method is very efficient for the solution of boundary value problems as well as initial value problems including coupled systems of fractional differential equations. One can easily extend the method for obtaining the solution of such types of problems with other kinds of boundary and initial conditions. Bernstein polynomials also give best approximate solutions to fractional order differential equations like Legendre wavelet method (LWM), approximation by Jacobi polynomial method (JPM), etc. The new operational matrices obtained in this method can easily be extended to two-dimensional and higher dimensional cases, which will help in the solution of fractional order partial differential equations. Also, we compare our result to that of approximate methods for different scale levels. We observed that the proposed method is also an accurate technique to handle numerical solutions.

Acknowledgements

This research work has been supported by Higher Education Department (HED) of Khyber Pakhtankhwa Government under grant No: HEREF-46 and Higher Education Commission of Pakistan under grant No: 10039.

Author details

Kamal Shah[1*], Thabet Abdeljawad[2,3], Hammad Khalil[4] and Rahmat Ali Khan[1]

1 Department of Mathematics, University of Malakand, Khyber Pakhtunkhwa, Pakistan

2 Department of Mathematics and General Sciences, Prince Sultan University, Riyadh, Saudi Arabia

3 Department of Medical Research, China Medical University, Taichung, Taiwan

4 Department of Mathematics, University of Education Lahore, Attock, Punjab, Pakistan

*Address all correspondence to: kamalshah408@gmail.com

References

[1] Lakshmikantham V, Leela S, Vasundhara J. Theory of Fractional Dynamic Systems. Cambridge, UK: Cambridge Academic Publishers; 2009

[2] EL-Sayed AMA, Bin-Taher EO. Positive solutions for a nonlocal multi-point boundary-value problem of fractional and second order. Electronic Journal of Differential Equations. 2013; **64**:1-8

[3] Magin RL. Fractional calculus in bioengineering—Part 2. Journal of Critical Reviews in Biomedical Engineering. 2004;**32**(2):105-193

[4] Magin RL. Fractional calculus in bioengineering. Journal of Critical Reviews in Biomedical Engineering. 2004;**32**(1):1-104

[5] Rossikhin YA, Shitikova MV. Applications of fractional calculus to dynamic problems of linear and nonlinear hereditary mechanics of solids. Journal of Applied Mechanics Reviews. 1997;**50**:15-67

[6] Ichise M, Nagayanagi Y, Kojima T. An analog simulation of non-integer order transfer functions for analysis of electrode processes. Journal of Electroanalytical Chemistry and Interfacial Electrochemistry. 1971;**33**: 253-265

[7] Inca M, Kiliça B. Classification of traveling wave solutions for time-fractional fifth-order KdV-like equation. Waves in Random and Complex Media. 2014:393-404

[8] He JH. Some applications of nonlinear fractional differential equations and their approximations. Bulletin of Science and Technology. 1999;**15**(2):86-90

[9] Rehman M, Khan RA. The Legender wavelet method for solving fractional differential equation. Communications in Nonlinear Science and Numerical Simulation. 2011;**10**:4163-4173

[10] Kilbas AA, Marichev OI, Samko SG. Fractional Integrals and Derivatives (Theory and Applications). Switzerland: Gordon and Breach; 1993

[11] Miller KS, Ross B. An Introduction to the Fractional Calculus and Fractional Differential Equations. New York: Wiley; 1993

[12] Podlubny I. Fractional Differential Equations, Mathematics in Science and Engineering. New York: Academic Press; 1999

[13] Hilfer R. Applications of Fractional Calculus in Physics. Singapore: World Scientific; 2000

[14] Kilbas AA, Srivastava HM, Trujillo JJ. Theory and Applications of Fractional Differential Equations, North-Holland Mathematics Studies. Vol. 204. Amsterdam: Elsevier; 2006

[15] Benchohra M, R. Graef J, Hamani S. Existence results for boundary value problems with nonlinear fractional differential equations. Journal of Applied Analysis. 2008;**87**:851-863

[16] Rehman M, Khan R. A note on boundary value problems for a coupled system of fractional differential equations. Computers & Mathematcs with Applications. 2011;**61**:2630-2637

[17] Su X. Boundary value problem for a coupled system of nonlinear fractional differential equations. Applied Mathematics Letters. 2009;**22**:64-69

[18] Idrees Bhattia M, Bracken P. Solutions of differential equations in a Bernstein polynomial basis. Journal of Computational and Applied Mathematics. 2007;**205**:272-280

[19] Henryk G, Jose' LP. On the approximation properties of Bernstein polynomials via probabilistic tools. Boletin de la Asociacion Matematica Venezolana. 2003;**X**(1):1

[20] Ordokhani Y, Davaei S. Approximate solutions of differential equations by using the Bernstein polynomials. Applied Mathematics. 2011. DOI: 10.5402/2011/787694. Article ID 787694

[21] Mandal BN, Bhattacharya S. Numerical solution of some classes of integral equations using Bernstein polynomials. Applied Mathematics and Computation. 2007;**190**(2):1707-1716

[22] Yousefi SA, Behroozifar M. Operational matrices of Bernstein polynomials and their applications. International Journal of Systems Science. 2010;**41**(6):709-716

[23] Blank L. Numerical treatment of differential equations of fractional order. In: Numerical Analysis Report 287. Manchester Centre for Computational Mathematics; 1996

[24] Sun HH, Onaral B, Tsao Y. Application of positive reality principle to metal electrode linear polarization phenomena. IEEE Transactions on Biomedical Engineering. 1984;**31**(10): 664-674

[25] Baillie RT. Long memory processes and fractional integration in econometrics. Journal of Econometrics. 1996;**73**:5-59

[26] Momani S, Odibat Z. Numerical approach to differential equations of fractional order. Journal of Computational and Applied Mathematics. 2007;**207**:96-110

[27] El-Wakil SA, Elhanbaly A, Abdou MA. Adomian decomposition method for solving fractional nonlinear differential equations. Applied Mathematics and Computation. 2006; **182**:313-324

[28] Odibat Z, Momani S. Numerical methods for nonlinear partial differential equations of fractional order. Applied Mathematical Modelling. 2008;**32**:28-39

[29] Erturk VS, Momani S. Solving systems of fractional differential equations using differential transform method. Journal of Computational and Applied Mathematics. 2008;**215**:142-151

[30] Khalil H, Khan RA. A new method based on legender polynomials for solution of system of fractional order partial differential equation. International Journal of Computer Mathematics. 2014;**91**(12):2554-2567

[31] Sweilam NH, Khader MM, Al-Bar RF. Numerical studies for a multi-order fractional differential equation. Physics Letters A. 2007;**371**:26-33

[32] Das S. Analytical solution of a fractional diffusion equation by variational iteration method. Computers & Mathematcs with Applications. 2009; **57**:483-487

[33] Momani S, Odibat Z. Homotopy perturbation method for nonlinear partial differential equations of fractional order. Physics Letters A. 2007;**365**:345-350

[34] Hashim I, Abdulaziz O, Momani S. Homotopy analysis method for fractional IVPs. Communications in Nonlinear Science and Numerical Simulation. 2009;**14**:674-684

[35] Arikoglu A, Ozkol I. Solution of fractional differential equations by using differential transform method. Chaos, Solitons & Fractals. 2007;**34**: 1473-1481

[36] Arikoglu A, Ozkol I. Solution of fractional integro-differential equations by using fractional differential transform method. Chaos, Solitons & Fractals. 2009;**40**:521-529

[37] Erturk VS, Momani S, Odibat Z. Application of generalized differential transform method to multi-order fractional differential equations. Communications in Nonlinear Science and Numerical Simulation. 2008;**13**: 1642-1654

[38] Odibat Z, Shawagfeh N. Generalized Taylor's formula. Applied Mathematics and Computation. 2007; **186**:286-293

[39] Saadatmandi A, Deghan M. A new operational matrix for solving fractional-order differential equation. Computers & Mathematics with Applications. 2010;**59**:1326-1336

[40] Kumar P, Agrawal OP. An approximate method for numerical solution of fractional differential equations. Signal Processing. 2006;**86**: 2602-2610

[41] Liua F, Anh V, Turner I. Numerical solution of the space fractional Fokker-Planck equation. Journal of Computational and Applied Mathematics. 2004;**146**:209-219

[42] Yang W. Positive solutions for a coupled system of nonlinear fractional differential equations with integral boundary conditions. Computers & Mathematcs with Applications. 2012;**63**: 288-297

[43] Diethelm K. An algorithm for the numerical solution of differential equations of fractional order. Electronic Transactions on Numerical Analysis. 1997;**5**:1-10

[44] Diethelm K, Walz G. Numerical solution of fractional order differential equations by extropolation. Numerical Algorithms. 1997;**16**:231-253

[45] Li Y, Shah K. Numerical solutions of coupled systems of fractional order partial differential equations. Advances in Mathematical Physics. 2017;**2017**:14

[46] Shah K, Wang J. A numerical scheme based on nondiscretization of data for boundary value problems of fractional order differential equations. RACSAM. 2018;**2018**:1-18

[47] Rehman M. Boundary value problems for fractional differential equations: Existence theory and numerical solutions [PhD dissertation]. Islamabad, Pakistan: NUST University; 2016

[48] Darania P, Ebadian A. A method for the numerical solution of the integro-differential equations. Applied Mathematics and Computation. 2007; **188**:657-668

Československá akademie věd

Sekce biologicko-lékařská

THE DENERVATED MUSCLE

The Denervated Muscle

Edited by E. Gutmann

Scientific Editors:

Zdeněk Servít and Otakar Poupa

Translated by:

Pavel Hník

Foreign language Editor:

Joseph Cort

THE DENERVATED MUSCLE

EDITED BY E. GUTMANN

List of Contributors

A. Bass, R. Beránek, Z. Drahota,

E. Gutmann, P. Hník,

O. Hudlická, V. Škorpil, L. Vyklický,

J. Zelená, R. Žák

PRAGUE 1962

Springer Science+Business Media, LLC

ISBN 978-1-4899-4856-4
DOI 10.1007/978-1-4899-4854-0

ISBN 978-1-4899-4854-0 (eBook)

Originally published by **Nakladatelství Československé akademie věd, Praha** in 1962

THE DENERVATED MUSCLE

Errata

Page 32, Line 2 from below: Ellenberg read Ellenbeck
Page 35, Line 6 from above: Horton and Smith read Horton-Smith
Page 127, Line 9 from above: methodical read methodological
Page 208, Line 13 from above: Ben Tor read Burton
Page 273, Line 20 from above: Huscles read Muscles
Page 273, Line 21 from above: mopkins read Hopkins
Page 321, Line 3 from above: Chagowets read Chagovets
Page 360, Line 11 from below: Nimmi read Nimni
Page 363, Line 15 from below: Waking read Wakim
Page 369, Line 3 from below: Nimmi read Nimni
Page 462, Line 24 from above: influence read influences

CONTENTS

Introduction

Problems concerning the fate of the skeletal muscle following peripheral nerve section and the mechanisms involved in denervation atrophy have attracted the attention of physiologists and pathophysiologists for more than a hundred years. This monograph presents the results of several years' united effort of physiologists, biochemists, morphologists and electrophysiologists in elucidating at least some of the questions which have hitherto remained unsolved in relation to the denervated muscle.

As an object of research, the denervated muscle is of considerable theoretical as well as of practical importance. The former is associated with questions of the trophic influence of the nervous system in neuromuscular relations and with the role of use and disuse in muscle atrophy. It appears in this connection that one of the main metabolic disturbances in the denervated muscle is due to the fact that section of the peripheral nerve results in the loss of postfunctional metabolic overshoot reactions which seem to be indispensible for the maintenance of normal muscle metabolism, structure and functional capacity. This aspect has particularly been studied from the biochemical point of view. However, it was necessary to supplement this approach to denervated muscle metabolism by further methodical approaches: physiological, morphological and electrophysiological. Different factors were studied which could influence denervation atrophy: vasomotor, humoral and hormonal influences and which could, perhaps, be used in the therapy and rehabilitation of peripheral nerve injuries and paralysis of different origin.

Chapter I

DENERVATION STUDIES IN RESEARCH OF NEUROTROPHIC RELATIONSHIPS

E. Gutmann, P. Hník

A. Definition of "trophic processes" and "nervous regulation of trophic processes" ("trophic influence of the nervous system")

Following section of the peripheral nerve muscle atrophy takes place with subsequent degeneration of muscle fibres innervated by this nerve. This is considered to be due to a disturbance in the trophic function of the nervous system. "Trophic functions" were originally used to designate changes described in patho-anatomical or clinical studies, characterized by either an increase (hypertrophy) or a decrease (atrophy) of these functions, without analysing or defining exactly the mechanisms subserving these "trophic functions". The efforts of physiological investigations should be aimed at giving these functions a concrete meaning. How does this trophic function manifest itself and in which way is it regulated?

Firstly, it will be necessary to define the "trophic functions" more precisely. This evidently includes a certain regulation of tissue (or cell) metabolism, which is dependent to a certain extent on the connection of these tissues with the nervous system. It is necessary to stress "to a certain extent", since it is well known that a denervated organ maintains its structure and its functional capacity, even though limited, and that the loss of specific tissue structure of the muscle is a relatively late symptom of denervation atrophy. Thus "trophic processes" take place even in the denervated tissues and, of course, in every living cell. During life cellular elements are continuously broken down and resynthetized again. These processes and their regulation have recently become more and more the object of biological research. Especially the evidence for a continuous turnover of proteins in the living organism have affected the development of modern biology (v. Schönheimer 1942). Of course, this continuous breaking down and rebuilding of structures, underlying the functional capacity of a tissue, is a phenomenon characteristic not only for individual tissues or organs, but also for the whole organism.

The organism develops and lives in unity with the surrounding environment and in this environment maintains its dynamic equilibrium. It is an equilibrium between catabolic and anabolic processes, and is under the constant influence of stimuli affecting this dynamic equilibrium. The organism maintains its integrity in spite of incessant changes in the environment by maintaining and renewing a constant "milieu interieur" on the one hand, and by maintaining the structure and functional capacity of the tissue on the other. The organism is thus an autoregulatory system in which the regulation of the milieu interieur (C. Bernard 1880) is realized by mechanisms of the so-called "homeostasis" (Cannon 1932), and by processes maintaining and renewing the structure and functional capacity of tissues — i. e. so-called "trophic processes" — which maintain the structure and integrity of the tissues in spite of the everchanging environmental conditions. Living organism is characterized by its continuous reception, transformation and storage of information on the basis of which tissue metabolism becomes adapted to continuous environmental changes. This adaptation provides the organism with capacities to react quickly to changes in the environment on the basis of stored information and to be prepared to act on the basis of the regulation of tissue metabolism. The latter processes are thus meant for the maintainance and renewal of tissues and their adaptation to stimuli arising in the environment. It is possible to say that the "organism acts", this being mediated primarily by specific functions, and "prepares to act", this being the result of "trophic functions" of the organism. It is unnecessary to point out that both these functions are closely interrelated. However, the aim of research is not limited to demonstrating the interrelations, since it should also analyse individual functional components in biological processes. Tissue excitability is, of course, a necessary presupposition for both these basic functions, i. e. the ability of living tissues to react to stimuli by reversible changes in the dynamic equilibrium between anabolic and catabolic processes. Cellular excitability has made the development of control regulatory systems possible, and ensures the stability of the organism in relation to environmental changes. Reactibility ensures a continual elaboration of information concerning the environment and the ability of reacting continually to these stimuli. Mechanisms regulating trophic processes enable the organism to attain again the state of equilibrium following the application of each stimulus changing this state. But this is only one aspect of these trophic functions. It is evident that these regulatory mechanisms also prepare the organism for subsequent stimuli. These regulatory mechanisms develop and become more effective with the evolution of living matter. They play an important role in all organisms, but in higher organisms the response to external stimuli is mediated by the nervous system. In these higher organisms the trophic function is represented, to a great extent, by nervous reflex regulation of maintainance and recovery processes of specific structure and functional capacity of the tissue and the nervous regulation of preparatory or adaptive metabolic processes. This aspect of the function of the nervous system in these processes may best be designated as the trophic function of the nervous system.

The concept of the organism as a regulating system maintaining the state of equilibrium by a continuous inflow and integration of information, is, to a great extent, a matter of course for the neurophysiologist in the case of reflex activity resulting in specific function — movement or animal behaviour. It is not so clear, however, in the case of metabolic processes maintaining the structure and functional capacity of the tissue, i. e. trophic functions which can be, with a certain danger of abstraction, differentiated from the so-called specific functions of the organism. Denervation experiments indicate disturbances of these regulatory mechanisms resulting from the loss of the trophic influence of the nervous system, namely disturbances of metabolic recovery processes which are mediated by the nervous system. These then are elimination experiments; this is their limitation, but at the same time their advantage, since they make the analytical approach possible.

The object of research of trophic functions is thus to study metabolic recovery processes following every functional activity (i. e. a study of so-called physiological regeneration), similarly as investigations of reparative regeneration (i. e. a study of structural renewal after injury to the organism). In both research is concerned with a study of metabolic recovery processes following the break down of living matter and is therefore centred around questions mainly concerned with anabolic phases of metabolism.

The significance of the study of recovery processes has been clearly stressed by Pavlov (1890): "It is possible to say that contemporary physiology is primarily the physiology underlying catabolic processes in living tissues. Almost the whole field of physiology is represented by the functions and general conditions of various parts of the organism and of the functional activity of organs connected with their partial breakdown, as has been shown up to the present time for almost all organs. Thanks to the results of the last decades, physiologists have successfully become acquainted in many organs with their centrifugal nerves — representing the actual levers of organ function — and have freely, according to their fantasy, altered this breaking down of living tissues in its duration, and various other conditions and the investigation of these is progressively more complete and profound. However, progress of knowledge of the other aspect of living processes, i. e. of the renewal of organs, is by comparison very incomplete. It seems that here the main questions have not yet even been formulated clearly enough or suitably for research." It appears that Pavlov very clearly stressed the importance of research of trophic functions of the organism which have been, in comparison with the study of the so-called specific functions, subjected relatively little to experimental research. It is not, perhaps, necessary to point out that trophic processes are not the expression of a static resting state, but are the result of the dynamic state of protoplasm. For this reason, expressions such as "resting metabolism" or "maintenançe metabolism" which are sometimes used when discussing the trophic functions of the organism are not quite satisfactory.

B. The study of trophic functions from the aspect of cellular metabolism

To what extent can the trophic functions be differentiated at the cellular level? The close interrelationship of fundamental cellular functions makes this differentiation extremely difficult. What are the energy requirements of trophic functions which, as has been mentioned above, preserve and renew the structure and functional capacity of tissues? The solution of these problems could, perhaps, aid in differentiating the basic processes of cellular metabolism. This question is especially complicated during development. According to Boell (1955) the developing organism requires energy for four basic processes, namely for maintenance of structure, growth, differentiation and specific function. Processes connected with structural maintenance are in action throughout life, the remaining three appear in temporal sequence during development. It is natural that, under normal conditions, these processes are closely interconnected and that their differentiation is very complicated and experimentally difficult to achieve (Needham 1942). Thus, according to Boell, trophic processes are those energy producing or energy consuming processes which maintain the integrity of the cell, renew its structure following each functional activity and thus ensure its functional capacity. During growth an exceptionally powerful accentuation of anabolic processes takes place as during regeneration or reinnervation of terminal organs following section of the nerve. The total balance between breakdown and renewal of protoplasmatic components shifts towards anabolic processes, and the rate of these processes is so fast that the differentiation of trophic processes in the sense of "maintenance metabolism" is, according to Boell, extremely difficult or even impossible. On the other hand, the rate of anabolic processes during ageing is very much slower and this can be shown, for example, in the regeneration processes following nerve section in old animals (Drahota and Gutmann 1960), where the restitution of proteins in reinnervated muscles is exceptionally slow as compared with the reinnervation process in young animals. Under these conditions we may speak of a reduced trophic function of nerve cells. Thus the rate of trophic processes, in the sense of maintenance and renewal of structure and functional capacity of tissues differs widely at different stages of development and this aspect must be kept in mind when attempts are made at their differentiation. Extreme accentuation of anabolic processes involves an increase of cellular components and can best be expressed in processes of proteosynthesis. Differentiation concerns the progressive specialization of cells structurally as well as functionally and finally, specific function requires energy necessary for specific functional activity of various cellular structures, such as contraction of the muscle cell and secretion of gland cells. The anabolic activity of protoplasm is connected primarily with the metabolism of nucleic acids and proteosynthesis. Protein metabolism is evidently particularly important during the process of development and Pollister (1954) also mentions "maintenance protein synthesis". However the breakdown of protoplasmatic components is constantly taking place and the catabolic activity of protoplasm is mainly due to enzymes of hydrolysis, respiration, fermentation, phosphorylation etc.

The energy requirements necessary for the continuous renewal of cells, for the "trophic functions", are undoubtedly considerable and can be greater than energy consumption needed for all other basic functions (see Boell 1955). The process of proteosynthesis, however, represents only one aspect — even though the most important — of the processes. The cell must maintain its structure against the environment, against osmotic gradients and here an important role is played by processes of active transport which regulate cellular permeability without which the cell could not maintain its structural and functional integrity. Knowledge of metabolic requirements of these processes is especially important for understanding neuron function (see Hodgkin 1959). The maintenance of bioelectric potentials, the regulation of cellular volume and ion composition requires energy and to-day special attention is devoted to questions of a "metabolically driven sodium pump" (Ussing 1959) in connection with the explanation of these processes which maintain the structure of the cell. The nerve fibre requires metabolic energy in order to build up a concentration gradient of Na and K across the membrane and following activity the original state is renewed by a "secretory" process which utilizes metabolic energy in order to extrude Na and absorb K against concentration gradients (Hodgkin 1959). Thus complex regulatory mechanisms come into play in maintainance processes and in the recovery of cellular structures, these being of a fundamental and universal nature taking place, of course, on the cellular level. It is possible to point out only some of the main aspects of this regulation of cellular metabolic processes. These are primitive control mechanisms (Stadie 1954) which adapt metabolic activity to changed requirements and demands and control the rate of metabolic processes. The rate of these regulating "trophic" mechanisms change, as has been pointed out, in the course of development and undergo very substantial changes in the tissues after denervation. Fundamental mechanisms affecting this rate are given by the system enzyme-substrate. However, other reactions exist besides those between the enzyme and substrate, evoked by the so-called metabolic "pacemakers" (see Krebs 1957) which are of considerable importance in the regulation of metabolism. Evidently higher regulatory mechanisms (humoral and nervous) are superimposed on these fundamental control mechanisms. Here we are concerned, however, with the fundamental control mechanisms of cellular metabolism which include processes maintaining considerable concentration gradients of certain substances, especially of K ions, between the cell and extracellular environment, and which require energy gained by oxidation processes. Of these, oxidative phosphorylations control, for example, the permeability of the membrane to glucose. Influence on the permeability of the cellular membrane is doubtless an important component of these fundamental control mechanisms.

At the present time considerable progress is being made in the study of these "trophic" regulations of cellular functions. Application of these facts to the study of regulations of tissue metabolism and to the nervous regulation of these processes — this being in the centre of our interest — has so far been very difficult.

It is then not at all surprising that relatively few reports deal with this aspect of the metabolic research of cells and tissues. It is difficult to separate manifestations of the trophic function from those of the specific cellular function. This is also the case when studying the function of nerve cells.

As in other cells we have to differentiate the trophic functions of the neuron, i. e. the processes of maintenance and recovery of structure and functional capacity of the neuron, disturbances of which lead to its atrophy and degeneration. As a matter of fact it was the study of Waller (1850, 1856) on the degeneration changes of the peripheral cell processes after cutting the nerve which demonstrated experimentally for the first time the "nutritive" influence of the cell body on its axon. The question of the "trophic function" of the cells was exemplified in the case of the neuron in the special trophic relations between cell body and axon. The axon is sometimes over 1 m long and has a volume that may be hundred times that of the cell body. The question of the "trophic function" of the neuron, essential for structural and functional integrity of the cell process was here first presented in a special and provocative way. Bouchard as early as 1866 pointed out that, as the distal portion of a sensory nerve degenerates, transmission of these impulses cannot constitute the trophic function of the neuron and Cook and Gerard (1931) showed that continued activity in severed nerves even accelerates failure of conduction. The nature of the "trophic" processes of the neuron, maintaining integrity of the cell processes has not been clarified, as evident from the many different concepts used, such as Held's "vis a tergo", Heidenhein's "histodynamic impulse" or Young's "intra-axonic pressure" (viz Gutmann 1959). On the other hand disproportionally more and certainly more succesful work was devoted to studying the specific functions of the neuron, i. e. synaptic transmission and nerve impulse conduction, than to trophic processes which maintain the structural and functional integrity of the neuron. It is evident, however, that this function represents an equally

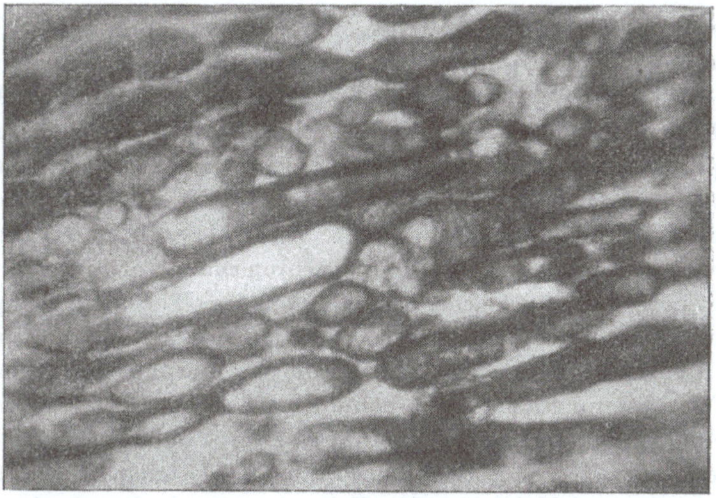

Fig. 1. Fragmentation of myelin sheath 27 hrs. after degeneration of the nerve in vitro (Gutmann E. and Holubář J. [1950]. J. Neur. Neurosurg. Psychiat. 13 : 89.)

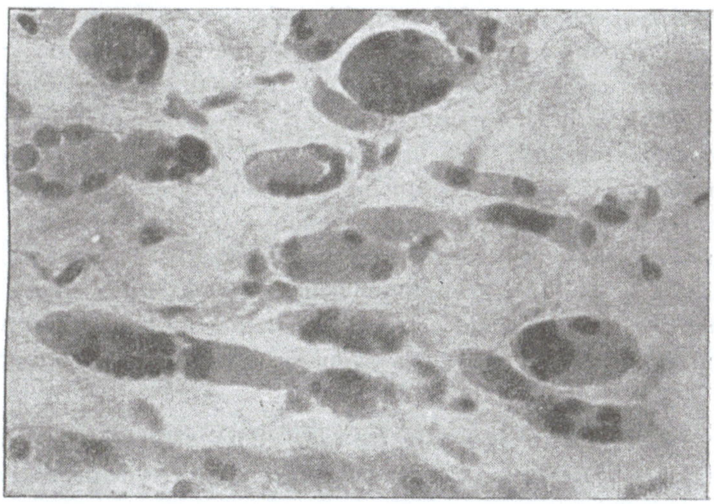

Fig. 2. Disintegration of the muscle into round and ovoid fragments 26 years after denervation (Bowden R. M. F. and Gutmann E. [1944]. Brain 67 : 273).

important aspect of neuronal function as nerve impulse conduction. Neurophysiology has so far paid attention primarily to those processes which are concerned with the production and propagation of nerve impulses and to their energy requirements. These investigations have already, to a considerable extent, made possible the understanding of synaptic activity of nerve cells, especially from the point of view of the properties of the nerve cell membrane. In comparison with the study of this "specific activity" of the nerve cell, research on metabolic processes maintaining the trophic function of the neuron has been neglected. Bodian (1947) has indicated that the energy requirements for this trophic function are probably greater than those for the specific activity of the neuron. Disturbances in the trophic activity can be deduced from the morphological picture of the nerve or muscle. After section of the nerve, the trophic activity of the neuron cannot be exerted on its cell process and consequently degeneration takes place. The action of proteolytic enzymes continues and catabolic processes are not followed by anabolic recovery processes, i. e. by protein resynthesis. From the morphological point of view the loss of trophic functions of the neuron become manifest as fragmentation of the axon into small particles. These processes are similar during degeneration in vitro and in vivo, so that it is not unsuitable to designate them as autolysis in vivo (Gutmann and Holubář 1950). Following muscle denervation a disturbance of the "trophic functions" of muscle cells occurs. The rate of the "trophic processes" is altered and processes for the "maintenance and recovery of cellular structure and functional capacity" are evidently retarded — "atrophy" of muscle cell takes place. Loss of the trophic influence of the nervous system on the muscle cell, which evidently plays a role in the regulation of "trophic processes" in muscle cells, is characterized by the same morphological picture after nerve section as in the peripheral stump, i. e. fragmentation and disintegration of muscle fibres into segments (Fig. 1 and 2).

This morphological picture convincingly demonstrates that cells in the organism are not isolated and that their metabolism is coordinated and regulated; various physiological mechanisms participate in this regulation and these mechanisms can influence tissue metabolism in various ways, under conditions of identical enzyme content and activity. What are these mechanisms affecting metabolic interrelations between the nervous system and innervated organs?

C. The trophic function of the nervous system and its place in the study of the physiology of neurotrophic relations

The foregoing section has shown that it is necessary to differentiate the trophic function of the neuron itself, (i. e. the general cellular activities concerned with the maintenance and recovery of the structure and functional capacity of the neuron) from the trophic functions, or better, influences of the neurons exerted on the peripheral innervated organs. As with the functions of the neuron itself we have to distinguish trophic and specific functions of the neuron in connection with this effect on the organs. Neurophysiology was primarily interested in studying fast processes connected with the reception and integration of information and their transmission to the effector organ. This means, in the first place, the study of changes accompanying the propagation of impulse activity, which is connected in the effector part of reflex mechanisms with specific function, e. g. muscular contraction and catabolic processes in the terminal organs connected with it. None the less important, even though considerably neglected, is the investigation of nervous control of processes maintaining the structural and functional integrity of organs, i. e. nervous regulation of trophic functions in the muscle. It is, of course, well-known that the nervous system reflexly regulates the specific function in the muscle, i. e. its contractions, mainly by the number and frequency of nerve impulses, this being associated with various degrees of catabolic processes in the muscle. These correlations were demonstrated, for example, in muscle glycogen by Yampolskaya and Yakovlev (1951). Furthermore, when the muscle was stimulated by the same number of stimuli applied at the same frequency, correlation was found between the initial glycogen content and the extent of glycogen breakdown. The amount of glycogen broken down was found to be dependent on the initial glycogen content. A direct correlation was also demonstrated after direct muscle stimulation (Bass et al. 1955) as well as following the administration of hormones, e. g. adrenaline. It is usually assumed that the main factor determining anabolic processes, which are of primary importance in trophic processes, is the amount of enzyme and substrate present. For example, Cori and Cori (1929) have shown that the content of liver glycogen depends mainly on blood glucose levels and a similar relationship in vitro has been found by Hechter et al. (1941). These autoregulatory mechanisms naturally also participate, but they are often considered to be the only important ones and according to this view (for example Soskin and Levin 1952) nervous regulation comes into play only under extraordinary conditions, e. g. stress.

However, the nervous regulations which are superimposed on the autoregulatory mechanisms cannot be underestimated. This is already indicated by experiments in which the rate of glycogen resynthesis was investigated in normal and denervated muscles, the experiments showing that, given the same initial content of glycogen in the muscles and the same blood glucose levels, a serious disturbance of postfunctional resynthesis, i. e. of metabolic recovery processes, occurs following the loss of neuromuscular excitability. These experiments demonstrate the significance of neurotrophic relations, i. e. the relations between nerve and innervated cell, for the metabolism of muscle. However, metabolic changes in nerve cells following disturbances of the peripheral organs are no less important.

These neurotrophic relations between the nerve cell and innervated organs exist and can be demonstrated throughout life, but they become most apparent when these connections are interrupted, e. g. after denervation, or when they are being renewed, e. g. in the process of nerve regeneration. It has been shown in several reports that the structural integrity of the neuron depends on its connections with the periphery, as the structural integrity of an innervated organ is dependent on its connections with the nerve cell. Of special interest for these interrelationships between the nervous system and innervated tissues are the results of morphogenetic studies (Zelená 1959, Studitskii 1959).

The mutual relations between nerve centres and innervated organs were widely studied from the point of view of nerve cell development following extirpation of the so-called peripheral field, e. g. an extremity (Detwiler 1936, Hamburger 1954, Hamburger and Levi Montalcini 1949) which leads to hypoplasia and atrophy of the corresponding nerve centres. On the other hand, an increase of the peripheral area by implantation of supernumerary limbs leads to the hyperplasia of neurons (e. g. Hamburger). The peripheral field exerts its influence not only on neurons innervating it under normal conditions, but also on more distant regions of the nervous system connected to these neurons. These "transneuronal" changes indicate that central synapses react sensitively to disturbances of normal "trophic signalization" and react by reorganization to new peripheral relations, so that it is possible to speak of peripheral regulation of central synapses (Sperry 1951). The peripheral tissue reacts equally sensitively to changes of neuronal function during development or regeneration. For example, the specialization of epithelial cells in taste buds occurs and is maintained only under the influence of the nervous system (May 1925). A clear indication of trophic interrelations between the nerve cell and the periphery are the compensatory growth processes demonstrated by the "extension" of axons from neighbouring nerves surrounding a denervated skin area (Weddell et al. 1941). This "collateral" regeneration of nerve fibres is apparently the result of a change of the metabolic equilibrium between different neurons initiated probably by changes in afferent signalisation. This equilibrium is apparently changed after section of neighbouring nerve fibres and the nerve cell reacts by increased metabolism which manifests itself by peripheral growth. Complex trophic adaptation mechanisms also participate during reinnervation of partially denervated muscles. The possibility of innervation of denervated muscle fibres by residual intact nerve fibres was first pointed out by Exner (1885) and later such "collateral regeneration" was demonstrated by Harreveld (1947), Hoffmann (1950) and Edds (1953). Hoffmann (1950, 1952) assumed chemical stimuli for the growth of neighbouring and residual fibres, the important point in this connection being the neurotrophic mechanisms which are evoked by changes in the periphery. On the other hand there exists considerable specificity of the trophic effect of nerve cells which has been demonstrated, for example, after cross-union of sensory and motor nerve fibres. The sensory fibres innervating a denervated muscle produce terminal plexiform nerve endings which are very similar to terminal nerve endings in the skin (Gutmann 1945). Of course, in this case reinnervation of the muscle is realized by sensitive fibres which are incapable of forming motor end-plates in the muscle or influencing muscle atrophy. The high specificity of neurons is also associated with the electrophysiological differentiation of the "fast" and "slow" muscles in the cat. If cross-union is made between the nerve innervating a fast muscle with a nerve originally innervating a slow muscle, then the fast motoneuron changes the slow muscle to a fast one (Buller et al. 1959). These changes could be evoked at early developmental stages, but neurotrophic relationships act throughout life. Nerve cells of the adult organism react by hypertrophy to increased activity, and by atrophy in case their axons are sectioned and regeneration of nerve endings is prevented (Edds 1949, 1950, Young 1950). Nerve regeneration experiments emphasize that neurons retain their capacity for growth and their reactibility to changes in the peripheral tissues throughout life (see Hamburger 1954), and questions how and to what extent these neurotrophic relations are linked with the processes of impulse transmission are important problems of neurophysiology.

These examples demonstrate that the investigation of neurotrophic interrelations represents an important aspect of physiological research reaching beyond the scope of embryology and morphogenesis, which has up-to-now received relatively little attention. Yet it is an independent branch of physiology completing traditional physiological research concerned mainly with the more rapid reactions of the organism provided for by the nervous system. The mechanisms of these neurotrophic relations between nerve cells and periphery should be demonstrated in their metabolic, morphological and reflex manifestations. Analysis must be made of the influence of nerve centres on the metabolism of the periphery and conversely the effect of signals from the peripheral organs on nerve centres. An important problem of the physiology of neurotrophic relations is the question by what mechanism are the mutual influences between the nerve cell, and the peripheral organ realized. The „specific" function of the neuron, i. e. the transmission of nerve impulses, is without doubt also important for the maintenance of trophic functions, but it cannot only be the impulse mechanisms which come into play. We must assume that the transmission of signals, their integration and storage is a pressupposition of nervous function and also of the initiation and realization of neurotrophic relations. All regulatory mechanisms are linked to "information" or signals, and reflex mechanisms must be assumed for both synaptic mechanisms as well as for trophic processes which maintain the structural and functional integrity of the tissue. The controlling mechanisms of the organism can function only on the basis of information obtained. However, these signals need not necessarily be of impulse character. Besides nerve impulses connected with the rapid synthesis and breakdown of specific mediators (especially acethylcholine), chemical signals producing slow changes of longer duration, i. e. hormones produced by endocrine glands and also products of neurosecretion are known. These chemical compounds may be important transmittors of the "trophic function" of the nerve cells. Evidence for the action of such chemical compounds of neurotrophic relations comes especially from two sources, i. e. from observations on the neurosecretory activity of nerve cells and from observations on the effect of tissue extracts on differentiation of nerve cells.

The first finding of secretory activity in nerve cells applies to "gland-like" cells in the spinal cord of sharks (Dahlgren 1914, Speidel 1919) and this activity was then conclusively demonstrated in special groups of neurons especially of the hypothalamic nuclei (Scharrer and Scharrer 1954). These cells are true neurons with axons and dendrites, Nissel's substance and neurofibrils and exhibit morphologically demonstrable evidence of secretory activity (viz Ortmann 1959). Neurosecretory material in the posterior lobe hormones is produced by nerve cells in the hypothalamic nuclei and these substances are transported along the axons in fibre tracts passing to the posterior lobe of the pituitary. Although the capacity of certain neurons both to conduct impulses and to carry on neurosecretion is still poorly understood (Ortmann 1959), the existence of neurosecretory processes acting as mediators between the nervous and endocrine system has been proved unequivocally. The fact that the streaming of cytoplasm from the region of the perikaryon to the periphery of the nerve cell can be demonstrated also in the peripheral nervous system, suggests that we are dealing with a phylogenetically old mechanism of nervous function having a general significance and possibly being an essential manifestation of its trophic function. If we assume a more general significance of "neurosecretory" processes concerned with trophic processes we are led to the question of the nature of the substance subserving the "non-impulse" relations between nerve cell and periphery. Some evidence for these processes is again found in morphogenetic studies. The nerve cell can react to extirpation of the peripheral field by degenerative changes even before growing axons approach the tissue which they should innervate (Hamburger and Levi Montalcini (1950). The neuron thus cannot be influenced by information connected with the conduction of nerve impulses via the axon. Neither can it be assumed that the compensatory growth of axons from neighbouring nerves into a denervated skin area (Weddell et al. 1941), or of motor nerve fibres (Edds 1953, Hoffmann 1952) into a partially denervated muscle are mediated by signals in the nature of nerve impulse conduction. These are evidently non-impulse in character and participate in the so-called repercussion phenomena as these neurotrophic relationships are sometimes called in Soviet literature (viz Golikov 1950). They include all the changes which occur in regions of the organism distant from the place of the original lesion. One of the best known phenomena, for example, are excitability changes (prolongation of chronaxy) on the healthy extremity following nerve injury in the contralateral limb (Uflyand 1950). What is the basis of these non-impulse mechanisms? It is probable that these are chemical "signals", but physical mechanisms, e. g. the "pumping action" of muscle contractions or changes of muscle tone, which might play a role in intracellular transport of substances, cannot be excluded (v.

Buller et al. 1959). A new approach to questions of the mechanism of these non-impulse effects of the nervous system is presented by the experiments of Levi Montalcini and Hamburger (1951, 1953) who found hypertrophy of sympathetic and spinal ganglia in chick embryos after implantation of mice sarcoma into the coelom cavity. The hypertrophy was not caused by direct innervation of the implanted sarcoma since a sarcoma implanted into the chorio-allantois, i. e. outside the reach of ganglion cells, also had this effect.

Fractionation of the sarcoma extract showed that the effective component of the neoplasm are nucleoproteins (Cohen et al. 1954). The agent causing the hyperplasia of the ganglia and which must have reached the embryo by direct diffusion or via the circulatory system is, or is at least associated with, a ribonucleoprotein (viz Hamburger 1954) and it may be related to substances which may also be concerned with the trophic influence of the nervous system on peripheral tissues. Thus impulse and non-impulse mechanisms play a role in neurotrophic processes. These non-impulse mechanisms can represent an integral part of the processes studied within the scope of physiology of neurotrophic relations and may thus explain some metabolic disturbances resulting from muscle denervation.

D. The trophic influence of the nervous system — an historical survey

As has been shown above, the physiology of neurotrophic interrelations requires a combined study of the regulation of fundamental metabolic processes in the cell and of the regulation of physiological functions. Both these aspects were not always and could not be synchronized and it is thus comprehensible that many questions, e. g. the question of the trophic influence of the nervous system, were postulated prematurely. Early investigations were primarily concerned with morphological changes occurring in peripheral organs following nerve section and it was not until later that this problem was approached from the biochemical aspect. Control mechanisms affecting cellular metabolism by way of the nerve cell functions could not be analyzed successfully until our knowledge of cellular regulating mechanisms was more complete. It was not until then, for example, that the above mentioned "pacemakers" of metabolism could be demonstrated as reactions which may mean alternate metabolic pathways or different metabolic routes which may be under the influence of the nervous system (Krebs 1957). This is fundamentally, without doubt, a question of metabolic adaptations to altered requirements and only the first steps have so far been made in elucidating the role of the nervous system in mediating this adaptation whatever its mechanism (by the activation or inhibition of enzymatic systems or by altered metabolic pathways).

Studies of denervated muscle are thus centred on physiological aspects of neurotrophic relations in the periphery which occur when these relations are impaired. Denervation studies approach the problem from the negative, eliminating aspect. Their aim is to analyse the trophic influence of the nervous system by means of experiments in which this function of the nervous system is excluded. This is only one aspect of the study of the trophic influence of the nervous system and cannot by itself be sufficient for the final solution of the basic questions. However, it was this aspect of the trophic function of the nervous system on which attention was first focused and its investigation has dominated the whole problem up to the present time. Although innumerable reports have been devoted to these problems, it cannot be stated that they have led to clear conclusions.

Research on the trophic influence of the nervous system was initiated by clinical observations long before fundamental questions concerning the regulation of trophic processes could be formulated in terms of the regulation of cellular metabolism. The question, whether nerves are important for tissue nutrition, for its "trophic state", was asked by physiologists long ago. This question was already formulated by Prochaska: "Are nerves necessary for nutrition?" and by this he understood the "maintenance" of the human organism. At that time the problem was discussed from the point of view of humoral physiology as is evident from the following quotation of Prochaska (1784): "As is known numerous physiologists, old and new, have maintained that nutrition is mediated by nerves. For example, Fr. Sylvius, Willis, Glisson and others assumed two kinds of nerve fluids: a dense, protein fluid which serves nutritional processes; the other vaporizing, more liquid, diffused in the former and destined for eliciting motion and sensation." Prochaska formulated the problem more precisely: "The question arises, however, whether nerves also participate in the substitution of lost particles and thus in the building up and renewal of our organism? Without doubt, the force which incorporates nutritional components into individual parts of the body, is admirable. This takes place with such perfection and wisdom that suitable and corresponding particles are incorporated in such a way as not to disturb, by perpetual assimilation of new molecules either the composition or character of the components, or the shape and structure of the organs which are to be sustained."

Trophic disturbances in peripheral tissues following section of the nerve necessarily drew the attention of clinical and experimental workers. For example, Romberg (1846) described hemiatrophy of the face („hemiatrophia faciei progressiva"), but did not find changes in sensitivity and motility, so that he considered this to be a trophoneurosis. Most important observations from the clinical point of view were those of neurosurgeons who encountered nerve injuries accompanied by considerable trophic disturbances, e. g. cases of causalgia leading to especially serious trophic changes (Mitchell et al. 1864). Soon the attention of experimental investigators was drawn to these "trophoneuroses". In 1824 Magendie sectioned the trigeminal nerve and demonstrated that the trigeminus is not only a sensory nerve but that it also has trophic functions. Questions of trophic changes after nerve section were then approached especially by Claude Bernard, namely after section of the trigeminal and vagus nerves. Claude Bernard observed vasodilatation and increased temperature after section of cervical sympathetic nerve fibres and this concentrated interest to the role of the vegetative nervous system in trophic processes. Magendie (1824) observed keratitis following section of the trigeminal nerve and in 1849 Brown-Séquard described trophic disturbances after section of the sciatic nerve. Similar observations were described by Novinski (1874) and Charcot (1876). At this time the character of the trophic influence of the nervous system on peripheral tissues was already being discussed and the existence of special trophic nerves was postulated. This assumption was formulated by Samuel (1860) in a monograph "The trophic nerves". Samuel attempted to prove experimentally that vasomotor nerves do not exert a direct influence on tissue metabolism. He therefore assumed that there is no other alternative than to admit the existence of trophic nerve fibres, but stressed that his conclusions were made by a process of elimination. According to this author the trophic influence of nerve fibres was defined in direct relation to metabolism of the cell. The results of Claude Bernard and Samuel founded the theory of trophoneuroses. However, the experimental method was primitive and the results were not in agreement with each other. Fortuitous clinical observations could not be satisfactory for the physiologist accustomed to evaluate results on the basis of analytical studies, so that Sechenov again put forward the question: "Do nerves influence nutrition?" (1858) and stressed the complexity of the whole problem of the trophic influence of the nervous system by stating (1866): "There are few questions in the physiology of the nervous system which have undergone such sudden changes as the questions concerning the nervous control of tissue nutrition." Problems of mechanisms of nervous trophic influence were formulated by him as follows: "These processes which ensure the anatomical and functional integrity of tissues during life are generally designated as nutritional processes. These functions enable muscles, nerves, bones and other tissues in the organism to remain unchanged as far as their physiological properties are concerned. Although the existence of these processes is still unknown, it is usually accepted that only two fastors take part in these processes: tissue elements and metabolites supplied by the blood". These considerations led him again to the question "whether metabolic processes which ensure the anatomical and physiological integrity of the tissue are not also dependent on other nervous influences (besides vasomotor)". Like Claude Bernard, Sechenov also goes back to the experimental evidence obtained by sectioning the vagus and trigeminal nerves. He rejected the effect of hypothetical trophic nerves on pulmonary changes following bilateral vagotomy, but was less certain about the trophic changes resulting from section of the trigeminal nerve. On the whole Sechenov appeared to be skeptical about trophic nerves similarly as Claude Bernard (1871) who wrote: "These nerves should directly regulate the metabolism of an organ, the assimilation processes and the dissimilation of elementary particles inside the organ. However, their anatomical existence has not been proved by anybody. The absolute necessity for their existence has been proved neither physiologically nor pathologically. Without doubt, the nerve fibres regulating the functional activity also regulate metabolic processes and can thus be considered, from this point of view, to be the trophic nerves. — As far as I am concerned, I presume that trophic nerves are actually vasomotor nerves, vasodilator in function, and this I have demonstrated after section of the trigeminal nerve" (1871). It can be seen that clinical interpretations and observations represented for a long time the leading aspect in the research of the trophic influence of the nervous system. Even though the main questions were already precisely formulated, exact physiological aims in the study of neurotrophic interrelations could not, at that time, be formulated accurately enough. The whole problem of nervous regulation of trophic processes could not be stated clearly. For this to be possible deeper knowledge of nervous regulation of physiological functions and of the regulation of fundamental metabolic cellular processes would have been necessary. For this reason the concept of "trophic nerves" was postulated prematurely, since it attributed trophic functions to special peripheral nerves, thus instituting a new, unfounded hypothesis. Clinical observations, however, played a very important role from the historical point of view and led to the extension of the reflex theory to trophic processes. This was mainly due to the work of Pavlov. Pavlov himself was inclined to acknowledge the existence of trophic nerves and his first ideas concerning specific trophic nerves were based on observations of cardiac function. He showed in his treatise on "Centrifugal cardial nerves" that there exist various cardiac nerves of different functional significance. Pavlov found that changes of the cardiac rhythm are not connected with changes in blood pressure. It was concluded that cardiac function is regulated by four centrifugal nerves. namely nerves retarding, accelerating, decreasing and increasing cardiac output. Following this Pavlov formulated the question of these nerves affecting cardiac output and further inquired whether they are vasomotor nerves or nerves regulating specific function.

"It is either a special trophic cardiac nerve, so long and so systematically sought for by physicians, or it is a simple vasodilatatory nerve innervating cardiac vessels." Pavlov returned to these questions in later work (1920). In a study on pathological reflexes evoked from the peritoneal cavity, he writes that during various operations performed with another aim in mind, various trophic disturbances, e. g. stomatitis, skin trophic ulcers etc. could be observed. This observation led him to the concept of so-called trophic reflexes: "It seems that all these phenomena described can best be explained if it is assumed that trophic reflexes come into play." Stimuli without effect on tissues under normal conditions, evoked stomatitis, inflammatory skin changes, etc. in operated animals by way of trophic reflex changes. Even though it is possible that some of the trophic changes observed were due to general nutritional disorders, e. g. avitaminosis, incompletely understood at that time, it can be said that these considerations were of primary importance for further development of the reflex theory especially from the point of view of extending it to longlasting processes, i. e. trophic processes. Reflex processes are generally defined as rapid processes and Pavlov pointed out that reflex mechanisms must also include longlasting processes. When analyzing this question of trophic innervation later (1920), Pavlov indicated that neurotrophic phenomena could not at that time be subjected to precise experimental investigation. In spite of this he wrote: "I have come to the conclusion — not experimentally, but gradually — that the clinical conception about the existence of special trophic nerves is acceptable." He mentions his own observations of trophic disturbances following surgical operations of the alimentary tract and continues: "... most of these facts have confirmed my assumption that the phenomena described could be explained as reflexes from pathologically excited centripetal nerves of the alimentary tract to special trophic nerves in different tissues." In a following report Pavlov and Rosenthal (1920) attempted to elicit trophic disturbances following various operations of the alimentary tract, but very inconsistent results did not make a precise analysis of nervous mechanisms possible. In spite of this Pavlov believed that the question would soon be solved experimentally: "It is possible that the trophic nerves are already in the hands of physiologists." Ideas analogous to those developed by Pavlov on the basis of observation of cardiac nerves were also held by Heidenhein (1886), who assumed trophic nerves to salivary glands, even though he used this term conditionally. Heidenhein himself differentiated secretory and trophic nerves and the secretion of saliva was subdivided by him into processes producing saliva and those producing and secreting organic constituents of saliva. The conclusions of Heidenhein and Pavlov are thus quite similar. Pavlov summarized his own opinion about trophic functions in the following way: "In this way every organ is under the influence of triple nervous control, i. e. functional nerves which either evoke or interrupt functional activity (muscle contraction, secretion in glands etc.), vasomotor nerves regulating roughly the supply of chemical material (and removal of metabolic products) by increasing or decreasing blood circulation through the organ and finally trophic nerves, which control in the interests of the organism the exact amount of final utilization of this material in each organ. This triple control we have demonstrated in the heart". From this survey of Pavlov's ideas concerning problems of the trophic function of the nervous system it is evident that Pavlov used the expression "trophic nerves" being aware that mechanisms of these trophic processes were not precisely defined. However, his conclusions about trophic reflexes have been very important for further development of views on the trophic functions of the nervous system by expanding the field of the reflex theory.

This conception of trophic reflexes was further studied most systematically by Speranskii and his coworkers. Speranskii stressed the reflex character of neurodystrophic disturbances resulting from section of peripheral nerves. The above author came to the conclusron (1935) that "the origin of trophic ulcers lies in reflex mechanisms ... excitation of the central stump of an injured nerve initiating the whole process". "The stimulation of any part of the complicated nerve net may evoke the excitation not only of neighbouring structures but also of more distant regions of the organism." A number of experiments inspired by these ideas investigated trophic disturbances following nerve section in relation to the excitation of the central nerve stump (e. g. Golub 1944, Borovški 1952 and others). These experiments corroborated the hypothesis of Speranski about the role of the nervous system in disease and recovery and led to the concept of "pathological reflexes". This author himself rejected an isolated study of the trophic function of the nervous system in the organism: "The trophic function of the nervous system in the organism does not act in an isolated manner, but in connection with all other functions ... every nervous process is followed by changes in the biophysical and biochemical state of the tissue, i. e. changes in its trophic state" (Speranskii 1952). In an earlier statement he says: „If a nerve is secretory, it also has a trophic effect. The one cannot exist without the other." Concepts of Speranskii were not concisely formulated and include mechanisms which could not yet be analyzed accurately. The "region" from which nerve processes spread and assumed mechanisms which determine the temporal sequence of processes taking place inside the nervous system are not formulated clearly. According to these ideas, excitation is non-specific in character with a certain latency and by this mechanism Speranski explains the initiation of trophic disturbances occurring after a certain latency period in the region of healed trophic lesions during intercurrent specific infectious processes. This has been termed the "second attack" by Speranskii. It is thus possible that the extinguished neurodystrophic process can be renewed during the second nerve trauma, which may be distant from the site of the first. These notions are thus closely related to Ukhtomski's conception of dominant processes (Speranskii 1937). Experimental observations of Speranskii and coworkers have attempted to show the manifestations of these "pathological reflexes". Generalization of the "laws" governing nerve functions in terms of mechanisms which had not been analyzed, could have only limited success. It is possible to agree with Speranskii that the role of the nervous system during various diseases is much greater than has hitherto been believed. From this aspect, Speranski's contribution has been important and extensive. Speranskii simplifies the problem by rejecting the possibility of differentiating trophic functions of the nervous system, even though he intends to carry on the research of disturbances of this trophic function. This procedure, however, cannot lead to a satisfactory approach to questions dealing with the trophic influence of the nervous system. It enables a phenomenological description of the reflex nature of neurodystrophic lesions to be made, but it cannot achieve a concrete analysis of the influence of the nervous system on the metabolism of peripheral tissues.

It is, of course, not possible to underestimate the significance of reflex mechanisms in trophic processes. A further contribution was the investigation of conditioned reflex mechanisms which led to research on cortical influences on trophic processes in the work of Bykov (1941). Other important results, e. g. the work of Petrova (1946) indicate the role played by Pavlov's "trophic reflexes" in trophic lesions in neuroses. These investigations deserve extension even though sometimes not too adequate approaches have been chosen, for example in experiments intended to demonstrate the influence of cerebral cortex on trophic processes, in which ablation experiments were performed (Bayandurov 1949). Questions of the trophic influence of nerves cannot, however, be elucidated by investigating only central regulations of trophic processes. The physiologist must return to the question of the mechanism by which the trophic influence of the nervous system is mediated. Here, denervation experiments can still be very helpful. Thus the main fundamental problem is as follows: By what mechanism is the trophic influence of the nervous system on muscle mediated? This problem can be concentrated on two questions:

1. What components of the nervous system participate in trophic processes?
2. By what mechanism is this trophic influence of the nervous system realized?

E. Mechanisms of the trophic influence of the nervous system

1. Components of innervation affecting trophic processes

a) The trophic function of the motoneuron

The conception of reflex regulation of trophic processes assumes that various components of the nervous system participate in this regulation, i. e. nervous centres as well as afferent and efferent reflex pathways. However, in muscle functions it can be pressumed that the principal role is played by the motoneuron which represents the "final common pathway" in trophic regulations, as in the integration of reflex functions (Sherrington 1906). To what extent and in what way do other peripheral components of the nervous system participate? The research on neurotrophic interrelations was in the first place based on observation of patients in whom marked dystrophic changes of the skin occurred after peripheral nerve lesions. Vasomotor disturbances of the denervated extremity were so evident that soon the decisive factor in the trophic influence of the nervous system on muscle metabolism was assumed and sought in vegetative nerve fibres (see also Chapter VII). Considerable time elapsed before the leading role of motor nerve cells for trophic processes in muscle was acknowledged. This question was decided by denervation experiments. It is, of course, difficult to differentiate individual components of innervation from each other after sectioning a mixed peripheral nerve, as far as their effect on muscle metabolism is concerned. It is possible, however, to interrupt individual nervous components contained in the peripheral nerve, i. e. somatic motor and afferent nerve fibres and sympathetic nerve fibres. This is made possible by the fact that in most animal species somatic innervation of the hind limbs originates in lower spinal segments than is the case for the sympathetic nerve fibres (viz Sheenan and Marrazzi 1941 in the cat and monkey, Siems and Kosman 1949 in the dog). A number of authors found in such experiments (Tower 1935, Siems et al. 1948, Slozhenikina 1957, Kutsenko and Nechayeva 1953 and others) that maximal metabolic changes take place only following section of motor nerve fibres, i. e. anterior rhizotomy. Tower (1939) summarizing her own results stated that "... all the results of total denervation of muscle were reproduced by ventral root section only ... The reaction to denervation in skeletal muscle therefore develops entirely as the result of motor denervation, the other innervations of the organ, skeletal muscle, making no contribution whatever". Tower's experiments (1935) demonstrated the significance of motor neuron lesions as the primary factor for the initiation of denervation atrophy and degeneration, and from this point of view it is possible to consider motor nerve fibres to be the trophic fibres of skeletal muscles.

b) Trophic function of vegetative nerve fibres

It has already been pointed out why in the original hypotheses about the mechanism of denervation muscle atrophy the main role was attributed to vasomotor disturbances. A trophic function of vegetative nerve fibres has also been used to explain metabolic disturbances in denervated muscle. According to this hypothesis muscle atrophy was caused by disturbances in blood circulation which resulted in insufficient substrate supply to the muscle. For example, Ricker and Ellenberg (1899) and Ricker (1901) considered venous hyperaemia to be of exceptional importance. This hyperaemia was caused by the paralysis of these blood vessels due to sectioning of vasomotor nerve fibres and the absence of muscle contractions. Vasoparalytic origin of muscle atrophies was also sought by Schiff and Zak (1912), while Brown Séquard (1872) considered neurotrophic skin disturbances (trophic ulcers) to be of ischemic origin. Vasomotor mechanisms in muscle atrophy were also assumed by Bechterew (1900), Lapinski (1904) and lately Beritov (1947). The trophic role was usually ascribed to sympathetic nerve fibres. Since most sympathetic nerve fibres innervating limb muscles pass along the peripheral nerve and not along blood vessels (Barcroft et al. 1943, Clonninger and Green 1955, Gilding 1952, Lehman 1930, Moore et al. 1933), section of the nerve is followed by circulatory changes similar to those after sympathectomy. The trophic influence of sympathetic nerve fibres could, of course, be realized by other effects besides vasomotor changes by which nutritional supplies are transported into the muscle, e. g. by influencing the muscle cells directly. The direct trophic influence of the sympathetic fibres on muscle would require direct innervation of muscle fibres by sympathetic nerve fibres. Boeke (1927), Agduhr (1920) and recently Coronini et al. (1957) published data offering histological evidence that sympathetic nerve fibres end directly on muscle fibres. However, the majority of authors did not find sympathetic nerve fibres directly innervating muscle fibres (Hinsey 1927, Wilkinson 1934, Tiegs 1953, Gurkov and Bast 1958 and others). The main evidence for the trophic influence of the sympathetic nerves was Orbeli's theory of the adaptative-trophic function of the sympathetic nervous system (Orbeli 1949). The basic evidence for this theory was provided by the finding of Ginetzinsky (1923, Ginetzinsky, Nekhoroshev and Tetyaieva 1927), who reported restitution of muscle contractions in a muscle exhausted by repeated twitches, when sympathetic nerve fibres were stimulated at tetanic frequencies. It seems that this phenomenon is not mediated directly by the sympathetic nerves but indirectly by way of blood circulation (Beritov 1947) and is missing following adrenalectomy (Belovintseva 1948). Aleksanian and Nasledov (1959) have demonstrated, however, that the effects of sympathetic stimulation and adrenaline on the muscle are fundamentally the same, i. e. both facilitate a more rapid mobilization of all muscle fibres during muscle contraction. This finding can perhaps indicate that sympathetic nerves, like the motor neuron, affect muscle fibres by way of mediators. But very little has been said concerning the trophic influence of the sympathetic nervous system. It is surprising, for example, that such wide differences and often contrary findings have been reported after sympathectomy. According to Dworkin et al. (1931), Popov and Mikhailova (1935), Baum et al. (1934), Hoffmann and Wertheimer (1928) and others the content of glycogen in the muscle following sympathectomy is higher, according to others (Britten 1930, Herrin and Meek 1931, Veselkina 1938, it is on the contrary decreased. Finally, in other reports, no differences were found (Beattie et al. 1930, Kutsenko and Nechaieva 1953, Slozhenikina 1957).

Hudlická (1954) also found no changes in the content of glycogen following sympathectomy in the rat (Tab. 1).

Table 1

The content of glycogen in muscles of rats following sympathectomy.

Experimental situation	No. of animals	Mg % glycogen		E/C in %
		Control	Experimental	
4 hrs. after periart. SE	7	293	257	89·0 ± 5·0
24 hrs. after periart. SE	10	801	775	97·2 ± 3·6
6 days after periart. SE.......	10	628	707	113 ± 15·3
6 days after periart. and unilateral lumb. SE	19	1004	988	98·7 ± 14·5
6 days after periart. and bilateral lumb. SE	15	1271 ± 79	1196 ± 48	

As can be seen, the content of glycogen did not change in any of the above groups, although a clear increase in blood flow occurred after sympathectomy (rise in intramuscular temperature 24 hours after periarterial sympathectomy was 1·6°C on the average). Similar differences after sympathectomy were found in oxygen consumption. Gilbertini et al. (1955) and Petkovic et al. (1958), for example, described an increase in oxygen consumption, Zubenko (1954), on the other hand, reported a decrease, while Nakamura (1921) and Slozhenikina (1957) found no changes in oxygen consumption after sympathectomy. These contrary findings are probably caused by species differences, functional characteristics of muscles, etc. Most convincing evidence against the trophic role of sympathetic nerves has been put forward by authors who investigated the trophic state of peripheral tissues after sympathectomy and found that there was no muscle atrophy (Nevin 1930, McCullych et al. 1930, Tower 1926, 1939). Following sympathectomy, not even changes in muscle tone occur (Ranson and Hinsey 1927), so that it is so far dubious, if sympathetic nerves in skeletal muscles have other functions besides vasomotor regulation. The function of sympathetic nerve fibres in the skin is, of course, different, since here the origin of skin trophic disturbances is much more dependent on the sympathetic nerves (Zellweger and Mort 1950).

A trophic influence on tissue metabolism has also been ascribed to parasympathetic nerve fibres. Ken Kuré and coworkers (Kuré et al. 1928a, Kuré et al. 1928b, Kuré et al. 1930) attempted to prove that it is this component of the vegetative nervous system which is responsible for trophic regulation. According to these authors the dorsal roots contain fine nerve fibres which are of intraspinal origin, are interpolated in the spinal ganglion and exert a trophic influence on skeletal muscle. The above authors even described changes following section of dorsal roots analogous to muscle changes in muscle dystrophy. A similar view on the trophic role of parasympathetic nerves is also held by Vrubel (1958).

This attractive hypothesis concerning the trophic role of parasympathetic nerve fibres is not, however, sufficiently founded either morphologically or functionally. Most authors studying this question came to the conclusion that dorsal roots contain nerve fibres of exclusively extraspinal origin (their trophic cell is localized in the spinal ganglia — Sherrington 1894, 1897) and that the dorsal roots do not fulfill an efferent function (Hinsey 1927, 1934, Horton and Smith 1897, Nevin 1930, Westbrook and Tower 1940 and others). The ratio between nerve fibres and ganglion cells is about 1 : 1 (Duncan and Keyser 1937, Davenport and Bothe 1934, Storey et al. 1936) and furthermore it seems that there is no synaptic interruption in the ganglion (e. g. Meshcheriakov 1954), even though this question does not appear to be quite clear (Dogiel 1908, Rose 1935 and others). Even degeneration studies were of no avail and did not help to decide this question conclusively. In spite of this, however, the non-appearance of degenerating nerve fibres in the peripheral stump following section of the dorsal roots proximally from the ganglion (Storey et al. 1936) speak against Kuré's theory. The small, finely myelinated nerve fibres in muscle nerves which were considered by Kuré to be parasympathetic, are probably fibres which were later found by Leksell (1945), Hunt and Kuffler (1951) and others to innervate intrafusal muscle spindles.

Vegetative nerves thus cannot exert any direct influence on skeletal muscle, since neither sympathetic nor parasympathetic nerves have been shown to end directly on muscle fibres. Following section of the peripheral nerve mainly vasoconstrictor nerve fibres are affected. In accord with this are findings that blood flow increases in consequence of denervation (see chapter VII), so that it is improbable that circulatory disturbances play a substantial role in denervation atrophy.

c) The trophic function of afferent nerve fibres

First concepts about the trophic function of afferent nerve fibres were formulated in connection with the work of Claude Bernard on trophic ulcers following transection of sensory nerves, as for example, after section of the trigeminal nerve. The occurrence of trophic ulcers after nerve section was explained either as the result of reflex excitation initiated in the central stump or of elimination of innervation (Khlopina 1957). The former theory was propagated first by Speranski (1935), who considered the excitatory process in afferent fibres in the central stump of the nerve to be the source of the reflex disturbance. Of course, trophic ulcers develop even without excitation of the central stump after section of the nerve (Speranski injected formaline or turpentine oil into the central stump) and this theory does not seem to be plausible. The main argument against this idea has been put forward by Khlopina (1957), who demonstrated that section of the dorsal roots proximally to the spinal ganglia does not lead to trophic disturbances and that the decisive factor is the connection with the trophic centre, in this case the spinal ganglion cell. Thus if afferent nerve fibres play a role in trophic disturbances, this is caused by the elimination of trophic influence of nerve cells in spinal ganglia. Section of dorsal roots has been reported to result in trophic disturbances by several authors (Pineles 1891, Mott and Sherrington 1895, Kopczynski 1907, Hering 1897, Ranson 1928 and others). Wyburn (1950) assumed a direct trophic effect of afferent nerve fibres (cholinergic unmyelinated C fibres) exerted antidromically, these being the fibres of small nerve cells localized in spinal ganglia. This contention of the trophic effect exerted by spinal ganglion cells is also held by Kutsenko and Nechaieva (1953), who reported that greater atrophy of the gastrocnemius muscle occurs in the cat and dog following extirpation of spinal ganglia than after section of the dorsal roots proximally to the ganglia. These authors also observed trophic changes of the skin after extirpation of spinal ganglia which were not observed in the second case. We were not able to corroborate this hypothesis in our experiments on rats, since we found that the atrophy of the gastrocnemius and soleus muscles a month after the extirpation of spinal ganglia was the same as after section of the dorsal roots proximally to the spinal ganglion (Hník 1956b). The main arguments against the direct role of afferent fibres in the regulation of trophic processes are to be seen in experiments indicating that section of the dorsal roots is not followed by trophic changes comparable in extent to those found after section of the peripheral nerve or anterior roots (Tower 1939). Atrophy of the deafferented soleus four to five months (120—150 days) after deafferentation is about 30%, while after denervation, muscle atrophy is about three times greater at this period (see Chapter II). After section of the dorsal roots the weight loss corresponds evidently to simple muscle atrophy, without degenerative changes, which is localized, however, only in deafferented muscles of the physiological extensor group (m. gastrocnemius and m. soleus) and not in flexor muscles (m. tibialis ant. and m. extensor dig. longus) — (Hník 1956b). Muscle atrophy following dorsal rhizotomy could also be explained by reflex motor disturbances (Sherrington 1900), i. e. by an indirect mechanism. It was also assumed that following section of the dorsal roots the protective role of afferent signalization from deafferented tissues is excluded, so that trophic changes occur in deafferented limbs, and namely on the skin (Corbin and Hinsey 1934). According to Grigorieva (1951) trophic disturbances in deafferented tissues are primarily caused by inflammatory infiltrates as the organism attempts to rid itself of the deafferented tissue as a foreign body. But inflammatory processes are to be found especially in those cases where trophic ulcers have developed in the skin. Inflammatory changes are then secondary and cannot thus be considered as evidence for the trophic significance of afferent nerve fibres (see Chapter II). From this it can be concluded that neither circulatory changes nor vasomotor vegetative nerves nor afferent nerve fibres play a major role in the regulation of trophic processes in the muscle. It is thus evident that the somatic motoneuron is the decisive factor in regulation of trophic processes, as has already been stressed by Tower (1939). The question then remains, by what mechanism is this trophic effect of the motoneuron realized.

2. The role of specific muscle function in regulation of trophic processes

The concept of the "trophic influence of the nervous system" assumes that besides specific nerve and muscle function, which is associated with impulse activity, there exist non-impulse, so far undefined, nerve mechanisms affecting muscle metabolism. If all aspects of the nervous regulation of muscle metabolism could be explained by mechanisms of impulse activity, then the problem of the "trophic influence" of the nervous system would lose its raison d'être. If denervation atrophy is only "atrophia ex inactivitate", then it is superfluous to assume and study trophic functions of the nervous system. This is why denervation studies are so important. The comparison of various muscle atrophies of different origin (denervation, immobilization, atrophy following nociceptive stimulation, section of tendon etc.) has, for more than a hundred years, instigated the question whether in all these cases the cause of atrophy is not solely inactivation. One of the first who formulated this hypothesis was Harless (1799), who discussed the possibility that denervation atrophy is the direct result of muscle inactivity. This necessarily led to the conclusion that the nervous system is able to affect muscle metabolism only via impulse activity, i. e. by means of specific function — muscle contraction. For this reason Sherrington (1900) considered all nerve fibres affecting in any way muscle function to be also, at the same time, trophic nerves, and similarly Speranski (1952), Piťha (1937) and Bykov and Vasyutochkin (1954) concluded that muscle metabolism is affected by specific function and that the assumption of an independent trophic influence of the nervous system is unnecessary. In terms of the electrical theory of synaptic transmission, all metabolic changes in the muscle could then be explained by activation of postsynaptic elements, the action potential of presynaptic fibres being the stimulus for electrically excitable elements. Of course, this theory cannot be applied in this form at the present time. This can be demonstrated in experiments with curare, which blocks neuromuscular transmission, without modifying the spike potential of presynaptic fibres and need not change electrical excitability of muscle fibres in order to elicit the block (Rosenblueth 1950). This theory of the neuromuscular "specific" function was completed soon by the chemical theory of synaptic transmission. It was first shown by Kibjakow (1933) that a chemical agent could serve as a "transmitter" at synapses, and Feldberg and Gaddum (1934) demonstrated that the chemical substance liberated is acetylcholine. Finally it was proved that motor nerve endings in skeletal muscles are cholinergic (Dale and Feldberg 1934). Evidence about the role of acetylcholine in synaptic transmission was then increasing. The chemical theory of synaptic transmission by acetylcholine ascribes the role of activation of postsynaptic cells to acetylcholine liberated on presynaptic fibres (Nachmansohn 1955). This "cholinergic transmitter" is then capable of causing local depolarization on the endplate and this potential can evoke either a local or a propagated reaction. It was the work of Fatt and Katz (1952) on the miniature endplate potentials initiated by subthreshold liberation of acetylcholine which opened up the possibility of explaining all or the main nervous mechanisms affecting muscle metabolism on the basis of acetylcholine production. The main changes in muscle metabolism would thus be the result of acetylcholine production and related muscle contraction and for explanation of other phenomena subthreshold liberation of quantal amounts of acetylcholine would be sufficient. Denervation atrophy would then be the result of absence of muscle contraction and electrotherapy then represents a direct projection of this explanation in the therapy of muscle atrophy, in which impulses to the denervated muscle are substituted by electric stimulation. In spite of negative findings, the majority of authors are agreed that electrotherapy retards the rate of atrophy, but only under certain conditions (see Chapter X). One of these is that the stimulated muscle must be in the stretched position and should contract under isometric conditions. This condition together with other observations, e. g. that electrotherapy of the muscle after tenotomy is without effect because it is not possible to obtain sufficient tension during contraction (Eccles 1944) have led to the conclusion that atrophy is caused by the inability of the muscle to exert effective tension. Loss of the effective tension mechanism is according to Hines (1942), Solandt et al. (1943), and Eccles (1944) the fundamental mechanism leading to muscle atrophy. In this connection an interesting observation was made by several authors that for muscle atrophy to be slowed down daily stimulation of several seconds, duration is sufficient (Eccles 1944, Hines 1950). According to Fischer, muscle proteins are most probably a thixotrophic gel which requires at least minimal contraction changes in order to prevent irreversible changes. Of course, if the trophic regulations could really be reduced to problems of muscle function and the tension mechanism only, then it remains to be explained why electrotherapy does not prevent muscle atrophy, but in the optimal case can only retard its rate. In spite of the effect upon the rate of muscle atrophy, electrotherapy is unable to arrest some of the changes characteristic for denervation. For example, denervation fibrillations (see Chapter V), the increase in DNA content and contractile strength of the muscle are unaffected by daily intensive electrical stimulation.

Besides theories attributing denervation atrophy to the absence of specific muscle function, a hypothesis was formulated on the other hand, not sufficiently well-founded, which considers muscle atrophy to be the result of hyperactivity, i. e. exhaustion of the muscle. According to this hypothesis first put forward by Langley (1916), metabolic exhaustion of the denervated muscle occurs due to fibrillation activity, the energy requirements of which cannot be covered by the denervated muscle (Tower 1939). A similar analogy between a denervated and an exhausted muscle was drawn by Ferdman (1953). Langley considered the increased consumption of oxygen in the denervated muscle to be in direct relationship with the fibrillation activity. However, Solandt and Magladery (1942) and Hines (1942) apparently disproved this hypothesis. The above authors demonstrated that the rate of denervation atrophy remained unchanged whether fibrillation activity was increased (KCl, physostigmine, etc.) or decreased (quinidine, quinine etc.). Furthermore, fibrillations occur only in muscles deprived of motor innervation, even though muscle atrophy of different origin (e. g. after section of the tendon, immobilization, nociceptive stimulation and others) can progress at a similar rate to denervation atrophy. For this reason Langley's hypothesis is not generally accepted. The question whether influence of the nervous system on muscle metabolism is realized only by impulse transmission and the trigger mechanism of muscle contraction, remains therefore open.

Following every functional activity of the muscle, energy reserves as well as muscle proteins are renewed. There is no doubt that these metabolic recovery processes, which we consider to be the principal mechanism of the trophic influence of the nervous system, occur without the transmission of nerve impulses (Gutmann 1955) Here then the nervous regulation of trophic processes is manifested as metabolic recovery processes. The question is, however, (and this would prove the existence of the "trophic" influence of the nervous system) if these mechanisms can exist independently of muscle specific function. The answer would provide indirect evidence about the independent existence of the trophic influence of the nervous system and with its help could enable the muscle to recover as soon as possible and resynthetize substances broken down during preceding functional activity. In this connection the question arises whether dissociation between the trophic and specific function is possible, since under normal conditions they are evidently so closely interlinked that they appear to be in one inseparable unity, and furthermore, whether there exists direct evidence about a separate and independent trophic influence of the nervous system on the muscle.

Indirect evidence indicating an independent existence of the trophic influence of the nervous system may be subdivided into two groups: firstly, observations on preserved muscle function in spite of which muscle atrophy sets in, and secondly, observations where muscle atrophy is not present, even though specific function is missing. In both these groups then, evidence is accumulated demonstrating that under certain, chiefly pathological conditions, it is possible to assume a dissociation between the specific and trophic functions.

Electrotherapy of the denervated muscle, for example, which substitutes nerve impulses evoking specific muscle function for direct electric stimulation of the denervated muscle slows down the rate of muscle atrophy under certain conditions. It is typical, however, that even the most intensive muscle stimulation never succeeds completely in preventing muscle atrophy, but only in retarding its rate. If the mechanism of maintaining structural and functional integrity were only a matter of its specific function, intensive stimulation of denervated muscle should, on the contrary, lead to muscle hypertrophy as in the case of direct stimulation of normally innervated muscle (e. g. Siebert 1929). The whole effect of electrotherapy, however, is restricted to retarding of atrophy. According to all authors interested in this question, with the exception of Ginetsinsky (1956), it is never possible to attain even partial recovery of muscle weight already lost by stimulating the muscle during the denervation period (Osborne 1951 and others). It is thus evident that anabolic recovery processes in the muscle are not automatically coupled with or evoked by muscle contractions, but by some other special mechanism requiring the presence of intact motor innervation of the muscle. It is interesting to note in this connection that the effect of electrotherapy is not proportional to the amount of work performed (10 sec. stimulation, for example, has been reported to have the same effect on muscle atrophy as 2 hrs. stimulation — Eccles 1944, Hines 1950) and there does not seem to be therefore a direct relationship between specific function and the "trophic" state of the muscle. But still other examples can be cited in which specific function is present but in spite of that muscle atrophy occurs. For example, deafferentation of the hind limb in rats causes a tendency to extension to develop together with other symptoms indicating that increased excitability of anterior motorhorn cells innervating these muscles has taken place (Hník 1956a). Muscles from the extensor groups are thus functionally more active, but despite this it is in these muscles (gastrocnemius and soleus) that gradual muscle atrophy occurs (Hník 1956b). Similarly, cases of muscle atrophy have been described due to chronic exhaustion (functional overloading) of musculature (Lorenz cit. Tower 1939, Piťha 1937). It is possible that increased metabolic demands of the muscle could not be covered, so that the trophic function of the nervous system was impaired in the sense of a disturbance of metabolic recovery processes.

The second set of observations concerns indirect evidence of "trophic" influences of the nervous system exerted in spite of lack of specific muscle function. It was found, for example, by Denny-Brown and Brenner (1944) and later confirmed by Sunderland (1951) and Gelfan and Tarlow (1956), that it is possible to cause such pressure blocks of the peripheral nerves which abolish nerve conduction, the anatomical continuity of the axons being preserved. Such a block may last for several hours or days and yet muscles do not atrophy. It thus seems, according to Denny-Brown and Brenner, that "atrophy of the paralyzed muscle appears to be prevented by anatomical connection with the motor horn cell in the absence of conducted nerve impulses". Muscle atrophy does not take place even though muscle contraction is abolished for a long time by the nerve block. According to Fischer (1956) these findings indicate that "since the nerve is unable to conduct an impulse through the pressure region and since nerve conduction is probably a function of the nerve membrane, it is apparently the axoplasm which is responsible for the trophic influence". The observation that during reinnervation following crushing of the nerve, muscle was already found to regain weight at a time when restitution of muscle function had not yet appeared (Fig. 3) can serve as further evidence of the fact that trophic mechanisms of the nervous system can act independently of specific muscle function (Gutmann 1942). Moreover, regenerating axons reinnervate the end-plates several days before indirect excitability is recovered (Gutmann and Young 1944) and spontaneous fibrillations in a denervated muscle may disappear before motor unit potentials can be evoked during the course of regeneration (Jasper, 1946, Feinstein et al. 1945). Evidence for the assumption, that the trophic control exerted by the nerve may be independent of its excitatory function is also supplied by the observations of Desmedt (1950, 1959), who showed that, during regeneration of a nerve, chronaxie of the muscle decreases slowly towards normal values at a stage where there is still no response of the muscle to nerve stimulation.

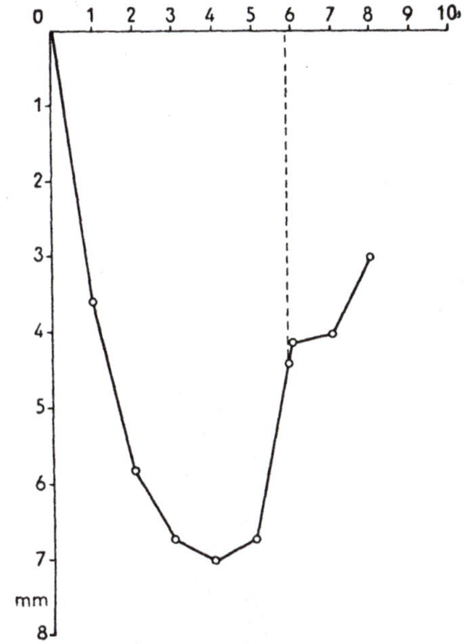

Fig. 3. Average values of the circumference of the calf after crushing the peroneal nerve in 12 animals. The dotted line represents the average time at which the spreading reflex of the toes first reappears following its absence during the denervation period (Gutmann E. [1942]. J. Neurol. Psychiat. 5 : 81). Abscissa: time after denervation in weeks. Ordinate: circumference changes in mm.

These findings point to the significance of mechanisms not connected with nerve impulse transmission, especially the transport of certain substances (besides specific mediators connected with impulse transmission) along the nerve, which may be effective in the regulation of trophic processes. These substances could evoke metabolic changes in the muscle and experiments demonstrating that the rate at which glycogen synthesis decreases following muscle denervation is dependent on the length of the peripheral stump of the sectioned nerve (Fig. 4) indicate that substances transported in the exoplasm are exhausted sooner the shorter the peripheral nerve stump after denervation (Gutmann et al. 1955). It has also been shown that the length of the peripheral stump also determines the rate at which neuromuscular endings degenerate (Gutmann et al. 1955) and at which denervation fibrillations appear (Luco and Eyzaguirre 1955). Jefferson et al. (1958) reported that no atrophy was found for three months after section and suture of the phrenic nerve, even though respiration function is absent. Similarly, no atrophy appeared following cross union between the vagal and phrenic nerves. These authors conclude that "therefore trophic effects may exist in the absence of functional activity and they may be transmitted by another nerve, independent of its original function". However, the authors did not take into account the denervation "hypertrophy" of the diaphragm (see Stewart 1955), so that their conclusions will have to be further experimentally analyzed. Similar conclusions were put forward in connection with the temporary increase of proteins in the denervated diaphragm. This phenomenon of transitory postdenervation hypertrophy was described by several authors (Martin and Sola 1948, Stewart 1955, Stewart and Martin 1956). It is evidently not just a matter of water content increase in the denervated diaphragm, as has been postulated by Thomson et al. (1951) and Thomson (1955), because Stewart (1955) and Stewart and Martin (1956) demonstrated an increase in the content of plasmatic as well as contractile proteins. The latter authors conclude that this phenomenon is probably not due to rhythmic passive movements and tension of the diaphragm, since at a later period the denervated half atrophies in any case. It is also known that passive manipulation is without effect on the course

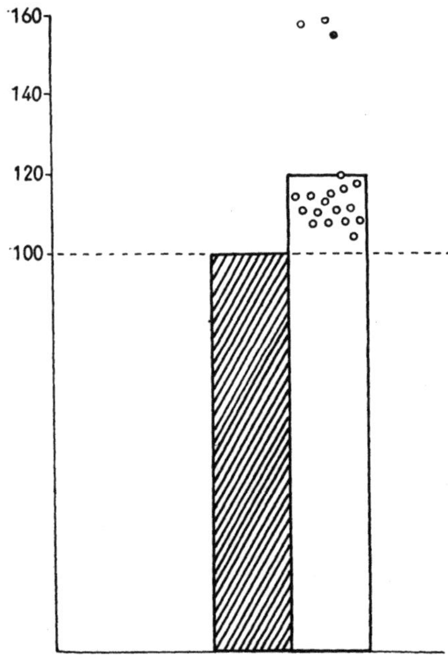

Fig. 4. The glycogen content in rat muscle with a long peripheral nerve stump expressed as percentage of muscles with a short peripheral stump. Black column — glycogen content in tibialis ant. with the short stump — 100 %. White column — glycogen content in tibialis ant. with the long nerve stump. The sciatic nerve was sectioned bilaterally (high section on one side, low section on the other) and the glycogen content was determined 24 hrs. after nerve section and 4 hrs. after peroral administration of glucose (50 % solution in amounts corresponding to 1·5 % body weight).

of muscle atrophy (Hajek and Hines 1947, Pollock et al. 1950). According to Stewart and Martin it is possible that "a substance is released by the nerve perhaps during the period of increased membrane permeability accompanying the nerve impulse. It is suggested that this material as well as activity is necessary for normal maintenance of this muscle" ..."release of substances from the nerve tissue is greater than normal due to degeneration of the fibres". — The influence of the peripheral nerve stump on the muscle has also been shown from the point of view of mechanical properties of the muscle during denervation atrophy (Thomson 1955). This author found that the progressive decrease of extensibility of the denervated muscle is preceded by a transitory period of increased extensibility 7—12 hours after nerve section. This transitory increase of extensibility is not caused by the abolition of nerve impulses, since 12 hrs. novocainization does not result in this phenomenon. — All these above data do not directly prove the existence of the trophic influence of the nervous system, but they indicate, however, that there are mechanisms participating in the regulation of muscle metabolism which are not or need not be coupled directly with muscle specific function, mechanisms mediating the "trophic influence of the nervous system".

3. The existence of an independent trophic mechanism of the nervous system

In the foregoing section examples have been given illustrating the fact that there is a relative independence of the tropic function of a neuron especially during growth or in the course of the regeneration process. The regenerating motor neuron recovers its proteins at a time when connection with muscle has not yet been realized and this renewal is not in any way connected with transmission of nerve impulses. Of course, if neuromuscular connection is not accomplished (e. g. regeneration of nerve fibres into the peripheral stump which had been separated from the muscle), then this process of proteosynthesis is disturbed (Aitken et al. 1947, Evans and Vizoso 1951). What is the nature of these "trophic" non-impulse signals which enable constant proteosynthesis in the neuron and which cause a reduction in proteosynthesis when contact with the peripheral organ is not realized? The progressing front of the regenerating nerve resembles a certain cellular pressure (vis a tergo — Held 1909, or formative turgor — Cajal 1928), or an intra-axonal hydrostatic pressure which maintains the longitudinal orientation of molecular structure inside the axon (Young 1945). The metabolic basis of these processes, of course, is proteosynthesis in the nerve cell which produces and maintains long, longitudinally oriented molecular protein structures in the axon. This proteosynthesis and its changes are manifested in the transient decrease of nerve fibre diameters in the central stump during nerve regeneration (Gutmann and Sanders 1943), the "proximodistal flow" of axoplasm following section of the nerve (Young 1937, 1944, Lubińska 1952, 1956) or the accumulation of axoplasm in the nerve above a ligature (Weiss and Hiscoe 1948). These proteosynthetic processes, so suggestive of continual axon growth, are a characteristic manifestation of the trophic function of the neuron continuously renewing and maintaining its structure. From this point of view the trophic function of the neuron is represented by a process of continual growth (see also Weiss 1954) which, of course, has its special energy requirements. Bodian (1947) wrote in this connection: "The morphologist and electro-physiologist often find it convenient to visualize the nerve cell as a relatively stable, if not rigid, system which serves as a connexion or a conduit between two distant points. This idea has its uses, but overlooks the possibility that the biggest job which the neuron may perform, in terms of biochemical work, is the job of continual replacement which is required to maintain its cytoplasmic dependency, the axon ... Perhaps this activity of the neuron is as characteristic as its work of conduction, and possibily the energy it requires is quantitatively greater. If this thesis holds, we must regard the embryonic outgrowth of the axon and the chemical activities which support it as processes which continue throughout life, as an axon replacement process." These "trophic processes" in the neuron are evidently realized by the proximo-distal transport of substances which move along the axons and these are also probably the carriers of intracellular "trophic signals" on the basis of which the rate of proteosynthesis in the cell and the reaction of the cell to injury of the axon is regulated. A number of authors have reported results indicating that there is a constant flow of axoplasm synthetized in the cell body from the cell to the periphery (Cook and Gerard 1931, Young 1942, Weiss 1944, Young 1956). Samuel et al. (1951) studied this transport with the aid of P^{32}. These authors investigated the flow of phosphoproteins in the sciatic nerve and found that the labelled protein was present in the proximal third of the nerve during the first ten days after injection, in the middle third during the next ten days and in the distal third of the sciatic nerve during the last ten days. Phosphoproteins thus travel along nerve fibres of the sciatic nerve at a velocity of 3 mm. per day, i. e. at a rate comparable with the "growth" of regenerating nerve fibres. Ochs and Burger (1958) have succeeded in demonstrating this transport in the anterior roots following intraspinal microinjections of P^{32}. These latter authors, however, have not determined the rate of "intra-axonal flow". The mechanism of intraaxonal transport of substances was originally visualized in the first outgrowth during development and in regenerating nerve fibres. However, it can be assumed that this "centrifugal supply stream" continues throughout the life of the mature neuron (see Weiss 1954). This rate of movement has been presumed to be similar to the actual rate of regeneration (i. e. about 3 mm/day). Nevertheless, it must be considerably increased after functional activity of the neuron (Vodička 1960,

Vodička and Gutmann 1958). This demands continuous synthesis of new protoplasm by the perikaryon. A rapid turnover of proteins and nucleic acids in the nerve cells is then to be expected and this has indeed been repeatedly shown to be the case (see Hyden 1960). It remains to be seen what relative contribution of this metabolic activity manifested by the continuous axonal transport process applies to the "tropic function" of the neuron. Morphological observations indicate that this transport concerns the products of nucleic acid metabolism synthetized in the nucleus of the nerve cell (see Young 1956) and there are related data on changes in cell nucleus, Nissl's substance, high RNA and protein production in nerve cells (see Hyden 1960) and the changes in nucleic acid metabolism during activity (Geiger 1957). Of course, transport of substances which should be associated with "trophic signals" may also concern acetylcholine, which plays such an important role in the initiation of nerve impulse transmission connected with synaptic function. According to De Robertis and Bennet (1954) acetylcholine and other chemical mediators could be associated with vesicles of submicroscopic dimensions which are present in presynaptic nerve endings and can move towards the presynaptic membrane. The size of these synaptic vesicles is reduced in retinal synapses in the dark and their numbers increase after electric stimulation of the splanchnic nerves (De Robertis 1959). Here we have a concept concerning the flow and discharge of synaptic vesicles which is related with the random discharge of isolated quantal units of acetylcholine which gives rise to spontaneous miniature potentials. The motor endplate is the place of this spontaneous subthreshold activity (Fatt and Katz 1952, 1953, Castillo and Katz 1956) and it is therefore possible that these synaptic vesicles may represent the quantal units of transmitter substance. The subthreshold liberation of acetylcholine and the related nervous control of chemical sensitivity of skeletal muscle (Thesleff 1960) appears to be an important factor transmitting the trophic function of the nervous system. Lack of transmitter release does actually reproduce some of the changes characteristic for denervation such as fibrillation and increase of sensitivity to acetylcholine (Thesleff 1960, Josefsson and Thesleff 1961). It remains to be seen whether other metabolic changes in the denervated muscle can be reproduced by lack of subthreshold transmitter discharge. There are chemical substances, of course, which are produced by the nerve cell and transported along the axon and which have no connection with the specific function of the neuron, i. e. with the initiation and conduction of nerve impulses, and with the rapid synthesis and breakdown of specific mediators (especially acetylcholine). They may represent or be analogous to the products of neurosecretion such as the well-known neurosecretory activity of cells of the hypothalamus. A nerve cell can thus serve as a secretory organ without losing all the typical properties of a neuron: a vesicular nucleus, Nissl's substance, axon and dendrites and capacity of impulse conduction. Such a nerve cell can produce secretory granules and transport them along the axon (Scharrer and Scharrer 1954). This function of the nerve cell can be classified from the phylogenetic aspect as a transitory function of endocrine glands and of impulse conduction by a differentiated nerve cell, so that these cells are

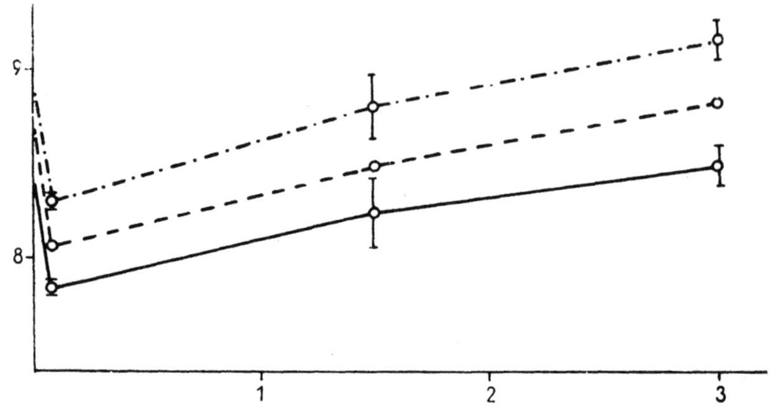

Fig. 5. Changes in total nitrogen expressed as percentage of dry weight in the distal (dashed-dotted) and proximal half (solid line) of the sciatic nerve before 5 min. swimming, immediately after and 1½ and 3 hrs. later. Ordinate % of total nitrogen in dry weight of the nerve. Abscissa: time in hours. Interrupted line — mean values in the whole nerve (Vodička 1960).

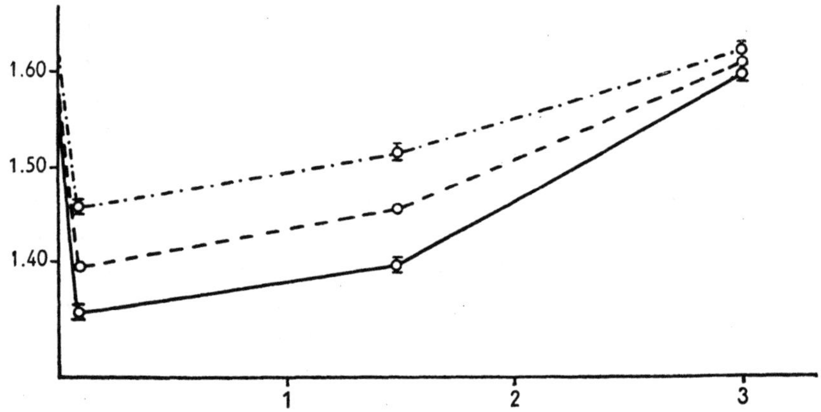

Fig. 6. Changes in total phosphorus expressed as percentage of dry weight in the distal (solid line) and proximal half (dashed dotted) of the sciatic nerve before 5 min. swimming, immediately after and 1½ and 3 hrs. later. Ordinate: % of total phosphorus in dry weight of the nerve. Abscissa: time in hours. Interrupted line: mean values in the whole nerve (Vodička 1960).

able to receive nerve impulses and to react by hormonal secretion (Scharrer and Scharrer 1954). The production of secretory granules which, for the morphologist serves as criterion of neurosecretory activity, has not been demonstrated in peripheral neurons innervating skeletal muscles. This does not mean, however, that such neurosecretory activity does not exist in peripheral motor neurons. On the contrary, it was pointed out by Vodička (1956) that the transport of substances in the peripheral nerve substantiates this possibility and new aspects of the physiological significance of this transport have been shown (Vodička 1956, 1959, 1960, Vodička and Gutmann 1958). It has been demonstrated that following brief experimental functional activity in rats (five minute's swimming), there is a transitory decrease of total nitrogen and total phosphorus in the sciatic nerve. The decrease in the content of nitrogen and phosphorus is evidently not the result of changes in the hydration of the nerve, since the dry weight of the nerve remains practically unchanged and the decrease is significant even if it is calculated per dry substance. During the three hours of recuperation following swimming, the content of nitrogen and phosphorus returns again to original values (see Fig. 5 and 6).

These results may indicate an accelerated axoplasmatic flow during functional activity. If it is assumed that this axoplasmatic flow of substances, which are probably the products of nucleic acid metabolism, is a manifestation of neurosecretory activity of the nerve cell, there remains, of course, the question as to the significance of this activity for trophic processes in skeletal muscle. In the following chapters material has been accumulated demonstrating that following every functional activity activation of metabolic recovery processes occurs in the muscle. For example, the resynthesis of glycogen is accelerated after direct muscle stimulation (Gutmann et al. 1954, Bass et al. 1955) and it reaches higher values than initially (Yampolskaya 1950). This activation of glycogen synthesis in innervated muscle occurs in two phases. The second phase of activation is absent in the denervated muscle (Bass et al. 1955). With the same initial values, glycogen does not surpass initial values and its content following direct stimulation returns slowly and asymptotically to values similar to those of the control, nonstimulated denervated muscle. Thus the overshoot reaction (supercompensation) in the content of glycogen does not occur under these experimental conditions. However, this over-shoot reaction does not disappear immediately after section of the nerve. It is still present 24 hours after section of the nerve and gradually disappears, being more pronounced in muscles with the longer peripheral nerve stump. All this speaks in favour of the fact that substances in the peripheral stump of the sectioned nerve are still able to exert certain trophic effects in the sense of regulating or activating metabolic recovery processes prior to nerve degeneration (Gutmann et al. 1955). This activation of metabolism has also been demonstrated in the diaphragm in vitro, so that it cannot be caused by vasomotor changes. What is the connection between the neurosecretory processes in the neuron and this activation of recovery processes? Is it possible that the proximo-distal transport of substances, which appears to be considerably increased during functional activity, could affect the regulation of trophic processes? It is true, of course, that during this activity acetylcholine is liberated in greater quantities and it is possible that following functional activity subthreshold liberation of microquantal amounts of acetylcholine is increased. The decrease of nitrogen in the nerve need not

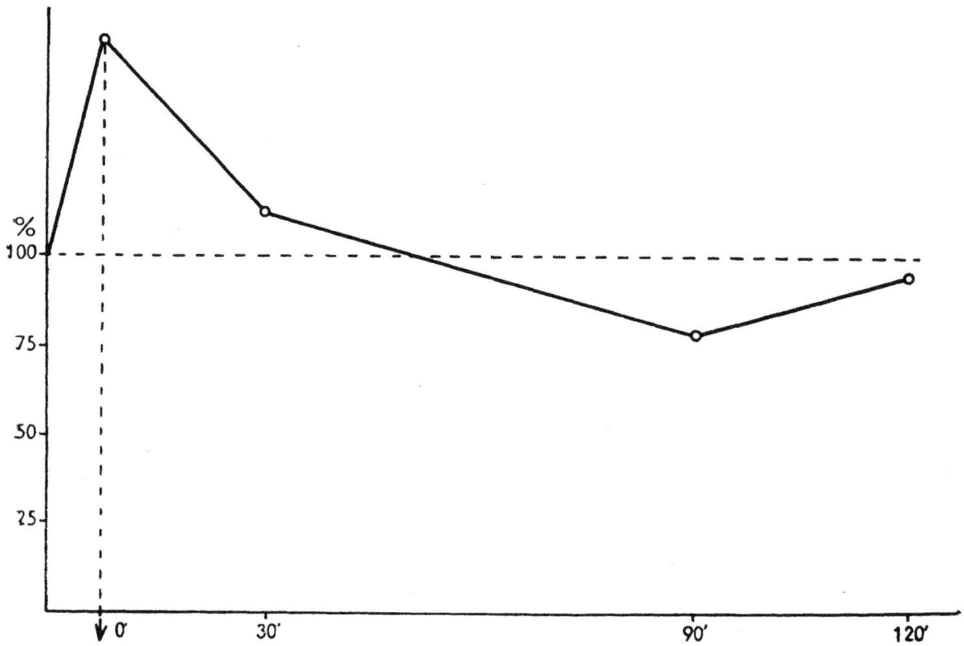

Fig. 7. Changes in the dimension of the end-plate region impregnated with $AuCl_3$ in the gastrocnemius muscle following indirect muscle stimulation. Abscissa: time after muscle stimulation in min. Vertical interrupted line — dimensions of the end-plate region immediately after stimulation. Ordinate: dimensions of the end-plate region in percentage of the control, non-stimulated side.

necessarily mean the transport of substances from the nerve into innervated structures. However, this possibility is indicated in experiments performed by Vodička (1958) in which the size of motor end-plates was determined after indirect muscle stimulation. End-plates impregnated with gold according to Kühne's method were found to increase in size during nerve stimulation and later returned to initial dimensions (Fig. 7). If a substance or some "activator" of metabolism penetrated into the muscle via the end-plates, then its effect should first appear in those regions of the muscle which are nearest to the end-plate zone — if the diffusion velocity of these substances from the end-plate region is sufficiently slow to make this differentiation of such a "metabolism activating wave" possible. These were the considerations which led Vodička to experiments in which the synthesis of glycogen was determined separately in proximal and distal halves of the anterior tibial muscle, utilizing the observation that the end-plate zone in this muscle was localized in the proximal half of the muscle (Vodička 1959). It was found that following the first phase of resynthesis which occurred simultaneously in both halves, activation of glycogen resynthesis took place sooner in the proximal than in the distal half (Fig. 8).

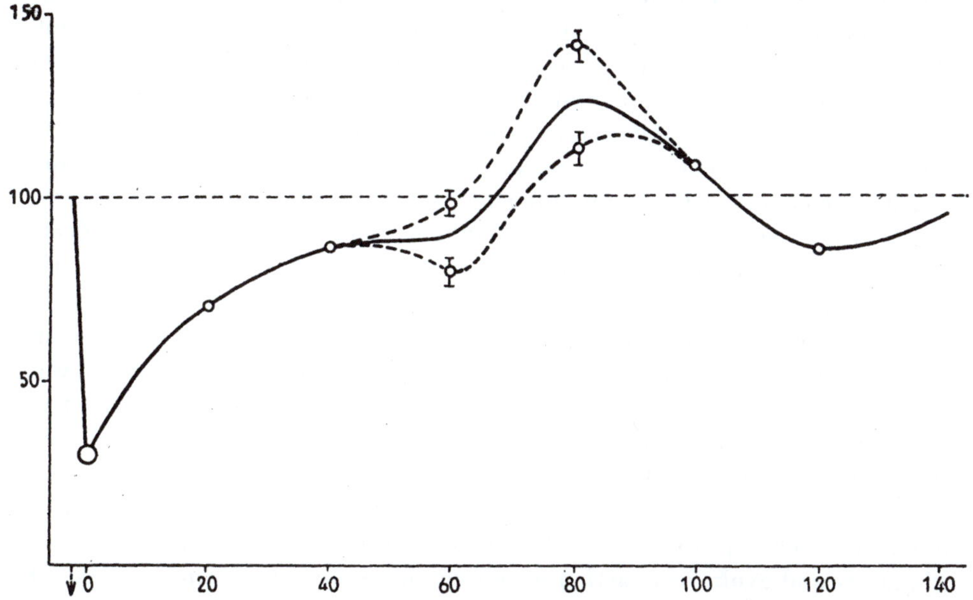

Fig. 8. Changes in glycogen content of the proximal (upper interrupted line) and distal (lower interrupted line) half of the anterior tibial following 3 min. stimulation with faradic current at tetanizing frequency. Abscissa: time in minutes after termination of stimulation. Arrow indicates onset of stimulation. Ordinate: glycogen content as percentage of control side.

Spatial and functional relations between transport processes of substances and metabolic recovery processes and the disturbance of these processes in denervated muscles are all so far only indirect approaches to problems of the fundamental mechanism of the "trophic influence of the nervous system". However, they have indicated that the function of a neuron also includes processes which, although correlated with the specific function of the neuron, are not identical and play an important role in the metabolism of innervated organs by regulating the trophic processes in peripheral tissues.

F. The significance and limitations of denervation studies

Denervation studies are part of the physiology of neurotrophic relations. This physiological field has extensive tasks and should analyze processes under conditions when these relationships are 1. formed during development, i. e. study of morphogenesis and the development of nervous regulations of metabolism, 2. disrupted, i. e. study of degeneration and denervation atrophy, 3. renewed, i. e. study of regeneration and reinnervation, and finally 4. adapted to new conditions of the environment, i. e. study of adaptations. The mutual relations between the nerve cell and peripheral innervated organ are gradually built up during development, adapted to new conditions and can be impaired and again renewed. Purposeful and efficient processes of this kind make a rapid and adequate reaction of the organism possible. They are an expression and the result of control systems which ensure the integrity of the organism in spite of constant changes in the environment.

We are thus dealing with mechanisms by which nervous processes are affected by receiving information and these mechanisms play a role in rapid reflex processes in response to changes in the environment as well as in slow metabolic processes of recovery in nerve cells and terminal organs in which they are manifested as neurotrophic regulations. In both cases the nerve cell constantly receives information and adapts itself to new environmental conditions on the basis of this information. Analogy with the theory of information which deals with the mechanisms by which the response is affected by incoming "information" is attractive and can be helpful in defining the relationship more precisely. While in fast reflex processes signals are conducted at a rapid rate, are impulse in character and evoke contraction of muscle fibres, signals informing about the trophic state of peripheral organs and mediating neurotrophic interrelations are non-impulse in character. This is already given by the fact that trophic processes or the nervous regulation of trophic processes manifests itself primarily in "metabolic recovery processes" which follow functional activity.

In fast reflex processes the signals are modulated primarily by changes of the impulse frequency, this being the main mechanism by which the nerve cell can react to increased or decreased activity. In trophic processes, however, it is changes in the rate of metabolism in the first place which follow preceding functional activity and renew the structural and functional integrity of the neuron and thus the stability of neurotrophic interrelations.

An attempt has been made to formulate a physiological definition of the so-called trophic influence of the nervous system. We have pointed out the significance of the neurosecretory activity of the nerve cell, of the proximo-distal transport of substances probably connected with the continued production and transport of the metabolic products of nucleic acids (see Young 1956) which accelerate or activate muscle metabolism following functional activity by a mechanism as yet unknown. It is therefore possible that a definition of the "signals" maintaining relations between the nerve cell and the peripheral organ will soon be made. In any case substances must be transported distally along the nerve and thus prevent its degeneration. This neurotrophic disturbance occurs when a part of the neuron is isolated. However, information non-impulsive in character proceeds also centripetally, as can be observed in active processes of proteosynthesis taking place in the nucleus of a neuron following section of its axon. Whatever the nature of this information, its absence evidently leads to serious disturbances in the trophic state of peripheral organs — and they cannot be caused by the elimination of impulse activity of the neuron only. This is where denervation studies are of paramount importance. But they are elimination experiments and this is at the same time their limitation. Denervation experiments have been, and still are, however, something like an "agent provocateur" in the research of neurotrophic relations and should contribute to the solution of the problem of the "trophic influence of the nervous system".

It is the aim of the present studies to contribute to the solution of the old problem already formulated by Prochaska: "Are nerves necessary for nutrition?"

References

Aitken I. T., Sharman M., Young J. Z.: J. Anat. 81 : 1, 1947.

Aleksanian A. M., Nasledov C. A.: XXI. Internat. Congress of Physiol. Sciences, p. 10, Buenos Aires 1959.

Barcroft H., Bonnar W. McK., Edholm O. G., Effron A. S.: J. Physiol. 102 : 21, 1943.

Bass A., Gutmann E., Vodička Z.: Physiol. bohemoslov. 4 : 267, 1955.

Baum I., Christen W., Pichler E.: Pflügers Arch. 233 : 43, 1934.

Bayandurov B. I.: Troficheskaya funktsiya golovnogo mozga. Medgiz, Moscow 1949.

Beattie F. J. R., Beattie M. K., Milroy T. H.: J. Physiol. 69 : 364, 1930.

Bechterew: Dtsch. Zeitsch. f. Nervenheilk. 17 : 221, 1900.

Belovintseva M. F.: Fiziol. zh. 34 : 361, 1948.

Beritov I. S.: Obshchaya fiziologiya myshechnoi i nervnoi sistem. Izdatelstvo AN SSSR, Moscow 1947.

Bernard C.: Leçons de Pathologie Expérimentale. Paris 1880.

— 1871, cit. Voronin V. V. Gedevanishvili I. D. Troficheskaya yazva, Tbilisi 1955.

Bodian D.: Symposia of the Soc. for experimental Biol. 1 : 163, 1947.

Boeke J.: Zeitschr. f. mikroskop. Anat. 7 : 561, 1927.

Boell J. E.: Energy exchange and enzyme development during embryogenesis. In: Analysis of development, Saunders Co. 1955.

Borovskii M. L.: Regeneracija nerva i trofika. Moskva 1952.

Bouchard C.: Arch. gen. de méd. 1 : 272, 1866, 2 : 273, 1866.

Britten S. W.: Am. J. Physiol. 93 : 213, 1930.

Brown-Séquard M.: Gaz. méd. de Paris 4 : 880, 1849.

— Leçons sur les nerves vaso-moteurs. Paris 1872.

Buller A. J., Eccles R. M., Eccles J. C.: XXI. Internat. Congress of Physiological Sciences. Abstracts p. 45, 1959.

Bykov K. M.: Kora golovnogo mozga i vnutrenniye organy. Medgiz, Moscow 1941.

Bykov K. M., Vasyutochkin V. M.: Fiziol. zh. SSSR, 40 : 555, 1954.

Cajal S. R. Y.: Degeneration and Regeneration of the Nervous System. Oxford 1928.

Cannon W. B.: The wisdom of the body. London 1932.

Castillo I. del, Katz B.: J. Physiol. 132 : 630, 1956.
Charcot L. M.: Leçons sur les maladies du systéme nerveux. Vol. 1, Paris 1876.
Clonninger G. L., Green N. D.: Am. J. Physiol. 181 : 258, 1955.
Cohen S., Levi Montalcini R., Hamburger V.: Proc. Amer. Ass. Cancer Res. 1 : 3, 1954.
Corbin K. B., Hinsey J. C.: Anat. Record 75 : 307, 1939.
Cori C. F., Cori T.: Am. J. Physiol., 90 : 317, 1929.
Cook D. D., Gerard R. W.: Am. J. Physiol. 97 : 412, 1931.
Coronini C., Kovac W., Lassman G., Smercker J.: Acta neuroveg. 15 : 446, 1957.
Dale H. H., Feldberg W.: J. Physiol. 81 : 39, 1934.
Dahlgren V.: Science 40 : 862, 1914.
Davenport R. A., Bothe R. T.: J. Comp. Neur. 59 : 167, 1934.
Denny-Brown D., Brenner Ch.: Arch. Neurol. Psychiat. 51 : 1, 1944.
De Robertis E.: XXI. Internat. Congress of Physiological Sciences. Symposia p. 211, 1959.
De Robertis E., Benett H. S.: Fed. Proc. 13 : 35, 1954.
Desmedt J. E.: Arch. Int. Physiol. 58 : 23, 1950a.
— Arch. Int. Physiol. 58 : 125, 1950b.
— Amer. J. Phys. Med. 38 : 248, 1959.
Detwiler S. R.: J. exp. Zool. 48 : 1, 1927.
— Neuroembryology, New York 1936.
Dogiel A. S.: Der Bau de Spinalganglien des Menschen und der Säugetiere. G. Fischer, Jena 1908.
Duncan D., Keyser L. L.: J. Comp. Neurol. 68 : 479, 1938.
Drahota Z., Gutmann E.: Gerontologia, 5 : 88, 1961.
Dworkin S., Bacq Z. M., Dill D. B.: Am. J. Physiol. 96 : 308, 1931.
Eccles J. C.: J. Physiol. 103 : 253, 1944.
Edds M. V.: J. Exp. Zool. 112 : 29, 1949.
— J. Comp. Neurol. 93 : 258, 1950.
— Quart. Rev. Biol. 28 : 260, 1953.
Evans D. H. L., Vizoso A. D.: J. Comp. Neurol. 93 : 429, 1951.
Fatt P., Katz B.: Acta physiol. scand. 29 : 117, 1953.
Exner S.: Pflügers Arch. ges. Physiol. 36 : 572, 1885.
— — J. Physiol. 117 : 109, 1952.
Feinstein B., Pattle R. E., Weddell G.: J. Neurol., Neurosurg. and Psychiat. 8 : 1, 1945.
Feldberg. W., Gaddum I. H.: J. Physiol. 81 : 305, 1934.
Ferdman D. L.: Biokhimiya zabolevanii myshts. Izd. AN SSSR, Kiyev 1953.
Fischer E.: in Le Muscle, Colloque tenu à Royaumont, France 1950.
— (1956) Symposium on Nervous Regulation of Metabolism and Active Transport of Ions. Prague 1958, p. 141.
Gelfan S., Tarlov I. M.: Am. J. Physiol. 185 : 217, 1956.
Gilbertini G., Malaguti C., Malsciani R., Vechiati R.: Riv. pat. clin. 10 : 581, 1955.
Gilding H. P.: J. Physiol. 74 : 34, 1932.
Ginetzinsky A. G., Nekhoroshev N. P., Tetyaieva M. B.: Russkij fiziol. zh. 10 : 483, 1927.
Ginetzinsky A. G.: Russ. fiziol. zh. 6 : 1923.
— Problems of the modern physiology of the nervous system and muscle system. Ac. Sc. of the Georgian SSR, Tbilisi 1956, p. 409—417.
Golikov N. V.: Fiziologicheskaja labilnost, Leningrad 1950.
Golub D. M.: Voprosy morfologii perifericheskoj nervnoj sistemy. Moskva 1949.
Grigorieva T. A.: DAN SSSR 78 : 387, 1951.
Gurkov H. J., Bast T. H.: Am. J. Physiol. Med. 37 : 269, 1958.
Gutmann E.: J. Neurol. and Psychiat. 5 : 81, 1942.

— J. Anat. 79 : 1, 1945.
— Čas. lék. čes. 94 : 452, 1955.
— Die Funktionelle Regeneration Peripherer Nerven. Berlin 1959.
Gutmann E., Holubář J.: J. Neurol. Neurosurg. Psychiat. 13 : 89, 1950.
Gutmann E., Sanders F. K.: J. Physiol. 101 : 489, 1943.
Gutmann E., Vodička Z., Vrbová G.: Physiol. bohemoslov. 3 : 182, 1954.
Gutmann E., Vodička Z., Zelená J.: Physiol. bohemoslov. 4 : 200, 1955.
Gutmann E., Young J. Z.: J. Anat. 68 : 15, 1944.
Hajek N. M., Hines H. M.: J. Iowa State Med. Soc. April issue: 2, 1947.
Hamburger V.: Trends in Experimental Neuroembryology. In: Biochemistry of the developing Nervous System, New York 1954.
Hamburger V., Levi Montalcini R.: J. exp. Zool. 111 : 457, 1949.
— — Some aspects of neuroembryology. In: Genetic Neurology, Chicago 1950.
Harless: Archiv f. d. Physiol. v. Reil. ,4 : 189, 1799.
Harreveld A.: Am. J. Physiol. 150 : 670, 1947.
Hechter O., Levine R., Soskin S.: Proc. Soc. exper. Biol. Med. 46 : 390, 1941.
Held H.: Die Entwicklung des Nervengewebes bei den Wirbeltieren. Leipzig 1909.
Herrin R. C., Meek W. J.: Am. J. Physiol. 97 : 57, 1931.
Hering H. E.: Neurol. Centralbl. 17 : 1077, 1897.
Hines H. M.: JAMA 120 : 515, 1942.
— Le Muscle, Colloque tenu à Royaumont, France, 1950.
Hines H. M., Knowlton G. C.: Am. J. Physiol. 128 : 97, 1939.
Hinsey J. C.: J. Comp. Neurol. 44 : 87, 1927.
— J. Comp. Neurol. 59 : 117, 1934.
Hník P.: Physiol. bohemoslov. 5 : 305, 1956a.
— Physiol. bohemoslov. 5 : 411, 1956b.
Hodgkin A. L.: XXI. Internat. Congress of Physiological Sciences, Abstracts p. 168, 1959.
Hoffmann H.: Australian J. Exp. Biol. Med. Sci.: 28 : 383, 1950.
— Australian J. Exp. Biol. Med. Sci.: 30 : 541, 1952.
Hoffmann A., Wertheimer E.: Pflügers Arch. 218 : 176, 1928.
Holub F. M.: Perifericheskij otrezok povreschdennogo nerva v razvitii nervnyk distrofii. Moskva 1944.
Horton-Smith R. J.: J. Physiol. 21 : 101, 1897.
Hudlická O.: Candidate's Thesis, Institute of Physiology, Czechoslovak Academy of Sciences, Prague 1954.
Hunt C. C., Kuffler S. W.: J. Physiol. 43 : 283, 1951.
Hyden H.: The Neuron, the Cell, vol. IV., 1960.
Josefsson I. O., Thesleff S.: Acta physiol. scand. 51 : 163, 1961.
Jasper H. H.: Trans. R. Soc. Canada 40 : 81, 1946.
Jefferson N. C., Ogawa T., Necheles H.: Am. J. Physiol. 193 : 563, 1958.
Khlopina I. D.: Morfologicheskiye izmeneniya denervirovannykh tkanei i razvitiye troficheskoi yazvy neurogennogo priskhozhdeniya. Medgiz, Leningrad 1957.
Kibjakow A. W.: Pflügers Arch. 232 : 432, 1933.
Kopczynski S.: Poln. Arch. f. biol. und med. Wissensch. 3 : 99, 1907.
Krebs H. A.: Endeavour 16 : 125, 1957.
Kutsenko N. A., Nechayeva G. A.: Fiziol. zh. SSSR 39 : 719, 1953.
Kuffler S. W., Hunt C. C., Quilliam J. P.: J. Neurophysiol. 14 : 29, 1951.
Kuré K., Saégusa O., Kawaguchi K., Shiraishi K.: Quart. J. of Exp. Physiol. 20 : 51, 1930.
Kuré K., Sakurasawa F., Tsuji M.: Zeitsch. f. d. ges. exp. Med. 60 : 233, 1928.
Kuré K., Nitta Y., Matsuura H., Tsuji M.: Zeitschr. f. d. ges. exp. Med. 60 : 250, 1928.

Langley J. N.: J. Physiol. 50 : 335, 1916.
Lapinskii M. N.: O degeneratsiji i regeneratsiji perifericheskikh nervov. Kiyev 1904.
Lehmann E. P.: Arch. Surg. 21 : 838, 1930.
Leksell L.: Acta physiol. scand. 10 (Suppl. 31): 1, 1945.
Levi Montalcini R., Hamburger V.: J. Exp. Zool. 116 : 321, 1951.
— — J. Exptl. Zool. 123 : 233, 1953.
Levine R., Hechter O., Soskin S.: Am. J. Physiol. 132 : 336, 1941.
Lubińska L.: Acta Biol. Exper. 16 : 73, 1952.
— Exper. Cell. Res. 10 : 40, 1956.
Luco J. V., Eyzaguirre C.: J. Neurophysiol. 18 : 65, 1955.
Magendie M.: Journ. de Physiol. exper. et pathol. 4 : 176, 1824.
Martin A. W., Sola O. M.: Fed. Proc. 7 : 78, 1948.
McCullych G. P., McFadden G. D., Milroy T. H.: J. Physiol. 69 : 353, 1930.
May R. M.: J. exp. Zool. 42 : 371, 1925.
Meshcheryakov A. M., Korotkov A. G.: Fiziol. zh. SSSR 39 : 443, 1953.
Mitchell S. W., Morehouse G. R., Keen W. W.: Gunshot Wounds and other Injuries of Nerves. Lippincott and Co., Philadelphia 1864.
Moore R. M., Williams J. H., Singleton A. C.: Arch. of Surg. 26 : 308, 1933.
Mott F. W., Sherrington C. S.: Proc. Roy. Soc. 57 : 481, 1895.
Nachmansohn D.: Am. J. Phys. Med. 34 : 33, 1955.
Nakamura H.: J. Physiol. 55 : 100, 1921.
Needham J.: Biochemistry and Morphogenesis. Cambridge 1942.
Nevin S.: Quart. J. of Exper. Physiol. 20 : 281, 1930.
Novinski M.: Arkh. vet. nauk 2 : 1874.
Ochs S., Burger E.: Am. J. Physiol. 194 : 499, 1958.
Orbeli L. A.: Fiziol. zh. SSSR 35 : 594, 1949.
Ortmann R.: Neurosecretion in: Handbook of Physiology. Section 1: Neurophysiology vol. II, Chapter XL.
Osborne S. L.: Arch. Phys. Med. 32 : 523, 1951.
Pavlov I. P.: O troficheskoi innervatsii 1920. In: I. P. Pavlov, A. N. SSSR, Moskva 1949.
— Laboratornyye nabljuddeniya nad patologicheskimi reflexami s brushnoi polosti. 1898. In: I. P. Pavlov AN SSSL, Moskva 1949.
— (1890): Sämtliche Werke, Akademie Verlag, Berlin 1954, Bd. II/1.
Petkovic D. S., Husni E. A., Simeone F. A.: Am. J. Physiol. 192 : 106, 1958.
Petrova M. K.: O roli funktsional'no oslblenoi kory golovnogo mozga v vozniknovlenii razlichnykh patologicheskikh protsessov v organizme. Medgiz, Moscow 1946.
Pineles F.: Arch. f. d. ges. Physiol. 48 : 17, 1891.
Piťha V.: Svalové atrofie a hypertrofie. Melantrich, Praha 1937.
Pollister A. W.: Cytochemical aspects of protein synthesis, In: Dynamics of Growth Processes, Princeton 1954.
Pollock L. J., Arieff A. J., Sherman I. C., Schiller M., Tigay E., Hiller F., Liebert E., Yacorzynski G.: Arch. Phys. Med. 31 : 265, 1950.
Popov N. F., Mikhailova N. M.: Arkh. Biol. Nauk 40 : 209, 1935.
Prochaska G.: De functionibus systematis nervosi, 1784. Adnot. acad. Fasc. III. sec. I. Prague 1954.
Ranson S. W.: Arch. Neurol. Psychiat. 19 : 201, 1928.
Ranson S. W., Hinsey J. C.: J. Comp. Neurol. 42 : 69, 1927.
Ricker G.: Arch. Path. Anat. Physiol. 165 : 263, 1901.
Ricker G., Ellenbeck J.: Virchows' Archiv 158 : 199, 1899.
Romberg (1846): Cit. Voronin V. V., Gedevanishvili I. D.: Troficheskaya yazva, Tbilisi 1955.

Rose M.: In Handbuch der Neurologie. Bumke O., Foerster O.: Berlin 1935, vol. I.
Rosenblueth A.: The Transmission of Nerve Impulses at Neuroeffector Junctions and Peripheral Synapses. Chapmann and Hall, London 1950.
Samuel A. T., Boyarsky L. L., Gerard R. W., Libet B., Brust M.: Amer. J. Physiol. 164 : 1, 1951.
Samuel S.: Die trophischen Nerven. Königsberg 1860.
Sechenov I. M.: Fiziologia nervnoi sistemy. 1866. In: Fiziologia nervnoi sistemy, vol. II. Moskva 1952.
Sheehan D., Marrazzi A. S.: J. Neurophysiol. 4 : 68, 1941.
Sherrington C. S.: J. Physiol. 17 : 211, 1894.
— J. Physiol. 21 : 209, 1897.
— In Schaffer E. A.: Text-book of Physiology vol. II, p. 785, 1900.
— Integrative Action of the Nervous System. Yale University Press, New Haven 1906.
Scharrer E., Scharrer B.: Handbuch der mikroskopischen Anatomie des Menschen. Springer Verlag. Berlin 1954.
Schiff, Zak: Münch. med. Woch. 1 : 669, 1912.
Schoenheimer R.: The dynamic state of body constituents. Cambridge (Mass). 1942.
Siebert W. W.: Zeitschr. f. klin. Med. 109 : 350, 1929.
Siems L. L., Kosman A. J.: Amer. J. Physiol. 156, 185, 1949.
Siems L. L., Kosman A. J., Osborne S. L.: Arch. Phys. Med. 29 : 759, 1948.
Slozhenikina L. V.: Vopr. med. khimii 3 : 137, 1957.
Solandt D. Y., Magladery J. W.: Brain 63 : 255, 1940.
Solandt D. Y., Patridge R. C., Hunter J.: J. Neurophysiol. 6 : 17, 1943.
Soskin S., Levine R.: Carbohydrate metabolism. The University of Chicago Press. 1952.
Speidel C. C.: Carnegie Inst. Wash. Publ. No. 281, 13 : 1, 1919.
Speranskii A. D.: Uchenie o nervnoj trofike kak put issledovatelskoj raboty v medicine. 1937. In: Izbranye trudy — Moskva 1955.
— Elementy postroyeniya teorii meditsiny, 1935.
— In: Zabolevaniye, lecheniye i vyzdorovleniye. AMN SSSR, Moskva 1952.
Sperry R. W.: Growth Symposium 10 : 63, 1951.
Stadie W. C.: Physiol. Rev. 34 : 52, 1954.
Stewart D. M.: Biochem. J. 59 : 553, 1955.
Stewart D. M., Martin A. W.: Am. J. Physiol. 186 : 497, 1956.
Storey M. H., Corbin K. B., Hinsey J. C.: Proc. Soc. Exp. Biol. and Med. 35 : 309, 1936.
Studitskii A. N.: Eksperimentalnaja chirurgija myshts. Moskva 1959.
Sunderland S.: Brain 74 : 491, 1951.
Thomson J. D.: Am. J. Physiol. 180 : 202, 1955.
Thomson J. D., Morgan J. A., Brodish A.: Am. J. Physiol. 167 : 832, 1951.
Tiegs O. W.: Physiol. Rev. 33 : 90, 1953.
Thesleff S.: Physiol. Rev. 40 : 734, 1960.
Thesleff S.: J. Physiol. 151 : 598, 1960.
Tower S. S.: Am. J. Physiol. 78 : 1926.
— Am. J. Anat. 46 : 1, 1935.
— Physiol. Rev. 19 : 1, 1939.
Uflyand U. M.: Voprosy fiziologii nervno i myshechnoi sistem. Trudy LSGM I, vol. 7, Leningrad 1950.
Ussing H. H.: XXI. Internat. Congress of Physiological Sciences, Abstracts p. 172, 1959.
Veselkina V. M.: Izv. inst. im. Lesgafta 21 : 33, 1938.
Vodička Z.: Physiol. bohemoslov. 5 (suppl.): 55, 1956.
Vodička Z.: Physiol. bohemoslov. 8 : 285, 1959.
Vodička Z.: Physiol. bohemoslov. 9 : 367, 1960.

Vodička Z., Gutmann E.: Symposium on Mechanism of Excitation. Berlin 1958.
Vrubel J.: Pathogenetická terapie. Kandidátská práce. Praha 1958.
Waller A.: Philos. Trans. Roy. Soc. London 1 : 423, 1850.
— Gaz. méd. Paris 11 : 218, 1856.
Weddel G., Feinstein B., Pattle R. E.: Brain 67 : 178, 1944.
Weddel G., Guttmann L., Gutmann E.: J. Neurol. Neurosurg. Psychiat. 4 : 206, 1941.
Weiss P.: Neurogenesis in Analysis of Development. Ed. by Williers. B. H. Weiss P., Hamburger V., 1954.
Weiss P., Edds M. V.: Am. J. Physiol. 145 : 587, 1946.
Weiss P., Hiscoe H. B.: J. exp. Zool. 107 : 315, 1948.
Weiss P., Taylor A. C.: J. exp. Zool. 95 : 233, 1944.
Westbrook W. H. L., Tower S. S.: J. Comp. Neurol. 72 : 383, 1940.
Wyburn-Mason R.: Trophic Nerves, London 1950.
Yampolskaya L. J.: Fiziol. zh. SSSR 36 : 749, 1950.
Yampolskaya L. J., Yakovlev N. N.: Fiziol. zh. SSSR 37 : 110, 1951.
Young J. Z.: Proc. Roy Soc. B. 121 : 319, 1937.
— Physiol. Rev. 22 : 318, 1942.
— Proc. Roy. Soc. Med. 37 : 351, 1944.
— The history of the shape of a nerve fibre. Assays on Growth and Form presented to D'Arcey W. Thompson. Oxford, p. 41, 1945.
— The determination of the specific characteristics of nerve fibres. In: Genetic Neurology, 1950.
— Endeavour 15 : 5, 1956.
Zelená J.: Vliv inervace na vývoj svalu. Babákova sbírka č. 12, Praha 1959.
Zellweger H., Mort H.: Helv. pediatr. acta 5 : 434, 1950.
Zubenko P. M.: Vopr. biokhimii myshts. Kiev p. 149, 1954.

Chapter II

MORPHOLOGICAL CHANGES IN THE DENERVATED MUSCLE

E. Gutmann, J. Zelená

A. Aims and limitations of morphological studies

The first histological studies of denervated muscles were clinical-pathological in character, and were associated with the rapid development of neurological diagnostic methods of peripheral nerve lesions, especially with electrodiagnostic changes developing during denervation, as studied by Erb (1868). In general there was agreement about the morphological changes in denervation which corresponded to the description given, for example, by Oppenheim (1894), who repeatedly investigated questions of muscle diseases: "From the histological aspect the muscle fibres are smaller in diameter, the content of muscle fibres disintegrates into granulated and lipid particles, which are resorbed, so that only sarcolemmal tubes containing nuclei remain." And further: "Primitive muscle fibres become thinner, lose their cross-striation, their content undergoes granular disintegration combined with waxlike degeneration and there is proliferation of subsarcolemmal nuclei and perimysium." The description of denervation changes is only very brief, and more attention was paid to degenerative changes in the peripheral nerve stump occurring, for example, after nerve injuries. This is quite natural when taking into account how important at this time were the findings of Waller (1852), concerning degeneration of the nerve, which led to the formulation of laws of neurotrophic relations (so-called "Waller's degeneration"). The main aim of authors at this time was evidently to show corresponding degenerative changes also in the muscle. It is probably for this reason that Erb (1868) used the term "reaction of degeneration" when describing excitability changes of denervated muscle to galvanic stimulation and Oppenheim similarly concluded in 1894: "Alterations in electrical excitability of the muscle, reaction of degeneration, the course of which have been studied in considerable detail by Erb, Ziemssen and Weiss, correspond to degenerative changes in the peripheral nerve" and further: "Where the reaction of degeneration is present, severe anatomical changes must also be expected, and the degree of these degenerative changes can be anticipated, to a great extent, by evaluation of the degenerative reaction" (see Kopits 1929). The description of degenerative changes in the atrophic muscle is usually stressed in pathological-anatomical literature, so that, for example, in the text-book by Beattie and Dickson (1943) it is stated that: "Cross-striation in the atrophic muscle is less evident or missing, fibres exhibiting vacuolization and fragmentation and having a granular or hyaline appearance."

However, for evaluation of the morphological denervation changes, especially from the point of view of their reversibility, the classification of degenerative changes without knowledge of their time relations is not sufficient. The first experimental studies which investigated denervation changes in muscle after section of the nerve more precisely (Ricker and Ellenbeck 1889, Gessler 1883, Stier 1896, Schmidtmann 1916, Willard and Graw 1924, Chor et al. 1937) were mostly restricted to relatively early denervation periods. Denervation changes could not be properly classified from the point of view of these time relations and development of degenerative changes, and of the ultimate fate of denervated muscle fibres, until experimental studies were performed on long term denervation models.

Muscle atrophy in the course of one year after denervation was described in the cat by Tower (1935). Tower differentiated three stages of denervation atrophy: atrophy, degeneration and fibrotic dedifferentation, and concluded from her experimental results that the decisive component of peripheral innervation responsible for denervation changes is the motor innervation. The second question, however, considered by Tower (1939) to be of fundamental importance, i. e. the problem of the mechanism and character of changes occuring after denervation, was more complicated. The relationship between "atrophic" and "degenerative" changes remained unsolved. It was stressed by Tower (1939) that for complete destruction of muscle fibres, which results in fibrosis, the degenerative processes are not so important as the process of "fibrotic metaplasia" or fibrotic dedifferentation which lead to loss of specific muscle structure and to its transformation into connective tissue. This process of muscle "metamorphosis" was already discussed by Durante in 1902, but neither Bowden and Gutmann (1944) nor Adams et al. (1954) could find any evidence confirming this conclusion. On the other hand, Altschul (1942), in agreement with Tower, considered fibrotic and adipose infiltration to be the result of muscle fibre metaplasia. This question is thus controversial. The problem of metaplasia in muscle fibres is, of course, closely related to questions concerning the course and significance of degenerative changes in muscle fibres. According to Tower (1939), atrophy and muscle degeneration after denervation is the most out-standing evidence of the loss of the trophic influence of the nervous system on peripheral tissue. The development and fate of degenerative changes after denervation thus gained in significance. Nevertheless, the degenerative changes as described by various authors differ substantially. A number of authors have not reported degenerative changes until late stages of denervation atrophy.

Stier (1896) observed only atrophic changes $3\frac{1}{2}$ months after denervation in the dog and rabbit, and Willard and Graw (1924) also did not find degenerative changes even 63 days after denervation in the rat. No degenerative changes, with few exceptions, could be found by Sunderland and Ray (1950) in the opossum as late as 485 days after denervation. However, Tower (1935) reported degenerative changes in cat muscle already one month after denervation. According to this author, degeneration starts by swelling of muscle fibres followed by vacuolization, which destroys the continuity of fibres, or the muscle fibres disintegrate into granules, their nuclei swell and disintegrate so that in the end the fibre is broken down into debris. However, Tower (1939) considers these acute degenerative changes as accidently determined and does not consider them to be the unavoidable result of denervation atrophy, as is the case in fibrotic metaplasia. Adams et al. (1954) found degenerative changes six months after denervation in the cat in all larger muscles; these authors described local vacuolization, loss of myofibrils, clumping and vacuolization of nuclei which may be located freely in the tissue, with mobilization of endomysial fibroblasts and histiocytes, which at later stages undergo regressive changes. Transitory stages between muscle cells and fibroblasts or adipose cells were not found. Inconsistencies in the descriptions are evidently caused by the fact that the occurrence and extent of the changes is different in various animal species and different muscles.

An important question from the practical point of view concerns the reversibility of denervation changes. Undoubtedly interspecies differences also play an important role in this connection. In man neither hyaline, amyloid nor adipose degeneration, nor fragmentation of muscle fibres was observed even three years after denervation (Bowden and Gutmann 1944). Changes, evidently irreversible after reinnervation, were not observed until later. The question of how long the denervated muscle retains its fundamental morphological properties is evidently extraordinarily important. Tower (1939) did not consider the question of how long the denervated muscle can survive denervation as solved. This problem was also considered especially important by Sunderland and Ray (1950). It is not possible to give an answer to this problem without the aid of reinnervation studies and some data are already at our disposal (Gutmann and Young 1944, Gutmann 1948, Gutmann 1958). Reinnervation represents the crucial test of the reversibility of denervation changes, and it is surprising that extreme changes in the muscle are reversible after reinnervation. For example, 20 months after denervation in the rabbit there are mostly only sarcolemmal tubes filled with chains of nuclei to be found (Fig. 47) and in spite of this endplates are formed after reinnervation and the muscle fibres are restored (Gutmann 1945), the muscle fibres showing normal cross striation again (Fig. 48). Prior to this, of course, restitution of muscle fibres is considerably affected by the fact that 1. regenerating nerve fibres do not come into contact with the original endplate, 2. spatial arrangement of Schwann sheaths is disorganized and 3. that part of the nerve fibres can no longer form new connections (Gutmann and Young 1944).

Are degenerative changes characteristic for denervation atrophy? It is important here to draw comparison with disuse muscle atrophy. For this various models of disuse atrophy have been used. Extensive immobilization was achieved by Tower (1937) by isolating spinal cord segments, i. e. by transection of the spinal cord above and below the segments investigated, combined with bilateral section of the dorsal roots. Only a gradual reduction in diameter of muscle fibres was found, without proliferation or enlargement of nuclei which can be observed after denervation, No acute vacuolar or granular degenerative changes were present. Similar results were reported by Reid (1941). If the differences between denervation and disuse atrophy are so marked, then histological differentiation can offer important evidence in favour of the trophic influence of the nervous system and changes after denervation could not be explained by muscle inactivity only. This question therefore deserves further investigation.

B. Denervation changes in human muscles

Atrophy in human muscles was investigated most after poliomyelitis (Kopits 1929), or after various forms of locomotor diseases, e. g. in amyotrophies, progressive muscle atrophy, amyotrophic lateral sclerosis, etc. It is a disadvantage of these studies that the duration of the pathological process and its localization are seldom known. Morphological muscle changes in various diseases leading to the destruction of spinal motoneurons or to lesions of peripheral nerves are very similar (see Adams et al. 1954): generally speaking they are characterized by atrophy of muscle fibres. Muscle atrophy is always secondary and its extent is dependent on the affection of motor nerve cells in the spinal cord. Atrophy affecting small groups of muscle fibres, which was first observed by Darkewitch (1905), indicates that it is due to the damage of individual motor units which evidently react as functional atrophic units (Wohlfart 1949). The degree of atrophy of muscle fibres in the affected groups may be very different. According to Wohlfart this serves as evidence that different muscle fibres exhibit a different resistance towards the pathological process. Fibres which retain an almost normal diameter are in Wohlfart's opinion probably the so-called "b fibres", which develop earlier during embryogenesis than other fibres and which retain in some muscles, for example, extraocular muscles, an exceptionally large diameter even in adults (Wohlfart 1938). To these fibres is ascribed a slower rate of atrophy (Wohlfart 1937). These differences in the rate of atrophy of muscle fibres are especially marked in poliomyelitis, and they can be caused by variations in the extent, localization and distribution of the pathological process in the spinal cord. It is, of course, possible that mesenchymal tissue is attacked by the virus directly, with different intensities. This could explain why only partial fibrosis is sometimes observed even with fragmentation of muscle fibres or when atrophy is far advanced (Adams et al. 1954), while in other cases the whole muscle is substituted by fibrous and adipose tissue.

In motoneuron diseases, similarly as in experimental denervation, nuclear changes are usually described in muscles. Nuclei are more spherical with enlarged nucleoli (Wohlfart 1949). Sometimes even migration of nuclei into the centre of a muscle fibre can be observed, but this is never so marked as in progressive muscle atrophy, this being an important criterion for differential diagnosis. Similarly as in denervation atrophy, it has not been conclusively proved whether subsarcolemmal nuclei are proliferated or not. Proliferation of nuclei was considered to be typical for muscle atrophy of nervous origin by Slauck (1921). However, Wohlfart (1949) concluded on the basis of quantitative evaluation of his material that the proliferation of nuclei is not a true one, the appearance of an increase being due to the decrease of diameter of muscle fibres.

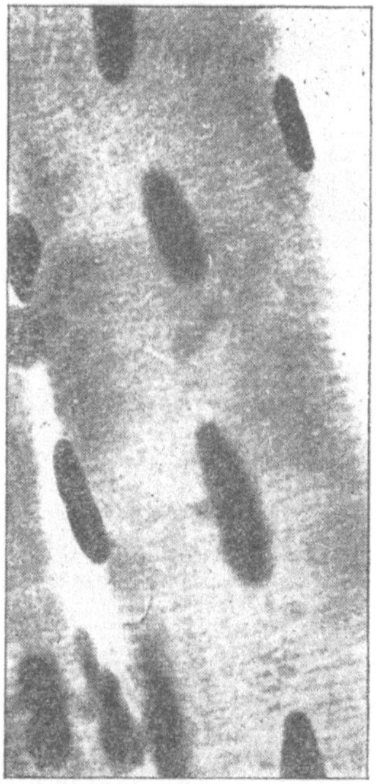

Fig. 9. Human muscle 145 days after denervation. The nuclei are in a central position inside the muscle fibre. (Glees-Bielschowsky; 350×.) (Bowden R. E. M., Gutmann E.: Brain 67 : 273, 1944).

Hyaline or vacuolar degeneration and phagocytosis of muscle fibres has been described in exceptional cases only. At later stages granular degeneration and fragmentation of muscle fibres can be observed (Adams et al. 1954).

More accurate information concerning changes in the denervated muscle have been obtained by muscle biopsies in peripheral nerve lesions. In these cases the state of the nerve, the type of lesion and the exact time of denervation are all usually known. This is hardly ever the case in other diseases of the lower motor neuron, so that it is usually not possible to correlate histological changes in the nerve with those of the muscle. Denervation changes following nerve trauma were studied by Bowden and Gutmann (1944). Their results will be briefly summarized here. No degenerative changes were found in muscle at early stages of denervation, i. e. 42–69 days after the nerve lesion. Chains of subsarcolemmal nuclei seem to indicate an increased number of nuclei. Nuclei are often localized centrally (Fig. 9) or in clumps. Chains of nuclei with hyperchromasia were also

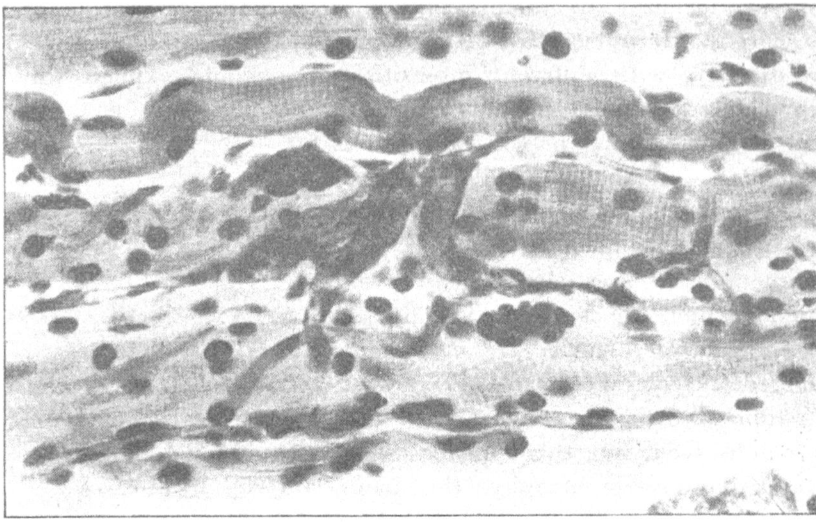

Fig. 10. Human muscle 69 days after denervation. The architecture of Schwann sheaths is well defined (Glees-Bielschowsky; 350×.) (Bowden R. E. M.: Gutmann E.: Brain 67 : 273, 1944).

described by Adams et al. (1954). The number of nucleoli inside the nuclei is increased. End-plate nuclei, however, are not altered at this early stage. The empty Schwann sheath structure is conspicuous (Fig. 10), endplates are preserved (Plate Ia,b) and enable differential diagnosis with progressive muscle dystrophy to be made (Bowden and Gutmann 1945). Arteries are thickened, with hypertrophic media, and it seems that the endothelial layer in capillaries is also thickened. Jamin found a proliferation of collagenous tissue around capillaries, and Kopits (1929) also described hyperplasia of the media in small arteries at late stages of denervation. There is only a little collagenous and adipose tissue at early stages after denervation, especially in the vicinity of blood vessels.

Four months to one year after the nerve lesion, muscle fibre diameter is already considerably reduced, some cases being characterized by a surprising variation in diameter. Cross striation is always preserved on observation in the polarizing microscope even though it is almost undiscernible in stained preparations. However, in some fibres cross-striation is already lost, even though longitudinal structure is clearly preserved. It is possible that the molecular organization of these fibres is damaged. The arrangement of nuclei into chains is more evident and there is also considerable proliferation of collagenous tissue, especially in perivascular spaces, where many fibroblasts and polymorphonuclear and monocytic cells can be found.

Degenerative changes were not observed until three years after nerve interruption. The protoplasm of muscle fibres is then often dedifferentiated (this

Fig. 11. Normal human muscle fibres. Granules oriented mostly on the border of cross-stration. (Glees-Bielschowsky; 900×.) (Bowden R. E. M., Gutmann E.: Brain 67 : 273, 1944).

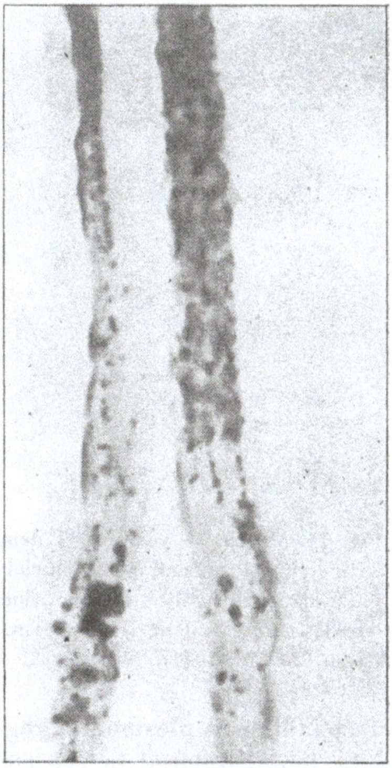

Fig. 12. Muscle fibre 25 years after denervation. Two muscle fibres considerably atrophied, with granules of various sizes irregularly distributed. (Glees-Bielschowsky; 900×.) (Bowden R. E. M., Gutmann E.: Brain 67 : 273, 1944).

Fig. 13. Muscle 26 years after denervation. Atrophied muscle fibre in the process of "indentation", which ultimately results in fragmentation of muscle fibres. (Glees-Bielschowsky; 900×.) (Bowden R. E. M., Gutmann E.: Brain 67 : 273, 1944).

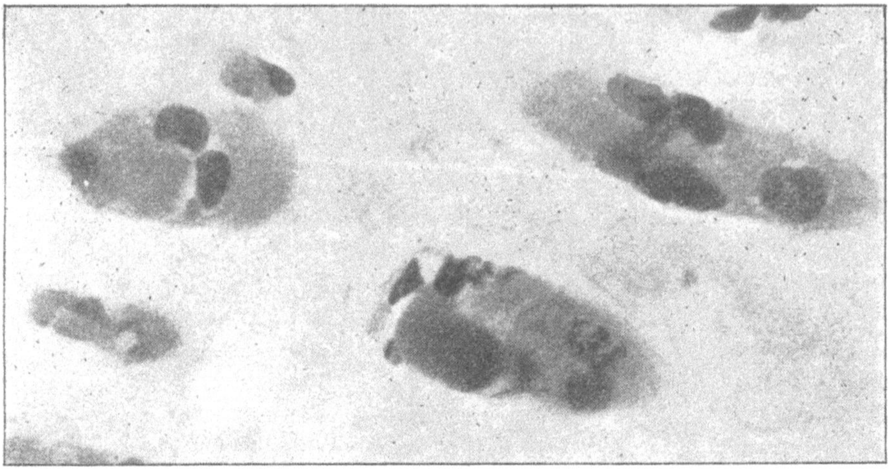

Fig. 14. Muscle 26 years after denervation. Fragments of atrophic muscle fibres filled with pyknotic nuclei; some nuclei arranged in aggregates; a large droplet of eosinophile material (hyaline degeneration). Occassional retraction of sarcoplasm, so that sarcolemma is more evident. (Haematoxyline-eosine; 500 ×.) (Bowden R. E. M., Gutmann E.: Brain 67 : 273, 1944).

corresponding to plasmodial regression of Durante 1902). Single muscle fibres can be found isolated in the collagenous tissue. The anatomical continuity of muscle fibres is lost by two mechanisms: either by longitudinal splitting into separate fibrils, or by fragmentation into oval segments. Fragments of individual muscle fibres resembling fibroblasts can be easily differentiated since they contain one or more pyknotic nuclei and granules which attain varying size and irregular distribution in the course of denervation. These granules correspond probably to sarcosomes and their changes in size and distribution in the course of denervation can be seen in Fig. 11, 12, Plate IIa.

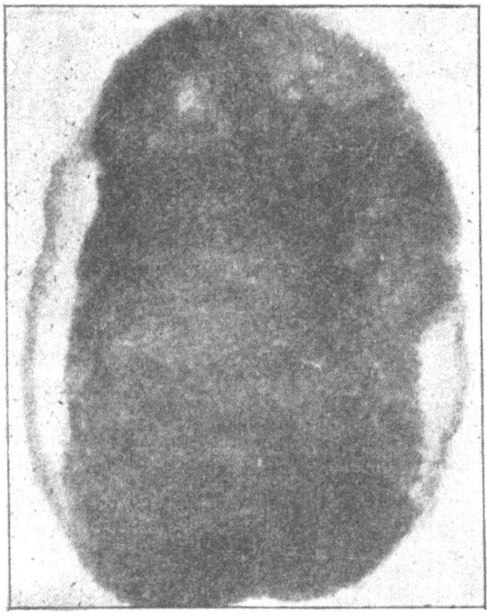

Fig. 15. A muscle fibre 26 years after denervation with the sarcoplasm retracting from the periphery leaving the sarcolemma clearly outlined. (Haematoxyline-eosine; 900 ×.)

During the latest stages of denervation spherical or oval fragments of muscle fibres are to be found surrounded by a layer of nuclei (Plate IIb). Occassionally fibres seem to be beaded in appearance (Fig. 13), and it is probable that this thickening precedes the formation

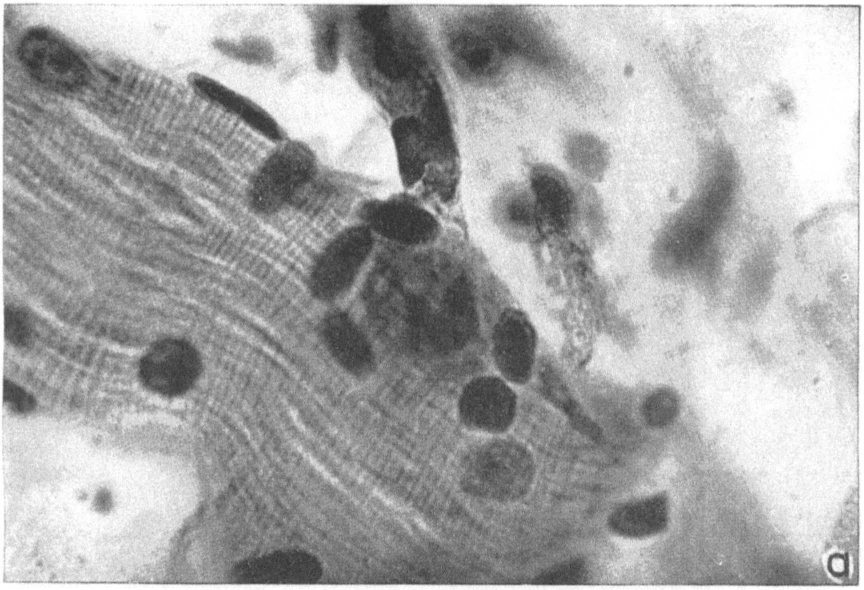

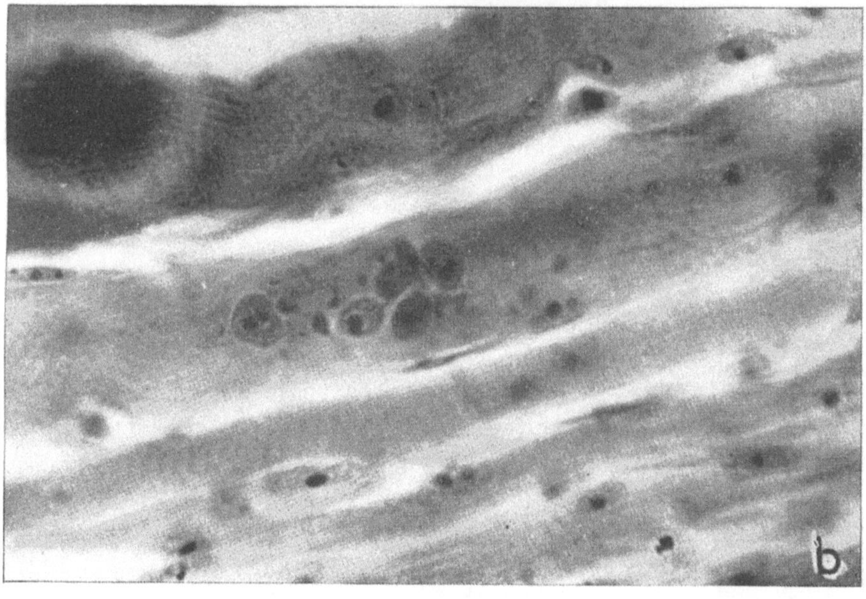

Fig. a. Human muscle fibre 69 days after denervation. End-plate sarcoplasm is still conspicuous and a single Schwann sheath may be traced to the end-plate. (Glees-Bielschowsky; 900×.)

Fig. b. Muscle 250 days after denervation. The end-plate is preserved. (Glees-Bielschowsky; 900×.) (Bowden R. E. M., Gutmann E.: Brain, 67 : 20, 1944.)

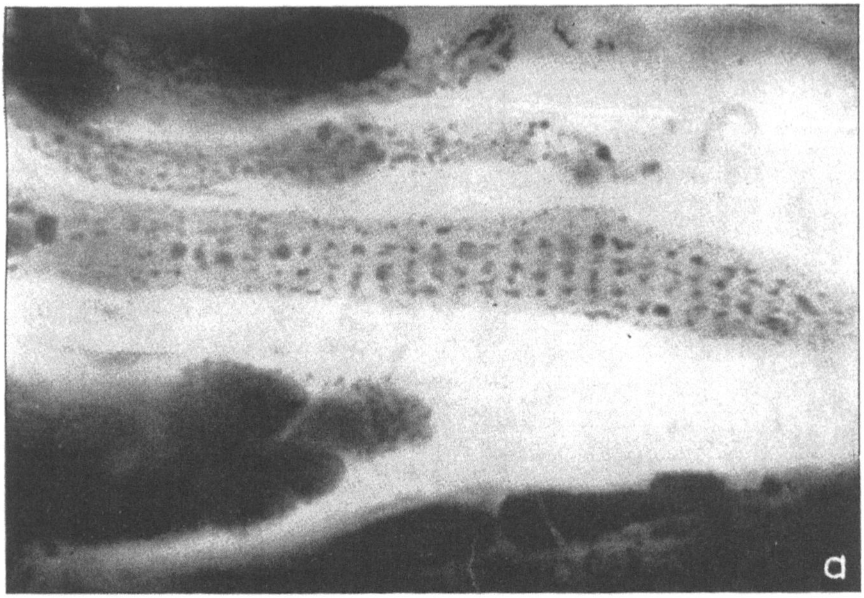

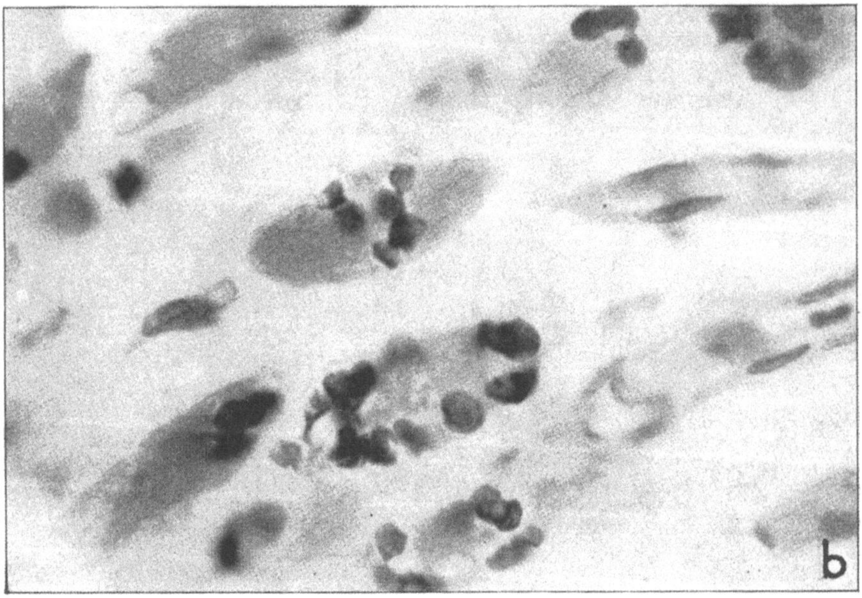

Fig. a. Human muscle fibre 20 years after denervation. Considerable atrophy of the fibre, granules of various sizes still oriented on the border of weakly defined cross-striation (Glees-Bielschowsky; 1800 ×).

Fig. b. Muscle 26 years after denervation. Spherical and oval fragments of muscle fibres with nuclear lining surrounded by connective tissue. (Haematoxyline-eosine; 500 ×.) (Bowden, R. E. M., Gutmann E., Brain 67 : 20, 1944.)

of spherical or oval fragments. In some fibres it is still possible to observe cross-striation even when the fibre diameter is very small. It is not present, however, in fragments where droplets of eosinophile material are described as the result of hyaline degeneration (Fig. 14). Homogenous sarcoplasm is sometimes retracted, so that the sarcolemma is then well defined (Fig. 15).

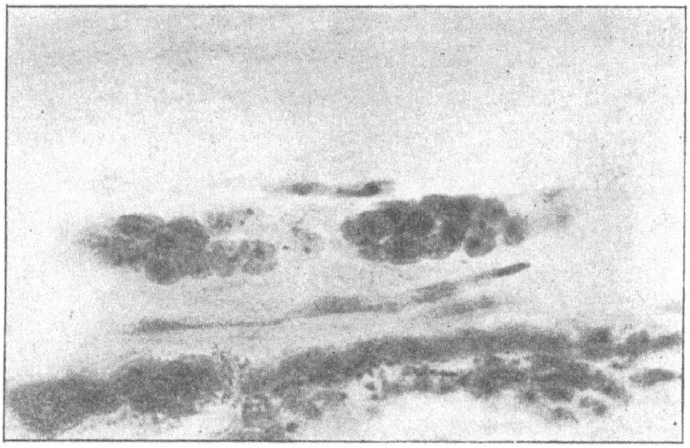

Fig. 16. Muscle 26 years after denervation. Nuclei arranged in groups inside the atrophic muscle fibre. (Haematoxyline-eosine; 900×.) (Bowden R. E. M., Gutmann E.: Brain 67 : 273, 1944).

Nuclear degenerative changes are characteristic. At late stages a large number of pycnotic nuclei can be seen, some of which are arranged in chains or clumps (Fig. 16). Often there is a light border around the nuclei which

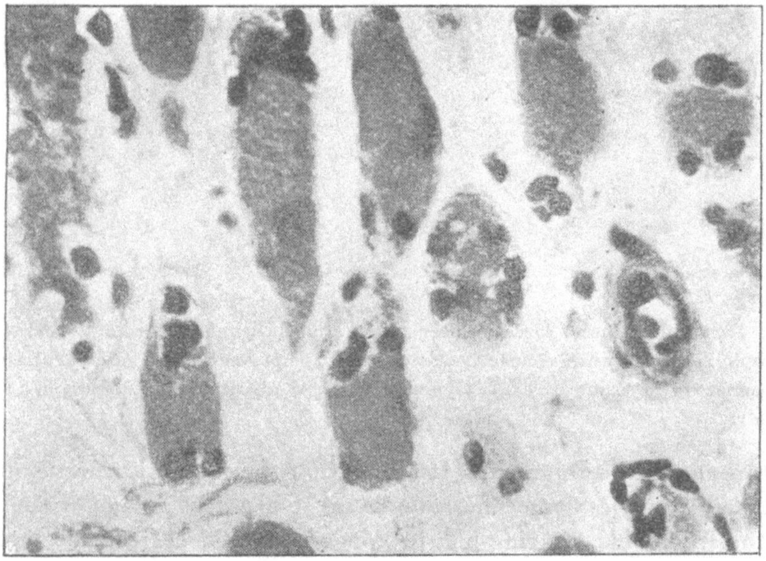

Fig. 17. Muscle 25 years after denervation. Nuclei exhibit various stages of pyknosis and disintegration. (Haematoxyline-eosine; 600×.) (Bowden R. E. M., Gutmann E.: Brain 67 : 273, 1944).

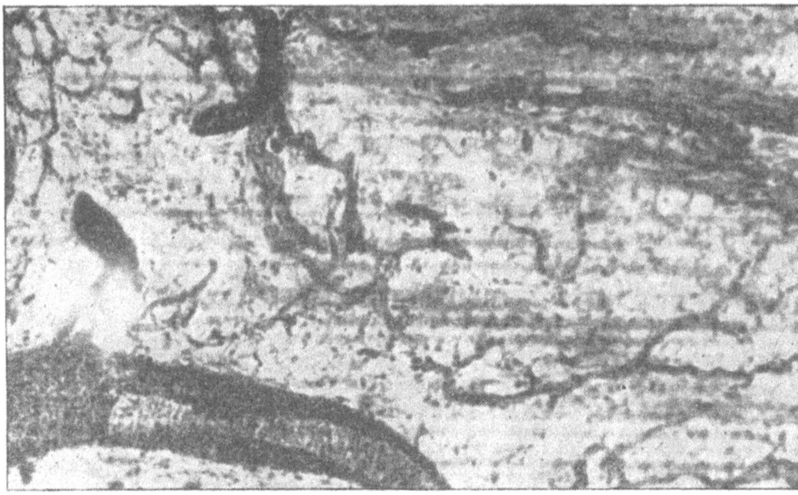

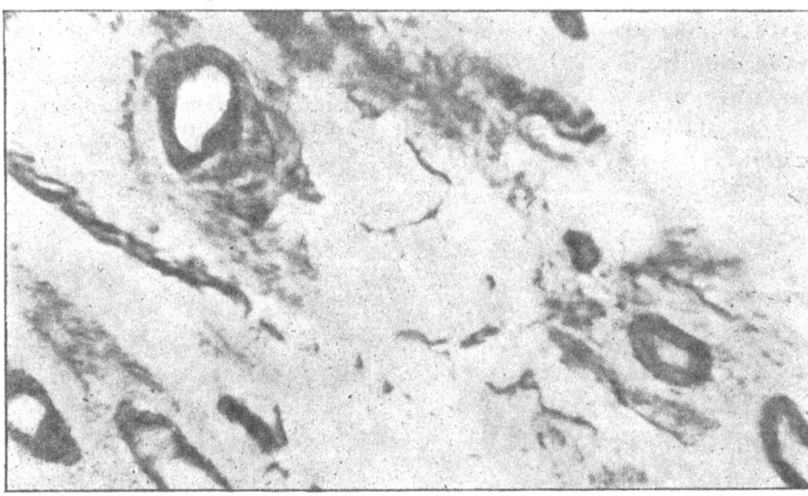

Fig. 18. and 19. Muscle 25 years after denervation. Ultimate result of denervation atrophy; only fatty and connective tissue is left; empty nerve and vascular strands. (Haematoxylin-eosine; 120 ×.) (Bowden R. E. M., Gutmann E.: Brain 67 : 273, 1944.)

may contain chromatin fragments. It seems that the pyknotic nucleus undergoes lysis and that vacuoles sometimes appear as a negative image of the nucleus. Some of the fragments are ring-like in appearance and correspond, perhaps, to remnants of the nuclear membrane (Plate IIb). Thus nuclei are being destroyed during late denervation, so that various stages of destruction, pyknosis, condensation, disintegration and chromatolysis can be observed (Fig. 17). Finally the muscle is substituted by fibrous and adipose tissue (Fig. 18 and 19).

Nuclear changes beginning with proliferation of nuclei directly after denervation and leading to progressing degeneration are evidently a very important part of denervation atrophy. Even though they appear to be typical for denervation, they can also be found in other types of muscle atrophy. Long chains of nuclei, often centrally situated, were also found in a case of motor paresis of central origin lasting for seven years. It may be more of a general metabolic reaction of the muscle fibre and it is interesting that this arrangement is to be found in smooth muscles as well as in intrafusal fibres of muscle spindles.

Direct phagocytosis of muscle fragments by neighbouring histiocytes and macrophages, as can be seen in experimental denervation studies, was not observed with certainty. It is rather a case of disintegration of fragments, the nuclei of which already exhibit irreversible metabolic disturbances. Direct metaplasia into collagenous fibrous tissue was never observed. Krösing (1892) assumed that this metaplasia does occur, namely as transformation of muscle fibres into collagenous connective tissue, and termed it "status fibrosus". Fibrous dedifferentation described by Tower (1935) is an analogous concept. According to Krösing, metaplasia of sarcolemmal cells into osteoblasts was reported, and is considered by this author to be the basis of myositis ossificans.

However, it appears that a direct transformation of muscle fibres into connective or adipose tissues is not possible. We are dealing rather with phagocytosis of muscle fibre remnants, the transformation into collagenous tissue beginning, most likely, by activation of mesenchymal elements, especially in the neighbourhood of blood vessels. Remnants of muscle fibres actually do sometimes resemble fibroblasts, but their nuclei are pyknotic (see also Adams et al. 1954). These authors were also unable to observe metaplasia of muscle fibres either into connective, adipose tissue, or into osteoblasts.

C. Experimental studies of the denervated muscle

Experimental studies of morphological changes occuring in the denervated muscle have concentrated on several problems of this pathological process— on questions of the initial changes in the muscle, on atrophy and degeneration of muscle fibres and on terminal stages of denervation atrophy, i. e. on the way in which muscle fibres are destroyed, and by what kind of tissue they are replaced. Although the description of these changes can hardly be expected to solve the basic mechanism of denervation atrophy, it could help in defining its course in considerable detail and give the key to the solution of several questions of its mechanism. Up to now a number of controversies exist concerning the description and interpretation of morphological findings, especially about questions of degeneration and disintegration of muscle fibres. These controversies are partly due to actual differences which are present in individual animal species as far as the course and extent of denervation changes are concerned, and partly due to the fact that in histological sections only a small portion of the muscle is evaluated, and since denervation changes need not be localized in the muscle homogeneously, this can cause considerable distortion and differences in conclusions. Many of these difficulties could, of course, be removed if the changes were quantitatively evaluated in whole muscles in the course of denervation. Reliable quantitative evaluation of morphological changes has so far been possible only in measuring muscle fibre diameters, and for this reason it is difficult to record a complex map of structural transformation during denervation, especially at late stages.

The development of denervation changes is different in various animal species and in various muscles, so that it is difficult to classify and compare material according to the duration of denervation. Attention was therefore paid to the main characteristic changes of this process, to their classification, time relation and role in determining the course of denervation atrophy.

1. Atrophy of muscle fibres

A reduction of sarcoplasm and myofibrils, which results in the decrease of muscle fibre diameter, represents the main process in the muscle after denervation. It runs in parallel with loss of weight of the denervated muscle. Changes in the fibre diameter are a more reliable indicator of loss of contractile mass than loss of muscle weight, since the results are not distorted by the presence of connective and adipose tissue.

The diameters of all muscle fibres in the anterior tibial muscle were measured one to eight months after denervation in the rabbit (Gutmann 1948, Fig. 20). During the first month the average muscle fibre diameter was decreased by one third, during the second and third months almost by 60 % and after eight months by 70 % as compared with average fibres of the control muscle (Fig. 21). The percentage decrease of fibre diameters was, with the exception of muscles six months after denervation, always lower than loss of weight (Fig. 21).

This may be explained by the fact that the loss of contractile mass grows in proportion to the square of the reduction in diameter. Similarly, there was a decrease in the weight and muscle fibre diameters in denervated muscles of the rat. Maximum loss of muscle weight was already observed after three months of denervation (70—85 % of the control muscles). At later periods the weight of denervated muscles was about 20 % of control muscles (Fig. 23). The decrease in diameters was smaller three months after denervation then the weight loss, being less than one half in the extensor digitorum longus and about two thirds in the soleus. During the following months there was a further decrease in diameter of 15—25% and oscillated, similarly as the muscle weight, at about 20% of control muscles (Fig. 22).

The rate of atrophy was faster in the soleus than in the extensor digitorum

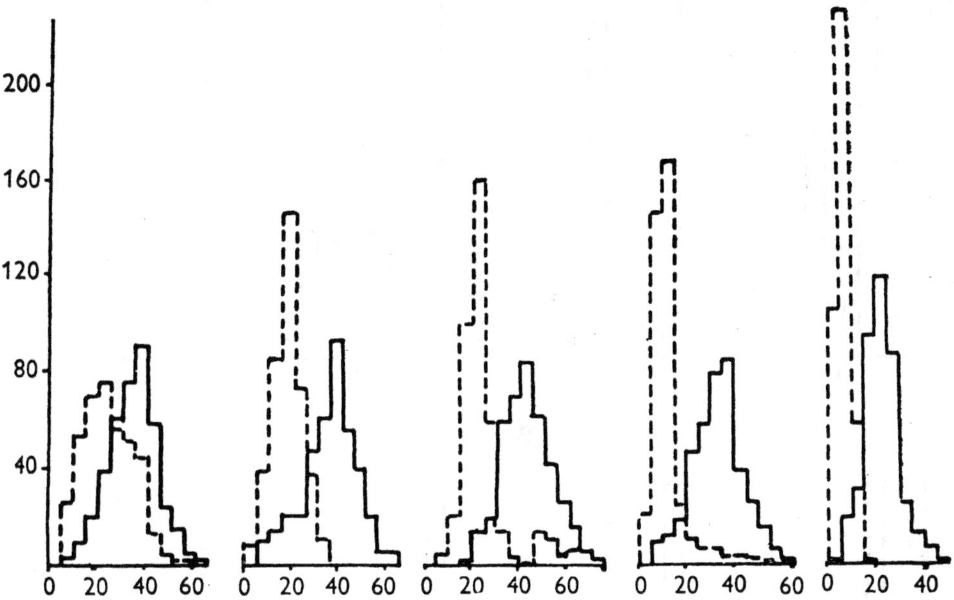

Fig. 20. Changes in the distribution of muscle fibre diameters in the tibialis anterior in the course of 1, 2, 3, 6 and 8 months after denervation (interrupted line) compared with the distribution of muscle fibres in the normal muscle (solid line). (Gutmann E.: J. Neurophysiol. 11 : 279, 1948).

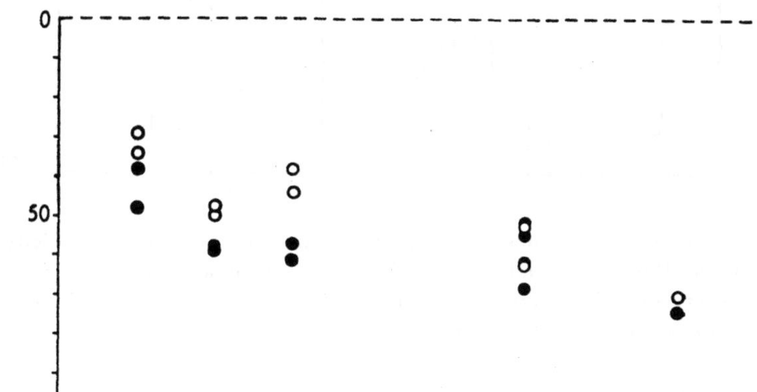

Fig. 21. Percentage loss of weight (black circles) and muscle fibre diameters (white circles) in the anterior tibialis (rabbit) as compared with the normal contralateral muscle in the course of 8 months' denervation. Abscissa: months after denervation. Ordinate: percentage decrease in weight and muscle fibre diameter. (Gutmann E.: J. Neurophysiol. 11 : 279, 1948).

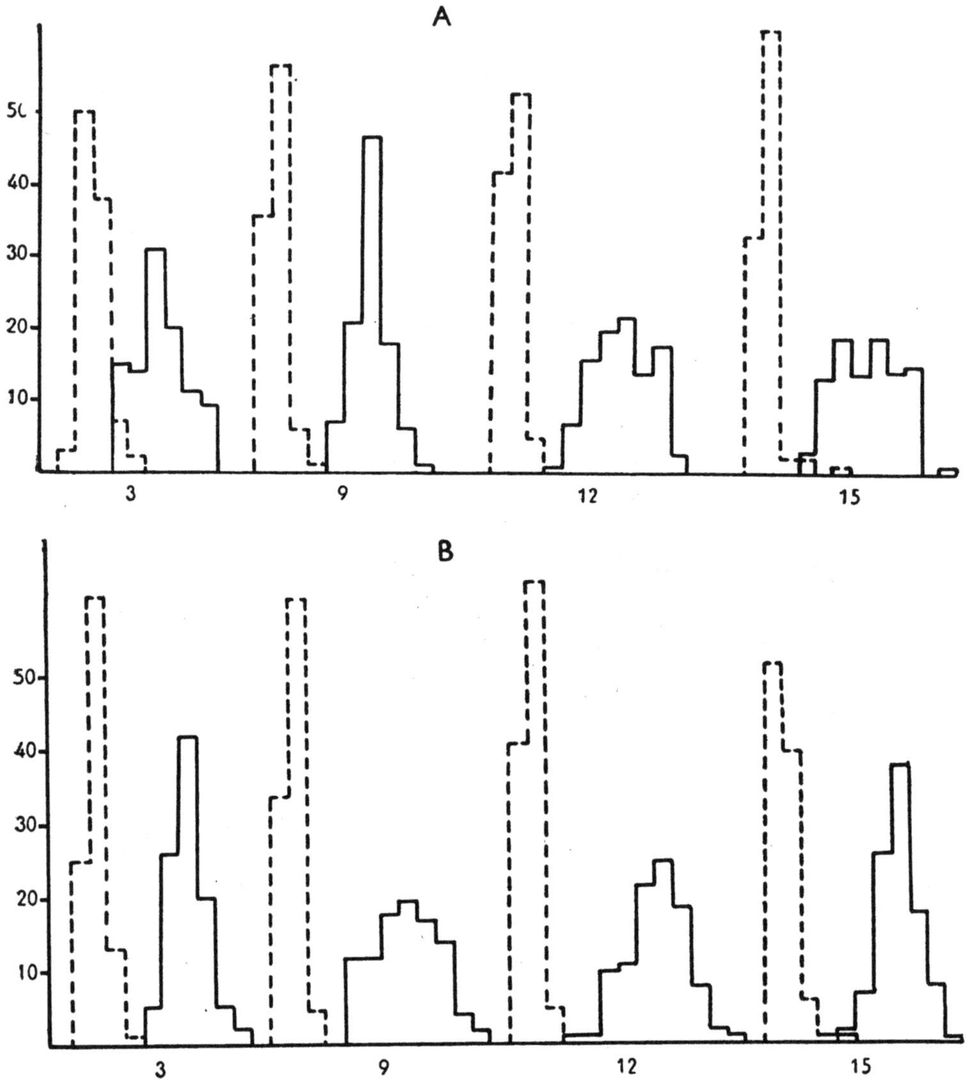

Fig. 22. Distribution of muscle fibres according to fibre diameter in the extensor digitorum longus (A) and the soleus (B) in normal and denervated muscles. Abscissa: number of fibres in individual groups (as percentage) 3, 9, 12 and 15 months after denervation. Ordinate: fibre diameters in microns. Fibre spectrum in 2μ steps. (Rat)

longus (Fig. 23). Mean fibre diameters in the denervated soleus were always smaller than in the extensor digitorum longus, although under normal conditions the diameters are larger in the former. It is thus apparent that the absolute loss of contractile substance is greater in the soleus than in the extensor digitorum longus, although the reduction in weight of the former 12 and 15 months after denervation was occasionally smaller than in the latter. The loss of the con-

contractile substance in the soleus is probably masked by the presence of adipose tissue which extensively infiltrates the muscle at these late stages of denervation (Plate VI, VII).

Observations on changes of weight during denervation cannot therefore give a clear picture of the essential processes going on in the muscle. Sunderland and Ray (1950), when investigating the loss of fibre diameters in the pronator teres and flexor carpi ulnaris in the opossum 9—485 days after denervation, found that the percentage loss of fibre diameter was always higher than loss of weight, although only moderate proliferation of connective tissue was observed in the above material and no increase of fat tissue was found. The most rapid loss of fibre diameter was registered in the course of the first two months. During the next two months the average diameter was reduced by a further 10%. From this time on until 16 months after denervation, the diameters were reduced to 80—90% of the control; the maximum values of the diameter in the pronator teres were reduced by 97% as compared with the control diameter. Willard and Graw (1924) observed a 68% decrease of diameters in mice 63 days after denervation. According to Tower's estimate (Tower 1935) the denervated muscle fibre diameters in the cat do not decrease until the end of the first month, this decrease attaining 30—50% loss by the end of the second month.

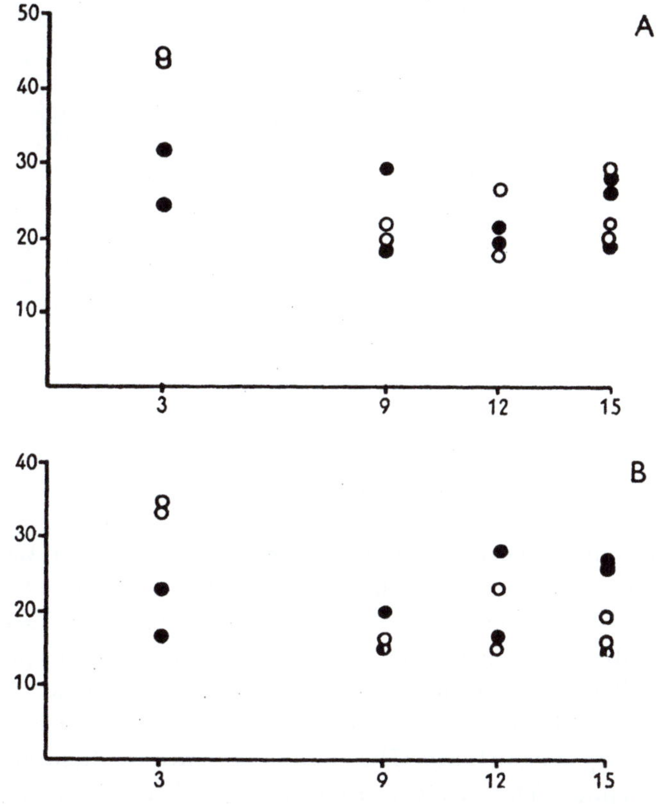

Fig. 23. Percentage weight loss (black circles) and diameters (white circles) of muscle fibres of the rat in the extensor digitorum longus (A) and the soleus (B) in the course of 15 months denervation. Abscissa: months after denervation. Ordinate: percentage of the control values.

Atrophy in the opossum results in uniformity of muscle fibre diameters (Sunderland and Ray 1950). In the rabbit, on the other hand, diameters are different in different fibres of a given denervated muscle. These differences are already apparent three months after denervation (Fig. 24). This difference can also be observed in the histogram where three to six months after denervation fibres with large diameters again appear, in spite of the fact that by two months after denervation they already disappeared. Perhaps the size differences which

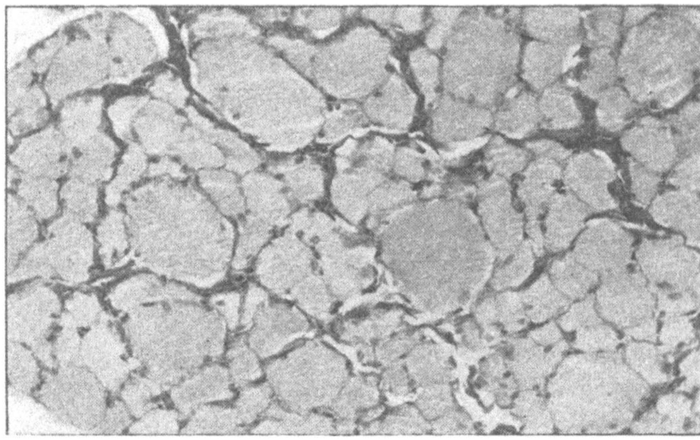

Fig. 24. Heterogenous atrophy in the soleus of the rabbit three months after denervation. Large fibres surrounded by groups of atrophic fibres. (Haematoxyline-eosine; 200 × .)

exist in the normal muscle are preserved to some extent during denervation and they may be even exaggerated by relatively slower atrophy in some fibres. A marked difference in muscle fibre diameters is already markedly developed one month after denervation in young rats, and it is conspicuous six months later in adult rats. Even nine to fifteen months after denervation these "giant" fibres are occasionally to be seen and they are 80% larger than the average neighbouring fibres (Plate IIIa,b). The presence of these fibres has not, as yet, been explained. They are very rare, so that it is not probable that these fibres are the resistent original "mother" fibres, as has been postulated in man by Wohlfart (1938). This could be the case in the rabbit since single large fibres are often present in one muscle fascicle, corresponding thus to the course of muscle development. These large fibres exhibit degenerative changes relatively more often than atrophic fibres and it cannot be excluded that their "hypertrophy" precedes their subsequent degeneration.

In many fibres myofibrils disappear completely after long periods of denervation, so that only narrow sarcolemmal tubes remain containing a chain of centrally placed nuclei. This process probably precedes the complete breakdown and disappearance of the fibre, which can also be accomplished by degeneration. In this connection it was of considerable interest to evaluate the loss of muscle fibres during the course of denervation atrophy. There was a small reduction in the number of muscle fibres in the denervated soleus of the rat three months after denervation and 15 months after denervation their number was reduced to about one half of the control muscle (see Table 2). In this case the reduction in the number of fibres was also less advanced in the extensor digitorum longus (Table 2). The process of elimination of fibres slowly progresses even at the time, when reduction of muscle weight and fibre diameters becomes stationary in denervated muscle at later stages of denervation. It is thus evident, that the loss of contractile substance is a continuous and progressive process in denervated muscle, although it did not result in complete disintegration of the muscle and total substitution by adipose or connective tissue in our experimental material. It is even probable, that loss of muscle fibers after denervation does not necessarily occur in all species, as Sunderland and Ray (1950) found no differences in the number of fibres between normal and denervated muscles in the opossum.

Table 2

Total number of muscle fibres in normal and denervated muscle 6, 12, 15 and 24 months following denervation

Months after denervation	M. soleus			M. extensor dig. long.		
	Normal	Denervated	Change in %	Normal	Denervated	Change in %
6	1806	1474	82			
9	2315	1800	77			
12	2318	1740	75	2617	2280	87
	1918	1237	64	2114	1786	85
15	2007	603	31	3464	2007	58
	2225	1225	55	2894	2214	76
24	1804	790	43	1865	736	39

As far as there are no degenerative changes in muscle fibres, and as long as they are not reduced to a sarcolemmal sack with nuclei only, cross-striation can usually be observed (Plate IV). In previous reports changes in sarcomere segments were often described. According to Tower (1935, 1939), the isotropic segment in the denervated muscle is longer than the anisotropic segment. Adams et al. (1954) reported greater variations in the length of sarcomeres in the denervated muscle. Birefringence was found to be decreased in denervated muscle fibres by Fischer (1940). Cross-striation is lost in some muscle fibres at later stages of denervation (Tower 1939). It was shown by Dempsey et al. (1946) that different staining changes the appearance of cross-striation, impregnation methods giving the opposite result from that of haematoxylin, for example. In our opinion, changes in cross-striation cannot be reliably differentiated by the normal microscope. A satisfactory solution of these questions can be expected from electron microscope studies which may represent an adequate approach to studying changes in the molecular orientation of muscle proteins probably substantially altered in the denervated muscle.

2. Changes in motor end-plates and muscle spindles

The first structural changes appearing in the muscle take place in the motor endplates. In most experimental animals degeneration of nerve endings occurs in the endplate one to three days after section of the nerve. A detailed description of the development of these changes is given on p. 91, and was correlated with the loss of indirect excitability. A muscle can react in many respects as an innervated muscle even after nerve section, as long as its nerve endings are preserved (Gutmann et al. 1955). It is therefore important, when investigating denervation changes in the course of the first few days after section of the nerve, to know the extent to which nerve endings are still preserved.

a) End-plates

Denervation changes do not appear until degeneration of nerve endings has taken place. It was observed, at high magnifications in the electron microscope, that a few days after the disappearance of nerve fibres the Schwann cells in the end-plates of frog muscles closely adhere to the postsynaptic membrane, to the so-called subneural apparatus (Katz and Miledi 1959, Reger 1959), and that they begin to contain vacuoles similar to those in terminal axon endings (Katz and Miledi 1959). Concomitant with the appearance of these vacuoles in the Schwann cell protoplasm, which are presumed to contain acetylcholine, miniature end-plate potentials reappear in the frog muscle after a short period of silence, at a frequency 10—100 times slower than in the normal muscle (Birks et al. 1959). Further changes in the end-plates affect their nuclei. Three kinds of nuclei are normally found in the end-plate region: muscle endplate nuclei ("Sohlenkerne" of Kühne 1862, "noyaux fondamentaux" of Ranvier 1878), i. e. oval or spherical nuclei with clear nucleoplasm and a distinct nucleolus, Schwann cell nuclei, the so-called teloglia, which were designated as "noyaux de l'arborisation" with a higher content of chromatin, and finally outer end-plate nuclei which correspond to "noyaux vaginaux" according to Ranvier and are mostly nuclei of fibrocytes with granular, often darkly staining nucleoplasm. Gutmann and Young (1944) described a proliferation of outer end-plate nuclei and reported that their number is doubled 12 days after nerve section. An increase in the size and number of muscle end-plate nuclei was found by Boeke (1916) with subsequent atrophy and disintegration of teloglial nuclei at later periods. Such degenerative changes, however, were not found by Tello (1907) or Gutmann and Young (1944). According to Tower (1939), end-plates cannot be discerned in the cat by six months after denervation, because their cytoplasm has disappeared and their nuclei can no longer be differentiated from the subsarcolemmal nuclei. Chor et al. (1937) maintained that after 14 days denervation endplates are to be found only very exceptionally. However, the granular endplate cytoplasm persists for a long time (Plate Ib) and enables the end-plate to be localized even at a late denervation period. According to Sunderland and Ray (1950) surviving muscle fibres of the opossum always contain an endplate irrespective of the duration of the denervation period; these authors were able to demonstrate end-plates even after 17 months' atrophy.

As far as the enzymatic activity of the end-plates is concerned, it was found that a high content of cholinesterase persists in the muscle following section of the nerve (Martini and Torda 1937, Marney and Nachmansohn 1937, Leibson 1939, Couteaux and Nachmansohn 1940, Meng 1940, Couteaux 1942, Stoerk and Morpeth 1944, Brooks and Myers 1952, Brooks and Chipman 1952) and that it is localized at the level of the subneural apparatus of the end-plates (Couteaux 1942, Couteaux and Nachmansohn 1940, 1946). Cholinesterase activity is decreased when it is determined histochemically (Waser, Hadorn 1961). Biochemically, both a decrease (Martini, Torda 1937, Feng, Ting 1938 and oth.) and an increase (Marney, Nachmansohn 1937, Sawyer, Davenport, Alexander 1950 and oth.) of its concentration were found following section of the nerve. This disagreement can probably be explained when concentration of cholinesterase is compared with the rate of atrophy. It is possible that during rapid atrophy the concentration of cholinesterase is increased for a short period, when the loss of weight progresses at a faster rate than the decrease in the total enzyme activity in the muscle (Couteux, Nachmansohn 1940, Brooks, Myers 1952). Anyhow the specialized endplate membrane, i. e. the subneural apparatus (Couteaux 1942) does not remain unaltered by denervation. Czillik and Sávay (1955) found "hypersegmentation" of this apparatus 14 days after denervation while its fragmentation was already observed two months after nerve section and continued progressively onward.

It cannot be excluded, however, that these changes are due only to a decrease of enzymatic cholinesterase activity and that the subsynaptic membrane itself remains unaltered. The disintegration of subneural apparatus after denervation conspicuous in young animals (see page 114), occurs in adults at least in some end-plates. For example, during reinnervation formation of new "gutters" was observed amidst the remnants of the old ones (Czillik and Sávay 1955). During reinnervation not only are new end-plates formed besides the old persisting end-plates, as was described by Gutmann and Young (1944); some components are differentiated de novo also in old end-plates after their reinnervation.

b) Muscle spindles

Muscle spindles appear to be unchanged for a long time after denervation and it was therefore assumed that intrafusal fibres, in contrast to extrafusal fibres, retain their structure after denervation (Sherrington 1894, Batten 1897, Horsley 1897, Willard and Graw 1924 and others). Tower (1932, 1935) first demonstrated atrophic changes and a reduction of nuclei in the nuclear bag and disintegration of intrafusal fibres one year after denervation. Chor et al. (1937) also agree that at late stages spindles are substituted by connective tissue. On the other hand, Adams et al. (1954) found intact spindles in cat muscle with preserved nuclear bag as long as eight months after denervation. Sunderland and Ray (1950) also observed simple atrophy of intrafusal fibres in the opossum, which was somewhat smaller than atrophy of extrafusal fibres, but did not find any changes in the connective tissue capsule or the perifascicular space.

In our experimental material we found a gradual decrease in the number of nuclei in the nuclear bag of intrafusal fibres one month after denervation (Plate V). Six months after denervation oedematous swelling was found in the equatorial zone together with a clear reduction in the number of nuclei in the nuclear bag (Plate Vd). The spindle capsule seemed to be thickened, and the perifascicular space was narrowed. Spindles could easily be located in both the muscles studied — the soleus and extensor digitorum longus even more than a year after denervation. The number of fibrous strands in the lymphatic space, connecting the capsule with intrafusal fibres, was increased (Fig. 25).

Spindles in rabbit muscles six months and one year after denervation were also very conspicuous. Atrophy of intrafusal fibres is evidently less extensive than that of extrafusal fibres, since the diameter of the former in the polar zone was the same as that of neighbouring fibres. Thus it seems beyond doubt that denervation changes concern mainly the equatorial zones, in which the special spindle structure — the nuclear bag — is dedifferentiated. The polar zones of intrafusal fibres are dependent for their trophic state on motor nerve fibres so that they atrophy after motor denervation and do not change after elimination of sensory fibres (Tower 1932).

It is difficult to evaluate degeneration and the extent of atrophy from changes in individual spindles. Some spindles may retain a nearly normal structure after denervation for a long time, whereas other spindles may completely disintegrate

Table 3

Number and size of spindles in control and denervated muscles of the rat.

Time after denervation	Number of spindles				Diameter of spindles			
	M. ext. dig. long.		M. soleus		M. ext. dig. long.		M. soleus	
	Control	Denerv.	Control	Denerv.	Control	Denerv.	Control	Denerv.
15 months	35	31	14	13	60,6	34,9	65,5	35,5
	30	29	13	15	60,7	33,5	71,8	50,8
	29	31	14	17	60,5	38,3	66,0	54,0
2 years	37	31	21	18	67,0	39,1	70,0	52,2
	—	—	16	16	—	—	64,7	56,3

after denervation as it was found in very young animals (see page 112). To what extent do these receptors disintegrate in adults?

To answer this question, the number of spindles was counted in serial sections of whole normal and denervated muscles 15 months and 2 years after denervation. All remnants of spindles were taken into account, in every case where a capsule surrounding intrafusal fibres could be discerned. No decrease of the number of spindles was found in the denervated muscles 15 months following denervation. Two years after denervation in spite of the extensive atrophy and fat infiltration of the muscle a normal number of spindles was still present. (Table 3). Evidently muscle spindles are more resistent to complete disintegration than extrafusal fibres, which are reduced considerabely in the soleus and in the extensor digitorum longus. As the remaining spindles are condensed into smaller volume, they are more conspicuous after denervation, when their atrophy has not yet advanced. To evaluate the degree of atrophy of muscle spindles in later stages the maximum diameter of each spindle was selected from serial sections and measured in denervated and control muscles. Maximal decrease of fifty percent of the control diameter was found in denervated spindles. This was due mainly to an evident reduction in the lymphatic space of some spindles, which was reduced to a narrow ring between the capsule and intrafusal fibers or was completely absent. In spite of this spindle atrophy the number of the intrafusal fibres was increased by about 50% at the average, this being apparently due to division of intrafusal fibres in the equatorial zone.

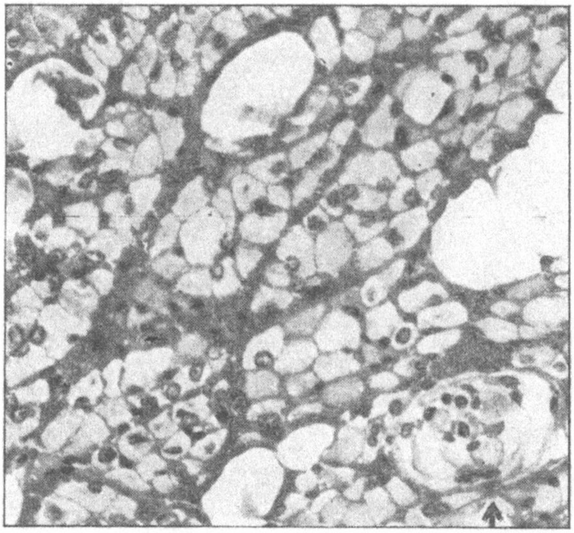

Fig. 25. Muscle spindle in the extensor digitorum longus a year after denervation. (Van Gienson; 400×.)

3. Changes in muscle nuclei

Changes in the size and shape of subsarcolemmal nuclei represent the earliest change in the muscle fibre following section of the nerve and the degeneration of nerve endings. Changes in muscle nuclei were described already two days after denervation by Krauss (1888) and Schmidtmann (1916), and in the course of the second week after denervation in the cat by Tower (1935, 1939) and Adams et al. (1954). According to Altschul (1947) the nuclear volume increases because of decreased pressure in the interstitial space. Smaller diameters of nuclei were only reported by Willard and Graw (1924) in mice soon after denervation.

Following denervation muscle nuclei in the rat cease to be elongated and become oval in shape. The karyoplasm becomes very light; the one or two nucleoli in the nucleus are very conspicuous because of the surrounding light karyoplasm (Fig. 26). It seems that the number of nucleoli does not change in the rat. On the other hand, in man, an increased number of nucleoli has been described by Bowden and Gutmann (1944), while a decrease was noted in the cat by Tower (1935).

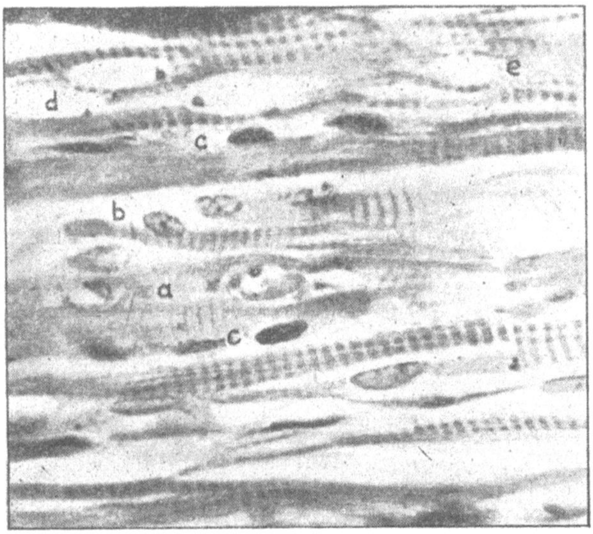

Fig. 26. Changes in muscle nuclei in rat muscle two months after denervation. a) spherical nuclei with conspicuous nucleoli; b) nuclei with shrunken nuclear membrane; c) pycnosis of nuclei; d) vacuoles inside muscle fibre with nuclear remnants at the poles; e) empty vacuole inside muscle fibre. (Heidenhain's haematoxyline; 900 ×.)

It is still a question whether the number of nuclei also increases in addition to changes in their size. Indication of amitosis can be found relatively often in the rat, two or more nuclei adhering to each other. Even the chains of nuclei, which are formed during the first month either beneath the cell membrane or inside the fibre between myofibrils, can rather be explained by amitotic division than by migration of nuclei. Sometimes even elongated bags filled with nuclei appear inside the fibre. This nuclear proliferation could be balanced by their equally rapid disintegration and need not necessarily result in an absolute increase of their numbers in the denervated muscle. According to the quantitative evaluation of Willard and Graw (1924) in mice up to 65 days and of Sunderland and Ray (1950) in the opossum, the increase of nuclei is only apparent and is actually only due to their condensation in a reduced space. In spite of this, Sunderland and Ray (1950) found proliferation of nuclei by 50% in two exceptional cases out of 36 (after 185 and 280 days respectively). According to the opinion of other authors, nuclear proliferation is a regular occurrence (Krauss 1888, Ricker and

Ellenbeck 1899, Jamin 1904, Schmidtmann 1916, Tower 1935, Chor et al. 1937, Altschul 1942 and others). These studies, however, are not based upon reliable data.

However, biochemical studies show an absolute increase of desoxyribonucleic acid (DNA) in the first stage of denervation. The absolute amount of DNA in the denervated muscle undergoes a temporary 40% increase during the first month of denervation, but from the second month onwards the content is the same as in the normal muscle (Gutmann 1960) — (Fig. 30). The increase in the number of nuclei in later stages of denervation is apparent, and their proliferation is balanced by their increased disintegration. In rat muscles beginning with the second month after denervation pycnotic nuclei are often present (Fig. 26). In others a central vacuole can be observed. Besides these two extreme cases, numerous transitory shapes exist which are suggestive of two mechanisms of nuclear degeneration. Nuclei shrink and disintegrate, or swell and undergo lysis. Nuclei with shrinkage of the surface and a wavy nuclear membrane can be observed, the karyoplasm of which is still clear and contains a distinct nucleolus (Fig. 26). Most of the round nuclei arranged in chains have their surface retracted in this way. The shrunken nuclei with concentrated chromatin content represent the next stage of this form of degeneration. The terminal stage evidently corresponds to pyknosis and karyorhexis, i. e. disintegration into free fragments of chromatin. The other form of degeneration is nuclear vacuolization, observed in the cat from the first month onwards (Tower 1935). In rats nuclei were found containing a central vacuole, with karyoplasm extruded towards the poles of the circumference of the nucleus by the second month after denervation. Karyolysis of these nuclei leaves behind empty vacuoles inside the muscle fibre with nuclear remnants at the poles (Fig. 26). Nuclear remnants also persist in the strands between vacuoles in muscle fibres undergoing vacuolar degeneration (Fig. 32, 33). Muscle fibre degeneration always leads to disintregation of nuclei after preceding pyknosis (Plate IVb). However, pyknosis and karyolysis of nuclei can also be found in fibres with preserved cross-striation. Nuclear destruction was already

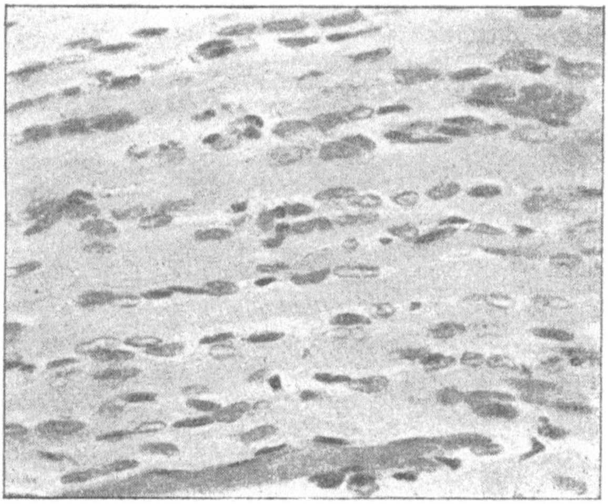

Fig. 27. Relative nuclear proliferation in the soleus muscle of the rat three months after denervation. (Haematoxyline-eosine; 400 ×.)

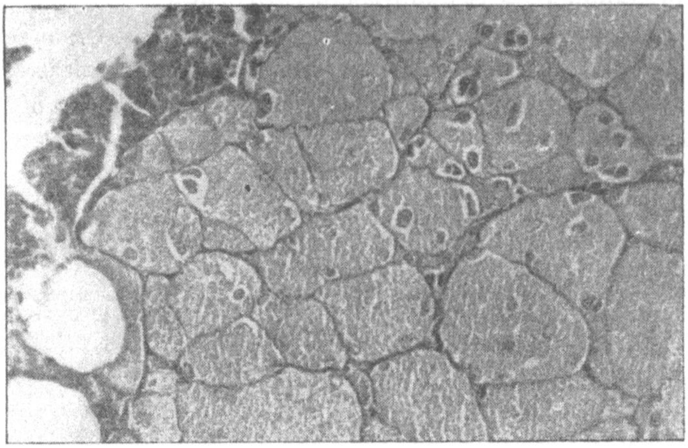

Fig. 28. Nuclear proliferation inside muscle fibres of the rabbit soleus muscle three months after denervation. Empty spaces around nuclei. (Van Gieson; 400 ×.)

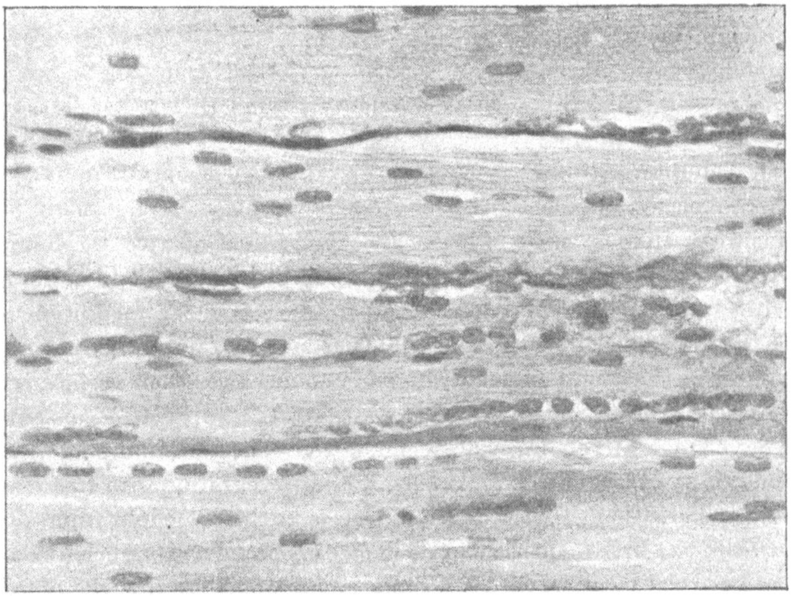

Fig. 29. The same muscle in longitudinal section. Proliferation of nuclei in chains on the surface and inside muscle fibres.

described by Stier (1896), but Sunderland and Ray (1950) never encountered this change in their extensive material.

In the rat, nuclei are localized mainly on the circumference of denervated muscle fibres (Fig. 27), less often inside the fibre. In the rabbit, however, a marked

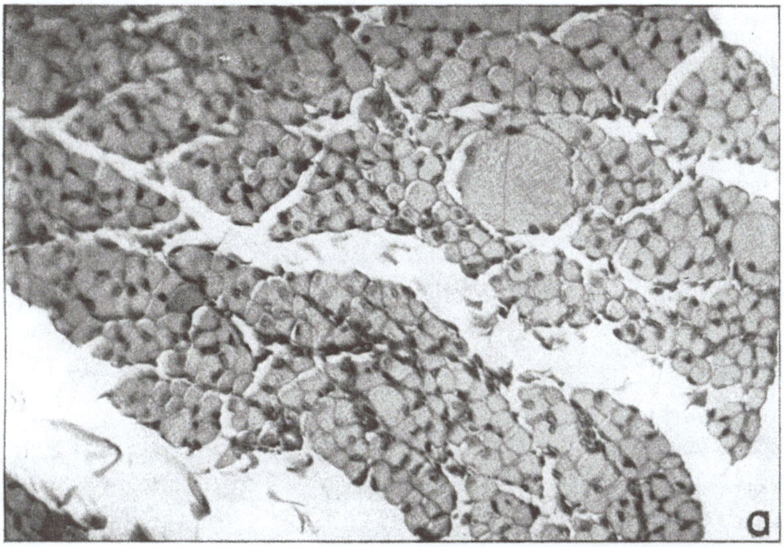

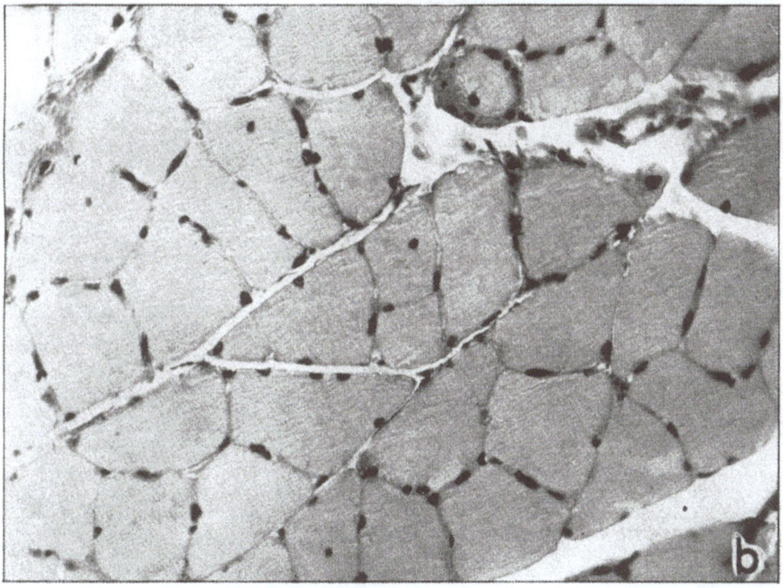

Fig. a. The soleus muscle in the rat nine months after denervation. A preserved large fibre surrounded by atrophic fibres.

Fig. b. Control soleus muscle in the rat. Centrally localized nuclei in several muscle fibres. (Van Gieson; 400×.)

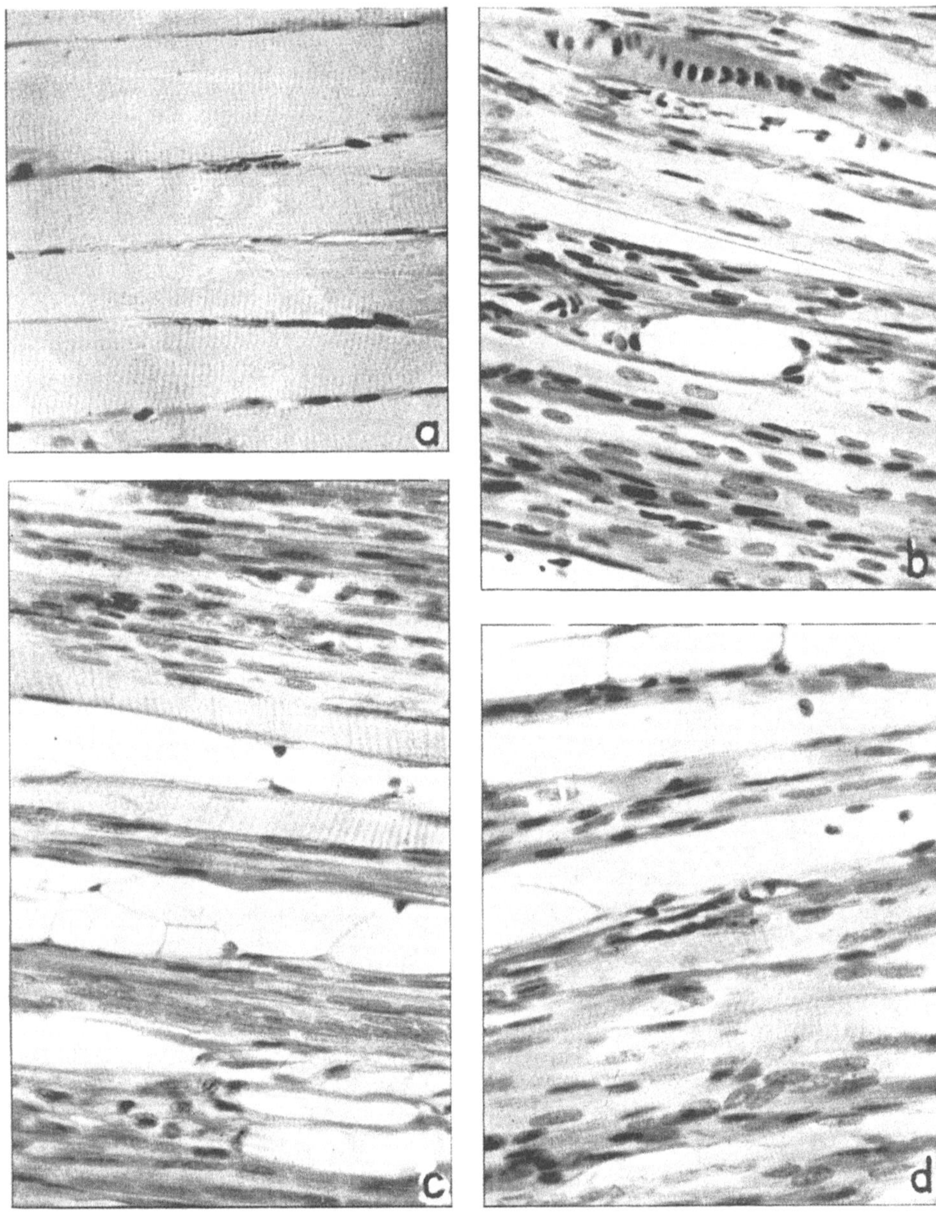

Fig. a. Normal soleus muscle in the rat. Fig. b. The soleus muscle of the rat six months after denervation. Fig. c. The soleus muscle of the rat nine months after denervation. Fig. d. The soleus muscle of the rat one year after denervation. (Van Gieson; 400 ×.)

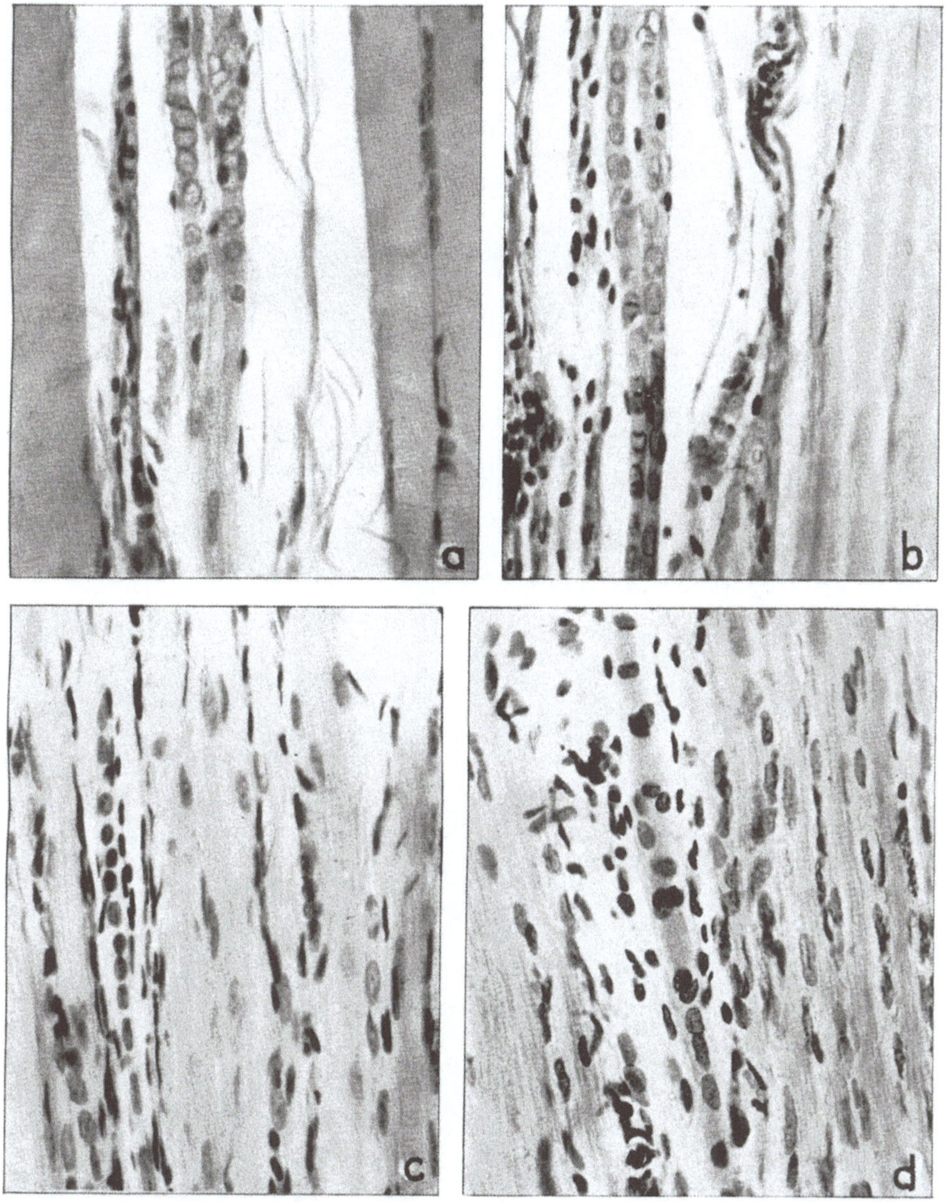

Fig. a. Normal muscle spindle in the rat.

Fig. b. Muscle spindle in the rat a month after denervation. Slight decrease in the number of nuclei in the nuclear bag on the left.

Fig. c. Muscle spindle in the rat two months after denervation. Dedifferentiation of the nuclear bag; marked reduction in the number of nuclei.

Fig. d. Muscle spindle six months after denervation. Swelling of an intrafusal fibre in the de-differentiated nuclear bag on the left and in the myotubal zone on the right. Pyknosis of nuclei in the nuclear bag. Cellular infiltration in the perilymphatic space. (Haematoxyline-eosine; 400 ×.)

a

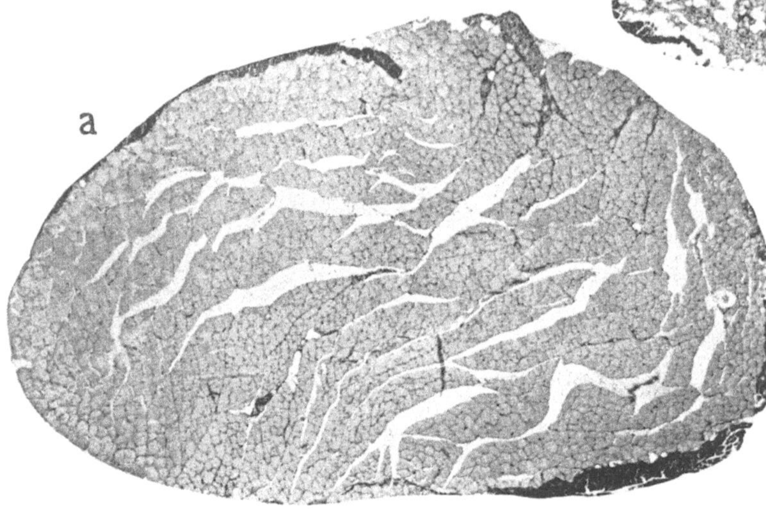

b

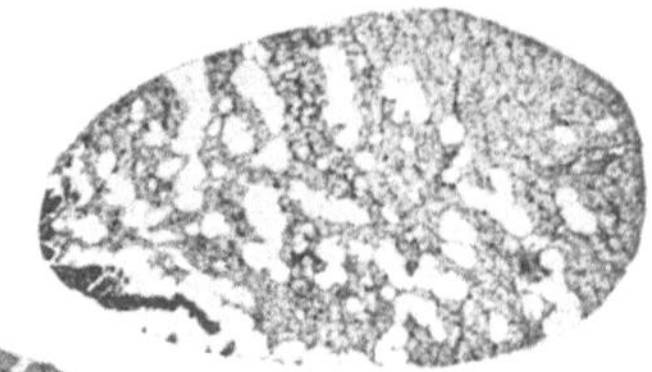

c

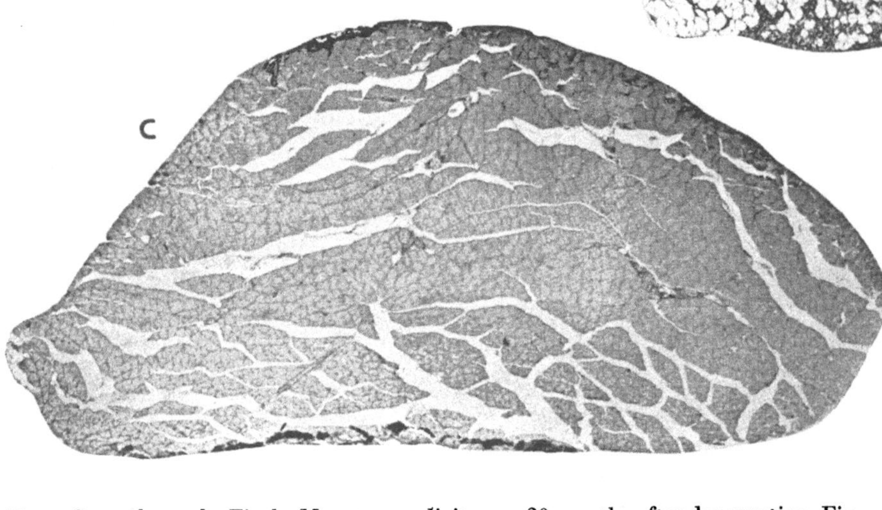

d

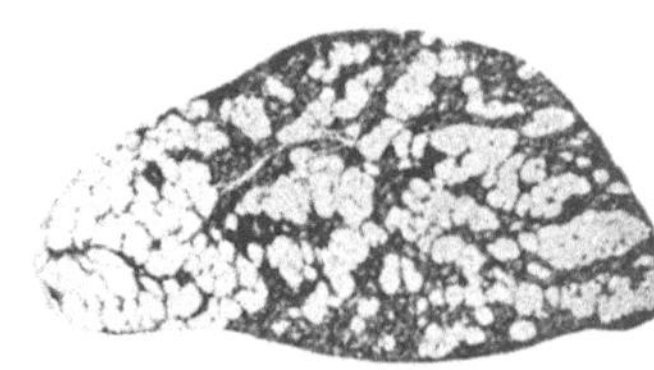

Fig. a. Control muscle. Fig. b. M. extensor digitorum 20 months after denervation. Fig. c. Control muscle. Fig. d. M. soleus 20 months after denervation.
Cross sections through the belly of the muscle. (Van Gieson; 35 ×.)

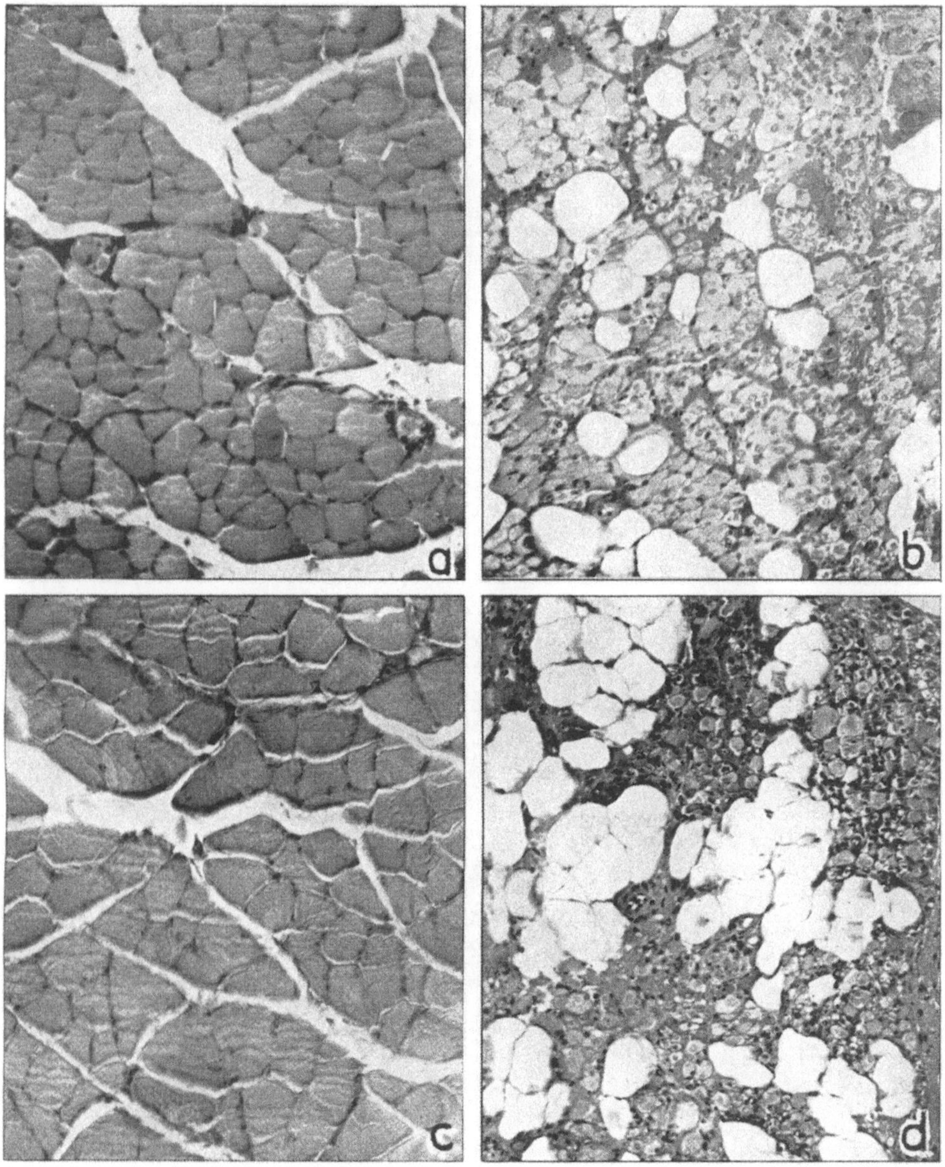

Same as preceding Plate VI.; details. (200 ×.)

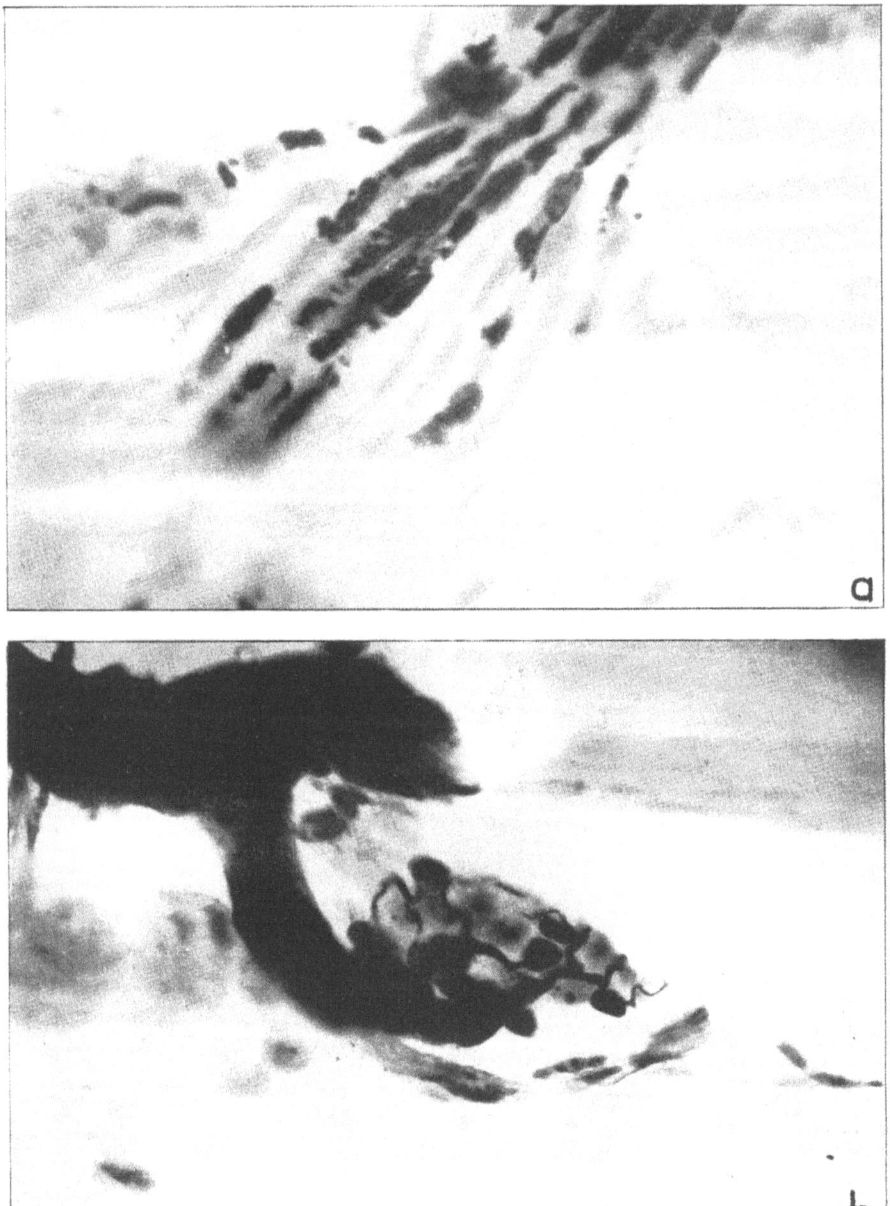

Fig. a. Empty end-plates 36 hours following section of the nerve in the rabbit.
Fig. b. Normal end-plate three days following the nerve blockade. (Bielschowsky-Gros; 1000×.)

proliferation of nuclei was found inside fibres of extensive regions of the soleus muscles three months after denervation (Fig. 28, 29).

Nuclei were usually surrounded by empty areas. Optically empty spaces were often observed around nuclei in the rat after the first month of denervation

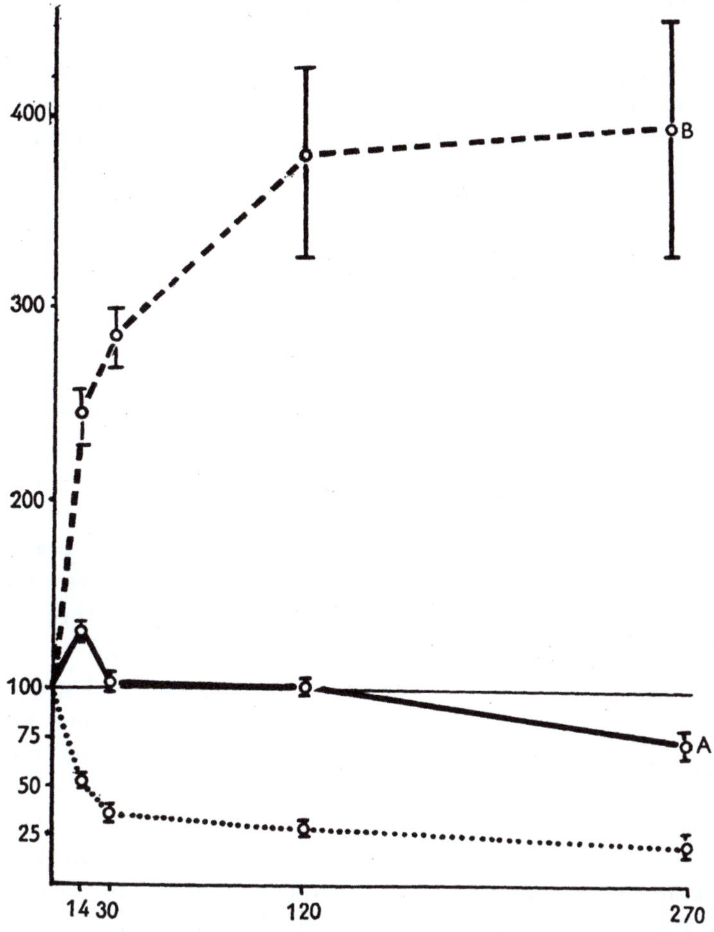

Fig. 30. Changes in the absolute (A) and relative (B) content of desoxyribonucleic acids and non collagenous protein nitrogen (dotted line) following section of the nerve. Abscissa: days after denervation. Ordinate: percentage changes in relation to control normal muscle. DNA determined in whole muscles in γ/P; the relative content of DNA determined in γ/P-DNA mg. N.

(Fig. 33). It often seems that nuclei lie in the space between fibres, however on cross-section they are clearly embedded in a groove-like space beneath the sarcolemmal membrane. The origin of these spaces, which are sometimes very extensive, is not clear. It is a question whether the pathologically altered nuclei can affect proteolysis and whether they can actively participate in the process of reduction of contractile tissue.

4. Degenerative changes in muscle fibres

When comparing the main symptoms of denervation atrophy according to their time relations after denervation, the initial changes affect endplates and sarcolemmal nuclei, and these are followed directly, and accompanied concomittantly, by atrophy of muscle fibres, while degeneration in muscle fibre does not begin until a certain time after denervation. The first period, in which no degenerative changes are present, is ususally designated as the stage of simple atrophy. This period can persist for a different length of time. Stier (1896) found degenerative changes in the rabbit as late as three and a half months after denervation. Hyaline degeneration of single muscle fibres were described one month after denervation by Tower (1935) in the cat. Chor et al. (1937) observed granular and vacuolar degeneration in the cat which was also demonstrated by Tower (1935) at later stages. Degenerative changes, mostly hyaline, were also found in the cat by Adams et al. (1954) from the sixth week onwards. The degenerative process in muscle culminates in the cat in the course of the fourth month after denervation, and later retreats (Tower 1935).

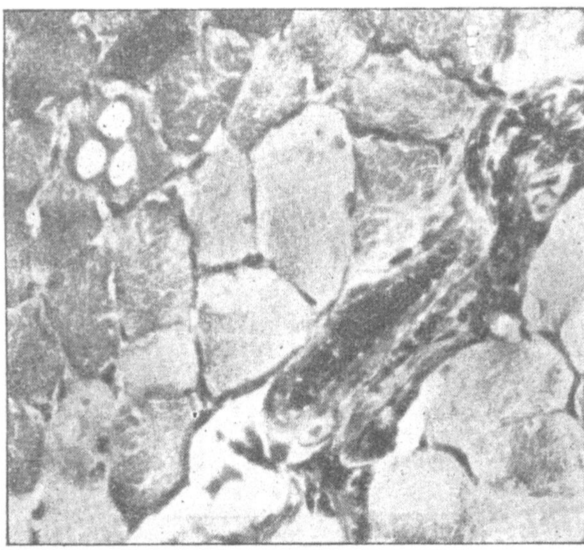

Fig. 31. Vacuolar degeneration of a muscle fibre in extensor digitorum longus a month after denervation. (Heidenhain's haematoxyline; 500 ×.)

Degeneration and phagocytosis of muscle fibres was described only in exceptional cases by Hines and Knowlton (1933) a month after denervation in the rat. We have also observed degenerative changes in the rat during the first month after denervation, but this does not seem to be an exceptional finding. Numerous muscle fibres are affected by the degenerative process in the soleus, so that sometimes continuous loci or bands of degeneration are formed across the muscle. In the extensor digitorum longus single fibres degenerate, but they can be found regularly.

Degeneration is mostly vacuolar, fibres containing either one central vacuole, or several larger vacuoles (Fig. 31, 33), in others small vacuoles are accumulated in a longer segment (Fig. 32). It is probable that centrally placed vacuoles develop by fusion of smaller vacuoles and it is often possible to find remnants of their septa in longitudinal sections on the circumference of the large elongated central vacuole. It is sometimes possible to observe remnants of disintegrated nuclei on the periphery of vacuoles or in the protoplasmatic strands between them.

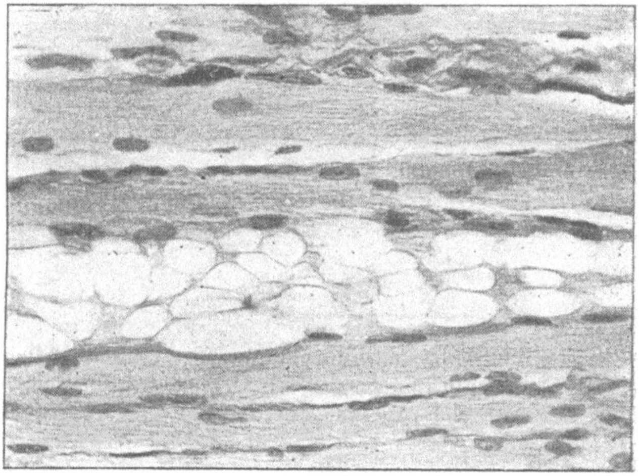

Fig. 32. Vacuolar degeneration in the soleus three months after denervation. Longitudinal section. (Van Gieson; 400 ×.)

In our experiments vacuolar degeneration was observed in the soleus of the rabbit three months (Fig. 36) and six months after denervation (Fig. 37), while it was absent in the extensor digitorum longus.

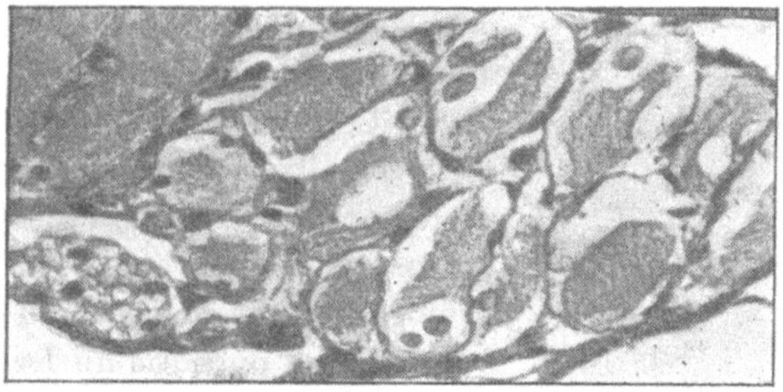

Fig. 33. A group of degenerating muscle fibres of the soleus in the rat a month after denervation. Proliferation of nuclei in empty peripheral spaces in fibres with reduced contractile substance. Vacuolar degeneration. (Van Gieson; 500 ×.)

At later periods degenerative changes in the rat appear in occasional fibres throughout the whole muscle. Hyaline degeneration predominates, although it was only exceptionally to be found in rats one month after denervation. Usually a short segment of the fibre degenerates, cross-striation is lost, and the staining affinity of homogenous cytoplasm is generally increased (Fig. 35). The continuity of the fibre is usually interrupted in the degenerated segment. Both stumps retract and their endings become club-shaped (Fig. 34). Sometimes the endomysial sheath between the fibres contains free, probably phagocyte-like cells,

Fig. 34. Hyaline degeneration in the extensor digitorum longus of the rat a month after denervation. Degenerated fibre interrupted and both stumps with club-like endings. (Haematoxyline-eosine; 400 ×.)

in other small loci with two or three adipose cells in the same locality. Activated hystiocytes filled with phagocytized fragments were observed in the close vicinity of muscle fibre remnants, the cross-striation of which was preserved.

Degenerating fibres are a rare finding even during late denervation periods. After three months, when the difference in the extent of atrophy between individual fibres begins to be conspicuous in the rat, degeneration is more common in fibres with large diameters. Degenerative changes similar to the physiological degeneration of myotubes during development (Couteaux 1941) were exceptionally found in the rat (Plate IVb). These muscle fibres are oedematous with homogenous sarcoplasm and a number of pycnotic nuclei centrally placed with their longitudinal axis lying transversely across the fibre.

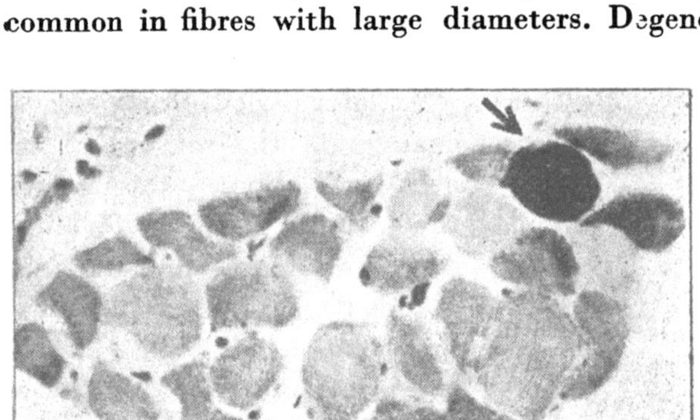

Fig. 35. Hyaline degeneration of a fibre on cross-section. The soleus muscle in the rat two months after denervation. (Heidenhain's haematoxyline; 400 ×.)

The border-line between atrophic and degenerative changes is usually sharply defined; atrophy results in a progressive reduction of diameters, degeneration alters the structure of the fibre and terminates by its desintegration. However, in denervated muscles of the rat and rabbit there is a change which can be classified as both atrophy and degeneration. Sarcoplasm with myofibrils is reduced and disappears from the space underneath the sarcolemma, so that an optically empty groove-like or horseshoe-shaped segment filled with a nucleus or with an aggregate of nuclei is thus formed. The fibre can thus be excavated from several sides, so that its remnants is star-shaped in cross-section. Muscle fibres with myofibrils reduced in this fashion, occur in groups. In some places these changes resemble vacuolar degeneration localized beneath the surface membrane. However, nuclei are usually present in the empty spaces of these excavated fibres, and since transitory forms can be found between nuclei surrounded by an empty space and these structures, it will be necessary to explain whether nuclei can affect the proteolytic process.

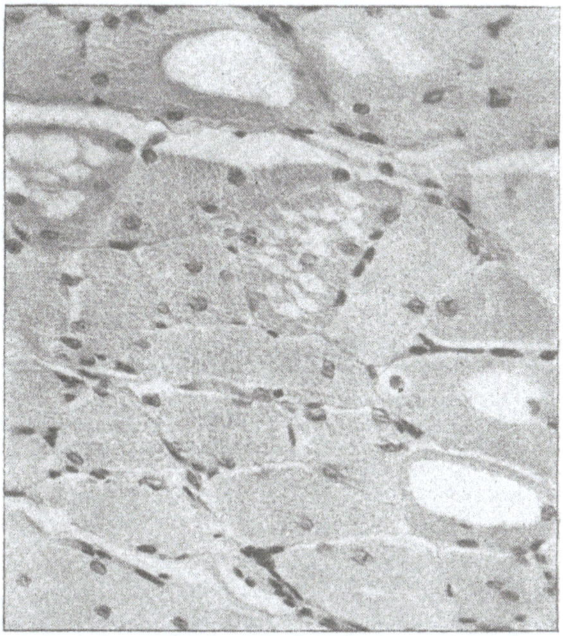

Fig. 36. Vacuolar degeneration and nuclear proliferation in the rabbit soleus three months after denervation. (Haematoxyline-eosine; 400 ×.)

Differences in degeneration changes can even be demonstrated in different muscles of the same animal species. For example, degenerative changes are rarely to be seen in insect coxal red muscles throughout the whole denervation period, and are restricted to fragmentation of muscle fibre segments into necrotic ovoids (Fig. 38). On the other hand, degenerative changes are extensive and very variable in pale muscles. Some of the muscle fibres succumb to vacuolar degeneration (Fig. 39), in others disintegration of central myofibrils occurs, so that elongated sarcolemmal bags develop.

It can be concluded that degeneration is a regular occurrence in the above experimental material (in rats and rabbits), but that it affects only a few fibres in the whole muscle cross-section, with the exception of the soleus in the rat a month after denervation and three to six months after denervation in the rabbit, in which whole muscle fibre groups and fascicles degenerated.

Numerous authors either found no, or almost no, degenerative changes in denervated muscles. Degeneration was completely absent in mouse muscle 63 days after denervation (Willard and Graw 1924). Degenerative changes were

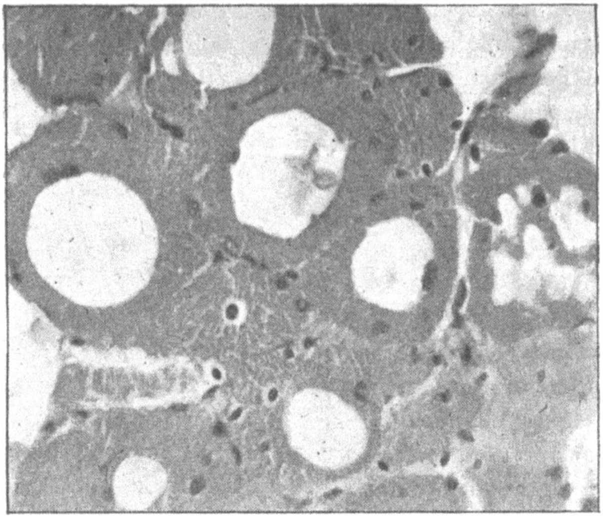

Fig. 37. Vacuolar degeneration in the rabbit soleus six months after denervation. Nuclei on the circumference of central vacuoles. (Haematoxyline-eosine; 400 ×.)

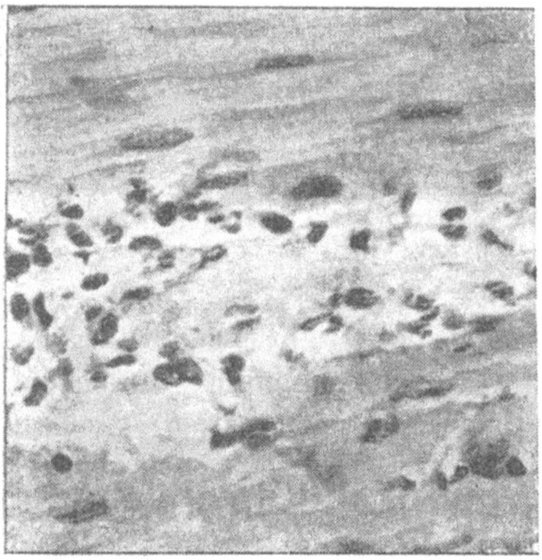

Fig. 38. Disintegration of muscle fibres into necrotic ovoids with lymphocyte infiltration. Red muscle of the cockroach (Periplanata americana) 80 days after denervation. (Haematoxyline-eosine; 500 ×.)

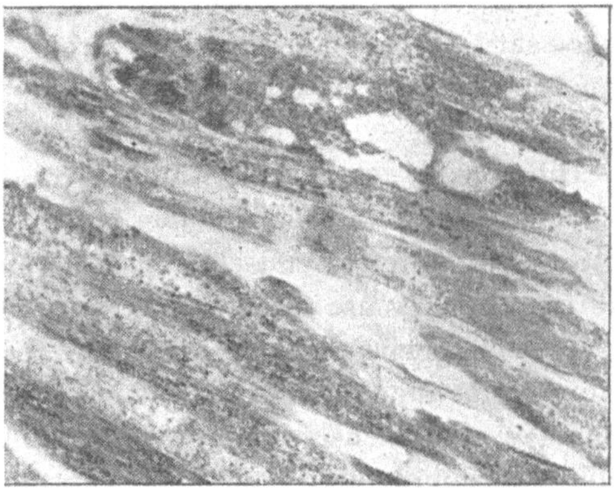

Fig. 39. Vacuolar degeneration (on the right) in a pale muscle of the cockroach (Periplanata am.) 80 days after denervation. (Stained for sarcosomata according to Altmann; 250 ×.)

found in a single case out of 36 in the opossum 49 days after nerve section (Sunderland and Ray 1950). In contrast to Tower (1935), Reid (1941) could find no signs of degeneration in the cat six weeks after denervation.

In some species, degenerative changes probably do not develop at all in denervated muscles. But even in those cases where they occur regularly, degenerative changes do not exhibit a progressive tendency in the course of the denervation period.

5. Connective and fatty tissues

It is usually assumed that a muscle after longlasting denervation ceases to exist as contractile tissue and is transformed into connective or fatty tissue.

According to Tower (1935), muscle fibres are transformed into connective tissue by direct metaplasia; progressive muscle fibre atrophy leads to an extreme thinning of diameter, to loss of cross-striation of myofibrils, and to their transformation into collagenous fibrils. In this way the border-line between fibres changed into connective tissue bands and proliferated endomysial connective tissue is difficult to detect. However, Tower (1935) did not find complete transformation into connective tissue in the interosseal muscle of digiti V. in the cat as long as one year after nerve section. The muscle was substituted by connective tissue in the manner described above only at occasional sites. In spite of this, Tower (1935) concluded that: "... the whole constitutes a tissue intermediate of scar tissue and tendon. A year did not suffice to complete this transformation except locally, but the end may be predicted, though not the time of its accomplishment."

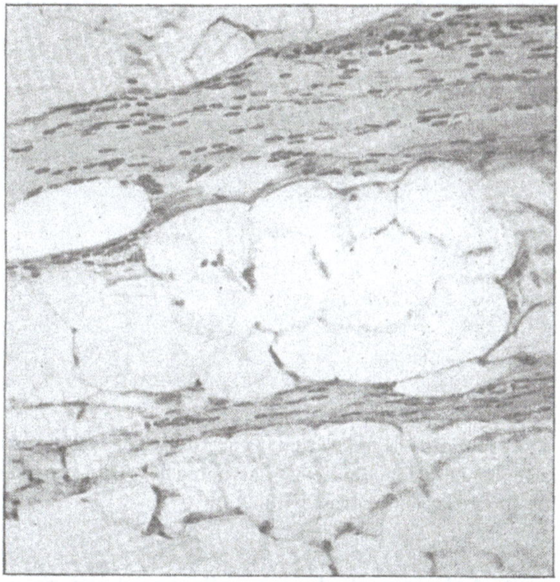

Fig. 40. Fatty degeneration in the rabbit soleus six months after denervation. Remaining fascicles of atrophic muscle fibres embedded in fatty tissue. (Haematoxylin-eosine; 200×.)

The problem of the ultimate fate of muscle fibres is the main research question put in the extensive report of Sunderland and Ray (1950). No transformation of muscle fibres was observed by these authors in muscles of the opossum, only an extreme reduction of 97% in diameter. Besides atrophy, only a slight proliferation of endomysial connective tissue was demonstrated, beginning at the end of the first month and culminating at the third month.

This proliferation of endomysium caused a thickening of the connective tissue capsule around individual fibres only, but was never invasive in character and did not replace muscle fibres.

We were also unable to find fibrous metaplasia in muscles of rats up to one year after denervation or in rabbits nine months after denervation. Perimysial and endomysial connective tissue was markedly thickened in rats and rabbits from the third post-denervation month onwards. However, connective tissue foci did not develop even as late as one year after denervation (Fig. 40), only endomysial sheaths, while perimysium internum and fibrous tissue of nerve and blood vessel bundles were several times more abundant. Nevertheless, no period of fibroblastic proliferation was observed. Biochemical analysis has shown that collagenous tissue does not increase in the denervated muscle of the rat (Kolínská and Vodička 1960). Connective tissue in the denervated muscle thus increases only relatively, by condensation in the smaller space of the atrophic muscle.

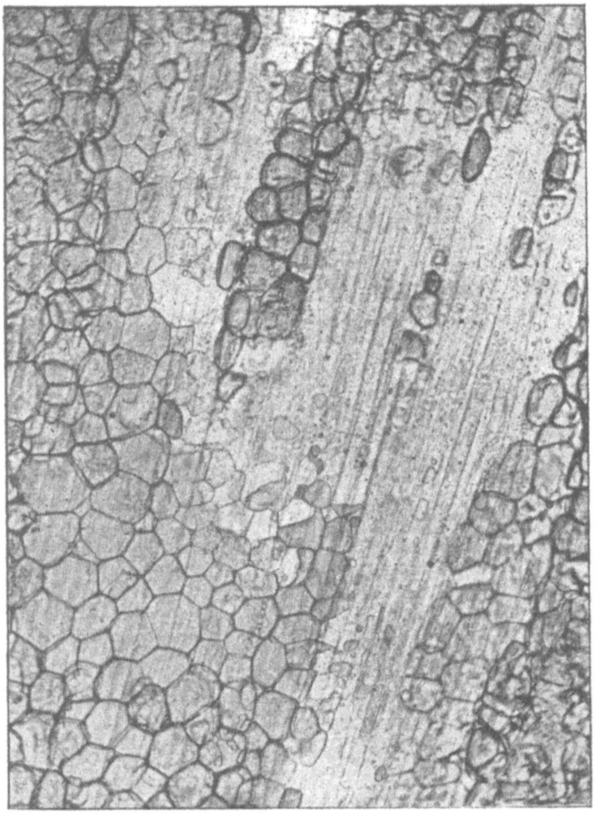

Fig. 41a. Fatty degeneration in the rabbit soleus nine months after denervation. Fascicles of atrophic fibres are surrounded by fatty tissue. (Scarlet R; 80 ×.)

On the other hand, the substitution of the major part of the muscle by fat cells has been convincingly documented in a number of reports (Krauss 1888, Grund 1912, Audova 1923, Chor et al. 1937, Altschul 1942, Gutmann and Young 1944, Adams et al. 1954, and Bowden and Gutmann 1944 in man). However, Sunderland and Ray (1950) do not consider the development of fat tissue to be typical for denervation, since in their material the appearance of fat tissue in denervated muscles was accidental. Its amount was independent of the duration of denervation, and did not even markedly predominate over the fat content in the normal muscle.

Slight fat infiltration was observed perivascularly and between muscle fibres in rats already a month after denervation. More extensive fat infiltration was found in the soleus and to a lesser degree in the extensor digitorum longus six months after denervation, but even 15 months after nerve section, interstitial columns between fascicules only are formed but fatty tissue never occupies the major part of the volume even in the soleus (Plate VI). Almost complete replacement by fat was found in the muscle 2 years after denervation.

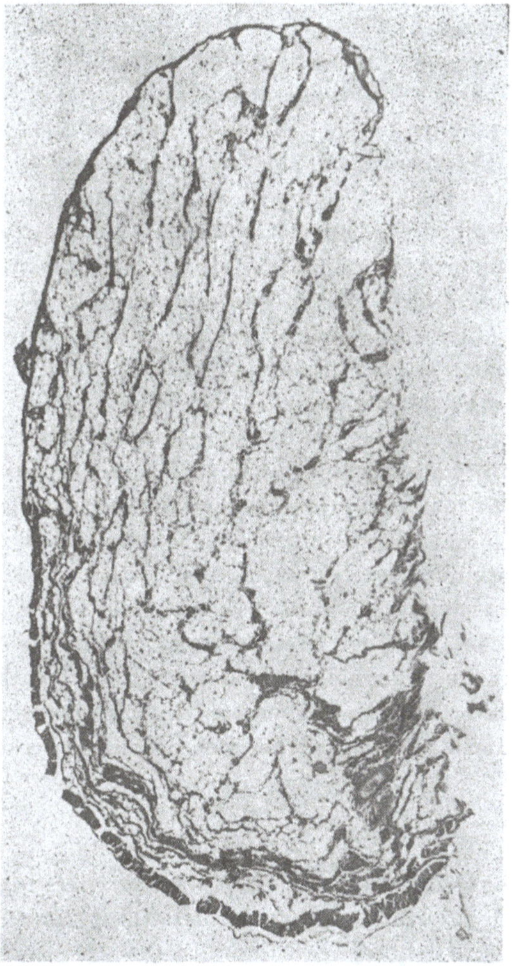

Fig. 41b. Cross section of the ant. tib. of the rabbit, replaced by fatty tissue, 16 months after denervation (Van Gieson 200 ×).

In the rabbit, on the contrary, where replacement by fat has already been described several times (Krauss 1888, Audova 1923, Altschul 1942, Gutmann and Young 1944), extensive fat infiltration was found in the extensor digitorum longus and soleus from the third post-denervation month onwards. Fat cells are mainly localized interfascicularly in the extensor digitorum longus, singly or in groups between muscle fibres, so that an intricate pattern can be seen in cross-section. This fatty degeneration is present in about one third of the muscle, in other parts connective tissue proliferation is marked and fat cells are missing. In the soleus muscle, lipids are accumulated only in the intrafascicular spaces. At later periods of denervation replacement by fat in the extensor digitorum longus is still restricted to a part of the muscle, while the soleus is affected by this process from the sixth month after denervation as a whole (Fig. 40). In the fat tissue only small groups of muscle fibres remain. The extent of fatty degeneration still increases up to the ninth month after denervation (Fig. 41a), but the fat tissue contains groups of muscle fibres which are only atrophic, without further impairment of structure. Almost complete transformation of muscle into fat tissue was described in the rabbit 11 months after denervation (Gutmann and Young 1944). Practically complete replacement by fat was found in the rabbit 16 months following denervation (Fig. 41b).

Extensive fatty degeneration was also found in pale coxal muscles of the cockroach (Periplaneta am.) 80 to 100 days after denervation, while there were only small areas of fat in red muscle along the tracheae (Gutmann and Zelená 1959). Fatty degeneration of a muscle is evidently dependent on its metabolic characteristics in different animal species and types of muscle.

The relationship between degeneration of muscle fibres and the appearance of fat cells will have to be elucidated. Usually groups of two or three cells are found at the end of the disrupted degenerated fibre. In the extensor digitorum longus of the rabbit columns of fat cells are evidently substituted for individual muscle fibres. Extensive degenerative changes precede massive fat infiltration in the soleus. In Periplaneta am. fat infiltration is also preceded by degenerative changes (Gutmann and Zelená 1959).

The development of fat cells has not been explained even under normal conditions, so that their origin in the denervated muscle still remains obscure. It is not probable, however, that muscle fibres could be transformed directly into fat cells with the persistence of nuclei and the cellular membrane of the original muscle fibre. Fat cells in denervated insect muscle are polyglobular with different cytoplasm filled with a greater or smaller content of lipid droplets. It is thus not possible to consider this as direct transformation of muscle cells into fat tissue. In mammals, mesenchymal elements are probably filled with fat in an analogous manner, by activation of either fibroblasts or cambial cells. Thus the development of fat tissue in the muscle and lipid accumulation in cells requires further more detailed investigation.

D. Correlation between physiological, biochemical and morphological changes accompanying denervation atrophy

Morphological studies accentuate the significance and necessity of correlation with physiological and metabolic findings. On the one hand, for example, muscle weight and protein nitrogen undergo small changes between the third and twelfth month after denervation. Yet, at this period, profound and extensive changes occur from the morphological aspect which result in complete destruction of groups of muscle fibres. Neither physiological nor metabolic data were able to discover these apparently irreversible changes. Metabolic research thus receives new tasks. The question of how long the muscle fibre "survives", and what changes lead to its destruction is a fundamental problem, and must be approached by correlating various findings. On the other hand, stages of morphological processes taking place in the muscle following section of the nerve should be interpreted from the aspect of physiological changes, i. e. primarily from the point of view of neuromuscular and muscle excitability.

Correlation of electrophysiological and histological data

Such correlation is of paramount importance for analysing the trophic influence of the nervous system. Changes in muscle excitability and the development of fibrillation activity in the denervated muscle are evidently chronologically closely interrelated, and require exact correlation with morphological changes. It has already been pointed out that the postulate of morphological and electrophysiological correlation led Erb to formulate his experimentally unfounded generalization concerning the "reaction of degeneration" from the functional and morphological aspect. The original finding of Erb (1868) about increased galvanic excitability was later confirmed by other authors, e. g. Pollock et al. (1945, 1947). Increased galvanic excitability occurs concomittantly with increased sensitivity of muscle fibres to chemical substances, especially of the acetylcholine type of mediator. Increased sensitivity and fibrillation activity are relatively precisely correlated in time (Luco and Eyzaguirre 1955, Axelsson and Thesleff 1959). These connections sufficiently clearly demonstrate the fundamental character of these findings for analysing the basic mechanism of the trophic influence of the nervous system. Morphological correlations are all the more important. It is not only a question of time correlations of complicated excitability changes or changes of electrical properties of the muscle fibre membrane, but rather of correlation during early stages of denervation atrophy

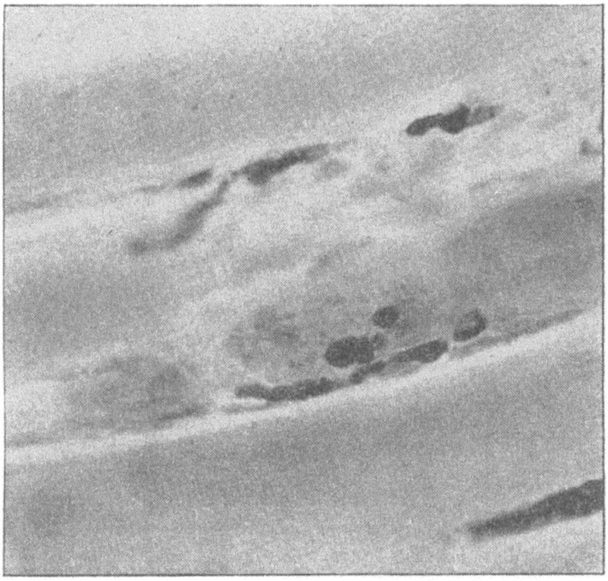

Fig. 42. M. peroneus longus 39 hrs. after section of the peroneal nerve. Irregular argentophile granules at the end-plates. (Glees-Bielschowsky; 700 ×.)

during which nerve excitability and indirect muscle excitability are lost. It is important to explain from the morphological aspect why indirect muscle excitability is lost relatively quickly (30 hrs. in the cat), while nerve excitability persists for as long as four days after nerve section (Titeca 1935, Holobuth and Jalowy 1936).

The correlation between physiological data, especially of muscle excitability and morphological changes at late stages of denervation atrophy is difficult. It can be assumed that, as long as cross-striation of muscle fibres is preserved.

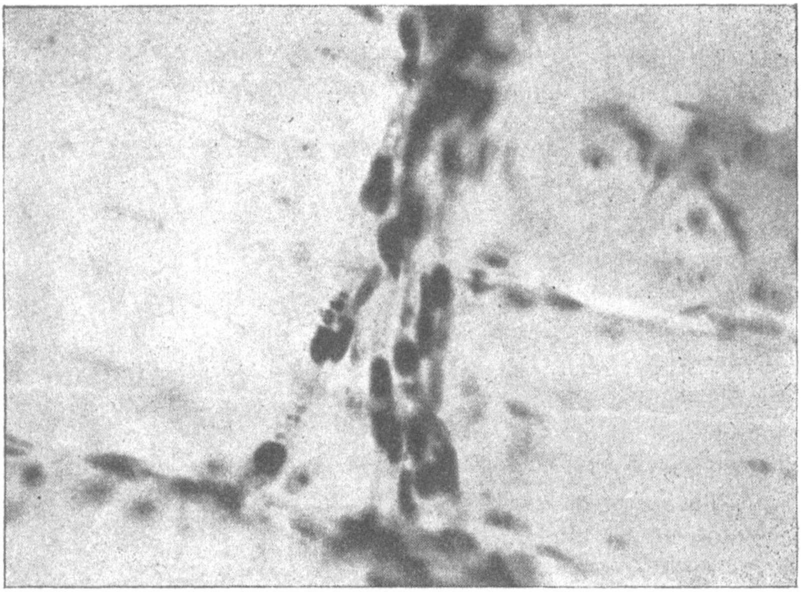

Fig. 43. M. peroneus longus 39 hrs. after section of the peroneal nerve. Disintegration of axons inside the Schwann sheaths. The end-plate is empty. (Glees-Bielschowsky; 700 ×.)

muscle excitability is intact. However, since the process of degeneration and destruction of muscle fibres does not progress at the same rate, it is necessary to differentiate between the destruction of individual fibres and destruction of the muscle as a whole. It is for this reason that muscle excitability is often reported in the clinical literature to be preserved a long time after denervation. Fibrillations can probably be demonstrated in the muscle as long as its contractile properties are not lost (Tower 1939) and since islets of muscle fibres can survive for long periods of time surrounded by fibrotic tissue, it is not surprising that fibrillations were reported in clinical cases even eleven (Proebster 1928) and eighteen (Weddell et al. 1943) years after denervation or poliomyelitis. It is, of course, necessary to exclude partial denervation.

The fundamental problem of investigating the reversibility of changes remains unsolved. This has, so far, been only possible by reinnervation. Of course, quantitative data must first be obtained and this is primarily the task of biochemical analysis.

Moreover, study of the initial changes in the denervated muscle, especially of the exact time relations of the loss of excitability and onset of degeneration of the axons in the endplates which may signalize the development of the trophic disturbance should give important clues for metabolic research. In the rabbit degeneration in nerve trunks and in end-plates were investigated during the first hours after cutting the nerve. Loss of continuity of axons which correlates in time with the loss of excitability and conductivity in the nerve fibres, could be defined as criterion of degeneration in the peripheral nerve stump (Gutmann and Holubář 1950). Similarly degeneration of axons in the end-plates can be defined and correlated as loss of continuity of the terminal arborizations of axons. This process takes place in the rabbit about 32 hours after cutting the nerve and at this time indirect excitability of the muscle is also lost. Thirty two hours after cutting the nerve almost all terminal axons in the end-plates exhibit fragmentation and myelin sheaths form ovoids. The fragmented axonal material is quickly removed and 39 hours after cutting the nerve only occassionally can irregular argentophile granules be found (Fig. 42). The majority of endplates, however, are empty. In the corresponding Schwann tubes there is considerable breakdown of axons (Fig. 43). The myelin sheaths are swollen and disintegrate into ovoids (Fig. 44). Seventy hours after cutting the nerve there is complete breakdown of axons in intramuscular nerve trunks, mostly

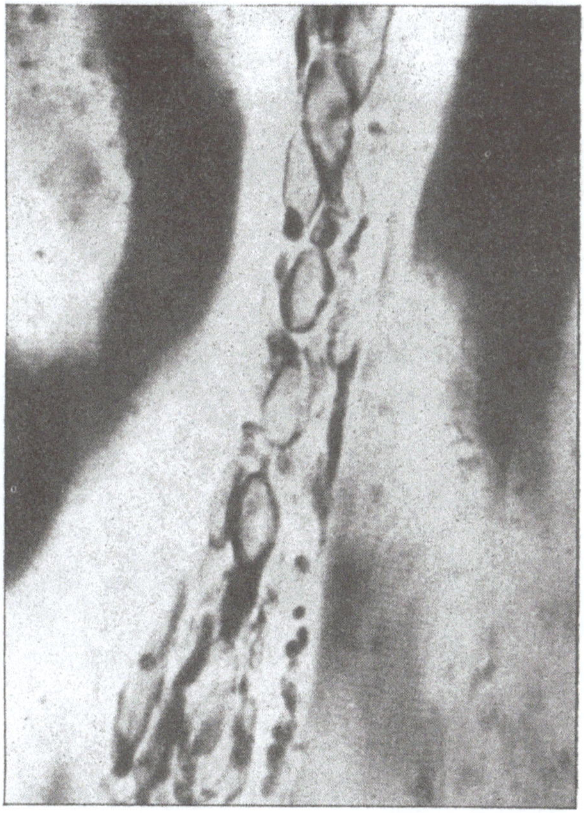

Fig. 44. M. peroneus longus 39 hrs. after section of the peroneal nerve. Desintegration of myeline sheaths into ovoids (Spielmayer; 500 × .)

only dust-like granules are found and the end-plates are empty (Fig. 45). At this time the continuity of the axons in the peripheral nerve stump and excitability of the nerve are mostly still intact (Gutmann and Holubář 1952). The first degenerative change thus occur in the end-plates and this explains the quick loss of indirect muscle excitability as suggested also by Titeca (1935) and Holobuth and

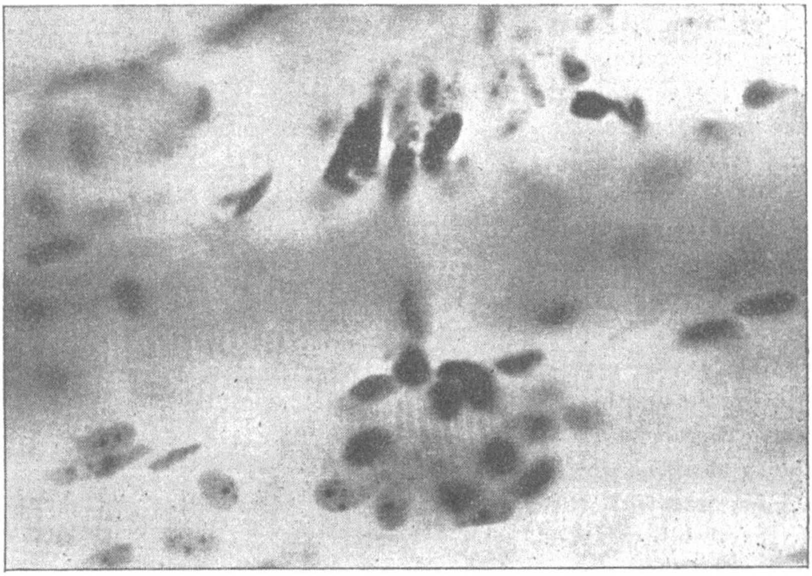

Fig. 45. M. peroneus longus 70 hrs. after section of the peroneal nerve. Empty end-plate with Schwann sheaths. (Glees-Bielschowsky; 700×.)

Jalowy (1936). Autolytic processes, which are morphologically correlated with the disintegration of axons and of myeline sheaths, progress at a faster rate in axons contained in the muscle than in axons of the nerve trunk. It is possible that metabolism in the nerve has smaller demands than that in the muscle and that increased muscle metabolism turnover could precipitate autolytic processes occuring during nerve degeneration.

Correlation of biochemical and histological data

Experimental morphological studies are mostly restricted to a description of histological sections, which is handicapped by a certain inaccuracy of observation since it is difficult, or even impossible, to obtain quantitative data concerning changes in structure of muscle fibres in the denervated muscle as a whole.

Morphological observations can be supplemented by quantitative biochemical determinations. Both methodological approaches stand in a complementary relation to each other. Biochemical determinations usually analyse the muscle as a whole, their disadvantage, however, lies in the fact that they cannot mostly differentiate which tissue structures are being affected. On the other hand, histological observations describe changes in individual components so that they can aid in interpreting correctly biochemical findings.

1. Protein changes

The most rapid loss of muscle weight in the course of denervation occurs during the first period after denervation (see p. 68), i. e. in most cases during the first two months after neurotomy. After the muscle weight had decreased to 20—30%, usually no further changes occur and it varies about these values in the course of the rest of the experiment (Fig. 46, see also 21, 23).

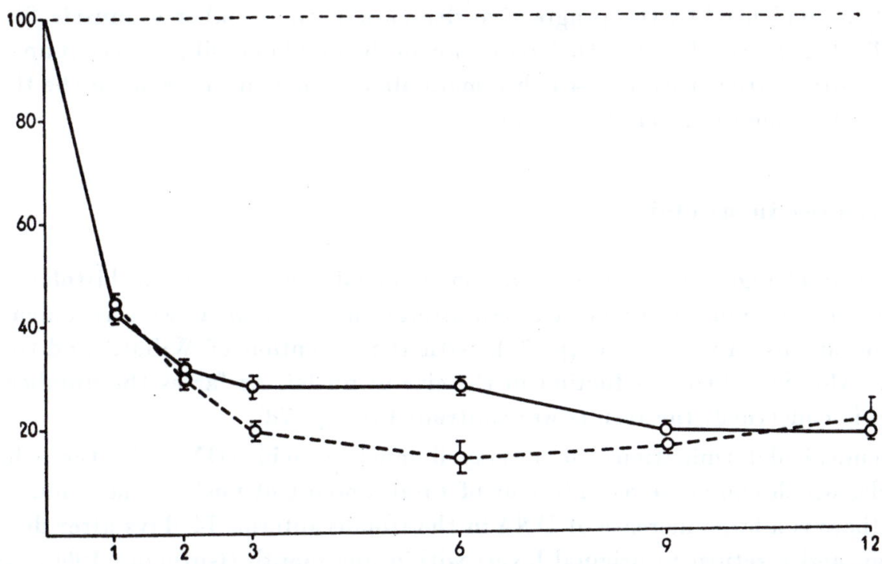

Fig. 46. Percentage loss of weight in the extensor digitorum longus (solid line) and the soleus (interrupted line) in the rat. Abscissa: months after denervation.

Non-collagenous nitrogen, which is a reliable indicator of changes in total muscle protein, alters in parallel with the weight of denervated muscle. A decrease to 20% of the original value was found in denervated muscle of the rat, which is analogous to muscle weight, and this value persists during the whole experimental period (Fig. 30). Thus, following the main loss of weight, no further reduction was noted in the total protein content of the muscle investigated.

The reduction in diameter corresponds approximately to the weight curve and the non-collagen protein curve, but within small limits it can be either greater (Sunderland and Ray 1950) or smaller (Gutmann 1948) than the loss of weight. Since the muscle — especially the denervated muscle — is not a homogenous tissue, the connective and faty tissues contribute to total weight in proportion to their content in the muscle. For example, Sunderland and Ray (1950) found a 10—20% greater reduction in diameter than would correspond to the loss of weight, and described proliferation and an increase of collagenous tissue, which evidently explains the discrepancy when the reduction in muscle fibre diameter is smaller than the weight loss (Gutmann 1948). It must be borne in mind that the diameter gives information about the loss of contractile substance only indirectly.

Changes in muscle weight at later stages of denervation atrophy may give misleading information. From the sixth month onwards, very little or no changes in muscle weight were recorded. However, this is at a time when irreversible changes in the muscle fibres take place and many muscle fibres are lost completely as evidenced by the progressive decrease in the number of muscle fibres (see Table 2, Plate VI and VII). A reduction in the number of fibres was not apparent in total nitrogen values, which remain almost unchanged between the third and twelfth month after denervation.

2. Changes in nuclei

Nuclear changes still represent a controversial question in the histology of the denervated muscle. In most cases, alterations in shape as well as increased volume are usually reported (p. 77), with the exception of Willard and Graw (1924) who described a reduction in the size of nuclei. As far as the number of nuclei is concerned, the results are contradictory (p. 78).

Chemical determinations of desoxyribonucleic acid (DNA) in the whole muscle, which can serve as indicator of total amount of nuclear mass, indicate that there is a large increase of DNA in the tibialis anterior 14 days after denervation, and a return to original levels within one month (Gutmann 1960). The content of DNA remains unaltered in the course of the next three months, but after nine months it decreases to 75% of the initial value (Fig. 30). At early stages nuclei in histological sections do actually appear to be larger, and it is thus possible that it is only the volume and not the number of nuclei which is increased. There is, however, as much as a 284% relative increase of DNA in the denervated muscle within a month, and it persists at this value up to the ninth month. This means that the number of nuclei increased about three times in relation to the protein mass. This immense relative increase of DNA can explain the apparently increased number of nuclei in denervated muscle. If it is taken into account that the absolute content in the whole muscle remains unchanged, and later even decreases, then it is evident that it is a question of condensation of approximately the same number of nuclei in the atrophic muscle. The late decrease of DNA can be caused by a small reduction in the absolute number of nuclei, or also perhaps by a decrease in their volume at late stages of atrophy. It can be demonstrated in histological sections that regressive nuclear changes take place at the same time. Thus the turnover of DNA is also probably increased in the denervated muscle.

In contrast to DNA, the absolute content of ribonucleic acid (RNA) does not increase, but its relative increase is considerable; this can probably be explained by a relative increase in the sarcoplasmatic component at the cost of the contractile component of he muscle fibre.

3. Changes in the content of lipids

There is a progressive increase in the lipid content of the muscle after denervation (see p. 87). Although a transient small decrease has been reported in the rat five days after denervation, lipids rise immediately afterwards and were increased by 20% fifteen days after denervation (Humoller et al. 1952). Changes in lipids were studied at late denervation periods in the rabbit by Audova (1923); the maximum increase of lipids was found in the gastrocnemius group, which attained 564% of the initial value by the 84th day. No further increase was noted until the fourth month. Thus the biochemical findings are in agreement with most histological observations that the content of lipids in the denervated muscle is increased. Exceptionally, no proliferation of adipose tissue was found after denervation, as compared with normal muscle, in the opossum by Sunderland and Ray (1950).

The mechanism of lipid formation and storage in the denervated muscle is, as yet, not clear. Some biochemical findings (Bass and Hudlická 1960) indicate that the denervated muscle forms lipids because of increased utilization of glucose for lipid synthesis, so that it is evidently not only a case of lipid formation from "imported" lipids "ex vacuo". According to these authors lipids are formed in excessive amounts in the muscle fibre so that lipids would be deposited in the fibre itself.

The formation of fat tissue has not been elucidated even under normal conditions, so that attempts to elucidate pathological deposition of fat in the denervated muscle encounter considerable difficulty. Direct transformation into fat tissue when muscle nuclei and the membrane are preserved is difficult to accept from the histological aspect. It is more probable that the fat deposited is phagocytized by free mesenchymal cells, and that these elements are then transformed into fat cells.

4. Changes in the content of collagen

It was the general opinion, based on histological observations, that the amount of collagen increased in the course of denervation, although it was not clear whether this was an absolute or only a relative increase. Biochemical determinations have shown that collagen remains unaltered during the early denervation phase, and that later it even decreases (Fischer 1950, Kolínská and Vodička 1958). An absolute increase in connective tissue takes place in denervated muscle of young animals only (Schapira and Dreyfus (1950), the collagen increment due to growth being the same as in the normal muscle in contrast to actomyosin which does not increase after denervation The loss of collagen in adult animals already occurs during the early denervation phase, in the soleus and extensor digitorum longus of the rat by one month after nerve section (Kolínská and Vodička 1958). The proliferation of connective tissue in histological sections is thus the result of its condensation, and is not due to an actual increase. It cannot, however, be excluded that at later periods activation of connective tissue, its proliferation and increase, does actually occur in comparison with initial values (Sunderland and Ray 1950). This would, of course, require determinations of the total collagen content at longer periods after denervation.

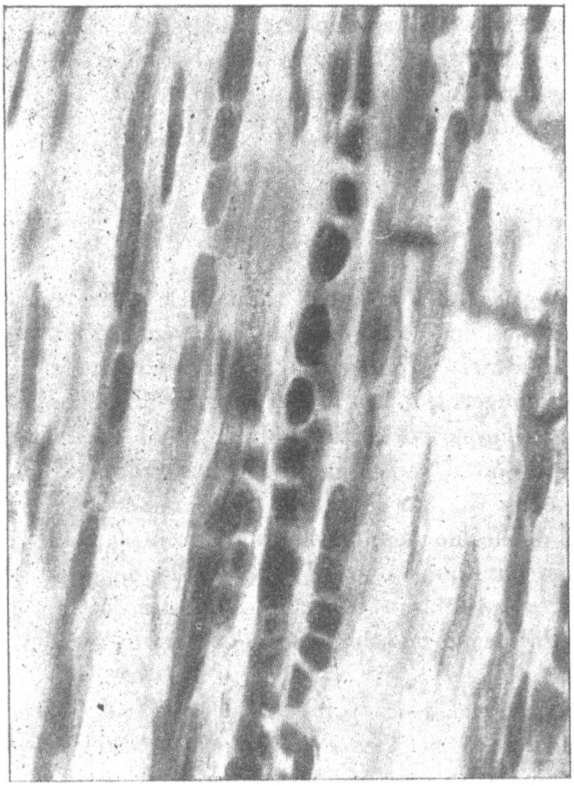

Fig. 47. Rabbit gastrocnemius 26 months following motor denervation. (Glees-Bielschowsky; 500×.) (Gutmann E.: J. Anat. 79 : 1, 1945).

It is, of course, theoretically possible that "collagenization" of the muscle occurs even though there is a decrease in net connective tissue. This possibility has not, however, been confirmed experimentally; complete transformation of muscle into connective tissue has not yet been reported. On the contrary, biochemical and histological data seem to indicate that maximum denervation changes occur in the course of the first stage after nerve section. Later, when the contractile substance in the muscle is reduced approximately to a fifth, further changes progress very slowly, and in some parameters there is hardly any change.

A survey of morphological changes in the denervated muscle does not, as yet, give adequate explanation of questions which morphological research is capable of answering. It would be necessary to answer the question whether there exist morphological changes specific for denervation, what is the ultimate fate of the denervated muscle, and how long can a muscle survive denervation. Many factors, such as span of life, basal metabolism and especially differences of metabolism in different muscles influence the rate of denervation atrophy. In the rabbit we observed practically complete destruction of muscle tissue 16 months following denervation. Such a state could not be found in the rat even 24 months following denervation. The finding of specific changes in the denervated muscle could help in formulating fundamental characteristics of the relationship between nervous and non-nervous tissues. Tower (1939) also evaluated the muscle reaction to denervation in this sense as a phenomenon of such general significance.

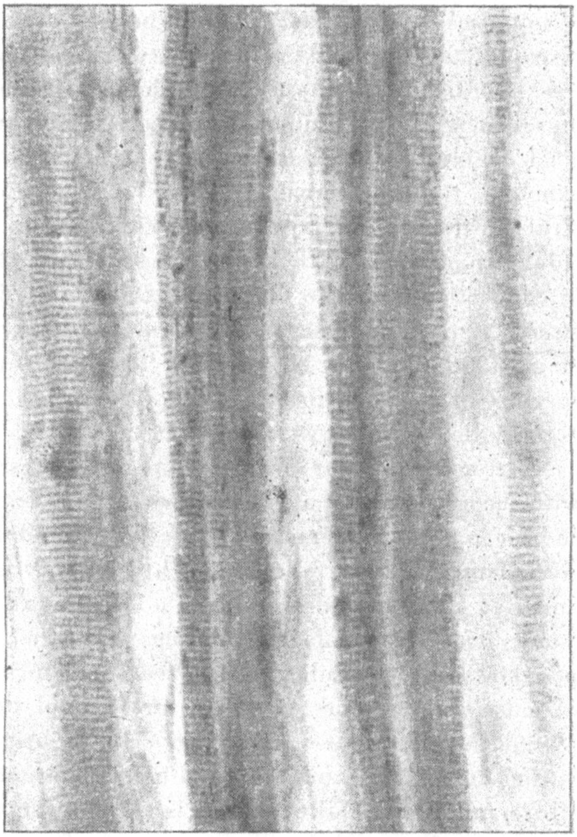

Fig. 48. Rabbit gastrocnemius 20 months after denervation and six months following reinnervation. (Glees-Bielschowsky; 500 ×.) (Gutmann E.: J. Anal. 79 : 1, 1945).

Up-to-date studies do not make a precise differentiation between denervation and disuse atrophy possible. Tower (1937) and Reid (1941) only found a reduction in muscle fibre diameters in disuse atrophy without, however, finding signs of degeneration or proliferation of nuclei. According to Tower (1939), fibrillation activity, which causes exhaustion of metabolic reserves and thus results in atrophy, is classified as the most important of the specific changes occuring in denervation. Extreme proliferation of muscle nuclei, could, perhaps, explain the exceptionally high demands laid especially on oxygen metabolism. Morphological studies, however, cannot contribute to solving this question.

Submicroscopic studies, which have not paid much attention to the denervated muscle structure, will doubtlessly bring a number of important findings. Investigation of the sarcoplasmatic reticulum "of the internal membrane", which can, perhaps, serve as the structure for intracellular conduction of impulses (see Ruska 1958) will be of special interest. The network of the sarcoplasmatic reticulum attached to sarcolemma, and placed between fibrils, was described by Bennett (1956). Changes in these structures in the denervated muscle may be very important for solving the question of specifity of morphological changes in the denervated muscle. It was found by Ruska (1957), for example, that the so-called "tonic" muscle fibres — fibres with a "fieldlike" structure (according to Krüger 1952) contain much less endoplasmatic reticulum than fibres with a fibrillar structure. This finding is of special interest in connection with the finding that following denervation the differences between "tonic" and "phasic" fibres are lost (Krüger and Günther 1956, Kuschinsky et al. 1956). It will be possible to solve further questions concerning the specifity of morphological changes with the aid of studies of the submicroscopic organization of denervated muscle.

More accessible for ordinary microscopic studies are questions associated with the ultimate fate of the denervated muscle. In this connection it is necessary to bear in mind the large differences between different animal species. Degenerative changes in man, for example, appear very late. In the opossum, Sunderland and Ray (1950) found practically no degenerative changes up to 485 days after denervation. Morphological findings indicate that in some muscle fibres the possibility of reversibility of denervation changes is very great indeed. Remnants of muscle fibres represented by sarcolemnal tubes filled with chains of nuclei 20 months after denervation (Fig. 47) evidently recover 6 months after reinnervation and again exhibit normal cross-striation and normal diameter (Fig. 48) (Gutmann 1945). However, this is certainly not the rule and does not apply to all muscle fibres. If the fibre spectrum of a muscle innervated immediately is compared with the fibre spectrum achieved after delayed reinnervation, the deficiency in recovery of muscle fibre diameters is considerable (Gutmann 1948). All these findings underline the importance of the evidence related to reversibility of the changes taking place during denervation.

The findings so far do not yet allow a generalization of the sequence of denervation changes to be made. A definite answer can be given by quantitative studies of terminal stages of denervation atrophy, the "terminal stage" being species specific. Further progress in morphological studies can thus be expected in the first place from submicroscopic studies and quantitative studies, especially of the initial and terminal stages of denervation atrophy.

References

Adams R. D., Denny-Brown D., Pearson C. M.: Diseases of muscle; a study in pathology, P. B. Hoeber, New York 1954.
Altschul R.: Arch. Path. 34 : 982, 1942.
— Rev. Cand. Biol. 6 : 485, 1947.
Audova A.: Skand. Arch. Physiol. 44 : 1, 1923.
Axelsson J., Thesleff S.: J. Physiol. 147 : 178, 1959.
Bass A., Hudlická O.: Physiol. bohemoslov. 9 : 401, 1960.
Batten F. E.: Brain 20 : 138, 1897.
Boeke J.: Verhandl. Akad. Wet. Amst. sec. 2, 18 : 1, 1916.
Beattie J. M., W. E. C., Dickson A.: Textbook of Pathology, London 1943.
Bennet H. S.: Amer. J. Phys. Med. 34 : 46, 1955.
Birks R., Katz B., Miledi R.: J. Physiol. 146 : 45, 1959.
Bowden R. E. M., Gutmann E.: Brain 67 : 20, 1944.
Bowden R. E. M., Gutmann E.: Lancet 2 : 768, 1945.
Brooks V. B., Chipman L. M.: J. Cell. Comp. Physiol. 40 : 1, 1952.
Brooks V. B., Myers D. K.: J. Physiol. 116 : 158, 1952.
Couteaux R.: Bull. Biol. 75 : 101, 1941.
— Bull. Biol. 76 : 14, 1942.
Couteaux R., Grundfest H., Nachmansohn D., Rothenberg M. A.: Science 104 : 317, 1946.
Couteaux R., Nachmansohn D.: Proc. Soc. exp. Biol. N. Y. 43 : 177, 1940.
Chor H., Dolkart R. E., Davenport H. A.: Am. J. Physiol. 118 : 580, 1937.
Czilik B., Sávay G.: Acta Physiol. Hung. 7 : 290, 1955.
Darkewitsch L.: Die pathologische Anatomie der Muskeln. Handb. der path. Anat. des Nervensystems, vol. 2, Berlin 1905.
Dempsey E. W., Wislocki G. B., Singer M.: Anal. Rev. 221, 1946.
Durante G.: Arch. de Méd. Exp. 14 : 658, 1902.
— Anatomie pathologique des muscles, in Cornil and Ranvier Manuel d'histologie pathologique, vol. II, pp. 1—477, Paris 1902.
Erb W.: Deutsch. Arch. klin. Med. 4 : 534, 1868.
Feng T. P., Ting Y. C.: Chin. J. Physiol. 13 : 141, 1938.
Fischer E.: Amer. J. Physiol. 131 : 156, 1940.
— In "The Muscle". Symposium Royaumont 1950.
Gessler H.: Deutsch. Arch. f. klin. Med. 33 : 42, 1883.
Grund G.: Arch. exp. path. Pharmak. 47 : 393, 1912.
Gutmann E.: J. Anat. 79 : 1, 1945.
Gutmann E., Young J. Z.: Journ. Anat. 78 : 15, 1944.
Gutmann E.: J. Neurophysiol. 11 : 279, 1948.
— Die funktionele Regeneration der peripheren Nerven. Berlin 1958.
— Čs. fysiol. 9 : 231, 1960.
Gutmann E., Holubář J.: Journ. Neurol., Neurosurg., Psychiat. 13 : 89, 1950.
Gutmann E., Holubář J.: Physiol. bohemoslov. 1 : 168, 1952.
Gutmann E., Vodička Z., Zelená J.: Physiol. bohemoslov. 4 : 200, 1955.
Gutmann E., Zelená J.: Comparative studies of muscle atrophy. 1959, Babákova sbírka, vol. 16, p. 147.
Hines H. M., Knowlton G. C.: Am. J. Physiol. 104 : 379, 1933.
Holobuth W. S., Jalowy B.: Ztschr. f. Zellenforschung mikr. Anat. 25 : 541, 1936.
Horsley V.: Brain 20 : 375, 1897.
Humoller F. L., Hatch D., McIntyre: Am. J. Physiol. 169 : 654, 1952.

Jamin F.: Experimentelle Untersuchungen zur Lehre von den Atrophien gelähmter Muskeln, Jena 1904.
Katz B., Miledi R.: J. Physiol. 146 : 44P, 1959.
Kolínská J., Vodička Z.: Čs. Fysiol. 7 : 223, 1958.
Kopits J.: Arch. f. Orthop. Unfallschirurgie 27 : 277, 1929.
Krauss E.: Virchows Arch. 113 : 315, 1888.
Krösing R.: Virchows Arch. path. Anat. 128 : 445, 1892.
Krüger P., Günther P. G.: Zeitschr. f. Biol. 109 : 41, 1956.
Krüger P.: Tetanus und Tonus der quergestreiften Skelettmuskeln der Wirbeltiere und des Menschen. 1952.
Kühne: Über die peripherischen Endorgane der motorischen Nerven, Engelmann, Leipzig 1862.
Kuschinsky G., Lüllmann H., Hoefke W., Muscholl E.: Anat. Auz. 103 : 116, 1956.
Leibson R.: Bull. Biol. Med. Exp. USSR 7 : 514, 1939.
Luco J. V., Eyzaguirre C.: J. Neurophysiol. 18 : 65, 1955.
Marney A., Nachmansohn D.: Compt. Rend. Biol. 126 : 785, 1937.
Martini F., Torda K.: Klin. Wchsch. 16 : 824, 1937.
Meng C. W.: Chin. J. Physiol. 15 : 143, 1940.
Oppenheim H.: Handbuch, Heilkunde der Nerven, Berlin 1894.
Pollock L. J., Golseth J. G., Arieff A. J.: Surg., Gyn., Obst. 84 : 1077, 1947.
Pollock L. J., Golseth J. G., Arieff A. J., Sherman I. C., Schiller M. A., Tigay E. L.: War Medicine 7 : 275, 1945.
Proebster R.: Klin. Wschr. 7 : 297, 1928.
Ranvier L.: Leçons sur l'histologie du système nerveux. Savy, Paris 1878.
Reger J. E.: Exp. Cell Research 2 : 269, 1959.
Reid G.: Med. J. Austr. 2 : 165, 1941.
Ricker G., Ellenbeck G.: Virchows Arch. path. Anat. 158 : 199, 1899.
Ruska H.: Exper. Cell Res. suppl. 5 : 560, 1958.
Sawyer C. H., Davenport C., Alexander L. M.: Anat. Rec. 106 : 287, 1950.
Schapira G., Dreyfus J. C.: The muscle, Symposium at Royaumont. 1950.
Sherrington C. S.: J. Physiol. 17 : 211, 1894.
Schmidtmann M.: Zbl. allg. Path., path. Anat. 27 : 337, 1916.
Slauk A.: Zschr. ges. Neurol. Psychiat. 71 : 352, 1921.
Stier S.: Arch. Psychiat. Nervenkrankh. 29 : 249, 1896.
Stoerk H. C., Morpeth E.: Proc. Soc. Exp. Biol. N. Y. 57 : 154, 1944.
Sunderland S., Ray L. J.: J. Neurol. Neurosurg. Psychiat. 13 : 159, 1950.
Tello J. E.: Trad. Lab. Invest. biol. Univ., Madr. 5 : 117, 1907.
Titeca J.: Arch. Int. Physiol. 41 : 1, 1935.
Tower S. S.: Brain 55 : 77, 1932.
— Amer. J. Anat. 56 : 1, 1935.
— J. Comp. Neurol. 67 : 241, 1937.
— Physiol. Rev. 19 : 1, 1939.
Waller A. C.: R. Soc. Biol., Paris 34 : 582, 1852.
Waser P. G., Hadorn I.: Histochemistry of cholinesterase. Symposium. 1960.
Weddell G., Feinstein B., Pattle R. E.: Lancet 1 : 236, 1943.
Willard W. A., Graw E. C.: Anat. Rev. 27 : 192, 1924.
Wohlfart G.: Acta psychiat. et neurol. Suppl. 12, 1, 1937.
— Zschr. f. mikr. anat. Forsch. 44 : 33, 1938.
— Arch. Neurol. Psychiat. 61 : 599, 1949.

Chapter III

THE EFFECT OF DENERVATION ON MUSCLE DEVELOPMENT

J. Zelená

Investigation of the effect of denervation upon processes of muscle development represents an important aspect of problems concerning the trophic influence of the nervous system. If it is assumed that the trophic influence of the nervous system affects muscle metabolism and primarily its anabolic processes, then ontogenetic and regeneration studies represent a suitable approach to studying this problem, since development is characterized by a marked predominancy of anabolic processes.

During development, the trophic relations between the nervous and muscular system are first established. What is the part played by the nervous system in ensuring normal development of muscle at individual stages of its development? What is its role in an analogous morphogenetic process, in regeneration? Elimination studies, i. e. elimination of the nervous system at different stages of development, can give at least a partial answer to these questions.

In the present chapter, a survey will be given of denervation studies performed during development and regeneration of vertebrate muscles, with the aim of elucidating the way in which denervation, i. e. elimination of nervous influences, affects the *development* of muscle, its further *differentiation, growth* and *regeneration.*

A. The effect of denervation upon the initial muscle differentiation

Mutual trophic relations between nerve cells and muscle are first formed during development. The question thus arises as to whether this relationship is already manifest at a time when these two tissues have not yet come into anatomical contact with each other. Is muscle differentiation from the undifferentiated blastema possible even in the absence of the nervous system?

The first answer to this question appeared long before the problem was formulated and investigated by experimental embryology. These were, of course, not experimentally founded, being mostly based on conclusions from teratological material, and rather speculative in nature.

1. Teratology

The first anatomical observations appeared to prove the decisive role of innervation in muscle differentiation. Alessandrini (1829, 1834) described malformations in domestic animals in which partial amyelia in dorsal and lumbal segments was accompanied by the absence of appropriate hind limb muscles, which were replaced by connective tissue only ("Zellgewebe"). Extensive fibrous changes also caused fixation and ankylosis of joints. Several years later, Neumann (1846) dissected a malformed human foetus with hemicephaly and partial amyelia and found, on the contrary, well-developed spinal roots and nerves together with well-developed muscles. This was accepted as positive evidence concerning the decisive role of peripheral nerves in muscle differentiation. Weber reported a similar case as Alessandrini in 1851. In a case of amyelia of the distal half of the spinal cord the corresponding region of the body was almost normally developed, with well-developed skin, normal skeleton and tendons well differentiated, yet musculature was completely missing. When interpreting these findings, Weber concluded that muscles have a closer association with nerves than other tissues, and that their differentiation during development is dependent on nerves, similarly as the formation of the eye is dependent on the optic nerve, so that muscles cannot be formed independently of the nervous system. This conclusion was accepted, with minor variations, even later (Manz 1870, Leonova 1893, Neumann 1903, 1904 and others).

Herbst (1901), under the influence of the first findings of experimental embryology and the fundamental discovery of sensory muscle innervation (which he attributes to Sachs, 1874), accepted Weber's opinion concerning the dependence of muscle development on peripheral nerves in his monograph on "Formative Influences in Ontogenesis". In contrast to Weber, however, he ascribed the influence of nerve fibres on muscle development exclusively to sensory nerve fibres. At the same time, Herbst rejected the description of this influence as "trophic" or "nutritive" as being insufficiently precise, and preferred to talk about the "formative" effect of sensory fibres, thus stressing their influence on the formation and differentiation of a structure. Motor nerve fibres can also affect the muscle by specific function, but according to Herbst, not until late stages of ontogeny. The formative influence of sensory nerve fibres was, in this authors' opinion, conclusively proved by the finding of differentiated muscles in amyelic foeti with intact spinal ganglia. Herbst considered the experiments of Schaper (1898), in which almost normal muscle development was reported in amphibia with spinal cord atrophy and intact spinal ganglia, as further evidence for his thesis, but further investigation proved that his conclusions were not correct.

None of the above findings and speculations, however, conclusively answered the question of whether muscles can develop in the complete absence of innervation. It was necessary to study this question further, and it was not until this problem had been investigated experimentally that Anders (1921) again reviewed the patho-anatomical data and interpreted them anew, in accord with experimental results. It was thus explained that in amyelic foeti the absence of muscles cannot be of primary origin, since muscles are formed independently of the nervous system, but that they undergo secondary atrophy. On the other hand, whenever muscles are developed in amyelia with intact spinal ganglia, this cannot be considered as evidence of the influence of spinal ganglia on their differentiation, but as a case of late disintegration or atrophy of motoneurons, in which neither advanced atrophy nor muscle degenerative changes could yet develop.

2. Experimental studies in amphibia

Experimental studies in amphibia disproved the above-mentioned speculations in a decisive manner. It was shown that muscles can develop even when the corresponding region of the nervous system is eliminated (Harrison 1903, Wintrebert 1903, 1905). Six days after extirpation of the neural tube in tadpoles of Rana sylvativa and Rana palustris at the 3—4 mm stage, differentiated muscle fibres with myofibrils were found by Harrison (1903) in muscles of the trunk and tail. Non-innervated fibres were only smaller in diameter than control fibres, and some of these were vacuolized. Since vacuolization was occasionally to be found also in the innervated musculature of the foetus, Harrison ascribed it to detrimental after-effects of the operation. In one case the development of a non-innervated hind limb was observed up to metamorphosis. The limb was atrophic with atrophic muscle fibres, which were otherwise normally differentiated. These findings were confirmed later by a number of further experiments (Harrison 1904, Wintrebert 1905, Braus 1905, Hooker 1911, Harrison 1918, Hamburger 1927, 1929, Hoadley 1925a, 1925b). At the same time, transplantations of these "aneurogenic" limbs began to be used for investigation of further problems (Braus 1911, Hamburger 1929, Piatt 1942, 1952). The most detailed histological muscle examinations in amphibian experiments have been described by Hamburger (1929). Unfortunately, in most of his cases reinnervation of the denervated extremities had taken place. According to Hamburger, the denervated muscles were atrophic and they were approximately 50% smaller than the control muscles. Occassionally hyaline degeneration was found in some fibres, in exceptional cases a portion of the muscle was undergoing disintegration. Sarcolysis of all muscles in both limbs was observed in a single case only. Most muscle fibres are differentiated normally, even though they are somewhat atrophic. It was thus conclusively proved that *muscle differentiation in amphibia begins indepedently of innervation* and that most muscle fibres also attain morphological maturity.

3. Experimental studies in birds

Denervation experiments in birds were begun approximately 20 years later than in amphibia. This delay was caused by methodical difficulties connected with questions concerning the relationship between innervation and muscle tissue development, Direct denervation at an early development stage is very difficult in homeotherms, and usually causes extensive secondary malformations. Instead of this, the technique of limb transplantation onto tissues in which reinnervation is least likely to occur, i. e. chorio-allantois and the coelomic cavity, was used with success. In the first reports these grafts were observed for a short time, but in spite of this it could be demonstrated that the initial *differentiation of muscles in birds is not dependent on innervation*. Fragments of blastema with or without portions of the neural tube were transplanted by Hoadly onto the chorio-allantois in 1 to 2-day-old chick embryos and the grafts were allowed to survive six days at the most. Somites became differentiated even in the absence of the spinal cord and spinal roots and cross-striated myofibrils were present in the muscle tissue. That muscle differentiation is independent of innervation in birds was further corroborated by Hunt (1932), Hamburger (1939), Eastlick (1943) Eastlick and Wortham (1947). Most of these authors investigated the further changes in these grafts.

Hunt (1932) transplanted limb buds of 3-day-old chick embryos onto chorio-allantois and investigated their development during the next 16 days. Ten to twelve days after transplantation the muscle tissue disintegrated and was replaced by fatty tissue in all grafts which remained without innervation. On the other hand, grafts which were innervated by chance, developed further and contained only occasional and insignificant degenerative changes. From this it was concluded by the author that the trophic influence of the nerve impulse is missing in non-innervated muscles around the tenth day of development.

Eastlick (1943) first observed that the development of non-innervated grafts is retarded not only as compared with muscle development in control extremities, but also in comparison with innervated grafts of the same age. The differentiation of myoblasts from embryonal blastema in non-innervated grafts begins with a delay of two days as compared with control muscles, i. e. as late as the eighth day of development. Cross-striation appears a day later than in the control muscle. On the next day, i. e. tenth day of development, similarly as in the experiments of Hunt (1932), disintegration and phagocytosis begin in the non-innervated muscles, while innervated grafts develop further and their muscles are well differentiated and divided into fascicles of fibres by the fifteenth day. According to this author, innervation is necessary for completion of differentiation, i. e. for division and growth of myofibrils and muscle fibres, and for maintenance of muscle fibres in a good functional state. Without innervation, muscles, following initial differentiation, undergo fatty degeneration (Eastlick and Wortham 1947).

B. The effect of denervation upon further muscle differentiation

It is already evident from the above material that not only initial muscle formation, but also further muscle differentiation, can take place in the absence of innervation in amphibians. In these, morphogenesis can even be completed in non-innervated muscle. On the other hand, in transplantation experiments on birds muscle disintegrates prior to attaining morphological maturity. The rapid onset of degeneration is evidently species specific. Some interpretations of this finding, however, suggested that in homeotherms innervation is indispensable during the last stage of muscle differentiation. Tissue culture experiments and denervation studies in mammals further contributed to elucidating of this problem.

1. Differentiation of muscle in tissue cultures

In tissue cultures, muscle tissue develops being removed from the organism and deprived of innervation. It would appear, according to several reports, that mature muscle fibres can, surprisingly enough, develope in explanted muscle of chick embryos (Szepsenwol 1946, 1947, Manina and Bystrov 1955, Grigoryev 1960). However, most authors consider differentiation to be completed when cross-striated myofibrils are formed, so that immature myotubes containing these myofibrils are already designated as muscle fibres. Manina and Bystrov (1955, Grigoryev 1960) themselves point out that they cannot decide to what extent these "muscle fibres" are fully histologically differentiated. On the other hand, it is evident from reports in which histological structure was studied in detail (Teodorovich 1953) that muscles in tissue cultures attain a stage of atypical myotubes at the most. Teodorovich (1953) cultivated muscles of human foeti at various stages of development in tissue cultures. Muscle tissue cultures grow differently after explantation, in connection with age of the donor. Myoblasts migrate out of explanted young foetal muscles which contain myotubes only, and these myoblasts develop into "myosymplasts", i. e. myotubes. In tissue cultures of older foeti, myotubes grow directly out of the remnants of muscle fibres; new muscle fibres are not formed. Thus, the differentiation of muscle tissue in tissue cultures is arrested at the myotubal stage or it dedifferentiates back to the myotubal stage.

2. Muscle differentiation in mammals following denervation at a late stage of development

Both in tissue cultures and transplantation experiments, muscle tissue is excluded from normal relations in the organism, and it is not possible to decide whether arrest of differentiation is caused by the absence of innervation only, or whether further factors are responsible. However, since the non-innervated muscles degenerate shortly after the time when they should have been innervated, it may be assumed that the muscle disintegration could be in direct connection with the absence of nerve fibres in the graft just at this period. It is possible that the last stage of muscle histogenesis, i. e. the transformation of myotubes into muscle fibres is dependent, in warm blooded animals, on the direct formative influence of the nerve endings at the developing synapses.

An attempt was made to answer these questions by studying denervation changes at a late stage of muscle development in mammalian foeti. Unilateral denervation of developing muscles of the leg was performed by sectioning the sciatic nerve in rat foeti in utero. It was thus possible to compare the development of normal and denervated muscles at a late stage of differentiation, at the time when myotubes change into mature muscle fibres.

During this transition period from the myotubal stage into that of a mature muscle, myotubes differentiate into two kinds of muscle fibres: fibres which form the major part of the muscle, the so-called extrafusal muscle fibres, and fibres which become the main component of muscle spindles, being arranged in bundles inside the connective tissue capsule, the so-called intrafusal fibres. At the primitive motor endings the postsynaptic region of the end-plate differentiates on the extrafusal fibres. The muscle component of the end-plate, the so-called sole with the characteristic postsynaptic membrane and an accumulation of sarcoplasm and nuclei underneath, is formed at this time.

In the following experiments the development and differentiation of:

a) muscle fibres,

b) muscle spindles,

c) the postsynaptic region of the end-plates

were studied in normal and denervated muscles.

a) Differentiation of myotubes into muscle fibres in the denervated muscle

Myotubes change into muscle fibres by increasing the number of myofibrils, which are originally localized around the circumference of the tube, and which then fill up the whole cross-section of the fibre. The diameter of the fibre becomes larger. At the same time, nuclei which were placed in the centre of the tube migrate towards the periphery of the fibre (Fig. 49). The change in the position

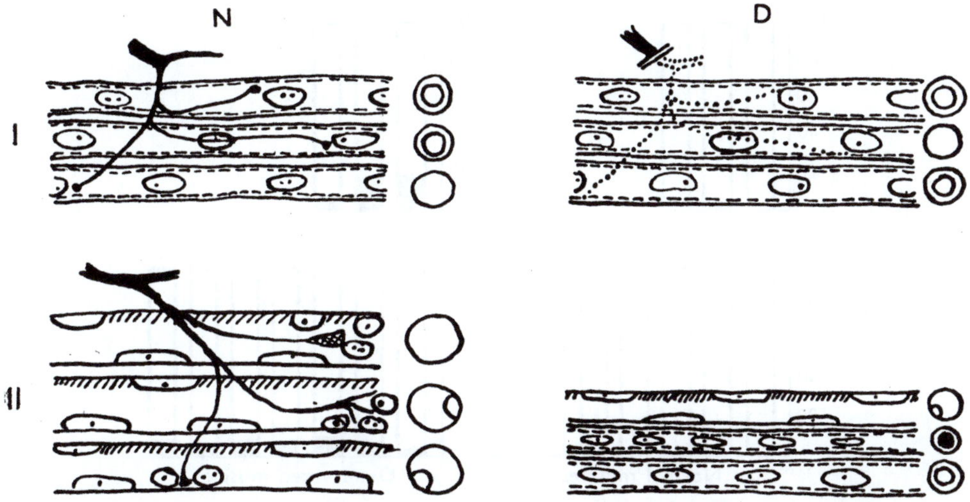

Fig. 49. Diagram of muscle fibre differentiation: I: Myotubal stage with normal innervation and following section of the nerve. II: N — mature muscle with normal innervation; III: D — denervated muscle with myotubes and muscle fibres. Longitudinal and transverse sections.

of the nuclei is the most characteristic sign of this late differentiation period, and for this reason it was taken as the criterion for evaluating the transformation of myotubes into muscle fibres.

The differentiation of myoblasts into myotubes in calf muscles of the rat already begins on the 16th day of intrauterine life, so that by the 19th day, i. e. three days before birth, at the time when the muscles were denervated, they contain fully developed myotubes. The last stage of development, during which myotubes are transformed into muscle fibres, begins in these muscles shortly before birth. Differentiation of the normal anterior tibial muscle is practically terminated on the third day after birth, the muscle containing only 3% of myotubes.

When the nerve is sectioned in foeti three days before birth, the differentiation of muscle tissue continues. Muscle fibres differentiate even in denervated muscles, but with a certain delay as compared with control muscles (Zelená 1959). In the denervated tibialis anterior six days after section of the nerve there is

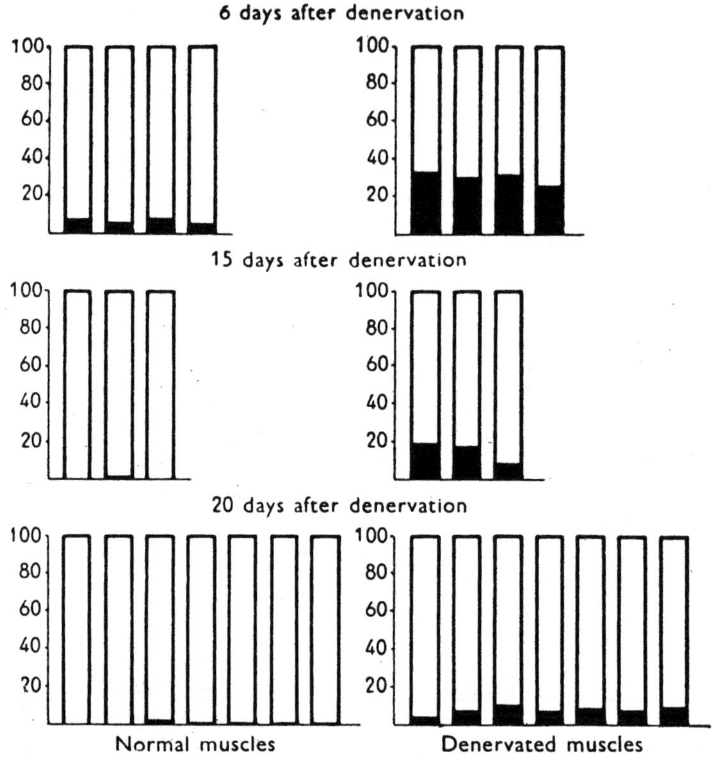

Fig. 50. Number of myotubes and muscle fibres (expressed as percentage) in the tibialis anterior 6, 15 and 20 days after intrauterine denervation. Abscissa — normal and denervated muscles. Ordinate — percentage of myotubes ■ and muscle fibres □.

approximately 30% of myotubes. At the same time the denervated muscle atrophies, and division of its fibres is slowed. Fifteen days after denervation a considerable percentage of myotubes (15%) is still preserved in the denervated muscle (Plate IX). However, in spite of considerable atrophy, differentiation continues, so that by the 20th day after nerve section the number of myotubes is reduced to 7% (Fig. 50).

Differentiation was also studied in the flexor digitorum brevis (Plate X), where it begins a day later and lasts for twice as long as the differentiation of the tibialis anterior. Denervation of this muscle was performed either three days before birth when myotubes are still immature or two days after birth when they are well developed. Following denervation the number of myotubes remains considerably high. From the 15th day onwards the number of remaining myotubes becomes relatively constant at about 30% (Fig. 51), so that 20 days after denervation it is approximately four times as high as in the denervated anterior tibial muscle.

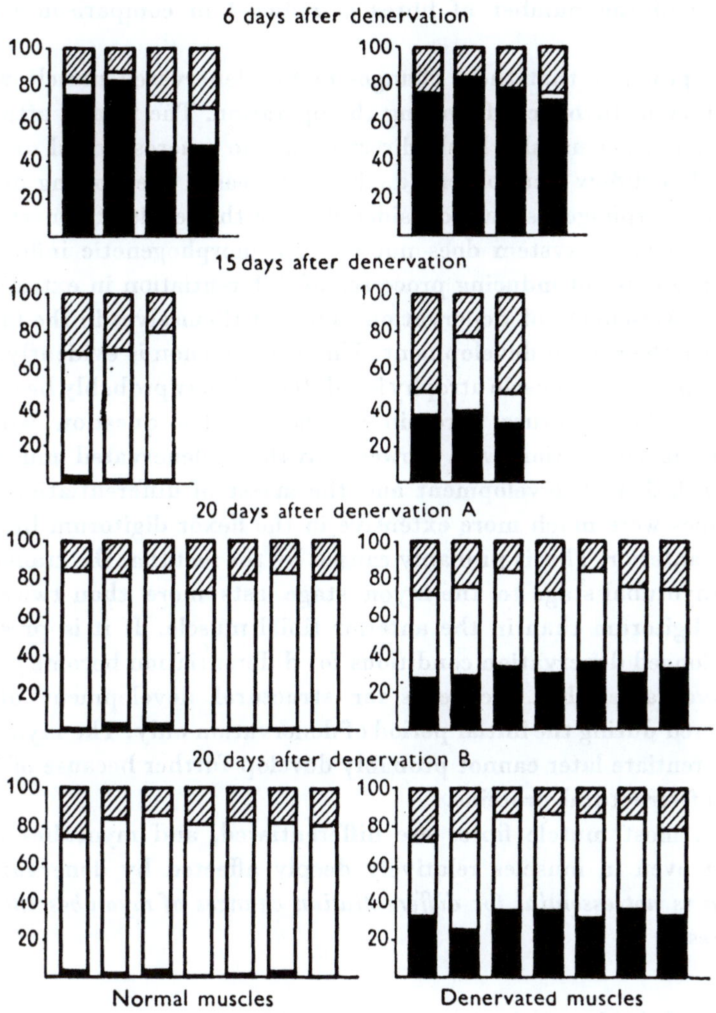

Fig. 51. Number of myotubes and muscle fibres (expressed as percentage) in the flexor digitorum brevis 6, 15 and 20 days after intrauterine denervation. Abscissa—normal and denervated muscles. Myotubes and muscle fibres denoted as in Fig. 50. Shaded parts indicate unclassified fibres (cross-sections containing no nuclei).

As can be concluded from our experiments, the differentiation of myotubes into mature muscle fibres takes place even in denervated muscles. Following denervation, however, the process of maturation is slowed down as compared with normal muscle, and in some fibres differentiation is arrested at the myotubal stage. It also seems that the division of myotubes and muscle fibres is reduced, and that some of these disintegrate. Denervated muscles are also markedly atrophied, and the number of fibres is reduced in comparison with normal muscles.

It is not probable that these changes in the denervated muscle were caused by circulatory disturbances following the operation. The wound after the intra-uterine denervation usually heals directly and no macroscopical or histological changes in blood flow were observed. For this reason the slowing down and retardation of morphogenesis are considered to be the result of denervation.

Thus the nervous system does not exert a morphogenetic influence on the muscle in the sense of inducing processes of differentiation in extrafusal muscle fibres. The after-effects of denervation manifest themselves in the first place by slowing down the rate of development. This phenomenon is evidently of the same nature as denervation muscle atrophy in adults and may probably be explained by a depression of structural protein synthesis. This question will again be brought up in connection with slower growth of denervated muscles.

The retardation of development and the arrest of differentiation in some of the myotubes were much more extensive in the flexor digitorum brevis than in the tibialis anterior. This is probably caused by the fact that the transition period from the myotubal stage to the myon stage lasts more than twice as long in the flexor digitorum than in the anterior tibial muscle. If it is presumed that during prolonged denervation conditions for differentiation become progressively more unfavourable, then the needs for structural development of myotubes can be covered during the initial period of denervation only. The myotubes which would differentiate later cannot probably develop further because of insufficient synthesis of structural proteins.

However, most muscle fibres are differentiated, and myotubes represent a minor part even in muscles relatively deeply affected by denervation. Thus, *innervation is not essential for differentiation changes of myotubes into extrafusal muscle fibres.*

b) Arrest of differentiation of myotubes into intrafusal fibres of muscle spindles in the denervated muscle

Myotubes differentiate in two directions. Most myotubes are transformed into extrafusal muscle fibres, a smaller portion, however, differentiates into intrafusal muscle fibres which represent the principal component of the muscle spindle. Is the transformation of myotubes into intrafusal fibres also independent of innervation, similarly as the differentiation of extrafusal fibres?

The myotubes in normally innervated muscle change into intrafusal fibres by nuclear proliferation in the vicinity of afferent nerve endings (Sutton 1915, Tello 1922, Cuajunco 1927, Couteaux 1941, Kalugina 1956 and others). These nuclei at first form a chain in the central part of the myotube (Plate XIa) later they become spherical and form a large aggregation which compresses

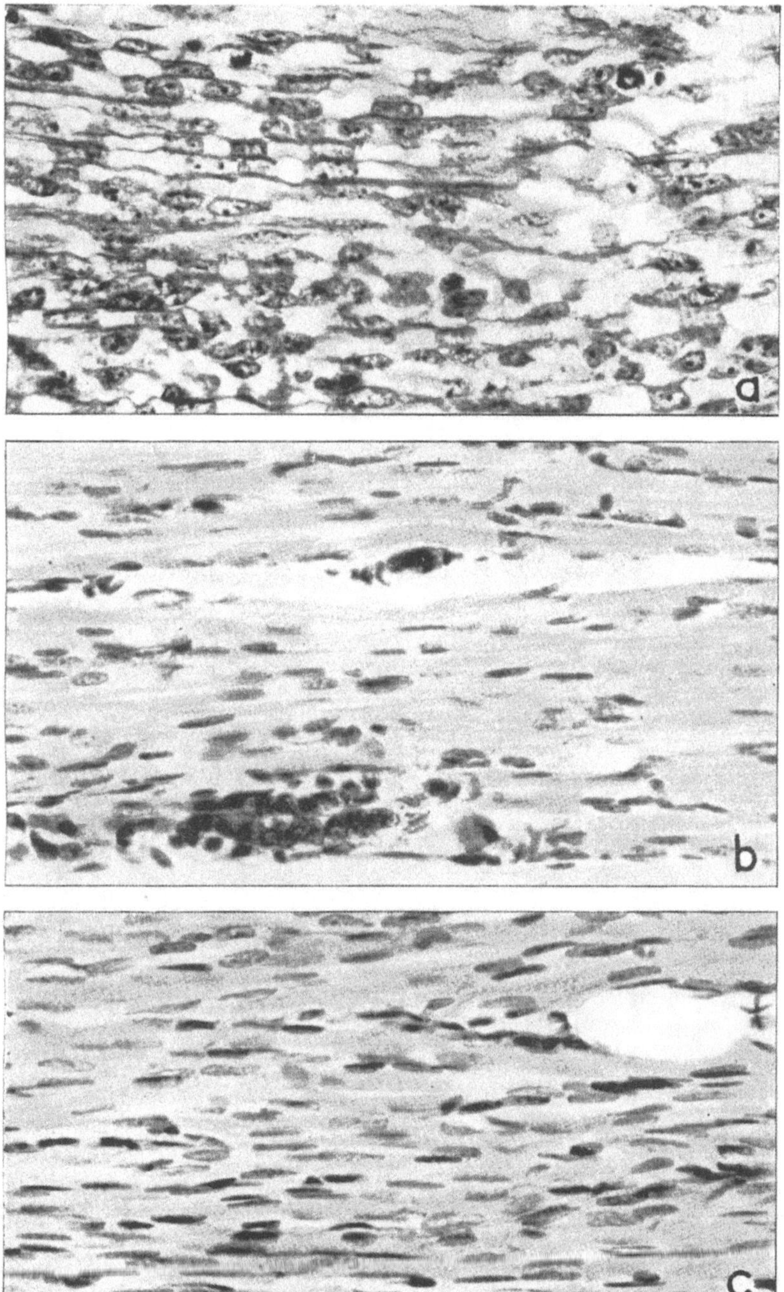

Differentiation of muscle fibres in normal and denervated m. tibialis anterior of the rat. — Fig. a. Myotubes with central nuclei and peripheral myofibrils three days before birth. — Fig. b. Normal muscle fibres and a muscle spindle in control muscle of a twelve-day-old rat. — Fig. c. Differentiated atrophic muscle fibres in denervated muscle 15 days following intrauterine denervation performed three days before birth. (Heidenhain's haematoxyline [a], Van Gieson [b, c]; 500 ×.)

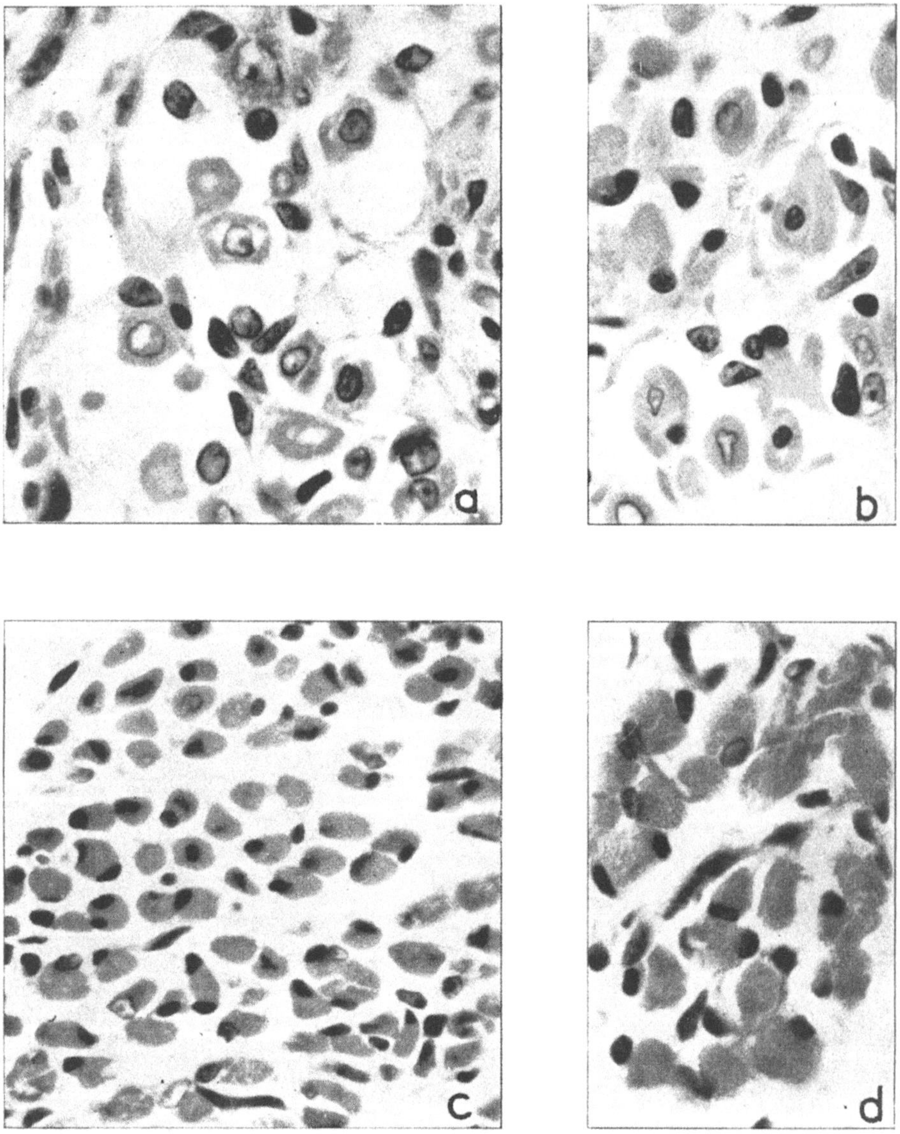

Muscle fibres in denervated and normal m. flexor digitorum brevis of the rat. Fig. a. Muscle six days after intrauterine denervation. Fig. b. Control muscle. Fig. c. Muscle 20 days after intrauterine denervation. Fig. d. Control muscle.
(Haematoxyline-eosine, transverse sections; 1000 ×.)

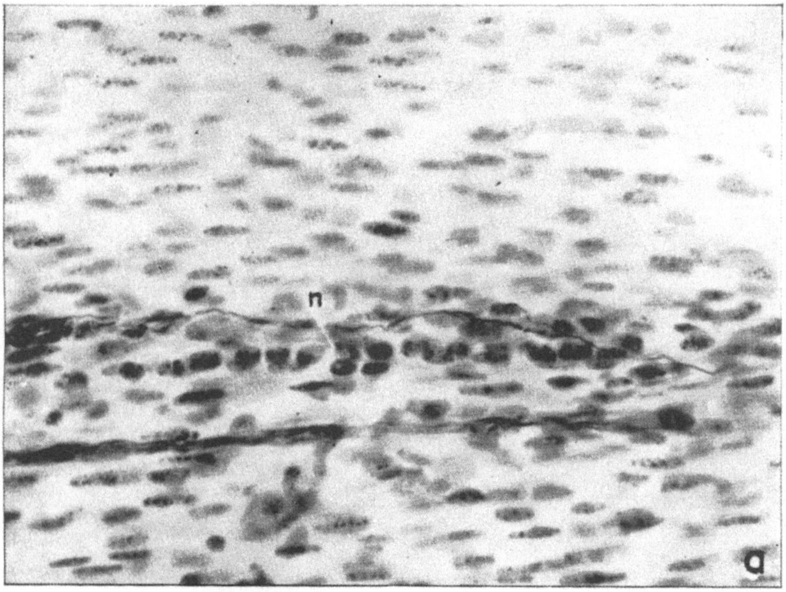

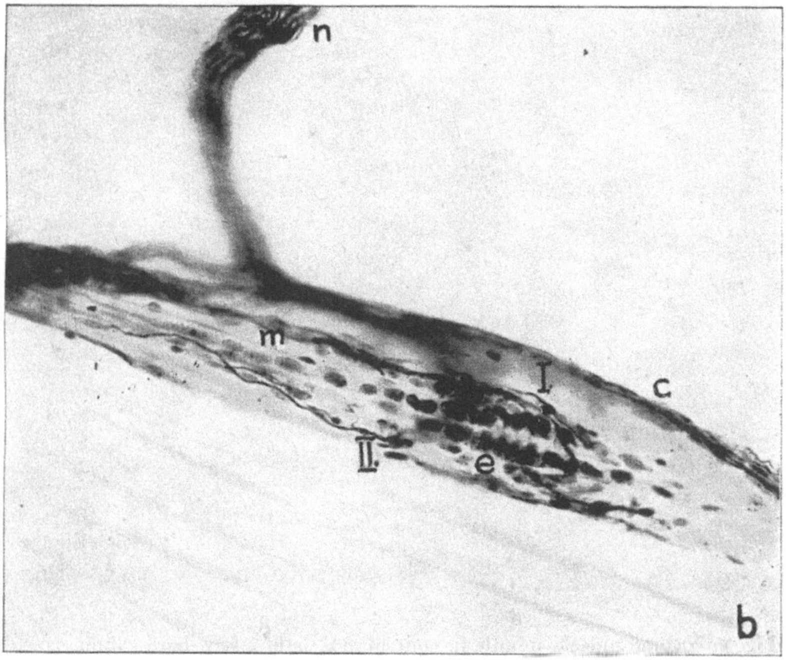

Fig. a. Beginning of differentiation of a muscle spindle at the sensory nerve ending (n) in the rat two days before birth.
Fig. b. Neuromuscular spindle of adult rat.
e — equatorial zone; m — myotubal zone; c — connective tissue capsule; n — nerve fibres; I — primary sensory endings; II — secondary sensory endings. (Bielschowsky-Gros; a — 500 ×, b — 300 ×.)

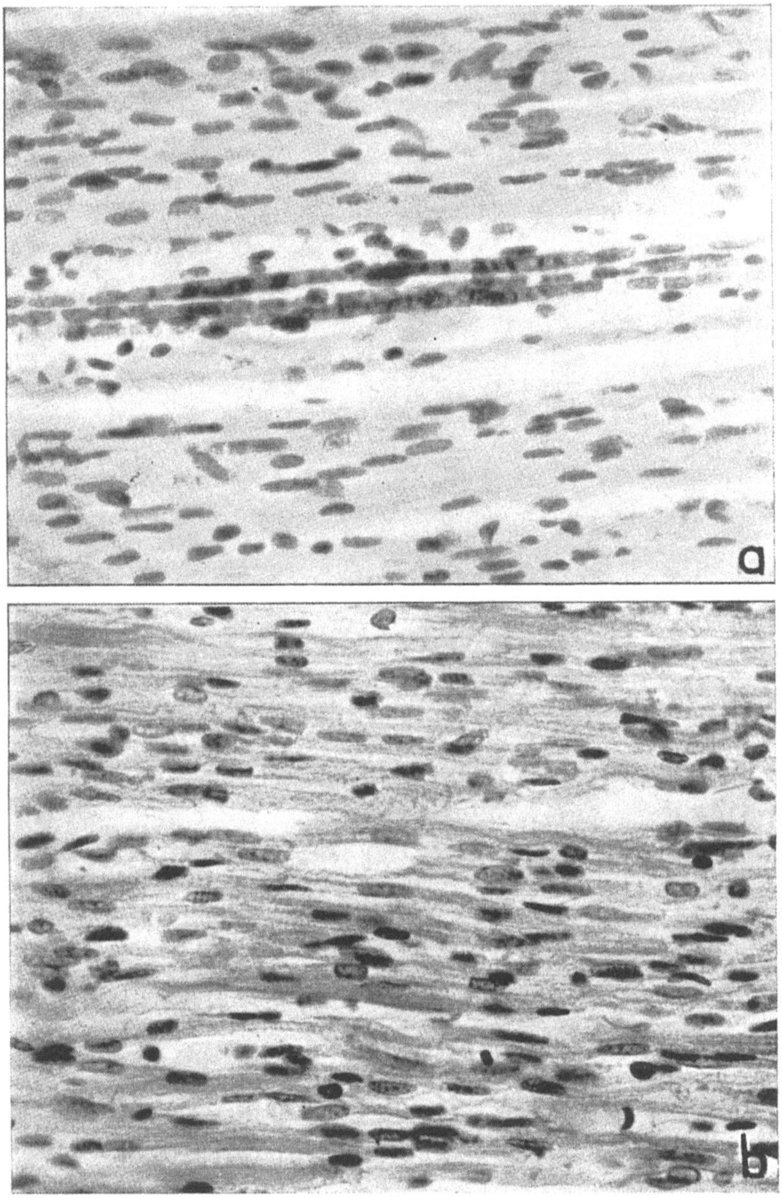

Fig. a. Normal muscle spindle in control muscle of a new-born rabbit.
Fig. b. Denervated muscle without spindles in a new-born rabbit seven days after intrauterine denervation. (Haematoxyline-eosine; 500 ×.)

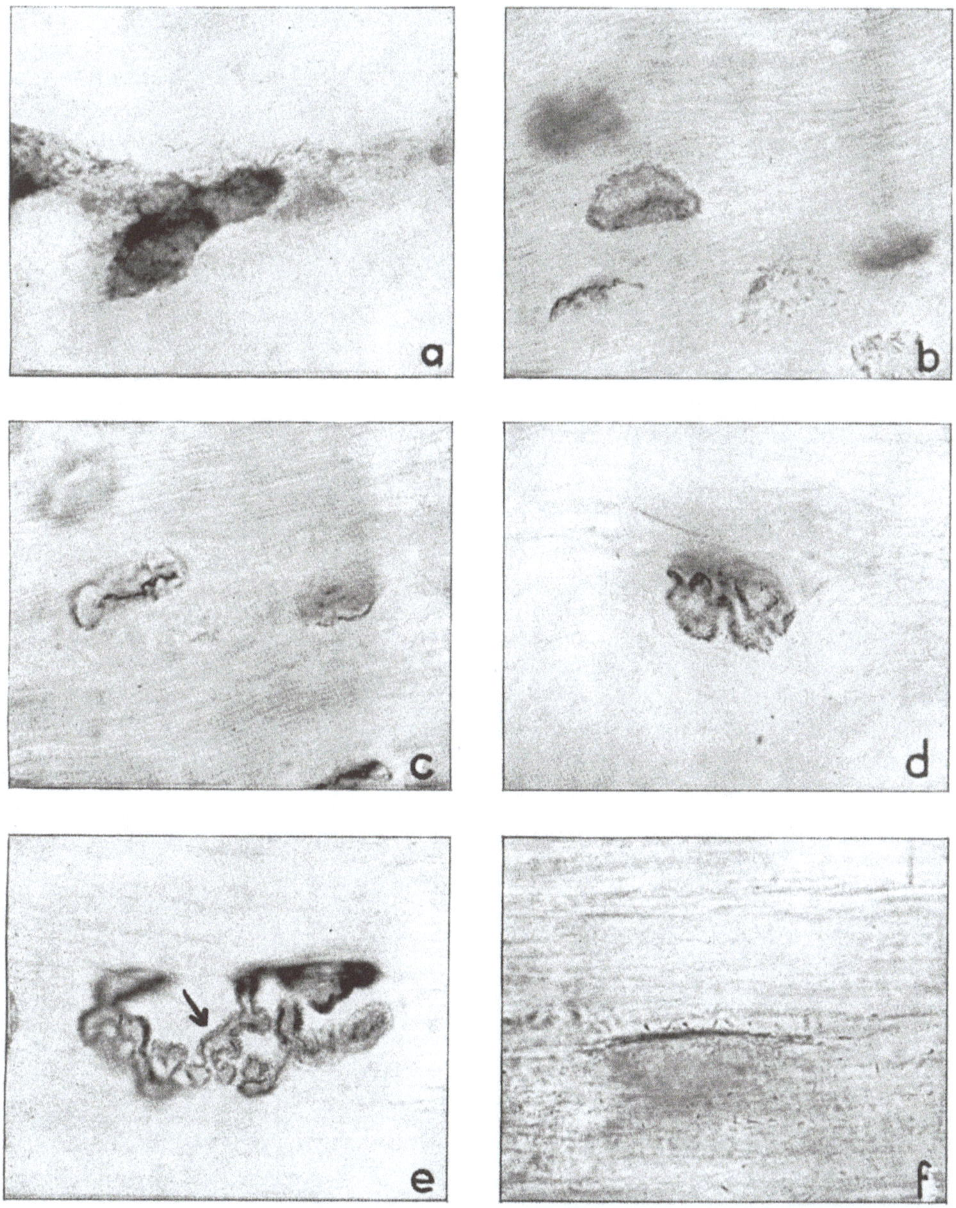

Differentiation of the subneural apparatus of motor end-plates in; Fig. a. three-day-old rat, Fig. b. fifteen-day-old rat, Fig. c, twenty day-old rat, Fig. d. thirty-day-old rat, Fig. e. adult rat, Fig. f. thirty days following intrauterine denervation.
(Method of Koelle and Friedenwald in Gerebtzoff's modification; 1100 ×.)

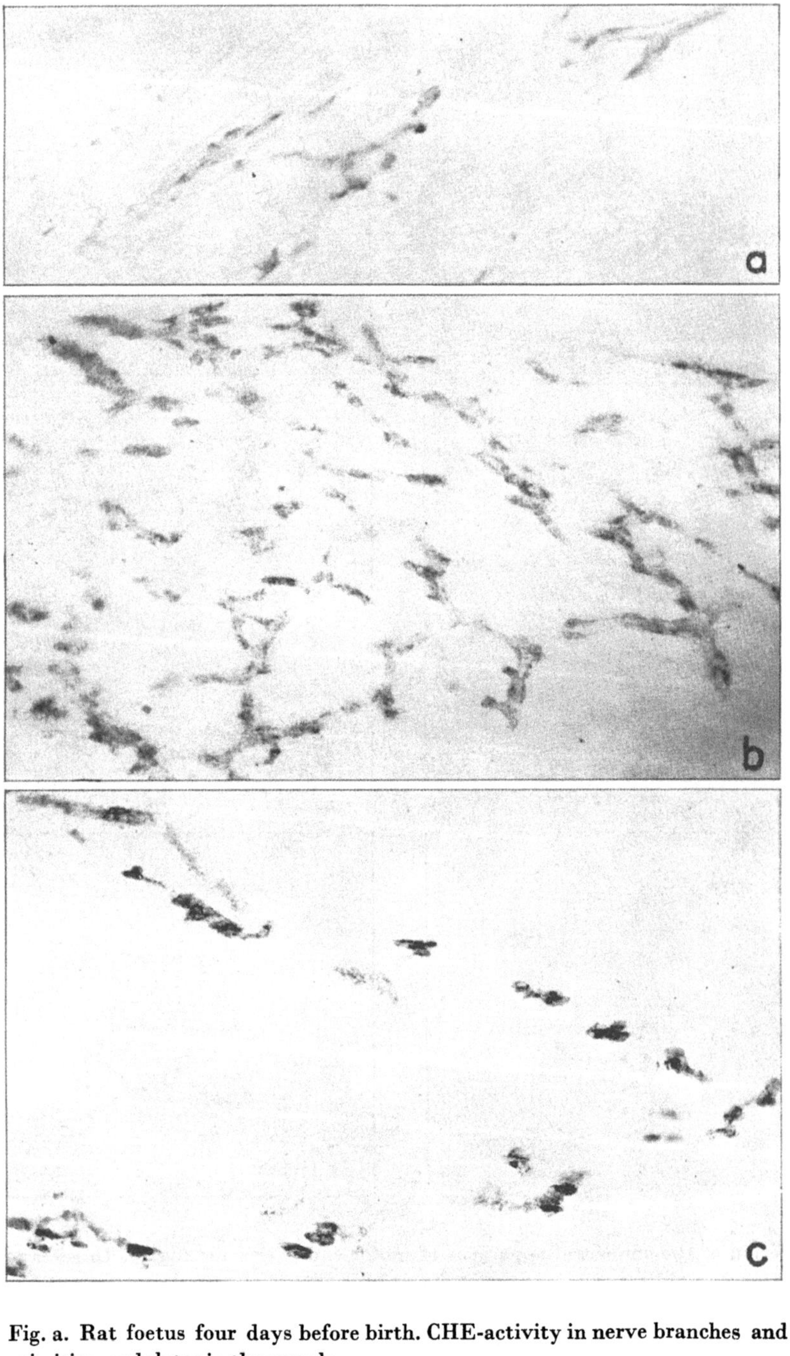

Fig. a. Rat foetus four days before birth. CHE-activity in nerve branches and primitive endplates in the muscle.
Fig. b. Rat foetus two days before birth. CHE-activity in nerve plexus with a more intensive reaction in primitive end-plates.
Fig. c. Three-day-old rat. Group of motor end-plates in characteristic V-shape arrangement. (Method of Koelle and Friedenwald in Gerebtzoff's modification; 200 ×.)

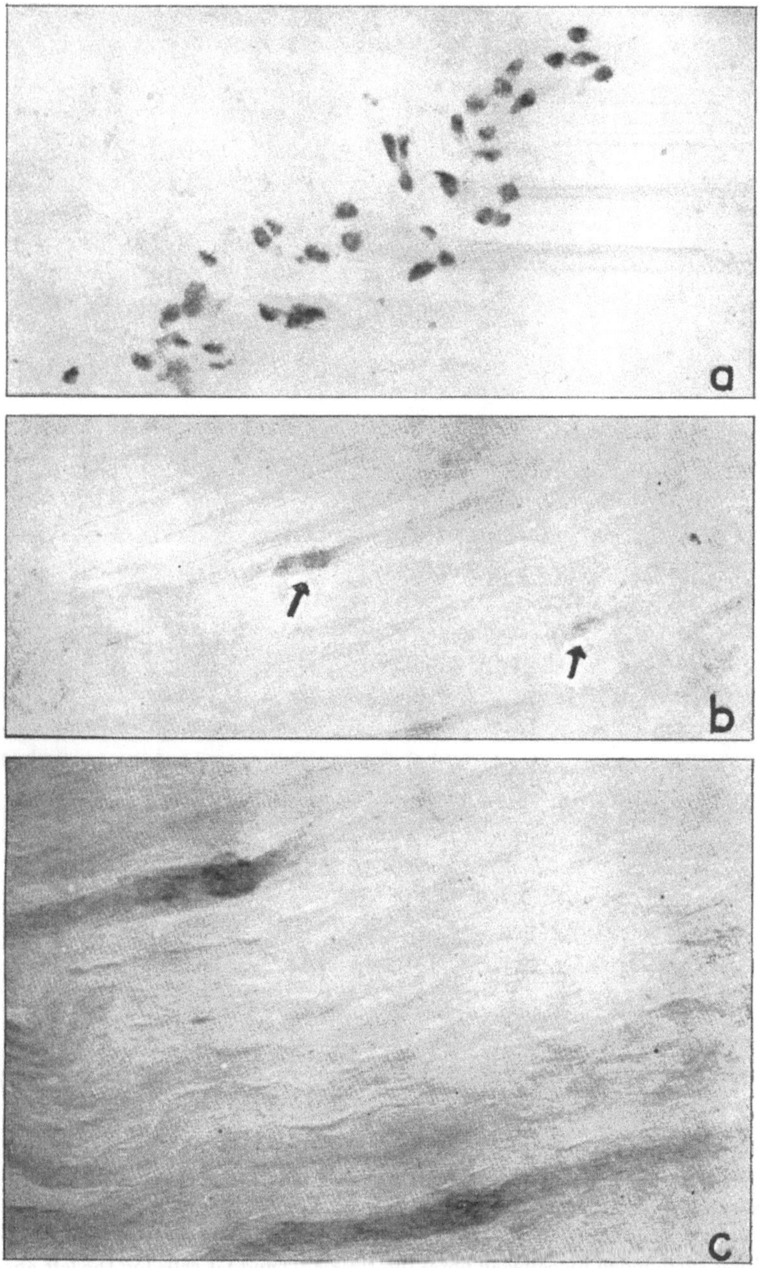

Fig. a. Cholinesterase in end-plates of control gastrocnemius muscle in a seven-day-old rat. (200 ×.)

Fig. b. Remnants of cholinesterase activity in dedifferentiated end-plates seven days after denervation. (200 ×.)

Fig. c. Same as preceding Fig. (400 ×.)

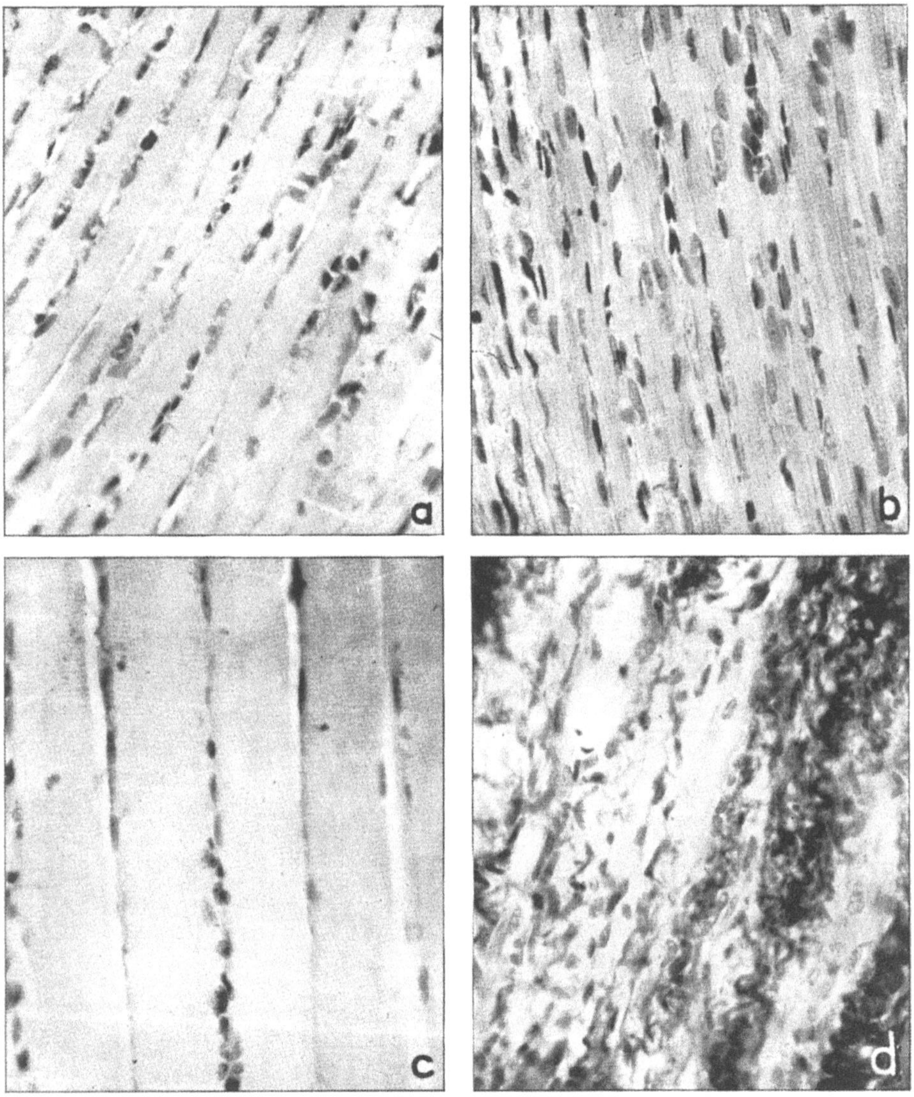

Fig. a. Normal m. soleus of a 14-day-old rat.
Fig. b. Denervated soleus muscle in a 14-day-old rat following section of the nerve at birth.
Fig. c. Normal m. soleus of a 2-month-old rat.
Fig. d. Denervated soleus in a 2-month-old rat following section of the nerve at birth. Atrophic muscle fibres surrounded by collagen fibres (on the right) and fat cells (on the left above).
(Van Gieson, longitudinal section; 400 ×.)

sarcoplasm to the periphery and even enlarges the lumen of the fibre at the time of birth in the rat (Fig. 52a). This characteristic central zone of the intrafusal fibres is called the "nuclear bag". It is not until the formation of the nuclear bag that the connective tissue capsule enclosing the spindle at the equatorial zone grows towards the polar zones, the lymphatic space enlarges between this and the intrafusal fibres and gamma motor endings form junctions in the polar zones.

However, *no spindles differentiate in muscles which had been denervated before onset of differentiation.* The nuclear bag does not develop, only occasionally is a short chain of nuclei formed in the central part of the myotube in the leg muscles of rat foeti denervated three days before birth (Fig. 52b,c). Further spindle components do not differentiate at all. A similar arrest of spindle differentiation was observed in muscles of rabbit foeti denervated at the myotubal stage seven days before birth (Zelená 1957) (Plate XII).

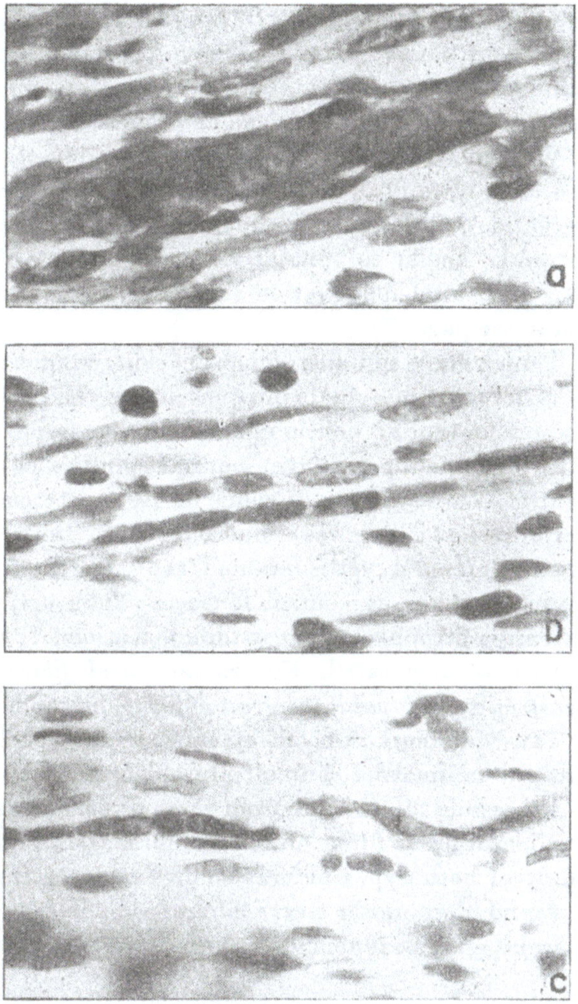

Fig. 52. a) Nuclear bag of muscle spindle in control muscle of a new-born rat; b) Alignment of nuclei in denervated muscle of a newborn rat three days after denervation; c) Chains of degenerating nuclei in denervated muscle of a new-born rat three days after denervation. (Haematoxylin-eosine; 1100 ×.)

At the period of early development, intrafusal fibres of spindles are innervated by sensory nerve fibres only. The motor gamma-fibres reach the muscle spindles later, and motor end-plates in the polar zones of the spindle cannot be observed in the rat until after birth (Zelená 1957). This means that although peripheral mixed nerves are sectioned in our experiments, spindles are deprived of their sensory nerve fibres by this procedure, so that the arrest of their differentiation is due to sensory denervation.

Denervation in newborn rats, in which the nuclear bag is already differentiated, arrests further differentiation of spindles and leads to disintegration of intrafusal muscle fibres. Spindles in calf muscles usually disappear 10 days after section of the sciatic nerve. Muscles remain permanently devoid of spindles even after reinnervation, although the muscles are reinnervated by a sufficient number of sensory nerve fibres (Zelená and Hník 1960a, b).

When denervation is performed in 20-day-old rats, in which muscle spindles are already fully developed, intrafusal fibres are only slightly affected by denervation. Fibres do not undergo rapid degeneration, but only slow general atrophy, similar to spindles of adult animals (Tower 1932). It is not until a long time after denervation that degenerative changes in intrafusal fibres take place (see page 75).

Thus, muscle spindles do not develop without innervation. The loss of spindles in denervated muscles is not just due to the rapid atrophy. Following section of the tendon during development, muscle atrophy progresses at a similar rate as after denervation, and yet normal spindles develop in this muscle, as in the control (Zelená 1960). Weight of the gastrocnemius muscles with tenotomy performed at birth was reduced by 70% three weeks later, as compared with control muscles, yet contained the same number of normally differentiated muscle spindles as control muscles. Thus neither loss of muscle tension nor extensive atrophy endanger differentiation of stretch receptors, as long as the muscle is innervated. Future *intrafusal fibres react selectively to the loss of innervation by immediate arrest of differentiation.*

Thus differentiation of extrafusal fibres continues after denervation even if atrophy is involved, but intrafusal fibres do not develop without innervation. Is it possible to conclude from these experiments that the influence of sensory and motor nerve fibres on muscle fibres is different? On the contrary, when the effect of both types of nerve fibres on specialized end-organs is compared, it is found that motor nerve fibres, similarly as sensory fibres, exert a direct morphogenetic influence.

a) Arrest of differentiation of the postsynaptic membrane of the motor end-plate after denervation

Motor nerve fibres exert a direct influence on the differentiation of extrafusal muscle fibres within a short segment in which the motor end-plate region — the sole — is being formed. The postsynaptic membrane, the so-called subneural apparatus exhibiting cholinesterase activity (Couteaux 1947, 1950, 1955, Couteaux and Taxi 1952) represents an important component of the neuromuscular junction. This subneural apparatus and its cholinesterase activity develop very early (Gerebtzoff 1955, Gerebtzoff et al. 1954, Kupfer and Koelle 1955, Portugalov 1955, Zelená and Szentágothai 1957, Beckett and Bourne 1958, Csillik 1960). It is the first component of the postsynaptic region which appears during development, much earlier than the accumulation of nuclei and sarcoplasm in the sole.

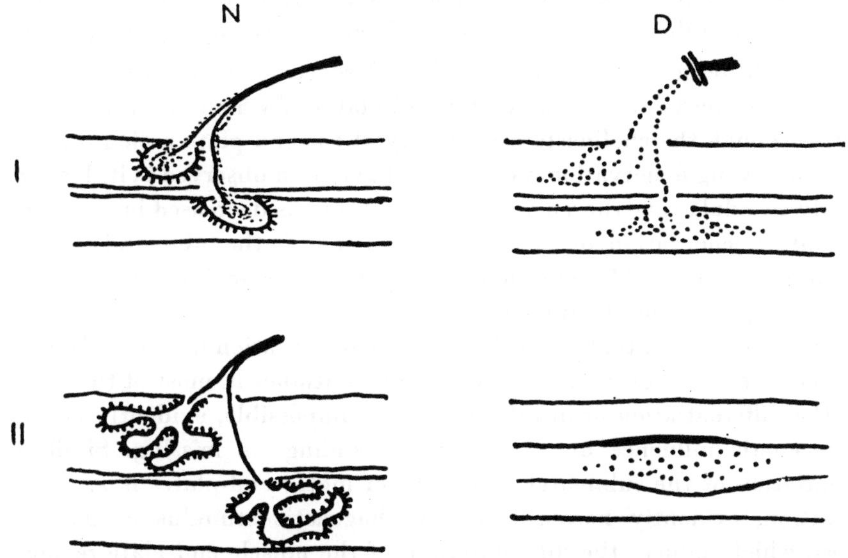

Fig. 53. Differentiation of the subneural apparatus of the motor endplates in normal and denervated muscles. I — beginning differentiation of the subneural apparatus at the myotubal stage. II — subneural apparatus of adult motor end-plate (N) and remnants of CHE-activity in denervated muscle (D) following intrauterine denervation.

At first, cholinesterase activity is localized on the surface of nerve branches in the muscle (Plate XIVa,b). Cholinesterase active segments of the muscle membrane differentiate in junctional areas only in direct relation to nerve fibres (Fig. 53, Plate XIV). This process already begins from the 17th day onwards in the calf muscles of the rat. Although we have not succeeded in denervating muscles prior to the onset of motor end-plate differentiation, denervation performed during their differentiation on the 19th day of intrauterine development leads to results proving that motor nerve fibres exert a morphogenetic influence on the differentiation of the subneural apparatus.

Following denervation, cholinesterase activity disappears from the nerve plexus, and in the muscle it remains at occasional primitive end-plates which were already differentiated prior to denervation. New cholinesterase-active loci on the muscle membrane are not formed after nerve section. The incompletely developed subneural apparatus of primitive motor end-plates does not develop further, and the intricate segmentation of the junctional gutters is not formed, as in the case of innervated muscle. The membrane is, on the contrary, slightly bulged and the cholinesterase activity quickly disappears from the subneural apparatus (Fig. 53, Plate XIII). In the course of several days, only insignificant traces of cholinesterase activity remain in the muscle at the places of the original endplates (Plate XV) (Zelená and Szentágothai 1957, Zelená 1959). Thus *the differentiation of the subneural apparatus of the end-plate is directly dependent on the influence of nerve endings* and cannot be completed in their absence.

This result represents a considerable difference in comparison with young animals three weeks of age, the motor end-plates of which are already differentiated. Although the cholinesterase activity of the end-plate region is somewhat reduced following denervation, no changes have been observed in its localization. Similarly in adult rats the cholinesterase activity is decreased in the end-plates shortly after denervation, but degenerative changes in the subneural apparatus — fragmentation and dedifferentation — do not appear until several months after denervation (Sávay and Csillik 1956).

It can be concluded that myotubes react to denervation in two different ways: the differentiation of extrafusal muscle fibres continues in most of the myotubes, while the differentiation of intrafusal fibres is impossible, evidently because the morphogenetic influence of afferent nerve endings is missing. Similarly the differentiation of the subneural apparatus of motor end-plates is arrested after denervation, evidently because the morphogenetic stimulus of motor nerve endings, which induces the differentiation of the muscle end-plate region under normal conditions, is eliminated. The differentiation of muscle fibres towards the end of the period of histogenesis is thus only indirectly affected by denervation, while the differentiation of terminal end-organs is completely dependent on innervation and is not realized in the denervated muscle.

C. The effect of denervation upon muscle growth

The denervated muscle lags behind normal muscle very markedly as far as its development and growth are concerned. This form of atrophy has not been studied quantitatively at an early period of development. It was, however, studied during the growth period following termination of differentiation, and these studies show interesting results.

Muscle atrophy during the growth period is not an absolute atrophy, i. e. is not accompanied by weight loss, but on the contrary, the muscles gain weight. (Dreyfus et al. 1949, Schapira et al. 1950, Leonard 1957, Zelená and Hník 1957, Alder et al. 1959 and others). For this reason growth was studied in denervated muscles at this period of intensive growth (Zelená and Hník 1957). Changes in muscle weight and fibre diameter were determined following unilateral section of the sciatic nerve in denervated and control soleus and extensor digitorum longus muscles. The weight of the denervated muscles increased about three times the initial values in the course of the first 14 days, five times by the 30th day and by two months underwent no further changes (Table 4, Fig. 54).

Control muscles, on the other hand, gained rapidly in weight, reaching a sixfold increase by 14 days, a twenty fold increase by one month, and by two

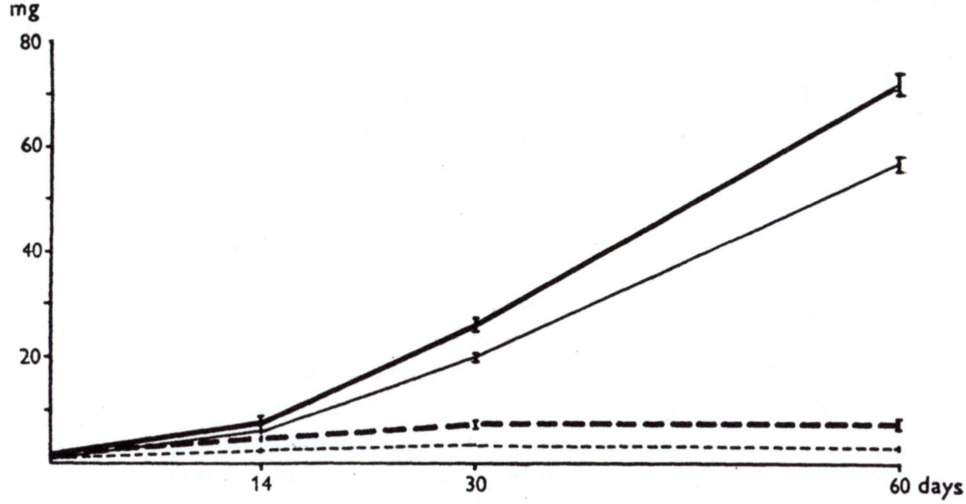

Fig. 54. Weight increment in normal and denervated muscles of young rats during the first two months of life following denervation at birth. Abscissa — time after birth in days. Ordinate — muscle weight in mg. ——— m. extensor digitorum longus, ——— m. soleus, - - - - denervated m. extensor digitorum longus, ------ denervated m. soleus.

Table 4

Muscle weight changes in denervated and control limbs 14 days, one and two months after birth in rats. Unilateral denervation was performed at birth (D), the contralateral limb serving as control (C). In the first line are average values found in new-born rats.

Time after denervation	Average body weight in g.	M. soleus			M. ext. dig. long.			n
		D mg	C mg	D/C %	D mg	C mg	D/C %	
At birth	5·0		0·75 ± 0·07 mg.			1·39 ± 0·06 mg.		12
14 days	23·4	2·3 ± 0·16	5·5 ± 0·28	40·5 ± 1·35	4·7 ± 0·26	7·4 ± 0·35	64·2 ± 1·94	27
1 month	51·3	3·6 ± 0·19	19·3 ± 0·98	19·3 ± 0·96	7·5 ± 0·44	25·6 ± 1·09	29·7 ± 1·19	27
2 months	131·3	3·4 ± 0·37	55·9 ± 2·98	6·0 ± 0·44	7·6 ± 0·66	70·1 ± 3·59	10·8 ± 0·76	20

D — mean muscle weight of denervated muscles; C — mean average weight of control muscles; D/C % — denervated weight expressed as percentage of control (mean value of individual values); n — number of animals.

months the extensor digitorum longus attained a weight which was fifty times greater than at birth, the soleus seventy-five times greater (Table 4, Fig. 54). Muscle fibre diameters in denervated muscles increased during the first 14 days without marked proliferation of interstitial tissue (Table 5, Plate XVIa,b). However, within a month the muscle fibre diameters decreased, and at 60 days after denervation muscle fibre diameters were on the average smaller than in new born rats (see Table 5, Fig. 55). Simple atrophy of the first period was now accompanied by degenerative changes and by considerable connective tissue proliferation (Plate XVIc,d).

Thus during the first period after denervation there is an increase in muscle weight and fibre diameter. A small decrease in longitudinal growth in young rabbits has been reyorted by Alder et al. (1959). It is therefore possible to assume that there is also an absolute increase in structural proteins. This conclusion is, of course, drawn on the basis of indirect evidence, by assuming that longitudinal and transverse growth of muscle fibers is associated with the synthesis of myosin. This assumption would have to be confirmed by biochemical analysis.

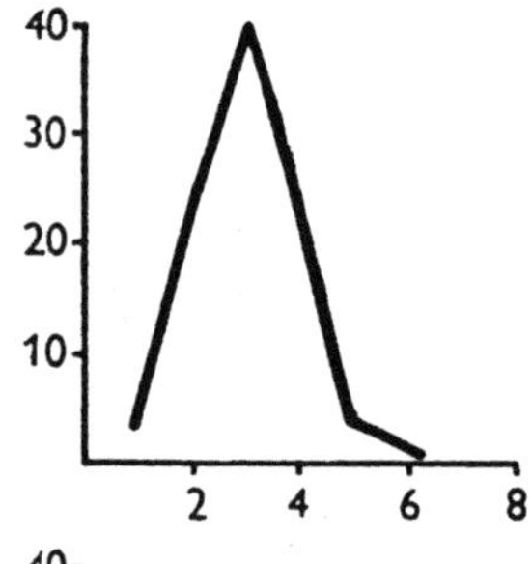

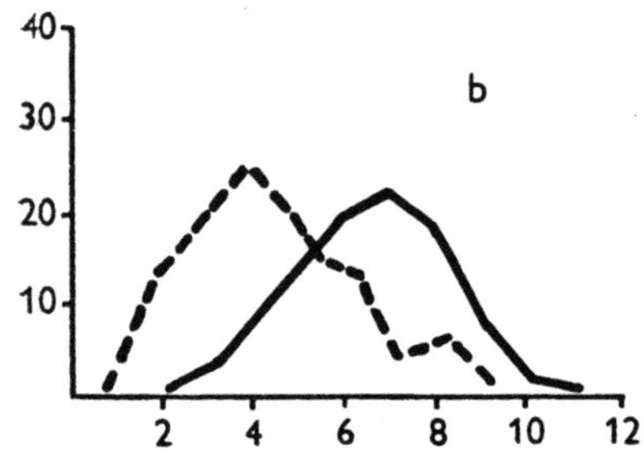

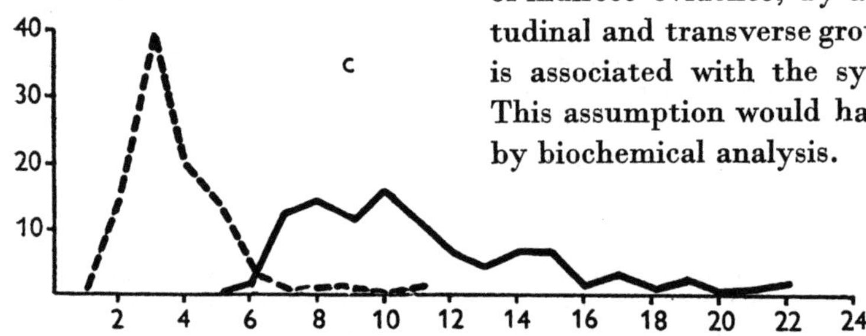

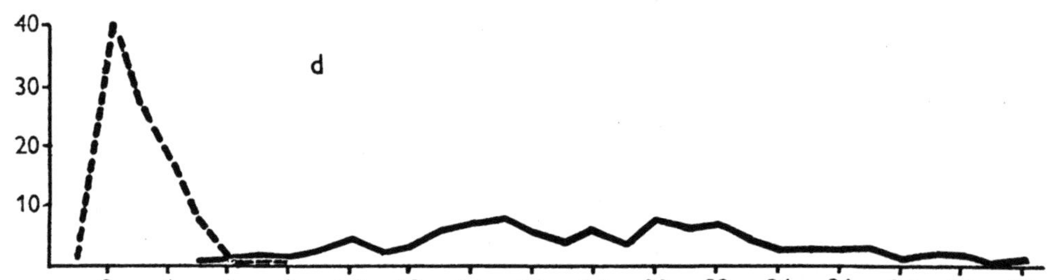

Fig. 55. Changes in the histogram of normal and denervated muscles in young rats in the course of the first two months after birth. a — newborn rat; b — 14 days; c — one month; d — two months following denervation. Abscissa — muscle fibre diameters in μ. Ordinate — percentage composition. ——— normal m. extensor digitorum longus, - - - - denervated m. extensor digitorum longus.

Schapira, Dreyfus and coworkers (1949, 1950a,b) investigated the growth of the denervated muscle from the biochemical and biometrical aspect. (For biometrical equations for rate of atrophy during growth, see Chapter X.) It was found (Schapira et al. 1950, Schapira and Dreyfus 1950) that following section

Table 5

Muscle fibre diameters (in μ) in denervated (D) and control (C) muscles of rats 14 days, one and two months after denervation at birth respectively.

Time after denervation	M. soleus			M. extensor dig. longus			n
	D μ	C μ	D/C %	D μ	C μ	D/C %	
At birth		4·81			4·04		
14 days	6·9 ± 0·17	10·1 ± 0·13	68·1 ± 0·89	6·3 ± 0·46	9·0 ± 0·84	71·1 ± 3·46	5
1 month	4·7 ± 0·42	17·9 ± 0·79	26·8 ± 0·86	5·3 ± 0·74	16·3 ± 0·85	32·4 ± 3·37	5
2 months	3·6 ± 0·30	24·2 ± 1·00	14·8 ± 1·33	3·9 ± 0·18	21·0 ± 1·27	18·7 ± 1·37	5

of the nerve in 10-day-old rabbits, the weight of the gastrocnemius muscle is approximately doubled in the course of the next ten days, while the normal muscle attains a weight which is three times the original weight. Total protein content is slightly increased in denervated muscles, actomyosin does not change, and when expressed as percentage of dry muscle weight, it even decreases (Fig. 56). Extractable proteins without actomyosin, on the other hand, are increased, even though this increase is not very marked. The considerable increase in total dry muscle weight is doubtlessly due, at least in part, to proliferation of collagen which increases about five times above its original content and equals the content in the normal, control muscle. Since the content of myosin remains unchanged or is decreased, it is concluded by these authors that it actually is muscle atrophy, even though it is not accompanied by muscle weight loss. The structural protein balance remains unchanged, since

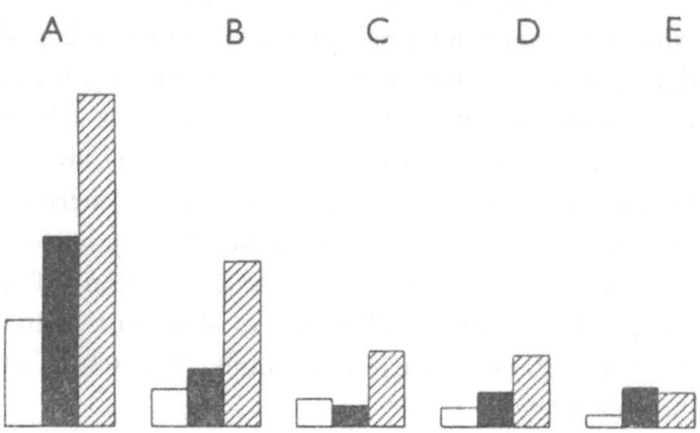

Fig. 56. Changes in protein components in the growing muscle of the rabbit following denervation. (From Schapira and Dreyfus, 1950). A — dry weight, B — soluble proteins, C — actomyosine, D — soluble proteins with the exception of actomyosine, E — collagene. [white box] — normal muscle of ten-day-old rabbits, [black box] — denervated muscle of 20-day-old rabbits 10 days following denervation, [hatched box] — control muscle of 20-day-old rabbits

the rate of myosin synthesis is the same as the breakdown rate. It is pointed out by the above authors that results of the experiment correspond to algebraic summation of growth and atrophy. The algebraic summation of both processes in the above experiment is nil, as both processes are of the same intensity. If protein synthesis due to growth was more rapid, and predominated over protein breakdown due to atrophy, the final balance would be positive. Such is probably the case in rat muscle denervated immediately after birth.

On the contrary, when growth is retarded, atrophy predominates in denervated muscle and muscle weight decreases. This loss of weight is less extensive, however, than in adult animals with terminated growth. A less extensive atrophy and slower reduction of muscle fibre diameter was found in 6-week-old rats than in animals over a year old (Vrbová 1952). There was a 30·14 $\pm$ 2·53% atrophy in the extensor digitorum and anterior tibial muscles in young rats, while a 45·30 $\pm$ 0·93% atrophy occurred in older rats. Muscle fibre diameters were reduced by 20% in young rats and by about 30% in old rats.

The rate of atrophy during the denervation period in growing animals probably undergoes changes similar to muscle atrophy at different stages of development from slowing down of the muscle weight increment to absolute loss of structural proteins. During the first stage when the growth curve is steep, proteosynthesis is partly preserved, later it is arrested, and finally protein degradation predominates and there is a decrease in the protein content, with the exception of collagen. If these probable changes are projected into the denervation period when the muscle is still undergoing differentiation, then it is comprehensible that differentiation is not affected directly by denervation. During the first phase it is slowed down, corresponding to the reduction in proteosynthesis, then it is arrested, and finally more or less rapid disintegration takes place with fibrous or fatty transformation of the muscle. The rate of these changes is evidently different in different animal species, similarly as muscle atrophy differs in various animals (Knowlton and Hines 1936 and others, see Chapter X).

So far, experimental data have been put forward indicating that, in contrast to the adult organism, *denervation atrophy in the developing and growing organism is a relative atrophy*, characterized by an absolute increment is muscle weight. It is, of course, necessary to stress that the delay in growth, as compared with normal muscle, is extreme. While denervated muscle in the above experiment increased in size to five times the initial weight, the normal muscle increased about seventy times. The elimination of the nervous system during the period of rapid growth causes only relative atrophy, but this atrophy is *of such extent*, in comparison with normal muscle, *that* equivalent atrophy *can never be observed in the adult organism.*

D. The effect of denervation on muscle regeneration

While the development of non-innervated limbs and limb muscles is not actually endangered in amphibians (Harrison 1904, Hooker 1911, Hamburger 1929 and others), regeneration of muscles without innervation is impossible (Weiss 1955, Schotté et al. 1941, Singer 1943a, b, 1945, 1946a, b, 1947, 1949, 1950, 1951, 1952, and others).

It is actually a process concerning the regeneration of the whole limb following amputation. Of course, regenerating muscles are as dependent on innervation as the whole limb bud. Limbs can regenerate independently of the nervous system only in organisms at the earliest developmental stage (Polezhayev 1946). At this early stage of development it is more suitable to consider this process as repetition of development rather than as regeneration. Limb regeneration is not possible at a later stage of development, if the amputation stump is denervated. This fact was already known to Todd in 1832, and was later confirmed by various experimental approaches. The relationship between regeneration and innervation was analysed by Singer (1943—1952) in adult Triturus. According to this author, successful regeneration can take place when the amputation stump contains something between a third and a half of normal nerve fibres. It was found that it is irrelevant whether these fibres are sensory, motor or vegetative. The influence of the nerve fibres is considered by Singer to be undoubtedly humoral, non-impulse in character. It is, of course, not clear whether innervation of a certain tissue is not if paramount importance in the process of regeneration. Thornton (1955), for example, points out that the epidermal cap of the regenerating limb bud must be innervated. If reinnervation of the epidermis is prevented, limb regeneration cannot take place. Limb regeneration is not only dependent on nervous influences at the very beginning of the process, but also at later stages at the beginning of muscle differentiation itself. If nerves are sectioned in the course of limb regeneration during the first thirteen days in the adult Triturus, the process does not progress further. Muscle differentiation is arrested immediately after denervation and the regenerated limb bud usually undergoes regression. During a later phase of regeneration, on the other hand, when myotubes have already differentiated in muscle fibres, morphogenesis progresses uninterruptedly even after denervation (Singer and Craven 1948). In contrast to this, Gorelova (1954) reported in a similar experiment, that only the initial differentiation of myoblasts is fully independent of innervation. This author studied regeneration in the tail of the tadpole following amputation, and observed that 20 days after denervation the differentiation of myoblasts in the regenerated limb was retarded. Within a period of four to six months, muscles disappeared not only from the regenerated part, but also from a small band above the margin of amputation. In analogous experiments, Zhinkin (1949) and Kryukova (1938) found only retardation of regeneration following denervation.

The fact that limb regeneration in tailed amphibians is dependent on innervation was further confirmed by an interesting experiment of Singer (1954) on the frog after metamorphosis. Under normal conditions limbs do not regenerate in adult frogs; however Singer provoked regeneration by doubling the density of innervation in the stump by transposition of the contralateral sciatic nerve into the amputation stump. The role of the nervous system for inducing regeneration was thus confirmed, not only by elimination experiments, but also by a positive effect of supernumerary innervation.

On the other hand, regeneration experiments in denervated muscles following trauma have not given unanimous results. According to Sanders and Sissons (1953), regeneration is not dependent on the presence of the nervous system. These authors confirmed the negative finding of Kirby (1894). They compared regeneration in normal and denervated muscles three weeks after section of the nerve, following trauma with a narrow forceps. A similar course of regeneration was found in normal and denervated muscles up to the fifth week after trauma. Regenerating fibres bridged the traumatized segment within a week. However, later they only attained the diameter of undamaged denervated fibres, while in the normal muscle they continued to increase in diameter. After the fifth week it was difficult to differentiate in the denervated muscle between the zone of trauma and regeneration, because of general atrophy and fibrosis. The exact location can be discerned from the irregularly arranged muscle fibres. This irregularity in arrangement is considered by the authors as the result of disorganization of sarcolemmal tubes. Where these tubes are preserved, the regenerating fibres grow into them. On the contrary, when they are considerably disorganized, regeneration is limited in extent.

Irregular arrangement of regenerated muscle fibres in the denervated muscle of the frog is considered by Vidik (1953) to be the main difference from normal muscles 50 days after injury by ligature. According to this author, a certain slowing down of the process of regeneration can be observed in denervated muscle.

Denny-Brown (1951), on the other hand, described a different course of regeneration in atrophic muscle of cats following trauma three weeks after section of the nerve. Normal muscle fibres did not appear in denervated muscle during regeneration, but only myotubes were formed. If the muscle was injured within three weeks after nerve section, the regeneration processes were normal. Thus denervation did not, in this case, affect the morphogenetic process until after a certain latent period.

According to Studitski and coworkers, denervation has an adverse effect upon regeneration of traumatized muscle in birds and mammals (Studitski and Striganova 1951, Studitski 1952, 1953, Zhenevskaya 1954, 1955, Studitski et al. 1956, Korytnyj 1957). In model experiments of Studitski and coworkers, extensive muscle regeneration is caused by excision of the whole muscle with the exception of a short remnant at the insertion. The empty fascia space is then filled with mashed muscle tissue and regeneration is studied in this region of the muscle. Innervation of the regenerating tissue is, according to these authors, indispensible for differentiation from the myoblastic stage into the myotubal stage. In case the regenerating tissue is not innervated, it does not develop further and disintegrates at the myoblastic stage. Studitski et al. (1956) consider the disintegration of heterotopic autografts and homografts, in which it is difficult to obtain innervation of the regenerating tissue, as an example of this. However, in these experiments, other more probable causes of disintegration have not been excluded, and the degree of innervation was also not controlled histologically. On the other hand, Zhenevskaya (1954, 1955) studied in detail the regeneration of the rat gastrocnemius following resection and crushing of the central part of the muscle with intact innervation, and following section of the tibial nerve. According to her results, denervated and innervated regenerating muscles develop similarly during the first week and attain the myoblastic stage. Further development in the denervated muscle is impossible, and the regenerating tissue is gradually substituted by connective tissue. However, at later periods the author found in all cases occasional muscle and nerve fibres scattered loosely in the connective tissue. These experiments cannot therefore be considered as conclusive evidence for the hypothesis that differentiation is arrested at the myoblastic stage after denervation, they indicate, however, that innervation is very important for successful regeneration after extensive damage to the muscle.

There are many controversies in the above experiments, and attempts at their analysis is further complicated by the great variability of methods used. The problem of regeneration of denervated muscle requires further investigation.

If the nervous system is excluded from participation in morphogenetic processes in the muscle, then formation and initial differentiation of the muscle can still take place. The nervous system does not exert a trophic influence on the muscle until nerve endings are formed in the muscle tissue. Thereafter, further development and growth of the muscle are deeply impaired by denervation, which causes a slowing down of the differentiation and a retardation of muscle growth. These changes are evidently the result of a reduction of proteosynthetic processes, and in the end by the prevalence of protein breakdown.

Differentiation of muscle fibres is not directly dependent on innervation. On the other hand, differentiation of muscle spindles and of the postsynaptic part of the motor end-plates is directly dependent on the formative influence of the nerve endings which is probably neurohumoral in character. The developing muscle thus represents a suitable model for analysing the basic mechanisms of the nervous influence on muscle tissue.

References

Alder A. B., Crawford G. N. C., Edwards R. G.: Proc. roy. Soc. B. 151 : 551, 1959.
Alessandrini: 1829, 1834, see Weber E. H.: 1851.
Anders H. E.: Roux Arch. Entw. Mech. Org. 47 : 452, 1921.
Beckett B. E., Bourne G. H.: Acta anat. 35 : 224, 1958.
Braus H.: Anat. Anz. 26 : 433, 1905.
— Ver. Ges. dtsch. Naturf. Ärzt. 83 : 114, 1911.
Couteaux R.: Bull. Biol. 75 : 101, 1941.
— Rev. canad. Biol. 6 : 563, 1947.
— Innervation du muscle strié et organisation du sarcoplasma au niveau des terminaisons motrices. Le Muscle: étude de biologie et de pathologie. L'expansion Scientifique Francaise, éd., p. 173, Paris 1950.
— Localization of cholinesterases at neuromuscular junction. Internat. Rev. of Cytology IV, ed. by Bourne G. H. and Danielli J. F. Academic Press, p. 335. New York 1955.
Couteaux R., Taxi J.: Arch. Anat. micr. 41 : 352, 1952.
Csillik B.: Z. Zellforsch. 52 : 150, 1960.
Cuajunco F.: Contr. Embryol. Carneg. Instn. 30 : 129, 1942.
Denny-Brown D.: J. Neuropath. 10 : 94, 1951.
Dreyfus J. C., Schapira G., Schapira F.: C. R. Soc. Biol. 143 : 681, 1949.
Eastlick H. L.: J. exp. Zool. 93 : 27, 1943.
Eastlick H. L., Wortham R. A.: J. Morphol. 80 : 369, 1947.
Gerebtzoff M. A., Biochemistry of the Dev. Nerv. System, Symp. Academic Press, p. 315, 1955.
Gerebtzoff M. A., Philippote, Dallemagne M. J.: Acta anat. 20 : 234, 1954.
Gorelova L. P.: DAN SSSR 96 : 415, 1954.
Grigoryev L. M.: Čs. Morfol. 8 : 5, 1960.
Hamburger V.: Naturwissenschaften 15 : 657, 677, 1927.
— Roux Arch. Entw. Mech. Org. 114 : 272, 1929.
— J. exp. Zool. 80 : 347, 1939.
Harrison R. G.: Amer. J. Anat. 2: Proc. Ass. Amer. Anat. 16, 1903.
— Amer. J. Anat. 3 : 117, 1904.
— J. exp. Zool. 4 : 239, 1918.
Herbst, C.: Formative Reize in der tierischen Ontogenese. A. Georgi, Leipzig 1901.
Hoadley L.: J. exp. Zool. 42 : 163, 1925.
— J. exp. Zool. 42 : 143, 1925.
Hooker D.: J. exp. Zool. 11 : 159, 1911.
Hunt E. A.: J. exp. Zool. 62 : 57, 1932.
Kalugina M. A.: Arkh. anat. 33 : 59, 1956.
Kirby: 1894, cit. Sanders, Sissons 1953.
Knowlton G. C., Hines H. M.: Proc. Soc. exp. Biol. 35 : 394, 1936.
Kryukova: 1938, see Gorelova 1954.
Korytnyj E. J.: Tezisy dokladov I. Beloruskoj konferencii ant. gist. embriol., 1957.
Kupfer C., Koelle G. B.: J. exp. Zool. 116 : 397, 1951.
Leonard S. L.: Proc. Soc. exp. Biol. 94 : 458, 1957.
Leonova O.: Neurol. Zbl. 12 : 218, 1893.
Manina A. A., Bystrov V. D.: DAN SSSR 103 : 499, 1955.
Manz W.: Virchows Arch. 51 : 313, 1870.
Neumann: 1846, see Weber E. H., 1851.
Neumann E.: Roux Arch. Entw. Mech. Org. 16 : 642, 1903.
— Roux Arch. Entw. Mech. Org. 18 : 296, 1904.

Piatt J.: J. exp. Zool. 91 : 71, 1942.
— J. exp. Zool. 120 : 247, 1952.
Polezhayev L. W.: Biol. Rev. 21 : 141, 1946.
Portugalov V. V.: Ocherki gistofiziologii nervnykh okonchanij. Medgiz, Moscow 1955.
Sachs C.: Arch. Anat. Physiol. 461 : 645, 1874.
Saunders J. H., Sissons H. A.: J. Bone Surg. 35-Bl : 113, 1953.
Sávay Gy., Csillik B.: Acta morph. Acad. Sci. Hung. 6 : 289, 1956.
Schaper A.: Roux Arch. Entw. Mech. Org. 6 : 157, 1898.
Schapira G., Dreyfus J.: Métabolismes du muscle atrophié. Le muscle, étude de biologie et de pathologie, L'Expansion, éd. 1950, 299.
Schapira G., Dreyfus J., Schapira F.: C. R. Soc. Biol. 144 : 829, 1950.
Schotté O. E., Butler, E. G.: J. exp. Zool. 87 : 279, 1941.
Singer M.: J. exp. Zool. 90 : 377, 1943.
— J. exp. Zool. 92 : 297, 1943.
— J. exp. Zool. 98 : 1, 1945.
— J. exp. Zool. 101 : 221, 1946.
— J. exp. Zool. 101 : 229, 1946.
— J. exp. Zool. 104 : 223, 1947.
— J. exp. Zool. 111 : 189, 1949.
— Anat. Rec. 108 : 518, 1950.
— Proc. Soc. exp. Biol., N. Y. 76 : 413, 1951.
— Quart. Rev. Biol. 27 : 169, 1952.
— J. exp. Zool. 126 : 419, 1954.
— Craven L.: J. exp. Zool. 108 : 279, 1948.
Studitski A. N.: Arkh. anat. gist. i embriol. 29 : 27, 1952.
— Arkh. anat. gist. i embryol. 30 : 10, 1953.
Studitski A. N., Striganova A. R.: Vosstanovitelnye processy v skeletnoi muskulature. Izd. AN SSSR, Moskva 1951.
Studitski A. N., Zhenevskya R. P.: Rumyavceva V. N.: Čsl. morfologie 4 : 331, 1956.
Sutton A. C.: Amer. J. Anat. 18 : 117, 1915.
Szepsenwol J.: Anat. Rec. 95 : 125, 1946.
— Anat. Rec. 98 : 67, 1947.
Tello J. F.: Z. Anat. 64 : 348, 1922.
Teodorovich V. J.: Arkh. anat. gistol. i embriol. 30 : 12, 1953.
Thornton Ch. S.: J. exp. Zool. 127 : 577, 1955.
Todd: see Singer 1952.
Tower S.: Brain 55 : 77, 1932.
Vrbová G.: Čs. Fysiol. 1 : 38, 1952.
Weiss P.: Nervous System (Neurogenesis). Analysis of Development. Ed. by Willier B. H., Weiss P. A., Hamburger V. — W. B. Saunders Comp. New York 346, 1955.
Vidik T. V.: Trudy Leningrad. Sanit. med. inst. 16 : 299, 1953.
Wintrebert P.: C. R. Acad. Sci. (Paris), 137 : 131, 1903.
Zelená J.: J. Embryol. exp. Morph. 5 : 283, 1957.
— Effect of Innervation on the Development of Skeletal Muscle (in Czech with English and Russian summary). (Babákova sbírka 12.) Státní zdravotnické nakladatelství, Praha 1959.
— Čs. fysiol. 9 : 195, 1960.
Zelená J., Hník P.: Physiol. bohemosl. 6 : 159, 1957.
— Physiol. Bohemoslov. 9 : 373, 1960.
— Nature 188 : 946, 1960.

Zelená J., Szentágothai J.: Acta histochem. 3 : 284, 1957.
Zhinkin: 1948 — see Gorelova 1954.
Zhenevskaya R. P.: DAN 96 : 217, 1954.
— Izvestija AN SSSR, Seria biol. 6 : 90, 1955.

Chapter IV

ELECTROPHYSIOLOGY OF DENERVATED SKELETAL MUSCLE

R. Beránek

Research on the electrical properties of denervated muscles was at first instigated by the practical interests of physicians in elaborating an electrodiagnostic method suitable for the evaluation of peripheral nerve lesions. This aspect together with positive results obtained by electrotherapy called forth extensive research work concerned with this question, especially during the two World Wars, when the occurrence of nerve injuries was considerably increased. At the present time, the denervated muscle is becoming more and more often the object of theoretical investigation aimed at elucidating neuromuscular trophic interrelations. Electrophysiology represents one methodical approach to this research. Its technical achievements have given investigations of the denervated muscle a progressively more analytical character. The following properties of the denervated muscle have been most extensively studied by electrophysiologists: 1. excitability and its time constants, 2. membrane potential and spontaneous activity, 3. passive electrical properties and constants of the muscle fibre membrane.

A. Excitability

Duchenne de Boulogne (1855) was the first to observe that the denervated muscle responds to electrical stimulation differently from the normal muscle. A systematic evaluation of electrical excitability of the paralysed muscle on the basis of clinical and experimental studies was performed by Erb (1868) who grouped the characteristic changes in denervated muscles under the name of the degeneration reaction (Entartungsreaktion): loss of nerve excitability, increased muscle excitability to galvanic current and lowered excitability to faradic current, reversal of polarity of the stimulating electrodes and mechanical excitability. Erb also explained these changes on the basis of then recent results of Waller (1850) who showed that the degeneration of the peripheral stump of an interrupted nerve is the result of disintegration of motor nerve fibres associated with loss of nerve conduction.

For determining excitability changes in the denervated muscle, clinical electrophysiology utilized general chacteristics of excitable tissues obtained by applying precisely defined electrical stimuli. Most of them had been defined on myelininated nerve fibres, they could, however, be applied generally also to muscle fibres.

The main outcome of clinical investigations based upon these physiological data was the finding that chronaxie of the denervated muscle is prolonged (from normal values of the order of tenths of a millisecond to tens of milliseconds) with a concomitant decrease of rheobase. The accomodation ability of denervated muscles is lowered. Following reinnervation the time constants of excitability tend to return to normal values. (For details see Chapter XII). At this period of clinical research the fibrillation potentials as correlate of denervation fibrillation were also described in electromyographic studies (see Chapter V).

All these findings have been successfully utilized in clinical diagnostics and therapy of peripheral nerve lesions (see Chapter XII), although a more precise electrophysiological analysis of denervation phenomena was lacking until it was performed under laboratory conditions.

The striated muscle is a complex organ containing nerve and muscle elements either of which can be endowed with different excitability. This has been experimentally proved for the frog skeletal muscle by Lucas (1907). Adrian (1917) utilized Lucas' data for explaining clinical findings of muscle excitability following nerve lesions. The slowing down of the process of excitation in denervated muscles represents, according to Adrian, the unmasking of the muscle component (substance α according to Lucas) after elimination of the nerve component (substance γ according to Lucas) following peripheral nerve degeneration. This conclusion was supported by the discontinuity of strength-duration curves found after denervation and during the reinnervation period.

It was not until later (during the II. World War period) that experimental investigations brought a more detailed account of muscle excitability changes following denervation (Pollock et al. 1945a, Pollock et al. 1945b, Desmedt 1950a, 1950b). These authors, utilizing the quantitative studies of functional and morphological changes by Young and his coworkers (Young 1942, Gutmann et al. 1942, Gutmann and Young 1944), classified, from the electrophysiological aspect, processes taking place in the muscle after nerve section into three phases: 1. degeneration — from nerve section to the loss of conduction in motor axons in the peripheral stump, 2. denervation and 3. reinnervation — from the first physiological symptoms of restitution to eventual recovery to normal. In his detailed and critical study, Desmedt (1950a, 1950b) differentiated two kinds of excitability in normal muscles of the hind limb in the rabbit: type I response is best evoked from the motor point, and its time constants are similar to those of the nerve. It is lost after curarization and evidently corresponds to the excitability of intramuscular motor axons. The type II response, best evoked outside the motor point, the time constants of which differ statistically from the first, does not change after curarization and represents the excitability of the muscle fibres themselves. The strength-duration curves are continual for each of these responses. When discontinual curves are obtained, this means that two different elements have been measured simultaneously and the discontinuity of the curves is of no physiological significance in the normal or in the denervated muscle.

According to Desmedt, the course of excitability changes in rabbit muscle after denervation is as follows:

At the end of the degeneration period motor axon disintegration occurs and the type I response, characterized by a short chronaxie (0·25 msec. in the normal anterior tibial muscle of the rabbit) disappears. The type II response (with average chronaxie 3 msec. and mean value 104 msec. for λ) may have a transient increase of rheobase during this temporary phase, together with a shortening of chronaxie and a decrease of λ. It can be observed, furthermore, that values of chronaxie and λ do not change in parallel, this serving as evidence that they represent two different processes.

Following complete loss of axon conduction 30—70 hrs. after nerve section (Gutmann and Holubář 1950), the muscle becomes a homogenous structure as far as its excitability is concerned and reacts only by the second type of response. Strength-duration curves are continuous and any discontinuity measured indicates that simultaneous excitation of a neighbouring non-denervated muscle had taken place. Thus after nerve degeneration "unmasking" of true muscle excitability occures, but the process does not terminate here, since during the next period slow continual denervation changes of excitability develop and these attain a plateau in the anterior tibial muscle of the rabbit on about the twentieth day after nerve section: rheobase decreases to about one fifth of its normal value and chronaxie is prolonged to 15 msec. and λ to 250—500 msec.

This state can last for several months until progressing fibrosis can increase the rheobase again up to and above its initial value. Chronaxie and λ, however, retain their original high denervation values. Changes of physical conditions in fibrotic muscles a long time after denervation evidently explain discrepancies in clinical reports concerning the rheobase changes (v. Richter and Katz 1943) after nerve injuries. The sequence of denervation changes is the same in different muscles, however, there can be differences as far as their time course is concerned. In Desmedt's experiments the gastrocnemius in the rabbit already attained the denervation plateau within ten days.

If reinnervation takes place, muscle excitability undergoes the opposite sequence of changes: rheobase increases, chronaxy and λ become shorter in duration. These changes distinctly appear sooner than indirect muscle excitability (Pollock et al. 1945a). During this transitory period rheobase can increase again above normal and chronaxie and λ can become shorter than their normal values. The excitability of the anterior tibial muscle in the rabbit is completely restored about 20 days after the first electrophysiological symptoms of reinnervation. All these data about the reinnervation period concern type II response. The type I response from regenerated intramuscular axons does not appear in the rabbit as a rule until several months after full restitution of muscle fibre excitability had taken place. This dissociation indicates the prolonged maturation of regenerated axons, their growth in diameter and their myelinization (Young, 1942). Inaccurate differentiation of both responses in clinical diagnosis led to much misunderstanding and scepticism concerning the significance of electrodiagnosis in nerve lesions, since short chronaxie values (several tenths of msec.) often are considered to be normal muscle chronaxie values, and yet it is known that they belong to the nerve component which cannot be normally detected even in cases when motor function has been restored. Values ten times higher (of the order of msec.) can be considered as representing fully restored muscle excitability. Another source of controversy about the value of electrodiagnosis were attempts at making conclusions on the basis of single chronaxie measurements. It was experimentally shown that even under optimal conditions for nerve regeneration, the onset of reinnervation is very unevenly distributed at different sites in the muscle, and this is the cause for the wide spectrum of various excitabilities which can be found in the course of this transitory period. It is therefore advisable to explore the muscle at various points and to attempt to obtain continuous strength—duration curves.

Desmedt (1950c) also pointed out the coincidence between the presence of fibrillations and prolongation of the accomodation constant after denervation.

B. Membrane potential and spontaneous activity

Recently the intracellular technique, first introduced into muscle physiology by Ling and Gerard (1949), has been used for studying the denervated muscle. Single denervated muscle fibres in their natural environment have become the object of research and thus a more detailed analysis of electrophysiological properties of their cellular membrane has been made possible.

Intracellular measurements have shown that there is considerable potential unrest (Migliaro et al. 1959) in the denervated muscle fibre membrane. Oscillations of the membrane potential are often rhythmic in character and when the depolarization reaches a critical value, fibrillation action potentials appear (Li et al. 1957 Luellmann 1957, 1960). By the intracellular technique Li et al. (1957) and Migliaro et al. (1959) found that the spikes of fibrillating muscle fibres are similar, as far as the amplitude is concerned, to those of an innervated fibre (they are characterized by an overshoot), their ascending part, however, is steeper and their descending part longer and their duration as a whole is greater.

At points corresponding to the endplates spikes with a prepotential are more often to be seen, this indicating the presence of a pacemaker. However, they can also be registered in regions outside the endplate zone which seems to corroborate the assumption that fibrillations may also originate at locations outside the original neuromuscular connection. Similar oscillations of the membrane potential accompanied by fibrillation potentials were also observed in embryonic muscle fibres in tissue cultures (Li 1960). The presence of nerve endings in the muscle prevents the onset of fibrillations, even when these nerve fibres do not mediate impulse activity (Tower 1937). It is pertinent to mention in this connection the findings of Ginetzinsky and Shamarina (1944) analysed by Axelsson and Thesleff (1959), according to whom increased sensitivity of the denervated muscle fibre to acetylcholine is not limited to the endplate region only, but occurs over the entire surface of the membrane.

The velocity at which the fibrillation action potential spreads along the muscle fibre was found to be reduced in the tenuissimus muscle of the cat after denervation (Jarcho et al. 1954). Similarly Buchthal and Rosenfalck (1958) reported slower conduction velocity of spikes evoked by electrical stimulation of the human denervated muscle. Whether slower impulse conduction along the muscle fibre membrane is fully sufficient to account for the slower contraction of the denervated muscle, or whether denervation slows down the reaction of the contractile substance itself seems not to have been analyzed.

As far as the mean value of the membrane patential is concerned, results of different authors do not agree. While Ware et al. (1954) reported a statistically significant decrease of the membrane potential in denervated muscles of the mouse, no changes were found between the membrane potentials of normal and denervated muscles by Nicholls (1956) in the frog and by Li et al. (1957) in the rat. Seeing that the technique used introduces some factors dificult to control, e. g. fibrosis and decrease in fibre diameter which can influence the selection of fibres measured, this question must be, as yet, considered undecided.

C. Passive electrical properties and membrane constants

Analysis of denervation changes in electrical properties of the muscle fibre membrane performed by Nicholls (1956) on the frog sartorius muscle by intracellular square pulse analysis, helped to elucidate some of the phenomena observed on the whole muscle. The most prominent feature of the denervated fibre was an increase (approximately double) in its specific resistance, together with a decrease of the critical depolarization value: i. e. depolarization necessary for initiating a spike. This result represents a sufficient explanation of increased electrical excitability of denervated muscles: a weaker current suffices to cause the critical decrease in membrane polarization if passed across a greater resistance. According to Lüllmann (1960) the increased resistance can be explained by lowered permeability for potassium ions. The decreased sensitivity to short stimuli (prolongation of chronaxie) observed in the mammalian denervated muscle would indicate an increase in the membrane capacity, but this value was found unchanged a month after denervation in Nicholls' experiments. The problem requires further analysis by the intracellular technique in homeothermic animals in which most of the known data about denervation excitability changes have been obtained. Denervation changes in homeothermic animals set in considerably sooner than in poikilothermic animals (Bremer and Gérard 1925).

No systematic research has yet been made concerning electrochemical mechanisms underlying the electrophysiological denervation phenomena in the skeletal muscle. Data concerning changes of ionic composition in striated muscles after denervation are discussed in Chapter IV..

D. Conclusions

Following section of the peripheral motor nerve fibres changes in excitability constants, the spike duration and conduction velocity of impulses occur. Spontaneous oscillations of the resting potential are present, giving rise to rhythmically produced spikes. Some of these changes can be accounted for by the values of electrical constants of the denervated muscle fibre membrane. These changes are not instantaneous but develop gradually with progressing time after denervation and they can thus be designated as "trophic". After reinnervation the electrical properties of the muscle return to normal. The process of restitution begins earlier than the recovery of the ability to evoke muscle contraction by neuromuscular transmission. This observation may serve as evidence (Desmedt 1950b) for a special trophic function of peripheral nerves which is different from the transmission function.

Electrophysiological phenomena in the denervated muscle, which are determined primarily by the physico-chemical state of its cellular membranes, go in parallel with metabolic changes studied predominantly in the muscle as a whole (see Chapters VIII and IX). At the present time, it is not possible to decide which of these changes represent the primary link in denervation atrophy. Only recently, have the interrelations between cellular metabolism and the ionic events in excitable membranes become the object of systematic research. It can be presumed that denervation studies will not only profit from the results of this investigation, but will themselves contribute some clues to this problem. Studies dealing with changes of chemical sensitivity of the denervated muscle membrane, discussed in more detail elsewhere (Chapter V), appear to be promising in this respect.

References

Adrian E. O.: Arch. of Radiol. Electrother. 21 : 379, 1917.
Axelsson J., Thesleff S.: J. Physiol. 147 : 177, 1959.
Bremer F., Gérard P.: C. R. Soc. Biol. 1327, 1925.
Buchthal F., Rosenfalck P.: EEG Clin. Neurophysiol. 10 : 521, 1958.
Desmedt J. E.: Arch. int. physiol. 58 : 23, 1950a.
— Arch. int. physiol. 58 : 125, 1950b.
— Acta neurol. psychiat. 179 : No. 4, 1950c.
Duchenne de Boulogne: De l'electrisation localisée et de son application à la pathologie et à la therapeutique. Paris 1855.
Erb W.: Dtsch. Arch. f. klin. Med. 4 : 535a, 5 : 42, 1868.
Ginetzinsky, A. G. and Shamarina, N. M.: Usp. sovr. biol., 1942.
Gutmann E., Guttmann L., Medawar P. B., Young J. Z.: J. of expl. biol. 19 : 14, 1942.
Gutmann E., Holubář J.: J. Neurol. Neurosurg. Psychiat. 13 : 39, 1950.
Gutmann E., Young J. Z.: J. Anat. 78 : 15, 1944.
Jarcho L. W., Berman B., Dowben R. M., Lilienthal Jh.: Am. J. Physiol. 178: 129, 1954.
Katz B.: Electric excitation of nerve. Oxford 1939.
Li Ch. L., Shy M., Wells J.: J. Physiol. 135 : 522, 1957.
Li, Ch. L.: Science, 132 : 1889, 1960.
Ling G. a Gerard R. W.: J. Cell. Comp. Physiol. 34 : 382, 1949.
Lucas K.: J. of. Physiol. 36 : 113, 1907.
Luellmann H.: Klin. Wschr. 38 : 1169, 1960.
Luellmann H. and Pracht W.: Experientia 13 : 288, 1957.
Migliaro E., Garcia-Austt E., Carlevaro P., Grünberg V., Eyzaguirre C.: XXI. Internat. congress of physiological sciences, Buenos Aires 1959, Abstracts p. 185.
Nicholls J. G.: J. Physiol. 131 : 1, 1956.
Pollock L. J., Golseth J. G., Arieff A. J.: Surg. Gynec. Obstefr. 81 : 451, 1945a.
Pollock L. J., Golseth J. G., Arieff A. J., Mayfield F.: Surg. Gynec. Obstetr. 81 : 192, 1945b.
Richter E. P., Katz D.: J. Amer. med. Ass. 122 : 648, 1943.
Tower S. S.: J. Comp. Neurol. 67 : 241, 1937.
Waller A.: Philos. Trans. 140 : 423, 1850.
Ware F. Jr., Bennett A. L., McIntyre A. R.: Am. J. Physiol. 177 : 115, 1954.
Young J. Z.: Physiol. Rev. 22 : 318, 1942.

Chapter V

FIBRILLATION ACTIVITY IN DENERVATED MUSCLE

Hník P., Škorpil Vl.

Denervation fibrillations are one of the most accessible electrical manifestations of the denervated muscle for experimental and clinical investigation. In spite of the extent to which they have been studied, however, their mechanism is still not clear. They represent one of the basic problems of denervation, because spontaneous activity following section of the peripheral nerve is characteristic for the denervated muscle and is not present in muscle atrophy of other origin. For this reason the registration of denervation fibrillations has become one of the diagnostic methods used for the examination of peripheral motoneuron lesions.

Denervation fibrillations are spontaneous rhythmical contractions of denervated muscle fibres which can be observed visually after removal of the skin and fascia (Schiff 1851, Langley and Kato 1914—15 and others) and which have their specific electrophysiological manifestation. On extracellular recording they appear in the EMG as mono-, bi-, or triphasic potentials, 10—100 μV in amplitude, of 1—3 msec. duration and 2—30/sec. frequency (Denny-Brown and Pennybacker 1938, Feinstein et al. 1945, Bergamini et al. 1955). These potentials can also be reproduced acoustically. They are characterized by a short, high tone.

Besides these typical denervation fibrillations there have also been described atypical denervation potentials (Lerique 1944, Kugelberg and Petersen 1949, Jasper and Ballem 1949, Buchthal 1957, Baumgartner 1957 and others). These potentials are of considerable duration, up to 100 msec. and often attain high amplitudes (up to several hundreds μV). They are less frequent than typical potentials, but are also considered to be a reliable sign of denervation. In our experiments on rats (Škorpil and Hník 1958) we also found different types of denervation potentials. Besides typical fibrillations, spike potentials of the usual duration (1—3 msec.) but of a noticeably high amplitude (500—700 μV) were observed. These potentials were evidently the same as those described by Landau (1951). They occurred rarely and appeared together with the other fibrillation potentials. They are probably initiated by an exceptionally close position of the recording electrode to the muscle fibre, as has been postulated by Landau. It is less probable that it is partial synchronization of several fibrillation potentials, because of their short duration. Besides these potentials, wide mono- or biphasic potentials were observed, about 30 msec. in duration and 50—100 μV in amplitude. Their frequency of discharge was lower than in typical potentials. These fibrillations are probably due to synchronized discharges of several typical fibrillation potentials. It is interesting that these potentials often first appeared following arrest of respiration and after arresting the blood flow through the muscle, at a time when the number of typical fibrillation potentials began to decline. If these potentials are the result of synchronization of several fibrillation potentials, then anoxia, which is the main factor in both experiments (arrest of respiration and arrest of blood circulation) could have a synchronizing effect at least during the first stages of anoxia. Other factors could also have a synchronizing effect (e. g. chloralhydrate anaesthesia, cold etc.) — (Škorpil and Hník 1958).

A. Conditions, onset and duration of denervation fibrillations

1. Conditions for appearance of denervation fibrillations

Denervation fibrillations appear exclusively after lesions of the peripheral motor neuron, independently of their localization. Fibrillations appear after lesions of the cell body, motor roots, motor nerve or nerve fibres branching to the neuromuscular endplate. They are not present in muscle atrophies of other origin. They have not been found after lesions of the first motor neuron, in atrophy following section of the tendon nor in immobilization atrophy (Tower et al. 1941, Solandt and Magladery 1942, Solandt et al. 1943, Johns and Thesleff 1961).

2. Time of onset of denervation fibrillations

Fibrillation activity does not appear immediately following the nerve lesion, but after a certain time which is dependent on a number of factors. Firstly a) on the animal species investigated. For example, in mice, fibrillations first appear 40 hrs. after nerve section, in albino rats 29—49 hrs. (Hník et al. 1958), in guinea-pigs 2—3 days, in rabbits 5 days, in the monkey 9 days and in man after 18 days (Feinstein et al. 1945). The onset in poikilothermic animals (frog) is especially slow. Reid (1941) observed fibrillation activity as late as 40—60 days after denervation. In the insect muscle (cockroach) they appear about 10—14 days after denervation and differ in size and shape from fibrillations in muscles of vertebrates (Beránek and Novotný 1959) — (Fig. 57).

Under identical conditions and in the same species the first appearance of denervation fibrillations was found to be dependent on b) body weight of the

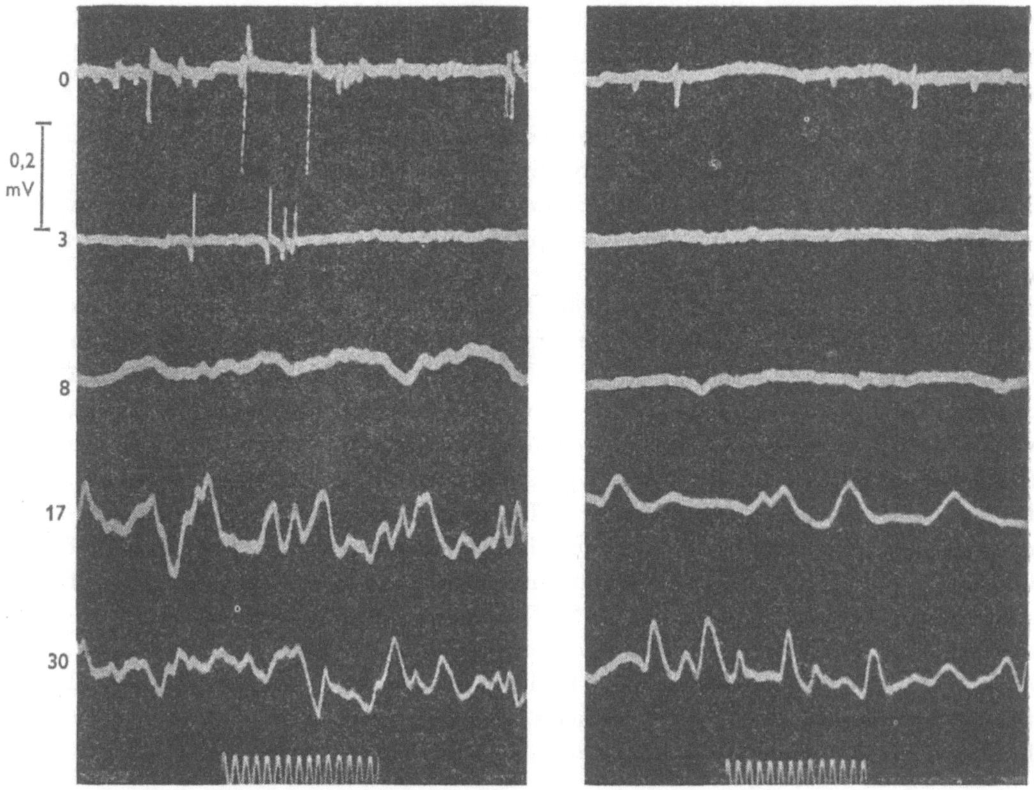

Fig. 57. The development of electrical activity in femoral (on the left) and coxal (on the right) muscles of the cockroach after denervation. Numerals on the left — time after denervation in days. Time 500 c./sec. (Beránek and Novotný 1959.)

animal. In heavier animals the onset of fibrillations is somewhat retarded. Besides this, denervation fibrillations are much more evident in small animals, e. g. rats and mice, and are also more numerous than in larger animals and man. Onset of fibrillations is also dependent on c) the type of muscle. Langley and Kato (1914—15) and Desmedt (1950) reported that fibrillations in the gastrocnemius muscle of the rabbit first appeared on the fifth day, while in the anterior tibial they appeared as late as on the tenth day after denervation.

Luco and Eyzaguirre (1955) showed in the cat that under otherwise identical conditions fibrillatory activity appeared later when the peripheral nerve stump innervating the muscle was longer and that fibrillations appeared earlier when the nerve was sectioned nearer to the muscle. The onset of muscle fibrillations is thus also dependent on d) the length of the degenerating peripheral stump. Some of these factors (*a*—c) can be brought to a common denominator, i. e. e) intensity of metabolism. Feinstein et al. (1945), for example, have reported that fibrillations appear later after thyreodectomy. On feeding animals with minced thyreoid extract, fibrillations following nerve section were more numerous and they appeared earlier.

f) If muscles are exhausted by extreme exercise immediately preceding section of the nerve, then fibrillation activity sets in earlier than in control animals. In our experiments (Hník et al. 1958) rats (30 animals) were kept running in a rotating drum for 6—8 hrs., until complete exhaustion. It was found electromyographically that fibrillations appeared earlier in these animals by about 3—5 hrs. in comparison with a control group. This result can be explained by assuming that the exhausted muscle enters the denervation period with a certain catabolic handicap and that this metabolic exhaustion accelerates denervation changes. It is possible that in these experiments, substances in the axoplasm, which according to some authors retard the onset of several denervation changes (Gutmann et al. 1955, Luco and Eyzaguirre 1955), are metabolized earlier. During these experiments it was observed that both in the control and experimental animals (exhausting motor activity prior to denervation) the onset of fibrillation activity is difficult to determine accurately, since the first signs of spontaneous activity appear about 29—36 hours after denervation and are characterized by single potentials of low amplitude (about 30 μV) and 2—3 msec. duration. In the control rats fifty-five to fifty-six hours after denervation typical fibrillations were present in all muscles. In the experimental rats, however, both these types of fibrillations appeared 3—5 hrs. earlier.

3. Duration of denervation fibrillations

If reinnervation does not take place, denervation fibrillations are still present in the muscle months and even years after denervation. However, there are no precise data available concerning this question. Tower (1939) still found denervation fibrillations in muscles of the cat a year after section of the nerve. The frequency of these potentials remained unchanged, but their amplitude decreased. According to Tower, fibrillation activity is present for as long as the muscle is capable of contracting. More is known from chance clinical examinations than from animal experiments, even though there are no detailed investigations of these questions. Marinacci (1955) stated that he registered denervation fibrillations in a muscle of a patient 54 years after acute poliomyelitis. Typical fibrillation activity was also found in paretic muscles of poliomyelitic patients 10—16 years after the onset of the disease by Janda et al. (1960). The presence of fibrillations in the denervated muscles is considered by some authors (Golseth and Fizzel 1947) as indicator of the fact that the muscle is still capable of reinnervation.

Fig. 58. Fibrillation activity in the gastrocnemius of a rat nine days after section of the sciatic nerve registered with coaxial needle from the same spot. A — immediately after insertion; B — 2 min. after insertion; C — 5 min., D — 10 min., E — 30 min. and F — 60 min. after insertion. Amplitude — 100 μV, time — 50 c./sec.

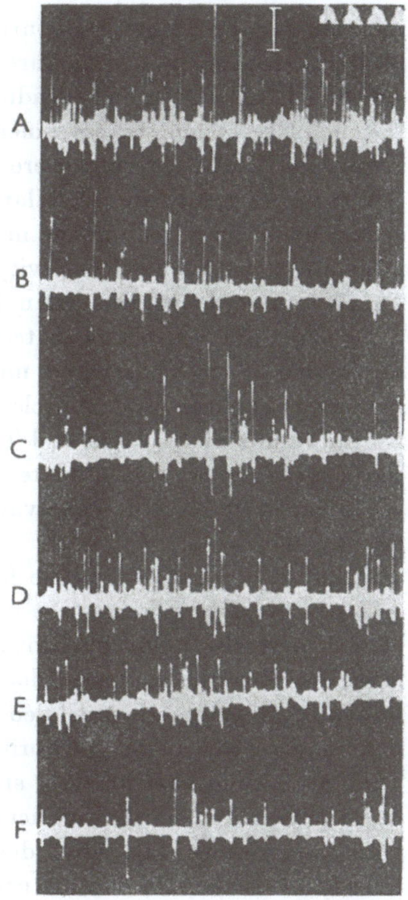

After reinnervation of the muscle fibrillation activity disappears about two weeks before action potentials reappear at this place on attempting voluntary contraction (Feinstein et al. 1945). They disappear first at the place where the nerve enters the muscle (motor point) and they gradually decrease in number at other places in the muscle. In some cases it is possible to record single fibrillation potentials in some places of the muscle even several months after reinnervation.

4. Number of fibrillating muscle fibres

According to Landau (1951), only a small percentage of fibres in the denervated muscle fibrillate. Clinical experience also indicates that in the denervated muscle there are many places where fibrillations cannot be found. Searching for fibrillation potentials is relatively simple in the rat and once found, fibrillation activity remains without substantial changes when recording with an EMG needle from the same place (Fig. 58).

B. Physiological mechanism of denervation fibrillations

The fibrillation potential is now known to be the expression of contraction of the whole muscle fibre. Langley and Kato (1914—15) and Langley and Hashimoto (1919) assumed that it is only a part of the muscle fibre which contracts. However, Eccles (1941) demonstrated that it is the whole fibre which contracts.

According to this author, the contraction starts from one focus localized anywhere on the surface of the muscle fibre. On the other hand, Hayes and Wolsey (1942) concluded that the impulse leading to the contraction of the fibre originates in the endplate region. In agreement with this were also the findings of Jarcho et al. (1954). These results were obtained by extracellular recording. Li et al. (1957) working with intracellular electrodes came to the conclusion that there are probably more foci on the muscle fibre from which fibrillation activity can originate, even though the activity of most is subthreshold at any given moment. This is in accord with the original findings of Ginetzinsky and Shamarina (1942) who, using the extracellular technique of recording, demonstrated that the sensitivity of the denervated muscle of the mouse to acetylcholine (ACh) is increased along the whole muscle fibre surface. This has lately been corroborated by the results of Axelsson and Thesleff (1959) who have shown that, although the sensitivity of the endplate region remains unchanged after denervation, the whole surface of the denervated muscle fibre acquires the sensitivity of the normal endplate.

Frank et al. (1922) were the first to point out that intravenous injection of ACh considerably increases the activity of denervation fibrillations. Small doses of ACh were also found by Brown (1937) to potentiate fibrillation activity, but large doses on the contrary blocked fibrillations. The above results were confirmed by Rosenblueth and Luco (1937) who furthermore observed that prostigmine and eserine increased fibrillations and sensitivity of the muscle to ACh. They also found that at early stages of denervation (5—6 days) ACh in large doses potentiated fibrillation activity, but did not lead to muscle contracture. Fifteen to twenty days after denervation, however, these same doses of ACh already evoke muscle contracture and block fibrillation activity. From this it was concluded by these authors that muscle contracture to ACh develops later than the response of denervation fibrillations to this substance. Relatively small doses of curare were found to decrease considerably the intravenous effect of ACh. Application of an ACh solution (1:500,000) onto the surface of the muscle with starting fibrillations leads to intense fibrillations lasting for several minutes which is preceded by a contraction of 12—20 sec. duration (Denny-Brown and Pennybacker 1938). However, it is very interesting that muscles atrophying from other causes have been found to manifest increased sensitivity to ACh even though fibrillations are not present. This has been reported in plegic muscles after section of the spinal cord (Solandt and Magladery 1942) and in immobilization atrophy (Solandt et al. 1943, Strelina 1956), although this increased sensitivity to ACh is not of the same magnitude as that after denervation.

Intravenous administration of small concentrations of KCl also increased fibrillation activity (Magladery and Solandt 1942). According to these authors it is ACh, KCl or both these aubstances which are responsible for evoking fibrillations. Denny-Brown and Pennybacker (1938) formulated a hypothesis according to which small amounts of ACh circulating in the blood cause fibrillation activity. However, free ACh is not present in the blood since it is hydrolyzed by the cholinesterase of erythrocytes and serum (Whittaker 1951). In case it is the ACh mechanism of fibrillations, it will be, most probably, endogenous ACh, i. e. ACh produced in the denervated muscle. A new aspect has been introduced by Desmedt (1950) who pointed out that for denervation fibrillations to be possible two factors are necessary: besides increased sensitivity to depolarizing substances it is the decreased accomodation ability of the denervated muscle fibre. On the other hand Mead (1947) came to the conclusion on the basis of negative results with cholinesterase that ACh is not the evoking factor of fibrillations. Recently Bergamini et al. (1955) investigated again the effect of ACh and reported a considerable increase in the frequency and amplitude of fibrillations in the guinea-pig after a dose of 0·8 γ/kg., but a block of fibrillation activity after a dose of 300 γ/kg. These authors also consider ACh as the main cause of denervation fibrillations. It is interesting to note that neither ACh nor eserine have any effect on spontaneous contractions of skeletal muscles in fish embryos prior to their innervation (Brinley 1954). In rats, however, the whole surface of the muscle fibre was found to be ACh-sensitive prior to birth (Diamond and Miledi 1959).

Although the membrane potential of a denervated frog muscle has not been found to change (Nicholls 1956), it has been reported by Lüllmann and Pracht (1957) to diminish gradually in the rat diaphragm. According to Lüllmann (1960) denervation fibrillation activity is associated with lowered K^+ permeability of the denervated muscle membrane and with the slow excursions of a labile membrane potential. The instability of the membrane potential is also considered to be the cause of fibrillation potentials by Li (1960).

C. Factors affecting fibrillation activity

1. Pharmacological

Decamethonium has a similar effect on denervation fibrillations to ACh (depolarization of the postsynaptic membrane). Small doses increase fibrillation activity, large doses cause muscle contractions and arrest of fibrillations. Succinylcholine at first increases fibrillation activity, later decreases it (Bergamini et al. 1955). Fibrillations are enchanced slightly by strychnine, veratrine, ATP and nicotine in the initial stage. Inhibition of denervation fibrillations is caused by quinine, quinidine and calcium. Histamine, papaverine, scopolamine, curare and chlorpromazine are reported to be without effect. Uncertain also was the effect of caffeine and atropine (Bergamini et al. 1955). Anaesthetics such as ether and nembutal did not affect fibrillation activity (Feinstein et al. 1945). Since ether anaesthesia which is commonly used for animal experiments has some disadvantages in longlasting experiments, some work was done with chloral hydrate anaesthesia in rats. It was found, however, that on recording from the same place in the muscle a marked diminution of fibrillation activity occurred within 10—15 min. after insertion of the needle. Generally atypical potentials also appeared. For this reason it appears to be less suitable for studying denervation fibrillations (Škorpil and Hník 1958). Langley was unable to influence fibrillation activity by injection of pituitry extract, while the application of adrenaline decreased fibrillations. On the other hand Bergamini et al. (1955) observed no effect of adrenaline. As far as desoxysterone or adrenalectomy is concerned, both were again without effect on denervation fibrillations.

2. Arrest of circulation

The effect of arrest of circulation on fibrillations was investigated by Langley and Kato (1915) and Tower (1939). Arrest of circulation caused, according to these authors, a decrease or disappearance of fibrillation activity. On occlusion of blood circulation for 10—15 min. fibrillations disappear (Tower 1939) and on renewal of circulation they recover incompletely or not at all. Mead (1947), on the contrary, denied that arrest of circulation was of any significance for denervation fibrillations.

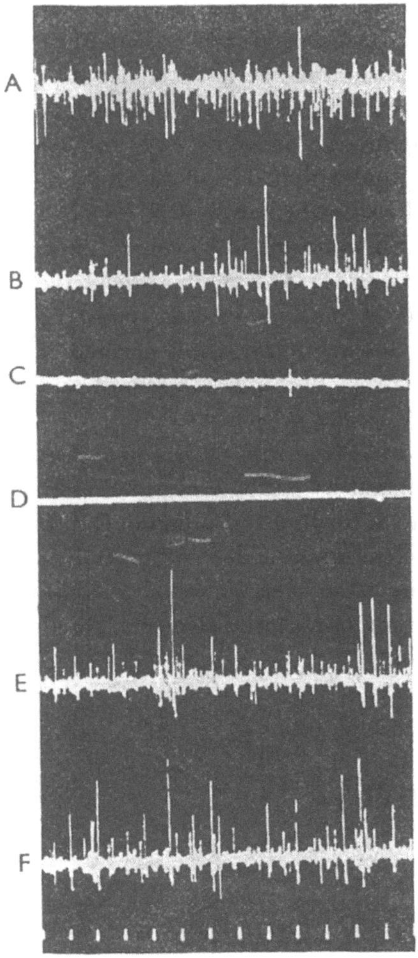

Fig. 59. The effect of arrest of circulation in the tibialis anterior on fibrillation activity in the rat 15 days after denervation. A — immediately after insertion; B — 5 min. after insertion; C — 2 min. after arrest of circulation by ligature of the abdominal aorta and v. cava caudalis; D — similar as C, 2 min. later; E — 45 secs. after release of ligature; F — 2 min. after release of ligature. Time — 50 c./sec.

When studying factors influencing denervation fibrillations by the electromyographic method, it is important to work under conditions which enable recording from the same place in the muscle over long periods of time. Otherwise there is danger that if a new recording site is sought, then this will either be without fibrillations, or, on the contrary, with higher fibrillation activity and this change in the EMG record could falsely be ascribed to the factor investigated. In the rat conditions for prolonged recording from the same place are relatively favourable (Fig. 58).

During our investigation of the effect of arresting circulation on fibrillation potentials in the rat, it was found that temporary ligation of the aorta abdominalis and v. cava caudalis leads to the complete disappearance of fibrillations within 3·3 min., and after restitution of circulation by loosening the ligature fibrillations reappear in about 20 sec. (Fig. 59, Tab. 6).

Table 6

Disappearance of fibrillation activity after ligature	n	Recovery of fibrillation after freeing of ligature	n
3·3 ± 0·33 min.	24	20·9 ± 2·82 sec.	16

n = number of measurements.

Fibrillation activity still returns after a ligature lasting for 2—3 hrs,. in a further 2—3 hrs. following renewal of circulation. Direct muscle excitability is lost about 1—2 hrs. after arrest of circulation.

If the ligation is performed at body temperatures lowered by 10°C, fibrillations disappear 2·8 times more slowly, i. e. after 7 min. (Tab. 7). Q_{10} is thus equal to 2·8, thus indicating that biochemical processes are concerned (Hník and Škorpil 1959).

Table 7

Normal body temperature			At lower body temperature		
Disappearance	Recovery	n	Disappearance	Recovery	n
2·7 ± 0·29 min.	28·0 ± 5·56 sec.	14	7·6 ± 0·91 min.	35·0 ± 10·9 sec.	12

In the human muscle fibrillations disappear reversibly during ischemia in 2½ to 4 min. and return after renewal of circulation in ½—1 min. (Janda et al. 1960).

3. Physical factors

a) Mechanical irritation due to insertion

When the EMG needle is inserted into a normal muscle, a short group of discharges appear in the EMG record which are caused by mechanical stimulation of muscle

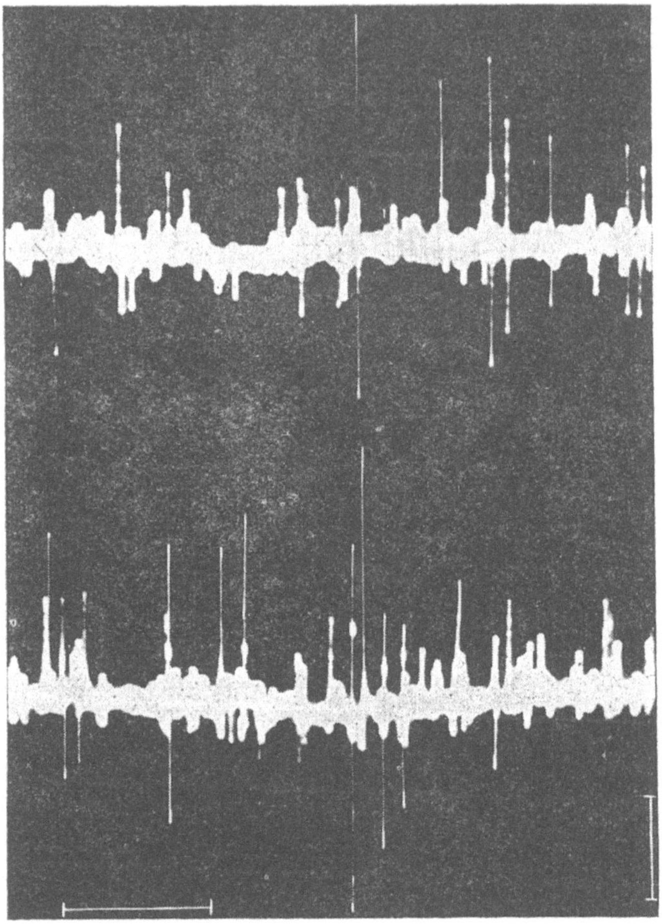

Fig. 60. Fibrillation activity in the denervated gastrocnemius of the rat following acute transection of the tendon (upper record) and during strong extension of the muscle with a 200 gr. weight (lower record). Time: 100 msec., amplitude: 100 μV.

fibres. These insertion potentials last longer in denervated muscles (even several seconds) — (Kugelberg and Peterson 1949).

b) The effect of muscle stretch

Langley and Kato (1915) as well as Tower (1939) reported that fibrillations disappear after stretching the muscle, however, these authors observed fibrillation activity visually. In our experiments it was found that denervation fibrillations were unaffected by muscle stretch or section of the tendon (Fig. 60). This possibility was already indicated by Tower (1939), i. e. that on stretching the muscle, contraction of fibres cannot be seen but electric manifestations could be normal. Thus the position of the extremity during EMG registration is of no significance (Škorpil and Hník 1958).

c) Pressure and muscle massage

According to Tower (1939) application of pressure on the muscle decreases fibrillation activity, but the dissociation of mechanical and electrical manifestations is probably similar as in the case of muscle stretch. Muscle massage is reported (Marinacci 1955) to enhance fibrillation activity.

d) Electric current

Stimulation by electric current either decreases or increases fibrillations in relation to the position of muscle fibres in the electric field (Kugelberg 1953, Landau 1951). Landau recommends the use of electric stimulation of denervated muscle in clinical work in order to aid diagnosis. Humbert et al. (1957, 1958) have described so-called provoked fibrillations during stimulation of denervated muscle which accompany muscle contraction but are independent of resting fibrillatory activity.

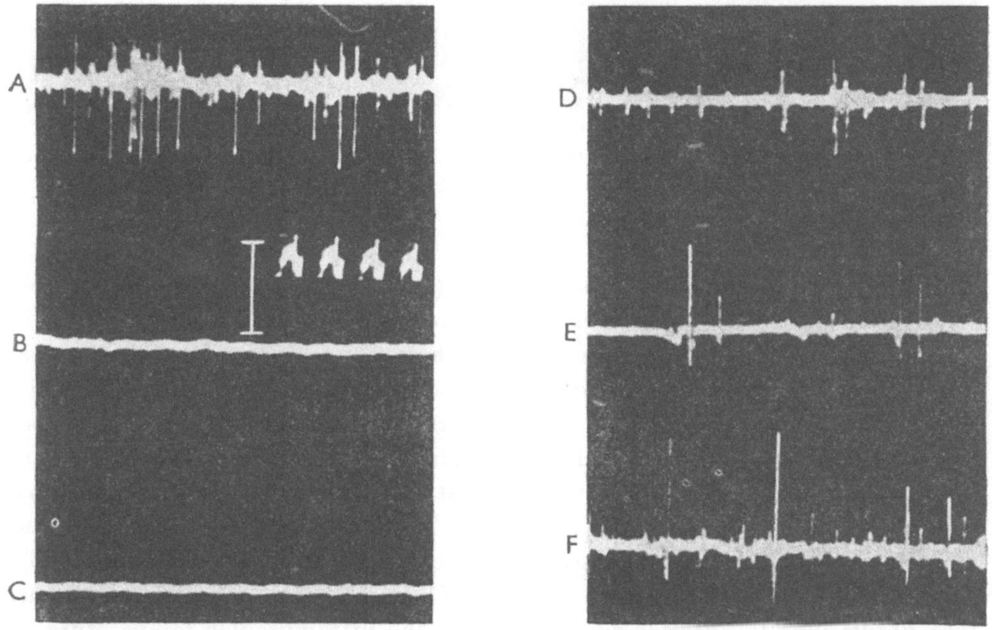

Fig. 61. The effect of lowered body temperature upon fibrillation activity. A — at 37°C, B — at 21°C, C — at 23°C, D — at 25°C, E — at 30°C, F — at 32°C. Amplitude — 100 μV. Time — 50 c./sec.

e) The effect of temperature

Langley and Kato (1915) found increased fibrillations on warming the muscle. Tower (1939) also demonstrated that on increasing muscle temperature to 45—50°C fibrillation activity was enhanced. It was not possible, however, to evoke fibrillations in a muscle which has not yet begun to fibrillate. When muscles were

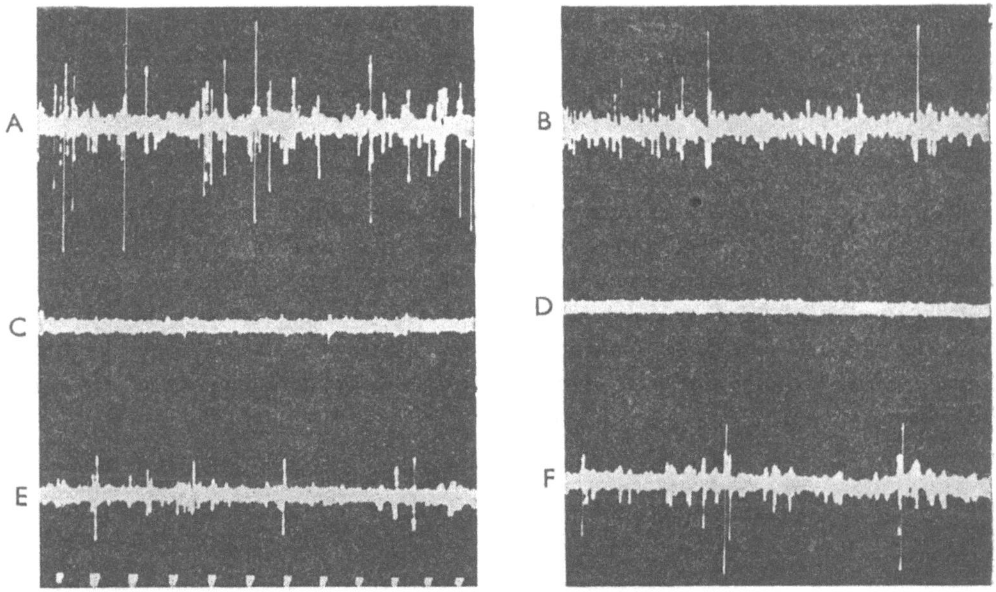

Fig. 62. The effect of increased muscle temperature on denervation fibrillations in the rat. A — at 31·5°C, B — at 38°C, C — at 45°C, D — at 46°C, E — at 33·5°C, F — at 31·5°C. Time — 50 c./sec.

heated to a higher temperature, Jarcho et al. (1958) found a reduction in atypical denervation potentials. Cooling the muscle reduced fibrillation activity until it disappeared, rewarming the muscle caused their reappearance. Muscle temperature in the rabbit lowered by 1½ to 2°C reduced denervation fibrillation (Feinstein et al. 1945). Thus according to these experiments a relatively small decrease in temperature caused considerable reduction of fibrillation activity. However, it was found in rats that fibrillations do not disappear until body temperature is lowered to 20—23°C (Fig. 61) and during heating to 41—44°C (Fig. 62). The temperature range in the rat is thus quite extensive (Škorpil and Hník 1958). Janda et al. (1959), during investigation of patients with poliomyelitis, found that fibrillations in anterior tibialis muscle, the intramuscular temperature of which is 31·7°C on the average, do not disappear until the temperature is lowered by 8°C. On warming the muscle by 3°C fibrillations are slightly enhanced, on further heating (5—7°C) denervation fibrillations return to initial values. It is thus apparent that the temperature range in man is also wider than reported by Feinstein et al. (1945).

Following death of the animal fibrillation activity disappears in a few minutes. Bergamini et al. (1955) observed arrest of fibrillations within three minutes after the heart ceased beating.

D. Relationship of denervation fibrillations to other symptoms of muscle denervation

1. Relation of fibrillations to muscle atrophy

This question was already investigated by Langley and Kato (1915—16) who postulated the hypothesis that fibrillations are responsible for denervation atrophy. In support of this view they pointed out that oxygen consumption in the denervated muscle rises (Langley and Igataki 1917). This explanation has also been accepted by Tower (1939). Levine et al. (1942) also found that administration of prostigmine increasing spontaneous denervation activity was associated with an increased degree of muscle atrophy by 47%. Atropine which has a depressant effect on fibrillations, slowed muscle atrophy by 37%. On the other hand Solandt and Magladery (1940) who depressed fibrillation activity by quinidine and did not arrest muscle atrophy, concluded that fibrillations are not the cause of atrophy. Feinstein at al. (1945) confirmed this latter conclusion, since they found atrophy in denervated muscles earlier than the first appearance of fibrillations. Neither Langley and Kato (1914—1915) nor Denny-Brown and Pennybacker (1938) found any difference between fibrillation activity in red and white muscles.

2. Relation of fibrillations to changes in muscle excitability

This question was studied by Desmedt (1950) who showed that muscle fibrillation activity is greatest at the period when symptoms of excitability changes characteristic for the denervated muscle first appear. It was on the tenth day after denervation for the gastrocnemius and 20th day for the anterior tibial in the rabbit. This was especially well expressed by the accomodation constant. This constant is about 20 in the gastrocnemius of the rabbit, and reaches values around 300 ten days after the operation. If reinnervation takes place, the accomodation ability increases again about ten days before neuromuscular transmission reappears and fibrillations disappear at about this time. Desmedt also found that chronaxie attained maximal values at the time of most abundant fibrillation activity.

3. Relation of increased oxygen consumption to denervation fibrillations

In most of the reports dealing with the metabolism of denervated muscle increased oxygen consumption was found as compared with the normal muscle (Langley and Igataki 1917, Knowlton and Hines 1934, Sato and Kasugai 1955, Sheves et al. 1956, Hudlická and Vrbová 1958 and others). Langley and Igataki consider this to be the result of fibrillation activity. However, oxygen consumption increases even in muscle atrophy of other origin in which fibrillations are not present (Harding 1926).

In conclusion it can be stated that fibrillation activity of the denervated muscle is evidently directly connected with the disintegration of motor nerve fibres in the muscle. This phenomenon is characteristic for denervation and represents a qualitative difference by which denervation muscle atrophy differs from atrophy of other origin (immobilization, tenotomy and others). Even though the explanation of spontaneous activity as the result of increased sensitivity to ACh is very attractive, several reports from the literature indicate that premature conclusions should not be made. Hines (1942), for example, reported that fibrillations disappear in close connection with reinnervation of the denervated muscle, while increased sensitivity of the reinnervated muscle to ACh remains high for another 20—30 days. Intravenous injections of cholinesterase in Mead's experiments (Mead 1947) were also found to be without effect on muscle fibrillation. Since curare does not abolish the extended receptor field to ACh in denervated muscle (Axelsson and Thesleff 1959), it seems improbable that supersensitivity to ACh is the basic mechanism of fibrillation activity (Thesleff 1960a, 1960b). On the other hand the increased sensitivity to ACh along the whole surface of a denervated muscle fibre indicates that changes in muscle excitability must be of paramount importance in determining functional characteristics of the denervated muscle, even though it is not specific for denervation only, as has been shown using botulinum toxin, where fibrillation activity was found similarly as after nerve section (Thesleff 1960b, Josefsson and Thesleff 1961). The relation between muscle excitability and supersensitivity to ACh, however, will not be a simple one, since during prolonged arrest of circulation in the denervated muscle, fibrillation activity disappears in the course of a few minutes and yet direct muscle excitability is still preserved for several hours (Hník and Škorpil 1959). This finding led to the conclusion that denervation fibrillations are dependent on the blood supply and hence on substances brought into the muscle by the blood. It is quite possible that it is primarily a question of the high oxygen consumption in denervated muscles. In that case it would be necessary to agree with Langley and Igataki (1917) that increased oxygen consumption is due to fibrillation activity. As has already been mentioned above, oxygen consumption is increased in other muscle atrophies, e. g. after nociceptive stimulation (Harding 1926) and yet no fibrillations are present. It is probable that elucidation of the fibrillation mechanism would substantially aid in solving the question of the trophic influence of the nervous system on skeletal muscle.

References

Axelsson J., Thesleff S.: J. Physiol. 147 : 178, 1959.

Baumgartner G.: Premier Congres Internat. des Sciences Neurologiques. Bruxelles 21—28 Jouillet 1957.

Beránek R., Novotný I.: Physiol. bohemoslov. 8 : 87, 1959.

Bergamini V., Ferro M. F., Ravizza L.: Arch. di Fisiologia 55 : 96, 1955.

Brinley F. J.: J. Pharm. and Exper. Ther. 112 : 257, 1954.

Brown G. L.: J. Physiol. 89 : 12P, 1937.

Buchthal F.: An Introduction to Electromyography. Scandinavian University Books, Gyldendal, København, 1957.

Denny-Brown D., Pennybacker J. B.: Brain 61 : 311, 1938.

Desmedt J. E.: Acta Neurol. Psychiat. Belgica 50 : 179, 1950.

Diamond J., Miledi R.: J. Physiol. 149 : 50P, 1959.

Eccles J. C.: Med. J. Australia 1 : 573, 1941.

Feinstein B., Pattle R. E., Weddell G.: J. Neurol. Neurosurg. Psychiat. 8 : 1, 1945.

Frank E., Nothmann M., Hirsch-Kaufmann H.: Pflügers Arch. f. d. ges. Physiol. 197 : 270, 1922.

Ginetzinsky A. G., Shamarina N. M.: Usp. sovr. biol. 15 : 283, 1942.

Golseth J. G., Fizzel J. A.: Am. J. Physiol. 150 : 558, 1947.

Gutmann E., Vodička Z., Zelená J.: Physiol. bohemoslov. 4 : 200, 1955.

Harding A. E. B.: J. Path. and Bacteriol. 29 : 189, 1926.

Hayes G. J., Woolsey C. N.: Fed Proc. 1 : 38, 1942.

Hines H. M.: JAMA 120 : 515, 1942.

Hines H. M., Thomson J. D., Lazere B.: Am. J. Physiol,. 137 : 527, 1942.

Hník P., Škorpil V.: Physiol. bohemoslov. 8 : 81, 1959.

Hník P., Škorpil V., Janda V.: Čs. fysiol. 7 : 468, 1958.

Hudli ká O., Gutmann E.: Naunyn-Schmiedebergs Arch. exp. Path. Pharmak. 234 : 501, 1958.

Hudlická O., Vrbová G.: Physiol. bohemoslov. 7 : 38, 1958.

Humbert R., Dehouve A., Laget P.: J. de Physiol., Paris 49 : 215, 1957.

— — — J. Radiol. électrol. 39 : 323, 1958.

Janda V., Škorpil V., Véle F.: Čas. lék. čes. 99 : 551, 1960.

Jarcho L. W., Berman B., Dowben R. M., Lilienthal J. L.: Am. J. Physiol. 178 : 129, 1954.

Jarcho L. W., Vera C. L., McCarthy C. G., Williams P. M.: EEG Clin. Neurophys. 10 : 527, 1958.

Jasper H., Ballem G.: J. Neurophysiol. 12 : 231, 1949.

Johns T. R., Thesleff S.: Acta physiol. scand. 51 : 136, 1961.

Josefsson J. O., Thesleff S.: Acta physiol. scand 51 : 163, 1961.

Knowlton G. C., Hines H. M.: Am. J. Physiol. 109 : 200, 1934.

Kugelberg E.: Clinical Electromyography. Progress in Neurol. Psychiat. 8 : 1, 1953.
Kugelberg E., Petersen J.: J. Neurol. Neurosurg. Psychiat. 12 : 268, 1949.
Landau W. M.: EEG Clin. Neurophysiol. 3 : 169, 1951.
Langley J. N., Hashimoto M.: J. Physiol. 52 : 15, 1919.
Langley J. N., Igataki M.: J. Physiol. 51 : 202, 1917.
Langley J. N., Kato T.: J. Physiol. 49 : 411, 1914.
Langley J. N., Kato T.: J. Physiol. 49 : 432, 1915.
Lerique J.: L'électro-myographie élémentaire chez l'homme. Thése, Paris 1944.
Levine R., Goodfriend J., Soskin S.: Am. J. Physiol. 135 : 747, 1942.
Li C., Shy G. M., Wells J.: J. Physiol. 135 : 522, 1957.
Li C.: Science 132 : 1889, 1960.
Luco J. V., Eyzaguirre C.: J. Neurophysiol. 18 : 65, 1955.
Lüllmann H.: Klin. Wschr. 38 : 1169, 1960.
Lüllmann H., Pracht W.: Experientia 13 : 288, 1957.
Magladery J. W., Solandt D. Y.: J. Neurophysiol. 5 : 357, 1942.
Marinacci A. A.: Clinical Electromyography. San Luca Press, Los Angeles 1955.
Mead S.: Arch. Phys. Med. 28 : 93, 1947.
Nicholls J. G.: J. Physiol. 131 : 1, 1956.
Reid G.: Australian J. Exp. Biol. and N. Science 19 : 199, 1941.
Rosenblueth A., Luco J. V.: Am. J. Physiol. 120 : 781, 1937.
Sato T., Kasugai F.: Tohoku J. Exp. Med. 26 : 336, 1935.
Schiff N.: Arch. f. Physiol. Heilkunde 10 : 579, 1851.
Sheves G. S., Epel'baum S. Je., Rjumina V. I.: Biokhimija 21 : 71, 1956.
Solandt D. Y., De Lury D. B., Hunter J.: Arch. Neurol. Psychiat. 49 : 802, 1943.
Solandt D. Y., Magladery J. W.: Brain 63 : 255, 1940.
Strelina A. V.: Fiziol. zh. SSSR 42 : 410, 1956.
Škorpil V., Hník P.: Plzeň. lék. sbor. 6 : 119, 1958.
Thesleff S.: Physiol. Rev. 40 : 734, 1960a.
— J. Physiol. 151 : 598, 1960b.
Tower S. S.: Arch. Neurol. Psychiat. 42 : 219, 1939.
Tower S., Howe H., Bodian D.: J. Neurophysiol. 4 : 398, 1941.
Whittaker V. P.: Physiol. Rev. 31 : 312, 1951.

Chapter VI

ELECTROLYTES IN DENERVATED MUSCLE

Drahota Z.

Electrolyte composition of denervated muscles

The mechanism of muscle atrophy is hitherto unknown. It is necessary to investigate the exact sequence of processes beginning with interruption of the neuromuscular connection and ending by disintegration of muscle fibres. This problem is difficult to solve. One of the first tasks is analysis of changes in various substances during the course of denervation atrophy which will provide fundamental data concerning the denervated muscle. Inorganic ions, especially sodium and potassium, belong to this class of substances characterizing the muscle tissue. Sodium and chloride ions can provide information concerning the extracellular fluid volume in muscle due to a primarily extracellular distribution (Hastings and Manery 1939). It is true that sodium ions diffuse into the cell. They are transported back into the extracellular fluid, however (see Ussing 1958, Conway 1958). Potassium, on the other hand, is retained primarily in the intracellular fluid and represents the osmotic medium of intracellular fluid, being the main intracellular ion. These ions play a very important rọle in muscle metabolism and its functional activity. Potassium ions can affect various enzymatic reactions in the metabolism of proteins (Iacobelis et al. 1956), carbohydrates (Hastings et al. 1956) as well as lipids (Geyer et al. 1953a, 1953b). According to Szent-Györgyi (1952) potassium ions are indispensible for maintaining myosin in colloidal state, and Hodgkin and Horowitz (1959) have shown the significance of equilibrium between K and Na ions for conduction of the electric stimulus along the muscle fibre.

Electrolytes are thus very important for metabolism and muscle cell function, and for this reason considerable attention has been paid to them in studies of denervation atrophy. The following questions were first studied: 1. How does the extracellular space and fluid change after denervation? 2. What are the osmotic conditions like inside the cell following denervation? 3. Whether after denervation, mechanisms maintaining concentration gradients of Na and K ions between the extra- and intracellular fluid are disturbed or not.

By answering the above questions it will be possible to interpret more precisely metabolic changes in the denervated muscle: 1. The relation between volume of extracellular fluid, which supplies the cell with all indispensible compounds and substrates, and intracellular fluid volume. 2. Whether osmolarity of intracellular fluid which could affect colloidal and chemical state of cellular proteins is not altered by denervation. 3. Whether denervation does not change the intracellular content of K and Na, or their ratio, since these changes could explain altered activity of various enzymatic systems.

A. Water and extracellular ions in the denervated muscle

1. Changes in water content after denervation

Electrolytes in animal tissues cannot be studied isolated from the total water content. In most cases it is more important to know the cellular concentration of ions than their total content.

There is considerable disagreement in literature concerning changes of water in the denervated muscle. Lipschütz and Audova (1921) reported a slight increase in the water content of denervated muscle. A small increase was also found by Hines and Knowlton (1937) and Fenn(1937). According to Fenn this increase in water is due to vasodilatation and increased capillary permeability. On the other hand Chen et al. (1924) did not find any changes in water content of rabbit muscles nine weeks after denervation. Similarly Chor et al. (1937) found no changes of water either in the rabbit or monkey even ten weeks following muscle denervation. No changes in denervated muscles of the cat could be found thrce weeks after section of the nerve by Westenbrink and Krabe (1936). In recent experiments, however, Humoller et al. (1950b) demonstrated that the content of water does increase in denervated rat muscles. Sutfin et al. (1954) also found a slight but significant increase of water in denervated anterior tibial, gastrocnemius, soleus and rectus femoris muscles of the rat. Increased content of water in denervated muscles of puppies was reported by Eichelberger et al. (1956) and in rats by Beránek et al. (1957) and Drahota (1959).

Differences between individual findings are evidently caused by the fact that the water content is not calculated per fat free dry muscle weight. Following denervation, an increased lipid content was demonstrated by Humoller et al. (1952) and Sutfin et al. (1953), so that changes in water, which are relatively small in comparison to the total water content, can be partially or even completely masked by increased lipid content per unit weight of denervated muscle. Controversy in the literature can be explained by the fact that the water content in various muscles of the same animal does not begin to rise at the same period after nerve section (Drahota et al. 1959).

Hydratation and dehydratation of muscles following osmotic load were investigated by Hofmann and Wertheimer (1927). These authors administered a hyper- or hypotonic solution intravenously and demonstrated that the denervated muscle loses water more quickly during dehydratation than the normal muscle. The absorption of water by the denervated muscle after adminostration of a hypotonic solution was found to be minimal in comparison with the normal muscle. On the other hand, Gutmann and Vrbová (1952) demonstrated that following an intravenous injection of distilled water the decrease of % of dry weight is the same in normal and denervated muscles.

2. Changes of sodium and chloride, and calculations of extra- and intracellular water

The determination of total water content is only a superficial indicator of water changes in muscle tissue. It is necessary to answer the question, whether the increased water content in the denervated muscle means an increase of intracellular water, or whether changes in water are not caused solely by an increase of extracellular volume. An increase of extracellular space in the denervated muscle can be deduced from increased content of extracellular Na and Cl ions. The increase of Na and Cl ions in the denervated muscle, however, could also be caused by entry into the cell. In order to answer the question of whether it is extra- or intracellular water which changes in the denervated muscle, it must be clear whether or not Cl ions do not enter the denervated muscle cell. Or it would be necessary to use another substance for extracellular space determinations, the extracellular distribution of which is not affected by metabolic disturbances after denervation, i. e. for example inulin or sucrose.

The rise in sodium from the fifth day after denervation was demonstrated by Humoller et al. (1950a). In our experiments we found an increase of Na ions in the anterior tibial of the rat commencing on the third day after nerve section (Drahota 1959). Fischer (1949) described analogous changes in the rabbit gastrocnemius six days after denervation and Eichelberger et al. (1956) found an increased Na content in denervated muscles of puppies. An increase of Na has also been demonstrated in muscle atrophy of other origin than denervation: after section of the tendon and immobilization (Fischer 1949, Eichelberger et al. 1958), in E avitaminosis (Fenn and Goetsch 1937, Morgulis and Osherhoff 1938) and in muscle dystrophy (Horwath et. al. 1955 Young et. al. 1959, Baker et. al. 1958).

An increase of Cl in denervated muscle was found by Hines and Knowlton (1937), who pointed out that the content of chloride and collagen rose in parallel. Fischer (1952) reported an increase of chloride in rabbit muscle, but presumed that chloride enters into the denervated muscle cell since extracellular sodium calculated from the chloride space was higher than would correspond to total sodium found in the muscle (v. Fischer 1955). In our experiments increased chloride content was present in the rat muscle three days after denervation (Drahota 1959, Drahota et al. 1959).

Chloride is one of the most commonly used substances for determining extracellular fluid in the muscle (see Manery 1954). However, chloride ion is not very suitable for measuring extracellular space in the denervated muscle, since the loss of connective tissue and intracellular muscle phase following denervation do not run in parallel. The decrease of the intracellular muscle phase is considerably faster than that of connective tissue, and in consequence there is a relative increase of intercellular connective tissue in the denervated muscle. This latter contains non-diffusible chloride (Manery and Danielson 1938) which means that the content of chloride in the denervated muscle per unit weight is higher than corresponds to extracellular fluid. In spite of this, however, chloride can give at least an approximate idea of extracellular fluid volume in the denervated muscleand is currently used (see Eichelberger et al. 1956).

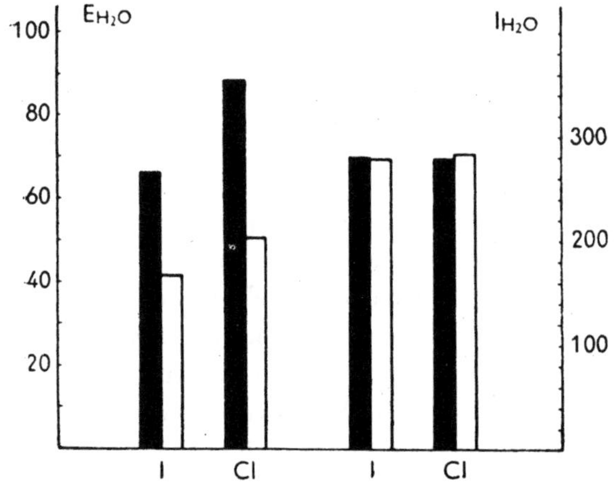

Fig. 63. Extra- and intracellular water in the normal tibialis anterior of the rat and in the muscle 10 days after denervation. Black columns — denervated, white columns — normal muscle. I (Cl) — extra- or intracellular water calculated with the aid of inulin (chlorides); see Drahota (1959).

More precise data concerning extra- and intracellular fluid in the denervated muscle can be obtained with inulin. Its content in muscle is directly proportional to extracellular water and this relation is not in any way affected by changes in the ratio muscle phase: connective tissue phase, as in the case of chloride. Another advantage of inulin lies in the fact that its extracellular position, due to its macromolecular structure, cannot be affected to the same extent by changes in the cellular membrane with chloride or sodium. For this reason extra- and intracellular water in the tibialis ant. ten days after denervation was determined with inulin. Changes in extra- and intracellular water were then compared with extra- and intracellular sodium and chloride water determined under the same experimental conditions. Extra- and intracellular water changes were studied under conditions when atrophy, measured as the decrease of fat-free dry muscle substance, was 40—45%. The results are presented in Fig. 63, in which the content of extra- and intracellular water in normal and denervated muscle is given. Absolute values of inulin and chloride extracellular water in normal muscle agree with published data (see Cotlove 1954). Inulin extracellular water is only equal to 80% of chloride extracellular water. These differences are primarily the result of the fact that intracellular collagen tissue contains nondiffusible Cl which is not in osmotic equilibrium with the extracellular fluid. These differences become even more marked after denervation. Extracellular inulin water in denervated muscle was increased by 52%, while extracellular chloride water was increased by 75% (Drahota 1959).

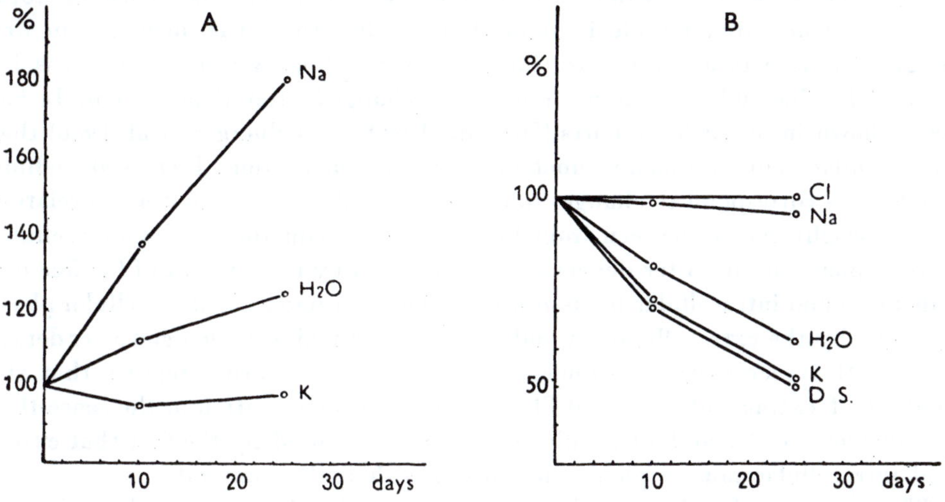

Fig. 64. Changes in the content of water and electrolytes in the extensor digitorum longus of the rat following denervation. A — the content of electrolytes and water expressed relatively, i. e. m. equiv. (ml.)/100 g. fat-free dry substance in the normal and denervated muscle; B — the content of electrolytes and water expressed absolutely, i. e. m. equiv. (ml.)/muscle; % — change expressed as percentage of the normal muscle.

Intracellular water calculated on an inulin basis, as well as with chloride, is quite the same in normal and denervated muscle (Fig. 63). The disadvantages of chloride extracellular water evidently do not come into play in this case since absolute values of total water in the muscle are very high. It can thus be concluded that changes of water which occur after denervation are changes in extracellular water.

The above results indicate that the content of extracellular Na, Cl and water rises in denervated muscle if m. equiv. (ml.) Na, $Cl(H_2O)$/100 g. fat-free dry substance in the normal muscle is compared with that of the denervated muscle. Evaluation of ion and water changes need not be carried out by this commonly used method, and the ion composition of normal and denervated muscle can be evaluated in absolute units. Comparison can be made between total Na, Cl and water of control and denervated muscle. When this is done, it will be found that only the intracellular muscle phase decreases after denervation, while the extracellular phase remains almost unchanged (Drahota 1960a). These results are summarized in Fig. 64, in which changes of ion composition in the extensor digitorum longus of the rat are given.On evaluation of results in relation to dry weight, data corresponding to reports in the literature are obtained (Fig. 64A). Changes of ions in absolute units, however, demonstrate that the total content of sodium and chloride (μ^1. equiv. Na, Cl per muscle) remains unchanged after denervation (Fig. 64B). Potassium in denervated muscle decreases at the same rate as fat-free dry solids. Decrease of total water in denervated muscle is somewhat smaller than the decrease in dry weight, because total water also includes extracellular water which, as is indicated by Na and Cl content, remains unchanged after denervation. It has been shown in an earlier reports (Drahota 1960b, Eichelberger et al. 1956) that intracellular sodium remains unchanged after denervation. Increased sodium would thus correspond to changes in extracellular fluid on evaluation in relation to dry weight. As can be seen from Fig. 64, only a reduction of the intracellular muscle phase occurs after denervation (evaluated by the amount of fat-free dry substance and intracellular potassium). On the other hand the extracellular phase (judging by the extracellular Na and Cl content) remains unchanged after denervation. It is necessary to supplement conclusions of earlier reports that the content of extracellular Na and Cl increases after denervation in the sense that the increase of Na and Cl is only relative and is caused by the fact that extra- and intracellular muscle phases do not diminish at the same rate.

The turnover of sodium in denervated muscle has been relatively little studied. Hudlická (1955) measured the uptake of Na^{24} into various skeletal muscles and found that the Na^{24} content 30 min. after equilibration is different is various muscles. Uptake of Na^{24} was higher in the soleus than in the extensor digitorum longus. Three days after denervation the uptake of Na^{24} in the soleus was lower, while it was increased in the extensor digitorum longus. The difference between normal muscles are evidently caused by the fact that the extracellular space in the soleus is greater than that of the extensor digitorum longus (Drahota et al. 1959). It was also calculated that the content of intracellular sodium is higher in the soleus (Drahota 1960c). Following denervation, the comparison of ion composition in these two muscles is still more complicated. Extracellular fluid increases in both muscles (measured by Cl content), i. e. including extracellular sodium, calculated per 100 g. fat-free dry weight, this increase, however, being considerably slower in the soleus than in the extensor digitorum (Drahota et al. 1959). The difference in Na^{24} uptake into various denervated muscles evidently corresponds to different course of changes in extra- and intracellular sodium in various denervated muscles.

These results demonstrate the importance of evaluating results obtained in different muscles carefully, particularly during the early denervation period. Denervation changes in different muscles can appear at different times after denervation and progress at different rates, even though the general character of changes is usually the same.

3. Changes of Na, Cl and water during reinnervation

It is very important to know the way in which the altered relation between extra- and intracellular muscle phases returns to normal after recovery of innervation and motor function. As yet, this question has not been dealt with in the literature. For this reason investigation was made of changes in sodium, chloride,

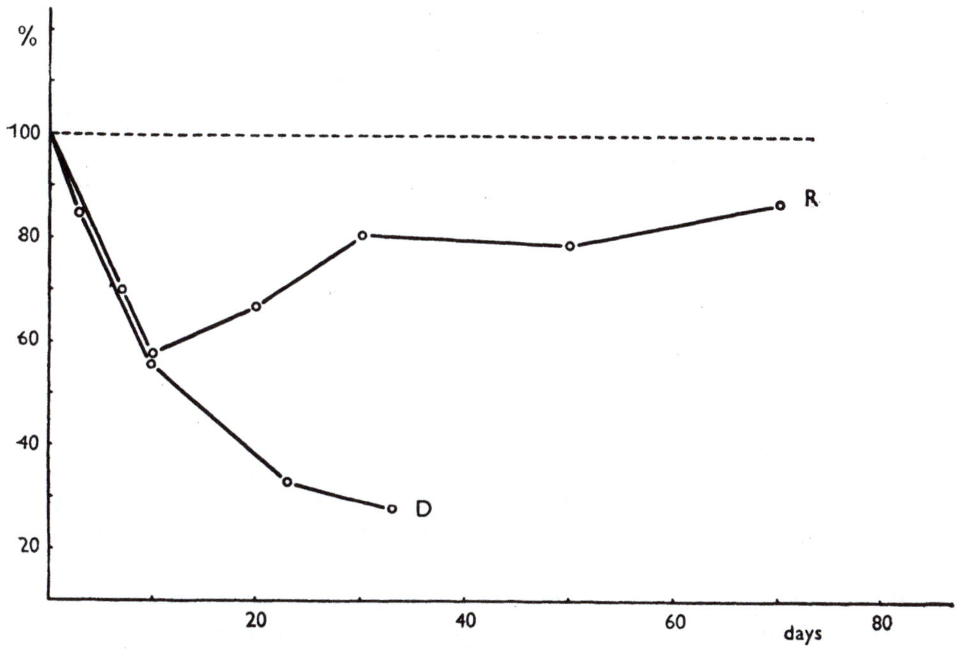

Fig. 65. Changes in fat-free dry substance in the tibialis anterior of the rat following section of the sciatic nerve (D) and after crushing of the sciatic nerve (R).

water, potassium and fat-free dry muscle substance at various time intervals after crushing the nerve (Drahota 1960b). Following crushing of the nerve, temporary interruption of the neuromuscular connection takes place. Innervation is renewed in the anterior tibial muscle about ten days after crushing the sciatic nerve (Gutmann and Kučerová 1949). The recovery of indirect excitability of the tibial muscle appears about the 14th day after nerve crushing. Muscle atrophy is characterized by changes in dry muscle weight (Fig. 65). The decrease in dry muscle weight following nerve crushing progresses at the same rate as after nerve section. This period can thus be considered as the denervation period. From the tenth day onwards, when the first nerve fibres begin to regenerate into the muscle, recovery of muscle weight begins. Weight of the reinnervated muscle increases very rapidly between 10 and 30 days after crushing the nerve. However from the thirtieth day on, the reinnervated muscles only change very slowly as compared with the weight of the contralateral control muscle. Not even 70 days after nerve crushing was muscle weight fully recovered. Hines (1942) also found only incomplete restitution of muscle weight 84 days after crushing the nerve, and Bajusz (1959) not even after five months.

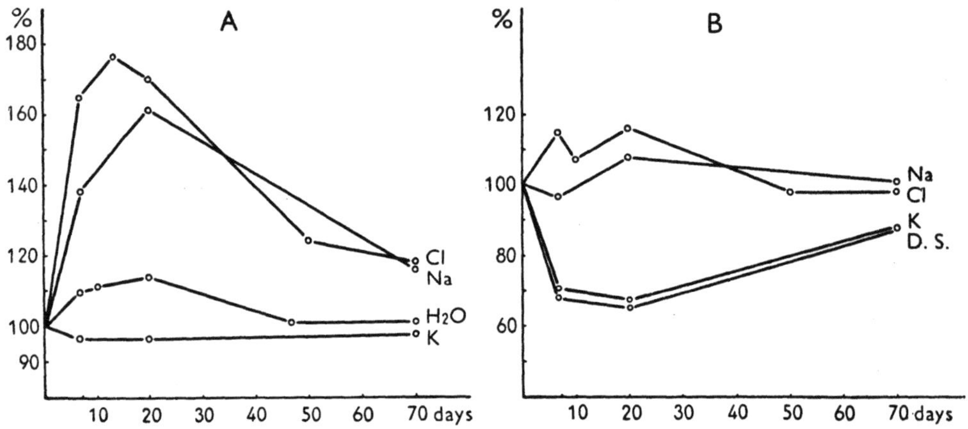

Fig. 66. Changes in the content of electrolytes and water in the tibialis anterior of the rat during reinnervation following crushing of the nerve. A — the content of electrolytes and water expressed relatively, i. e. m. equiv. (ml.) /100 g. fat-free dry substance in the normal and denervated muscle; B — the content of electrolytes and fat-free dry substance (D. S.) is expressed absolutely, i. e. m. equiv. (ml., mg)/muscle; % — change expressed as percentage of the normal muscle.

Results ofo in and water estimations (Fig. 66A) were evaluated in relation to dry weight (m. equiv. ml./100 g. dry weight) and in absolute units (μ.equiv./muscle). The former calculation leads to the conclusion that Na and Cl and water per 100 g. fat-free dry substance are increased during the denervation period after nerve crushing, which is analogous to that after nerve section. From the twentieth day after nerve crushing, sodium and chlorides begin to return towards normal values, but do not attain values of the normal muscles even 70 days after crushing of the nerve. The calculation of extra- and intracellular water using the chloride space (Drahota 1960b) indicate that, as after denervation, the increased content of sodium during reinnervation is due to changes in extracellular fluid.

On evaluation of ion changes in the reinnervated muscle in absolute amounts, it can be seen that, as after denervation, the increase of Na and Cl ions is only relative. Extracellular space, calculated from Na and Cl contents (m. equiv. Na, Cl/muscle) during reinnervation is quite the same as that of the contralateral control. Intracellular phase calculated from the fat-free dry substance and the content of potassium (mg. or m. equiv./muscle) first decreases and after renewal of innervation again increases (Fig. 66B). Na and Cl changes calculated in relation to dry weight (m. equiv./100 g. dry weight) therefore exhibit a relative increase of the extracellular muscle phase as compared with the intracellular phase during denervation period, and relative diminution of extracellular space in the period following reinnervation when the muscle regains its weight.

B. Intracellular ions in the denervated muscle

1. Changes of Mg and Ca ions in the denervated muscle

Magnesium and calcium ions play an important role in the regulation of cellular metabolism. Mg ions are indispensible for normal oxidative phosphorylations and can, like Ca ions, affect ATP-ase activity. Ca ions have an antagonist effect to potassium in various reactions, e. g. potassium contractions. Evidently these ions are very important for muscle metabolism and have therefore also been studied in denervated muscle.

Calcium was investigated in the denervated muscle by Hines and Knowlton (1933, 1937) and Shimizu and Kuriaki (1960). They found that the content of calcium in the denervated rat muscle increases. Similarly Humoller et al. (1950a) demonstrated that, although Ca ions in rat muscles are unchanged ten days after denervation, there is already a considerable increase by the 20th day. Increased calcium was also reported by Eichelberger et al. (1956) in denervated muscles of puppies.

It has not yet been demonstrated, however, whether it is intracellular calcium or Ca in intercellular connective tissue which increases after denervation. Usually the increase of calcium in denervated muscles is discussed in connection with the results of Fenn and Goetsch (1937) and Morgulis and Osterhoff (1938) who found a considerable rise of calcium in dystrophic muscles. These authors correlated the increase of calcium with histologically observed calcification and formation of fibrous tissue.

Even though it has not yet been decided whether calcium increases in the muscle cell or the connective tissue, it is obvious that the increase of calcium is not the cause but the result of denervation, since calcium changes do not occur until later, when denervation atrophy is already considerable advanced.

Changes of magnesium ions after denervation have also been investigated in several reports. All these come to the same conclusion that magnesium per unit weight of denervated muscle remains unchanged in the course of denervation (Humoller et al. 1950a, Fischer et al. 1952, Eichelberger et al. 1956). It can thus be concluded that magnesium apparently does not participate in the initiation of muscle atrophy.

2. Changes in potassium content immediately after denervation

Potassium changes in the course of denervation atrophy can be divided into two phases: 1. changes which occur shortly after denervation, i. e. up to the third or sixth day after section of the nerve; 2. changes which occur at a later period of advanced atrophy, i. e. about 10 to 30 days after denervation, when atrophy reaches about 70% (Fig. 65). Longer denervation periods are not usually studied experimentally, since at later stages already considerable morphological changes occur, thus making accurate determinations more difficult. Classification of potassium changes into two phases is justified, since changes in potassium content shortly after denervation are opposite in character to changes found in late denervation.

Fenn (1937) reported an increase of potassium in the rat gastrocnemius shortly after denervation. In other muscles investigated (anterior tibial, biceps femoris, semimembranosus), however, Fenn either did not find any increase in potassium at all or only a small increase. Later this finding of a transient increase of potassium was confirmed by Fischer (1949). One day after section of the nerve, the potassium content in denervated rabbit gastrocnemius was increased by 6% on the average. This temporary increase disappeared about the sixth day after denervation.

However, other authors (Hines and Knowlton 1933, 1937, Humoller et al. 1950a) did not find this increased potassium content after denervation even though they used the same muscle for investigation as Fenn. Then differences between the results of Fenn and Humoller et al. could perhaps be explained by the fact that while the former author calculated potassium on the basis of dry muscle weight, Humoller et al. used wet weight. Since the water content in the muscle increases, a rise in potassium in the experiments of Humoller et al. could be masked by this increase of water. However, Hines and Knowlton (1933) calculated potassium in wet muscle substance as well as in dry substance (1937) and yet no increase in the content of potassium was found in the rat gastrocnemius three days after denervation.

In our experiments the question of a temporary potassium increase after denervation was also investigated (Drahota 1960c). An attempt was made to determine whether the transient rise of potassium occurs only in some muscles or whether it is a more general phenomenon the course of which differs in various muscles. Changes of potassium were therefore investigated in various types of muscle: in physiological flexor muscles (anterior tibial and extensor digitorum longus), in an extensor muscle (soleus) and in the diaphragm. The diaphragm was included because it is a special kind of muscle which, because of its continual rhythmic activity, is more like heart muscle than skeletal muscles. As can be seen from the results in Fig. 67, the transient increase of potassium shortly after denervation is a general phenomenon, since it is present in all the muscles studied.

The temporary increase of potassium, however, differs in individual muscles in amplitude and duration. This increase is longest in duration in the denervated diaphragm. Approximately the same magnitude of change, but of shorter duration was found in the soleus. The smallest increase of shortest duration was found

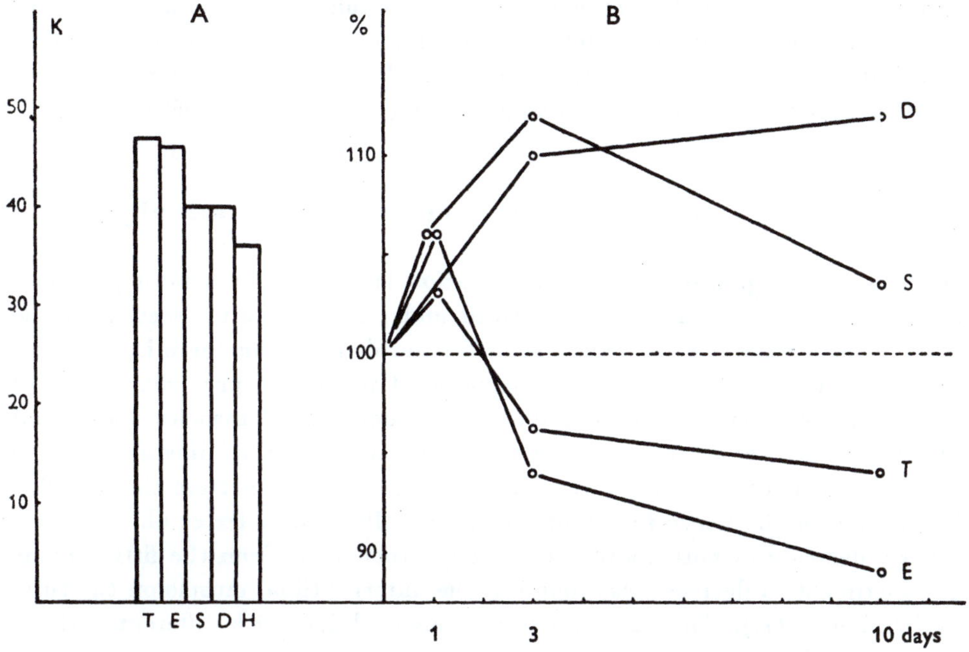

Fig. 67. The content and changes in potassium in different muscles after denervation.
A — the content of K (mequiv./100 g. fat-free dry substance) in the tibialis anterior (T), extensor digitorum longus (E), diaphragm (D) and heart muscle (H); B — changes in K ions following denervation; % — change expressed as percentage of the normal muscle (see Drahota 1960c).

in the denervated anterior tibial and extensor digitorum longus. Values of total potassium in anterior tibial, extensor digitorum longus, soleus and the diaphragm from 12 normal rats is given for the sake of comparison. These muscles were further supplemented with values of potassium content in heart muscle (Fig. 67A). It can be seen that the potassium content is different in various muscles under normal conditions. The highest potassium content was found in physiological flexor muscles (anterior tibial and extensor digitorum longus), it is somewhat lower in the soleus (physiological extensor) and in the diaphragm, and lowest potassium content was found in the heart muscle.

The increase of potassium is thus higher and of longer duration in muscles which contain less potassium under normal conditions. It is not yet possible, however, to answer conclusively the question of the mechanism underlying these differences. The differences in the duration and extent of the increase are evidently not caused by a different rate of atrophy, since extensor digitorum longus and soleus, which differ as far as the increase and the duration of potassium increase is concerned, atrophy at a similar rate (Hník 1955). Our experiments rather indicate that differences which exist between various muscles under normal conditions tend to level out after denervation. The transient increase of potassium is higher in muscles which contain less potassium under normal conditions. A similar tendency was also observed during investigation of Na^{24} and P^{32} uptake into various denervated muscles (Hudlická 1955) and also of changes in chlorides, water and glycogen (Drahota et al. 1959).

3. Potassium changes in the course of denervation atrophy

The increase of potassium after denervation lasts only for a brief period of time. Relatively soon, between the third and tenth day after section of the nerve, it disappears and in the course of further denervation there is, in agreement with most authors, a gradual decrease of potassium per unit weight of the denervated muscle. Loss of potassium in denervated muscles was found in 1932 by Leurier et al. and can be demonstrated according to Hines and Knowlton (1937) from the third day after denervation. It is evident from the fifth day after section of the nerve in rabbits (Fischer 1949). Humoller et al. (1950) reported a decrease of potassium in the rat gastrocnemius from the fifth day, in our experiments a decrease was found in the anterior tibial muscle on the third day (Drahota 1957). The exact course of changes differs with different authors, but all agree that the content of potassium per unit weight of denervated muscle is lower than that in the normal muscle. The only exception is Bogatzki et al. (1952) who did not find any decrease of potassium in denervated rat muscles 40 days after nerve section. These authors investigated the content of potassium in group experiments in the left normal muscle and left denervated muscle. It was their intention to eliminate any possible error which could originate by considering the non-denervated extremity as normal. However, these group experiments are not advantageous since the biological dispersion of potassium values between individual animals is considerable (the mean for 50 animals was $360{\cdot}6 \pm 39{\cdot}56$ mg%, from 296·8—458·9 mg.%), while changes after denervation are relatively small, being about 10—20%.

The content of potassium in denervated muscles was also studied by several authors in the frog. Cicardo and Guererich (1942) reported a 28% decrease in potassium in the denervated frog gastrocnemius (calculated per wet muscle weight). According to Bogatzki and Fischer (1951) potassium in denervated frog muscles (Oberschenkelmuskulatur) decreases within five days after denervation. On the other hand, Harris and Nichols (1956) did not find any changes in potassium in the sartorius muscle of the frog even 30 days after denervation.

Thus potassium in the denervated muscle decreases in the course of denervation. In spite of this, many authors doubt whether the content of intracellular potassium actually decreases in muscle cells (Hines and Knowlton 1937, Tower 1939, Fischer 1949, Humoller et al. 1950a). The fall of potassium in the denervated muscle could only be apparent and actually be the result of a relative increase of collagenous tissue and a decrease of the cellular phase itself per unit weight of the denervated muscle. It was Fischer and Ramsey (1946) who found that following denervation the loss of muscle phase takes place at a considerably faster rate than that of the intercellular connective tissue, so that a relative accumulation of collagen takes place in the muscle. Since connective tissue contains practically no potassium ions (Manery and Hastings 1939), the decrease of potassium per unit weight of denervated muscle could be explained by these changes. A unit weight of denervated muscle contains less muscle cells rich in potassium and more connective tissue cells without potassium, than normal muscle. Potassium concentration in dry and wet muscle weight cannot thus be compared without reserve. Fischer (1955) is the only author who has made this correction when determining K in the denervated muscle, by calculating the concentration of potassium in normal and denervated muscles in the noncollagenous muscle phase. According to Fischer, the concentration of potassium in the denervated rabbit gastrocnemius decreases after the 20th—30th day.

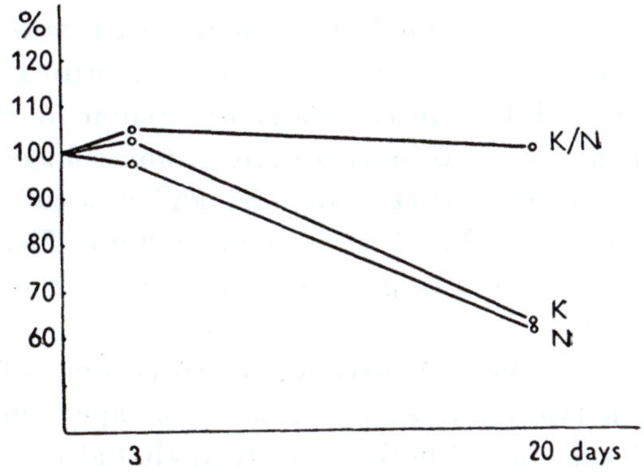

Fig. 68. K/N ratio in the rat extensor digitorum longus muscle. K — mequiv. K/muscle; N — non-collagenous protein nitrogen in mg.; % — change expressed as percentage of the normal muscle.

Decrease in the concentration of potassium in rat muscles, however, can be explained by the relative increase of collagenous tissue, since potassium changes are very small as compared with those in the rabbit (Fischer 1949).

In our experiments the potassium content in normal and denervated muscles was also corrected on the basis of the assumption that the absolute amount of collagen remains unchanged after denervation (Fischer and Ramsey 1946). It was found that the decrease of potassium in the denervated muscle can be explained by the relative increase of connective tissue up to the thirtieth day after denervation (Drahota 1957). Not even after calculation of the potassium concentration in non-collagenous muscle phase does the result represent accurately the true situation, however. This correction is based on the assumption that the total collagenous tissue remains unchanged after denervation. Fischer and

Ramsey (1946) pointed out that the increase in collagen per unit weight of denervated muscle is somewhat smaller than would correspond to the assumption that the amount of collagen in the denervated muscle remains unchanged. This was later confirmed by the work of Kolínská (1958) who showed that the total collagen content in denervated muscle is slightly reduced. For this reason changes in potassium in denervated muscles were determined together with non-collagenous protein nitrogen which gives information concerning the intracellular muscle mass. Potassium in normal and denervated muscles was then expressed in μ.equiv.K/g. non-collagenous protein nitrogen. From Fig. 68 it is evident that potassium and non-collagenous protein nitrogen in absolute units are reduced to a similar extent. The concentration of potassium in non-collagenous nitrogen thus remains unchanged during denervation. (Drahota 1960a).

Since the concentration of potassium in the muscle cell does not change after denervation, it is not necessary to expect any changes of K during reinnervation. Results in Fig. 66B demonstrate that after crushing the nerve no changes in potassium occur in the muscle either during the denervation period or after reinnervation. The decrease of potassium in absolute units corresponds exactly to the reduction of fat-free dry weight. Similarly, after renewal of innervation, muscle weight recovers in parallel with potassium.

4. Potassium turnover in denervated and reinnervated muscles

Potassium and sodium can diffuse freely across the cell membrane in both directions. The high potassium and low sodium content in intracellular fluid must therefore be maintained against a concentration gradient. The total content of both ions inside the cell is thus the result of a dynamic equilibrium of K and Na along and against the concentration gradient. The rate at which K and Na ions move passively, i. e. by physico-chemical diffusion, is dependent on the concentration difference of both ions on either side of the membrane and on its permeability. The rate at which K and Na ions move actively, i. e. against the concentration gradient is dependent primarily on the activity of transport mechanisms. There are principally two possibilities how an animals cell transports K and Na ions against concentration gradients: 1. mechanisms of the so-called sodium pump (v. Ussing 1948, v. Conway 1948) which "actively" transport Na ions only out of the cell the movement of K ions being the result of electrochemical forces; 2. mechanism by which both ions are "actively" transported either by a common system for the transport of both ions, or by separate systems by which each ion is transported separately against the concentration gradient (Glynn 1956, Scott and Hayward 1954).

As yet, little is known about the function and still less about the enzymatic basis of mechanisms by which ion concentration gradients are maintained. It is also surprising that one of the fundamental properties common to practically all animal cells — ability to maintain the concentration gradients — would be realized in different tissues of the same organism by different systems. Harris (1957) concluded his report on transport through biological membranes by saying that for interpretation of ion kinetics more fundamental knowledge will have to be accumulated concerning their localization in the cell, as well as their exchange.

Investigation of ion exchange in the denervated muscle is very difficult, because as yet the central question of ion exchange have not been solved in normal muscle. Even though it is not yet quite clear whether it is only sodium which is actively transported across the muscle cell membrane, it is generally agreed that the sodium concentration inside the cell is minimal, since sodium which diffuses into the cell is immediately extruded out of the cell. For this reason, the chief interest in denervated muscle has centred on potassium turnover. The movement of potassium into the cell, active or passive, can serve as indicator of the cell's ability to maintain the ion concentration gradient.

In the literature, attention has focussed primarily on the following questions: 1. whether free and bound potassium undergoes any changes after denervation; 2. what is the uptake of K^{42} into denervated muscles; 3. what is the capacity of denervated muscle to maintain the potassium concentration under different conditions (e. g. osmotic dialysis, effect of drugs, anoxia etc.); 4. whether the denervated muscle is able to renew a reduced content of potassium.

Practically all potassium in the normal muscle is exchangeable for the radioactive isotope (Noonan et al. 1941b). Recently Fischer (1950) and Lenan (1957) have demonstrated that about 14% of potassium in the muscle can be considered as bound.

It is not yet known whether the amount of exchangeable potassium changes after denervation, or not. Following perfusion of a muscle with a potassium-free solution, Neuschlosz and Trelles (1924) found that the amount of unexchangeable (bound) potassium increases after denervation. On the other hand Fischer (1950), using isotopic K^{42} for determination of free and bound potassium concluded that "a small fraction of potassium which does not participate in K^{42} exchange decreases somewhat". Lyman (1942) also came to the conclusion that following denervation there is a decrease of non-exchangeable, i. e. bound potassium. According to this author the uptake of K^{42} into the rat gastrocnemius is increased 8—30 days after denervation. This increased uptake is caused by the fact that the denervated muscle contains less bound potassium, so that the rate of potassium exchange, calculated per total muscle potassium, is higher. This increased uptake, however, can be demonstrated only when the time interval between administration of isotope and autopsy does not exceed nine minutes. According to Lyman, no difference could be found between the activity of normal and denervated muscles four hours after isotope administration. In two experiments which were reported by this author, the denervated muscle had 79 and 89% activity of the normal muscle respectively. Increased uptake was also found by Noonan et al. (1941b) in rat muscles 24 hrs. after denervation. This increase, however, was explained not due to the effect of denervation on cellular membrane permeability, but to increased vascularization which occurs shortly after denervation, leading to increased perfusion of K^{42} in the capillary bed.

Potassium exchange in denervated rabbit muscle was also studied by Fischer (1950). He pointed out that the investigation of potassium turnover in the denervated muscle is very difficult in "in vivo" experiments, because of the wide biological variability of uptake values both after intravenous and still more so after intraperitoneal application of the isotope. Fischer summarizes the results of a considerable number of experiments as follows: 1. In experiments in which equilibration of the isotope was longer than three hours no difference could be observed between the turnover of potassium in normal and denervated muscles. 2. The turnover of K^{42} was found to be decreased in the denervated muscle in experiments where animals were sacrificed sooner than three hours after isotope administration, if denervation lasted for more than three days. 3. If the denervation period was shorter than three days then the initial turnover of potassium in the denervated muscle is higher than in the normal muscle.

Harris and Nichols (1956) measured the uptake of K^{42} into normal and denervated frog muscle in vitro and found that the uptake of K^{42} into the denervated muscle is lower by 23% and this difference remained unchanged throughout the incubation. These authors explain the differences between their own results and those of Lyman by the fact that they were obtained in quite different muscles of different animal species. The frog muscle does not atrophy even 30 days after denervation (wet muscle weight), while the rat muscle in Lyman's experiments is already considerably atrophied. Lyman performed in vivo experiments and used brief isotope equilibration (max. 9 min.), while the experiments of Harris and Nichols were in vitro under conditions of long equilibration (1—5 hrs.). The differences between these findings could therefore be caused either by different experimental conditions or by the use of different muscles of different animal species. In order to elucidate this point we measured the uptake of radioactive potassium in vitro into rat diaphragm five days after denervation. The uptake was thus measured in mammalian muscle which had not yet atrophied under the same experimental conditions as in frog muscle in the experiments of Harris and Nichols. Sola and Martin (1954) found that muscle weight in the diaphragm is not reduced until 20 to 40 days after denervation. In our experiments it was found (Drahota and Hudlická 1960) that the uptake of K^{42} after denervation of the hemidiaphragm is lower than that of the normal half. A decrease was, however, demonstrable as in experiments of Harris and Nichols following longer incubation periods. During the first ten minutes of incubation of the diaphragm in a medium with radioactive potassium, the uptake into normal and denervated hemidiaphragm did not differ. After 60 min. of incubation the uptake of K^{42} was 16% lower, after 120 min. incubation 20% lower.

In the in vivo experiments the isotope was administered by the intraperitoneal route and the time interval between this administration and sacrificing the animal was 30 min. It was found under these experimental conditions that the uptake of K^{42} into various muscles three days after denervation is different. It was higher into extensor digitorum longus, lower into the soleus and unchanged into the anterior tibial (Drahota et al. 1959).

It is evident from this survey that data concerning potassium cannot yet give a definite picture about the exchange of K^{42} in the denervated muscle. The differences between individual muscles are caused by the fact that various authors worked under different experimental conditions on different muscles.

The metabolism of ions could also be studied in experiments in which ion metabolism is loaded in different ways. Kraup et al. (1955) found, for example, that the denervated cat muscle liberates more potassium after administration of acetylcholine into the perfusion fluid than normal muscle. Differences between the normal and denervated muscles were observed from the eighth to eleventh day after denervation. On using low doses of acetylcholine (0·02—0·59 μM), potassium was liberated from the denervated muscle only. After higher doses the total potassium liberated from the denervated muscle was higher than that from the normal muscle.

Werner et al. (1955) administered tetramethylammonium bromide under similar experimental conditions. On using lower concentrations, liberation of potassium from the denervated muscle was found only in some of the cases investigated. After higher concentrations (3·11—6·65 μM) the liberation of potassium from the denervated muscle was higher.

Potassium ions can also be extruded from the muscle by osmotic dialysis, i. e. by intraperitoneal application of potassium-free Tyrode's solution (Drahota et al. 1957). If the intensity of dialysis is low, the content of potassium remains unchanged and potassium decreases in the denervated muscle only. When an isotonic solution of KCl is administered after termination of dialysis, then the decreased content of potassium in the normal muscle returns to normal values within four hours, while in the denervated muscle the lower potassium content remains unchanged. A reinnervated muscle, i. e. muscle in which restitution of innervation and motor function has taken place, behaves partly as a normal muscle, and partly as a denervated muscle. It loses potassium after moderate dialysis as the denervated muscle, but this is renewed at a rate similar to normal muscle.

After partial occlusion of blood circulation by ligature of abdominal aorta and the v. cava caudalis, as after partial exsanguination, the uptake of K^{42}

into normal rat muscle and the denervated muscle is reduced three days after denervation. Following both these procedures the return to normal uptake values occurs considerably more rapidly in the normal muscle than in the denervated muscle (Drahota and Hudlická 1958).

It is evident from the above experiments that, even though no changes in total potassium occur in the course of denervation, the dynamics of ion movement are changed after denervation. The denervated muscle can neither maintain a constant potassium content under different loads, nor restore a reduction of potassium.

C. Conclusions

It can thus be concluded that the intracellular ion composition of the muscle remains unchanged after denervation. This conclusion is based on the following facts: 1. The amount of intracellular water in the normal and denervated muscle is the same (Eichelberger et al. 1956, Drahota 1959); 2. The concentration of potassium per mg. of non-collagenous protein nitrogen is the same in normal and denervated muscles (Drahota 1960a); 3. Intracellular sodium remains unchanged after denervation (Eichelberger et al. 1956, Drahota 1960b). Thus the osmolarity of the denervated muscle cell is maintained as long as the cell exists as a morphological entity, and even under conditions, when considerable metabolic disturbances occur inside the cell (v. Gutmann 1959a).

When evaluating potassium changes, it is necessary to pay special attention to changes occuring shortly after denervation. The course of muscle atrophy begins by the accumulation of various intermediary products, substances and substrates in the muscle cell after section of the nerve (see also Chapter VIII). At this early denervation period, when there is a predominance of anabolic processes, accumulation not only of proteins (Gutmann 1959) and glycogen (Bass et al. 1957) takes place, but also of potassium ions (v. Drahota 1960c). This transient rise in potassium, which is rather the result of immobilization than of denervation itself disappears relatively soon and the intracellular ion composition remains unchanged during the further denervation period.

The total potassium content in the muscle remains unchanged, but the dynamics of ion metabolism are impaired. Although the denervated muscle, under normal conditions, maintains the content of potassium unchanged, it is more sensitive to various stimuli which disturb the ionic gradients. After osmotic dialysis, administration of acetylcholine tetramethylammonium bromide, the decrease of potassium ions in the denervated muscle is more promiment than in the normal muscle. However, it is not only the ability to maintain the concentration gradient which is impaired, but also the restitution of changes in ion relations. This is borne out by the fact that the content of potassium, decreased by osmotic dialysis or the uptake of K^{42} reduced by anoxaemia or hypotension, are restored at a slower rate in the denervated muscle than in the normal muscle.

Investigation of extracellular ions Na and Cl, and of extracellular water calculated from these and the inulin space indicate that metabolic disturbances in the denervated muscle cannot be caused by an inadequate supply of substrates or other substances indispensible for the cell. Absolute values of extracellular space in the muscle demonstrate that following denervation the volume of the extracellular space remains unaltered, while the intracellular muscle phase decrease. It is therefore necessary to supplement the conclusions of preceding authors who maintained that extracellular fluid increases after denervation. This increase, however, is only relative and is caused by the fact that following denervation extra- and intracellular muscle phases do not decrease at the same rate. The extracellular phase remains unchanged even under conditions where the loss of protein mass reaches almost 40% (Fig. 68). Thus, in the denervated muscle, more extracellular phase is present per unit of muscle proteins than in the normal muscle. These findings together with the results included in Chapter VII. about circulation changes in denervated muscle, indicate that metabolic disturbances in denervated muscle cannot be caused by an insufficient supply of substrates or other indispensible materials at the cell membrane.

References

Bajusz E.: Endocrinology 64 : 262, 1959.
Baker N., Blahd W. H., Hart P.: Am. J. Physiol. 193 : 530, 1958.
Beránek R., Hník P., Vrbová G.: Physiol. Bohemoslov. 6 : 200, 1957.
Bogatzki M.: Z. ges. exptl. Med. 119 : 457, 1952.
Bogatzki M., Fischer L.: Z. ges. exptl. Med. 118 : 143, 1951.
Chen K. K., Meek W., Bradley H. C.: J. Biol. Chem. 61 : 807, 1924.
Chor H., Dolkart R. E., Davenport H. A.: Amer. J. Physiol. 118 : 580, 1937.
Cicardo H., Guereich E.: Rev. Soc. Argentina Biol. 18 : 425, 1942.
Conway E. S.: Symp. soc. exp. biol. 8 : 297, 1954.
Cotlove E.: Am. J. Physiol. 176 : 396, 1954.
Drahota Z.: Candidatẽs Thesis, Institute of Physiology ČSAV, Prague 1957.
— Physiol. Bohemoslov. 3 : 501, 1959.
— Proc. Soc. Exp. Biol. Med. 103 : 849, 1960a.
— Physiol. Bohemoslov. 9 : 240, 1960b.
— Physiol. Bohemoslov. 9 : 1, 1960c.
Drahota Z., Gutmann E., Vrbová G.: Physiol. Bohemoslov. 6 : 77, 1957.
Drahota Z., Hník P., Hudlická O.: Konference o srovnávací fysiologii výměny látek, Biol. fak. KU 23. 4. 1959 Prague, in press.
Drahota Z., Hudlická O.: Physiol. Bohemosl. 7 : 489, 1958.
— — Nature 186 : 4722, 1960.

Eichelberger L., Akeson W. H., Roma M.: Am. J. Physiol. 185 : 287, 1956.
— — — J. Appl. Physiol. 12 : 42, 1958.
Fenn W. O.: Am. J. Physiol. 120 : 675, 1937.
Fenn W. O., Goetsch C.: J. Biol. Chem. 120 : 41, 1937.
Fischer E.: Arch. Phys. Med. 30 : 375, 1949.
— Am. J. Phys. Med. 34 : 212, 1955.
Fischer E., Chevalier B. A., Bratton M. J.: Pflügers Arch. 255 : 1, 1952.
Fischer E., Moore J. W., Skowlund H. V., Ryland N. J., Copenhaver N. J.: Arch. Phys. Med. 31 : 429, 1950.
Fischer E., Ramsey V. W.: Am. J. Physiol. 145 : 571, 1946.
Geyer R. P., Meadows M. F., Marschall M. D., Gongaware M. S.: J. Biol. Chem. 205 : 81, 1953a.
— — — — J. Biol. Chem. 203 : 625, 1953b.
Glynn J. M.: J. Physiol. 134 : 783, 1956.
Gutmann E.: Am. J. Phys. Med. 38 : 104, 1959.
— in press 1960.
Gutmann E., Kučerová M.: Věstník čs. zool. spol. 11 : 145, 1947.
Harris E.: An. Rev. Physiol. 19 : 13, 1957.
Harris E. J., Nichols J. G.: J. Physiol. 131 : 473, 1956.
Hastings A. B., Ashrnore J. A., Cohil G. F.: Arch. Bioch. Biophys. 65 : 78, 1956.
Hines H. M.: J. Am. Med. Assoc. 120 : 515, 1942.
Hines H. M., Knowlton G. C.: Am. J. Physiol. 104 : 397, 1933.
— — Am. J. Physiol. 120 : 719, 1937.
Hník P.: Candidate's Thesis. Institute of Physiology, ČSAV, Prague 1955.
Hodgkin A. L., Horowitz P.: J. Physiol. 145 : 405, 1959.
Hoffman A., Wertheimer E.: Arch. ges. Physiol. 217 : 138, 1927.
Horvath B., Berg L., Cummings D. J., Shy G.: J. Appl. Physiol. 8 : 22, 1955.
Hudlická O.: Physiol. Bohemosl. 4 : 276, 1955.
Humoller F. L., Griswold B., McIntyre A. R.: J. Gen. Physiol. 33 : 723, 1950a.
— — — Am. J. Physiol. 161 : 406, 1950b.
Iacobelis M., Muntwy er E., Dodgen G. L.: Am. J. Physiol. 185 : 275, 1956.
Kolínská J. Čs. Fysiologie 7 : 223, 1958.
Kraupp O., Werner G., Kobinger W., Stormann H.: Arch. exper. Pathol. Pharmakol. 266 : 301, 1955.
Lenan: Biochim. Biophys. Acta 16 : 87, 1955.
Lenrier A., Bernard I., Richard A.: C. R. Soc. Biol. 110: 848, 1932.
Lipschutz A., Audova A.: J. Physiol. 55 : 300, 1921.
Lymann Ch. P.: Am. J. Physiol. 137 : 392, 1942.
Manery J. F.: Physiol. Rev. 34 : 334, 1954.
Manery J. F., Danielson I. S.: J. Biol. Chem. 124 : 359, 1938.
Manery J. F., Hastings A. B.: J. Biol. Chem. 127 : 657, 1939.
Morgulis S., Osherhoff W.: J. Biol. Chem. 124 : 767, 1938.
Neuschlosz S. M., Trelles R. A.: Arch. ges. Physiol. 204 : 374, 1924.
Noonan T. R., Fenn W. O., Haege L.: Am. J. Pnysiol. 132 : 312, 1941a.
— — — Am. J. Physiol. 132 : 474, 1941b.
Scott G. T., Hayward H. R.: J. Gen. Physiol. 37 : 601, 1954.
Shimizu S., Kuriaki K.: Am. J. Physiol. 198 : 943, 1960.
Sola O. M., Martin A. W.: Am. J. Physiol. 172 : 324, 1954.
Sutfin D. C., Thomson J. D., Hines H. M.: Am. J. Physiol. 178 : 535, 1954.
Szent-Gyorgyi A.: Chemical physiology in body heart muscle. Acad. Press 1952.
Tower S. S.: Physiol. Rev. 19 : 1, 1939.

Ussing H.: Symp. Soc. Exp. Biol. 8 : 407, 1954.
Vodička Z., Gutmann E., Bass A.: Čs. fysiol. 6 : 354, 1957.
Werner G., Kraupp O., Stormann H.: Arch. exper. Pathol. Pharmakol. 227 : 1, 1955.
Westenbrink H. G. K., Krabe H.: Arch. Neerl. Physiol. 21 : 55, 1936.
Young H. L., Young W., Edelman I. S.: Am. J. Physiol. 197 : 487, 1959.

Chapter VII

VASOMOTOR MECHANISMS IN GENESIS OF DENERVATION ATROPHY

Hudlická O.

The maintenance of structural and functional integrity is ensured by nervous influences (as has been shown in preceding chapters) and by sufficient blood supply to the muscle. It is not possible to expect normal muscle function without an intact vascular supply. Functional, morphological and metabolic disturbances occur not only in trauma of blood vessels and Volkmann's paresis, which are extreme cases, but also in Bürger's disease, arteriosclerosis etc. But, of course, the limit below which blood flow must not drop if metabolic disturbances are not to occur, has never been determined. It is not even possible since such a value would vary between individuals (blood flow through human limbs varies by as much as 100%, as has been shown by Golenhofen and Hildebrandt in 1957) and in relation to the work performed.

On the other hand, an increase in blood flow can lead to marked improvement of metabolism and function of the extremity previously impaired as has been demonstrated, for example, after administration of vasodilating drugs in gangrene due to arteriosclerosis of peripheral blood vessels. But the relationship between circulation and metabolism is reciprocal: vasodilating drugs, e. g. derivatives of adenylic acid, can directly affect muscle metabolism. The increase of metabolism can consequently lead to adaptation of circulation, as has already been pointed out by Rein and Mertens (1936) and as is universally known about hyperaemia following functional activity. Such adaptation of circulation is usually evoked by local accumulation of metabolites-lactic acid or changes of pH (e. g. Rein 1931, Kester et al. 1952, Stillwell 1957) — and mediated by axon reflexes, the afferent fibres of which pass in or along blood vessels (Fleisch 1956). Prolonged denervation, which leads to the loss of axon reflexes (Celander and Folkow 1953a, Stillwell 1957) in the muscle, also interferes with this mutual interplay between metabolites and blood flow. Insufficient adaptation of circulation in the denervated muscle could then be one of the causes of denervation atrophy. This insufficiency can manifest itself either by disproportionate increase or decrease of blood circulation. Exaggerated increase of blood flow will not affect muscle metabolism as long as it is itself sufficiently high.

Schoedel (1936) did not succeed, by increasing the blood flow through muscle by administration of adenylic acid, histamine or acetylcholine, in increasing the efficiency of muscle work. On the other hand, a decrease in blood flow can lead to grave metabolic disturbances, since a sufficient supply of substances brought by the blood is one of the fundamental conditions necessary for maintaining the normal trophic state of a tissue. Supply of substances, however, is not given by a sufficiently large blood flow through large vessels and capillaries only (and these two values need not always be correlated: it was found by Barcroft and Donhorst in 1954 that on heating the body total blood flow is increased but is not exhibited in the capillary bed). A sufficient supply of nutrients is also dependent on adequate intracellular uptake.

When studying denervation atrophy it is thus very important to know whether and how the inflow of substances into the muscle changes, i. e. what are the changes in

a) blood flow,
b) uptake of substances into the muscle after section of the peripheral nerve.

This question is all the more urgent, since cold and cyanotic extremities were usually described in clinical cases of peripheral nerve injuries. Such observations have led not only the skin, but also muscle trophic disturbances to be explained by insufficient nutrient supply and have inspired the theory about vasomotor mechanisms of trophic changes (e. g. Cassierer and Hirschfeld 1924, Leriche 1952 and others). To what extent is this theory tenable?

Measurements on a large number of cases have shown that blood flow through the skin is actually decreased following lesions of the peripheral nerve, even though it is not true in every case. Following complete lesions, the skin of the upper limb was cold in 50% of cases, in the lower limb in 40% of cases (Richards 1954). The reduction in blood flow need not always result in trophic changes. Basic metabolic requirements of the skin are so small that relatively insignificant skin circulation is sufficient to maintain normal function (Hertzmann 1959) and even after denervation decreased blood flow in the fingers is enough to cover metabolic needs, which are decreased (Doupe 1943). On the other hand, Bilisoby et al. (1954) even assumed that trophic ulcers are caused by lowered threshold of sensitivity to pain due to vasodilatation. Even though at least some trophic disturbances in the skin could be explained by vasomotor changes, the question remains, whether this is also the case in the muscle. Is the blood flow through the muscle decreased after section of the peripheral nerve, or do the changes occur only at the stage of cellular uptake? Or is this influx of metabolites into the cell sufficient, and the disturbance occurs when intracellular utilization of these substances should take place?

These questions have not been answered unequivocally in the literature. Contrary findings, especially those concerning blood flow, are mainly due to the fact that various methods were used for measurements and the results obtained were not always correctly evaluated from the point of view of methodological adequacy.

What methodological errors are most likely to occur when measuring blood flow through the muscle?

Muscle circulation is two-fold: at rest part of the blood by-passes the capillary bed and is shunted directly through the arterio-venous anostomoses (Zweifach et al. 1949, 1955); during work and also after application of adrenaline, and possibly also under other conditions, blood flows to a marked extent through capillaries. Measurements of total blood flow cannot differentiate these routes. On the other hand, some methods, e. g. clearance of Na^{24}, measure blood flow through the capilary bed only (and only under certain conditions). Intramuscular temperature measurements, often used as indicator of circulation, cannot give objective results, especially in muscle, since changes in temperature are caused primarily by metabolic changes. It has been shown by Kemp et al. (1947, 1948) that blood flow can be decreased even though temperature is increased at the same time, and vice versa. Heated thermocouples are a relatively good indicator of relative changes in blood flow, even though the results are dependent on the location of their tips in the muscle: for correlation with the direct method it was necessary to change the position of couples in the muscle 10—20 times. Values obtained of small or fast changes do not correspond to those given by the direct method (Graf and Rosell 1958). Direct methods, on the other hand, have one drawback, i. e. it is necessary to work on anaesthetized animals. A number of vasodilating substances are liberated as a result of tissue injury. The most accurate method employs measurement of venous outflow (Folkow 1953b). Plethysmography makes work possible without anaesthesia, however, its disadvantage lies in the fact that it registers changes not only of muscle circulation but also of the skin. In the human forearm there is 15—43% of other tissues besides muscles (skin, bones, fat, tendons) and skin circulation was eliminated with the aid of adrenaline iontophoresis only in a very small number of cases. For this reason changes in blood flow through the denervated muscle must be evaluated very critically.

A. Blood flow through denervated muscle

The blood flow through denervated muscles was already investigated during the last century, partly by direct methods especially by means of histology (Levashiev 1880, Ricker and Ellenbeck 1894, Lapinsky 1906, Chen et al. 1926), and partly by determining circulation through the whole extremity (Goltz 1874). It was shown by both methods that blood flow is increased during the initial period after denervation (1—5 months) and decreased later after periods longer than five months. However, no one of these methods could gain information about the actual blood flow, since dilatation of capillaries demonstrated histologically can mean an increased content of blood in the muscle, but it also indicates venous congestion and consequently a decreased volume of blood flowing through in a given unit of time. Besides, more recent histological observations have shown that there are no changes in the capillary bed (Sunderland and Ray 1950, Richards 1954), or that the number of capillaries is increased (Chor et al. 1937). Such findings are perhaps suggestive of potential possibilities of increased blood flow, but do not conclusively prove this. The thermometric method used by Goltz also does not measure the rate of blood flow. It measures the intensity of muscle metabolism and it is possible that the rise of temperature is caused by increased catabolism in the denervated muscle, rather than by increased blood flow. Methods for measuring blood flow directly in the muscle were used for a long time for investigating changes in blood flow immediately after sectioning the nerve. It was found by various methods, e. g. measurements of blood flowing out of a vessel into a calibrated tube (Gaskell 1877, Verzár 1912), measurements performed with an anemometer or Rein's flowmeter (Keller et al. 1930, Anrep et al. 1934, Anrep and Saalfeld 1935) and with other equipment (Seeling 1909, Mulikov 1937, Kramer and Quensel 1938 and others), that the amount of blood flowing through a muscle in a given unit of time is increased immediately after section of the nerve.

This increase of blood flow lasts for different periods of time according to various authors. Issekutz et al. (1956) reported an increase after 3—4 weeks, Goltz (1874) and Griffin et al. (1954) after 3—4 days. No changes in blood flow were found by Kemp et al. (1947) 12—14 days after denervation, while Randall et al. (1952) described a decrease at the same period. One to two months later Kemp et al. found a decrease. However, Doupe et al. (1943) and Duff and Shepherd (1953) reported increased blood flow in the human forearm several months and even years after section of the peripheral nerve. No changes have been reported more recently by histological methods, as long as blood vessels were not injured at the same time, with the exception of Chor et al. (1937) who described an increased number of capillaries after denervation. Obliteration of blood vessels in denervated limbs of the rabbit found by Khlopina (1953) was evidently due to inflammatory complications.

It is thus possible to conclude from studies of circulatory changes in the muscle following section of the peripheral nerve that decreased blood flow found histologically (Ricker 1901, Ricker and Ellenbeck 1894, Kopits 1929) cannot explain the cause of muscle atrophy, since first signs of the decrease are described at a time when muscle atrophy is already far advanced (after five month's denervation). It is possible rather to agree with Bowden (1954) that "this altered vascular pattern reflected the decreasing functional demands of the muscle". Different findings of blood flow through the extremity by Duff and Shepherd on the one hand, and Randall and Kemp on the other, could be explained either by the fact that different methods were used by these authors (plethysmography and electromagnetic flowmeter respectively) or the object of investigation. Doupe and Duff worked on the human forearm where there is relatively more musculature than on the leg of the dog, which was studied by Kemp and Randall in the femoral artery. Kemp and coworkers themselves pointed out that their results could be distorted by decreased blood flow through the skin.

Are there differences between blood flow through the whole extremities and the muscle only? To what extent does the method of measurement affect the results? On determination of blood flow through the limb in rats and dogs (Hudlická 1958, Hudlická and Vrbová 1958) and muscle itself (Bass and Hudlická 1960), it was found that in principle there was no difference between measurements in the whole limb and the muscle.

Blood flow measurements were performed in the extremity of rats by weighing samples of blood flowing out of the femoral vein and in dogs with a rotameter introduced into the femoral vein. For measuring blood flow through the gastrocnemius in dogs a photoelectric drop counter was used. Blood flow was increased in dogs under all conditions, there were, however, no differences found in rats three days after section of the peripheral nerve (Table 8).

Table 8

Blood flow through the whole extremity and muscle respectively following section of the sciatic nerve.

Organ and animal animal	Time after denervation days	Number of measurements	Changes in blood flow (D/N in %)
Rat-extremity	3	17	92 ± 5·4
Dog-extremity	3—4	6	305 ± 58·2
Dog-gastrocnenius	3	2	172
Dog-gastrocnemius	7—10	16	153 ± 31
Dog-gastrocnemius	30—35	20	194 ± 27·6

Blood flow through the muscle is increased even 30—35 days after denervation as is shown in Fig. 69. What is the reason for this increase?

Immediately after nerve section, it is evident that the effect is due to interruption of sympathetic nerve fibres which are present in peripheral nerves in considerable numbers (Langley 1923, Barcroft and Edholm 1946, Randall et al. 1950). According to Clonninger and Green (1955), the sciatic nerve contains 89% of all sympathetic nerve fibres which innervate the hind limb of the dog. The question whether the increased blood flow through the muscle after denervation is caused by elimination of sympathetic nerve fibres should be answered by comparing this with blood flow after sympathectomy. However,

Fig. 69. Blood flow through the denervated (D) and normal (N) gastrocnemius of the dog 35 days after section of the sciatic nerve. The record in number of drops in 10 sec. intervals.

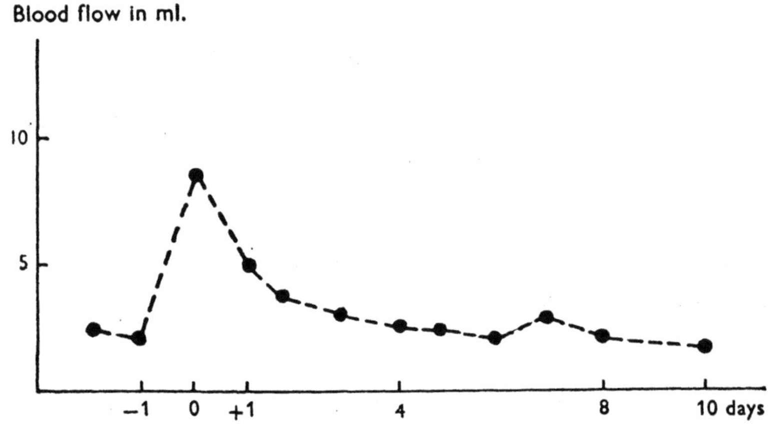

Fig. 70. Blood flow through the human fore-arm following sympathectomy (after Barcroft and Swan 1953). Abscissa: in days (sympathectomy performed at 0). Ordinate: blood flow in ml./ /100 g. tisues/min.

increased blood flow through the muscle after sympathectomy is very short-lasting (Fig. 70 and 71): between 12—14 hours (Lambert 1957c, Duff 1951a, 1951b, 1954) and 2—4 days (Barcroft and Swan 1953, Stillwell 1957) after sympathectomy blood flow values returned to initial values. At longer periods after sympathectomy only reflex vasodilatation is missing on warming (as long as six years — Barcroft and Swan 1953) and there is a prolongation of reactive hyperaemia (Donhorst and Sharpey-Schafer 1951).

The rapid restitution of vasomotor tone after sympathectomy may be caused by incomplete sympathectomy, sprouting of sympathic fibres from neighbouring innervated regions, or by increased sensitivity to mediators circulating in the blood, the vasoconstictor effect of which can manifest itself primarily on denervated blood vessels.

The question whether sympathectomy was complete or not is certain to play an important role. Following section of preganglionic fibres the effect of sympathectomy is more permanent (Cannon and Rosenblueth 1949). It is difficult to eliminate all vegetative nerve fibres by sympathectomy, because some sympathetic ganglia are located directly in the anterior roots (Randall et al. 1950). High frequency stimulation of the remaining nerve fibres leads to the liberation of mediators which can also diffuse to blood vessels in the denervated region and evoke their constriction (Randall et al. 1953).

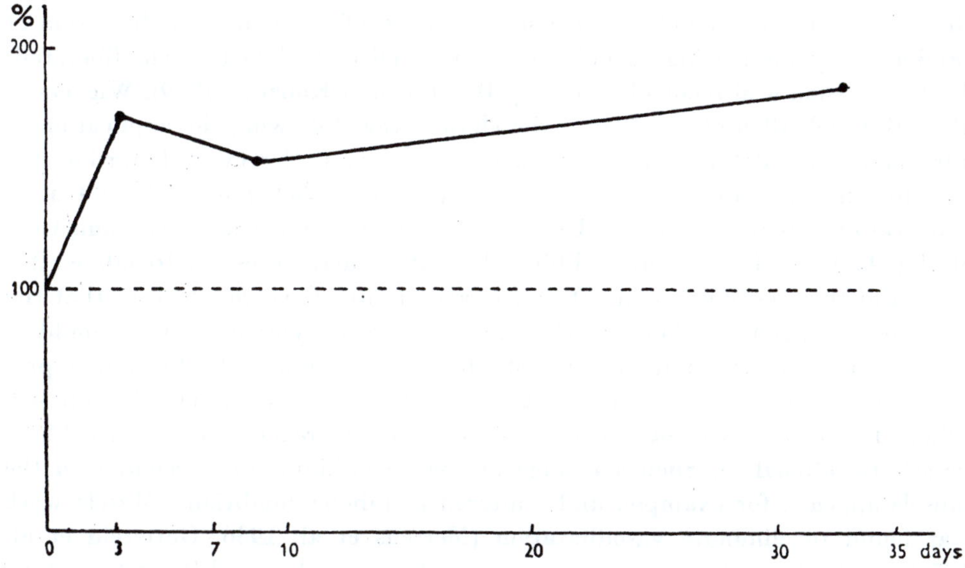

Fig. 71. Blood flow through the denervated gastrocnemius muscle of the dog expressed as percentage of values obtained on the control side. Abscissa: days after nerve section.

The sprouting of sympathetic nerve fibres from neighbouring innervated regions, as considered by Barcroft (1952), and which has been demonstrated in the skin by Murray and Thompson (1957) probably does not participate in the rapid recovery of vasomotor tone following sympathectomy, since it takes place much later than the return of blood flow to initial values.

One of the mediators to which the denervated blood vessels are hypersensitive is adrenaline. This hypersensitivity of blood vessels in the denervated extremity of the frog to adrenaline was described by Okuyana (1926), Schneider (1926) and Cannon and Rosenblueth (1949) which was found to cause vasoconstriction in these experiments. However, in the muscle adrenaline leads to vasodilatation (Rein and Schneider 1930, Grant and Pearson 1938, Friedlander et al. 1940, Wiggers et al. 1942, Allen et al. 1946, Barcroft and Konzett 1949, Harpunder et al. 1947, Issekutz et al. 1956, Imig et al. 1952, Quidot et al. 1953, Whelan 1952, Dörner and Kuschke 1955, Gottstein et al. 1955, Dörner 1956, Hensel 1956, Went et al. 1956 and others) which, on intravenous administration, passes through two phases — the first, short, of nervous origin and second — of blood vessel origin which is similar to the dilatation after intraarterial administration of adrenaline. The mechanism of this vasodilatation is not quite clear. According to Dörner and Kuschke (1955) vasodilator fibres are activated (since vasodilatation is smaller after injection of novocaine into the blood vessel and it is sometimes diminished by atropine). It was demonstrated by Dörner in experiments on dogs with crossed circulation that adrenaline acts via the nervous system. On the other hand Stillwell (1957) stresses the effect of adrenaline on dilatation of capillaries by influencing carbohydrate metabolism and lactic acid liberated. This conclusion is also corroborated by Barcroft and Konzett (1949), Wiggers et al. (1942) and Allen et al. (1946), who showed that following the application of adrenaline vasodilatation of the capillary bed occurs. However, dilatation due to adrenaline is absent in chronic sympathectomy (Whelan 1952) and after denervation (Went et al. 1956, Rein and Schneider 1931). Duff (1954) and Imig et al. (1952) even found increased blood flow after small doses of adrenaline after section of the sciatic nerve or after sympathectomy. This would mean that the increased sensitivity of blood vessels to mediators after sympathectomy should — in case this hypersensitivity was mainly due to adrenaline — lead rather to prolongation of dilatation after sympathectomy. It is, of course, possible that still other effects are involved which could modify the reaction of muscle blood vessels to adrenaline. Such a change in the reactibility to adrenaline in the muscle appears, for example, under altered metabolic conditions. Muscle work was found to eliminate vasodilatation (Mertens et al. 1936, Gottstein et al. 1955), as did hypoxia or traumatization in the surgical wound (Gottstein et al. 1955). Increased temperature even causes constriction after application of adrenaline (Rein and Schneider 1930), similarly as excessively large doses of adrenaline (Harpunder et al. 1947).

However, the above data indicate that it is not very probable that the early return of vasomotor tone after sympathectomy is the result of increased sensitivity to adrenaline. It would be more plausible to expect increased sensitivity of muscle blood vessels to noradrenaline, which is normally liberated at the endings of vasoconstrictor nerve fibres (Folkow 1956), its effect having been shown to be vasoconstrictor (Barcroft and Konzett 1949, Barcroft and Swan 1953, Hensel 1956, Stillwell 1957 and others). In older reports increased vasoconstriction of denervated blood vessels was described to adrenaline and this was evidently due to the action of noradrenaline which was yet unknown. Hypersensitivity to noradrenaline is explained (Burn and Rand 1959) by the fact that there is a considerable decrease of noradrenaline content in denervated structures (iris, aorta and spleen in the rabbit); administration of very small doses then evokes contractions of smooth muculature in these structures. Grant (1954) also explains the restitution of vasomotor tone to normal after sympathectomy by changes in the wall of blood vessels. According to this author, increased sensitivity to vasoconstrictor stimuli is caused by the disappearance of acetylcholine from the blood vessels; pseudocholinesterase activity is decreased following sympathectomy, while monoaminooxidase remains unchanged. Keyssler and Schmier (1951) assume that after sympathectomy, cholinergic organs play a role in regulation of blood flow and help to adapt it to metabolic requirements of the tissue.

It is thus apparent that rapid restitution of blood flow through the muscle to normal values after sympathectomy is probably due to greater effect of the remaining vegetative nerve fibres as well as to increased influence of humoral factors — either to higher sensitivity to noradrenaline or to a lack of acetylcholine.

If this is the case, why is increased blood flow in the denervated muscle of longer duration than after sympathectomy? It is unlikely that section of the peripheral nerve would eliminate more sympathetic nerve fibres than sympathectomy itself. However, complete sympathectomy, besides eliminating the effect of vasoconstrictor fibres, also removes vasodilator fibres which play a conspicuous role, expecially in skeletal muscles (Dale and Gaddum 1930, Rein 1944, Mulikov 1937, Barcroft an Edholm 1946, Barcroft and Swan 1953, Celander and Folkow 1953, Lindgren and Uvnäs 1953, 1955, Uvnäs 1954). It is possible that section of the peripheral nerve meinly eliminates vasoconstrictor fibres, while vasodilatating fibres remain intact. Their effect would in that case predominate and cause prolonged vasodilatation.

Following sympathectomy vasodilatation is of short duration since the sensitivity of blood vessels to humoral factors is increased. Is this sensitivity of blood vessels less marked in the denervated extremity? It appears that there is some evidence for this assumption. Rein (1931), for example, found that carbon dioxide does not evoke an increase of blood flow during muscular work of a denervated extremity. In addition to this, acetylcholine is not liberated in blood vessels following injection of adrenaline (Went et al. 1956) and dilatation does not occur after adrenaline (Went et al. 1956, Rein and Schneider 1930). Similarly, dilatation after d-tubocurarine is smaller after section of the sciatic nerve (Bean and Ehwell 1951). It was demonstrated by Grant (1954) that on simultaneous section of peripheral nerves and sympathectomy cholinesterase disappears completely from blood vessels; this means that following section of the peripheral nerve no acetylcholine is broken down in the wall of blood vessels (while after sympathectomy it is according to Grant) and thus ACh could be the cause of a longlasting increase of blood flow.

After interruption of the peripheral nerve, however, other factors besides those in sympathectomy can participate in changes of circulation. The section of motor nerve fibres leads to decreased muscle tone. It could be assumed that in a muscle the fibres of which do not contract, dilatation of blood vessels would occur as the result of diminished total resistance. Barcroft and Edholm (1946) attempted to verify this assumption experimentally. These authors found increased blood flow through the human forearm after blockade of the median, ulnar and radial nerves, similarly as after sympathectomy. However, if the blockade was performed in subjects in whom the forearm was deprived of sympathetic innervation, no increase of blood flow was observed after blockade of the nerves mentioned. This would indicate that the increase of blood flow after elimination of peripheral nerves is due exclusively to sympathetic nerve fibres contained in these nerves. When they were not present, the exclusion of motor nerve fibres was without effect. According to the above authors, increased blood flow following denervation is caused by the elimination of sympathetic nerve fibres. Their experiment, however, confirms this assumption only a short time after section of the nerve, i. e. at a time when there is no difference as compared with sympathectomy. It is also possible that the decreased resistance against blood flow through the limb is affected by longlasting decrease of muscle tone and that the increase of blood flow after denervation is a form of adaptation to changes in mechanical conditions.

Investigation of muscle circulation in poliomyelitis should help to elucidate the role played by motor nerve fibres in regulation of blood flow, since lesions of the anterior spinal horns are assumed in this disease. The blood flow in poliomyelitic patients has been found to be lower in the extremity more paralyzed (Kessel and McPherson 1955). The number of capillaries in atrophic muscles is increased at first, but 3·5—5·5 years after onset of the disease their lumen decreases (Kopits 1929). Zellweger and Mort (1950) found a smaller calibre in the arterial part of the capillary loop and Bloch (1953) described changes in the whole capillary bed in monkeys, mice and men: aggregation of erythrocytes, diminution of plasma and anemia. All these findings indicate that in poliomyelitis the pathological process involves not only anterior spinal horns, but also the lateral horns and this is confirmed by Weinstein (1957), Goetz et al. (1955) and deGroot (1956) who reported changes in arterial lumen and the reactibility of skin vessels to increased temperarature and pharmacological stimuli. It was then shown by Whitten and Kramer (1953) that skin circulation in poliomyelitis can be favourably affected by dibenamine. All the above data indicate that it is not possible to make conclusions concerning the role of the motor neuron in regulation of circulation in skeletal muscle from changes of blood flow in poliomyelitis.

A satisfactory answer has not yet been given to the question why section of the peripheral nerve results in increased blood flow of longer duration. The whole problem is all the more complicated since increased blood flow has been observed not only after nerve section, but also after its stimulation and even after stimulation of the ventral roots (Krogh et al. 1922). These results, however, obtained in frogs, have not been confirmed in other animals and they cannot therefore be considered of general significance. It is possible that increased blood flow after denervation is to a certain extent due to elimination of afferent innervation conducting impulses which lead to reflex vasoconstriction. Moore and Moore (1933) demonstrated that pain and application of various solutions in the vicinity of blood vessels do not evoke vasoconstriction if the dorsal roots had been cut.

The most plausible explanation seems to be the larger quantities of acetylcholine liberated in denervated vessels, which is corroborated by Grant's finding of the disappearance of cholinesterase from blood vessels after section of the peripheral nerve.

It is possible to state with certainty that the increase of blood flow in the denervated muscle is of longer duration than that after sympathectomy. This finding itself contrasts with clinical observations that in peripheral nerve lesions and in poliomyelitis there are often markedly cold extremities. What is the cause of this discrepancy?

Clinical observations are based either on direct inspection of the extremity or on plethysmographic records. Direct observation, of course, gives information about changes in the skin only. However, in the forearm or calf (where blood flow was found to be increased after denervation), muscle bulk predominates: according to Cooper et al. (1954) muscles represent 63·5% of the whole volume (skin 8·5%, fat 14·5%, bones 13·5%), according to Grant (1938) as much as 85%. Abramson and Ferris (1940) reported 57% for muscles, 12% for skin and subcutaneous tissue, 31% for fat, bones and tendons. Exact plethysmographic measurements of changes in denervated skin have shown that there occurs a prolonged decrease of blood flow (Ederstrom et al. 1956, Griffin et al. 1954, Doupe 1943). However, even in complete peripheral nerve lesions skin was colder only in one-half of cases in the upper extremities and in 40% of cases in the lower extremities. Decreased blood flow appeared after a transient vasodilatation sometimes lasting for as long as three weeks (Richards 1954). Vasomotor changes in denervated skin are doubtlesly caused by elimination of the sympathetic nerves: sympathectomy is followed by practically the same changes as section of the peripheral nerve — transient vasodilatation, which is of much longer duration than in the muscle, is followed by vasoconstriction (Doupe 1943, Griffin et al. 1954, Ederstrom et al. 1956). This latter is evidently caused by increased sensitivity to adrenaline which in contrast to muscle always evokes vasoconstriction in the skin (see, for example, Friedlander et al. 1940, Quidot et al. 1953, Barcroft and Swan 1953 and others). In the denervated skin then vasoconstriction takes place after much smaller doses of adrenaline than in the normal skin (Van Dobben-Brockema and Dirken 1950, Barcroft and Swan 1953). The reactibility of skin vessels is different from muscle blood vessels in other respects as well. For example, hyperventilation decreases the blood flow through the skin and increases it in the muscle (Page 1952). Following interruption of the arterial supply to a limb there is prolonged vasodilatation in the skin, in the muscle, however, brief transient vasodilatation is followed by vasoconstriction (Lambert et al. 1951). Inhalation of CO_2 causes vasoconstriction in the muscle (Rein 1944) and vasodilatation in the whole limb (Ukolova and Bobryushkov 1957). Administration of Priscol results in increased blood flow in the skin and decreased blood flow through the muscle (Brück et al 1952). Vasomotor postdenervation changes in the skin are affected not only by sensitivity to adrenaline, but also by cold, which also leads to vasoconstriction (Hertzmann 1953). In contrast to this the application of heat leads either directly or by the reflex route to vasodilatation. For this reason therapy of trophic changes in the skin is justified in applying measures increasing either blood flow directly or temperature of the skin.

From the aspect of muscle atrophy, this question becomes much more complicated. It is true that direct application of heat leads in most cases to an increase of blood flow in the muscle (Feucht et al. 1949, Richardson et al. 1950, Quidot et al. 1953, Imig et al. 1954, Richardson 1954, Paul and Imig 1955) with the exception of diathermy which, while increasing temperature, decreases muscle blood flow (Kemp, Paul and Hines 1948). But already indirect heating reduces blood flow (Rein and Schneider 1930, Rein 1931, Barcroft et al. 1955, Hensel 1956). In addition to this the application of heat probably increases the intensity of metabolism, so that increased blood flow is insufficient to cover increased metabolic demands (Randall, Imig and Hines 1952). For this reason it would perhaps be better to use either ultrasonic radiation (Paul and Imig 1955), electromagnetic radiation (Richardson et al. 1950), or indirect reflex cooling which induces increased blood flow in the muscle (Rein and Schneider 1930, Rein 1931, Barcroft et al. 1955, Hensel 1956, Imig et al. 1956) for improving muscle circulation. There are doubtlessly several possibilities of increasing the blood flow in the muscle. The physiologist, however, is primarily interested in the question whether increasing blood flow in the denervated muscle is rational or not, since blood flow is already increased and in spite of this the muscle atrophies. Is it because, even though blood flow is increased, blood does not pass through the capillary bed in the muscle?

Such a question logically follows from views expressed by Rein (1931) and Stillwell (1957) who assumed that muscle circulation is under nutritional control (local metabolites) which involves capillaries, as well as nervous control which involves arteriovenous anastomoses. Folkow (1953b) and Golenhofen and Hildebrandt (1957b) stressed the role of myogenic reactions — automacity of smooth muscles in the vessel walls — in the regulation of muscle blood flow. Elimination of the nervous system could result, in consequence of exclusion of muscle contractions, in reduction of local nutritive regulation factors in the capillaries and to increased blood flow through A—V anastomoses. This is confirmed by the work of Rein (1931, 1944) and Rein and Mertens (1936) which indicates that the decisive factor for capillary circulation is intensity of metabolism. If the intensity decreases after denervation as the result of acute immobilization, blood flow through capillaries can be reduced together with a lower supply of substances which could, in the chronic stage, lead to atrophy. This possibility — decreased capillary blood flow with simultaneously increased total blood flow — must be taken into account, since it has been shown that there is double the number of A—V anastomoses after denervation (Piiper and Schoedel 1954). This increase could, of course, be attributed to changes in skin A—V anastomoses the existence of which is indispntable. The existence of such connections in the muscle has long been disputed.

The presence of A—V anastomoses can be investigated by means of indirect methods such as observation of pulsations, changes in the colour of venous blood, changes in capillary pressure, angiography, changes in the volume or injection of pellets of various sizes (Schroeder 1952, Luckner 1955, Lambert 1957) or by direct histological methods of observation in vivo (Zweifach and Chambers 1947). Histological proof of A—V anastomoses in the muscle was not forthcoming for a long time. For this reason the presence of A—V anastomcses was not acknowledged by a number of authors until Zweifach's description of A—V connections in the trapezius muscle of the rat (Zweifach 1955), even though indirect evidence obtained with various methods had already been reported earlier (Zweifach 1949, Hyman 1957, Schroeder 1953, 1954, 1955, Mottram 1954, Bostroem and Schneider 1953, Barcroft 1952, Lambert 1957 and others). Today it seems to be conclusively proved that A—V anastomoses do exist in the muscle. It is also approximately known under what conditions blood flows mainly through the anastomoses and when through capillary bed. For example, muscular work, reactive hyperemia, arrest of circulation, barbiturate anaesthesia, loss of blood and lack of oxygen are followed by an increase of blood flow through the capillaries (Barcroft 1952) with simultaneous closing up of A—V anastomes (Schroeder et al. 1954) Cooling also decreases blood flow through the anastomoses (Bostroem and Schoedel 1953). On blockade of sympathetic nerve fibres (Hyman 1957), heating of the body (Bostroem and Schoedel), the administration of adrenaline (Barcroft 1952) and acetylcholine (Schoedel 1952) there is on the contrary an increased blood flow without an increase flow through the capillaries. It would thus be possible that interruption of peripheral nerves causes, similarly as sympathetic blockade, increased blood flow through A—V anastomoses and that circulation through the capillary bed could even be diminished, so that substances necessary for normal metabolism could not come into contact with muscle cells. Increased number of A—V anatomoses was reported not only in the denervated limb, but also in a limb with the paw amputated when the region with most numerous A—V anastomoses is eliminated (Piiper and Schoedel 1954). An increased number of A—V anastomoses was also found in the denervated muscle (Bostroem and Schoedel 1953, Dieter 1954), namely 3—60 min. after section of the nerve, i. e. at a time when blood flow is increased due to elimination of sympathetic vasoconstrictor tone. Preobrazhanskaya (1956) described the formation of A—V-anastomoses in the diaphragm even longer periods of time after denervation.

An investigation was made of the number of open A—V anastomoses in calf muscles of dogs 3—4 days after section of the sciatic nerve. In these experiments an indirect method was used: agar pellets 17—23 μ in diameter were injected into the femoral artery and their number was determined in the femoral vein. Agar pellets are compressible and do not thus traumatize the blood vessel walls. For this reason relatively small numbers of these pellets were used (about 6 000), in order to prevent more extensive embolization of the capillary bed. The number of open A—V anastomoses in denervated calf muscles of the dog is almost three times as large as that on the control side (Hudlická et al. 1959).

Increased blood flow in the denervated muscle is thus partly caused by increased flow through A—V anastomoses. However, for muscle nutrition the blood flow through the muscle capillary bed is decisive. It was thus necessary to find out whether capillary blood flow in the denervated muscle is not decreased.

The determination of capillary blood flow is methodically very difficult. Besides histological methods (e. g. Mchedlishvili 1957), which give a picture about the number of capillaries, but not about the amount of blood flowing through them, it is possible to obtain an idea about capillary circulation by measuring capillary pressure (Schroeder et al. 1953), or by plethysmography during which blood flow through capillaries is blocked by a pressure of 35 mm Hg (Schroeder 1955). The most commonly used method recently is measurement of Na^{24} clearance. A number of objections have been raised against this method, first described by Kety (1949) and Elkin et al. (1948), in which the decrease of activity is measured above the place of injection of physiological saline with Na^{24} into the muscle. Decreased activity could be due not only to blood circulation, but also to circulation of lymph or various degrees of capillary permeability (Burton 1953). It was convincingly demonstrated by Stone and Miller (1949) that the efflux of Na^{24} into the lymph is insignificant, since three hours after intramuscular injection only 1 % of the total amount of Na^{24} injected appears in the chylus; on the other hand the decreased activity in the muscle correlated very accurately with the rise in activity of venous blood. Walder (1953, 1955) found good correlation in man between values of clearance Na^{24} and plethysmography. A linear relationship between venous outflow from the muscle and clearance was also reported by Prentice et al. (1955). These authors also showed that diffusion and permeability factors are not important in this connection. Warner et al. (1953) pointed out that it is important to inject small amounts (the smaller the volume of injected solution, the larger the clearance and vice versa). Hyman et al. (1959) reported a good correlation between clearance values and direct venous outflow during constriction, but not during vasodilatation (excercise, adrenaline etc.), which is in agreement with Barcroft (1952). This can, however, be explained by the fact that during vasodilatation A—V anastomoses are usually opened up and there

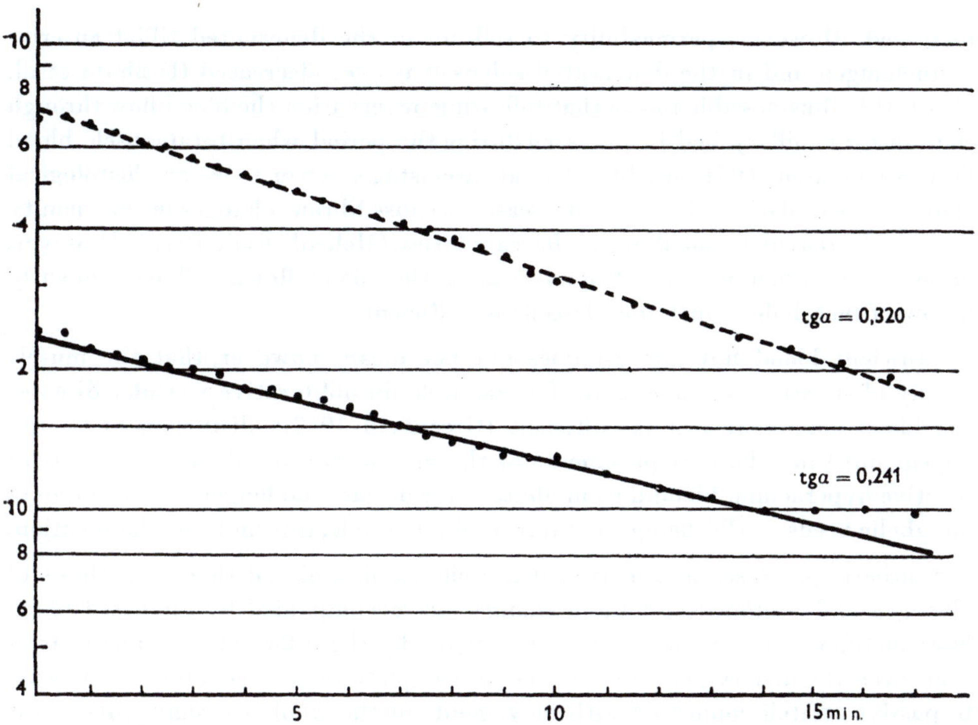

Fig. 72. Na^{24} clearance in the normal (solid line) and denervated (interrupted line) gastrocnemius of the dog. Abscissa: time in minutes. Ordinate: logarithm of the number of impulses.

is no correlation because part of the blood does not pass through capillaries where clearance measures blood flow. Walder (1953) and Hyman et al. (1957) consider clearance as a good method for measuring capillary blood flow. In our opinion, clearance of Na^{24} is a suitable method for measuring capillary blood flow if permeability of capillaries to Na^{24} is not changed. The clearance of Na^{24} was measured in the anterior tibial and gastrocnemius mucles of the dog three to four days after section of the sciatic nerve. Activity measured expressed as the logarithm of the number of impulses is in a linear relation to time. The values in different muscles can be compared when a comparison of some function of the curve is made. Instead of the constants of Kety, we compared tan α of the slope since always a relative comparison of two cases was required without necessity of calculating the absolute blood flow values. In normal animals (rabbits) the dispersion was $99{\cdot}3 \pm 6{\cdot}16\,\%$, while in the denervated muscles the ratio D/N was $165{\cdot}5 + 25{\cdot}6\,\%$. This means that the clearance of Na^{24} in the denervated muscle is significantly increased (Hudlická et al. 1959) — (Fig. 72). In order to be able to conclude that blood flow in the capillary bed is increased, it would be necessary to know that the permeability of capillaries for sodium is not increased. However, permeability to sodium in the denervated tibial anterior is unchanged and in the denervated soleus it is even decreased (Drahota et al. 1960). It is thus possible to say that following denervation the blood flow through the muscle capillary bed is increased during the period when total muscle blood flow is increased. It is possible that at later stages when there are histological changes in capillaries, blood flow decreases because fibrous changes in the vicinity lead to decreased blood flow in the capillaries (Mchedlishvili 1957). However, at the time when muscle atrophy begins, the blood flow, and consequently inflow of metabolites into muscle cells is sufficient.

Sufficient blood flow at rest does not yet mean, however, that the muscle supply of substances is adequate. For example, in obliterative vascular diseases the blood supply at rest is sufficient (Greenfield 1957); disturbances do not appear until metabolic requirements of the muscle are raised. After exercise or reactive hyperaemia, blood flow in obliterative diseases no longer suffices to cover metabolic needs — this being the cause of, for example, intermittent claudication.

Synthetic processes in denervated muscle are diminished (lower synthesis of glycogen and proteins) and these changes are accompanied by increased blood flow; metabolic requirements are thus covered by the inflow of substances. However, even the denervated muscle is not at complete rest. It is subjected at least to passive stretch connected with movements of the whole extremity, its metabolism changes in relation to more extensive changes of surrounding temperature, is sometimes stimulated during electrotherapy, at other times circulation can be restricted for transient periods of time by bandages, etc. It would then be possible that under such exceptional conditions the blood supply is inadequate and that metabolic over-shoot reactions, which are present in the normal muscle after muscle contraction, are impaired.

Blood flow in the normal muscle increases considerably after functional activity (Gaskell 1877, Verzár 1912, Keller et al. 1930, Anrep et al. 1934, Mertens et al. 1936, Grant 1938, Fleisch and Weger 1938, Schretzenmeyer 1933, Jalavisto et al. 1947. Barcroft and others 1952, Gollwitzer-Meier 1950, Donhorst and Whelan 1953, Barcroft and Swan 1953, Garbulinski et al. 1957). This increase lasts a certain time after termination of functional activity (Kramer and Quensel 1938, Barcroft and Swan 1953, Hilton 1953 and others) and spreads even to muscles which did not perform any work, on either the ipsilateral or contralateral extremity (Rein and Schneider 1930, Keller et al. 1930, Rein 1931, Anrep et al. 1934, Garbulinski et al. 1957 and others). Shortly after section of the peripheral nerve, the increase of blood flow after muscle stimulation is either the same (Keller et al. 1930, Rein 1931, Schretzenmeyer 1933) or smaller (Anrep et al. 1934) than in the control muscle. Since the resting values of blood flow in the denervated muscle are high, blood flow immediately after section of the nerve is never smaller than that in the normal muscle.

Three to four days after section of the nerve blood flow increases in the extremity following functional activity (Hudlická and Vrbová 1958). Even though the increase is smaller than in normal muscle, blood flow in the denervated limb reaches values analogous to those in the normal extremity (Fig. 73). The rate with which it returns to initial values after direct stimulation of the muscle was also the same as in the control limb. However, a late increase of blood flow was observed 60—90 min. after stimulation in the normal extremity, and this increase sometimes attained values found during functional activity. This late increase lasted for about 5—20 min. and was never observed in the denervated limb. The

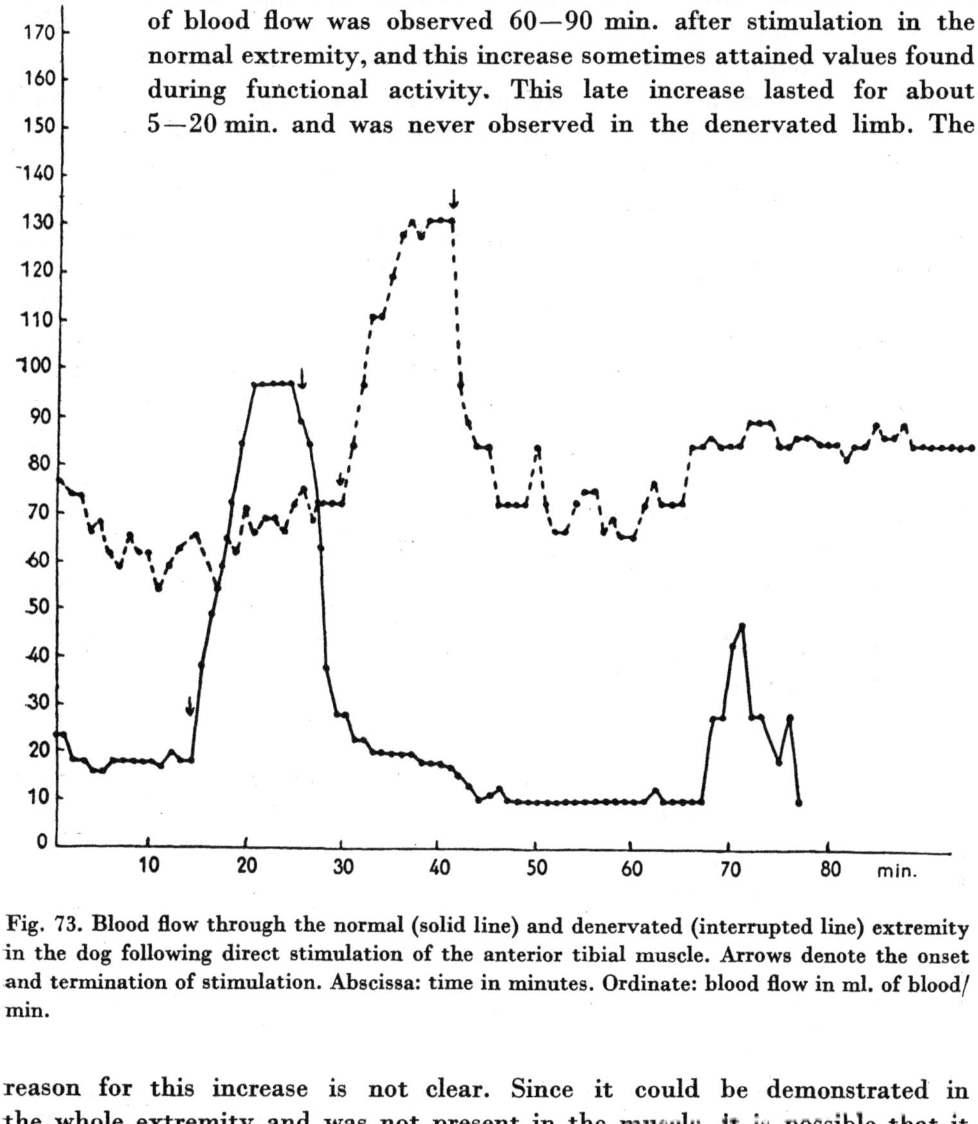

Fig. 73. Blood flow through the normal (solid line) and denervated (interrupted line) extremity in the dog following direct stimulation of the anterior tibial muscle. Arrows denote the onset and termination of stimulation. Abscissa: time in minutes. Ordinate: blood flow in ml. of blood/min.

reason for this increase is not clear. Since it could be demonstrated in the whole extremity and was not present in the muscle, it is possible that it involves blood flow in the skin — perhaps opening up of A—V anastomoses. It is therefore possible to say that blood flow in the denervated muscle is sufficient even after functional activity. After denervation, however, the increase of blood flow due to the activity of other muscles or of the contralateral extremity is missing (Keller et al. 1930, Anrep et al. 1934, Rein 1944). Seeing that such reflex vasodilatation is also absent after sympathectomy (Schretzenmeyer 1933), it is not probable that it is mediated by peripheral nerves.

A further question is related to the problem of whether reactive hyperaemia is possible in denervated muscle. This question is important from the practical point of view, since after denervation the blood supply of the muscle is often temporarily interrupted either because the main blood vessels had been injured, because bandages had been too tightly applied or because of unfavourable position of the limb. It is then important to renew a normal blood supply as soon as possible and to repay the debt which might have been made in the mean time.

Trauma to blood vessels causes disturbances of metabolism which, if extensive, can even lead to impairment of muscle function (Makins 1919). The blood supply is affected more, if vasospasm of the injured blood vessel spreads to collateral branches (Kinmonth et al. 1949, deSuoville 1956). Sometimes vasospasm can appear even on the contralateral extremity (Barnes and Trueta 1942) and the damage is greater when blood flow is occluded simultaneously in the artery and vein (Shepherd 1952). However, if collateral circulation is formed quickly, then blood vessel injury need not manifest itself in a disturbance of muscle metabolism and function. This is why, for example, Griffiths (1945) states that ligature of the iliac artery and vein does not lead to serious anoxaemia of the limb. Sympathectomy accelerates the formation of collateral circulation (Lambert 1957b, Beaconsfield 1954), but after damage of the central nervous system a smaller number of anastomoses is formed after arterial occlusion (Bardina 1956). Similarly, the development of collateral lymphatic circulation is retarded after denervation (Ljan 1957).

Table 9

Changes of blood flow in normal and denervated extremity of a dog during ligature of the abdominal aorta and after its release (in ml./min.).

Initial values		At the end of ligature		15 min. after loosening		One hour after		Four hours after	
N	D	N	D	N	D	N	D	N	D
8·2	9·9	5·2	6·9	8·4	9·0	7·1	9·1	8·3	7·9

After removal of the obstruction, a marked increase of muscle circulation occurs — reactive hyperaemia (Rein 1944, Schoop 1956, McArdle and Verel 1956, Lambert 1957c, 1957d and others) which is sometimes even greater than that after functional activity. Sympathectomy does not affect the extent of this hyperemia (Malméjac and Haimovici 1936, Lambert 1957), except, perhaps, prolonging its duration (Beaconsfield 1954). It is not clear, whether reactive hypaeremia changes or not after denervation. According to Malméjac and Haimovici it is the same as in normal muscles, while Mercker and Urtig (1942) found no reactive hyperaemia in the denervated extremity. In our experiments, we were unable to find any basic difference in the course of circulatory changes during ligature of the abdominal aorta at various intervals after releasing the ligature in the normal and denervated limb of the dog (Table 9).

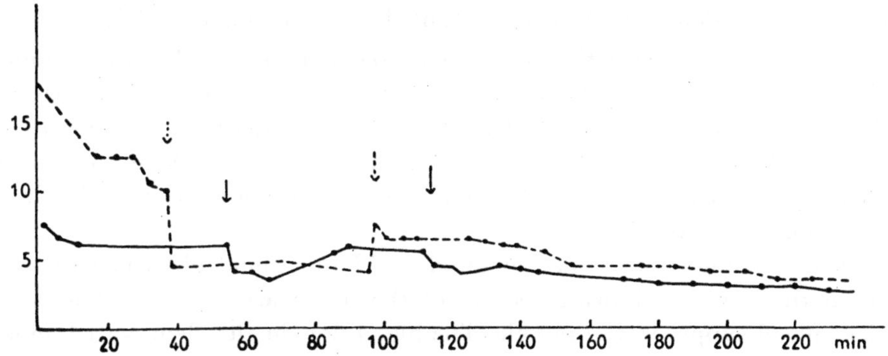

Fig. 74. Blood flow through the normal (solid line) and denervated (interrupted line) gastrocnemius of the dog during ligature of the abdominal aorta. Arrows denote the beginning and release of ligature. Abscissa: time in minutes. Ordinate: blood flow in ml./100 g. of tissue/min.

It seems, however, that no collateral circulation was being formed in the denervated gastrocnemius in the course of a 60 min. ligature (Fig. 74); after releasing the ligature no differences in blood flow were found between the denervated and normal muscle. In rats greater vasodilatation was found histologically in denervated muscle 24 hours after loosening the ligature (Hudlická and Gutmann 1958).

An insufficient blood supply leads to serious changes in the muscle. Prolonged arrest of circulation causes paresis (Makins 1919) with either massive necrosis or extensive interstitial fibrosis (Bowden and Gutmann 1949, Scully and Hughes 1956). Tourniquet or ligature of blood vessels is accompanied by a decrease of muscle glycogen, ATP, creatinephosphate and an increase of lactic acid (Bollman and Flock 1944, Harman and Gwinn 1949, Fuhrman and Crismon 1951, Fetisova 1954, Hudlická and Gutmann 1958 and others): after tourniquet of longer duration (five hours) the metabolic changes are practically irreversible — ATP is broken down and histological changes set in (McPherlane and Spooner 1946, Harmann and Gwin 1949). There are also considerable changes in the ion composition: extrusion of potassium from the muscle and an influx of sodium into the muscle takes place, with an increase in the water content (McPherlane and Spooner 1946, Fox and Baer 1947, Fuhrmann and Crismon 1951, Mc Phee 1955 and others) and decreased uptake of K^{42} (Drahota and Hudlická 1958). If insufficient blood supply was the reason for metabolic disturbances in the denervated muscle, then further damage should be caused in such a muscle by limiting the blood supply further. However, there were no differences registered during ligature of the abdominal aorta and the caudal vena cava between the normal and denervated muscle in either the breakdown of glycogen or creatinephosphate — no substantial metabolic disturbance occurred in the denervated muscle. On the contrary, after releasing the ligature the restitution of glycogen was accelerated in the denervated muscle (Hudlická and Gutmann 1958).

It can thus be concluded that the denervated muscle is sufficiently supplied with blood even under circumstances when the muscle is functionally stressed, i. e. by contractions and reactive hyperaemia. In addition to this, restriction of blood flow does not cause more serious consequences than in the normal muscle and, as far as glycogen metabolism is concerned, a transient arrest of circulation rather stimulates its synthesis.

It is thus possible to say that metabolic and functional disturbances which appear in the muscle following section of the nerve are not caused by an insufficient blood supply in the muscle. The blood flow in the denervated muscle is increased, and this increase includes the capillary bed, i. e. the region most important for metabolism.

If the blood flow in the denervated muscle is sufficient, it does not necessarily mean that sufficient quantities of substances essential for covering muscle metabolic requirements enter into the muscle. Not only must substances be brought into muscle capillaries by the blood, but they must obtain access into the muscle extracellular space and hence into the inside of muscle cells. This means that besides knowing the blood flow, the uptake of substances in the muscle must be determined, i. e. the amounts of substances which pass through the capillary walls into the extracellular space and then through the muscle cell membrane to the muscle cells in order to be able to talk about sufficient nutrient supply to muscles.

B. Uptake of substances in the muscle

The uptake of substances by muscle cells depends on the permeability of both capillaries and muscle cells. The permeability of capillaries was considered for a long time to be primarily a physico-chemical phenomenon (e. g. Landis 1946). Kühnau (1955), however, demonstrated how the uptake of individual substances is influenced by enzyme activity of the capillary walls. The permeability of the muscle cell membrane is clearly dependent upon metabolism (e. g. Teorell 1949, Steinbach 1954). It would thus be logical to expect that changes in metabolism occuring after denervation would be reflected in changes of permeability. At the same time it is necessary to realize that some substances can pass through capillaries only and others can enter into the muscle cell itself. For example, labelled albumins or dyes can pass only through the capillary wall. In addition they usually have the disadvantage of relatively large molecular size and are thus more suitable for investigation of capillary permeability only when the capillary wall is already pathologically damaged. On the contrary the uptake of sodium can, under certain conditions, give accurate information only about the permeability of capillaries. Sodium enters the inside of muscle cells only relatively little (Hevesy 1948, Morell 1950), while it passes quite freely through the capillary wall under normal conditions. The influx of sodium into the muscle cell can be restricted by administering a relatively large dose of radioactive sodium which produces a cosiderable concentration gradient while minimizing the time between administration of the isotope and autopsy (Hudlická 1955).

Investigation of muscle cell permeability is far more difficult, since it is not possible to talk about permeability of the muscle cell as such. The muscle cell is almost impermeable to some substances (chlorides), for others it is permeable only in one direction (glucose). It is therefore necessary when studying the muscle cell permeability to refer to a concrete substance. Phosphorus, for example, enters into the cell only in the form of organic compounds and P^{32} found in organic compounds in the muscle means that phosphorus has entered the cell. Similarly K^{42} uptake can be considered as a certain criterion of muscle cell permeability. By measuring of uptake of different radioactive ions in the muscle the changes in both capillary and muscle cell membrane permeability can be determined.

What are the nervous factors determining the permeability of capillaries and muscle cell membranes?

Sympathectomy increases the permeability of muscle capillaries (Gabbe 1926, Hesselmann 1932). As far as the effect of peripheral nerves on the permeability of muscle cells is concerned, Behrendt (1922) described differences in the permeability to phosphorus and sucrose in muscles in relation to their function: semimembranosus muscle of the frog exhibited higher permeability for sucrose and phosphorus than the gastrocnemius. According to Bogatzki (1952) there is an increased permeability in the denervated frog sartorius (tested with the aid of the demarcation potential). It seems that the nervous system influences the permeability of both the capillaries and muscle cell membranes. It is uncertain whether this influence can be of significance in the mechanism of metabolic disturbances in the denervated muscle. This possibility was already postulated by Tschermak-Seysenegg (1933), who assumed that trophic disturbances occurring after denervation could be caused by decreased permeability. The findings mentioned above, on the contrary, indicate quite the reverse.

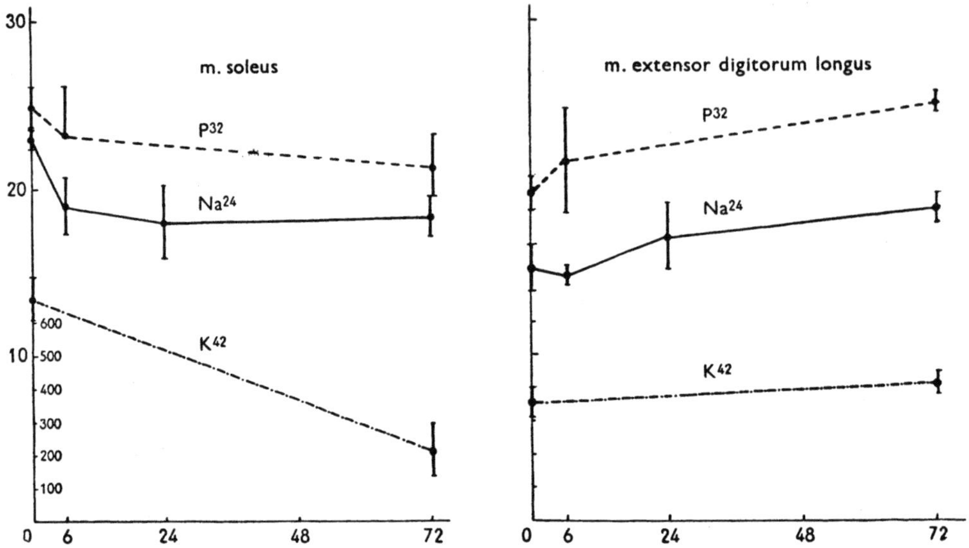

Fig. 75. Uptake of Na^{24}, P^{32} and K^{42} in the rat soleus muscle (left half of figure) and extensor digitorum longus (right half) after denervation. Abscissa: time in hours. Ordinate: activity in the muscle expressed as percentage of activity in the plasm (for Na^{24} and P^{32}) and as number of impulses g of tissue/min. (for K^{42}).

In order to elucidate this question changes in the uptake of Na^{24}, P^{32} and K^{42} were investigated in rat muscles at various intervals after nerve section (Hudlická 1955, Hudlická et al. 1956, Drahota and Hudlická 1960).

In agreement with Behrendt, different uptake of sodium, phosphorus and potassium was found in muscles belonging to different functional groups. The uptake of all three ions was found to be higher in m. soleus than in the extensor digitorum longus (Fig. 75). These differences could be explained on the basis of differences in blood circulation: Smith and Giovacchini (1956) demonstrated that in red muscles, such as the soleus muscle in the rat, there are 1·5 capillaries per muscle fibre, while this ratio is 0·2—0·9 capillaries in white muscles (e. g.

Table 10

Uptake of P^{32} in muscles of the rat after sympathectomy (expressed in % of plasma activity).

No. of animals	Activity in soleus		Activity in extensor dig. longus	
	Experimental	Control	Experimental	Control
17	21·9 ± 1·5	20·9 ± 1·7	13·9 ± 1·2	14·2 ± 1·08

extensor digitorum longus or the anterior tibialis muscle in the rat). Of course, the structure of capillaries only indicates the potential possibilities of circulation. The actual capillary blood flow measured by the Na^{24} clearance was found to be the same for soleus and extensor digitorum longus (in the former tan $\alpha = 0{\cdot}238 \pm 0{,}04$ and for the latter tan $\alpha = 0{\cdot}213 \pm 0{,}068$). If these differences of uptake in different muscles were given by differences of blood flow, then they should disappear after sympathectomy which increases blood flow in the soleus as well as in the extensor digitorum. Howerer, the uptake of P^{32} after sympathectomy remains unchanged in both these muscles (Hudlická 1955) (table 10).

Differences in uptake are thus evidently not given by different blood flow, but are more likely to be connected with different functional characteristics of individual muscles. In favour of this functional relation is the finding that following section of the nerve the uptake in the soleus diminishes while in the extensor digitorum longus it increases (v. Fig. 75). These changes appear gradually. Six hours after section of the nerve the uptake of Na^{24} and P^{32} is still higher in the soleus than in the extensor digitorum. This difference levels out three days after denervation. Differences in uptake also level out after transec-

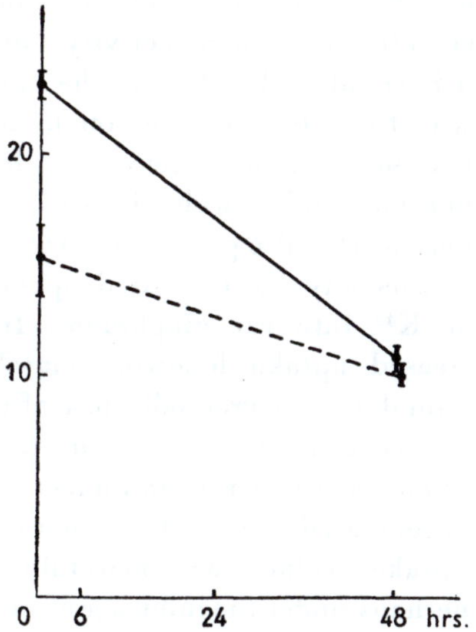

Fig. 76. Na^{24} uptake into the rat soleus muscle (solid line) and extensor digitorum longus (interrupted line) 48 hrs. after transection of the spinal cord (as percentage of activity in the plasm).

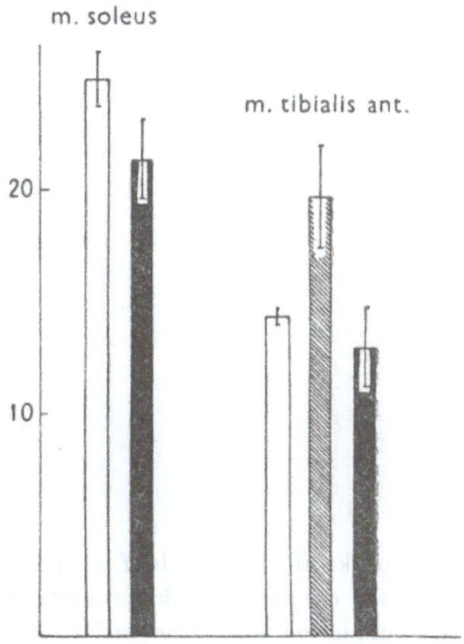

Fig. 77. Uptake of P^{32} in the rat soleus muscle and tibialis anterior expressed as percentage of activity in the plasm. White columns — normal muscle.
Hatched column — normal muscle 4 hrs. after direct stimulation. Black columns — denervated muscle (m. tibialis ant.) after direct stimulation.

tion of the spinal cord, but here uptake was reduced considerably more than after denervation in both muscles (Fig. 76). It is possible that differences in uptake are maintained by higher nerve centres (Hudlická 1955).

What is the reason that decreased uptake after section of the peripheral nerve is found only in some muscles? Physiological extensor muscles, the uptake of which (sodium and phosphorus) is higher, are functionally active practically all the time, which means that they will have higher metabolic requirements. On the other hand physiological flexors were found to have lower uptake, which increased considerably after functional activity (Fig. 77). However, increased uptake after direct muscle stimulation is absent in the denervated anterior tibial muscle (Fig. 78). This means that in muscles requiring increased influx of substances after functional activity (and it is possible that this is the case with the soleus all the time), the nervous system is necessary for normal uptake mechanisms in the muscle. It is also possible to explain in this way the decreased uptake of K^{42} into rat diaphragm. Increased uptake following muscle stimulation is gradually lost after denervation, so that it seems that as long as the nerve stump has not degenerated completely, normal uptake values are maintained, perhaps under the influence of substances contained in this stump, as has been demonstrated for the glycogen content by Gutmann et al. (1954).

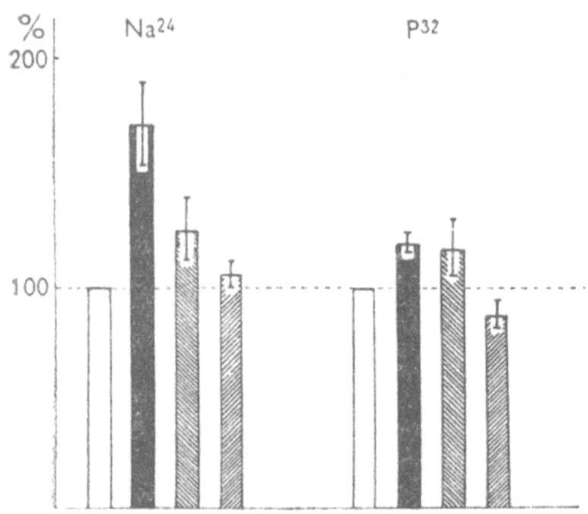

Fig. 78. Uptake of Na^{24} (left half of figure) and P^{32} (right half) in the anterior tibial muscle after direct muscle stimulation and denervation expressed as percentage of the activity in the control muscle. White columns — control muscle. Black columns — normal muscle 4 hrs. after stimulation ▨ 4 hrs. after stimulation, and 36 hrs. after section of the sciatic nerve, ▨ 4 hrs. after direct stimulation and 72 hrs. after nerve section. Interrupted line — 100 % (activity of control side).

The uptake in sympathectomized animals is not decreased after muscle stimulation. It would thus appear that the sympathetic nervous system does not participate in the regulation of uptake in the muscle, and this is true under normal conditions without load (Hudlická 1955). The differences in uptake between muscles with different functional activity are then maintained by the somatic neuron. It is questionable whether decreased uptake following section of the nerve into some muscles can be of such magnitude as to explain the appearance of denervation atrophy. This is not very probable since no fundamental differences in the rate of atrophy have been found between the soleus and extensor digitorum longus. Moreover, the rat diaphragm hypertrophies at the same time that uptake of K^{42} is decreased. Changes in the uptake of sodium can rather be explained by changes in extracellular space which are not important from the point of view of intensity of muscle metabolism. On the other hand, it has been reported in papers dealing with the content of radioactive phosphorus in organic compounds of phosphorus in the muscle — and which determine the uptake of P^{42} inside the muscle cells — that the permeability of the muscle fibre is increased in denervated muscles, including physiological extensor muscles (Bass 1958).

When summarizing the role of vasomotor mechanisms during denervation atrophy, it may be concluded that blood flow in the denervated muscle is increased not only immediately after section of the nerve, but also at a time when marked atrophy occurs. Although the number of open a—v anastomoses increases, blood-flow in the muscle capillary bed is also increased. The uptake of some inorganic substances is decreased in some muscles after denervation, probably in connection with changes in the extracellular space (see Chapter VI). The uptake of substances into muscle cells is increased. In spite of this denervation atrophy sets in. It is therefore probable that an insufficient blood or nutritional supply cannot be the cause of denervation atrophy. However, it is necessary to know, whether a denervated muscle is able to utilize the substances supplied. This will be discussed in the following chapter.

References

Abramson D. J., Ferris E. B.: Amer. Heart J. 19 : 541, 1940.
Allen N. J., Barcroft H., Edholm O. G.: J. Physiol. 105 : 255, 1946.
Anrep G. V., Blalock A., Saaman A.: Proc. Roy Soc. B 114 : 223, 1934.
Anrep G. V., Cerqua S., Saaman A.: Proc. Roy. Soc. B 114 : 245, 1934.
Anrep G. V., Saalfeld E.: J. Physiol. 85 : 375, 1935.
Barcroft H.: Visceral circulation. Ciba Symposium, London 1952.
Barcroft H., Bock K. D., Hensel H., Kitchin A. H.: Pflügers Archiv 261 : 199, 1955.
Barcroft H., Donhorst Q. C.: J. Physiol. 109 : 402, 1949.
Barcroft H., Donhorst A. C.: Peripheral Circulation in man. Ciba Symposium. London 1954.
Barcroft H., Donhorst A. C., Mc Clatchey H. M., Tanner J. M.: J. Physiol. 117 : 391, 1952.
Barcroft H., Edholm O. G.: Lancet 513, 1946.
Barcroft H., Konzett H.: J. Physiol. 110 : 194, 1949.
Barcroft H., Swan H. J. C.: Sympathetic control of human blood vessels. London 1953.
Bardina R. A.: Arkhiv Anat. gist i embriol. 33 : 55, 1956.
Barnes J. M., Trueta J.: Brit. J. Surg. 30 : 74, 1942.
Bass A., Hudlická O.: Physiol. bohemoslov. 9 : 408, 1960.
Bass A.: Čs. Fysiol. 7 : 1958.
Beaconsfield P.: Ann. Surg. 140 : 786, 1954.
Bean J. W., Elwell L. H.: Am. J. Physiol. 164 : 734, 1951.
Behrendt H.: Hoppe-Selyers Ztschr. 118 : 123, 1922.

Bilisoby F. M., Goodell H., Wolff H. G.: AMA Arch. Int. Med. 94 : 759, 1954.
Bloch E. H.: Am. J. Med. Sci 226 : 24, 1953.
Bogatzki M.: Zeitschr. ges. exper. Med. 118 : 544, 1952.
Bollmann J. L., Flock E. V.: Am. J. Physiol. 142 : 290, 1944.
Bostroem B., Schneider P. W.: Pflügers Archiv 257 : 241, 1953.
Bostroem B., Schoedel W.: Pflügers Arch. 256 : 371, 1953.
Bowden R. E.: Peripheral nerve injuries. London 1954.
Bowden R. E., Gutmann E.: J. Bone and Joint Surg. 31 B : 356, 1949.
Boyd J. D.: Visceral C rculation. Ciba Symposium, London 1952.
Brück K., Hensel H., Poche R., Rotzler A., Spang K.: Klin. Wschr. 30 : 889, 1952.
Burn J. H., Rand M. J.: Abstracts of the XXI. Congress of Physiol. Sci., p. 47, Buenos Aires 1959.
Burton A. C.: Am. Rev. Physiol. 15 : 213, 1953.
Cannon W. B., Rosenblueth A.: The supersensitivity of denervated structures, New York 1949.
Cassierer, Hirschfeld: Vasomotorisch-trophische Erkvankungen in Kraus-Brugsch: Spez. Pathologie und Therapie innerer Krankheiten vol. *9*, 1924.
Celander O., Folkow B.: Acta physiol. scand. 29 : 359, 1953a.
— — Acta physiol. scand. 29 : 241, 1953b.
Chambers R., Zweifach B. W.: Phys. Rev. 27 : 456, 1947.
Chen K. K., Meek W., Bradley H. C.: J. Biol. Chem. 61 : 807, 1926.
Chor H., Dolkart R. E., Davenport H. A.: Am. J. Physiol. 118 : 580, 1937.
Cloninger G. L., Green H. D.: Am. J. Physiol. 181 : 258, 1955.
Cooper K. E., Edholm O. G., Mottram R. F.: J. Physiol. 123 : 33P, 1954.
Dale H. H., Gaddum J. H.: J. Physiol. 70 : 109, 1930.
Dieter E.: Pflügers Archiv 258 : 470, 1954.
Van Dobben-Broekema M., Dirken M. N. J.: Acta physiol. pharmacol. Neerl. 1 : 584, 1950.
Donhorst A. C., Sharpey-Schafer E. P.: Clin. Sci. 10 : 371, 1951.
Donhorst A. C., Whelan R. F.: Clin. Sci. 12 : 33, 1953.
Dörner J.: Pflügers Archiv 262: 265, 1956.
Dörner J.: Kuschke H. J.: Arch. exper. Pathol. u. Pharmacol. 224 : 368, 1955.
Doupe J.: J. Neurol. Neurosurg. Psych. 6 : 97, 1943.
Doupe J., Barnes R., Kerr A. S.: J. Neurol. Neurosurg. Psych. 6 : 136, 1943.
Drahota Z., Hník P., Hudlická O.: Babákova sbírka 16 : 152, 1960.
Drahota Z., Hudlická O.: Physiol. bohemoslov. 7 : 489, 1958.
Drahota Z., Hudlická O.: Nature 185 : 396, 1960.
Duff R. S.: Clin. Sci. 10 : 529, 1951a.
— J. Physiol. 113 : 9P, 1951b.
— Amer. J. Phys. Med. 33 : 109, 1954.
Duff P., Shepherd J. T.: J. Physiol. 122 : 26P, 1953.
Ederstrom H. E., Vergeer T., Rohde R. A., Ahlness P.: Am. J. Physiol. 187 : 461, 1956.
Elkin D. C., Cooper R. K., Rohrer R. K. B., Miller W. B., Shea P. C., Dannis E. W.: Surgery, Obst. Gyn. 87 : 1, 1948.
Fetisova T. B.: Voprosy biochimiji myshts, Kiev, 1954.
Feucht B. L., Richardson A. W., Hines H. M.: Arch. Phys. Med. 30 : 687, 1949.
Fleisch A.: Acta neuroveget. 14 : 88, 1956.
Fleisch A., Weger P.: Pflügers Archiv 240 : 553, 1938.
Folkow B.: Ann. Rev. Physiol. 18 : 159, 1956.
— Acta physiol. scand. 27 : 118, 1953 a.
— Acta physiol. scand. 27 : 99, 1953b.
Fox C. L., Baer H.: Am. J. Physiol. 151 : 155, 1947.

Friedlander M. Silbert S., Bierman W.: Am. J. Med. Sci. 199: 257, 1940.
Fuhrman F. A., Crismon J. M.: Am. J. Physiol. 167 : 289, 1951.
Gabbe E.: Ztschr. ges. exper. Med. 51 : 728, 1926.
Garbulinski T., Popandink L., Bula B.: Polski tugod. lek. 12 : 381, 1957.
Gaskell W. H.: J. Anat. Lond. 11 : 360, 1877.
Goetz R. H., Du Toit J. G., Sewart B. H.: Acta med. scand. supply 306 : 56, 1955.
Golenhofen K. Hildebrandt G.: Ztschr. Kreislaufforsch. 46 : 257, 1957.
— — Pflügers Archiv 264 : 492, 1957b.
Gollwitzer-Meier K.: Lancet 371 : 1950.
Goltz F.: Pflügers Archiv 9 : 17, 1874.
Gottstein V., Hille H., Oberdorf A.: Pflügers Archiv 261 : 78, 1955.
Graf K., Rosell S.: Acta physiol. scand. 42 : 51, 1958.
Grant R. T.: Clin. Sci. 3 : 157, 1938.
— Vascular reactivity following sympatectomy. Peripheral Circulation in man. Ciba, Symp. London 1954.
Grant R. T., Pearson R. S. B.: Clin. Sci. 3 : 119, 1938.
Greenfield A. D.: Am. J. Med. 23 : 675, 1957.
Griffin P. D., Green H. D., Youmans P. L., Johnson H. D.: J. Pharmacol. exper. Therapeutics 110 : 95, 1954.
Griffiths D. L.: Brit. J. Surg. 28 : 239, 1945.
De Groot A. L.: IInd Congress of Physiol. Medicine, Kobenhavn 1956.
Gutmann E., Vodička Z., Vrbová G.: Physiol. bohemoslov. 3 : 182, 1954.
Harman J. W., Gwinn R. P.: Am. J. Path. 25 : 741, 1949.
Harpunder K., Byer J., Stein I.: Am. K. Physiol. 150 : 181, 1947.
Hensel H.: Klinische Wochenschrift 34 : 1223, 1956.
Hesselman J.: Ztschr. f. Biol. 92 : 287, 1932.
Hertzman A. B.: Am. J. Phys. Med. 32 : 233, 1953.
— Physiol. Rev. 39 : 280, 1959.
Hevesy G.: Radioactive indicators, New York 1948.
Hilton S. M.: J. Physiol,. 120 : 250, 1955.
Hudlická O.: Physiol. bohemoslov. 4 : 276, 1955.
— Sborník prací konference o nervovém řízení metabolismu. Praha 1958.
Hudlická O., Bass A., Vodička Z.: Physiol. bohemoslov. 5 : 63, 1956.
Hudlická O., Gutmann A.: Arch. exp. Path. u. Pharmak. 234 : 501, 1958.
Hudlická O., Kovách A. G., Boloday-Varga A.: Acta physiol. hung. 16 : 51, 1959.
Hudlická O., Vrbová G.: Physiol. bohemoslov. 7 : 38, 1958.
Hyman C.: Angielogie 9 : 25, 1957.
Hyman C., Rossel S., Rosen A., Sonnenschein R. R., Uvnäs B.: Acta physiol, scand. 46 : 358, 1959.
Hyman C., Stone A., Sosnow M.: Proc. Soc. exp. biol. 83 : 643, 1953.
Imig C. J., Randall B., Hines H. M.: Am. J. Phys. Med. 33 : 100, 1954.
— — — Am. J. Physiol. 169 : 22, 1952.
Imig C. J., Roberson W. J., Hines H. M.: Am. J. Physiol. 186 : 35, 1956.
Issekutz B. J., Hetenyi G. J., Winter M., Mosonyi M., Forbath M.: Acta physiol. hung. 10 : 327, 1956.
Jalavisto E., Mertens O., Schoedel W.: Pflügers Archiv 249 : 167, 1947.
Keller Ch. J., Loesser A., Rein H.: Ztschr. f. Biol. 90 : 260, 1930.
Kemp C. R., Paul D., Hines H. M.: Arch. Phys. Med. 29 : 12, 1948.
Kemp C. R., Tuttle W. W., Hines H. M.: Am. J. Physiol. 150 : 705, 1947.
Kessel A. W. L., McPhearson A.: Clin. Sci. 14 : 361, 1955.

Kester N., Richardson A. V., Green H. D.: Am. J. Physiol. 169 : 687, 1952.
Kety S. S.: Am. Heart J. 38 : 321, 1949.
Keyssler H., Schmier J.: Pflügers Archiv 253 : 301, 1951.
Khlopina I. D.: DAN 92 : 839, 1953.
Kinmonth J. B., Simeone F. A., Perlow V.: Surgery 26 : 452, 1949.
Kopits I.: Arch. Orthop. u. Unfall-Chir. 27 : 277, 1929.
Kramer K., Quensel W.: Pflügers Archiv 239 : 620, 1938.
Krogh A., Haropp C. A., Rehberg P. B.: J. Physiol. 56 : 179, 1922.
Kühnau J.: Kapillaren und Interstitium. Hamburger Symposium 1955.
Lambert J.: Angeioloie 9 : 11, 1957a.
— Arch. Int. Physiol. Biochem. 45 : 46, 1957b.
— Arch. Int. Physiol. Biochem. 45 : 504, 1957c.
— Medizinische 45 : 1666, 1957d.
Lambert J., Neven C., Stevens A.: Arch. Int. Physiol. 59 : 451, 1951.
Landis E. M: Ann. of the N. Y. Academy of Sci. 46 : 713, 1946.
Langley J. N.: J. Physiol. 58 : 70, 1923.
Lapinsky M.: Virchows Arch. 183 : 1, 1906.
Levashiev: 1880. Cit. by Lapinsky.
Leriche R.: Angiology 3 : 222, 1952.
Lindgren P., Uvnäs B.: Acta physiol. sand. 29 : 137, 1953.
— — Acta physiol. scand. 33 : 108, 1955.
Ljan, J. M.: Arkh. anat. 34 : 70, 1957.
Luckner H.: Kapillaren und Interstitium, Hamburger Symposium 1955.
Mc Ardle B., Verel D.: Clin. Sci. 15 : 305, 1956.
Mc Phee: J. Clin. Sci. 14 : 451, 1955.
Mc Pherlane M. G., Spooner S. J. L.: Brit. J. exp. Pathol. 27 : 339, 1946.
Makins G. H.: On gunshot injuries to the blood vessels. John Wright et al., Bristol 1919.
Malméjac J., Haimovici H.: C. r. Soc. biol. 88 : 23, 1936.
Mchedlishvili G. I.: Usp. sov. biol. 43 : 82, 1957.
Mercker H., Urtig G.: Pflügers Arch. 245 : 756, 1942.
Mertens O., Rein H., Valdecasas F. G.: Pflügers Archiv 237 : 454, 1936.
Moore R. M., Moore R. E.: Am. J. Physiol. 104 : 267, 1933.
Morel: Acta physiol. helv. 8 : 146, 1950.
Mottram R. F.: J. Physiol. 125 : 57, 1954.
Mulikov A. I.: Arkiv biolog. nauk 46 : 79, 1937.
Murray J. L., Thompson J. W.: J. Physiol. 135 : 133, 1957.
Okuyama Y.: Biochem. Z. 175 : 18, 1926.
Page I. H.: Bull. New York Acad. Med. 28 : 131, 1952.
Paul W. D., Imig. C. J.: Am. J. Phys. Med. 34 : 370, 1955.
Piiper J., Schoedel W.: Pflüg. Archiv. 258 : 489, 1954.
Prentice J. C., Stahl R. R., Dial N. Q., Ponterio F. V.: J. Clin. Invest. 34 : 545, 1955.
Preobrazhanskaya I. N.: Bjul. exper. biol. med. 41 : 72, 1956.
Quidot J., Notali J.: Presse Méd. 61 : 868, 1953.
Randall W. C., Alexander W. F., Cox J. W., Hertzman A. B.: Circulation. Res. 1 : 16, 1953.
Randall W. C., Alexander W. T., Hertzman A. B., Cox J. W., Henderson W. P.: Am. J. Physiol. 160 : 441, 1950.
Randall B. F., Imig C. J., Hines M. H.: Arch. Phys. Med. 33 : 73, 1952.
Rein H.: Pflügers Archiv 248 : 100, 1944a.
— Pflügers Archiv 248 : 111, 1944b.

Rein H.: Ergebnisse Physiol. 32 : 28, 1931.
— Schneider M. Z.: Biol. 91 : 13, 1930.
Rein H., Mertens O.: Pflügers Archiv 237 : 231, 1936.
Richards R. L.: Peripheral nerve injuries. London 1954.
Richardson A. W.: Am. J. Phys. Med. 33 : 103, 1954.
Richardson A. W., Imig Ch., Feucht B., Hines H. M.: Arch. Phys. Med. 31 : 19, 1950.
Ricker G.: Arch. Path. Anat. Physiol. 165 : 263, 1901.
Ricker G., Ellebeck: Virchows' Archiv 158 : 199, 1894.
Schneider W.: Biochem. Z. 173 : 11, 1926.
Schoedel W.: Pflügers Archiv 237 : 190, 1936.
Schretzenmayer A.: Pflügers Archiv 232 : 743, 1933.
Schroeder W.: Verh. dtsch. ges. Kreislauf. 289 : 1952.
— Kapillaren und Interstitium. Symposium 1954. Thieme Verlag, Stutg. 1955.
Schroeder W., Schoop W., Stein E.: Pflügers Archiv 259 : 124, 1954.
Schroeder W., Stein E., Schoop W.: Pflügers Archiv 258 : 8, 1953.
Scully R. E., Hughes C. W.: Am. J. Path. 32 : 805, 1956.
Seeling M. G., Lyon E. P.: JAMA 52 : 45, 1909.
Shepherd J. T.: Circulation 6 : 281, 1952.
Shoop W.: Klin. Wchschr. 34 : 12, 1956.
Siems L. L., Kosman A. J.: Am. J. Physiol. 156 : 2, 1949.
Smith D. R., Giovacchini R. P.: Acta anat. 28 : 342, 1956.
Staubesand J.: Kapillaren und Interstitium. Hamburger Symposium. Stuttgart 1955.
Steinbach B.: Symp. Soc. Exper. Biol. VIII : 438, 1954.
Stillwell G. K.: Arch. Phys. Med. 38 : 682, 1957.
Stone P. W., Miller W. B.: Proc. Soc. Exper. Biol. Med. 71 : 529, 1949.
Sunderland S., Ray L. J.: J. Neurol. Neurosurg. Psychiat. 13 : 159, 1950.
de Suoville A.: Arch. Int. Physiol. Bioch. 64 : 213, 1956.
Teorell H.: Ann. Rev. Physiol. XI: 545, 1949.
Tschermak-Seysenegg A.: Medizinische Klinik 29 : 213, 1933.
Ukolova M. A., Bobryushkov J. N.: Bjul. exper. biol. med. 44 : 24, 1957.
Uvnäs B.: Physiol. Rev. 34 : 608, 1954.
Verzar F.: J. Physiol. 44 : 243, 1912.
Walder D. N.: Clin. Sci. 12 : 153, 1953.
— Clin. Sci. 14 : 303, 1955.
Warner C. F., Dobson E. L., Pace M., Johnston M. E., Finney C. R.: Circulation 8 : 732, 1953.
Weinstein L.: Circulation 15 : 735, 1957.
Went I., Szücz E., Hetényi E.: Acta physiol. hung. 9 : 193, 1956.
Whelan R. F.: J. Physiol. 118 : 575, 1952.
Whitten W. L., Kramer J. G.: Arch. Phys. Med. a. Rehab. 34 : 552, 1953.
Wiggers H. C., Duschatko A., Kory R. C.: Am. J. Physiol. 136 : 86, 1942.
Zellweger H., Mort H.: Helv. pediat. acta 5 : 434, 1950.
Zweifach B. W.: Josiah Macy Foundation, Caldwell 1949.
Zweifach B. W., Metz D. B.: Angiology 6 : 282, 1955.

Chapter VIII

ENERGY METABOLISM IN DENERVATED MUSCLE

Bass A.

A. Energy metabolism in normal muscle

1. Introduction

Specific function of the muscle is represented by muscular work, or, more accurately, by muscle contraction. This work is mechanical in character and the energy needed is obtained from biochemical reactions. It is thus possible to say that specific muscle function is based on the transfer of chemical energy to mechanical energy. The muscle utilizes this energy for its specific function and also for the production, maintenance and renewal structure and chemical composition and for adaptation to new stimuli, in other words for trophic processes which ensure its specific function. Trophic processes also take place in the muscle at rest, but they are especially accentuated after functional activity.

Liberation of energy which the muscle can utilize is associated with the oxidation of substrates. Even though muscles represent about two thirds of the active body mass in adult mammals and their participation in respiration is considerable even at rest, the oxygen consumption per unit weight of tissue is quite small as compared with other organs. Besides the relatively low oxygen consumption values at rest, there are still other metabolic characteristics resulting from adaptation of the muscle to its specific function during the evolutionary process. These metabolic differences are characteristic in the way the muscle utilizes various sources of energy from substrates of energy metabolism for its basal resting consumption, for the production of stores covering energy requirements during activity and finally for recuperation following functional activity.

2. Sources of energy in muscle at rest and their utilization

The best known sources of energy are blood glucose and muscle glycogen, i. e. carbohydrates. Fletcher and Hopkins (1907), Meyerhof (1930) and Hill (1950) maintained, mainly from their *in vitro* experiments, that carbohydrates are almost the sole, and certainly the most important source of energy in muscle. This conclusion was based on the investigation of glycogen and glucose consumption, intermediate metabolites of glycolysis and the respiratory quotient (RQ) in isolated muscle. Thus, according to this theory, carbohydrates were

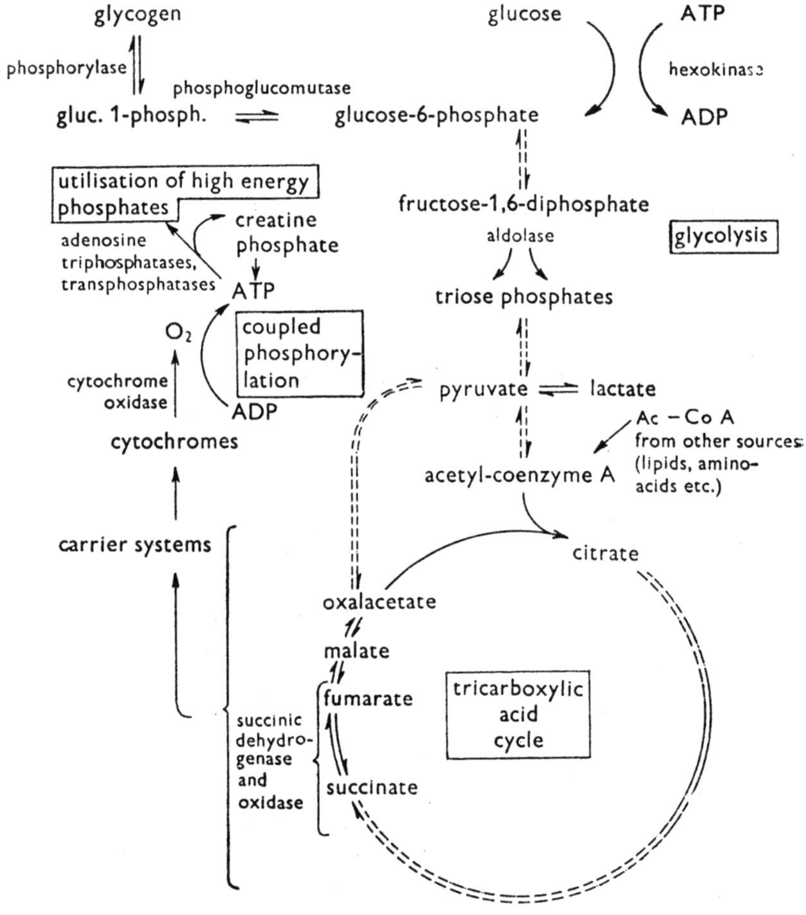

Fig. 79. Diagramatic representation of carbohydrate, lipid and energy metabolism in the muscle.

considered to be the main source of energy in the muscle, and this theory has also been accepted by most text-books.

In muscle, carbohydrates serve as substrate for glycolysis, which has already been extensively studied from the beginning of the century. A simplified diagram of glycolysis and oxidative muscle metabolism is appended in as far as it is useful for easier comprehension of some of the biochemical processes. Enzymes are also included if mention of them is made in the present chapter (see diagram of metabolism — Fig. 79). As can be seen from the diagram, glucose or glycogen serve as substrates for the glycolytic formation of lactic acid, a part of the energy from glucose being liberated for ATP synthesis. Glycolysis does not require oxygen, but under anaerobic conditions only a small part of energy is liberated. Glycolysis, however, still plays another role. Under aerobic conditions it prepares the substrate for further oxidation in the so-called Krebs'citric acid or tricarboxylic acid cycle, pyruvic acid serving as substrate. During this oxidation the remaining carbohydrate energy is liberated and this is bound to ATP or to other high energy phosphates. The tricarboxylic acid cycle also oxidizes other substrates, such as oxidation products of fatty or amino acids.

In several tissues yet another route of glucose catabolism has been found, namely direct oxidation of glucose or the pentose cycle. During this process the first atom of C in glucose is split off and the pentose thus formed is further broken down. Such direct oxidative breakdown of glucoso-6-phosphate has been found, for example, in the liver, but not in the muscle (Bloom et al. 1953, Katz et al. 1954, Bloom 1955, Bloom and Stetten 1955, Bloom et al. 1956).

In contrast to anaerobic glycolysis, where the high energy phosphate is transferred directly from the intermediate product to ADP giving rise to ATP plus a dephosphorylated intermediary product, the transfer of energy in oxidative metabolism from the substrate to high energy phosphate is indirect. Enzymes, the so-called dehydrogenases, transfer hydrogen from the substrate onto hydrogen or electron carriers which form the respiratory chain. Hydrogen and electrons are transferred by this chain onto oxygen, water being formed. The individual reactions are oxidoreductions in character and are coupled with the phosphorylation of ADP to ATP. This linkage, however, is not too firm or indispensible. Although phosphorylation cannot be realized without oxidation, oxidation, on the other hand, is not dependent on phosphorylation, so that the phosphorylation of ADP to ATP can be inhibited without the mechanism of oxidation itself being impaired. Consumption of substrate and oxygen is thus not a reliable indicator for obtaining information about energy utilization. Rate of synthesis of high energy phosphates is a better criterion, but this cannot, as yet, be reliably determined in the living organism. An important role is possibly played by a mechanism suggested by Kaplan et al. (1956). This mechanism involves concentrations of DPN and TPN and the enzyme pyridinenucleotide dehydrogenase which may regulate the effectiveness of oxidative reactions in the organism.

As is evident from the diagram, either *carbohydrates, lipids or proteins* can serve as the substrate for muscle energy metabolism. No reliable method is available, as yet, for determining which of these serves as substrate and the extent to which it is being utilized. The investigation of substrate consumption is not sufficient by itself, since it does not enable differentiation to be made between substances being used as sources of energy, those which are stored in the original or altered form and stores which are being utilized. Work with radioisotopes is probably more reliable, but even this has its difficulties and disadvantages. Certain information concerning the character of the substrate oxidized can be obtained from RQ determinations. Carbohydrates (RQ = 1), lipids (RQ = 0·7) or protein (RQ = 0·84) can serve as substrates of muscle energy metabolism. It was concluded from experiments in vitro, i. e. on more or less damaged tissues, that the main substrate of energy metabolism at rest is glucose. However, according to some in vitro experiments (Gemill 1942, Stadie and Zapp 1943, Gruber et al. 1952) it is doubtful whether RQ in the muscle at rest is always approximately 1·0. Doubts concerning the principal role of carbohydrates were already formulated in 1913 by Lafon, who demonstrated that the muscle at rest utilizes lipids. It was shown later that only about 20% oxygen utilized at rest is used for the oxidation of glucose (Holling 1949, Somogyi 1948, Bell et al. 1952, Baker et al. 1954. Feyer et al. 1955 and others). Although these experiments were not performed on isolated muscles, they were at least done on the whole extremity, the main metabolically active part of which is represented by musculature. This has been confirmed by Andres et al. (1955—56) in experiments with isolated circulation through the muscle. Other authors, on the basis of their experimental results, also appear to doubt that glucose is the main energy substrate in the muscle: Harrop (1919), Himwich and Castle (1927), Brünner and Grosse-Brockoff (1939), Geyer (1949), Goldmann et al. (1950), Weinhouse et al. (1950), Hansen and Rutter (1952), Wertheimer and Ben Tor (1952), Gordon (1957), Lifson (1957) and Fritz et al. (1958, 1959). Neptune et al. (1959 a,b). Thus another major substrate is being oxidized besides glucose. This is also corroborated by the fact that RQ of a more or less intact muscle is usually lower than one. According to data in the literature RQ values are about 0·6—0·9, in our experiments with arterio-venous differences in isolated gastrocnemius muscle of the dog an RQ of $0{\cdot}74 \pm 0{\cdot}06$ was found.

It does not seem probable that resting muscle would oxidize amino acids and proteins to any great extent. This would be disadvantageous for the organism from the biological aspect. Furthermore, experimental data refute this possibility, since non-protein nitrogen released from the muscle is extremely low.

Ketone products of lipid metabolism in the liver are unlikely to cause the low respiratory quotient, since their oxidation results in RQ around one.

Thus lipids remain as the major energy substrate in the muscle at rest. It was shown by Lafon (1913) using the method of arterio-venous differences, that lipids are utilized in the levators of the upper lip in the horse. Lately, a number of authors have demonstrated (e. g. Gordon 1957) that the major role in the transport of lipids is probably played by unesterified fatty acids (UFA), the transport of which from adipose tissue to other tissues and especially their uptake through the cellular membrane depends in part on their concentration and combination with albumins.

It thus seems that the resting muscle could utilize energy primarily from UFA, less from glucose and to a very small extent from proteins and amino acids. However, several findings are in contradiction with this conclusion. For example, according to Roheim and Spitzer (1958), UFA are, on the contrary, liberated from hind extremities and the abdominal region. They could be liberated from subcutaneous or other adipose depots in the extremity, or directly from muscle lipids present inside or outside muscle fibres. It has been shown that in dog gastrocnemius (Bass and Hudlická 1960) the A—V difference of UFA in the muscle in situ is negative, i. e. that the muscle liberates UFA at rest (under Nembutal anaesthesia) at a rate of $1{\cdot}52 \pm 0{\cdot}50\ \mu$ equiv./min./100 g. tissue. At the same time RQ was determined and was found to be near 0·7, this indicating that lipids are being oxidized. UFA can perhaps be liberated from stores in the muscle or by lipolysis of blood lipids in the capillary bed of the muscle.

3. Sources of energy in different muscles during and after work

There are a large number of different muscles in higher animals. Each of these has its special mode of function. As a result of this function certain types of skeletal muscles have developed a characteristic metabolism during evolution. The Muscle "with a capital M", as Szent-Györgyi (1958) has stated, does not exist.

Differences between *various types of muscles* have been studied by a number of authors: Needham (1926), Kashpur (1952), Yakovlev and Yakovleva (1953), Lawrie (1953). Sheves (1955) and others. This question was recently investigated in insect muscles by Kubišta (1959). Some muscles are active almost all the time Their function is more static in character, as, for example, in antigravity muscles, or tonic, as in the diaphragm. Others are in action only from time to time in a phasic manner, often almost "explosively", but they usually work against a small resistance. These muscles have been designated as dynamic muscles by Yakovlev and Yakovleva (1953) or phasic muscles by others in contrast to tonic muscles. The former authors themselves pointed out, however, that especially in mammals, there are virtually no sharply defined muscle types, either dynamic or static. It is only possible to say that a given muscle is either predominantly one or the other type. These differences are also defined histologically and histochemically (George and Naik 1958a, 1958b) according to their appearance, orientation of muscle fibres, their diameter, the shape of nerve endings and differences in the blood supply (see Chapter X). They are also characterised by their biochemical properties not so much of contractile proteins, excepting perhaps their net content, as of their energy metabolism.

The functional characteristics and the type of muscle are determined to a large extent by the direct influence of the nervous system. It has been shown by Bach (1948), Yakovlev and Yakovleva (1953), Yakovlev (1956) and Buller et al. (1959)

that it is possible to transform the characteristics of muscle predominantly of one type into those predominantly of the other type by experimental training, or cross-unions of tendons or nerves.

Oxygen consumption in muscles at rest is relatively small as compared with some other tissues. Poor blood supply corresponds to this low oxygen consumption in the muscle. There is sufficient substrate and oxygen in the muscle at rest for its metabolism, even though the greater part of the capillary bed is closed. When the muscle in this situation begins to perform work suddenly, it takes some time before the capillary bed opens and before the supply of substrates and oxygen is increased to such a level so as to ensure a sufficient amount of energy. In static muscles and during prolonged tetanic contractions, the contraction mechanically obstructs blood flow and the blood supply is decreased (Gasskell 1877, Ben Tor and Opitz 1903, Anrep et al. 1934 and others). When the supply of metabolites is inadequate, the muscle draws energy from its endogenous stores. The energy from lipids and proteins can be utilized only under conditions of sufficient oxygen supply. It then depends on the character of work performed, on the type of muscle, its blood reservoirs and their content, on the content of myoglobin (as long as the low partial pressure of oxygen which is liberated by the latter is sufficient for the oxidation of lipids) etc., whether the muscle in this situation has or has not enough oxygen for the oxidation of substrates. Under hypoxic conditions the muscle can make use of carbohydrates only, especially of glycogen, which are broken down to lactic acid. The latter is liberated from the muscle into the blood and can be oxidized in the liver, and a part of it can be resynthetized in the liver into carbohydrates — and as glucose can consequently again serve as substrate for various organs, including muscle (Cori's cycle — Cori 1926). When work is performed at a rapid rate so that the organism is unable to obtain enough oxygen for oxidations by respiration, lactic acid accumulates in muscles, liver and blood. This so-called oxygen debt is subsequently „repaid" after finishing work by increased oxygen consumption.

If the organism is able to supply itself with sufficient amounts of oxygen, the oxygen debt is more local in character and is limited by the capacity of the contracting muscle to liberate lactic acid into the blood stream. The main, or at least the most important role for liberating energy in the muscle in this situation is played by glycolysis. Oxidative metabolism and especially utilization of non-saccharide substrates are less important. Since different muscles are differently equipped with aerobic and glycolytic enzymes and also differ in their blood supply, more energy is derived from either carbohydrates or lipids in relation to the type of muscle in question.

The *content of glycogen* decreases during work, especially at the beginning — this being most probably due to insufficient supply of oxygen and other substrates from the blood — later its cotnent does not change further. At the beginning of work the RQ of the whole organism increases to values around 1·0, later it decreases again. It is thus evident that at the beginning of work the organism utilizes primarily carbohydrates and later compounds less rich in oxygen — *lipids and proteins* — as substrates for obtaining energy. Main attention will be paid to lipids, even though proteins are also utilized, but it is not clear to what extent. However, protein metabolism is discussed separately in the following chapter. The utilization of endogenous lipids during functional activity has been shown by Palazzolo (1913) and Buchwald and Cori (1931) who demonstrated that in winter frogs and in fasting summer frogs lipids are broken down during muscle stimulation. This has not been found in mammals, either because methods for lipid determination are not sensitive enough to detect slight changes in their content, or because endogenous lipids are quickly recovered or substituted by lipids supplied from the blood. Lipids are probably utilized as energy substrates especially during prolonged work at a slower rate, or after muscle exhaustion when the content of ketone bodies and urea in the blood is increased, indicating an increased utilization of lipids and ketogenic amino acids. (Issekutz and Spitzer 1960). It has been shown by Žák et al. (1957) that there is a decrease in the protein content of muscles following stimulation. These are possibly utilized as a source of energy in the muscle or their fragments (amino acids or peptides) reach the liver by way of the blood stream and are oxidized there.

Muscle activity is followed by a metabolic *recovery phase*. The muscle is still well supplied with blood and receives a large amount of substrates. Later on, circulation gradually returns to normal (see Chapter VII). An analogous situation can be observed as far as the utilization of metabolites and the formation of energy stores, especially of glycogen, is concerned—(Flock and Bollmann 1940 and others). Changes in the RQ after stimulation or functional activity indicates that lipids are probably also synthetized. The restitution of metabolites to normal values, however, does not occur gradually by returning smoothly to initial values, as has been presumed by older authors. It was not until 1952 that Yampolskaya demonstrated that following direct muscle stimulation, an over-shoot reaction takes place during which glycogen rises above initial levels. Such changes have also been shown in creatinephosphate (Yakovlev 1955, 1956), proteins (Žák et al. 1957) and nucleic acids (Žák 1958).

According to Yakovleva, the over-shoot is present only within a certain definite range of frequencies used for stimulation, i. e. between a minimum threshold frequency, and a maximum frequency. In our experiments it was found (Bass et al 1955) that the threshold frequency, which will evoke the over-shoot of glycogen after stimulation, is dependent on the functional and metabolic type of muscle. Three different muscles were stimulated in the rat — tibialis anterior, gastrocnemius and the diaphragm (the latter muscle in vitro) always with the same number of impulses, but of different frequencies. In the predominantly dynamic muscle, the tibialis anterior, it was already possible to obtain an overshoot reaction of glycogen by stimulating with a frequency of 30 impulses per minute. In the gastrocnemius, an antigravity muscle of a static type, no overshoot reaction was found even when stimulating with a frequency 60 impulses/min. and it was not present until the frequency used for stimulation reached 120 impulses/min. For the diaphragm, a tonic muscle adapted to practically uninterrupted functional activity, the frequencies and number of impulses used were insufficient to evoke an overshoot reaction. The over-shoot was not observed until the stimulation caused a tetanus and complete fatigue of the muscle (20—25 min.). Changes in these over-shoot reactions have also been studied following section of the peripheral nerve and a comparison was made between the normal and denervated muscle. These results will be discussed later in connection with denervation changes. It can be concluded from the above that glycogen as a source of energy plays an different role in different muscles. However, muscles have other sources of energy besides glycogen. What are these sources and how are they utilized and stored in the muscles?

4. Formation of energy stores in the muscle

The best known of these are *carbohydrates* stored predominantly as glycogen. Glycogen is synthetized from blood glucose via intermediary metabolites glucose-6-phosphate and glucose-1-phosphate. In muscles of poikilotherm animals glycogen is also formed from lactic acid produced by glycolysis (Meyerhof 1930, Wilhelmi 1955). This ability is completely or almost completely lost in skeletal muscles of homeothermic animals, as has been shown by Omachi and Lifson (1956), Cheosakul et al. (1956), Villee et al. (1952). Practically all muscle glycogen in mammals takes its origin from blood glucose or its six-carbon (C_6) derivatives. Glucogenesis cannot, however, be excluded, but this would have to take place by another route than glycolysis.

Glucose which has entered the muscle fibre from the blood is not usually released into the blood as free glucose or as its six-carbon derivatives. It is first transformed to lactic or pyruvic acid. This is borne out by following findings:

a) Arteriovenous differences of glucose are usually positive. Negative values commonly occur under pathological conditions (Kreienberg 1942).
b) The content of free glucose in muscle elements is insignificant according to calculation (Gey 1956).
c) In contrast to liver and some other tissues, the enzyme which enables the liver, for example, to liberate glucose from cells (glucoso-6-phosphatase) has not been found in the muscle (Chiquoine 1955).

Muscle contains amylases which can split low-molecular oligo-, di- and monosaccharides from glycogen. The biological significance of these enzymes is hitherto unknown, but it is possible that they do not come into play until pathological states (severe hypoxia) occur, in which the muscle liberates glucose into the blood. Glucoso-6-phosphate is the only intermediary product of saccharide metabolism which is present in the muscle in major quantities; the others constitute only minimal amounts, so that they do not represent a substantial source of energy. Similarly, intermediary products of the tricarboxylic acid cycle are not sufficiently represented to form energy stores since they are broken down very quickly during oxydation of saccharides, lipids, amino acids and proteins. Their content in the muscle is usually very low. A more important role could be played by *lipids*, the content of which is quite high in the muscle, even though different in various types of muscles. Their utilization during work has already been discussed, blood lipids being probably the more important source in mammals. The low RQ *in situ* values, however, which are sometimes found in muscle at rest indicate that it is, perhaps, partial oxidation of lipids which can serve as the substrate of glycogenesis in the muscle.

According to Gemill (1942), muscle RQ does not usually differ from RQ of of the whole organism. Muscle RQ varies in different conditions. In our experiments (Bass, Hudlická 1960) it was found to be 0·74 in the gastrocnemius of dogs and varied from 0·3 to 0·9. As far as the values of RQ below 0·7 are concerned, it is possible that these muscles synthetize saccharides from fatty acids. However, experiments with C^{14} labelled lipids and fatty acids do not corroborate this assumption.

Muscle *proteins* are, for the major part, metabolically inert and their halftime is relatively very long. Of course, there is a protein fraction which has a relatively short halftime and which probably plays the role of a "general precursor of specific proteins" (Fritz 1957). This fraction could, under certain conditions, for example during muscular activity, serve as a further source of energy. The question, whether addional proteins are utilized during work, and which proteins, has not yet been solved (see Chapter IX).

A relatively large amount of energy which the muscle can utilize even under hypoxic or anoxic conditions is stored in the form of *high energy phosphates* (ATP and others). This energy reserve is in constant motion at rest and during work, since it is being continuously broken down and resynthetized and yet is maintained at relatively stable level or stationary state. During muscular activity, especially under hypoxic conditions, the breakdown of creatinephosphate predominates over this resynthesis and its content decreases. It is thus being used up as an energetic reserve. The content of ATP is maintained for a longer time and it does not decrease until the main part of creatinephosphate has been exhausted (Flock et al. 1939).

The normal muscle thus utilizes lipids, carbohydrates as well as proteins for obtaining energy. It stores these substances and utilizes them according to the type of functional adaptation of the muscle, in connection with the supply of substrates and oxygen and in relation to the type of work which it performs.

This is associated with further regulatory mechanisms besides autoregulation at the cellular level, i. e. humoral and direct nervous regulation. From the present point of view nervous regulation of trophic processes and particularly, the changes which occur after loss of this regulation following nerve section are most important in this connection.

B. Energy metabolism in the denervated muscle

1. A survey of disturbances in muscle energy metabolism following denervation

Section of the somatic nerve excludes the muscle from function by interrupting its connection with the central nervous system. After a time indirect muscle excitability is lost, nerve endings in the muscle degenerate and almost simultaneously first signs of atrophy appear, the latter being associated with loss of weight due to a decreased content of contractile proteins. This simple atrophy gradually changes into degeneration, i. e. the disintegration of muscle elements and their exchange for non-specific adipose and connective tissue.

Normal trophic function, which ensures the morphological and chemical composition of muscle and its ability to adapt itself to various stimuli, requires an intact nervous system. For maintaining a normal trophic state, muscle needs energy which it obtains from metabolism. If the trophic influence of the nervous system on muscle is impaired, a disturbance of energy metabolism seems to take place, either because enough energy cannot be obtained by normal procedures from usual sources for maintaining the normal trophic state, or because, although sufficient energy is being produced, its utilization for trophic processes is impaired; perhaps both these alternatives occur. In other words, in the first case a disturbance of energy liberation from normal sources occurs causing the muscle to utilize other sources of energy, in the second sufficient energy is at its disposal, but this energy would be accumulated in vain since the muscle is unable to make use of it for normal trophic processes. Evidence has been presented indicating both these types of disturbance, so that it is not clear which of them is primary, and whether it is at all possible to assume that changes in energy metabolism represent the primary link causing trophic changes after section of the peripheral nerve. It is necessary to discuss in more detail changes in muscle energy metabolism following denervation, especially changes in intermediary metabolism.

Discussion concerning energy metabolism in the denervated muscle is somewhat misleading, since its main function in the muscle — mobilization of energy stores for functional activity — ceases. Denervation, however is accompanied by trophic disturbances. Energy metabolism remains as an integral part of metabolism and only its specific character, acquired during development of adaptation to muscle function is lost gradually. As the contractility of muscle is not lost immediately, utilization of energy sources and the formation of energy stores are not lost at once. Energy liberated is utilized for covering the needs of general cell metabolism and possibly for incoordinated muscle contractions — denervation fibrillations.

a) Historical survey

Physiologists have studied changes in energy metabolism following denervation for more than 100 years. These investigations with their methodological approaches are closely related to development of research methods and knowledge of energy metabolism in normal muscle.

The first period of research was associated with *analysis of individual compounds* present in the muscle. At this stage it was mainly the content of glycogen which was determined (Cl. Bernard 1859), together with changes in its structure studied with the aid of its iodine staining properties (Cl. Bernard 1877). These estimations were at first made at random intervals after denervation, but later the time course of these changes was studied. Quantitative determinations of the glycogen content developed gradually, but experiments led to contradictory results, which have only recently been elucidated. — Similarly lipids were at first estimated at random intervals after denervation (Rumpf and Schumm 1901), mostly in human pathological material, later, in experimental animals (Grund 1912, Audova 1923 and others).

With deeper knowledge of the role of other metabolites, their changes were also studied after denervation; the content of creatinephosphate (Muschini 1928 and others), ATP, and ADP (Hines and Knowlton 1933) etc.

The method of random determinations of metabolites are rather static in character, they can give information concerning changes in the steady state at a given moment, but cannot elucidate the more subtle metabolic changes following denervation. Nevertheless, they represented basic findings for further research.

Hand in hand with more dynamic concepts on processes in living organisms, other approaches of studying changes in the denervated muscle have gained in significance. Before the chief reactions of energy metabolism were discovered, i. e. before fundamental knowledge had been obtained concerning intermediary metabolism, changes in metabolic processes could only be demonstrated by a single approach. This was accomplished by measuring the input and output of substances by means of *arterio-venous differences* and of blood flow. According to Langley and Igataki (1917), the consumption of oxygen is increased after denervation. This question was studied later by other authors and in 1935 Sato and Kasugai reported an increased consumption of glucose and O_2 and higher utilization of lactic acid in the denervated muscle. Later the consumption of substrates and oxygen and the release of metabolic products was also studied in denervated *muscle preparations in vitro* (Knowlton and Hines 1934 and others), with various results.

In vitro experiments have been particularly valuable in studying intermediary metabolism of muscle. During these investigations several enzymes have been discovered and a number of methods for investigating enzymatic activities have been elaborated. These methods have also been used, and still are being used, for studying energy metabolism in denervated muscle. Hines and Knowlton (1934), for example, investigated changes in succinic dehydrogenase after denervation and later various authors studied other enzymes, so that at the present time changes in the activity of a number of enzymes engaged in energy metabolism have been studied.

However, contradictory findings were often obtained between *in vivo* and *in vitro* experiments, especially as far enzyme activity in vitro and the metabolic rate in vivo was concerned. For example, at a time when the activity of glycolytic and oxidative enzymes is reduced, the consumption of glucose and oxygen is increased. Today it is known that this is not necessarily in contradiction, since there is usually a surplus of enzymes and their activity is generally not the limiting factor. The limiting factors are evidently of a different nature — they may either be associated with the supply of substrate and the removal of metabolic products or with the concentration of other substances, the so-called pacemakers, which may either accelerate or retard the rate of individual intermediary biochemical processes, with humoral factors or with the localization of metabolic processes on structural cellular elements and with the diffusion of substrates to these elements. *In vitro* enzymological studies only provide information about the maximal metabolic capacity. This capacity is only rarely fully exploited in the muscle, perhaps during maximal functional activity, and also after activity or excursions of metabolism from the steady state by other stimuli. For this reason changes in metabolism have been studied after denervation following procedures which evoke these excursions, at least in the normal muscle. Energy metabolism in the muscle was studied under various experimental conditions. For example, Mansfeld and Müller (1913) and Wertheimer (1927) studied the effect of *starvation*, other worked with *alimentation*, *muscle stimulation*, artificial *changes in blood flow* etc. It was shown that, with the exception of muscle stimulation, the metabolic reactibility of the denervated muscle is decreased in comparison with the normal muscle. However, the denervated muscle is hypersensitive to most humoral agents (e. g. acetylcholine — Cannon and Rosenblueth 1949) — see Chapter XI.

Using the above-mentioned approach of altering metabolic conditions, it has been found that the reactibility of the denervated muscle is changed. However, changes in intermediary metabolism and in the rate and direction of individual reactions could not be demonstrated.

It is thus not possible to rely on enzymological *in vitro* experiments only, and the *in vivo* approach does not yet have adequate methods at its disposal. It is therefore necessary to seek methods which would make the investigation of intermediary metabolism possible *in vivo* and with the aid of which the rate of *in vivo* metabolism could be measured, (at least the differences in relation to the controls). Such a method could, perhaps, be represented by studies with *labelled compounds*. Isotope studies connected with questions of metabolic regulations are only beginning. So far it has been very difficult to differentiate indirect influences, especially the effects of different blood flows, etc. This would perhaps be possible by studying incorporation gradients in precursors and products in a metabolic chain. With the aid of this method a number of difficulties could be overcome, namely difficulties with the experimental procedure and their interpretation.

Results obtained by each methodological approach concerning questions of energy metabolism in the denervated muscle will now be discussed in more detail. An attempt will then be made to utilize these results for elucidating the question of the main link in disturbances of metabolism in the denervated muscle and to answer the question to what extent can energy metabolism play a fundamental role in trophic processes in the muscle.

b) Content of metabolites following denervation

The first report on changes in metabolite content after denervation already appeared 100 years ago. Cl. Bernard showed in 1859 that the content of muscle *glycogen* in the rabbit increases following denervation. These results were later confirmed in other animals and are being confirmed by contemporary authors, who work with more accurate and reliable methods of chemical glycogen determination. Confirmatory reports were published by McDonnel (1863), Ogle (1868), Chandelon (1876), Manché (1899), Varg (1894), Avelone and Di Macco (1925), Wertheimer (1927), Masayama (1932), Cendragolo (1937), Gutmann and Vrbová (1952, 1952b), Ferdman (1953, 1957), Vrbová (1954,1955), Martínek (1955, 1956) and others. However, contradictory findings appeared in the thirties when rats and mice became more widely used for experimental work. It was shown that the content of glycogen in these and some other animals decreases after denervation. This has been shown by Wertheimer (1927-V, VI), Britton (1930), Hines and Knowlton (1933, 1935b, Knowlton and Hines (1934), Fainshmidt (1939), Lazere et al. (1942, 1943), Levine et al. 1941, Humoller et al. (1950), Yakovlev (1951), Bass et al. (1955), Beránek et al. (1953), Vrbová (1955), Zubenko (1956).

When studying the time course of glycogen changes in muscle following section of the nerve, Martínek (1956) showed that the decrease of glycogen occurs not only in rats and mice, but also in guinea-pigs and rabbits, in which an increase had previously been described. It was demonstrated by Martínek that during the first phase after section of the nerve there is always an increase in the content of glycogen which differs, however, in duration in different species. For example, in the rat the content of glycogen returns to initial values in about two days, in the guinea-pig in five days and in the rabbit in 13 days. The contradictory findings in the literature were evidently caused by these differences. It thus appears that the rate of glycogen break down and synthesis differs in various species and also perhaps in different muscles (Fig. 80).

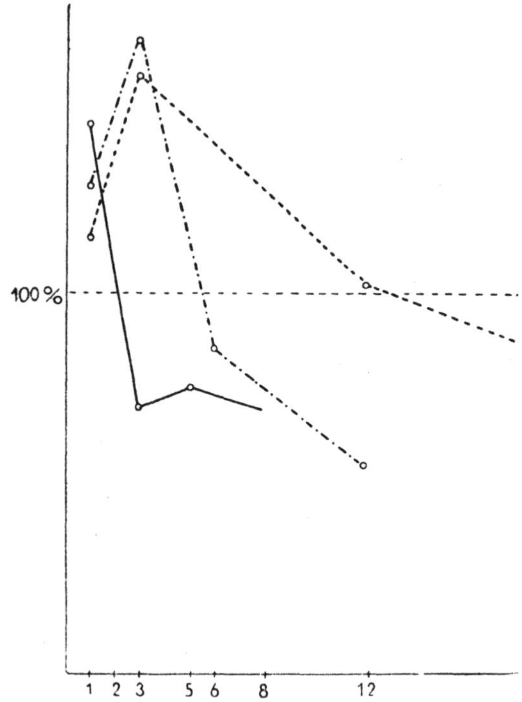

Fig. 80. Changes in glycogen content in the gastrocnemius after denervation in the rat (———), guinea-pig (- - - -), and rabbit —.—.—.). Abscissa: time after section of the sciatic nerve in days (Martínek 1956). Ordinate: content of glycogen expressed as percentuage of the control muscle.

Besides changes in glycogen, other substrates of energy metabolism have been investigated after denervation. The content of intermediary metabolites of glycolysis was studied in several papers: according to Hines and Knowlton (1933), the content of *acid soluble phosphate* decreases in parallel with the loss weight, while Fainshmidt (1939) described an increase of *lactic acid* and Bieth et al. (1956) reported a decreased content of organic *phosphate and glucoso-1-phosphate.*

The content of *creatinephosphate* after denervation was studied by the following authors: Muschini (1928), Zanghi (1929), Martino (1929), Masayama (1932), Hines and Knowlton (1933), Paladin and Sigalova (1933), Wehrmacher et al. (1945), Humoller et al. (1950), Grigorieva (1953, 1954), Zubenko (1956). Its content corresponds, on the whole, to changes in glycogen. i. e. an increase followed by a subsequent decrease (Fig. 81).

The content of *ATP* and *ADP* was determined by Hines and Knowlton (1933),

Humoller et al. (1951), Ferdman (1953), Grigorieva (1954, 1958). The ATP content was found to be slightly increased at first, then, after some oscillation, it decreased.

Changes in the content of *lipids* following denervation were studied by Audova (1923), Wertheimer (1927), Chor et al. (1937), Dreyfus and Schapira (1948), Fischer et al. (1952), Humoller et al. (1952), Kanashenok and Klyutchayeva

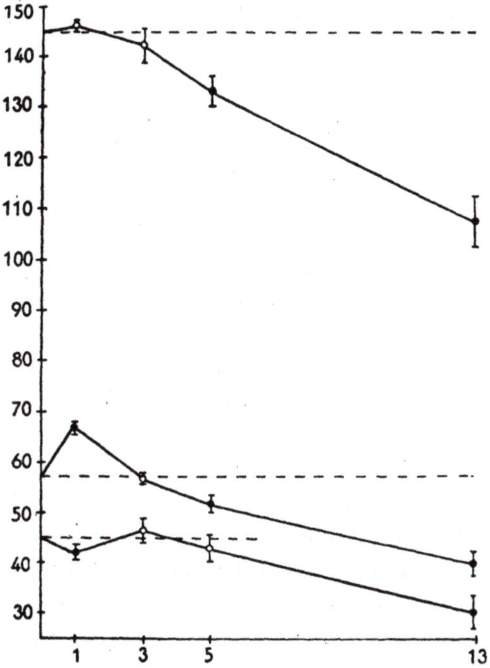

Fig. 81. Changes in total phosphorus (top curve), creatinephosphate (middle curve) and ATP (bottom curve) after denervation in the gastrocnemius of the rat. Ordinate: content of the above compounds in mg% P. Abscissa: time after section of the sciatic nerve in days. (Modified from results of Humoller et all. 1950a.)

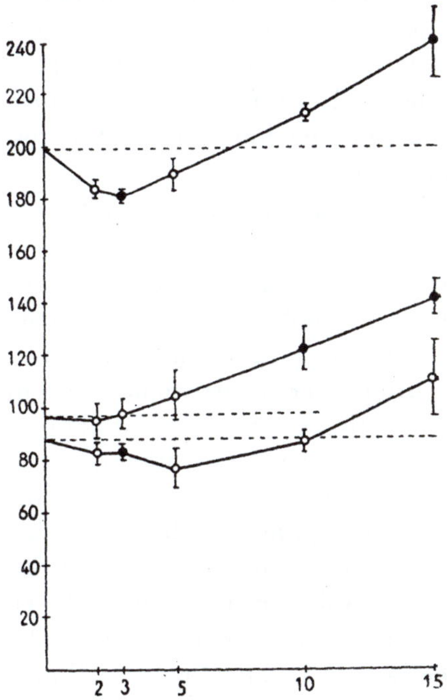

Fig. 82. Changes in lipid content after section of the somatic nerve. Top-curve: total lipids. Middle curve: cholesterol. Bottom curve: phospholipids. Ordinate: the content in mg%. Abscissa: time after denervation in days. (Modified from the results of Humoller et al. 1952.)

(1956), Sutfin et al. (1954). *Phospholipids* and *steroids* were investigated by Grund (1912), Artom (1941), Humoller et al. (1952), Hines and Thomson (1956). Following section of the nerve, lipids, after brief stabilization, exhibit a tendency to rise. However, lipids are neither a homogenous chemical compound nor a group of chemically closely related substances, and they behave quite differently from the chemical and biochemical aspect. The determination of lipids in biological material is very difficult and inacurate since lipids must first be isolated, lipoproteins must either be quantitatively removed or split and their lipid component determined together with other lipids.

Similar difficulties and inaccuracies complicate the determination of individual lipid compounds and only rarely is it possible to attain the necessary accuracy, as e. g. in cholesterol and lipid phosphate determinations. For this reason changes in their content following denervation have been studied most. However, in older reports, the net content of lipids was studied, i. e. lipids extractable with fat-solvents, and an increase was found. Humoller et al. (1952) demonstrated a decreased content of total lipids and especially of lipid phosphate five days after denervation. Fifteen days after denervation an increase of net lipids was found by these authors, and this increase was most prominent in the phospholipid and cholesterol fraction, but was less defined in glycerides (Fig. 82).

It can thus be concluded from the above that the content of metabolites investigated decreases during denervation atrophy, with the exception of lipids. However, it is not yet clear to what extent further conclusions can be drawn about metabolic processes after denervation from these changes. Determinations of metabolites and their changes are not reliable indicators of metabolic processes occuring after denervation. The content of a metabolite is the result of the balance between rate of its formation and degradation. Its changes after denervation, as well as under other conditions, represent changes in the balance of these two antagonistic processes, since the content of a metabolite is directly associated with the equilibrium between synthesis and degradation. The stable concentration of metabolites expresses the so-called metabolic steady state. However, since metabolism is not just the sum of individual enzymatic reactions, but a complexly integrated system, a change in the rate of one reaction can change the rate of a whole "pathway" and direction of other reactions, or on the contrary, it may be compensated for by other reactions, so that the content of metabolites can return to a stationary state which may differ but little from the original state.

Which factors could be responsible for changes in the rate of metabolic reactions following section of the somatic nerve?

In the first place, it is the enzyme activity in metabolic reactions. However, even the most active enzyme, under otherwise optimal conditions, cannot catalyze a reaction unless sufficient substrate is present. Thus the supply of substrate represents a further regulating factor and in the case of enzymes catalyzing reversible reactions, also the removal of products of the reaction. But these two examples do not exhaust all the possible regulatory mechanisms.

The activity of a number of enzymes, besides requiring substrate, is also dependent on coenzymes or activators which may be products of other reactions. The concentration of these often determines the rate of reactions and their direction. These substances are especially effective in certain metabolic reactions

Krebs's pacemakers (1957). This concerns reactions, the rate of which regulates further metabolism. This rate is dependent on the concentration of substances entering a given reaction in addition to the main substrate, being formed by a reaction other than the substrate. For example, the content of oxalacetic acid regulates the rate of oxidation in the tricarboxylic acid cycle. Its lower concentration retards the rate of the tricarboxylic acid cycle. Oxalacetic acid is formed either by carboxylation of pyruvic acid, i. e. during oxidation of carbohydrates, of from glutamic acid, aspartic acid or from alamine, i. e. as a product of glycolysis or of protein catabolism. It cannot, however, originate from lipids. For this reason, besides regulating the rate of the tricarboxylic cycle, it also affects the ratio and interaction of carbohydrates, lipids and proteins as substrates of oxidative metabolism, especially the rate of lipid catabolism. There are, as yet, only few of these "pacemakers" known in metabolism. Best known are those glycolysis and the tricarboxylic acid cycle. They can not only regulate the metabolic rate, but in places where alternate metabolic pathways are possible, they may also change the whole character and direction of metabolism. The theoretical explanation of such a mechanism at a metabolic cross-roads has been given by Pasynskiy (1957).

The physico-chemical state of the cell, altered pH due to changes in the content of metabolites or other factors which can alter enzymatic activity, their localization on structures etc. can represent further factors affecting regulation of metabolism.

Four main groups of regulation mechanisms can thus affect the rate of metabolic reactions:

1. Enzyme activity, which can limit the rate only under conditions where no other limiting factor comes into play.

2. Passage of metabolic substrates from the blood into interstitial spaces and from these, and or directly from the blood, into the muscle fibre itself.

3. Intracellular regulations of metabolic reactions by activators and inhibitors, especially by the so-called pacemakers.

4. The physico-chemical character of the environment and external influences, nervous and humoral, which modify them.

Of these, changes in enzyme activity and in the supply and removal of substances in the blood after denervation have been most extensively studied.

c) Changes in enzyme activity associated with energy metabolism following denervation

A number of authors have investigated postdenervation changes in the activity of enzymes, especially of those associated with glycolysis, oxidative metabolism and the transfer of high energy phosphates. The determination of enzyme acticity presumes that optimal conditions for utilizing total potential activity of enzymes

Table 11
Changes in enzyme activity after denervation

Enzyme	Increase (+) or decrease (—)						Species	Authors
	1 day	2—3 days	5 day	1 week	2—3 weeks	4 w. and more		
Phosphorylase (phosphorolytic activity)					—		rat	Levine et al. 1941
				—	—		rabbit	Mirski and Wertheimer 1942
					—		cat	Veselkin, Veselkina 1942
				norm.	—	0	cat	Varga 1950
a		—	+	+			rat	Hummoller et al. 1952
a + *b*		+	norm.	—	—		rat	Hummoller et al. 1952
				—	—	—	rabbit	Eppelbaum, Kantor 1954
					—		rat, cat	Slozhenikina 1957
					—	—	rabbit	Telepneva 1955
Phosphorylase (synthetic activity)					norm.		rat	Matsui, Kuriaki 1959
		+	+	+	— (a) norm. (a+b)		rat	Humoller et al. 1952
Hexokinase		+	+	+			rat	Humoller et al. 1951
					—		rat	Gerritsen 1952
	norm.	norm.					rat	Bass, Vodička 1956
						—	rabbit	Telepneva 1955
Aldolase		In relation to atrophy					rabbit	Fischer 1949
			—		—	—	rabbit	Epelbaum, Kantor 1954
Phosphoglucomutase		In relation to atrophy					rabbit	Fischer 1948
Creatine phosphokinase					—		rabbit	Telepneva 1955
					norm.		rabbit	Slozhenikina 1957
Creatine synthetizing enzyme					—		rat	Matsui, Kuriaki 1959
ATP-ase	—						guinea-pig	Michelazzi et al. 1957
					—		rat	Nachmias, Padykula 1958
					—	—	rabbit	Sheves 1953
						—	frog, cat, guinea-pig	Zubenko et al. 1950
Succinic dehydrogenase		+					rat	Humoller et al. 1951
					+		rat	Nachmias, Padykula 1958
					—		rat	Knowlton, Hines 1934
						—	rat	Schmidt, Schlief 1956
					—		rat, rabbit	Mirski, Wertheimer 1942

Table 11 continued

Enzyme	Increase (+) or decrease (—)						Species	Authors
	1 day	2—3 day	5 day	1 week	2—3 weeks	4 w. and more		
Choline esterase					—		guinea-pig	Brooks, Myers 1952
					—		rat	Couteaux, Nachmansohn 1951
						—	rat	Dreyfus, Schapira 1948
						norm.	guinea-pig	
						norm.	nerve endings	Sávay, Csillik 1954, 1956
		—		—		norm.	rat	Schwarzacher 1957
					—	norm.		Snell, Mc Intyre 1956
						norm.		Strelina 1956
				—			dog (myosin)	Varga et al. 1957
				norm.			dog nerve endings	Varga et al. 1957
		—					rabbit (myosin)	Varga et al. 1957
						—	rabbit nerve endings	Varga et al. 1957
Succinic oxidase		+					rat	Humoller et al. 1951
				+			rat	Levine et al. 1941
						—	rat	Schmidt, Schlief 1956
Cytochrome oxidase				norm.			rabbit	Ferdman 1953
						—	rat	Schmidt, Schlief 1956

are known, that they can be reproduced in vitro, and that all of the enzyme is isolated or is contained in a preparation quantitatively without losses. The ideal case would be to determine the activity of an enzyme isolated in the pure state. However, during purification losses cannot be avoided and often the enzyme is also partly inactivated, so that the accuracy of total enzyme activity determinations is thus adversely affected. As long as the enzyme is not inactivated during isolation, it is possible to express its activity only in relation to its net protein. For this reason most authors have ceased to use this method and they usually determine the total enzyme activity in tissue slices and homogenates, or in subcellular fractions, the main advantage of which lies in the fact that they contain a simpler enzymatic system and some of the interfering enzymes can be removed with other fractions. Interfering enzymes can sometimes be inhibited by more or less specific inhibitors. In this way it was found possible to demonstrate changes in enzyme activity following denervation. A survey of these changes together with the time after denervation are given in Table 11.

As can be seen from the Table, the first changes were found in ATP-ase activity of sarcosomes and in *enzymes associated whith glycolysis and glycogenolysis*

(Michelazzi et al. 1957). Humoller et al. (1951) reported increased phosphorylase and hexokinase activity from the third day after denervation onwards in the rat gastrocnemius. On the third day after denervation synthetic activity of phosphorylase was found to be increased in phosphorylase *a* and *b*. The glycogenolytic activity of phosphorylase is decreased at first, and then increases on the fifth day after denervation. Ten to fifteen days after denervation phosphorylase activities are decreased. The above authors also found increased hexokinase activity five to ten days after denervation, but did not investigate later periods. They also found an increased activity by three days after neurotomy, but this is not statistically significant (see Bass and Vodička 1956). Other authors have shown that there is a reduction in phosphorylase and hexokinase ten days after denervation in the rabbit and cat (Mirski and Wertheimer 1942, Eppelbaum and Kantor 1954, Varga 1950). The activity of phosphorylase then decrease together with atrophy. A similar decrease of hexokinase was found in the rat by Gerritsen (1952) and in the rabbit by Telepneva (1955). The activity of aldolase — another glycolytic enzyme — decreases from the fifth day onwards (Eppelbaum and Kantor 1954) in parallel with denervation atrophy (see Fischer 1949) and similar changes were described in the case of phosphoglucomutase. As far as *oxidative enzymes* are concerned, the activity of succinic acid dehydrogenase increases early after neurotomy (Humoller et al. 1951), but this enzyme is not associated with oxidations directly. After this, its activity decreases, so that two to six weeks later it is below original values (Knowlton and Hines 1934, Mirski and Wertheimer 1942, Schmidt and Schlief 1956, Nachmias and Padykula 1958). Changes in the activity of the enzyme systems of succinoxidase and cytochromoxidase were not found until later (Levine et al. 1941, Ferdman 1953, Schmidt and Schlief 1956).

The activity of sarcosomatic ATPase, one of the enzymes associated with transfer of energy, is increased very soon, however, within 12—16 hours after section of the nerve it decreases (Michelazzi et al. 1957). It is then maintained at a reduced level (Zubenko et al. 1950, Sheves 1953, Nachmias and Padykula 1958). The ratio between soluble and insoluble myosin-bound ATPase is also changed (Fischer 1946). The content of creatinephosphokinase does not change until three weeks after neurotomy (Telepneva 1955, Slozhenikina 1957, Feuer 1957) and at the same time decreased activity was found in the creatine synthetizing enzyme (Matsui and Kuriaki 1959).

Changes in the activity of enzymes associated with energy transfer do not become apparent until changes in energy metabolism are quite evident and when atrophy is pronounced. An exception to this has been found in the activity of sarcosomatic ATPase and ATPase in myofibrils, but this has only been found in one report so far and has not been, as yet, confirmed.

As has already been mentioned above, the enzyme content and activity are not usually the limiting factor in energy metabolism in the normal muscle or other tissues, especially at rest or under conditions of submaximal functional activity. It seems that this is not the case soon after denervation, since enzyme activity seems rather to increase, even though temporarily. This would correspond to faster metabolic turnover or at least to an increased metabolic capacity. In addition to this, the activity of enzymes of energy metabolism is decreased at a time when the consumption of glucose and oxygen is increased. Changes in enzymatic activities — as far as they are known — cannot thus be the limiting factor in energy metabolism after denervation.

Enzyme activity measured *in vitro* is, of course, an artificial value, since it is determined under the most optimal artificial conditions possible, which can differ substantially from conditions *in vivo*. With this reservation, it is possible to say that changes in the activity of enzymes are probably not the primary cause of postdenervation atrophy, but rather its result. Enzyme activity *in vitro* is, in the optimum case, a measure of the maximum capacity of performing a certain reaction, a capacity which is usually never fully utilized. Its full utilization is prevented, among others, by the limited supply of substrate. The supply of substrate for individual reactions in metabolic chains or cycles is given by the rate at which the substrate is formed in the preceding reaction, i. e. by the rate at which the energy substrate penetrates into the organ from the blood. Less often, under very exceptional conditions, enzyme activity can be hindered by accumulation of its reaction products. Accumulation of intermediary metabolites has not been found after denervation, so that it does not seem plausible to suppose that this could be the cause of the metabolic disturbance.

As a further possibility, the supply or utilization of substrates of energy metabolism or their utilization can be changed. This can be determined by investigating the consumption of substrates and the release of metabolic products after denervation.

d) The consumption of substrates and formation of products of energy metabolism in the denervated muscle

Thus determinations of the content of metabolites and enzyme activity cannot, at present, give sufficient information about metabolic changes after denervation. The maximal capacity to utilize substrates for metabolic purposes is greater than corresponds to the consumption, especially the resting consumption, and metabolism can progress at a different rate without apparent changes in the content of intermediary compounds. Several authors were aware of this, since they attempted to determine changes in the consumption of substrates and formation of their products after denervation. According to the methodological approach used these experiments can be divided into three groups: *in vitro* experiments on muscle preparations with a damaged cellular membrane, *in vitro* experiments with isolated muscles or muscle strips, i. e. with the cellular structure more or less intact, and lastly experiments in which the consumption and output are determined in situ with a perfusion technique or by measuring A—V differences and blood flow.

The muscle can utilize *oxygen and glucose as well as fatty acids* (Fritz 1958, Spitzer and Issekutz 1960) and lactic acid (Omachi and Lifson 1956, Cheosakul et al. 1956, Drury and Wick 1956) as sources of energy from the blood.

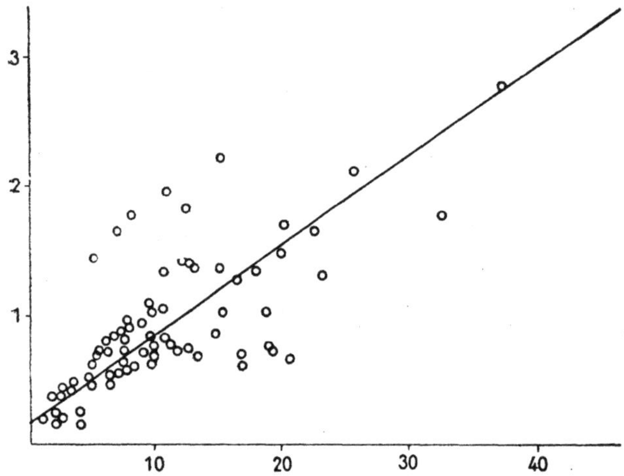

Fig. 83. Correlation between muscle blood flow and oxygen consumption. Individual values — O. Straight line — calculated regression line. Ordinate: oxygen consumption in ml./min./ /100 g. of tissue. Abscissa: blood flow in ml./min./100 g. of tissue.

As has already been stated, the consumption of these substrates can be investigated either in vitro or by means of A—V differences and the blood flow in vivo. Both methods have their advantages and disadvantages: *in vitro* exact amounts of individual substrates can be added, while extramuscular influences such as of humoral agents can be excluded to a great extent. Nevertheless, the fact that a considerably damaged tissue is being investigated cannot be underestimated, since tissue structure is impaired in many cases and it is thus no wonder that the results obtained often differ from those *in vivo*. The main disadvantage of the use of A—V differences lies in the fact that in most cases measurements are carried out on anaesthetized animals and that direct venous outflow determinations necessitate manipulation in the traumatized region. Experiments in man or unanaesthetized animals would be ideal from this aspect, but they encounter considerable methodological difficulties and in addition to this, determinations of arterio-venous differences are complicated by a natural variation of the level of metabolites in arterial blood. Such variation, although unpleasant for the experimentator, is nevertheless physiological and the method of studying A—V differences gives for this reason, especially when performed in unanaesthetized animals, the least distorted picture of metabolism. When investigating the regulation of muscle metabolism, the situation is made more complicated by the difficulty of differentiating nervous, humoral and other regulating factors affecting individual intermediary metabolic processes.

Oxygen may be considered as the most important component of energy metabolism. Its consumption in the muscle varies greatly under different conditions. Is it governed by metabolic requirements only, or is it dependent on the oxygen supplied by the blood? How are these factors affected by denervation?

Oxygen consumption in muscle has been studied by a number of authors. According to Ludwig and Schmiedt (1861), oxygen consumption is proportional to the blood flow. However, it was shown by Pflüger and Finkler (1875) that although blood flow is reduced during partial exsanguination, oxygen consumption remains unchanged. Harding (1926) demonstrated an increase in oxygen consumption, while the blood flow remained unaltered, and Issekutz et al. (1955) even found a decreased consumption during increased blood flow. Blood flow is considered to be the decisive factor in determining the consumption of oxygen by Holling (1938), Lilienthal et al. (1956) and Gibertini et al. (1955), however, Mottram (1954, 1955, 1958) has demonstrated the reverse in several reports.

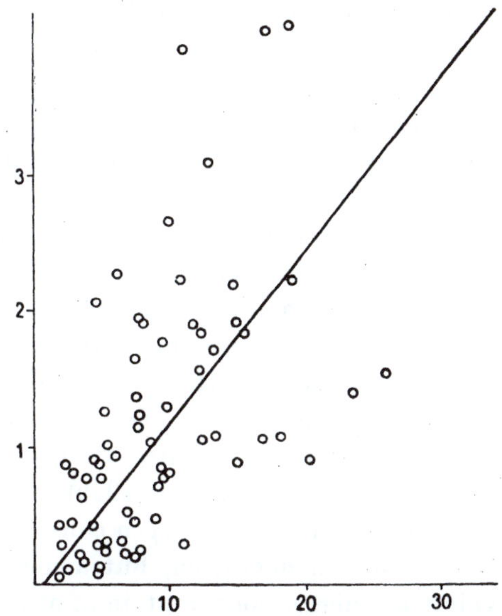

Fig. 84. Correlation between muscle blood flow and glucose consumption. Individual values — O. Ordinate: glucose consumption in mg./min./100 g. of tissue. Abscissa: blood flow in ml./min./100 g. of tissue.

In our experiments performed on the isolated gastrocnemius of the dog under nembutal anaesthesia, it was found that there is a direct correlation between blood flow and oxygen consumption at rest and after muscle stimulation (see Fig. 83).

Blood glucose is another important substrate of muscle energy metabolism. Its consumption, according to our experiments, correlates less closely to the blood flow through the muscle (Fig. 84). It is not affected, at least within normal physiological limits, by the content of glucose in arterial blood. It can be assumed that the consumption of both metabolites, i. e. oxygen and glucose, can also be influenced by other factors, especially humoral agents, as can be deduced from contradictory results in literature. Nevertheless, blood flow appears to be a very important regulating factor. Furthermore, insulin administration increases oxygen consumption in muscle (Hall et al. 1954) similarly as adrenaline (Shatenshteyn 1935, Issekutz et al. 1950). Thyroxin causes a decrease in oxygen consumption (Radsma et al. 1955). The nervous system represents a very important regulating factor for consumption of oxygen. Oxygen consumption in muscle is increased after sympathectomy (Petkowitz et al. 1958, Gibertini et al. 1955).

According to most authors, the *consumption of oxygen* is altered following section of the peripheral nerve. An increase has also been described in vitro (Chepinoga 1937, Fischer 1939, Hoagland 1946, Telepneva 1955, Sheves and Ryumina 1956, Sheves et al. 1956). Increased oxygen consumption was reported in vivo by Langley and Igataki (1917), Ijima (1924), Kimura and Takashi (1928), Ichimi (1929), Sato and Kasugai (1935), Hirohashi (1941), Zubenko (1954). Several authors, however, found no changes in oxygen consumption in brei of denervated muscle (Knowlton and Hines 1934, Levine et al. 1941) or in A—V differences (Nakamura 1921, Jansen 1927, Hines et al. 1931). In several cases, even decreased oxygen consumption was reported after denervation (Califano 1927, Shatenshteyn et al. 1935, Belicer 1937), but these investigations were either methodically inadequate (Shatenshteyn, for example, measured oxygen consumption 40 min. after excision of the muscle) or interspecies differences played an important role (Califano and Belicer determined oxygen consumption in frog muscles). It is probable that there are differences in oxygen consumption following denervation in various animal species, all the more so since Sheves and Ryumina (1956) reported differences between muscles in the same species (increased oxygen consumption in vitro in the rabbit gastrocnemius after somatic denervation, but a decrease in the soleus muscle — the soleus exhibited a higher consumption of oxygen prior to denervation than the gastrocnemius and this difference was anulled following denervation). Other contradictory findings can be explained by the fact that various time interval elapsed between nerve section and the beginning of the experiments. For example, Chepinoga found an increase of oxygen consumption on the fifth day after denervation, but a decrease on the second day after denervation. Similarly, Fischer reported a maximum increase in oxygen consumption fifteen days after section of the nerve in the rat. Our experiments with A—V differences were performed on dogs one week and four to five weeks after denervation. A survey of differences in the consumption and output gradients in various animals from literary data are shown in Table 12. A number of experiments were not performed on isolated muscles, but on the whole limb. However, there was no qualitative difference found between the results obtained by the two methods. The consumption of oxygen in the whole limb three to four days after section of the sciatic nerve is double that of the control extremity. Seven to ten days after denervation, oxygen consumption in the gastrocnemius was increased only in some cases. At thirty to thirty-five days it is increased in every case.

Thus consumption of oxygen is not limited by the quantity of enzymes. as is evident from the fact, for example, that cytochrome c present in the muscle is sufficient for Q_{O_2} 900, while actually it reaches values around 115 at the most (Schmidt, 1956). That holds even when the content of cytochrome c, succinic dehydrogenase and cytochromeoxidase is decreased in denervated muscle, and yet oxygen consumption is increased.

The increased consumption of oxygen appears to be paradoxical from the point of view of the biologist; why should muscles undergoing atrophy consume more oxygen and liberate more energy than normal muscles in which trophic processes are normal?

Table 12

Changes in oxygen consumption in muscles of different species after neurotomy.

Species and experimental object	Time after neurotomy	O_2 consumption	Authors
Rat diaphragm	11 days	+	Gruber et al. 1952
Rat diaphragm	5 and more days	—	Humoller et al. 1953
Rat muscle brei	14 days	+	Fischer 1939
Rat muscle brei	21—28 days	normal	Knowlton, Hines 1934
Rat muscle brei	14 days	normal	Levine et al. 1941
Rat muscle brei		—	Schmidt 1952
Frog muscle brei	60 days	—	Califano et al. 1927
Frog muscle brei		—	Beliceva 1937
Rabbit muscle brei	5 and more days	+	Chepinoga 1938
Rabbit muscle brei		+	Telepneva 1953
Rabbit muscle brei	15—23 days	+	Sheves, Ryumina 1956
Rabbit muscle brei	15 and more days	+	Sheves et al. 1956
Dog muscle brei	8—10 days	—	Shatenshtein 1935
Dog A—V differences		+	Ijima 1924
Dog A—V differences	24—26 days	+	Kimura, Takahashi 1928
Dog A—V differences	21 days	+	Ichimi 1929
Dog A—V differences	23—70 days	+	Sato, Kasugai 1935
Dog A—V differences		+	Hirohashi 1941
Dog A—V differences	3 days	+	Hudlická, Vrbová 1958
Cat A—V differences	17—56 days	+	Langley, Igataki 1917

According to Langley (1917), the denervated muscle must cover increased energy expenditure due to muscle fibrillations. However, oxygen consumption is also increased after tenotomy (Ijima 1924) and arthritis (Harding 1926), where fibrillation activity is not present. Furthermore, a significant increase in oxygen consumption was not always found in isolated muscles even seven to ten days after denervation in dogs, although fibrillations appear earlier.

It could be presumed from other findings given below that the increased oxygen consumption is associated with transformation of energy metabolism in denervated muscle, and probably also with the formation of lipids (Fig. 85).

There exists only scanty information concerning the *consumption of glucose* in denervated muscle. It was shown by Hines and Knowlton (1941) that the consumption of glucose remains unaltered in the denervated muscle, but Sato and Kasugai (1935) reported an increase. In our experiments on dogs an increase was found 30—35 days after denervation, but there were no significant changes in glucose consumption to be found 7—10 days after denervation. Hines and

Knowlton investigated O_2 consumption a week after denervation and Sato and Kasugai three weeks after denervation. Thus it seems that increased consumption of substrates does not appear until atrophy is fully developed. However in our experiments formation of carbon dioxide was found to be increased 7—10 days after denervation, but the increase in RQ was insignificant. RQ values increased

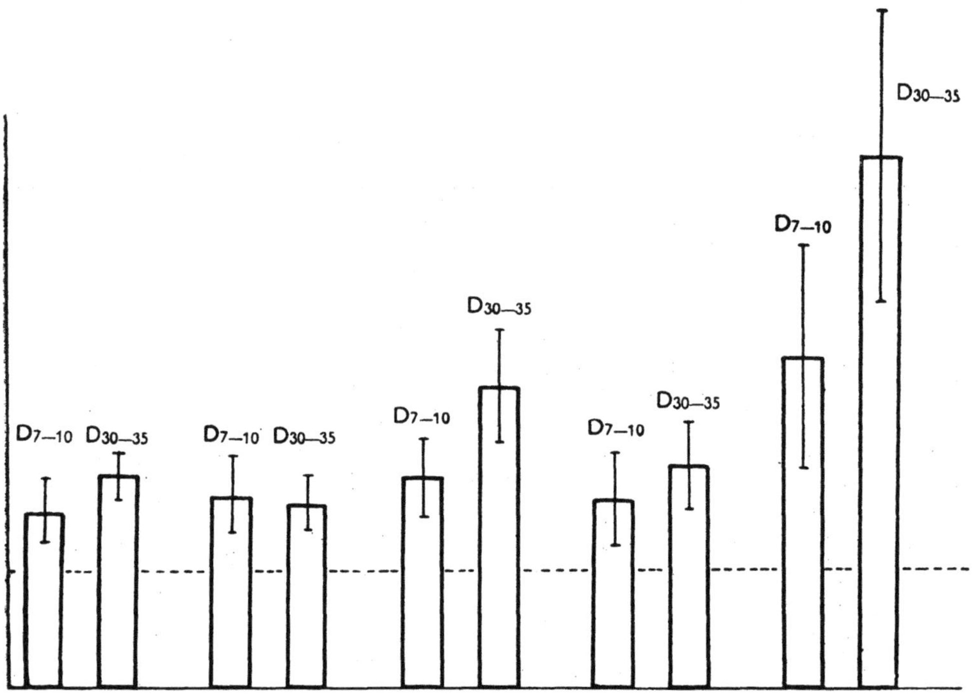

Fig. 85. Changes in blood flow, the consumption and release of metabolites and RQ after denervation in the gastrocnemius of the dog. Expressed as percentage of the corresponding values in the control muscle.
Pairs from the left to right: muscle blood flow, oxygen consumption, CO_2 liberation, RQ, glucose consumption.
D_{7-10}: 7 to 10 days after section of the sciatic nerve; D_{30-35}: 30 to 35 days after denervation. Interrupted line: 100 %, i. e. value of the control muscle.

a little above 1·0 thirty to thirty-five days after denervation and this, together with increased oxygen and glucose consumption, indicates the possibility that lipids are being formed from carbohydrates. This would mean that in addition to the usual oxygen consumption for resting energy requirements, oxygen would be needed in denervated muscle for obtaining energy for lipid synthesis from carbohydrates. This could at least partly explain the increased oxygen consumption as enhanced synthesis of lipids from carbohydrates following denervation. The content of lipids in denervated muscle was found to be increased. It is, of course, also possible that these lipids originate from blood lipids, which can serve as an important source of energy for the normal muscle and which could be stored in the denervated muscle.

Lactic acid, either brought to the muscle by the blood or originating from glycolysis in the muscle, could also serve as an energy source. It is known that the level of lactic acid in denervated muscle is increased (Knowlton and Hines 1934, Levine et al. 1941, Tower 1939) and the possibility that it is being utilized for glycogen synthesis (muscle glyconeogenesis) cannot be excluded. Sato and Kasugai (1935) demonstrated an increase in lactic acid utilization during infusion. Gutmann and Hudlická (1958) and Bass et al. (1960) suggested the possibility that lactic acid can be utilized by the denervated muscle for glycogen synthesis following hypoxaemia or repeated muscle stimulation.

It thus seems that increased oxygen consumption in the denervated muscle is also accompanied by increased glucose utilization, not only as a source of energy but also, perhaps, for lipid synthesis. This is in accord with Ferdman (1953, 1957) according to whom the denervated muscle utilizes more glucose from the blood and less glycogen. These conclusions of Ferdman were based on the results of Fainshmidt (1939), who reported that glycogen is not broken down in the denervated muscle following direct stimulation and who assumed that blood glucose serves as the source of energy. These experiments, however, were not sufficiently documented by initial values, so that they cannot be considered as conclusive.

It can thus be presumed from the above-mentioned results that, following denervation, *changes in metabolism* take place consisting of increased utilization of blood glucose as energy substrate and perhaps for increased lipid synthesis.

When the results of experiments on utilization and output *in vivo* and *in vitro* are compared, often contradictory values are encountered. It is probable that the latter are liable to a similar error as experiments in which enzymatic activity is being determined — they are inadequate for discovering changes associated with functional integration of a given organ into the whole organism, which manifest themselves *in vivo* only. It can thus be seen that for studying changes of metabolic regulations following denervation the *in vivo* approach is more suitable. *In vitro* experiments give information about progress of atrophy and about major pathological changes of metabolism. It has not been possible to demonstrate important changes in blood flow, but alterations in the permeability of muscle fibres have been found by this method.

A study of substrate consumption and formation of products yields information about the net changes of muscle metabolism following denervation. Of special significance are balance studies which demonstrate changes in the utilization of various substrates and their mutual interrelations. They thus make a more detailed biochemical analysis of certain links of intermediary metabolism possible and help in formulating the main questions for this analysis. This analysis should then lead to the localization of postdenervation changes in energy metabolism *in vivo*. Today two main approaches to this analysis seem to be practical: firstly, methods investigating changes in the reactibility of the denervated muscle (see Chapter XI which is dedicated to this question), and methods enabling the determination of changes in intermediary metabolism. Furthermore, experiments studying the reactibility of denervated muscle can also help in localizing changes of intermediary metabolism.

e) Metabolic changes in the denervated muscle under conditions of "metabolic loads"

At rest metabolism is a complex chain-like process which progresses at a more or less constant rate. For this reason its individual reactions are more or less equilibrated and the result of this equilibrium is a stable, so-called stationary concentration of intermediary metabolites.

This balance can be disturbed in various ways by changing the amount of substrate offered, by decreasing the concentration metabolites (for example, creatinephosphate and ATP during work), by their accumulation, or by retarding or accelerating an intermediary reaction. Most of such intervention (from the metabolic aspect it is possible to talk about metabolic loads) are complex in character, but some aspects may predominate. For example, during muscular work creatinephosphate and ATP are broken down, but at the same time there is often lack of oxygen and hexosophosphates and lactic acid are accumulated.

In experiments with these loads, changes in the balance can be observed both during the load as well as during restitution to the normal steady state. During this restitution an over-shoot recovery reaction takes place in the content of metabolites. These over-shoot reactions can be studied by determining the content of metabolites, metabolic stores changes in the activity of enzymes, or the consumption of substrates and formation of metabolic products at different time-intervals.

The main emphasis is necessarily laid on investigations carried out *in vivo*, although some experiments can also be performed in vitro. In experiments performed *in vivo*, regulating mechanisms of the whole organism participate in these reaction, i. e. nervous regulation, humoral factors, content of substrates etc.

In this sections the following three kinds of metabolic "loads" will mainly be discussed: alimentary load, "work" load, i. e. muscle stimulation, and metabolic loads resulting from artificial disturbances of blood circulation.

The most important factors belonging to the alimentary group are starvation and realimentation, especially by glucose administration.

aa) Starvation

During starvation the content of glycogen in the normal muscle is reduced. This decrease, however, occurs much later than that in the liver and is smaller in extent. Next, lipids are broken down and lastly proteins are utilized. Nevertheless, even when death due to starvation occurs, not all muscle glycogen and lipids are completely exhausted, and neither are the muscle proteins, of course, even though the total weight of the muscle is reduced.

The situation in the denervated muscle is somewhat different. In 1913 Mansfeld and Müller performed an interesting experiment of this kind. These authors determined the content of substances soluble in ether, i. e. crude *muscle lipids* in normal and denervated muscles of guinea-pigs dying of starvation. It was found that the denervated muscles contained considerably more lipids than normal muscles. From this it was concluded that the mobilization of lipids is considerably reduced after denervation and that the organism cannot utilize lipids present in denervated muscle for covering its energy requirements. Decreased mobilization of lipids was observed within four days after section of the nerve. Similar conclusions were drawn by Audova (1923) and Wertheimer (1927).

The *glycogen* content was studied much more extensively. According to Wertheimer (1927), Hines et al. (1931) and Hines and Knowlton (1935), the breakdown of glycogen in the denervated muscle during starvation is slowed down. Recently, a number of more detailed studies have been made concerning changes of glycogen in the denervated muscle in the course of starvation. It is pertinent to recall that glycogen content is different in different muscles and varies considerably in homonymous normal or denervated muscles (Riesser 1947) within a relatively large range. In our experiments on rat skeletal muscles, for example, between 300 to 950 mg%. The content of glycogen also varies during the day in relation to the intake of food (Forsgren 1928, 1935, Fábry 1955), and differences were found even from day to day (Riesser 1947). However, muscle glycogen was found to be fairly constant in animals from the same animal house, especially when animals from the same litter, of the same sex and same age were used. It is important to note that the ratio between glycogen content in normal and denervated muscle is the same under analogous nutritive conditions, irrespectively of the absolute values. This fact makes it possible to use the contralateral non-denervated muscle as control. Comparison of glycogen levels between different groups of animals is far more difficult. Such a procedure is possible only under the assumption that the experimental animals are absolutely homogenous and that all determinations are carried out not only under identical nutritive conditions, but also simultaneously. Most authors when comparing glycogen levels in the normal and denervated muscle performed unilateral nerve section and compared this with the homonymous contralateral muscle. Even though unilateral denervation can, to a certain extent, affect the control side (Baum 1931 — see also Chapter X), it cannot be expected that this would alter substantially the results obtained. On the contrary, the higher dispersion of results in so-called group experiments can mask any actual changes. It can be concluded from the results of Wertheimer (1927), Hines and Knowlton (1933, 1934, 1935). Gutmann and Vrbová (1952), Martínek (1956), Bass and Vodička (1958) and Vodička et al. (1957) that the rate of glycogen breakdown in the denervated muscle during starvation is reduced. The glycogen content in normal and denervated muscle was found to be lowered, but the ratio between glycogen content in denervated muscle to glycogen in control muscle (D : C) was found to have increased. Glycogen levels in denervated rat muscles are lower than in normal muscles (D : C = 0·51 ± 0·026) and this ratio D : C increases to 0·89 ± ± 0·018 in the course of 48 hrs. starvation. Likewise, in rabbits three days after denervation the content of glycogen in the denervated muscle of satiated animals is about 50% higher than in the controls, while after 48 hrs. of starvation it is about 200% higher (see Fig. 86). It is thus evident that the glycogen content in the control muscle decreases more rapidly than in the denervated muscle, independently of whether the initial values in the denervated muscle are higher or lower than in the control non-denervated muscle. Glycogen levels in denervated muscles, however, are most likely also reduced, so that glycogen values are lower than those usually found in denervated muscles of fed animals. This indicates that although glycogen breakdown in the denervated muscle is slowed down. it is, nevertheless, not lost completely. This is further borne out by the fact that the glycogen levels vary in the denervated muscle itself in parallel to its content in the control muscle and must therefore be not only synthetized but also broken down. It may thus be concluded that during starvation glycogen and lipids are broken down more slowly in denervated muscles than in the normal.

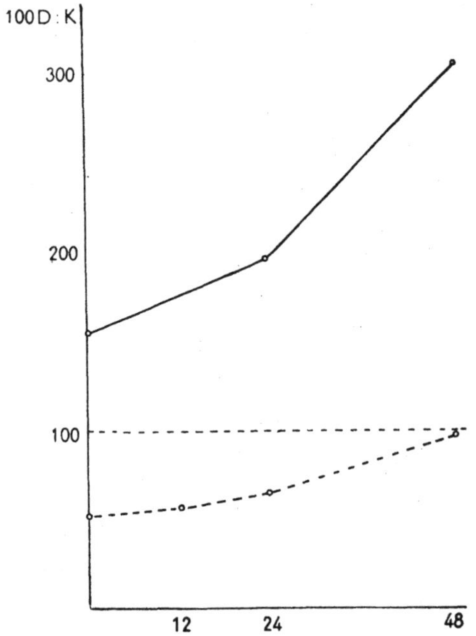

Fig. 86. Changes in glycogen in the denervated muscle as percentage of the control muscle in the rabbit (upper curve) and the rat (lower curve) during fasting. Ordinate: D:C . 100. Abscissa: duration of fasting in hours. Interrupted line: 100 %.

bb) Food intake

The most common and best defined method of accentuating muscle metabolism by offering surplus substrate is by *administration of glucose*. Administration of lipids or proteins is far more complicated since they are stored in the liver or in other tissues, and are liberated from the liver in altered form (for example as glucose).

Prior to glucose administration food is usually withheld for some time, so that glucose is given under conditions of fasting or starvation. According to Hines and Knowlton (1935), *glycogen* in denervated muscle is resynthesized from glucose, but never surpasses the level usually present in the denervated muscle at a given period of denervation. This would mean that there is some sort of maximum glycogen value in the denervated muscle for each denervation period, above which glycogen cannot rise. This conclusion is not in accord with the undulatory changes in glycogen levels mentioned above and also with the finding that, for example, in the rabbit, glycogen increases at a time when in the rat it was found to be decreased (Hines and Knowlton used rats for their experiments). It is probable that these findings can be explained by alteration in the rate of breakdown and synthesis of glycogen and their mutual interrelations.

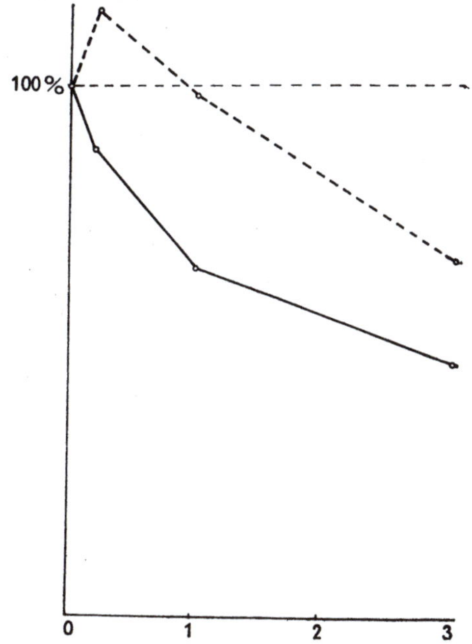

Fig. 87. Changes in glycogen content in the denervated muscle of the rat after 24 hours' starvation (interrupted line) and 4 hrs. after glucose administration (solid line). Abscissa: time after nerve section in days. Ordinate: D:C . 100.

The rate of glycogen breakdown is reduced in denervated muscle during starvation. There might also be changes in glycogen resynthesis after administration of glucose.

Glycogen resynthesis after glucose administration (glucose solution by stomach tube) was investigated in animals following 24 hrs. fasting. Three days before determination of the content of muscle glycogen, the sciatic or peroneal nerve was sectioned unilaterally. Contralateral non-denervated muscles and muscles of normal animals served as control. As can be seen from Fig. 87 and 88, glycogen synthesis following glucose administration is decreased in both rats and rabbits. It seems, however, that the experiments on rabbits are not as conclusive, in this case, as those on rats, perhaps because the glycogen content is higher in denervated muscles than in the normal muscles, so that it is difficult for these muscles to accumulate further amounts of glycogen. This might reduce the difference between glycogen levels in normal and denervated muscle after glucose administration and lead to incorrect conclusions. Glycogen levels are lower three days after denervation in the rat, although at shorter time intervals after denervation they are higher (Martínek 1956, Vodička et al. 1957). This fact was utilized in experiments in which glycogen levels were determined four, twenty-four and seventy-two hours after section of the nerve in rats which received, following 24 hrs. fasting, a solution of glucose by stomach tube four hours before excising the muscle (see Fig. 87).

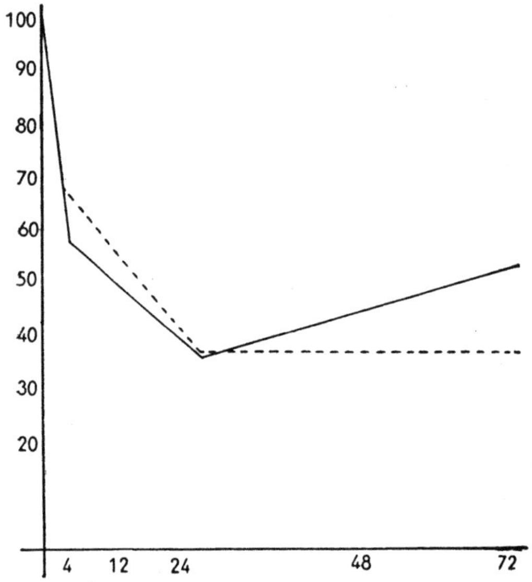

Fig. 88. Changes in the rate of glycogen synthesis (interrupted line) and breakdown (solid line) in the rat after denervation. Ordinate: the rate of glycogen synthesis and breakdown as percentage of control. Abscissa: time after nerve section in hours.

Within four hours after neurotomy the amount of glycogen synthetized in the denervated muscle was reduced to 68% of that synthetized by the normal muscle.

Twenty-four and seventy-two hours after denervation, i. e. at a time when the glycogen level is considerably reduced as compared with controls, the amount of glycogen synthetized amounted to 37% of the normal muscle. In normal rabbit muscles the glycogen content increases by 110% four hours after glucose administration (to 210% of the original value) and by 90% in denervated muscles. The D : C ratio has decreased from 1·94 to 1·75.

In the rat, glycogen synthesis from glucose administered is already reduced four hours after denervation. Glycogen synthesis is further reduced in the course of the first three days. When relative changes in the rate of glycogen synthesis four hours after denervation are compared with those during starvation, it is evident that both are reduced, but the course of these changes differs from each other. Four hours after neurotomy the breakdown is depressed more, so that synthesis predominates. Up to about twenty-four hours after nerve section, both processes are approximately balanced and at 72 hrs. glycogen synthesis is reduced more than its breakdown (Fig. 88). This is in agreement with detailed investigations of changes in muscle glycogen levels as reported in fed animals by Martínek (1956) and in fasting animals by Vodička et al. (1957). Where the rate of glycogen synthesis is higher than that of its breakdown, glycogen levels increase in comparison with the control muscle, even when both processes are retarded. This is the case, for example, four hours after denervation in fasting rats and still twenty-four hours after denervation when the rates of glycogen breakdown and synthesis are about equal. Seventy-two hours after denervation breakdown predominates and glycogen levels are decreased both in fed and fasted animals.

It seems that a single dose of glucose is relatively the most advantageous form of application, even though humoral factors, difficult to control, come into play. Starvation is a much more complicated stimulus, since the supply of metabolites does not cease rapidly but is maintained for a long time at more or less normal levels. This applies, for example, to glucose, which is liberated in the liver where increased gluconeogenesis takes place from reserve lipids and proteins. It is not clear whether such glucogenesis also takes place in the muscle.

Slowing down of glycogen synthesis and breakdown following these "loads" cannot be explained by changes in substrate supply in the muscle, since the consumption of neither O_2 nor glucose has been found to be decreased. The change in glycogen metabolism cannot evidently be explained by a lowered substrate supply, but has to be looked for exclusively in the metabolism of the muscle cell itself. Experiments with elimentary "loads" demonstrate one aspect of the metabolic reactibility of denervated muscle, i. e. its response to stimuli represented by reduced or increased offer of metabolites at rest, or at least under conditions of relative inactivity. Nevertheless the capacity of the muscle to synthetize glycogen need not be fully utilized here. The ability of the muscle to break down glycogen can be more fully exploited if the muscle is subjected to function, which in the case of denervated muscles has to be evoked by direct stimulation.

cc) ***Direct muscle stimulation***

Stimulation of normal muscle results in the reduction of glycogen and creatinephosphate and prolonged stimulation also in the decrease of ATP. The content of inorganic phosphate, intermediary products of glycolysis and lactic acid increases. The situation is the same in the denervated muscle (Levine et al. 1941). In our experiments a more detailed analysis of *glycogen* changes was attempted. Glycogen was investigated, because of its major role during muscle activity as an important source of energy. Muscles were stimulated directly with series of galvanic impulses of two frequencies: either 30 or 120 impulses per minute, in a similar way as Yampolskaya and Yakovlev (1951). These authors used two frequencies of stimulation — a slow and a fast frequency. The total number of galvanic pulses was the same in all their experiments. It was reported

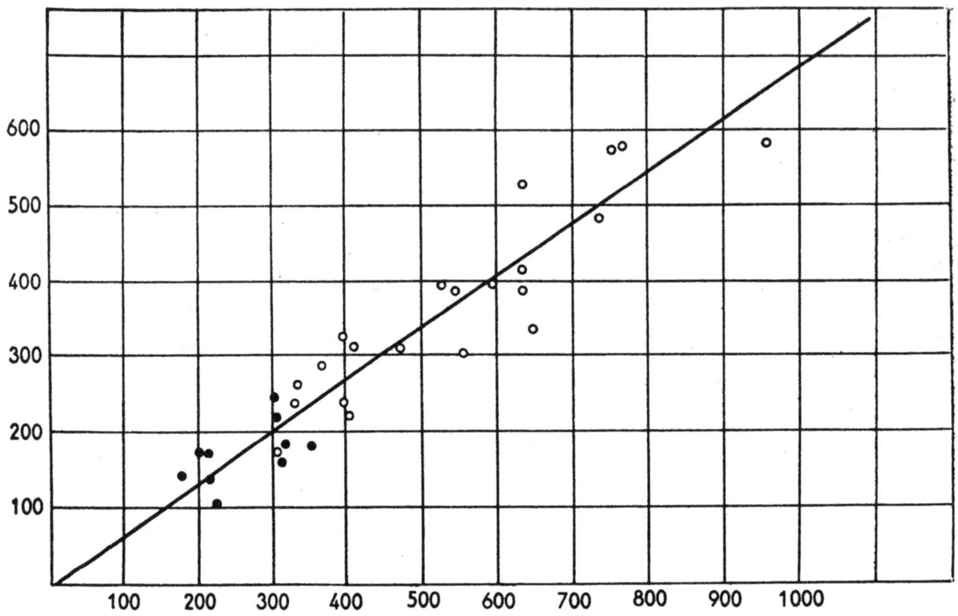

Fig. 89. The correlation between the amount of glycogen broken down after stimulation 120 imp./min. and its content in the muscle. Ordinate: glycogen content in the muscle after stimulation. Abscissa: content of glycogen in the control muscle. ○ — normal muscles, ● — denervated muscles.

that by suitable alternation of these frequencies resynthesis of at least a part of glycogen already occurs during the slow frequency of stimulation. These findings were obtained in normal muscles. According to Fainshmidt (1939) glycogen is not broken down during direct stimulation of denervated muscle. However, this result cannot be considered fully conclusive (see page 229). For this reason, denervated muscle was studied after stimulation by either frequency (Bass et al. 1954, 1955). Since it had already been shown that resynthesis occurs in normal muscle, it was not necessary to work under such complicated experimental conditions as Yampolskaya and Yakovlev. For this reason the reduction in glycogen content was calculated by comparing the glycogen decrease after stimulation by the "low" and "high" frequency in different animals. However, the content of glycogen even in normal muscles is not always the same and varied from 300 to 950 mg % in our experimental animals (rats). It was found, however, that following muscle stimulation glycogen is always reduced by the same proportion, independently of the initial value, but dependent on the state of nutrition of the animal and on the frequency of stimulation. In animals after 12—24 hrs. fasting, glycogen was reduced by $69{\cdot}4 \pm 4{\cdot}3\%$ of the content in contralateral muscle after stimulation at a frequency 120 imp. per min., and by 56·0 ± 5·4% on stimulation with the lower frequency. A similar result was obtained in the denervated muscle 72 hrs. after section of the nerve: glycogen content was reduced to 67·0 ± 4·8% at the higher frequency and 58 ± 8.7% at the lower. It thus appears that there is no difference in the breakdown of glycogen following stimulation between normal and denervated muscles, even though contractions of the denervated muscle are somewhat weaker. The content of glycogen in the denevated muscle three days after neurotomy is lowered, so that the proportional amount broken down during stimulation actually corresponds to an absolutely smaller glycogen amount. The correlation between the initial content of glycogen and the glycogen broken down during stimulation is given in Fig. 89. The lowering of glycogen content in mg% is expressed as a function of its original value. The correlation between the amount of glycogen broken down during stimulation and the content of glycogen in the control limb is statistically highly significant, the correlation coefficient being 0·84 for 32 pairs of values, this being significant for $p < 0{\cdot}001$. The reason for this relationship is unknown. It is not in any way affected by denervation. In this sense the thesis of Levine et al. (1941) that the breakdown of glycogen remains unchanged following stimulation of denervated muscle, can be formulated more precisely. It can thus be concluded that the percentage of glycogen broken down during direct muscle stimulation is constant under identical conditions and is not altered after denervation.

That the rate of glycogen resynthesis is similar in both cases was indirectly demonstrated by the fact that the same percentage of glycogen was broken down when stimulating with both frequencies. This postulate could be erroneous if the rate of glycogen breakdown by direct stimulation was altered as well. For this reason, the resynthesis and its rate was also studied directly, i. e. in the course of recovery after stimulation.

Recently it has been shown (Hudlická, Bass, unpublished results) that the role of carbohydrate oxidation in the denervated muscle is also enhanced during stimulation. The consumption of "oxidizable glucose" (= glucose consumption after subtracting lactic acid liberated) in the normal muscle decreases to such negative values, that carbohydrate oxidation can account only for a little part of the oxygen consumed. In the denervated muscle, however, there is the same amount of glucose, or even more, being consumed than there is lactic acid liberated, the consumption of "oxidizable glucose" being even slightly positive on the average. Since glycogen is broken down at the same time, there is more carbohydrate being oxidized. The role of carbohydrate oxidation is thus greater in the denervated muscle than in the normal muscle both at rest and during stimulation.

dd) Recovery processes after stimulation

Levine et al. (1941) studied both the breakdown and the resynthesis of substances important for energy metabolism, i. e. glycogen. ATP, creatinephosphate, inorganic phosphate, lactic acid, hexosomonophosphate and hexosodiphosphate, and found no differences between normal and denervated muscles. In our experiments (Bass et al. 1955), no differences in glycogen synthesis were found in rat muscles three days after denervation during the first phase of resynthesis, i. e. within the first 45 min. after stimulation of the tibialis anterior

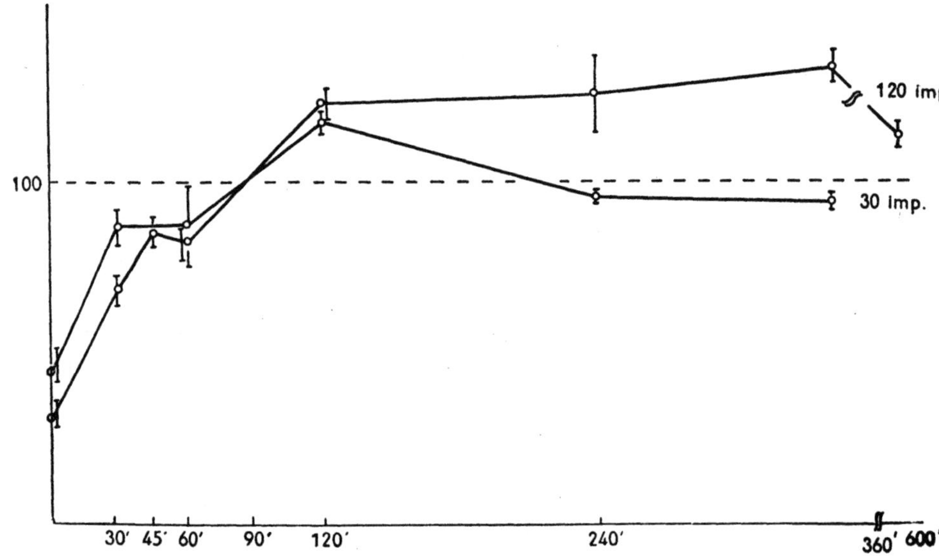

Fig. 90. Changes in glycogen content in the tibialis anterior of the rat following stimulation at a frequency of 120 and 30 imp./min. respectively. Ordinate: content of glycogen in percentage of control. Abscissa: time after termination of stimulation.

muscle. Up to this point it is thus possible to agree with Levine et al. that the rate of glycogen resynthesis immediately after stimulation is the same in normal and denervated muscles. Values compared are, of course, only relative, being expressed in relation to the glycogen content in the contralateral, non-stimulated control, under otherwise identical experimental conditions as on the stimulated side. After this phase (the duration of which varies in different muscles) the content of glycogen temporarily reaches higher values than those in the control muscle. This *over-shoot reaction* was first described by Yampolskaya (1950) and was designated as supercompensation. Changes in *glycogen* after stimulation were also investigated by other authors (e. g. Flock and Bollmann 1940), but either too short a time was allowed after stimulation or an unsuitable frequency of stimulation was used. Similar over shoot reactions following muscle stimulation were later found in the *creatinephosphate* content (Yakovlev 1956). This reaction and time course of glycogen changes was studied in detail by Yampolskaya (1952) and later by Bass et al. (1956), who reported that the over-shoot reaction is dependent on the character of stimulation; the excursion lasts longer the greater the breakdown of glycogen after stimulation (Fig. 90) and does not occur after slow frequency stimulation nor, in most cases, after tetanic contraction. The threshold frequency which evokes this over-shoot reaction also

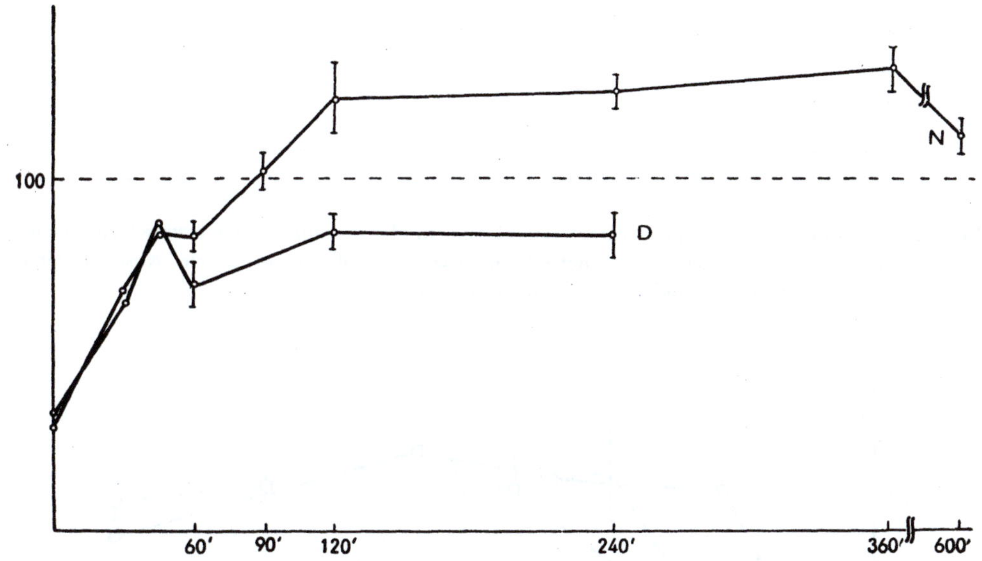

Fig. 91. Changes in glycogen content in the normal (N) and denervated (D) tibialis anterior muscle of the rat following stimulation at a frequency of 120 imp./min. Ordinate: glycogen content as percentage of control. Abscissa: time after termination of stimulation.

depends on the type of muscle — in fast muscles usually a low frequency of stimulation is sufficient, in slow, tonic muscles a higher frequency is necessary and for an over-shoot reaction in the diaphragm (*in vitro*), a prolonged tetanic contraction is necessary. Changes in this over-shoot reaction have been studied by Gutmann et al. (1954) following denervation and transection of the spinal cord. It was found that following denervation it gradually decreases and disappears within 24 hrs. The over-shoot reaction in tibialis anterior is also gradually lost after transection of the spinal cord (Th_{8-10}). When comparing the glycogen content and its time course in the normal and denervated muscle, it was observed that during the first 45 min. of resynthesis there are no differences in the glycogen resynthesis between the normal and denervated m. tibialis anterior (the resynthesis being expressed as percentage of glycogen broken down during

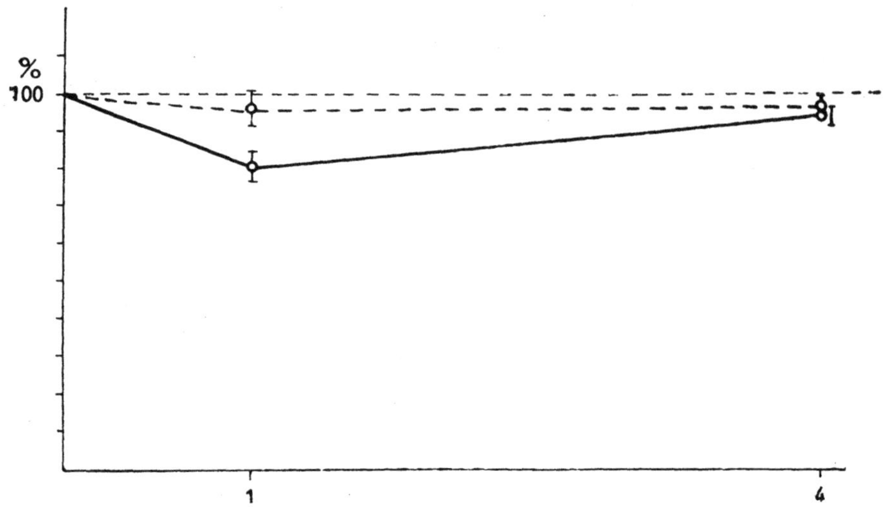

Fig. 92. Hexokinase activity after direct stimulation in the normal and denervated gastrocnemius of the rat. Ordinate: hexokinase activity in percentage of control muscle. Abscissa: time after stimulation in hours. Straight line — control muscle.

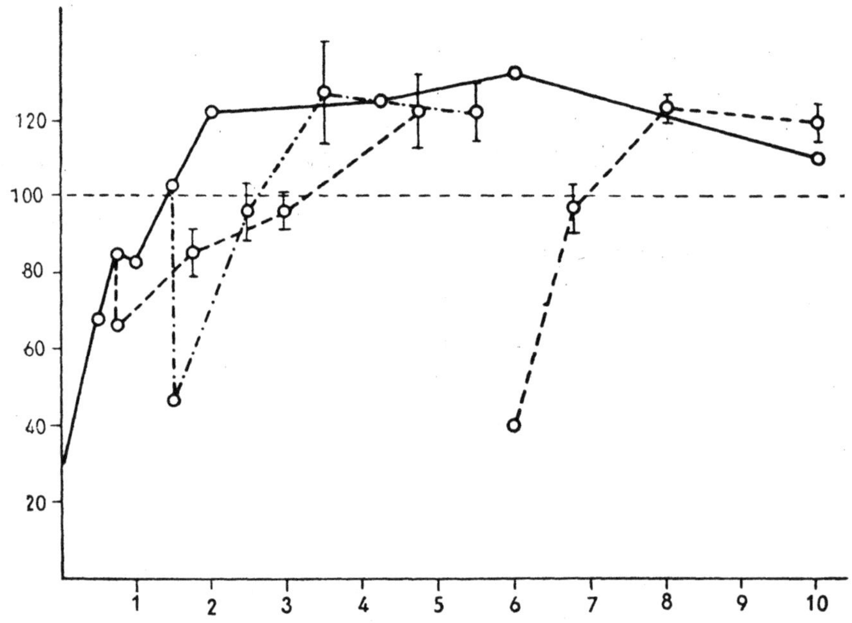

Fig. 93. Changes in glycogen content in the normal tibialis anterior of the rat following first and second stimulation at different time intervals. Ordinate: glycogen content in percentage of the control. Abscissa: time after first stimulation in minutes.

stimulation). After this, however, the normal muscle maintains the content of glycogen at the same level up to approximately 60 min. Between min. 60 and 90 after stimulation it begins to rise again (Fig. 91). Maximum values of glycogen content (20 to 30% higher than in the contralateral limb) are attained after about 120 min. and remain at this peak for a certain time which is dependent on the frequency of stimulation. In the denervated muscle the over-shoot reaction is quite different in character. The glycogen content reached a maximum

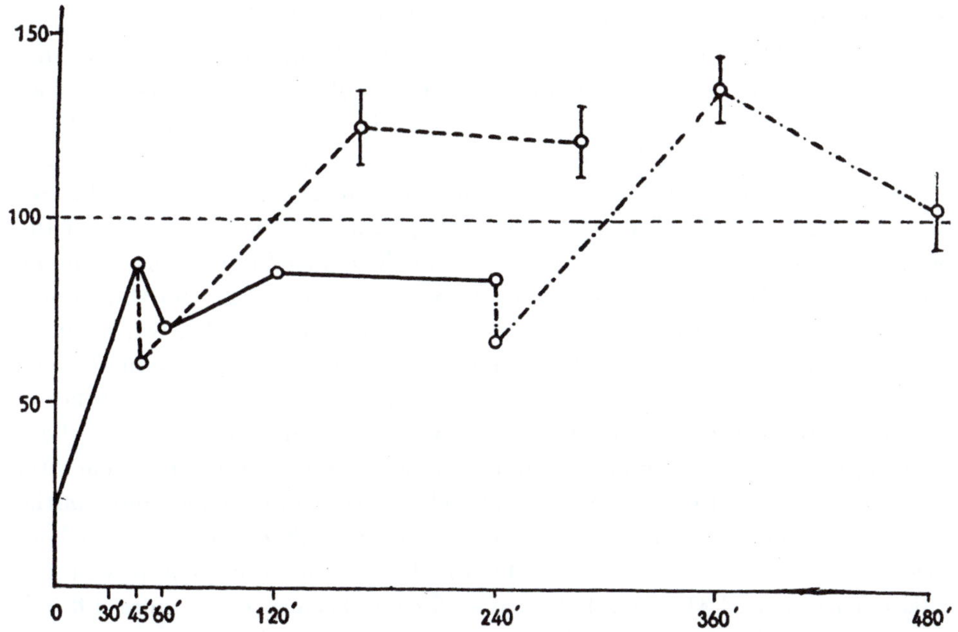

Fig. 94. Changes in glycogen content in the denervated tibialis anterior of the rat following first and second stimulation at different time intervals. Ordinate: glycogen content in percentage of the control. Abscissa: time after first stimulation in minutes. Straight line after first stimulation

value 45 min. after stimulation at which time an identical amount of glycogen was resynthetized, expressed as percentage of glycogen broken down, as in the normal muscle and amounted to about 85% of the level of the control muscle At 60 min. it decreased somewhat and after that gradually attained values of the control muscle. It did not, however, reach higher levels (Fig. 91).

Yakovlev (1956) demonstrated that stimulation increases the activity of several *enzymes* e. g. hexokinase. We assumed that such an increase can occur a second time in normal muscle at the period when glycogen content rises rapidly. It could be presumed that at this time the muscle assimilates large quantities of glucose. However, it was shown on the contrary that hexokinase activity in the normal muscle is reduced one and four hours after stimulation (Fig. 92).

Hexokinase activity in these experiments was determined in muscle extract by the method of Collowick and Kalckar (1943). The possibility cannot be excluded that hexokinase is being adsorbed at this period onto cellular structures — possibly the membrane — so that its content in the cytoplasm is thus reduced. It is also possible, however, that the factor which determines the rate of glucose penetration into muscle fibres is of another character, e. g. increased blood flow (glucose consumption in normal muscle correlates with the rate of blood flow) or an increased ADP concentration. However this change was not observed in the muscle three days after denervation (Bass and Vodička 1956).

It is also possible that the normal "trophic state" is maintained by continual over-shoot reactions similar in character as those observed in glycogen. Loss of these over-shoot reactions might result in muscle atrophy. On the other hand it is known from clinical and experimental reports (Hoffmann and Wertheimer 1927, Solandt and Magladery 1940, Gutmann and Guttmann 1944, Wertheimer 1942, 1944, Osborne 1951, Zubenko 1954, 1956) that it is possible to retard the progress of atrophy by repeated stimulation of denervated muscle, even though the normal state. If these over-shoot reactions are of significance for the trophic state of a muscle, then it can be assumed that repeated stimulation will elicit an analogous over-shoot reaction in the denervated muscle. This has actually been found to be the case. We have shown (Bass et al. 1959) that on second stimulation performed in the rat 45 min. or 240 min. after first stimulation, an over-shoot reaction appears which is analogous in character to that in normal muscle after low frequency stimulation (Fig. 93 and 94). The glycogen excursion above initial level is also of short duration, and its course does not differ from the over-shoot reaction observed in normal muscle after the first stimulation. The mechanism of this excursion and the metabolic cause for its absence, or its more difficult elaboration after denervation, will have to be studied further.

ee) *Changes in energy metabolism during reduced blood flow*

Muscle hypoxia represents another type of metabolic load, which forces the muscle to utilize more energy from glycolysis, but does not imply muscle activity. This approach may indicate the extent to which energy metabolism in the denervated muscle is capable of adaptation to altered conditions. Lack of oxygen can either be evoked by high altitude anoxia, inhalation of air poor in oxygen or of nitrogen, by poisoning some of the oxidation enzymes and finally by restricting muscle blood supply with tourniquet or ligature of blood vessels. This latter approach is also important for denervated muscle since it is also possible that the sensitivity of the denervated muscle to lowered oxygen supply and other blood constituents is altered. If this is the case, then it could be expected that such restriction of blood flow can play a role in the metabolic disturbances after denervation. If it is found, however, that muscle metabolism is not seriously impaired by transient reduction of blood flow, then it can be presumed that if in some cases insufficient supply of substrates is brought into the denervated muscle via the blood, it cannot reduce the intensity of metabolism and cause muscle atrophy.

Restriction of blood flow in normal muscle leads to a decrease in the content of glycogen, creatinephosphate and ATP, to an increase of lactic acid and to lowered activity of dehydrogenase and ATPase (Chandelon 1876, Manche 1889, Bollmann and Flock 1944, McPherlane and Spooner 1946, Harmann and Gwinn 1949, Fuhrmann and Crismon 1951, Bücherl and Schwab 1952, Gollwitzer-Meyer 1953, Fetisova 1954, Issekutz et al. 1955, Thorn et al. 1957 and others). The consumption of oxygen is decreased during constriction of vessels (Bücherl and Schwab 1952, Breuer and Ricker 1958), and increases after renewal of blood flow (Rein and Schneider 1938, Takácz 1957). Glucose consumption decreases (Takácz 1957) and during deeper hypoxia arterio-venous differences of glucose are even negative (Kreienberg 1942). Evidently the utilization of energy substrates is altered, since RQ rises to values above 1·0 (Breuer and Ricker 1958). Following release of the ligature, the *consumption of oxygen* in a muscle denervated immediately before the ligature (femoral artery) is increased similarly as in the normal muscle (Rein and Schneider 1938). The changes in energy metabolism under these conditions were compared in normal and denervated muscles three days after neurotomy by Hudlická and Gutmann (1958). Following ligature of the abdominal aorta and vena cava caudalis muscle excitability was lost during the ligature with rapid restitution after its release. This testifies that sufficient arrest of blood flow was achieved.

Firstly the abdominal aorta was ligated. After half an hours' ligature no changes in the content of *glycogen* were observed in both normal and denervated muscle; glycogen content decreased in the normal muscle one hour after application of the ligature, but remained unchanged in the denervated muscle. Four hours after releasing the ligature glycogen level in the denervated muscle rose. It thus seems that denervated muscle is, as far as can be judged from the glycogen content, more resistent to ischaemia than normal muscle.

However, the decrease of glycogen content in the normal muscle was only slight. Consequently in further experiments the v. cava caudalis was ligated in addition to the abdominal aorta, so that reduction of blood flow was more marked (Shepherd 1949). Reduction in the content of glycogen was more evident in the normal muscle than after ligature of the aorta only and there was even a decrease in the denervated muscle (Fig. 95). Following release of the ligature, glycogen returned to initial values within 24 hrs. in the normal muscle, while the original levels in the denervated muscle were found already two hours after release. Four hours after restitution of circulation, glycogen even attained values higher than prior to the ligature; this increase lasted throughout the experiment (further four hours).

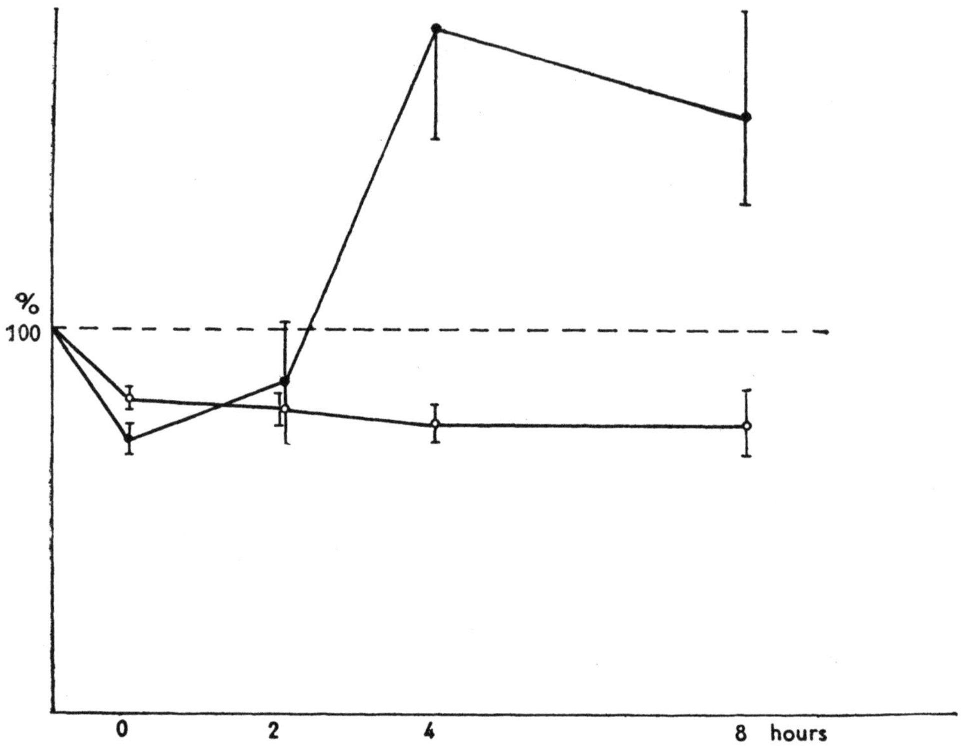

Fig. 95. Glycogen resynthesis in the normal (○) and denervated muscle (●) following ligature of the abdominal aorta and v. cava caudalis. Ordinate: glycogen content in percentage of the control nonligated muscle. Abscissa: time after ligature.

Prolonged decrease in the content of glycogen is not due to shock resulting from laporotomy, since it was not found following control operation. Neither can it be explained by vasomotor spasm which could perhaps be evoked by manipulation with blood vessels. The decrease in glycogen in normal animals lasted for as long as in animals following sympathectomy. Muscle temperature, which was lowered during the ligature, returned to normal values within four hours after releasing the ligature.

It is possible that ischaemia causes a greater disturbance of energy metabolism in normal muscle than in denervated muscle. Such a disturbance would become evident in the first place as a change in the content of creatinephosphate, which is rapidly lost after arrest of circulation (Fetisova 1954). However, the decrease in *creatinephosphate* during the ligature was found to be the same in both the normal and denervated muscle and the recovery of creatinephosphate following release of the ligature was faster in the normal muscle. Four hours after release of the ligature the content of lactic acid in normal and denervated muscle also attained the initial values. Nevertheless, the content of *lactic acid* during the ligature was higher in denervated muscle than in normal muscle. It is possible that the high glycogen content in the denervated muscle following release of the ligature might, at least in part, be due to the fact that the muscle is able to utilize lactic acid for glycogen resynthesis — as is known to be the case in frog muscle and as is indicated by experiments of Sato and Kasugai in denervated muscles of dogs. This is also borne out by experiments in which the content of lactic acid following direct stimulation of muscle has been found to

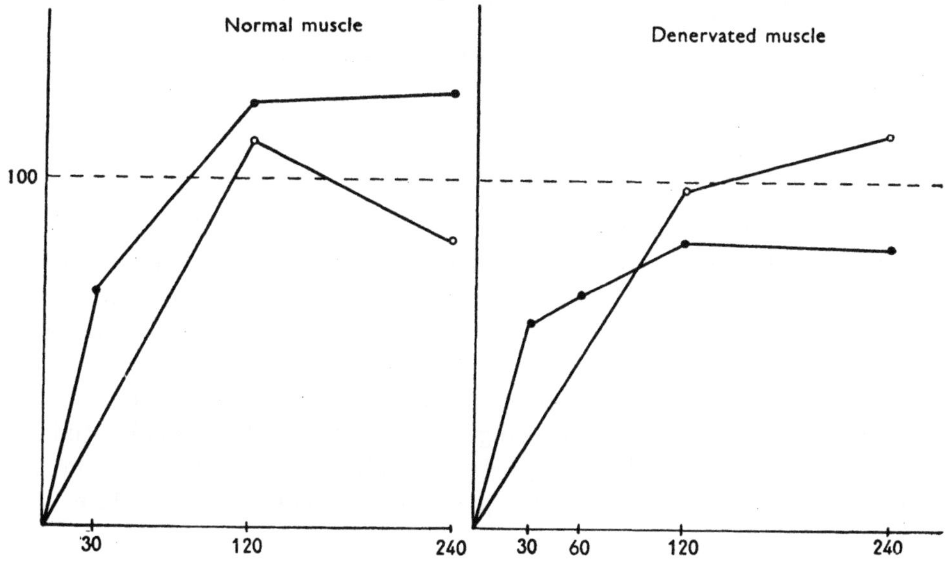

Fig. 96. Glycogen resynthesis after stimulation (●) and after stimulation during ligature (○). Ordinate: content of glycogen in the control muscle. Abscissa: duration of resynthesis in minutes.

be increased still further. In these experiments muscle stimulation was combined with 30 min. ligature; four hours after stimulation glycogen was found to rise considerably above initial values in denervated muscle (and above values obtained by stimulation only), while in control muscle values attained were still below those found after stimulation only. Following a transient rise the glycogen content decreases four hours after stimulation (Fig. 96). It is thus apparent that while the normal muscle is relatively highly sensitive to lack of oxygen, no substantial decrease of energy metabolism occurs in the denervated muscle under the influence of reduced circulation. On the contrary, it is possible to state that there is a certain stimulation of glycogen synthesis. This is, to some extent, corroborated by experiments demonstrating that accentuation of glycogen resynthesis occurs in denervated muscle following partial exsanguination (corresponding to 1·5% body weight). The level of hemoglobin was decreased by

10% on the average by this procedure and erythrocytes were decreased by half a million. One hour after exsanguination the glycogen content was reduced by 20 to 30% in the soleus and anterior tibial muscles. Four hours after the procedure glycogen resynthesis attained initial values in normal muscles and reached even higher values than before in the denervated soleus (Drahota and Hudlická 1958). The denervated muscle thus exhibited a higher resistance to lack of oxygen obtained by other means than ligature and manifested a tendency to accentuation of glycogen synthesis.

It is possible that the increased glycogen synthesis in denervated muscle is actually the result of higher utilization of lactic acid, as has been indicated by experiments with repeated muscle stimulation. This hypothesis, however, requires further experimental corroboration.

ff) Conclusions concerning metabolic loads

As can be concluded from above, metabolic loads can be useful in determining changes in muscle reactibility rather than in analysing details concerning metabolic processes and reactions taking place after denervation. In order to obtain such an insight into these metabolic processes, it is necessary to study metabolic processes *in vivo* in such a way as to penetrate into individual processes of intermediary metabolism approaching, as far as possible, resting conditions. It is not until then that changes following various metabolic loads can be studied. The most important method, and perhaps the only adequate method is the use of labelled compounds and their investigation in the course of metabolism.

f) The use of labelled compounds

Reports dealing with labelled compounds have often given very important results concerning intermediary metabolism, especially is investigations of isolated enzymes or enzymatic systems *in vitro*. However, these experiments can easily lead to misleading results even *in vitro*, as has been shown, for example, by Ogston (1950) in studies of the tricarboxylic cycle (biological asymmetry of citric acid) and in other similar instances. The probability of error is still greater in the organism as a whole, especially when intermediary metabolism of a normal organ is investigated. A similar source of error is also possible, of course, when metabolic changes are studied in pathologically altered organs. The situation in this case is somewhat simpler, however, when symmetric organs are investigated, so that conparison is made between a pathological and a normal organ. Such an approach is possible in denervation studies in skeletal muscles.

Labelled compounds have so far been used only to a small extent in investigations of denervated muscle metabolism and have still less been utilized for studying its energy metabolism. Upto date only a few findings have been obtained by this method. In some of these investigations increased metabolism was reported, in other this finding was repudiated.

However, these studies represent only the beginning of investigations concerned with metabolic regulations and their disturbances with the aid of labelled compounds, and it is therefore not surprising that first attempts of making use of this technique are not conclusive. A number of *interfering factors* make their interpretation difficult and place high demands on the correct and exact formulation of the experimental questions.

What are the main interfering factors which must be excluded, in order to ensure the reliability of experimental results with labelled compounds *in vivo* and to avoid incorrect conclusions?

Labelled compounds administered usually reach the muscle via the blood stream. Their concentration in the blood is not usually constant and its maintenance would be very difficult unless a perfusion technique was used. However, it is not easy to evaluate perfusion experiments since capillaries and arterio-venous anastomoses may be open to varying degrees, interfering thus with the uptake of the labelled compound. Labelled compounds are usually administered by a single injection. Following this, the content of the labelled compound in the blood rapidly increases, especially after direct administration into the blood stream. This is followed by a decrease of its content, the rate of which depends on the rate of uptake and output of the labelled substance in the organ (which may even be transformed into another labelled compound), until an equilibrium is reached between the organ and blood concentration. After this, labelled substances escape back into the blood, and as long as they are excreted, their concentration in tissue fluids is higher than in the blood until they are completely excreted from the organism. When symmetrical organs are investigated and the blood flow in one of these is increased, as may be the case in denervated muscle, the labelled substance can penetrate more quickly inside and in this way simulate more intensive metabolism, its intracellular concentration sooner attains its equilibrium with the blood and so increases the apparent turnover rate.

The time elapsed between administration of the labelled compound and excision of muscle tissue for analysis is also important. When using substances which penetrate slowly into muscle fibres, determinations of samples taken a short time after injection of the labelled compound, i. e. at a time when the uptake into muscle fibres may be ignored, can give information concerning the capillary volume and uptake into the extracellular space. The work of Hudlická et al. (1955), for example, with *labelled phosphate* was performed along these lines. It was found that the uptake in some muscles after denervation is increased (extensor digitorum longus), in others (soleus) it is decreased, as in the case of oxygen consumption (Sheves and Ryumina 1956). It is possible to investigate in this way changes in blood flow and uptake into extracellular spaces after denervation. Attempts have already been made to use radioactive compounds for studying energy metabolism in denervated muscle. Thus Bourguignon (1952), who studied the incorporation of radioactive phosphorus into denervated muscle at long intervals after injection, did not differentiate the individual compounds into which P^{32} was incorporated. On the other hand, Grigorieva (1954, 1958) investigated metabolism of the muscle fibre in the rabbit with the aid of labelled phosphorus and did not restrict her attention to the uptake only, but also determined the specific activity and the relative specific activity of further metabolic products: ATP, creatinephosphate, total acid soluble and inorganic phosphates.

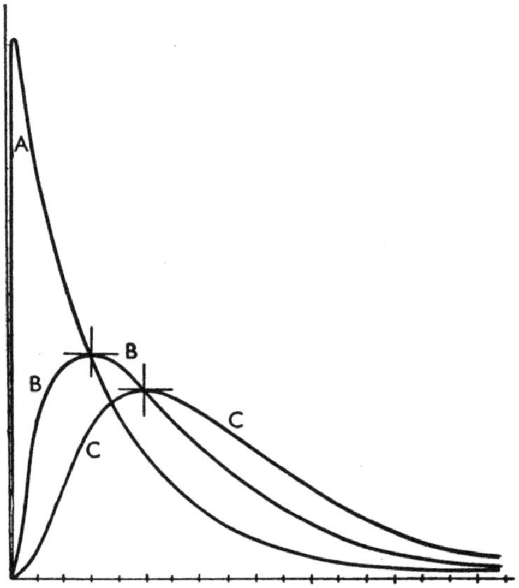

Fig. 97a. An illustration of specific activity in a three-member system of precursor (A), intermediary product-precursor (B), and the final product (C). Ordinate: specific activity arbitrary units. Abscissa: time after addition of labelled compound.

By this approach, i. e. by comparing the rate of formation of compounds from their precursors under different conditions, it is possible to determine changes in the rate of formation of intermediary metabolites and consequently of changes in metabolism as a whole. It will, however, be necessary to improve this method from the technical (technique of isolation and sample preparation etc.), as well as the biological aspect. In the first place, only small quantities of labelled compounds must be administered, in order not to alter the original concentration in the blood to any great extent and to avoid undesirable experimental conditions from increased levels in the blood. Danger of this was already pointed out by Sacks and Altschuler (1942) who found that after administration of large doses of labelled phosphorus there was a transient marked increase in the specific activity of hexosomonophosphates. This increase, however, was not observed when smaller doses of P^{32} — phosphate of higher specific activity — were used.

In addition it is necessary to investigate more than one or two points of the incorporation gradient, in order to be able to determine in which phase of the gradient a particular compound is at a given moment (Fig. 97a). During the phase of rapid rise in the activity of an organ the product (C) accepts the labelled isotope from its precursor (B), labelling of which also increases. When the concentrations of labelled precursor and the labelled compound (A) from which it is formed are balanced or is in equilibrium with the labelled precursor (A) in the blood, the specific activity of the precursor (B) suddenly begins to decrease, while in the product (C) the rate of incorporation is gradually reduced until the concentration of the labelled compound of the product attains equilibrium with the concentration of the precursor. At this moment specific activity reaches its maximum. In the following phase the concentration of the labelled component is reduced in both the product (C) and precursor (B), but that of the product is higher. It is therefore necessary to investigate the course of incorporation for longer periods of time after a single injection dose of the labelled compound, in order to avoid any error due, for example, to the fact that a higher concentration of the labelled product, after reaching the maximum, would be considered to be the precursor. This would mean that the direction of the reaction would be incorrectly deduced. As many intermediary metabolites as possible should be investigated at the same time, so that relative rates of their formation and breakdown may be determined. When investigating muscle energy metabolism with the aid of labelled compounds, the following substrates seem to be the most important: intermediary metabolites of glycolysis and the tricarboxylic acid cycle, and then intermediary metabolites of lipid metabolism and of protein catabolism. Essential are substances transferring energy, especially ATP and creatinephosphate.

Of these compounds, *phosphorylated intermediary metabolites* are most easily open to investigation. Their study makes it possible to compare the rate of energy transfer (with the aid of ATP and creatinephosphate) and the rate of glycolysis. Grigorieva (1958) compared the rate of formation of labile phosphorus in ATP and creatinephosphate by comparing their activity with the specific activity of inorganic phosphate corrected for inorganic phosphate present in the blood. According to Grigorieva, this function, i. e. relative specific activity — remains unchanged after denervation. However, specific, activity not expressed in relation to inorganic phosphate is increased after denervation. The interval between isotope administration and excision of the muscle being short (10 min.), this result indicates increased blood flow. Since oxygen consumption is increased at the same time (Sheves and Ryumina 1956, Bass, Hudlická 1960), this could mean that the effectiveness of oxidative phosphorylation is reduced in the muscle in vivo. The phosphorylation of ADP and creatine to ATP and creatinephosphate remains unchanged, but the same degree of phosphorylation requires more oxygen. The P : O quotient is thus smaller than in normal muscle. There is still a number of open questions to anwer. In the first place, differentiation has not been made between the beta- and gamma-phosphate group of ATP which would show possible changes, for example, in myokinase activity which could simulate or alter an apparent difference between the P : O quotient in normal and denervated muscle (see also p. 259). Secondly, evaluation of the specific activity of compounds investigated in terms of the specific activity of inorganic phosphorus does not seem to be most suitable, since, even after correction for capillary blood, it is calculated in relation to intracellular and extracellular (intercellular) phosphate, which cannot be differentiated by methods used at the present time (Ennor and Rosenberg 1954). This means that changes in extracellular space or uptake into it, may change the relative specific activity of substances investigated independently of changes in metabolism.

It thus seems more advantageous to express the results in relation to other compounds, which are not complicated by changes in extra- and intracellular space, for example, by comparing the specific activity of gamma-phosphorus of

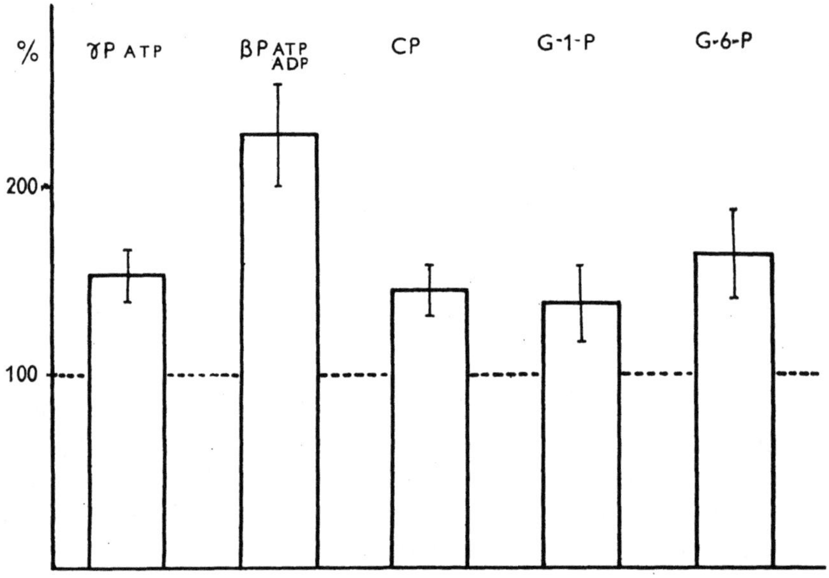

97b. Specific activities of phosphorus in phosphate compounds in denervated muscle, 15—45 after labelled phosphate injection Broken line-specific activity of same compounds in normal muscle as 100%.

ATP and the phosphorus of intermediary products of glycolysis in a complex analysis of a large number of metabolites with special reference to the gradient, the course of which can indicate the relations between precursors and products, and their changes after denervation.

More recently such an experiment was performed by Bass and Vítek (unpublished results). The specific activities of phosphorus in glucose-1-phosphate (G-1-P), glucose-6-phosphate (G-6-P), creatinephosphate (CP) and β- and γ- of ADP and ATP were compared in the normal and denervated gastrocnemius muscle. The values found in the denervated muscles were expressed in % of the specific activities of same compound in normal muscle. In the denervated muscle the specific activities of phosphorus were increased by about 50%, and this may be due to increased blood flow as can be deducted from the experiments of Grigoryeva (1954, 1958) and Bass (1958). But the specific activity of β-phosphate of ADP and ATP rises to a higher level in the denervated muscle — to about twice the value in the normal muscle (Fig. 97L). The turnover of β-phosphate in denervated muscle is thus considerably higher. The increased turnover is most probably the result of increased velocity of the myokinase reaction, since the content of ADP and ATP was not changed. No changes in myokinase activity, however, are known 3 days after denervation. It is thus probable that more substrate (ADP) of the myokinase reaction (2 ADP = AMP + ATP) is available. The results imply also an immediate resynthesis of ADP and ATP from AMP by both the intensification of the myokinase reaction and higher consumption of oxygen.The greater availability of ADP for the myokinase reaction would suggest a disturbance (uncoupling?) in oxidative phosphorylation. It thus seems that such a partial disturbance of oxidative phosphorylation represents one of the possible causes of metabolic disturbances after denervation.

Experiments with isotopes are furthermore complicated by the fact that in muscle, as in other organs and cells, various reactions take place simultaneously in different parts of the cell or subcellular structures. Even if this makes the evaluation of the results still more complicated, the isotope method may be able to elucidate some important points in changes of metabolic rates in denervated muscle, as compared with normal muscle.

g) Special methods for investigating energy metabolism in denervated muscle

Some of these methods cannot be classified into any of the categories mentioned above [a to f]. Such methods usually aim at solving a detailed question and it does not seem necessary to classify each into a special group.

One of the methods used in our laboratory for investigation of muscle energy metabolism after denervation was the method of structural analysis of glycogen (Bass and Vodička 1955). An attempt was made to evaluate changes in glycogen synthesis and breakdown by analysing its structural changes following denervation. Estimations were made of enzyme functions (not of their activity!) associated with the breakdown and synthesis of the alpha-1,4-glycosidic bond and of enzymes attacking the alpha-1,6-bonds. Changes were found in the relation between enzymes synthetizing and breaking down these two bonds in the periphery of the glycogen molecule, which is formed by glycosidic chains linked by alpha1,4-bonds and enzymes attacking the alpha-1,6-bond in the central part of the molecule. The net amount of glucose remnants in the periphery of the glycogen molecule is increased, so that there is a predominancy of alpha-1,4-bonds (Fig. 98). This work is in fact an extension of the investigation of Cl. Bernard (1877), who isolated glycogen exhibiting a blue reaction with iodine after denervation, and of Koshtoyants and Janson (1950) who demonstrated that glycogen from denervated muscle is more homogenous and gives a predominantly blue reaction with iodine. Both these results suggest longer glucosidic chains in the periphery of the glycogen molecule, which is in agreement with our results.

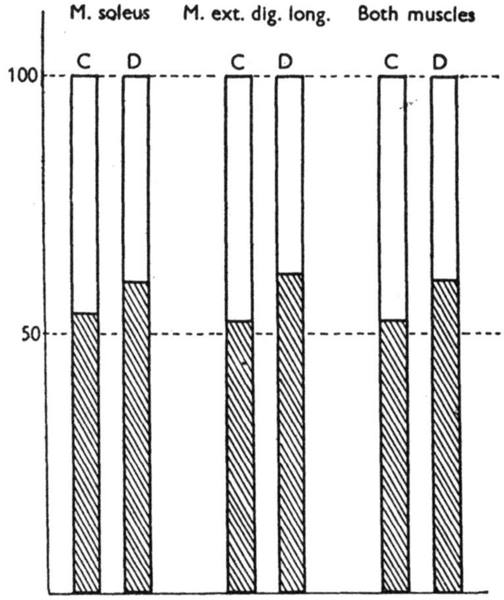

Fig. 98. Changes in the periphery: nucleus ratio in glycogen molecules in rabbit muscles after denervation (mm. soleus and extensor digitorum longus). Hatched field: periphery. (split by beta-amylase). White field: "nucleus" (not split by beta-amylase). 50% and 100% denoted by interrupted lines.

Besides chemical and physicochemical methods, muscle energy metabolism has also been studied with physical methods, namely thermometric methods (Kemp et al. 1947). If the consumption of energy substrates is increased after denervation (as has been found to be the case for oxygen and glucose), and their energy is not utilized for any other purpose, the surplus chemical energy mustbe transformed into thermal energy. Such a surplus could manifest itself by increased intramuscular temperature. However, in the experiments of Kemp et al. the temperature in denervated muscle was found to be lower than in control, contralateral muscle. Although this finding does not corroborate the possibility that heat formation is increased in denervated muscle, it does not definitely exclude this. Possible sources of error cannot be eliminated by this method. For example, introduction of a thermocouple into the normal control muscle may increase circulation by reflex mechanisms and consequently its temperature, while in the denervated muscle deprived of afferent signalization no increase of temperature would occur in response to nociceptive stimulation. This would mean that by this method comparison might be made between the temperature of the denervated muscle and the thermal reaction of a normal muscle to a nociceptive stimulus, and not its resting temperature. It is true, on the other hand, that the consumption of oxygen in some muscles after denervation is decreased (Sheves and Ryumina 1956), so that higher temperature could not even be expected.

h) Conclusions of the survey

As can be concluded from the above data, research on changes in carbohydrate, lipid and energy metabolism can be divided into several stages in which biochemical, physiological and physical methods are closely intermingled. The first stage, concerned with gathering information about the content of various metabolites associated with energy metabolism, at first at random, later in the time course after denervation. Gradually with deeper knowledge of the dynamics of metabolism and of the role of enzymes, the main interest of research was directed towards investigation of changes in substrate consumption and the removal of metabolic products, and in a more detailed inspection of enzymatic activity following denervation.

However, when intracellular autoregulation of metabolism was studied, it was shown that activity of enzymes, i. e. their maximal capacity to catalyze a reaction, is usually neither the only nor the main factor in determining the rate of individual reactions *in vivo*. This indicates that investigations of enzymatic activities in denervated muscle is a less suitable approach to discovering metabolic changes; it is important in those cases only where activity of an enzyme is the limiting factor in metabolism, i. e. where its activity in vivo is more or less fully exploited, as for example, during work (or muscle stimulation) and similar metabolic loads.

Reports dealing with metabolic loads and studying the input and output of energy substrates and products of their metabolism have shown that:

1. The role of carbohydrates as the substrate of energy oxidative metabolism increases after denervation.

2. The rate of glycogen synthesis is retarded as the reduction in its breakdown during starvation.

3. Glycogen is being broken down during stimulation in the same way as in normal muscle.

4. Glycogen is resynthetized to a similar extent during the first phase after stimulation.

5. It is not until the next phase of resynthesis that a difference between normal and denervated muscle becomes evident. This difference can be ascribed to altered reactibility of the denervated muscle (see Chapter XI). The metabolic basis of this altered reactibility is, as yet, unknown.

Further methods are being used for more detailed investigations such as methods, using radioisotopes, fractionation of cellular structures, histochemical methods, etc. These methods are aimed at elucidating changes in the localization of metabolic processes on intracellular structure or in individual metabolic reactions. However, they are not satisfactory as yet and reports in which changes in denervated muscle were investigated are not yet conclusive.

The main problem as to which link in energy metabolism — or metabolism as a whole — is primarily affected by denervation, has not so far been elucidated. Still less clear is the question of whether this primary disturbance is the cause of further disturbances leading to atrophy. For this reason it is necessary to summarize recent knowledge of this question in order to be able to formulate further problems.

2. Character of changes of energy metabolism of muscle after denervation

The main problem of the denervated muscle is to find the primary metabolic link affected by section of the nerve. This primary link must, of course, precede other disturbances if they are to be considered to be their cause (Dreyfus and Schapira 1950). This primary link is not necessarily located in energy metabolism. First changes were discovered elsewhere by Dreyfus et al. (1953), namely in the physical structure of myosin. These physical changes can, of course, be associated with metabolic changes. However, the first chemical changes after denervation were found in the content of glycogen, in decreased rate of its synthesis and breakdown.

Which metabolic changes after denervation can be considered as primary and what is their mechanism of action?

a) Disturbances in substrate supply

The decreased rate of synthetic processes could be caused by insufficient supply of substrate for energy metabolism from the blood. It is known, however. that such is not the case in a number of muscles after denervation. On the contrary, blood flow through denervated muscle is often greater than that in the control and the consumption of substrates is not reduced but, on the contrary, often increased. In addition to this, the denervated muscle is at relative rest and has also at its disposal the energy released from protein degradation, so that it is possible rather to consider that there is surplus energy after denervation.

However, in spite of an abundant supply of substrate, denervated muscle undergoes atrophy. The disturbance therefore cannot be caused by reduced supply. But if the supply of the muscle with energy substrates is sufficient, or even increased, then the disturbance has to be caused by their utilization.

b) Disturbances of substrate utilization

Utilization of energy substrates can be disturbed principally in two ways: either utilization of energy already acquired in utilizable form, or a disturbance in production of such energy. In the first case energy accumulated could be utilized for fibrillations, for example, or for surplus energy storage. In the second case other substrates than the normal ones could be utilized, usual substrates could be used via metabolic pathways less economic for the muscle, or a disturbance in the energy transfer of oxidative processes into a form easily utilized by the organ (ATP, creatinephosphate etc.) could develop. Most of these possibilities are necessarily associated with increased heat formation. It is only, perhaps, when energy is used for synthetic processes that this increase is not markedly conspicuous. However, the experiments of Kemp et al. (1947) in which the temperature in denervated and normal muscle was measured, indicate rather that heat formation is reduced. As has already been pointed out, these experiments do not seem conclusive enough, and do not exclude the possibility that heat formation is increased.

aa) *Fibrillation activity*

Fibrillation activity is discussed in Chapter V, so that it will be mentioned here only in connection with energy metabolism of denervated muscle. Fibrillations appear simultaneously with the degeneration of neuromuscular connections. According to some authors, fibrillations are the actual cause of denervation muscle atrophy (Langley 1916, 1917, Tower 1939), and this atrophy is due to exhaustion of the muscle. This conclusion does not appear probable, chiefly because electrotherapy, i. e. a certain type of "training" of denervated muscle, retards atrophy, but is without any effect on muscle fibrillations. If denervation atrophy was caused by excessive activity, repeated stimulation of the muscle should lead to increased atrophy. In addition, energy stores are not depleted by fibrillations, since at a time when fibrillation activity is already fully developed, the content of some of these acturally increases (e. g. glycogen, lipids, ATP and creatinephosphate etc.). The relation between fibrillation activity and atrophy is not yet clear. It could, perhaps, enhance denervation atrophy. There are other kinds of atrophy, however, without fibrillations, which progress at a faster rate even than denervation atrophy. An example of this is reflex atrophy following nociceptive stimulation, which attains the same degree of atrophy after 4 hrs., as in denervation after 3 days, i. e. approximately 20 hrs. after onset of fibrillation activity.

Considering these facts it does not seem probable that fibrillation activity is the cause of atrophy after denervation. It is not clear, at present, what amount of energy is consumed by fibrillations. It has not yet been measured since present methods are not adequate. For this reason only very approximate estimations can be made. However, since fibrillations are not manifested by increased tone or measurable twitches, the work done and energy consumed cannot be expected to be very great, even though increased oxygen consumption could, perhaps, accompany this activity. On the other hand, even a normal muscle continuously receives impulses from nerve centres which do not always lead to contractions but may increase catabolic and consequently anabolic processes. It is possible that normal muscle does not consume less energy for this resting activity than denervated muscle for fibrillations. It is thus an open problem whether fibrillation activity at all increases energy consumption in comparison with the normal muscle at rest.

It is also possible that surplus energy can also be utilized in the denervated muscle for the formation of energy stores which would be accumulated, but which are not, or cannot be, utilized.

bb) Formation of surplus energy stores

The glycogen content is increased almost immediately after neurotomy, the rate of its synthesis from glucose being decreased and its breakdown during inanition slowed down. In some animal species the content of glycogen increases and remains higher for some time, even when the rate of synthesis and breakdown is reduced. The uptake of glucose and oxygen, however, is not decreased but rather increased and similarly there is an increased output of carbon dioxide It seems as if the role of glycogen was diminished in resting metabolism and as if the muscle utilized substances circulating in the blood as substrate for energy metabolism, these substances containing progressively more oxygen than usually (change to carbohydrates as substrate from lipids). If the normal muscle utilizes according to recent findings, mainly lipids for obtain energy at rest, the denervated muscle gradually utilizes other sources of energy. This is also borne out by the fact that RQ gradually rises and is often higher than 1·0, as well as by the findings that oxidizable glucose consumption is increased in denervated muscle at rest and during direct stimulation of muscle. Normal muscles can also make use of various energy sources; lipids, proteins and carbohydrates (Meyerhof 1930, Werthermeir and Ben Tor 1952), but *in vivo* at rest it utilizes chiefly lipids. It is not until lack of lipids, or when they are not present in suitable form (in the presence of proteins), as in most *in vitro* experiments, that carbohydrates are used. For this reason the preference of denervated muscle for carbohydrates and others was difficult to detect *in vitro*. However, *in vivo* results concerning the gradual increase in RQ and the consumption of glucose following denervation indicate that the muscle transfers its preference from lipids as the main source of energy to carbohydrates, and possibly the utilization of carbohydrates as plastic material for lipid synthesis. Part of the energy is thus stored in denervated muscle in the form of lipids. So far, it is not yet possible to state with certainty in which way the consumption of lipids is altered. The increasing RQ speaks against the possibility that lipids are being oxidized. If they were stored to a greater extent, the mucle would be transformed rapidly into adipose tissue by formation of lipids from glucose and possibly also by their uptake from the blood. Increased content of lipids appears just at the time when RQ markedly rises. Proteins can also serve as source of the lipid carbon, but in denervated muscle only to a limited extent, if at all.

However, lipids can also be released from muscle. Unesterified fatty acids are released from the gastrocnemius of the dog approximately to the same extent in normal muscle and muscle 3—4 weeks after denervation. According to Mansfeld and Müller (1913), Audova (1923) and Wertheimer (1927—VI), lipids in denervated muscle are not utilized during starvation for covering energy requirements of the organism, or only to a small extent. It thus seems that the breakdown of lipids is reduced or even arrested completely. Glucose becomes the main substrate of energy metabolism in the muscle and may be partially used for the formation of lipids. The formation of this surplus energy store could at least partly explain the increased consumption of oxygen and glucose in the denervated muscle. It is not clear, however, whether this process suffices to use up all the "surplus" energy in muscles with increased oxygen and glucose consumption.

The degradation of proteins, a catabolic process, is a further factor which could, perhaps, consume energy.

cc) Degradation of proteins (see also Chapter IX)

In some tissues, e. g. the liver, degradation of proteins is a process requiring energy. Žák and Drahota (1960) have demonstrated, however, that this process in the muscle is purely exergonic, i. e. it does not require energy from oxidative processes, but itself releases energy. Degradation of proteins occurs in the muscle during atrophy, in fact it is identical with atrophy, and cannot thus be considered as its cause.

It is not known to what extent denervated muscle is capable of utilizing energy liberated from protein degradation, or whether it is released in the form of heat. Moreover it does not explain increased energy requirements of the denervated muscle, but rather indicates that this increase must be even higher than corresponds to the increase of oxygen consumption.

dd) Possible alternate metabolic pathways

The utilization of energy from substrates after section of the somatic nerve can be altered in still another way. The character of metabolism could, for example, change in the sense of Orbeli's hypothesis that a denervated organ becomes similar to an organ phylogenetically younger. It is possible that, perhaps, some less effective mechanisms corresponding to phylogenetically older tissues participate in metabolism. For example, the possible utilization of lactic acid for synthesis of glycogen corroborates Orbeli's assumption, as has been shown by Hudlická and Gutmann (1958). This is also borne out by the fact that denervated muscle consumes more lactic acid, administered intravenously, than normal muscle (Sato and Kasugai 1935). The possibility cannot be excluded then that the denervated muscle in mammals utilizes lactic for glycogen synthesis, like normal frog muscle. This is practically impossible in normal mammalian muscle, since lactic acid is either completely oxidized, or enters the blood stream and is not utilized for glycogen synthesis and for energy formation needed for this synthesis until it reaches the liver (Cori's cycle). In the frog this cycle is not the only possible mechanism by which carbohydrates are resynthetized from lactic acid. Muscle glycogen in the frog is resynthetized from lactic acid. It is possible that the mammalian muscle regains this ability after denervation.

It is evident that there may be many other alternate metabolic pathways in the denervated muscle, which do not take place in innervated muscle, or only to a limited extent. It is not clear whether such processes actually do exist and what is their possible significance.

The increased oxidation of glucose instead of lipids and the production of lipids from glucose cannot be considered as such an alternate metabolic pathway but rather as a quantitative change in the relation between processes occurring normally.

It is, of course, possible that the alternate metabolic pathways would be of the "shunt" type. However, changes in the various processes and their respective roles is given by the acceleration of one or the slowing down of an other metabolic process — possibly of a single reaction. Such changes, as long as they are not energetically favourable, mean that heat production is increased and that the indicator of utilization of energy substrate — oxidative phosphorylation — is reduced. An important example is the uncoupling of these coupled reactions which represent an universal mechanism for the utilization of energy in oxidative processes.

ee) Increased heat production and disturbances of oxidative phosphorylations

The energy formed by the oxidation of glucose may be utilized by denervated muscle for fibrillation activity, and for the synthesis of lipids (during which carbohydrates probably supply carbon atoms, as well as energy), and also perhaps for the formation of glycogen from lactic acid. Fibrillation activity was found to appear sooner than atrophy or increased lipid synthesis, but both of these develop with progressing atrophy. It is therefore not possible to state unequivocally that the denervated muscle has "surplus energy", with the exception of the first stage after denervation, before fibrillations set in. It is rather possible to say that the muscle does not utilize energy for trophic processes in the usual way, but for the formation of surplus stores. On the other hand, these changes must lead to increased heat production. In order to confirm this, determinations of muscle temperature are not sufficient since they can give misleading information.

At the present time, undisputable results cannot be expected from studies of energy exchange for methodical reasons, since it is not possible, as yet, to determine the energy needed for fibrillation activity and a number of other factors, e. g. changes in the permeability of cell membranes and the energy needed for active transport of ions under altered conditions.

If denervated muscle liberates surplus energy in the form of heat, energy can be lost in two phases of energy metabolism. Firstly when energy-rich bonds are not formed the mechanism of their synthesis being ineffective, oxidation of substrates occurs without formation of energy-rich phosphate bonds during the transfer of oxygen, as is the case under normal conditions. This could be realized, for example, by regulation with DPN, TPN and pyridine nucleotide dehydrogenase (PNDH), as was postulated by Kaplan et al. (1956). The transfer of electrons by the TPN — TPNH system is not accompanied by synthesis of high energy phosphates, as long as hydrogen is not transferred onto DPN with the aid of PNDH enzymes. If this transfer is impaired, high energy phosphates are not formed and hydrogen serves to a greater extent for the synthesis of fatty acids. Secondly, it is possible that these high energy phosphates are being formed but their splitting is "ineffective", i. e. by simple hydrolysis. These mechanisms can be projected into ATP synthesis during oxidative phosphorylation. This means that, in the first case, ATP is not being formed, or only to a lesser extent, in the second case the store of chemical energy in ATP is being broken down ineffectively by hydrolysis, so that the store of chemical energy formed is again transformed into heat energy. This latter possibility would presume an increased turnover of ATP, especially of its terminal (γ) phosphate, but also of its other (β) labile phosphate. Both these phosphates attain equilibrium with the aid of myokinase, which catalyses the dismutation reaction during which one molecule of ATP and AMP are formed from two molecules of ADP. Thus the beta-phosphate of one molecule of ADP becomes a gamma-phosphate of ATP, which has a greater turnover than the beta-phosphate (Fleckenstein et al. 1956, Janke et al. 1956). AMP is either deaminated and leaves the muscle, or is complemented from ADP and ATP. For this reason a more detailed analysis by means of specific activities of all three phosphates of ATP is necessary. Such a differentiation was not made by Grigorieva (1958). According to this author, however, the turnover of ATP remains unchanged after denervation. As has been shown (Bass, Vítek, unpublished results), the turnover of β-phosphate in ADP and ATP increases after denervation, indicating a disturbance (probably uncoupling) of oxidative phosphorylation mechanisms. If we accept the conclusion as correct, is it true that oxidative phosphorylation are unchanged? The so-called P : O-quotient represents a measure of oxidative phosphorylations and thus of the effectiveness of oxidative processes, and expresses the number of molecules of high energy phosphate of ATP, which is produced on consumption of one atom of oxygen. This P : O quotient equals approximately three in the normal muscle.

If the rate of ATP formation after denervation remains unaltered and if we at the same time, accept, the conclusions of Sheves and Ryumina (1954) and Bass and Hudlická (1960) that the consumption of oxygen is increased in denervated gastrocnemius, then this means that the P : O quotient here is reduced. If the P : O quotient of normal muscle is equal to three and the consumption of oxygen in our experiments was increased by 60%, the P : O quotient in denervated muscle would be decreased approximately to two, i. e. the effectiveness of oxidative metabolism would be lowered by about a third.

Data concerning changes of the P : O quotient after denervation consist of contradictory reports. Huf and Fischer (1949) have pointed out the conspicuous similarity between muscle metabolism after denervation and hyperthyroidism. The well-known uncoupling effect of thyroxin led to the experimental investigation of changes in P : O quotient in the denervated muscle. According to a number of investigators (Ferdman 1953, 1955,, Slozhenikina 1957, Telepneva 1957) the P : O quotient is lowered after denervation, while Severin (1957) did not find this effect. All the above results were obtained in muscle homogenates Experiments with muscle homogenates are not adequate for studying the P : O quotient since besides oxidative phosphorylation, anaerobic phosphorylation, glycolysis and glycogenolysis are also registered together with their changes. These processes may sum with oxidative phosphorylation, so that P : O values obtained on this material do not correspond to actual oxidative phosphorylation, but only to the ratio between total phosphorylating capacity and oxygen consumption This value is more complex and its biological significance is considerably smaller than that of oxidative phosphorylation, as can be determined *in vitro* in mitochondria or *in vivo* with the aid of oxygen consumption and turnover of phosphorylated high energy compounds. Although experiments *in vitro* are technically less complicated and exclude interfering influences from other structures, it is possible that changes in the P : O quotient are manifested mainly *in vivo*. This could, for example, be an analogy of reversible uncoupled oxidative phosphorylation as has been shown *in vivo* by Ernster (1955a, 1955b, 1956). The regulation of the P : O quotient could, perhaps, serve as the factor regulating metabolism even in normal muscle, since the P : O quotient in muscle is different at rest and during functional activity, as was discussed by Kubišta (1959).

It is also possible that the utilization of energy during normal oxidative phosphorylation is reduced by increased hydrolytic splitting of ATP, but this is not probable judging from experiments *in vitro*. According to Michelazzi et al. (1957) the activity of ATPase, an enzyme hydrolysing ATP, is already reduced 12—16 hours after denervation. The reduction of this activity, however could be the possible reason for the decrease of P : O quotient by insufficient amounts of free ADP which could accept high energy phosphates during oxidation. However, since lack of ADP also slows down glycolysis, and this has been found to occur at a faster rate in the denervated muscle, this possibility is also not probable.

It is thus quite probable that the P : O quotient is reduced after denervation i. e. that energy metabolism becomes less effective.

ff) Conclusions concerning the utilization of substrates

The question of whether denervated muscle has surplus energy and whether this surplus energy and adaptation to it causes atrophy, remains open.

However, whatever the answer, the consumption of energy in denervated muscle is increased for fibrillation activity and synthesis of lipids, the utilization of substrates with a lower energy content (carbohydrates instead of lipids) and probably also reduced utilization of energy from oxidative processes, i. e. to a decreased P : O quotient, would have to become evident by increased heat formation. Previous findings do not corroborate this, but not being quite conclusive, they do not exclude the possibility of increased formation of heat.

However, most experiments, on the basis of which these conclusions are being made, were performed at a period after denervation when atrophy was already advanced, so that changes found might be the result rather than the cause of metabolic changes after denervation. In consequence thereof it is necessary to seek the primary link or links which lead to denervation atrophy.

c) Questions of the primary link

As has been shown above, the utilization of energy substrates is impaired in denervated muscle. This disturbance should have a primary link or links which cause other changes. The primary link in the metabolic disturbance must therefore also be the first to appear in the sequence of events and secondary events cannot become manifested until a certain time after the primary events.

Following section of the nerve, the metabolic disturbance does not become apparent immediately to the full extent, but only after a certain time, which roughly corresponds to the time when indirect muscle excitability is lost and first degenerative changes in the terminal fibres of the neuromuscular junction appear.

However, the content of glycogen and other metabolites already changes earlier. It seems that this change could often rather be ascribed to immobilization than to disturbances of the trophic influence of the somatic nerve. Control non-denervated muscles — and normal muscles in general — receive impulses from the central nervous system, which lead to an accentuation of catabolic reactions. As has been demonstrated by Hoffmann and Wertheimer (1927-VII), glycogen is not synthetized without being broken down at the same time. The balance between its synthesis and breakdown is continuously being disturbed by impulses from the central nervous system and renewed. Decreased breakdown results in reduced synthesis.

This relationship, of course, expresses the actual situation very crudely. Later investigations concerning the over-shoot reaction following stimulation and physiological functional activity showed that it could be a mechanism which would stand in close relation to breakdown and synthesis (Folbort 1941, Yakovlev 1956). Over-shoot reactions might represent the element of muscle adaptation to work and perhaps also the element of other adaptation processes. However, this mechanism is quickly affected after denervation. Six hours after denervation in the rat the over-shoot reaction is still present, but by 24 hours after denervation — about the time when degeneration of neuromuscular endings occurs — it is lost. In 1939 Tower postulated that proximo-distal transfer of substances in the nerve could be the basis for the trophic function of peripheral nerves. This hypothesis has been corroborated by a number of findings, such as, for example, the persistence of the trophic influence some time after section of the nerve, the influence of the peripheral stump after nerve section on the onset of fibrillations (Luco and Eyzaguirre 1955) and on the time when the muscle loses its indirect excitability, when degeneration of motor endplates occurs and glycogen synthesis from glucose is decreased (Gutmann et al. 1955). It seems that during functional activity or indirect stimulation of muscle this transport is accelerated (Vodička 1956a, 1956b, Vodička 1957). This transport would be dependent, after section of the nerve, on the amount of substances contained in the peripheral nerve stump and which maintain normal trophic processes in the muscle. It would also be possible to explain in this way the fact that a short time after nerve section the breakdown of glycogen is slowed down more than its synthesis (since impulses from nerve centres do not arrive in the muscle). Glycogen synthesis is not slowed down to any great extent until later (see p. 234). This seems to be the reason why the content of glycogen (4—24 hours after neurotomy in the rat) and also of proteins (see Chapter IX) and potassium (see Chapter VI) is higher.

From this and experiments dealing with over-shoot reactions it can be concluded that the somatic nerve, which regulates catabolic processes by way of its impulse activity, may regulate synthetic, anabolic processes and consequently also processes in the muscle with the aid of a transport of substances along the nerve into the muscle. This transport of substance following section of the nerve probably lasts for some time in the peripheral stump until degeneration of terminal nerve fibres occurs at the motor nerve endings.

In a number of muscle atrophies, however, the neuromuscular junction is preserved. One of these is the reflex atrophy following nociceptive stimulation in which the neuromuscular transmission of nervous impulses is not inhibited, but trophic functions are blocked immediately. It was shown by Vodička (1957) that this block is located in the periphery, so that section of the nerve immediately after the application of nociceptive stimulation is followed in four hours by atrophy comparable with that observed three days after denervation. If the nerve is sectioned immediately before nociceptive stimulation, atrophy is similar to denervation atrophy, i. e. it is slower. It can be concluded from the results of Vodička that the inhibitory process is mediated by the central nervous system and that the inhibitory block, once evoked, becomes "fixed" in some way in the peripheral nerve endings in the muscle. The "trophic transport of substances" into the muscle would then appear to be blocked in some way, although action potentials are transmitted normally and the muscle can be stimulated indirectly.

However, since it is possible to evoke reflex atrophy of nervous origin in such a short time, and denervation atrophy, on the other hand, appears later in relation to the length of the peripheral stump, it is probable that the primary link in the initiation of the disturbance after denervation is to be found at a time when the transport of substances from the nerve into the muscle is arrested, i. e. at a time when axons are emptied or perhaps when indirect muscle excitability is lost.

No changes in the activity of muscle enzymes have been found at this period, with the exception of ATPase activity in sarcosomata and myofibrils, which was found to be decreased (Michelazzi et al. 1957). But enzyme activity may become the limiting factor in the muscle only under conditions of extreme metabolic loads. The relationship between enzyme and substrate could be a more important factor. This could play a more important role in individual reactions even under resting conditions, as in the reaction of oxalacetic acid with acetyl-coenzyme A is the relation of oxalacetic acid, acetyl-coenzyme A and the condensating enzyme. This relation can also play a role in other "pacemakers".

The inflow of metabolites into denervated muscle is not impaired, and so often even increased. It does not seem to be the fundamental mechanism causing the metabolic disturbance after denervation, or by the means of which the intact nervous system increases metabolism of normal muscle at rest. This has been demonstrated in balance studies performed by the method of comparing the intake and output of metabolites using arteriovenous differences multiplied by the blood flow through the muscle. By this method it was shown that the main role in energy metabolism is transferred from lipids to carbohydrates. However, this process does not take place to any measurable extent until atrophy occurs.

In order to get nearer to the cause of disturbances of muscle metabolism following denervation, more research will have to be done concerning changes in intermediary metabolism. The nearest change so far investigated is the reduction in synthesis of glycogen and loss of the ability to react to functional activity by the over-shoot reaction of glycogen proteins, etc. above initial levels.

C. Main conclusions, problems and aims of research on energy metabolism in the nervous regulation of muscle metabolism

Energy metabolism is undoubtably of paramount importance for muscle and is associated, as has already been demonstrated, with its biological role of transforming chemical energy into mechanical. It has been found possible in some tissues to differentiate, at least roughly, between energy necessary for specific function and that for trophic processes. In the muscle, however, the situation is more complicated since there are widely differing functional types of muscle. An extreme case of a muscle which works incessantly in a certain, fairly constant rhythm is the heart muscle, but our investigations do not deal with this muscle. The diaphragm is similar as far as uninterrupted functional activity is concerned. In contrast to these muscles, other skeletal muscles are mobilized into function according to need, often suddenly, while at other times there is relative rest. Even the types of function are different in different muscles and according to these are divided into static and dynamic, tonic and phasic, slow and fast, etc. However, there is a gradual transition between these two extreme types, so that it is not possible to study only one or the other, but only muscles with either a preponderance of one or the other type. From this a further difficulty arises in expressing quantitatively the energy consumption for trophic functions. Different types of muscle also synthetize energy stores for future activity to a different extent, so that even the consumption of oxygen and other metabolites at rest is not usually constant, even in the same muscle.

Energy stores are not formed, of course, only at rest, but also during functional activity, especially during contractions performed at slow frequencies. It is glycogen, in the first place, which can be utilized by muscle during activity under conditions of insufficient inflow of substances and oxygen from the blood, and also, perhaps, lipids which are utilized during work performed at a slow frequency or during prolonged activity, and finally proteins, which the muscle evidently utilizes during activity, mainly under extreme conditions. The muscle is equipped with an enzymatic system which is able to oxidize any one of these constitutents.

To what extent muscle is able to transform one these constituents into another remains an open question. It is probable that carbohydrates can be transformed into lipids, as is corroborated by a high RQ following work (Bass and Hudlická 1960) and experiments with the oxidation of C^{14} — fatty acids (Fritz et al. 1958). However, sometimes RQ drops below 0·7 at rest, this indicating that perhaps carbohydrates are being synthetized from lipids or proteins.

In muscle at rest after performing work synthesis of energy stores for subsequent functional activity takes place. Glycogen and evidently lipid synthesis is accelerated (as can be judged from RQ following work). Synthetic processes are intensified during training, i. e. during repeated functional activity, the muscle becoming adapted to work and namely to the type of activity to which it is trained — to short bursts of activity performed at high speed, or prolonged activity at slow speed, and various combinations of these. In the course of training there is not only an increase in the net content of contractile muscle proteins, but also in the percentage content of energy stores, for example, of glycogen. Less glycogen is then broken down during activity and the overshoot reaction is smaller in extent (Gutmann et al. 1954). This need not necessarily be the result of metabolic adaptation, but can be due to improved economy of movement.

As has been demonstrated in Chapter I, the trophic influence of the peripheral nerve affects the regulation of metabolism which ensures morphological and chemical composition of the muscle and its capacity of adaptation to external stimuli. This means that it affects the regulation of energy metabolism almost in its full range, i. e. with the exception of metabolic processes associated with specific muscle function, muscle contraction.

The trophic function of the nerve is not lost after its section. At first, immobilization evidently comes into play, this being the cause of an increased content of glycogen and further constituents of energy metabolism, including potassium (this being probably due to sudden exclusion of nerve impulses arriving from nerve centres). After this the rate of glycogen synthesis and breakdown is decreased, this being a further possible after-effect of a complete arrest of inflow of nerve impulses, as well as by the gradual loss of the trophic influence of the peripheral nerve stump. This effect is lost only after the degeneration of terminal nerve endings. The over-shoot reaction after muscle stimulation is then lost as well and the utilization of metabolic substrates at rest is probably also altered. The over-shoot reaction can then be evoked by repeated stimulation only and is shorter in duration.

Muscle atrophies of other origin are similar in character, the rate of muscle weight loss, however, is usually slower and atrophy is smaller in extent (e. g. tenotomy, immobilization). Muscle atrophy following nociceptive stimulation, which is reflex on origin (Vodička 1956, 1957), progresses rapidly at first, so that after four hours muscle weight loss corresponds to the atrophy three days after section of the nerve. Muscle motor function, however, is not affected. The trophic function can thus be impaired by a reflex mechanism and independently of motor function. It is thus possible that other muscle atrophies of neurogenic origin in which the motor nerve and its terminal endings are not anatomically affected, are also governed by a reflex component.

As can be concluded from Chapter I., the motor neurons play a decisive role in the regulation of trophic processes in muscle. The question of the mechanism by which the motor neuron affects muscle metabolism and the trophic functions are of paramount importance. A likely hypothesis, as yet not sufficiently founded, presumes that substances are transported along the nerve into the muscle, these substances in some way regulating and activating muscle metabolism. This hypothesis was tentatively postulated by Tower (1939) and was later further investigated by Vodička (1956).

In order to be able to answer which reaction or metabolic process is primarily affected after section of the nerve, it will be necessary to investigate on more detail metabolic changes after denervation. *In vitro* only relatively late denervation changes can be determined. It was shown that in denervation atrophy changes occur in muscle enzyme systems and that after a temporary increase, the activity of glycolytic enzymes and later of oxidative metabolism is decreased. *In vivo* it was demonstrated that muscle utilizes in the first place blood glucose as the energy substrate this being probably utilized for the production of lipids and their accumulation in the muscle, and finally that the P : O quotient is reduced, i. e. that the energy effectivenes of oxidative metabolism is decreased.

However, this has not been found to be the case during the first stage after denervation. In vitro experiments performed at this time indicate rather that energy metabolism is improved with a markedly increased capacity of glycolysis.

In order to investigate the individual reactions of energy metabolism *in vivo*, new techniques will have to be elaborated, probably with the aid of radioisotopes, which will enable the rate of formation of individual compounds from their precursors to be studied.

There is considerable controversy among physiologists whether it is correct to study changes in metabolism after denervation, since denervation atrophy does not differ from the metabolic aspect from other muscle atrophies of nervous origin and differs only from the electrophysiological (fibrillations) and physiological (increased sensitivity to acetylcholine) aspect. Fibrillations do not appear to be the only specific manifestation of denervation atrophy, fibrillation activity and postdenervation atrophy having probably a common genesis. While electrophysiological and physiological reports can indicate the direction of future metabolic studies, investigation of metabolic changes after denervation should explain the metabolic basis and consequences of these processes. For this reason both types of studies seem to be equally useful in research on denervated muscles. If various muscle atrophies of nervous origin are otherwise similar, it is possible that they have a common etiology, i. e. a disturbance of nervous trophic influence on the muscle, either complete loss of the trophic influence of neurons as after neurotomy, or perhaps a reflex depression of this influence in the case of other atrophies. Research on primary links in the metabolic disturbances in these atrophies is thus of paramount importance in studying the trophic influence of the nervous system.

Concomittantly with denervation studies, it is also necessary to investigate muscle hypertrophy, such as work hypertrophy, to study metabolism during functional activity, adaptation to work and the influence of the somatic nerve on this adaptation. This means that investigations of the trophic influence of the nervous system should not be limited to denervation studies, but should also be directed to questions of muscle hypertrophy.

References

Andres R., Cader G., Zierler K. L.: Amer. J. Phys. Med. 34 : 286, 1955a.
— — — J. Clin. Invest. 34 : 920, 1955b.
Anrep G. V., Blalock A., Saaman A.: Proc. Roy Soc. B 114 : 223, 1934.
Artom C.: J. Biol. Chem. 139 : 953, 1941.
Audova A.: Skand. Arch. Physiol. 44 : 1, 1923.
Avelonet L., Di Macco G.: Ann. Clin. Med. Sperim. 15 : 39, 1925.
Bach L. M. N.: Proc. Soc. Exptl. Biol. Med. 67 : 268, 1948.
Baker N., Shreeve W. W., Shipley R. A., Incefy G. E., Miller M.: J. Biol. Chem. 211 : 575, 1954.
Bass A.: Physiol. Bohemoslov. 7 : 220, 1958.
Bass A., Gutmann E., Hudlická O., Vrbová G.: Physiol. Bohemoslov. 8 : 93, 1959.
Bass A., Gutmann E., Vodička Z.: III. Czechosl. Congres of Physiologists, Pharmacologists and Biochemists, Brno 1954.
— — — Physiol. Bohemoslov. 4 : 267, 1955.
Bass A., Hudlická O.: Physiol. bohemoslov 9 : 408, 1960.
Bass A., Vítek V.: Unpublished results, 1961.
Bass A., Vodička Z.: Physiol. Bohemoslov. 4 : 409, 1955.
— — Physiol. Bohemoslov. 5 : 58, 1956.
Baum J.: Klin. Wochenschr. 10 : 987, 1931.
Baum J., Pichler E.: Pflügers Arch. 233 : 35, 1934.
Bell D. M.: J. Lab. Clin. Med. 40 : 337, 1952.
Beránek R., Gutmann E., Vrbová G.: Physiol. Bohemoslov. 3 : 39, 1954.
Bernard Cl.: C. R. Soc. Biol. 48 : 673, 1859.
— "Leçons sur la diabète". Paris 1877.
Bieth R., Weill J. P., Mandel P.: J. de Physiol. 46 : 245, 1954.
Bloom B.: Proc. Soc. Exptl. Biol. Med. 88 : 317, 1955.
Bloom B., Eisenberg F. Jr., Stetten D. W. Jr.: J. Biol. Chem. 222 : 301, 1956.
Bloom B., Stetten D. W. Jr.: J. Biol. Chem. 212 : 555, 1955.
Bloom B., Stetten M. R., Stetten D. W. Jr.: J. Biol. Chem. 204 : 681, 1953.
Bogatzki M.: Ch. Z. 126 : 145, 1955.
Bogatzki M., Schmidt C. C., Fischer L.: Z. Ges. Exptl. Med. 119 : 457, 1952.
Bollmann J. L., Flock E. V.: J. Biol. Chem. 147 : 155, 1943.
Bourguignon: C. r. Soc. Biol. 146 : 1083, 1952.
Breuer H., Richter W.: Langenbecks Arch. u. Dtsch. Z. Chirurg. 288 : 78, 1958.
Britton S. W.: Amer. J. Physiol. 93 : 21, 1930.
Bruner H., Grosse-Brockoff F.: Pflügers Arch. 238 : 361, 1938.
Buchwald K. W., Cori C. F.: Proc. Soc. Exptl. Biol. N. Y. 28 : 737, 1931.
Bücherl E., Schwab M.: Pflügers Arch. 254 : 337, 1952.
Burton-Opitz R.: Amer. J. Physiol. 9 : 161, 1903.

Califano L. J.: Riv. di Pat. Sperim. 48 : 2, 1927.
Cannon W. B., Rosenblueth A.: The Supersensitivity of Denervated Structures. New York 1949.
Cendragolo F.: Arch. Sci. Biol. 23 : 377, 1937.
Chandelon Th.: Pflügers Arch. 13 : 676, 1876.
Cheosakul P., Blide L. W., Vanko M., Knudson A.: Science 123 : 375, 1956.
Chepinoga O. P.: Ukr. Biokhim. Zh. 10 : 823, 1937.
Chiquione: J. Histochem Cytochem. C : 471, 1956.
Chor H., Davenport H. A., Dolkart R. E.: Amer. J. Physiol. 118 : 589, 1937.
Christensen E. H., Hansen O.: Scand. Arch. Physiol. 81 : 152, 1939.
Collowick S. P., Kalckar H. M.: J. Biol. Chem. 148 : 117, 1943.
Curtice F. C., Douglas C. G.: Proc. Roy Soc. Lond. B 119 : 338, 1936.
Dreyfus J. C., Joly M., Schapira G., Raeber L. (1953): C. R. Acad. Sci. 236 : 2351, 1953.
Dreyfus J. C., Schapira G.: Sem. Hop. Paris 24 : 1999, 1948.
— — "Le Muscle", Colloque tenu à Royamount, France 1950.
Drury D. R., Wick A. N.: Amer. J. Physiol. 284 : 304, 1956.
Ennor A. H., Rosenberg H.: Biochem. J. 56 : 302, 1954.
Epelbaum S. E. Kantor L. F.: Biokhimiya 19 : 660, 1954.
Ernster L.: Exptl. Cell. Res. 10 : 704, 1957a.
— Exptl. Cell. Res. 10 : 721, 1957b.
Fábry P.: Physiol. Bohemoslov. 4 : 145, 1955.
Fainshmidt O.: Biokhimiya 4 : 411, 1939.
Ferdman D. L.: "Biokhimiya Zabolevaniya Mishts". Kiev. 1953.
— Vopr. Med. Khim. 3 : 351, 1957.
Ferdman D. L., Grigoryeva V. A.: Ukr. Biokh. Zh. 22 : 41, 1956.
Fetisova: Voprosi biokhimiyi Mishts. Kiev 1954.
Feuer G.: Acta Physiol. Acad. Sci. Hung. 11 : 1, 1957.
Feyer J. H., Moore N. S., Williams H. H., Young L. M.: J. Lab. Clin. Med. 45 : 684, 1955.
Fischer E.: Amer. J. Physiol. 127 : 605, 1939.
— Amer. J. Physiol. 145 : 571, 1946.
— Arch. Phys. Med. 29 : 291, 1948.
Fischer E., Chevalier B. A., Bratton M. J., White E. A.: Pflügers Arch. 255 : 1, 1952.
Fleckenstein A., Janke J., Davies R. E.: Arch. Exptl. Pathol. Pharmacol. 228 : 596. 1956.
Fletcher W. M., Hopkins F. G.: Proc. Roy. Soc. 89 B : 444, 1907.
Flock E. V., Bollman J. L.: J. Biol. Chem. 136 : 469, 1940.
Flock E. V., Ingle D. J., Bollman J. L.: J. Biol. Chem. 129 : 99, 1939.
Folbort G. V.: "Fiziologiya processov istoschcheniya i vosstanovleniya". Kharkow.
Forsgren E.: "Über die Rhytmik der Leberfunktion, des Stoffwechsels und des Schlafes" Stockholm. 1935.
Fritz I. B.: Endocrinology 58 : 493, 1956.
— Science 129 : 334, 1959.
Fritz I. B., Davis D. G., Holtrop R. H., Dundee H.: Amer. J. Physiol. 194 : 379, 1958.
Fuhrman F. A., Crismon J. M.: Amer. J. Physiol. 167 : 289, 1951.
Gasskell N. H.: J. Anat. 11 : 360, 1877.
Gemill C. L.: Physiol. Revs. 22 : 32, 1942.
George J. C., Scaria K. S.: J. Anim. Morph. Physiol. 3 : 91, 1956.
— — Nature 181 : 709, 782, 1958a,b.
Gerritsen T.: Biochim. et Biophys. Acta 8 : 466 : 1952.
Gey K.: Biochem. J. 69 : 145, 1956.
Geyer R. P., Mathews L. W., Stare F. J.: J. Biol. Chem. 180 : 1037, 1949.

Gibertini G., Malagati C., Marsciani R., Vecchiti R.: Riv. Pat. Clin. 10 : 581, 1955.
Goldmann D. S., Chaikoff I. L., Reinhardt W. O., Entemann C., Danken W. G.: J. Biol. Chem. 184 : 719, 1950.
Gollwitzer-Meyer K.: Pflügers Arch. 256 : 228, 1953.
Gordon R. S. Jr.: J. Clin. Invest. 36 : 810, 1957.
Grigoryeva V. A.: "Voprosi Biokhimiyi Mishts". Kiev. 1954.
— Ukr. Biokh. Zh. 30 : 356, 1958.
Grund G.: Arch. Exptl. Pathol. Pharmacol. 67 : 393, 1912.
Gruner Ch. M., Kwang S. L., Lashichenko Z.: Proc. Soc. Exptl. Biol. Med. 80 : 549, 1952.
Gutmann E.: Čas. lék. čes. 94 : 452, 1955.
Gutmann E., Bass A., Vodička Z., Vrbová G.: Physiol. bohemoslov. 5 : (suppl.) 14, 1956a.
— — — — J. de Physiol., 1956b.
Gutmann E., Guttmann L.: Lancet 1 : 169, 1942.
— — J. Neurol., Neurochir., Psych. 7 : 7, 1944.
Gutmann E., Holubář J.: Physiol. bohemoslov. 1 : 221, 1952.
Gutmann E., Sobolová V., Vrbová G.: Theor. a praxe těl. a vých. 1 : 213, 1953.
Gutmann E., Vodička Z., Vrbová G.: Physiol. bohemoslov. 3 : 182, 1954.
Gutmann E., Vodička Z., Zelená J.: Physiol. bohemoslov. 4 : 200, 1955.
Gutmann E., Vrbová G.: Physiol. bohemoslov. 1 : 38, 1952a.
— — Physiol. bohemoslov. 1 : 205, 1952b.
Hall J. C.: Science 813 : 3101, 1954.
Hansen R. G., Rutter W. J.: J. Biol. Chem. 195 : 121, 1952.
Harding A. E. B.: J. Pathol. Bacteriol. 29 : 189, 1926.
Harmann J. W., Gwinn R. P.: Amer. J. Pathol. 25 : 741, 1949.
Harrop G. A. Jr.: J. Exptl. Med. 30 : 241, 1919.
Hill A. V.: Proc. Roy. Soc. B 137, 40, 1950.
Himwich H. E., Castle W. B.: Amer. J. Physiol. 83 : 93, 1927.
Hines H. M., Knowlton G. C.: Amer. J. Physiol. 104 : 379, 1933.
— — Amer. J. Physiol. 110 : 8, 1934.
— — Amer. J. Physiol. 111 : 243, 1935.
Hines H. M., Leese C. F., Knowlton G. C.: Amer. J. Physiol. 98 : 50, 1931.
Hines H. M., Thompson J. D.: Amer. J. Phys. Med. 35 : 35, 1956.
Hirohashi: Jap. J. Med., Sec. 3. Biophys. 7 : 76, 1941.
Hoagland C. H.: Adv. Enzymol. 6 : 193, 1946.
Hoffmann A., Wertheimer E.: Pflügers Arch. 216 : 337, 1927-VII.
Holling H. E.: Clin. Sci. 4 : 1949.
Hudlická O.: Physiol. bohemoslov. 4 : 433, 1955.
Hudlická O., Gutmann E.: Arch. Exper. Pathol. Pharmacol. 234 : 501, 1958.
Hudlická O., Vrbová G.: Physiol. bohemoslov. 7 : 38, 1958.
Huf E. G., Fischer R.: Amer. J. Physiol. 159 : 6, 1949.
Humoller F. L., Griswold B., Mc Intyre A. R.: Amer. J. Physiol. 164 : 742, 1951a.
Humoller F. L., Hatch D., Mc Intyre A. R.: Amer. J. Physiol. 167 : 656, 1951b.
— — — Am. J. Physiol. 169 : 654, 1952.
— — — Proc. Soc. Exptl. Biol. N. Y. 380 : 84, 1953.
Ichimi: Tohoku J. Exptl. Med. 13 : 100, 1929.
Ijima: Nippon. Geka Hokan 1 : 503, 1924.
Issekutz B. Jr., Hetényi G., Winter M.: Acta Physiol. Acad. Sci. Hung. 7 : 45, 1955.
Issekutz B. Jr., Spitzer J. J.: Proc. Soc. Exper. Biol. Med. 105 : 21, 1960.
Janke J.: XX-e Congrès International de Physiologie: 470, Bruxelles, 1956.
Jansen S.: Arch. Exptl. Pathol. Pharmacol. 119 : 31, 1927.

Kanashenok P. S.: Bjul Eksper. Biol. Med. 38 : 45, 1954.
Kanashenok P. S., Klyutchayeva M. M.: Bjul. Esper. Biol. Med. 42 : 36, 1956.
Kaplan N. O., Swarta M. N., Frech M. E., Cioti M. M.: Proc. Nat. Acad. Sci.: 42 : 481, 1956.
Kashpur A. M.: Zh. Obshch. Biol. 13 : 50, 1952.
Katz I., Abraham S., Hill R., Chaikoff I. L.: J. Amer. Chem. Soc. 76 : 2277, 1954.
Kemp C. K., Tuttle W. W., Hines H. M.: Amer. J. Physiol. 150 : 705, 1947.
Kimura K., Takashi H.: J. Exper. Med. 10 : 215, 1928.
Knowlton G. C., Hines H. M.: Amer. J. Physiol. 109 : 200, 1934.
Koshtoyants Ch. S., Janson Z. A.: Dokl. Akad. Nauk, USSR 75 : 81, 1950.
Krebs H. A.: Endevour 16 : 125, 1957.
Kreienberg W.: Pflügers Arch. 246 : 171, 1942.
Kutsenko N. A., Nechyeva G. A.: Fiziol. Zh. SSSR 39 : 374, 1953.
Lafon G.: C. R. Acad. Sci., Paris 156 : 1248, 1913.
Langley J. N.: J. Physiol. 50 : 355, 1916.
Langley J. N., Igataki M. J.: Physiol. 51 : 202, 1917.
Lawrie R. A.: Biochem. J. 55 : 298, 1953a.
— Biochem. J. 55 : 305, 1953b.
Lazére B., Thomson J. D., Hines H. M.: Ann. Rev. Physiol. 1942.
— — — Amer. J. Physiol. 138 : 357, 1943.
Levine R., Hechter O., Soskin S.: Am. J. Physiol. 132 : 336, 1941.
Lifson N.: Amer. J. Physiol. 188 : 227, 1957.
Ludwig, Schmidt: 1869. see Pflüger, 1875.
Manché E.: Zeitschr. f. Biol. 163 : 25(7), 1889.
Mansfeld G., Müller F.: Pflügers Arch. 152 : 61, 1913.
Martínek J.: Physiol. bohemoslov. 5 : 7,8, 1956.
Martino: 1929.
Masayama T.: Arch. Exptl. Pathol. 163 : 563, 1932.
Matsui T., Kuriaki K.: Amer. J. Physiol. 196 : 461, 1959.
Mc Dounel: Amer. J. Med. Sci. 46 : 523, 1863.
Mc Pherlane M. G., Spooner S. J. L.: Brit. J. Exptl. Pathol. 27 : 339, 1946.
Meyerhof O.: "Die Chemischen Vorgänge im Muskel". (Berlin.) 1930.
Michelazzi L., Mor M. A., Dianzani M. U.: Experientia 13.: 117, 1957.
Mottram R. F.: J. Physiol. 123 : 34 P, 1954.
— J. Physiol. 128 : 268, 1955a.
— J. Physiol. 130 : 42P, 1955b.
— J. Physiol. 142 : 314, 1958.
Muschini: C. R. Soc. Biol. 99 : 1889, 1928.
Nachmias V. T., Padykula H.: J. Biophys. Biochem. Cytol. 4 : 47, 1958.
Nakamura H.: J. Physiol. 55 : 100, 1921.
Needham J.: Physiol. Rev. 6 : 1, 1926.
Neptune E. M., Foreman D. R.: J. Biol. Chem. 234 : 1942, 1959a.
Neptune E. M., Sudduth H. C., Fash F. J., Foreman D. R.: Amer. J. Physiol. 196 : 269, 1959b.
Neptune E. M., Sudduth H. C., Foreman D. R.: J. Biol. Chem. 234 : 1659, 1959a.
Ogle M.: St. George Hospital Repts. 3 : 149, 1868.
Ogston A. T.: Nature 162 : 963, 1950.
Omachi A., Lifson N.: Amer. J. Physiol. 185 : 35, 1956.
Ochs S., Burger E.: Am. J. Physiol. 194 : 499, 1958.
Osborne S. L.: Arch. Phys. Med. 32 : 523, 1951.
Paladin A. V., Sigalova: Bull. Soc. Chim. Biol. Paris 15 : 1263, 1931.
Palazzolo G.: Arch. Fisiol. 11 : 558, 1913.

Pasynskiy A. G.: Usp. Sovr. Biol. 43 : 263, 1957.
Petkowitz D. S., Husni A. E., Simeone F. A.: Brit. J. Exptl. Pathol. 27 : 339, 1946.
Pflüger E.: Pflügers Arch. 10 : 346, 1875.
Radsma W., Kok N. H., Hoske J. O.: Acta Physiol. Pharmacol. Neerl. 4 : 438, 1955.
Rapoport S.: In "Symposium on Nervous Regulation of Metabolism and Active Transport of Ions", (in Prague, 1956). 1958.
Rein H.: Schneider M.: Pflügers Arch. 239 : 451, 1938.
Riesser O.: Erg. Physiol. 38 : 133, 1936.
— Biochim. et Biophys. Acta 1 : 208, 1947.
Roheim P. S., Spitzer J. J.: Amer. J. Physiol. 195 : 288, 1958.
Rumpf Th., Schumm O.: Dtzsche Z. Nervenheilk. 20 : 445, 1901.
Sacks J., Altschuler C. H.: Amer. J. Physiol. 137 : 750, 1942.
Samuel A.: Amer. J. Physiol. 36 : 78, 1957.
Sato T., Kasugai F.: Tokohu J. Exptl. Med. 26 : 336, 1935.
Schmidt C. G.: Biochem. Z. 323 : 266, 1952.
Schmidt C. G., Schlief H.: Z. Ges. Exptl. Med. 127 : 53, 1956.
Severin S. Y.: Biokhimiya 22 : 259, 1957.
Shatenshteyn D. Y., Zyukova M. A.: Arch. Biol. Nauk 40 : 63, 1935a.
— — Arch. Biol. Nauk 40 : 71, 1935b.
Shatenshteyn D. Y., Tsyrlina D. L.: Arch. Biol. Nauk 40 : 77, 1935a.
— — Arch. Biol. Nauk 40 : 87, 1935b.
Shepherd T. J.: Circulation 6 : 281, 1952.
Sheves G. S.: Biokhmiya 18 : 63, 1953.
— Biokhimiya 20 : 152, 1955.
Sheves G. S., Epelbaum S. Y., Ryumina V. Y.: Biokhmiya 21 : 385, 1956.
Sheves G. S., Ryumina V. Y.: Biokhmiya 21 : 385, 1956.
Slozhenikina L. V.: Vopr. Med. Khim. 3 : 137, 1957.
Sola O. M., Martin A. W.: Amer. J. Physiol. 172 : 324, 1953.
Solandt D. U., Magladery J. W.: Brain 63 : 255, 1940.
Somogyi M.: J. Biol. Chem. 174 : 189, 1949.
Stadie W. C., Zapp J. A.: J. Biol. Chem. 148 : 669, 1943.
Stainsby W. N., Fales J. T., Lilienthal J. L.: Bull. J. Hopkins Hosp. 249 : 99, 1956.
Stewart D. M.: Biochem. J. 59 : 553, 1955.
Stewart D. M., Martin A. W.: Amer. J. Physiol. 186 : 497, 1954.
Sutfin D. C., Thompson D. J., Hines H. M.: Amer. J. Physiol. 178 : 535, 1954.
Szent-Györgyi A.: Neurology 8 : 65, 1958.
Takács L.: Acta Physiol. Hung. 11 : 197, 1957.
Telepneva V. Y.: Biokhmiya 20 : 212, 1955.
— Vopr. Med. Khim. 3 : 456, 1957.
Thesleff S.: Acta Physiol. Scand. 34 : 386, 1955.
Thomson J. D.: Amer. J. Physiol. 180 : 202, 1955.
Thomson J. D., Hines H. M.: 1945.
Thorn W., Heimann I., Müldener B., Gercken G.: Pflügers Arch. 265 : 34, 1957.
Tower S. S.: Amer. J. Anat. 56 : 1, 1935.
— J. Comp. Neurol. 67 : 109, 1937.
— Phys. Rev. 19 : 1, 1939.
Varg F.: Arch. Exptl. Pathol. Pharmacol. 34 : 45, 1894.
Varga E., Kostya K., Szabó E., Aszodi L., Kesztus L.: Arch. Exptl. Pathol. Pharmacol. 210 : : 214, 1950.
Veselkin N. V., Veselkina V. M.: Fiziol. Zh. SSSR 33 : 345, 1947.

Villee C. A., Deane H. W., Hastings A. B.: J. Cell. Comp. Physiol. 34 : 159, 1949.
Villee C. A., White V. K., Hastings A. B.: J. Biol. Chem. 195 : 287, 1952.
Vodička Zd.: Physiol. bohemoslov. 5 (suppl.) : 55, 1956.
— Physiol. bohemoslov. 6 : 62, 1957.
Vodička Zd., Gutmann E., Bass A.: Physiol. bohemoslov. 6 : 343, 1957.
Vrbová G.: Physiol. bohemoslov. 4 : 62, 1955.
Wehrmacher W. E., Thomson J. D., Hines H. M.: Arch. Phys. Med. 26 : 261, 1945.
Weinhouse S., Millington R. H., Volk M. E.: J. Biol. Chem. 185 : 191, 1950.
Wertheimer E.: Pflügers Arch. 213 : 282, 1926.
— Pflügers Arch. 215 : 779, 1927-V.
— Pflügers Arch. 215 : 796, 1927-VI.
Wertheimer E., Ben Tor V.: Biochem. J. 50 : 573, 1952.
Westerbrink H. G., Krabbe H.: Arch. Neerl Physiol. 21 : 456, 1936.
Wilhelmi A. E.: In "Fulton's Textbook of physiology". Philadelphia 1955.
Yakovlev N. N.: Fiziol. Zh. SSSR 36 : 744, 1950.
— Dokl. Akad. Nauk, USSR 81 : 709, 1951.
— Ukr. Biokhim. Zh. 27 : 444, 1955.
— "Ocherki po biokhymiyi sporta", Moskva, Leningrad 1956.
— In "Symposium on Nervous Regulation of Metabolism and Active Transport of Ions". Prague 1956.
Yakovlev N. N., Yakovleva I. S.: Usp. Sovr. Biol. 35 : 134, 1953.
Yampolskaya L. J.: Fiziol. Zh. SSSR 36 : 749, 1950.
Yampolskaya L. J., Yakovlev N. N.: Fiziol. Zh. SSSR 37 : 110, 1951.
Yudayev N. A., Smirnov M. J., Razina P. G., Dobbert N. N.: Biokhimiya 18 : 732, 1953.
Zanghi: 1929.
Zubenko P. M.: In "Voprosy Biokhmiyi Mishts". Kiev. 1954.
— Ukr. Biokh. Zh. 28 : 223, 1955.
Zubenko P. M., Reva A. D., Plakhotishina Y. T.: Biokhimiya 15 : 79, 1950.
Žák R., Drahota Zd.: Physiol. bohemoslov. 9 : 329, 1960.
Žák R.: Fysiol. čs., 7 : 33, 1958.
Žák R., Gutmann E., Vrbová G.: Experientia 13 : 80, 1957.

Chapter IX

PROTEINS IN THE DENERVATED MUSCLE, CHANGES IN THEIR QUANTITY, PROPERTIES AND METABOLISM

Žák R.

Loss of muscle weight and finally complete degenerative dedifferentation of muscle tissue are the most prominent results of interrupting the connection between the muscle and the nervous system. Nevertheless, chemical investigations performed towards the end of the 19th Century analyzing quantitative changes in the denervated muscle were concentrated on study of glycogen, creatine and creatinephosphate. This is all the more surprising since, at that time, it was believed that muscle proteins play a substantial role in muscular work (Pflüger). First reports dealing with protein changes in denervated muscles were by Steyrer (1903) and Grund (1912). The former described changes in the myosin: myogen ratio, while the latter pointed out the marked decrease of phosphorus-free protein. The methods of investigating properties of muscle proteins were at that time very imperfect and it is therefore difficult to evaluate these two findings from the point of view of contemporary knowledge. The first exact data were published from a physiological laboratory, when Langley and Kato (1915) determined the gradient of denervation atrophy on the basis of changes in wet muscle weight. By 1921 Lipschütz and Audova showed that tenotomy as well as denervation leads to a gradual loss of muscle proteins. In spite of effort continuing up to the present it has not been possible to differentiate exactly between these two kinds of atrophy. The above investigations of atrophic muscles were made at the time of fundamental discoveries by Fletscher and Hopkins in 1907 and Meyerhof in 1919. These authors reported an increase of lactic acid in muscle following contraction in a nitrogen atmosphere and concluded that glycolysis serves as source of energy for contraction. During the course of the following years, therefore the interest of biochemists should focus on muscle contraction energetics, an area of study concerning the basic function of muscle. Systematic investigation of the denervated muscle was thus not undertaken until 20 years after the report of Langley and Kato, by Hines and Knowlton (1933). These authors studied rather the physiological characteristics of the muscle undergoing atrophy, such as loss of protein mass, the course of atrophy in various types of muscles, changes in different animals species, etc.

Biochemical knowledge of the mechanism of muscle contraction was enriched during the following years (1939—1945) by the work of Szent-Györgyi, Straub, Weber, Engelgart, D. Needham and others. Actomyosin was found to be the unit of muscle contraction and the reaction of ATP with myosin to be the triggering mechanism of contraction. Accurate methods were elaborated for precise determinations of muscle proteins and their properties. On the basis of these findings, Fischer and coworkers investigated changes in muscle proteins in the course of denervation atrophy on a biochemical level, beginning in 1940. Between 1940 and 1955 not only changes in the protein spectrum were studied in the denervated muscle, but also properties of individual proteins. Since 1949 more data have accumulated concerning proteins in the denervated muscle due to the work of Crepax with use of modern electrophoretic methods.

Incorporation studies of Schoenheimer and collaborators in 1939, which led to the concept of the dynamic state of body constituents, were followed by intensive study of proteosynthesis. In recent years the disturbance of proteosynthesis has also been investigated in the denervated muscle with the aid of incorporation of labelled amino acids. Although very extensive studies have been carried out especially by Schapira and Dreyfus in Paris and Ferdman in Kiev, it has not been possible to specify the primary link in disturbance of proteosynthesis in denervated muscle.

Reviews of protein changes in denervated muscles have been published by Fischer, Schapira and Dreyfus in the symposium at Royaumont (1950) and by Ferdman and Zubenko in a monograph "Questions of muscle biochemistry" published in 1954 in Kiev. The present chapter deals with denervation changes of muscle proteins from the functional aspect. The basic function of a muscle under the integrated control of the nervous system is muscle contraction. This process is not only accompanied by changes in energy metabolism. Following single or even repeated contractions qualitative and quantitative changes in muscle proteins have been described. It is possible that the reversal of these changes, restitution of the resting state, is the basis of the trophic influence of the nervous system. It is for this reason that in this chapter denervation changes are always compared with changes occurring during contraction. Comparison with other types of muscle atrophy will be made only when it is necessary for the sake of clarifying denervation changes.

A. Changes in protein composition of the denervated muscle

When studying denervation changes in muscle proteins, the results are usually expressed either in absolute or relative values. Absolute values give information about the amount of proteins per whole denervated muscle, while relative ones, on the other hand, give information only about changes in the composition of the denervated muscle. In both cases either 1. net changes in muscle proteins (which represent the sum of independent changes in individual substances) or 2. changes of specific proteins may be studied.

Study of net changes represents a methodically simpler approach, making a large number of determinations possible. This is an important aspect especially during investigation of physiological changes in the denervated muscle. However, the great functional differences between individual muscle proteins, allows an investigation of net changes to provide only information that is approximate. To obtain an accurate picture of denervation changes from the functional aspect, is possible only by the determination of certain specific proteins. Separation and determination of individual proteins has been extensively investigated, although it is methodically very difficult. For this reason often only several points of the time gradient are given in the literature and results are generally expressed in relation either to total muscle proteins or to a certain specific protein.

1. Net changes

a) Changes per whole muscle

Wet weight.

Muscle weight is the most approximate, but also the most evident indication of changes in muscle proteins. Although dry substances represents only 25 % of this weight, the increment of water in the denervated muscle is not so extensive as to mask the loss of muscle mass. Occasionally loss of dry substance has been found without changes in wet weight, but this is rather the exception than the rule (see Chapter X). However, generally the change in water content is not great (Audova 1923, Chor et al. 1937, Hines and Knowlton 1933, Fischer and Ramsey 1946, Zubenko 1954, Drahota 1958), or has not been observed at all (Peters and Guenther 1921, Chen et al. 1924, Westenbrink and Krabbe 1936, Chepinoga 1938). If the water content of denervated muscles is not substantially altered, wet weight can serve as indication of absolute loss of muscle substance.

The course of denervation changes in muscle weight is different in various types of muscles and in different animal species (for further details see chapter X). Nevertheless it is possible to generalize that loss of weight first appears three days after section of the nerve and that atrophy progresses faster during the first stage (loss of weight during the first 14 days is about 30—50 %) and later is somewhat slower (30 days after denervation weight loss is about 50—70 %). Individual results are given in Table 13.

Table 13

The loss of wet weight in the denervated muscle (percentage).

Experimental animal	Muscle	Time after denervation*			Author
		I. stage	14 days	II. stage	
Rat	Gastroc.	-8·4	-48·3	-72·3 (28)	Hines and Knowlton (1933)
	Gastroc.	-7·4	-49·9	—	Hník (unpubl.)
	Ext. dig. l.	0 (3), -17·5 (7)	-37·7	—	Hník (unpubl.)
	Tib. ant.	-7·6(3)	-46·0	—	Hník (unpubl.)
	Soleus	-4·9 (3)	-36·0	—	Hník (unpubl.)
Rabb t	Gastroc.	-10 (5)	-37·0	-53 (30)	Fischer and Ramsey (1947)
	Gastroc.	Decrease begins on 2nd—3rd day	-30	-50 (35)	
	Plantaris		-30	-50 (35)	Langley and Kato (1915)
	Soleus		-30	-50 (35)	
	Extensor		-30	-50 (35)	Peters and Guenther (1921)
Cat	Gastroc.	—	—	-50 (25)	Westenbrink and Krabbe (1937)

* Numbers in brackets denote days after operation. (I. stage: before, II. stage: after 14 days of denervation).

The gradual retardation of denervation weight loss is especially evident in the rat in which the weight loss of the gastrocnemius decreases about six times between the third and fourteenth day after section of the nerve, and only about 1·6 times between the fifteenth and thirtieth day (Hines and Knowlton 1933). Analogous results have been reported by Hník (Chapter X). Muscle weight loss in the rabbit during the first stage is approximately the same as in the rat, from the fourteenth days onwards, however, it is considerably smaller (by about 30%), as has been reported by Langley and Kato (1915), Chen et al. (1924), and Fischer and Ramsey (1946) — v. Table 13. The general course of weight changes in the rabbit is given in Fig. 99. Loss of muscle weight has also been studied in the denervated gastrocnemius of the cat and has been found to be roughly the same in extent as in the rabbit, if conclusions can be drawn from a single point 25 days after denervation (Westenbrink and Krabbe 1936).

Besides denervation, loss of muscle weight has also been observed, for example, after tenotomy. However, it is difficult to decide what is the difference between these two types of atrophy. Muscle atrophy was found to be approximately the same after denervation and tenotomy by Lipschütz and Audova (1921) and Fischer and Ramsey (1946), while according to Lippmann and Selig (1928) atrophy after tenotomy progresses at a faster rate. Immobilization also results in weight loss, the changes, however, are about 20% smaller than those after denervation (Lippmann and Selig 1928, Fischer and Ramsey 1946).

Dry weight.

A more direct index of losses in muscle weight after denervation is the estimation of dry weight. Even though the changes in the water content are not great, they can, at a certain period, as has already been mentioned, mask the loss of wet weight. An example of this is given in Chapter X: three days after denervation no changes were found in the weight of extensor digitorum longus of the rat, even though muscle dry weight was already decreased by 6%. The loss of dry muscle weight was further studied up to the fourteenth day after denervation at which time it was found to be about 50% (Chapter X).

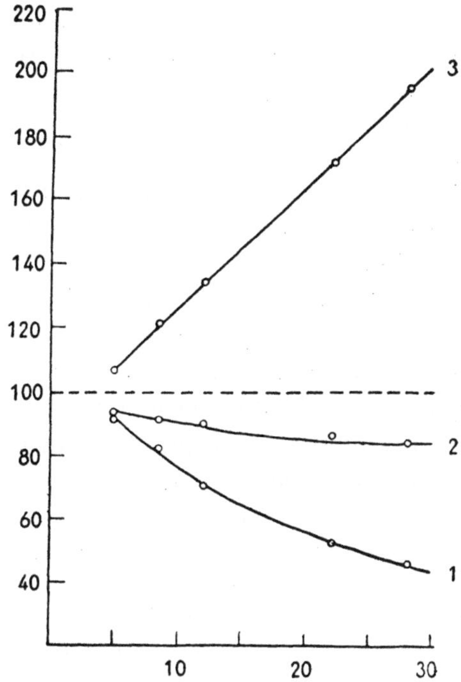

Fig. 99. The effect of denervation on the loss of wet muscle weight and on the content of total proteins and collagen. Ordinate: percentages. Abscissa: time after denervation (days). 1 — wet weight, 2 — total proteins, 3 — collagen. Changes expressed as percentage of corresponding value of normal muscle. (After Fischer and Ramsey 1946.)

Questions of the rate at which dry weight is lost in various types of muscles has not yet been clarified. However, it seems that the gastrocnemius in the rat atrophies at a somewhat faster rate than either extensor digitorum longus, soleus or the anterior tibial muscles. The difference is especially well marked 14 days after denervation at which time gastrocnemius loses 51%, soleus only 36·7%. These differences in the rate of muscle atrophy are evident even when only wet weights have been determined only.

Protein nitrogen.

Total protein nitrogen is decreased in the anterior tibial muscle of the rat by three days after denervation. Seven days after denervation this loss represents 23% of protein nitrogen in the control muscle and increases to 45% loss after 14 days (Gutmann 1960). According to an other reference three weeks denervated rat muscle lost 72% total proteins. (Hearn 1959).

On summarizing the above net index of muscle weight loss following denervation, it can be concluded that wet and dry weight, as well as protein nitrogen give similar results, with the exception, perhaps, of the first denervation stage (third day) when changes in wet weight are somewhat smaller than those in dry weight and total protein nitrogen.

b) Relative changes

Total nitrogen (TN) and total proteins (TP) (Tab. 14).

First information about muscle composition can be obtained by determination of TN and TP in unit wet weight. It is a disadvantage, however, that these indicators also include changes in water content and independent changes in individual proteins, as has already been pointed out above. This is probably the reason why no unity of results has been attained in studies of denervation changes by total nitrogen and total proteins determinations. Differences have expecially been found between different types of experimental animals (see Tab. 14).

The TN content has been found to decrease by about 10% in the soleus and extensor digitorum longus of the rabbit 21 days after denervation (Peters and Guenther 1921). A similar decrease was described for total proteins in the rabbit gastrocnemius by Fischer and Ramsey (1946) and by Helander (1957).

Course of these changes is shown in Fig. 99, from which it is evident that maximum loss occurs in the course of the first week after denervation. Later, changes are very slow. In the dog decrease in the total protein content similar to rabbit muscles was found (Avellone and Di Macco 1925, Cahn 1927). (Grund reported in an earlier paper in 1912 that the decrease of proteins three months after denervation in the dog affects only phosphorus-free proteins, this experiments, however, can hardly be taken into account from the modern point of view.)

Similarly as in muscle weight, a decrease of TP in the muscle has been described after tenotomy which is similar to that after denervation (Fischer and Ramsey 1946, Nechiporenko 1955). In E-avitaminosis similar decrease in total proteins (Spencer et al. 1937) and total nitrogen (Bender et al. 1959) was described as after denervation. Imobilization according to Helander (1957) leads to very small change only.

On the other hand, in experiments of other authors no changes have been found in muscle composition. For example, Hines and Knowlton (1933) did not find changes of TN in the rat gastrocnemius from the seventh to twentieth day after denervation. No changes were also found in the gastrocnemius of the cat 21 days after denervation by Westenbrink and Krabbe (1936) and neither were any changes of total nitrogen encountered in muscles of Maccacus rhesus one to twelve weeks after section of the sciatic and femoral nerves (Chor et al. 1937).

Table 14

Changes in protein composition of atrophic muscle (%): NET CHANGES

Indicator	Animal	Muscle	Content in normal muscle	Time after denervation: I. stage	Time after denervation: 14 days	Time after denervation: II. stage	Tenotomy	Immobilization	E-avitaminosis	Notes	Author
Total nitrogen	Rabbit	Ext. dig. l.	3·08	—	—	-13·5 (21)	—	—	—	—	Peters and Guenther (1921)
% wet weight	Rabbit	Soleus	3·36	—	—	-8·5 (21)	—	—	—	—	Peters and Guenther (1921)
% dry weight	Rabbit	Gastroc.	13·92	—	—	—	—	—	-4·0	—	Spencer et al. (1937)
% wet weight	Rat	Gastroc.	3·65	0 (7)	0	0 (20)	—	—	—	—	Hines and Knowlton (1933)
% dry weight	Monkey	***	13·60	0 (7)	0	0 (50)	—	—	—	—	Chor et al. (1937)
% wet weight	Cat	Gastroc.	3·10	—	—	0 (25)	—	—	—	—	Westenbrink and Krabbe (1936)
Total proteins % wet weight	Rabbit	Gastroc.	18·08	-8 (7)	-10	-14 (30)	-14 (30)	-4 (30)	—	—	Fischer and Ramsey (1946)
Non-collag. protein % wet weight	Rabbit	Gastroc.	14·88	—	—	-34 (25)	-34 (25)	-12 (25)	—	—	Fischer and Ramsey (1946)
Extract. protein	Rabbit	Gastroc.	60·40	—	—	-37 (30)	-35 (30)	+10 (30)	—	0·5M LiCl	Fischer and Ramsey (1946)
% non.-coll. protein	Rabbit	Gastroc.	9·00	0 (5)	-20	-30 (30)	—	—	—	0·5M LiCl	Fischer (1948)
% wet weight	Rabbit	***	***	—	-14**	-38 (40)	—	—	—	KCl	Crepax (1949)
Sarcoplasm. protein % dry weihgt	Rat			—	—	+16 (40)	—	—	—	0·1 μ	Schmidt and Schlief (1956)
% wet weight	Rabbit	Gastroc.	6·71	—	—	-22 (20)	0 (20)	+20 (20)	—	0·1 μ	Helander (1957)
mg./g. w. wt.	Rabbit	Calf muscl.	35·00	—	—	—	-12 (30)	—	—	H_2O	Nechiporenko (1955)
***	Rabbit	***	***	—	—	—	—	—	0	0·1 μ	Crepax (1952a)

* Numbers in brackets denote days after denervation (I. Stage before, II. stage after 14 days of denervation).
** 16—24 days after denervation.
*** Values not given by author.

Non-collagenous proteins (Tab. 14).

Following section of the nerve there is a relative increase in the content of collagen in muscle (see p. 294). This can perhaps explain why changes in total proteins (and TN) are relatively small after denervation, if present at all. The true decrease of protein content should thus be demonstrated after removal of collagen. This was actually the case as reported by Fischer and Ramsey (1946) who found that non-collagenous proteins were decreased by 34% in the denervated gastrocnemius of the rabbit 22—30 days after denervation, as compared with the 14% loss of total proteins.

In agreement with all the above mentioned indexes, section of the tendon evokes changes similar to denervation, while immobilization has a smaller effect — only 12% (Fischer and Ramsey 1946).

Extractable proteins (Tab. 14).

The extraction with 0·1N sodium hydroxyde generally used for the determination of non-collagenous proteins is very drastic and cannot be used for further separation of specific proteins. This disadvantage is removed when a further basic indicator is used, for example, determination of proteins extractable by salt solutions of ionic strength 0·3. By this method most structural muscle proteins are extracted (50—70%). Fischer, using extraction with lithium chloride, found that 22—30 days after denervation the content of extractable proteins expressed in percentage of non-collagenous proteins decreases by about 40% (Fischer and Ramsey 1946). The character of these changes is mainly determined by actin. It can be seen on comparison of time course of percentage loss of actin with LiCl-extractable proteins that the curves are identical (Fischer 1948). A smaller decrease, especially during the early denervation period, was found by Crepax (1949a) and Helander (1957). who, like Fischer, worked on the rabbit. Quantitative differences between the results of these authors can be caused by the fact that Crepax expressed the extractable proteins in percentage of wet weight which decreases at a faster rate than non-collagenous proteins, in relation to which the results of Fischer were calculated. Furthermore, the extracting media used by these authors were different: Crepax used KCl instead of LiCl. Lithium chloride used in Fischer's experiments, has somewhat more potent extracting properties.

Again, no substantial differences were reported between denervation and tenotomy, although immobilization again resulted in considerably smaller changes (Fischer and Ramsey 1946 and Helander 1957).

Extraction with a salt solution of ionic strength 0·1 μ is a good indicator of non-structural (sarcoplasmatic) proteins. A 16% decrease of these has been found in rat muscles at a late stage of atrophy (40 days after section of the nerve) by Schmidt and Schlief (1956), while 20 days after denervation a decrease of 22%

is reported by another worker (Helander 1957). No differences were found in late stages of E-avitaminosis as compared with muscles of control animals (Crepax 1952a). Similarly after tenotomy only a very small change of non-structural proteins has been described by Nechiporenko (1955) and Helander (1957) — see also note on p. 294. Immobilization on the other hand leads according to the later worker to 15% increase in sarcoplasmatic proteins.

It can thus be concluded from the above material that with the use of net indexes a precise gradient of denervation changes has been determined. Loss of muscle weight differs in various types of muscles. From the animals studied it can be stated that the rate of atrophy in cat and rabbit is approximately the same, while in the rat atrophy progresses somewhat faster.

There is a slight decrease in the content of total nitrogen and total proteins after denervation (about 10% in 14 days), or no change at all. The decrease can be due to increased water content and relative increase of non-protein compounds (e. g. lipids). Furthermore, the decrease in non-collagenous proteins and proteins extractable with solution of ionic strength 0·3 μ is approximately double the decrease in total proteins. This is due to the fact that there is a relative increase of collagen in atrophic muscles. Proteins extractable with solutions of ionic strength 0·1μ are decreased by about 20% at late stages of atrophy. As far as is known, this change is evident only after denervation, and has not been observed either after tenotomy or in dystrophic muscles of animals with E-avitaminosis. Changes in all the remaining indicators are not specific for denervation: section of the tendon leads to changes approximately of the same extent, while immobilization results in changes which are about one third smaller.

2. Specific proteins

Hines and Knowlton differentiated muscle into a muscle phase and an interstitial phase. From the point of view of protein composition, the muscle phase consists of structural proteins (about 60%) which represent the basic structure for muscle contraction, and of sarcoplasmatic proteins (about 40%) which represent the bulk of muscle enzymatic systems. The interstitial phase is formed by connective tissue the main constituent of which is collagen (about 20% of muscle proteins). According to Hines and Knowlton denervation changes affect the muscle phase only, without affecting the interstitial phase. However, as will be demonstrated below, an absolute decrease in the content of collagen has been found at late stages of muscle atrophy. Despite this, we shall adhere to this classification when discussing denervation changes in muscle proteins.

General information about the muscle phase, i. e. about intracellular proteins, can be obtained by determining non-collagenous proteins (representing 14%

wet muscle weight) and extractable proteins (representing 60% non-collagenous proteins). Changes in these during denervation have already been discussed above (see p. 280).

a) Structural proteins

Structural proteins are the main component of the myofibrils. Up to now the following proteins have been isolated: actomyosin, tropomyosin, Y-protein, N-protein extra-protein, contraction and metamyosin. From the methodical aspect these proteins are defined by their solubility. It is possible to extract fractions from the muscle homogenate with solutions up to 0·3 ionic strength. For example, approximately 60–75% structural proteins are extracted with 0·5M-LiCl. This lithium extract can be used as the first index of structural proteins. The decrease of proteins in the LiCl-extract of the denervated muscle was already mentioned among basic indicators on p. 280.

Myosin.

(Myosin-A according to Szent-Györgyi, myosin-L according to Weber, classified as myosin β on the electrophoretograms (Tab. 15).

Determination of myosin is very difficult even though it is the protein known longest and has been most extensively studied. Muscle extraction as well as the subsequent precipitation are dependent on the functional state of the muscle (v. section C, p. 311) and the method used. It is therefore not possible to expect that there would be complete agreement between results of individual authors. Nevertheless, it can be concluded from these results (Tab. 15) that myosin is the protein which is reduced most in the muscle after denervation (reduction of actomyosin is of similar extent since myosin forms 70% of this complex). The decrease begins very soon according to Fischer (1948), being 20% on the fifth day. As can be seen from Fig. 100, the decrease gradually slows down in the course of denervation, so that between the fourteenth and thirtieth day the difference is only 10%. These values have been obtained on the gastrocnemius of the rabbit as in the report of Ferdman and Mestechkina (1951) who have described analogous results. Similarly in the experiments of Crepax (1949a, 1949b, 1953a, 1953b), with the use of the Tiselius electrophoretic method of separation of structural proteins, a decrease of myosin was found after denervation. Changes of the electrophoretic pattern were studied between the sixteenth and eighty-fourth day after section of the nerve. In contrast to Fischer, only a small change was found in the content of myosin 16 days after denervation. Later, however, this decrease gradually accelerates to the forty-fifth day, when it again slows down. The decrease of myosin is also smaller at this late period than has been reported by Fischer who worked with the precipitation method. Smaller changes of myosin have also been reported by Gligliotti and Lamedica (1956), who separated myosin by paper

Table 15

Changes to protein composition of denervated muscle (%): SPECIFIC PROTEINS

Protein	Animal	Muscle	Content in normal muscle	Time after Denervation			Tenotomy	Immobilization	E-avit	Notes	Author
				I. stage	14 days	II. stage					
Actomyosin											
% w. wt.	Rabbit	Gastroc.	4·75	-5 (5)	-50	-76 (30)	-68 (30)	-10 (30)	—	Called myosin	Fischer and Ramsey (1946)
% non-coll. protein	Rabbit	Gastroc.	33·90	—	—	-68 (30)	-59 (30)	—	—	Called myosin	Fischer and Ramsey (1946)
mg N/100 g. dry wt.	Rabbit	Gastroc.	576	—	—	—	—	—	-76	—	Bonetti et al. (1952)
Myosin											
% w. wt.	Rabbit	Gastroc.	2·50	-20 (5)	-70	-80 (30)	—	—	—	Precipitated	Fischer (1948)
% protein	Rabbit	Ext. dig. l.	57·10	—	—	-12·5**	—	—	—	Paper electrophoresis	Gligliotti and Lamedica (1956)
***	Rabbit	Gastroc.	***	—	—	—	-50 (9)	—	—	KCl extract	Fischer (1950)
mg./g. w. wt.	Rabbit	Calf musc.	5·80	—	—	—	-55 (20—45)	—	—	Precipitated	Nechiporenko (1955)
mg. N/100 g. dry w.	Rabbit		931	—	—	—	—	—	—	Myosin peak disappeared (cuvette el.)	Bonetti et al. (1952)
Actin											
% w. wt.	Rabbit	Gastroc.	1·30	0 (5)	-20	-30 (30)	—	—	—	—	Fischer (1948)
mg. N/100 g. dry. wt.	Rabbit	Gastroc.	352	—	—	—	—	—	0	—	Bonetti et al. (52)
mg./g. w. wt.	Rabbit	Calf muscles	5·60	—	—	—	-12 (20—45)	—	—	—	Nechiporenko (1955)
Myogen											
% protein	Rabbit	Ext. dig. l.	40·30	—	—	-3·0	—	—	—	Paper electrophoresis	Gligliotti and Lamedica (1956)
***	Rabbit	Gastroc.	***	0 (5—10)	—20	-55 (30)	-20 (9)	0 (9)	—	Aldolase	Fischer (1950)
% protein	Cat	Gastroc.	2·70	—	—	0 (25)	—	—	—	Myogen + globulin -x	Westenbrink and Krabbe (1946)
Myoalbumin											
% protein	Rabbit	Ext. dig. l.	2·60	—	—	+420	—	—	—	—	Gligliotti and Lamedica (1956)
***	Rabbit	Gastroc.	***	+20 (5)	+120	+500 (30)	+125 (9)	+5 (9)	—	Cuvette electrophoresis	Fischer (1950)

* Days after denervation in brackets (I. stage before, II. stage after 14 days of denervation).
** Late stage of atrophy, without exact date of denervation. *** Values not given by the author.

electrophoresis, and Westenbrink and Krabbe (1936) who determined myosin in cat gastrocnemius by precipitation.

The increase of myosin after denervation is an exceptional finding in literature and deserves, therefore, further consideration. Using the method of heat fractionation of von Fürth, Steyrer (1903) found a rise in the ratio myosin: myogen in the denervated rabbit gastrocnemius. This result is thus in contradiction to the above mentioned, more recent findings. It should be pointed out that the myosin determined by von Fürth's method of heat coagulation represents only 25% of intracellular proteins, while that determined by contemporary methods represents 60%. A further difference between Steyrer's results and more recent reports lies in the fact that myosin was estimated in press juice, while in contemporary determinations it is estimated in an extract. Hines and Knowlton (1933) also criticized the work of Steyrer because non-buffered solutions were used, while the coagulation of myosin is dependent on the pH of the medium. Altered production of acids (glycolysis) by the denervated muscle can thus be responsible for the difference found above in the myosin: myogen ration. From the above remarks it is evident that the results of Steyrer cannot be compared with more recent results.

Denervation is again not the only type of muscle atrophy which leads to a decrease in myosin. The changes due to tenotomy are more extensive and appear earlier than after denervation. Later, the extent of these changes corresponds to those of denervation (Fischer 1950, Nechiporenko 1955). A very large decrease of myosin was also found in dystrophic muscles of animals with avitaminosis-E (Feuer and Frigyier 1952). Several authors have even reported complete disappearance of myosin from the electrophoretograms (Crepax 1952a, Bonetti et al. 1952). The effect of immobilization was studied for only short periods of time: no changes were found by 8—10 days, while a decrease of myosin was already found at this time after denervation or tenotomy (Fisher 1955).

All the above results concerning changes of myosin were expressed in relation to wet weight. However, in the course of atrophy there is a decrease in this weight, as has already been pointed out on p. 277. If then the relative content of myosin in the denervated muscle decreases, it can be expected that its absolute decrease will be still greater. On the other hand, it must be kept in mind that denervation changes the physico-chemical state of myosin (v. paragraph C, p. 312) which results in a lowering of its extractibility. The decrease of myosin found is thus greater than is actually the case, since less myosin is extracted from the denervated than from the normal muscle. This fact is especially clearly demonstrated in the paper of Bonetti et al. (1952). Although these authors found a decrease of actomyosin in the muscles of animals with avitaminosis-E, it was shown to be present in sufficient amounts to be determined. On the other hand myosin completely disappeared in the same experiment. Since myosin is a part of actomyosin (it forms 70% of this complex) it must, therefore, be present in the dystrophic muscle. The binding forces preventing myosin from being extracted have evidently become more potent due to the atrophy. However, it is known that the extractibility of actomyosin is not altered by denervation (v. section B, p.). The decrease of actomyosin following denervation can serve at the same time as evidence that myosin is also decreased, even though to a smaller extent than is usually believed. This is also in accord with histological observations (v. Chapter III).

Actin (Tab. 15).

Changes of actin during denervation have one characteristic difference from those of other structural proteins. The actin content remains unchanged for nine to ten days, then during a short period of about five days there

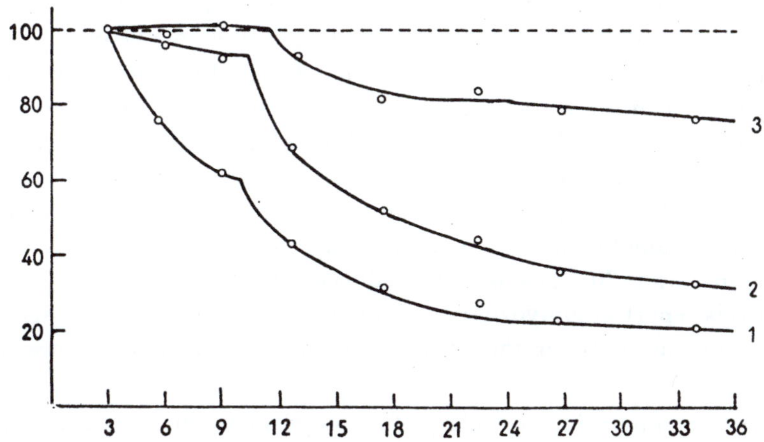

Fig. 100. The effect of denervation on the content of actin, myosin and actomyosin in rabbit gastrocnemius. Ordinate: percentages. Abscissa: time after denervation (days) 1 — myosin, 2 — actomyosin, 3 — actin. Changes are expressed as percentage of corresponding values of normal muscle. (After Fischer 1950.)

is a rapid decrease. During the following 15 days actin diminishes at the same rate as the other muscle proteins (v. Fig. 100) — (Fischer 1948, 1950, 1955).

As far as other types of atrophy are concerned, actin changes were studied in animals with avitaminosis-E and after section of the tendon. In the former case no differences were found from the content of actin in muscles of normal animals (Bonetti et al. 1952) and even in muscles with sectioned tendon the difference is not marked (Nechiporenko 1955) — (v. Tab. 15).

Actomyosin (myosin-B according to Szent-Györgyi, myosin-S according to Weber, myosin-α in electrophoreograms) — (Tab. 15).

Actomyosin is a complex of actin and myosin in the ratio 2 : 5. So far changes of actin and myosin have been discussed independently of each other.

Theoretically the mean of these two values should determine the changes of actomyosin. It can be seen in Fig. 100, which gives Fischer's results (Fischer 1948), that this is actually the case. However, during the first stage, changes in actomyosin are determined more by actin which remains unchanged. In the second stage, from the tenth day onwards, the character of the curve is determined more by the drop of myosin content. This variation from the mean value in given by the fact that actin decreases less than total actomyosin, while myosin decreases more. If this is actually the case (i. e. if it is not due to solubility changes of myosin) then this must mean that the ratio actin: myosin in the actomyosin complex increases and this could be responsible for some of the changes of its physicochemical properties (for details v. section C, p. 311). In addition to Fischer's studies determining actomyosin by the precipitation method, a decrease was also found by Crepax (1953a, 1953b) with the aid of quite a different method, i. e. electrophoretic separation.

Denervation, however, is not the only type of atrophy leading to a reduction of actomyosin. Avitaminosis-E has been found to evoke changes similar to denervation (Bonetti et al. 1952). Thirty days after section of the tendon the content of actomyosin is reduced by 68% as compared with the 76% decrease in denervation. There is again no fundamental difference between denervation and tenotomy. Immobilization has a markedly smaller effect, since the content of actomyosin after 30 days is only reduced by 10% (Fischer and Ramsey 1946 — in this report actomyosin is designated as myosin). This is thus the most prominent difference between these types of atrophy, as far as structural proteins are concerned.

As in the case of myosin, the values of actomyosin are given as relative values, i. e. calculated per wet muscle weight. In contrast to myosin it is not presumed that the decrease of actomyosin after denervation is higher than is actually the case. As will be stated in section C. on p. 312, the solubility and extractibility of actomyosin are not affected nearly as much by the functional state of the muscle as myosin. It is therefore not probable that less actomyosin is extracted from the denervated muscle than from the normal muscle. On the other hand the precipitating properties of actomyosin are lowered, so that not all actomyosin can be obtained from the denervated muscle by this method. However, a decrease of actomyosin after denervation has also been found after electrophoretic separation, which does not depend on precipitation.

Y-protein.

This protein was found in the muscle by Duboisson in 1950, and according to Crepax it is depolymerized actin. In contrast to all the other structural proteins in the denervated muscle it was shown to be increased, however, until a late stage — 30 days after section of the nerve (Crepax 1953a, 1953b). Thus, according to Crepax, there is a certain reciprocity of changes between Y-protein and actin. It must be noted, however, that reduction of actin was already observed 15 days after denervation, this being 14 days earlier than the increase in the content of Y-protein.

As after denervation, Y-protein is higher even in dystrophic muscles of animals with avitaminosis-E (Crepax 1952a). Changes in immobilized muscles, or muscles after section of the tendon have so far not been reported.

Other structural proteins (tropomyosin, N-protein, extraprotein, contractin and metamyosin) have not yet been studied.

b) Sarcoplasmatic proteins

Sarcoplasmatic proteins consist chiefly of enzyme of glycolytic cycle. These proteins represent 40% of muscle proteins. From the methodical aspect they are defined by solubility in solutions of ionic strength 0·1. Denervation changes of this extract have been discussed among net index on p. 280. The quantitative determination of individual extracted protein fractions is very difficult. As far as denervated muscle is concerned, either determinations of enzymatic activity have been made or protein separation has been performed by electrophoresis. The identification of individual proteins on the electrophoreogram is difficult and because of this individual peaks are often designated by numbers which, of course, apply only for a given method. Denervation in the rabbit has been found by this approach to cause a proportional increase of the faster and to reduce the slower fractions (Haan 1953). — (Fig. 101 and 102). A protein spectrum similar to that found by Haan, who used the electrophoretic method of Tiselius, was obtained by Hetényi (1956) with paper electrophoresis. This author found an absolute, but not a relative drop in an unidentified fraction I and II and a percentage rise of fraction III (almost by 100%) 22—29 days after denervation in the rat and 6—7 weeks after nerve section in the rabbit.

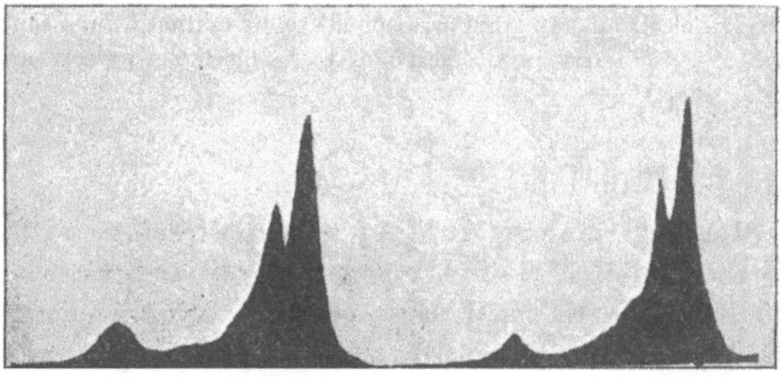

Fig. 101. Changes in electrophoretograms of sarcoplasmatic proteins of denervated muscles during the early period of atrophy.
The left electrophoretogram corresponds to the extract from a rabbit gastroenemius five days after denervation. On the right electrophoretogram from the control muscle. (After Fischer et al. 1949.)

The changes described are not, of course, characteristic for denervation only. A rise in faster and a decrease of the slower fractions has been reported in atrophic muscle after immobilization by Haan (1953) and in avitaminosis-E by Haan (1953) and Baechtel et al. (1957). Changes in spectrum of soluble proteins in myopathie are reported by Dreyfus and Schapira (1959). Of the other non-structural proteins the following have been further specified and studied.

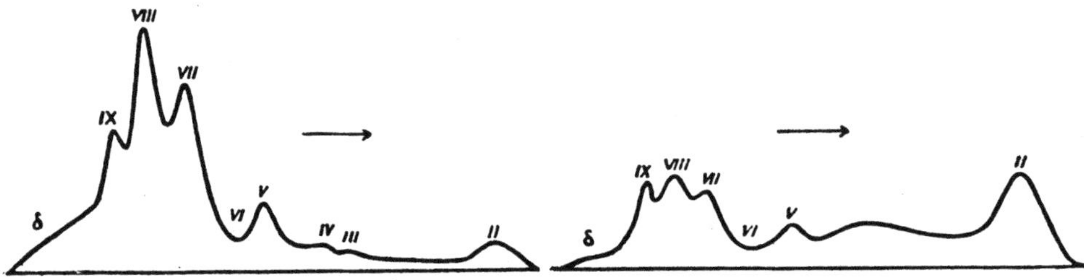

Fig. 102. Changes in electrophoretogram of sarcoplasmatic proteins from the denervated muscle at a late period of atrophy.
The left electrophoretogram corresponds to the extract from a control skeletal rabbit muscle. The right electrophoretogram corresponds to a muscle 2—5 weeks after denervation. (According to Haan 1952.)

Myogen (Tab. 15).

Myogen represents 20 % of muscle proteins and by crystalization has been further subdivided into myogen A, B and C, which correspond to individual enzymatic activities of the myogen complex such as: aldolase, d-3-glyceraldehyde dehydrogenase, phosphoglucomutase, phosphorylase, etc. Myogen as a whole decreases after denervation: this was shown by Crepax (1949a, 1949b) in the rabbit gastrocnemius 16—85 days after the operation using Tiselius' electrophoresis. This change is not very pronounced 16 days after denervation, after this, however, a rapid drop occurs lasting till about the 45th day. Later the changes again become more gradual. The late onset of these changes was also observed by Fischer et al. (1949) in the same muscle, however, the first signs of a decrease were reported by these authors by ten days after denervation, i. e. six days earlier than described by Crepax (Fig. 103). These marked changes are in contrast to the findings of insignificant changes of myogen in atrophic extensor digitorum longus of the rabbit (Gligliotti and Lamedica 1956) — (Tab. 15). How can these contradictory results be explained? It was already mentioned on p. 277 that the rate of atrophy in the gastrocnemius is greater than in the extensor digitorum longus. The difference, however, is far less pronounced than in myogen. The reason for this is apparently the different method used for separation of myogen.

The group of myogen proteins in the denervated muscle was further fractionated in several investigations. The main credit is due to Fischer and his coworkers. In papers from 1948—1949 (Fischer 1948, Fischer et al. 1949) these authors studied denervation changes of aldolase (zymohexase), the enzyme responsible for the breakdown and synthesis of hexosodiphosphate. This enzyme is part of myogen A. By determining activity of the enzyme and the amount of protein in the appropriate electrophoretic fraction, these authors found that, as in total myogen, the decrease does not begin until the tenth day after denervation. After this delay, however, changes set in rapidly, so that 30 days after nerve section the aldolase content is 55% lower (Fig. 103).

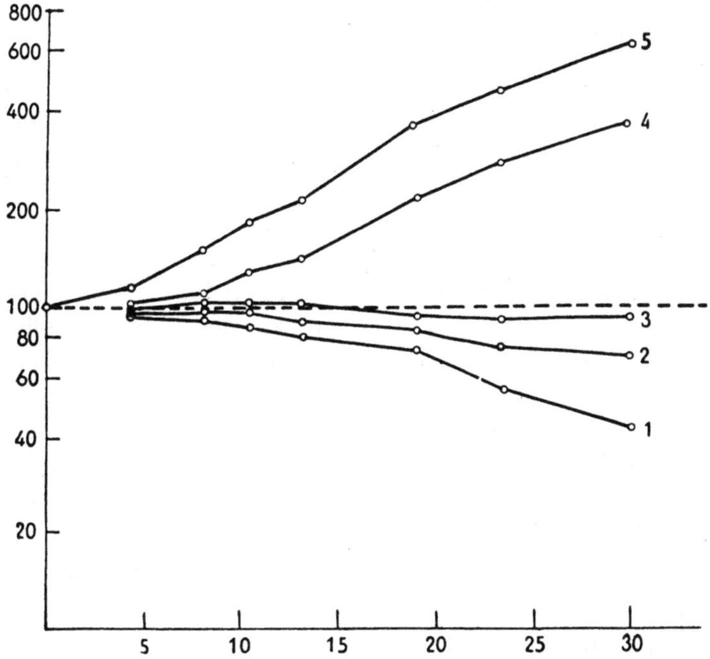

Fig. 103. Quantitative changes in myogen and myoalbumin in the denervated muscle. Ordinate: percentage. Abscissa: time after denervation (days). 1 — aldolase, 2 — fraction II, 3 — fraction III, 4 — unidentified fraction (perhaps the slowest subfraction of myoalbumin), 5 — myoalbumin. Changes are expressed as percentage of corresponding values of normal muscle. (After Fischer 1955.)

As far as differences between various types of muscle atrophy are concerned, it is important to note that immobilization does not lead to a decrease of aldolase, while changes after tendon section appear approximately two days earlier and are more pronounced than those after denervation (Fischer 1955). This, of course, applies to the first period after section of the nerve only. At later stages of atrophy Haan (1953) found the same change in the electrophoretogram after denervation as after immobilization, i. e. decrease of the slow fraction which corresponds to the migration properties of aldolase.

In the non-structural protein spectrum a decrease of the fraction *l* and *k* (Crepax 1949a) and fraction II and III (Fischer et al. 1949, Fischer 1950) was furthermore described after denervation. Judging from their respective migration velocities, the fraction *l* and II, and *k* and III are identical and correspond to further enzyme systems of myogen, evidently to phosphoglucomutase and phosphorylase. The general character of changes in these fractions is the same as for aldolase, i. e. changes first appear with a latency of ten days. On quantitative comparison, changes of fraction II and III are smaller in extent than those found for aldolase: a 20% decrease after 30 days of denervation, while aldolase is decreased by 55% (Fig. 103).

Myoalbumin (Tab. 15).

Myoalbumin represents 7% of sarcoplasmatic proteins and belongs to the fastest electrophoretic fractions of the non-structural proteins. It has already been mentioned above that Haan found a percentage rise of the fast fraction not only after denervation but also after immobilization and in avitaminosis-E. This corresponds to the increase of myoalbumin demonstrated by other authors with various methods (Crepax 1949a, 1949b, Fischer et al. 1949, Gligliotti and Lamedica 1956). According to Fischer et al. myoalbumin rises very soon after section of the nerve: a 20% increase was already found five days after denervation. With progressing atrophy the rate of this increment increases, so that 30 days after nerve section it attains 600% (Fig. 103). The onset of these changes as reported by Crepax was later — as late as on the sixteenth day after denervation. As far as the extent of these changes is concerned, however, the results of different authors are in agreement.

As in the case of aldolase, there were smaller changes in myoalbumin after immobilization during the early stage of atrophy and greater changes with earlier onset after section of the tendon (Fischer 1950).

Cytochrome C.

Approximately a 40% decrease in the content of cytochrome C was found in rabbit gastrocnemius 40 days after denervation by Schmidt and Schlief (1956). Changes in the content of cytochrome C are discussed in more detail in Chapter VIII, p.: 222, in connection with its role in biological oxidation systems.

Myoglobin.

Changes of myoglobin in denervated muscle have been shown at later stages of muscle atrophy: a 50% loss was described seven weeks after section of the sciatic nerve of the dog as reported by Whipple (1926). These results were later confirmed by Verbolovich and Aleksandrovich (1950). The early denervation period was studied by Ginetsinsky et al. (1945), but no changes were found. Dreyfus et al. (1959) found three fractions of myoglobin differing in electrophoretic modilities. The slow fraction (most abundant in the muscle) decreases after denervation, while the second and third fraction increases.

As in the case of structural proteins, the values of sarcoplasmatic proteins are relative. They express changes in the protein muscle composition only, but not absolute increments or decrements. Only Hetényi (1956) gives data about the absolute decrease of sarcoplasmatic proteins. The electrophoretic fractions II and III, corresponding evidently to proteins of the myogen group, were decreased by 50% in rat muscles 22 days after denervation and in rabbit muscles 4—7 weeks after denervation.

c) Interstitial muscle phase

The complex of proteins contained in the interstitial muscle phase is known as the stroma. It contains sarcolemmal proteins, insoluble parts of nuclei and cytoplasmatic granules, remnants of myofibrils (2—4%) and proteins of connective tissue and of blood vessels. The stroma represents 20% of proteins in pale muscle and 27% in red muscle. Collagen is the main stromaprotein. An 50% increase in stroma protein content of the denervated muscle is reported by Helander (1957). Tenotomy and immobilization led to only half such a change.

Collagen (20% of total muscle protein).

Increase in the content of collagen is one of the most pronounced changes accompanying muscle atrophy. A relative increase of collagen in denervated muscle was found chemically already five days after section of the nerve, when it attains 5% (Fischer and Ramsey 1946). Its content in the muscle increases linearly with progressing atrophy, so that 30 days after nerve section it is already increased by 100% (Fig. 99). It is interesting that the rise in collagen is smaller than would correspond to the theoretical value calculated on the assumption that no changes in collagen had taken place. This is considered by Fischer and Ramsay (1946) as indirect evidence that following denervation there is an absolute decrease of collagen.

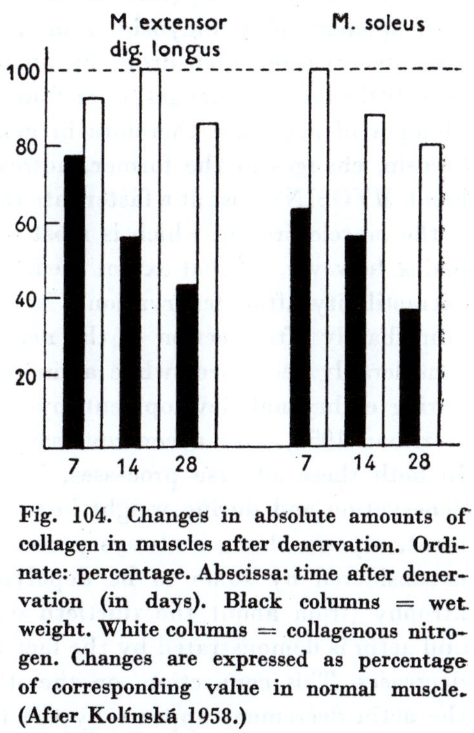

Fig. 104. Changes in absolute amounts of collagen in muscles after denervation. Ordinate: percentage. Abscissa: time after denervation (in days). Black columns = wet weight. White columns = collagenous nitrogen. Changes are expressed as percentage of corresponding value in normal muscle. (After Kolínská 1958.)

This was actually proved by direct experiment in which the content of collagen was expressed per muscle. These changes were investigated by Kolínská (1958), 7, 14 and 28 days after denervation. In the soleus the decrease was already about 10% 14 days after denervation, and 15—20% on the 28th day. However, in the extensor digitorum longus a 10% decrease did not appear until the 28th day of denervation (Fig. 104).

From among the various types of atrophy, avitaminosis-E causes the same relative increase of collagen as denervation (Spencer et al. 1937, Bender et al. 1959), while changes in tenotomy are 10% smaller and in immobilization 16% smaller (Fischer and Ramsey 1946). It can be concluded from the discrepancy between the theoretical and actual increase that there occurs an absolute loss of collagen in tenotomy as well as after immobilization (Fischer and Ramsey 1946).

What conclusions can be drawn about the general changes of denervated muscle proteins from the above review? The percentage composition of denervated muscle changes in the sense that specific proteins as a whole decrease, while non-specific (sarcoplasmatic) proteins increase. After denervation the main structural protein ACTOMYOSIN decreases very slowly at first (during 8—10 days). Then suddenly a rapid decrease in its content occurs and during a short period (5—8 days) almost 50% of this content is lost from the muscle. In the course of the next period the rate of this loss gradually slows down. ACTIN, a component of actomyosin, remains completely unchanged for 8—10 days after denervation, then drops by about 20% in the course of five days. Later very little further changes were found (a decrease of about 10% in 15 days). Changes of actin are therefore in general outline the same as those of actomyosin, changes in the former, however, are somewhat smaller. On the other hand, MYOSIN is lost at a faster rate than actomyosin. It even seems that myosin is the muscle protein which is most affected by denervation. (The question remains, however, to what extent the loss of myosin is associated with its decreased extractibility after denervation). In contrast to actin, myosin changes appear immediately after section of the nerve, so that five days later its content is considerably decreased, while actin remains unchanged. It is interesting that during embryonal development myosin appears earlier than actin (Csapo and Herrman 1951). But a controversary result were also reported (Ogata 1960). In both these adverse processes, i. e. during loss of muscle substance after denervation and during weight increment in the course of embryonal development, myosin changes always precede changes of actin. Y-PROTEIN, which is considered by some to be depolymerized actin, increases at later stages of atrophy (from about the thirtieth day). The relationship between Y-protein and actin is demonstrated by the fact that while the former increases, the latter decreases. This connection, on the other hand, is contradicted by the fact that the actin decrement appears by the tenth day, while the Y-protein increment does not appear until the thirtieth day.

Sarcoplasmatic proteins as a whole are decreased after denervation. Individual proteins change in such a way that highly specialized proteins (carriers of enzymatic activities) are lost, while myoalbumin, a protein without a specific function, rises. This increment of myoalbumin is smaller than the loss of specialized proteins, so that on the whole, the content of non-structural proteins decreases.

These changes are demonstrated on the electrophoreogram as percentage increase of fast and decrease of slower fractions. From the identified proteins a characteristic course of changes occurs in enzymes of the MYOGEN group: following section of the nerve their content in the muscle remains unchanged for about ten days. After this latency, however, a rapid decrease follows: the content of aldolase drops by 50% (myogen A) between the fifth and eight day, while Fischer's fraction II and III (which are identical with the fraction *l* and *k* of Crepax and correspond most probably to phosphoglucomutase and phosphorylase respectively) decrease less — only by about 20%. In the following period a general slowing down of enzyme loss takes place. The latency of denervation changes described is analogous to that of actin. A more detailed explanation of this phenomenon has not been given. A further non-structural protein. Myoalbumin, is increased after denervation. The percentage increase of myoalbumin appears practically immediately after section of the nerve and rapidly rises linearly up to the thirtieth day, when it attains a 600% increase. It is thus the most conspicuous change in the denervated muscle.

According to the interpretation of changes in non-structural proteins, denervation leads to the deterioration of highly specialized proteins of the myogen group. The breakdown products of these proteins then serve as basis for the rise of less specialized myoalbumin (whether this change requires complete reconstruction of the molecule, or only certain structural transformation). Since Hetényi (1956) did not find an absolute increase in the fraction corresponding to the migration velocity to myoalbumin, this hypothesis will require further experimental corroboration. Phylogenetic and ontogenetic development is accompanied by quite opposite changes. While the contractile capacity of the muscle increases, myoalbumin decreases and proteins of the myogen group increase. This was found, for example, by Renards (1952) in a phylogenetic line starting from fish and terminating with mammals. Similarly in one animal species myoalbumin gradually decreases and the content of myogen increases in muscles with increasing speed of contraction in the sequence: fetal muscle, uterus, myocardium, red muscle, white muscle (Crepax 1952b). The same interrelations have been found between smooth muscles when their contractile properties were compared with their content of myogen and myoalbumin, e. g. uterus at various stages of pregnancy (Crepax 1951a). It can thus be seen that the content of myoalbumin decreases in muscles in proportion to their functional differentiation. From this point of view denervation represents a dedifferentiation of proteins with their return to lower evolutionary forms.

The content of COLLAGEN in denervated muscle increases relatively, but absolutely decreases. Collagen loss in 14 days represents about 10%.

Which of these changes accompanying muscle atrophy are characteristic only for section of the nerve? Atrophy due to AVITAMINOSIS-E is accompanied by the same changes as denervation as far as myosin, actomyosin and Y-protein are concerned. The main difference lies in the fact that neither loss of actin nor decreased non-structural proteins were found in avitaminosis-E. Of course, the electrophoretic pattern of non-structural proteins does not differ from denervation. The loss of actomyosin after tenotomy is almost the same as after denervation, while changes in actin, as in avitaminosis-E, are inconspicuous. A marked difference from denervation was observed in that the decrease of myosin and myogen as well as the increase of myoalbumin appear sooner in tenotomy than in denervation, and that these changes are more extensive. At an advanced stage of atrophy non-structural proteins in general are slightly decreased in the muscle after tenotomy, while in denervation their content decreases by 20 %. Immobilization, on the other hand, does not, at first, cause any changes in myogen and myoalbumin. At late stages of atrophy, however, similar changes have been found as after denervation. After immobilization actomyosin decreases 80 % less than after denervation. It can thus be seen that decreased content of sarcoplasmatic proteins (about the validity of this finding see p. 287) and the loss of actin (changes after immobilization have, of course, not been studied) are changes typical for denervation.

3. Denervation hypertrophy

While investigating the time gradient of denervation changes in the rat diaphragm, Martin, Sola and Stewart found that weight loss was preceded by a period during which wet and dry weight of the denervated hemidiaphragm increased (Martin and Sola 1948, Sola and Martin 1953). Later it was demonstrated that this denervation hypertrophy includes sarcoplasmatic and structural proteins. Connective tissue proteins, however, remain unchanged. From the fourth day onwards both groups of proteins increase in the denervated hemidiaphragm to reach a maximum seven days after denervation: sarcoplasmatic proteins are increased by 36 %, structural proteins by 27 % (expressed in percentage of total proteins). Both these values return to normal between the 15th to 30th day and 40 days after denervation muscle atrophy is already under way and represents a loss of 45 % of sarcoplasmatic and 54 % of structural proteins (Stewart 1955) — (Fig. 105).

In a subsequent report hypertrophy was also demonstrated in the mouse and rabbit (Stewart and Martin 1956). Maximum weight attained in the denervated hemidiaphragm was found five to eight days after denervation in the mouse, after seven days in the rat and after ten days in the rabbit. While the time necessary for attaining the maximum weight differs in various animal species, the maximum increase is always the same and equals approximately 30%. During the following period initial weight of the denervated hemidiaphragm is reached in all the above mentioned species at the same period, i. e. 20 days after the operation. It thus seems that the faster the rate at which the maximum is attained, the slower the subsequent weight loss. This

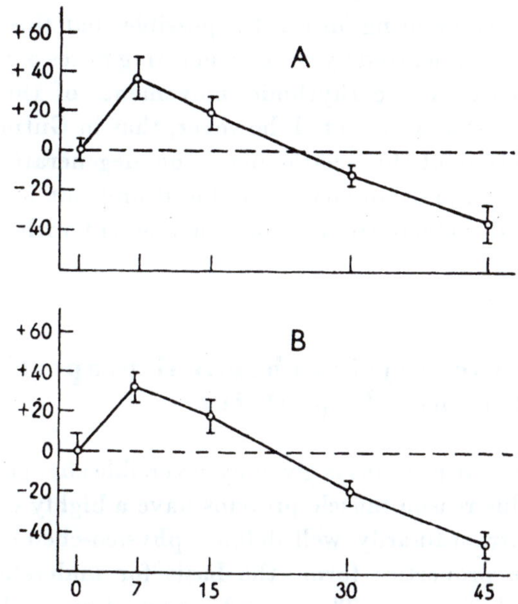

Fig. 105. Denervation hypertrophy of the diaphragm. Ordinate: percentage. Abscissa: time after denervation (in days). A — sarcoplasmatic proteins, B — structural proteins. Changes are expressed as percentage of the theoretical initial value. (After Stewart 1955).

is also evident in the period after the 20th day, when the weight of the denervated hemidiaphragm is lower than that of the control side: in order to obtain 40—50% loss of weight, this lasts 64 days in the mouse (where the maximum is attained sooner), but only 40 days in the rat.

However, Thomson et al. (1951) and Thomson (1955) were able to find only increased wet weight in the denervated diaphragm. According to these authors this hypertrophy is due to tissue edema only.

Up to now denervation hypertrophy has been considered to be a special reaction of the diaphragm to nerve section. Gutmann (1960), however, has succeeded in demonstrating this phenomenon in skeletal muscle. The maximum increase of dry weight and non-collagenous proteins is attained considerably earlier than in the diaphragm. Weight of the anterior tibial muscle four hours after section of the nerve is about 10% higher than in the control muscle. This increase persists for 24 hours, but there is evident a tendency to a quick return to normal values. It thus appears that the transient increase of weight in the denervated muscle is a general phenomenon and occurs in different muscles at different time intervals after section of the nerve. Stewart and Martin (1956) interpret their own results with the effect of neurocletin. This substance is liberated by degenerating distal nerve fibres and stimulates the growth of intact axons to branch to degenerating fibres. It is possible that this substance or some similar material can be liberated by the degenerating fibres of the hemidiaphragm and can, together with passive rhythmic movements of the diaphragm cause this hypertrophy. It should be noted, however, that in Gutmann's experiments four hours after section of the sciatic nerve no degeneration takes place and passive movements of the extremity are reduced and not increased. The hypothesis of Stewart and collaborators is not in agreement with this finding.

B. Changes in the physico-chemical properties of denervated muscle proteins

The function of muscle proteins is given by reversible changes of their molecular arrangement. For this reason muscle proteins have a highly crystalline structure which results in extraordinarily well defined physico-chemical properties. The knowledge of these properties forms the basis for understanding all changes accompanying muscle function. The second aspect of molecular transformations accompanying muscle contraction is that the properties of muscle proteins depend on the functional state of the muscle. This causes great methodical difficulties especially while studying the physiological properties of the muscle. The first report dealing with physico-chemical characteristics of muscle proteins after denervation is therefore associated with methodical difficulties. Chor et al. (1937) reported that when determining the dry muscle weight, acetone extracts more water from denervated muscle proteins than from normal. Since then extensive evidence has accumulated demonstrating changes in physico-chemical properties of proteins, especially after functional activity (see section C, p. 311), avitaminosis-E and denervation.

1. Solubility and extractionability

Solubility is the fundamental property expressing the state of individual muscle proteins inside the cell. The effect of denervation on myosin solubility was studied by Fischer and Ramsey (1944). According to these authors the solubility of myosin is considerably decreased in rabbit gastrocnemius 28 days after denervation. The solubility or extrability of myosin from muscle brei is dependent on pH with maximum extrability in the normal muscle at pH 6·9. Solubility is smaller at lower as well as higher pH (Fig. 106). On the other hand, the extractability of proteins in denervated muscle is not affected by pH. This means that myosin of denervated muscle loses its sharply defined isoelectric point. It is interesting that at the beginning of denervation atrophy — on the fifth day after nerve section — the changes are reversed: the extractability of myosin is increased in the whole range of pH studied with the exception of pH 6·9 when it is lowered. This pH corresponds to the isoelectric point of myosin in the normal muscle. It can thus be assumed that atrophic changes in solubility begin to be exhibited from this critical pH value. Immobilization (Fischer and Ramsey 1944) and avitaminosis-E (Azzone and Aloisi 1955) have also been found to decrease solubility. Immobilization evokes smaller changes than denervation at a stage of atrophy when losses of weight are the same.

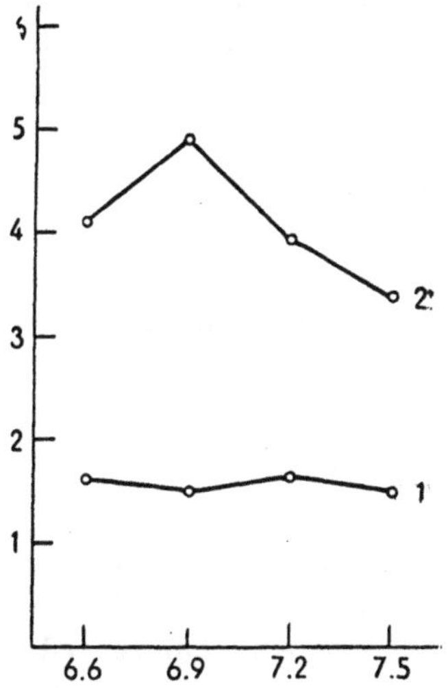

Fig. 106. The influence of denervation on extractibility of myosin. Ordinate: g. myosin/100 g. muscle. Abscissa: pH. 1 — denervated muscle, 2 — normal muscle. Myosin was extracted from the muscle with 0·5M LiCl and then fractioned by precipitation at a given pH. (After Fischer and Ramsey 1944.)

The capacity of myosin and actomyosin to precipitate from the extract is decreased after denervation (Fischer 1948). It seems that changes in actomyosin are greater than in myosin itself. With progressing atrophy from the fifth till the thirtieth day this lowering of precipitation becomes more evident. At the same time the density of the precipitate is greater in the myosin from the denervated muscle (Fischer and Ramsey 1944).

This is, of course, true only at a late stage of atrophy. Five days after denervation the reverse is true. The contradictory changes in the initial and late stages of atrophy are thus analogous to changes in solubility. Immobilization of five and also thirty days' duration leads to higher density of the precipitate, i. e. to changes analogous to late stages of denervation atrophy (Fischer and Ramsey 1944). Explanation of these changes has not been given.

2. Double refraction of flow (DRF)

As far as the optical properties are concerned, double refraction of flow characterizing the shape of the protein molecule, especially of fibrous proteins, has been studied after denervation. The main credit for this is due to Schapira and Dreyfus. According to these authors the length of the most frequent particle l_f which is 2850 Å in myosin of the normal muscle diminishes, so that the ratio between l_f of myosin in the atrophic and normal muscle is 0·94. The mean afferent anisotropy of myosin particles, on the orther hand, increases after denervation as could be calculated from DRF intensity. It is important to note that neither of these changes is dependent on the time of nerve section. They have been found already seven minutes after section of the nerve in rabbit gastrocnemius and remained without further changes 1½, 2, 5, 48, 56 hours and 10—30 days after denervation (Schapira et al. 1952, Dreyfus et al. 1953). Similar changes were described in myosin isolated from atrophic muscles in avitaminosis-E (Aloisi et al. 1953). This can be deduced from the fact that DRF intensity of the myosin solution in normal muscle increases after adding KCl, while the DRF of dystrophic muscles decreases under the same conditions. The above authors explain these findings by lost ability of myosin in the dystrophic muscle to form K-myosinates after addition of KCl.

3. Viscosity

Further changes of fibrous proteins in the denervated muscle have been demonstrated by Fischer (1955), according to whom the viscosity of actomyosin isolated from denervated muscle has lower values than found for normal muscle. At the same time there is a smaller drop of viscosity of actomyosin from denervated muscle when ATP is added. In these experiments, the ratio between actin and myosin in the actomyosin molecules is the same in the normal and denervated muscle. On the contrary it was found that the content of myosin decreases at a faster rate than actin, so that it would be expected that the ratio actin: myosin in the actomyosin complex undergoes a change (Fischer 1948). However, Fischer (1955) himself, having described this finding, states that the change can be due to a larger decrease in the extractibility or precipitating properties of myosin in the course of denervation than that of actin. Thus a larger decrease of myosin is found than is actually the case and the ratio actin-myosin remains unchanged. It is interesting that the viscosity of actomyosin in embryonal muscles or in new-born rats and mice is lowered by adding ATP to a much lesser degree than in adult animals (guinea-pigs are born with actomyosin corresponding to adult animals — Ivanov and Kasavina 1948). Denervation thus also represents a regression to a lower evolutionary stage from this aspect.

What is the basis of these viscosity changes? It seems that the ability of myosin and actin in the atrophic muscle to polymerize is lowered. However, experimental material was mainly obtained in dystrophic muscles of animals with avitaminosis-E. As has already been mentioned, KCl does not increase the myosin DRF in dystrophic muscles, since the ability to form K-myosinates is reduced

(Aloisi et al. 1953). Similarly, actin has a diminished capacity for linear association during the change of G-actin to the F-form (Bonetti et al. 1952) and there is also a reduction in the capacity of myosin to interact with actin to form actomyosin (Aloisi et al. 1953). This seems to be different from denervation changes; Crepax (1953a, 1953b) reported a conspicuously faster actomyosin formation 30 days after denervation. This difference between dystrophic and denervated muscles is especially interesting because viscosity changes of actomyosin are similar in both cases.

4. Changes under electron microscope

In the preceding section changes in the ability of myosin from dystrophic muscles to polymerize have been described. How are the changes manifested under electron microscope? Actin-F isolated from normal muscles appears as single filaments with a tendency to formation of aggregates by side-by-side associations. Individual filaments then appear as linear series of small globular elements lying end- to-end. This capacity for linear association of these globulins is diminished in the dystrophic muscle and especially the side-by-side aggregation of individual filaments is substantially reduced.

Actin-G can be observed as irregular globules of different sizes. The addition of KCl or LiCl results in the linear aggregation of small globules, and filaments formed by this aggregation exhibit a tendency to lateral aggregation. In dystrophic muscles the globules of actin-G are larger and their structure is coarser than usually observed in normal muscles (Aloisi et al. 1952).

The general symptom of these protein molecular changes in the atrophic muscle is the decrease of birefringence of the whole muscle. Contractile strength (tension per gram muscle weight) decreases in parallel with the decrease of birefringence and this is accompanied by thinning of muscle fibres (Fischer 1940). In constrast to this, no changes of birefrigence or tension have been found after tenotomy or immobilization.

5. Enzymatic properties

Changes of physico-chemical properties of myosin must necessarily become manifested in its enzymatic properties. As far as is known, two fractions have been separated, myosin β_1 and β_2 with the aid of electrophoresis. The former is the carrier of ATPase activity, the latter of the deaminase of adenylic acid. Myosin isolated from the denervated muscle has been found to have decreased ATPase activity by 30% from the fifth denervation day onwards (expressed per mg. of protein) — (Fischer 1948). ATPase activity progressively decrease with denervation, until it is reduced by 90% 35 days after denervation (Fig. 107). The relation of enzymatic activity to pH and $CaCl_2$ concentration remains unchanged after denervation. Extremely important is the fact that with progressing atrophy more ATPase remains in the solution following myosin precipitation. While in the normal muscle soluble ATPase is to be found in the supernatant only after the first precipitation of the rough extract, in the denervated muscle ATPase remains in the supernatant even after subsequent precipitation (Fischer 1950,

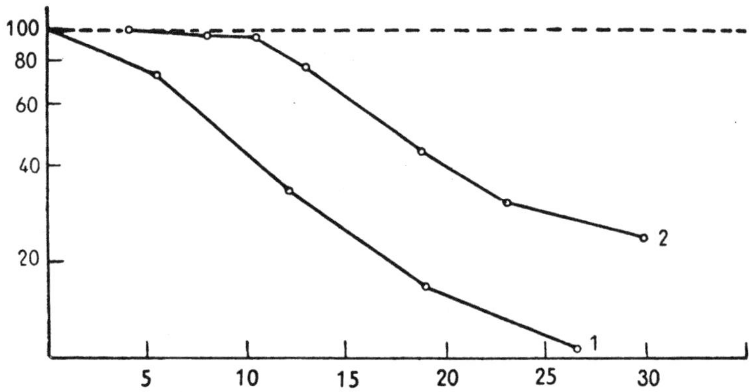

Fig. 107. Changes in enzymatic properties of muscle proteins. Ordinate: percentage. Abscissa: time after denervation (in days). 1 — ATPase, 2 — aldolase. Changes are expressed as percentage of corresponding values in normal muscle. (After Fischer 1950.)

1955). With progressing atrophy, there is a decrease of ATPase firmly bound to myosin, which is accompanied by a corresponding increase in soluble ATPase.

A contradictory finding has been reported by Beznák (1949), according to whom ATPase activity is increased after denervation. It is probable, however, that, as has been pointed out by Fischer (1950), this author only determined soluble ATPase or ATPase weakly bound to myosin. Muscle brei was not extracted in her experiments by solution of sufficient ionic strength necessary for extracting all of myosin. Her results are thus in agreement with Fischer's findings.

It is interesting that during embryonal development the ratio between ATPase bound to myosin and total ATPase rises with age of the embryo (Herrman et al. 1949). Similarly in invertebrates only a small part of ATPase is firmly bound to myosin (Humphrey 1949).

This means then that during denervation atrophy the ATPase system changes in the reverse direction as compared with development and evolution — i. e. returns to a lower evolutionary stage. Similar results as in the case of ATPase have been described by Nechiporenko and Ferdman (1953) for the adenylic acid deaminase. With progressing immobilization atrophy the linkage between enzyme and myosin weakens — the amount of deaminase which can be separated from myosin increases.

Besides enzymatic changes in myosin, changes have also been described in other proteins. For example, the percentage content of aldolase in denervated muscle extract decreases, as has already been cited above on p. 289. What is more important, however, is the fact that aldolase activity is also reduced when calculated per mg. of aldolase protein. Besides loss of the enzyme itself, there evidently also occurs a change in the molecular arrangement resulting in a decrease of the specific enzyme activity. The character of these changes is similar to the changes in aldolase content. This means that reduction in activity appears with a latency of 10—12 days (Fig. 107). After this period a reduction of specific activity as well as the content of the enzyme is observed, the former being lowered at a faster rate, so that 30 days after denervation the specific aldolase activity is reduced to one quarter of its normal value, while its content has been reduced by only 55% (Fig. 103 and 107).

It can thus be concluded that following denervation physicochemical properties of myosin undergo changes which are manifested by the loss of a sharply defined isoelectric pH value and by decreased solubility. The shape of the molecule changes in such a way that the viscosity of the actomyosin solution decreases and the decrement of viscosity after addition of ATP is smaller in the "denervated" actomyosin as compared with the "normal". It can be concluded from DRF studies that the length of myosin particles shortens after denervation and that anisotropy increases. The bond of ATPase and adenylic acid deaminase with myosin becomes weaker with progress of atrophy. All these changes indicate that denervation atrophy changes myosin properties in such a way that it forms a more stable complex with actin and actomyosin thus formed exhibits a reduced capacity for contraction. Linkage between the protein skeleton of myosin and active centres of ATPase and deaminase molecules becomes less firm. A number of denervation changes (such as viscosity and ATPase bonding with myosin) are contrary to changes occuring during evolution and development. Similarly as was the case with differentiated and non-differentiated proteins (see p. 293), physico-chemical properties after denervation also represent a regression to a lower evolutionary stage.

What is the difference between various types of atrophy? The decrease of myosin solubility also appears after immobilization and avitaminosis-E. Changes after immobilization are often smaller than after denervation. Changes in the molecular shape are evidently the same after denervation and avitaminosis-E. The formation of actomyosin, however, is unaffected in the denervated muscle or is at least less marked than in dystrophic muscles. The difference between denervation and tenotomy was observed when birefringence was studied in the whole muscle: tenotomy does not cause a reduction in birefringence.

C. Disturbances of protein metabolism in the denervated muscle

The loss of proteins in the atrophic muscle can be the result of two processes: 1. autolysis of muscle tissue, 2. change in the dynamic equilibrium between protein breakdown and synthesis.

The first possibility — the *in vivo* autolysis of muscle tissue-is improbable from the biochemical point of view. Autolysis is disintegration of the tissue (and generally of the whole organism) which begins post mortem. It is not certain whether a biochemicaly similar process can occur in the living cell. In the second case, the changed balance between degradation of proteins and their synthesis may result in muscle weight loss. This result can be due either to increased degradation which then predominates over synthesis, or by a decrease of synthesis which is then dominated by degradation which remains unaltered.

Under normal conditions degradation of proteins is balanced by their synthesis. Both these simultaneous processes form the basis of protein turnover. However, the question of turnover of animal proteins was lately subjected to criticism. Hogness et al. (see Žák 1959) came to the conclusion based on investigation of proteosynthesis in bacteria that proteins, once synthetized, are not subjected to degradation and for this reason the turnover is not measurable. Their results on bacteria were confirmed many times by various experimental approaches (see Žák 1959). Earlier work done on animal material and proving the turnover of proteins is criticized by Hogness et al. in the sense that it expresses the dynamic state of tissues rather than of protein molecules within the cell. The dynamic state of a tissue is given by death and lysis of cells (including all their components) together with secretion of specific proteins by the cells. In experiments performed on bacteria it is possible to work with a homogenous system of cells, which neither die nor secret any proteins. There is thus neither replacement of dead cells nor of proteins secreted. Consequently no turnover of proteins can be found. This criticism led to investigations of the turnover of animals tissues under new experimental conditions. In spite of this, however, turnover of proteins was found even in homogenous systems of animal cells. One of the numerous reports demonstrating the turnover of these tissues was by Heimberg and Velick (1954) and Simpson (1955) in the muscle. These authors found that various enzymes (e. g. aldolase, dehydrogenase of 3-phosphoglyceral-aldehyde and phosphorylase) have a different turnover in the muscle. Since these enzymes are present in one cell, the above result serves as evidence that the turnover is directly associated with the protein molecule and is not given by cellular turnover (living cycle). In that case the turnover of enzymes present in one cell would have to be the same, since it would be the result of the life cycle of the cell.

If protein turnover does exist in animal tissues, then theoretically it is plausible to assume that loss of muscle substance in atrophying muscles is caused by a change in the balance between degradation of proteins and proteosynthesis.

1. Synthesis and degradation of proteins in the normal muscle

It has been indicated several times in the literature that both synthesis and degradation of proteins in the muscle take place by a different mechanism than in other tissues. Mihalyi et al. (1957) have found, for example, that muscle RNA is bound almost exclusively to myosin. The incorporation of labelled amino acids

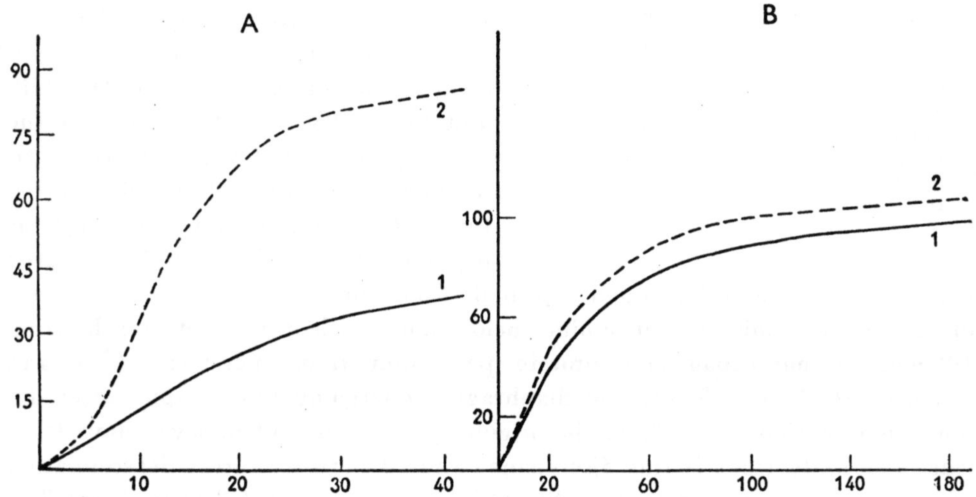

Fig. 108. Course of incorporation of labelled amino acids into mitochondria and microsomata of muscle and liver. A — liver. Ordinate: cpm/mg. proteins. Abscissa: min. 1 — mitochondria, 2 — microsomata. C^{14}-alanine was used (see Siekevitz 1952).
B — muscle. Ordinate: standard specific activity. Abscissa: min. 1 — mitochondria, 2 — microsomata. C^{14}-leucine was used (see Simpson and McLean 1955).

into myosin, however, is one of the lowest as compared with other muscle proteins. It can be seen from comparison in Tab. 17 that the turnover of myosin is 30% lower than the average turnover of all muscle proteins. In contrast to this, it is known that in other tissues a positive correlation exists between the level of RNA and the rate of incorporation of labelled amino acids. The time course of incorporation of amino acids into mitochondria and microsomata is also different in the muscle. In the liver, the incorporation of labelled amino acids precedes and quantitatively surpasses the incorporation into mitochondria. This difference, however, has not been found in the muscle (Fig. 108) — (Simpson and McLean 1955, McLean et al. 1958).

a) Proteosynthesis

What do we know about proteosynthesis in the muscle? The incorporation of labelled amino acids into muscle is, with the exeption of the brain, one of the lowest in the body (Tab. 16). Nevertheless, it was found by determining the net protein increase under certain conditions that muscle has the ability of proteosynthesis of the same order of magnitude as tissues producing proteins in the course of secretory function. It was reported, for example, that one gram of pancreatic tissue synthetizes 3—5 mg. of amylase per hour (Rychlík and Šorm 1957). In the course of the recovery phase following electrical stimulation one gram of muscle synthetizes 2·9 mg. of non-collagenous proteins per hour (Žák et al. 1957). The isolated diaphragm is capable, under optimal conditions (i. e. in the presence of a complete mixture of amino acids and nucleotides), of synthetizing 1·9 mg. of non-collagenous protein per gram tissue per hour (Žák and Klicpera 1958) and finally the diaphragm undergoing hypertrophy after denervation synthetizes 30 % of the original protein content in seven days (Stewart 1955). By what is this discrepancy caused between direct determination of proteins and incorporation studies of labelled amino acids? There is qualitative agreement between both methods: changes in incorporation are accompanied by confirmatory changes in the net protein increase or decrease. For example, addition of a complete mixture of amino acids into the medium in which diaphragm of the rat is incubated causes an increment of 1·9 mg. of non-collagenous protein. The incorporation of S^{35}-methionine is simultaneously increased by 50 %. The reversion of proteosynthesis to degradation (caused by administration of p-fluorophenylalanine, omission of an essential amino acid or by high concentration of amino acids) results in a reduced incorporation (see Fig. 109) — (Žák and Babický 1958). Quantitative evaluation of these relations has not been given. It is not yet possible to explain how the muscle is able to synthetize a marked amount of proteins although turnover is low (it must, however, be pointed out that incorporation was not measured under the above conditions of increased proteosynthesis). Very important is the work of Schapira and Dreyfus and their coworkers (Schapira et al. 1959, Dreyfus et al. 1960). According to their results the inertness of muscle structural proteins is only relative. The specific activity of free labeled amino acids stays high enough to maintain the

Table 16

Incorporation of S^{35} menthionine into different tissues of the rat.
(according to Niklas et al. 1958).

Tissue	Half-time (days)
Skeletal muscle	22·2
Brain	8·9
Heart	8·2
Liver	2·8
Spleen	2·6
Stomach	2·4
Lungs	2·4
Kidney	2·1
Intestine	1·8

radioactivity of myosin on constant level. If the specific activity was diluted by feeding a protein-rich diet a molecular turnover of myosin was seen.

It is interesting to compare the course of incorporation of labelled amino acids into tissues with a high turnover such as, for example, the liver, with muscle tissue. It is evident from Fig. 110 that the rates of growth of the specific tissue activity is the same in both cases. Thus the rate of incorporation is also

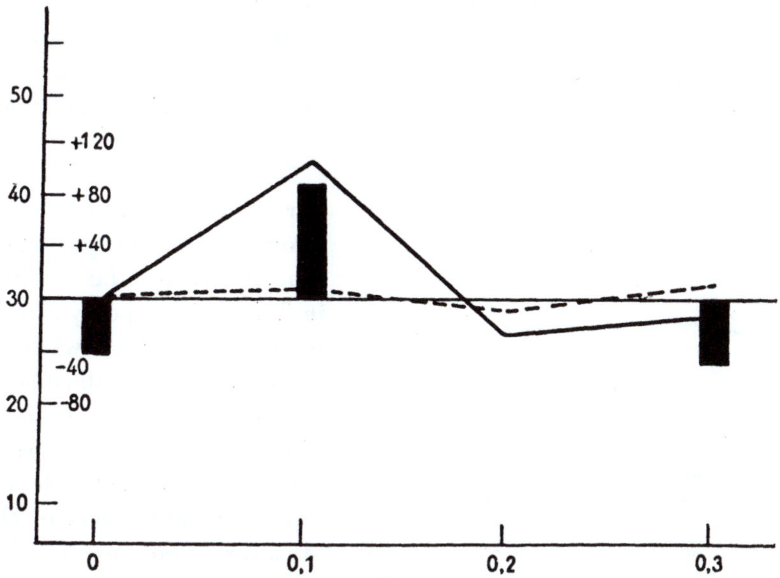

Fig. 109. Incorporation of S^{35}-methionine into isolated rat diaphragm under conditions of inhibited or enhanced proteosynthesis. Ordinate: to the left (i. e. 10—50) = specific activity; to the right (i. e. —80 to +120) = non-collagenous proteins N. Abscissa: concentration of amino acids (in %). Columns: non-collagenous proteins (decrease or increase during three hours incubation in Krebs-Ringer bicarbonate medium containing complete mixture of amino acids in the appropriate concentration). Solid line: incorporation of S^{35}-methionine during three hours' incubation in Krebs-Ringer bicarbonate containing complete mixture of amino acids in the appropriate concentration. Interrupted line: the same as expressed by solid line, tryptophane, however, was missing in the mixture (after Žák and Klicpera 1958).

the same. However, incorporation in muscle tissue stops at a final value which is lower, by an order of magnitude, than in the liver. How can one explain this initial rapid growth in the muscle specific activity? It was shown *in vitro* on the isolated rat diaphragm (Žák et al. 1959) that there exist two fractions of incorporated radioactivity. One fraction is labile and its radioactivity can be "washed out" by incubation with a non-labelled amino acid. The second fraction corresponds to firmly bound radioactivity. During the course of further incubation the second fraction increases at the expense of the first. As a result of this a transient radioactivity quickly reaches high values and then again drops. The sum of both radioactivities then gives a curve with a plateau characteristic for that found in vivo. It is possible that transient radioactivity corresponds to a protein precursor as has also been found in other cellular systems. A typical precursor product curve was also observed by Winnick and Winnick (1960) when comparing on one hand the incorporation of labeled amino acids into mitochondria and microsomes and soluble and myofibrilar proteins on the other hand. It is possible that such a protein precursor is responsible for the rapid rise in the specific tissue activity, as it is found in the muscle *in vivo* during the first period after injection of the labelled amino acid.

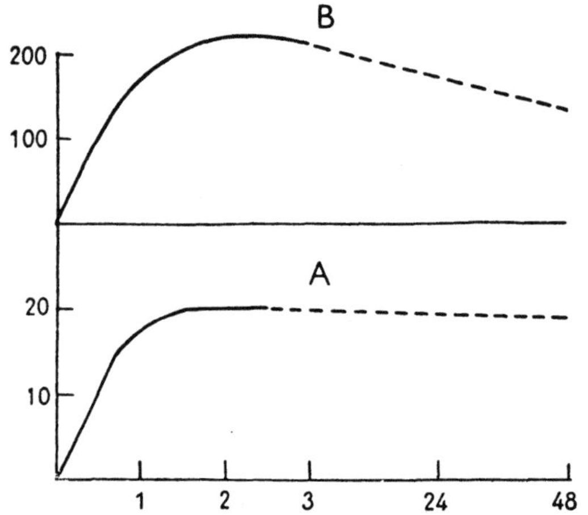

Fig. 110. Course of incorporation of S^{35}-methionine into muscle and liver. Ordinate: specific activity. Abscissa: time after intravenous injection of S^{35}-methionine (in hours). A — muscle (m. tib. ant. of the rat), B — rat liver. (After Žák 1959.)

What is known about the turnover of individual muscle proteins? From among the proteins in the muscle phase so far studied given in Tab. 17, tropomyosin has the highest turnover, while the dehydrogenase of 3-phosphoglyceraldehyde has the lowest turnover (Velick 1956). Proteins of the myogen group as whole are metabolically more active than the structural proteins (Ussing 1941). This is evidently due to the high activity of aldolase (Dreyfus et al. 1956). From the soluble proteins the following sequence of specific activities was found, phosphorylase, phosphofructokinase, dehydrogenase of d-glyceraldehyde (Dvornikova and Grigoryeva 1960). Molecular turnover of aldolase was studied by Schapira et al. (1960). As far as the structural proteins are concerned, it is interesting that the incorporation of labelled amino acids into actomyosin is faster than that into myosin itself (Bidinost 1951). From individual components of myosin meromyosin-H has a higher acticity than L (Brown et al. 1956, Velick 1956). As far as stroma proteins of the muscle are concerned, incorporation was studied into collagen, which appears to be metabolically inert (Neuberger and Perrone 1951). From different structural elements (as was mentioned above) there is not any substantial difference between the rate of incorporation into microsomes and mitochondria (McLean et al. 1958, Winnick and Winnick 1960). Incorporation into both of these particles far exceeds that one which was found with soluble and structural proteins. As far as RNA content is concerned, the richest are microsomes (10·4%), followed by mitochondria (4·8%), myofibrils (0·59%) and soluble proteins (0·21%) (Winnick and Winnick 1960).

As has already been mentioned on p. 303, the intracellular localization of the maximum RNA does not correspond to maximum incorporation in the muscle.

Table 17

Incorporation of labelled amino acids into various muscles.

Author	Velick 1956	Bidinost 1951	Dreyfus et al. 1956	Braun et al. 1956	Neuberger and Perrone 1951
Whole muscle	—	0·044	—	—	698
Actomyosin	—	0·085	—	—	—
Myosin	—	0·031	30	—	—
H-myosin	80	—	—	1310	—
L-myosin	20	—	—	2020	—
Tropomyosin	27	—	—	—	—
Actin	67	0·028	—	—	—
Dehydrogenase	100	—	—	—	—
Aldolase	50	—	18	—	—
Phosphorylase	50	—	—	—	—
Collagen	—	—	—	—	46
Expression of radio-activity	Half-time in days	Relative specific activity	Biol. time	Specific activity	cmp/cm²
Labelled amino acid	tyrosine C^{14} C^{14}-phenyl-alanine	C^{41}-glycine	C^{14}-glycine	C^{14}-glycine	C^{14}-glycine
Remarks		Per os, fed for two days		5 days after injection	Per os, fed for one day, rat 203—7 g. body wt.

In the muscle as a whole, as in other tissues, the level of RNA is decisive for the ability to synthetize proteins (Žák and Klicpera 1958). Preincubation of the diaphragm with ribonuclease leads to a decrease of RNA by 45% and this is accompanied by a loss of the ability to synthetize proteins. However, if the tissue thus preincubated is offered ribonucleotides together with a mixture of amino acids, the level of RNA recovers and proteosynthesis proceeds again (Fig. 111). An analogous result was also obtained in experiments *in vivo*. Muscle stimulation leads to a decrease of RNA and proteins. During the restitution phase, the resynthesis of proteins follows after preceding recovery of RNA (Žák and Gutmann 1959, 1960).

A general aspect of muscle proteosynthesis cannot, as yet, be accurately described similarly as in other tissues. It is worthwhile noting that the enzymatic mechanism activating amino acids prior to the formation of a peptide bond has also been demonstrated in muscle (Cole et al. 1957, Pennington 1960).

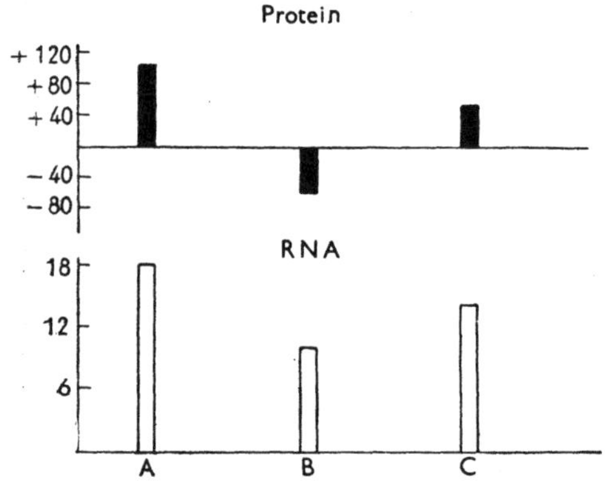

Fig. 111. Relationship between the content of RNA and protein synthesis in the muscle. Ordinate: non-collagenous protein N (in γ); RNA-P/100 mg. w. wt. (in γ). A — normal isolated rat diaphragm incubated for 120 min. in a medium containing complete mixture of amino acids in 0·1 % concentration. B — isolated diaphragm preincubated for 90 min. in a medium containing ribonuclease (1 mg./ml.), then incubated as in A. C — isolated diaphragm preincubated as in B, then incubated for 120 min. in a medium containing a complete mixture of amino acids (0·1%) and ribonucleotides in concentration 1 mg./ml. of medium.
The results concerning proteins are expressed as an absolute decrement or increment, those concerning RNA represent the content in the tissue. (After Žák and Klicpera 1958.)

b) Protein degradation

It can be concluded from the above facts that certain stages of proteosynthesis in the muscle differ from other tissues. What is known about degradation of proteins? This process has up to now been studied by determining the activity of proteolytic enzymes and by measuring the intracellular degradation of proteins. The former approach consists of the following methodological possibilities:

1. Autolysis: the tissue either remains in the organism and after a certain time the increase in non-protein nitrogen is determined (autolysis *in situ*), or the tissue homogenate is incubated in an unbuffered solution (autolysis *in vitro*). The rate of degradation is then given by three factors: by the activity of tissue cathepsins, the availability of the endogenous proteins to enzyme action and finally by the formation of acid products in the tissue resulting in a change of pH (i. e. glycolysis). By this procedure only very approximate information is obtained concerning proteolytic enzymes.

2. Proteolytic activity: this approach is similar to autolysis *in vitro* with the exception that the homogenate is incubated in a solution with a pH optimal for tissue cathepsins. The production of acids by the tissue is thus eliminated. Activity measured gives a survey of tissue cathepsins and of the availability of endogenous proteins to enzyme action.

3. Cathepsin activity: the activity of an enzyme extract splitting an exogenous substrate, for example, haemoglobin or casein is measured. This method offers information about activity of an enzyme only.

Table 18

Intracellular degradation of muscle proteins.

DIAPHRAGM

Incubation conditions	Oxygen	Nitrogen	DNP* (10^{-4}M)	Azide-Na (5×10^{-3}M)	p-FPA** (10^{-2}M)	MSI*** (2×10^{-3}M)	Mixture of amino acids (0·1 %)†
cpm liberated	460 ± 10	1110 ± 20	1190 ± 50	1220 ± 80	1290 ± 80	600 ± 50	370 ± 40
% cpm liberated††	1·40 ± 0·05	3·64 ± 0·44	—	—	4·14 ± 1·05	—	0·96 ± 0·12
change in %	100 %	+140	+160	+165	+180	+30	—31

* 2·4-dinitrophenol,
** p-fluorophenylalanin,
*** methioninesulphoximine,

† enzymatic hydrolysate of casein (pronutrin,)

†† $\frac{\text{cpm liberated}}{\text{cpm of proteins}} \times 100$.

MITOCHONDRIA

(% S^{35}-methionine liberated)

Origin in mitochondria	Rat		Pigeon	Cockroach	Rat
	Skeletal muscle	Heart	Pectoral muscle	Thoracic muscle	Liver
Oxygen	0·615	0·930	1·090	1·830	5·28
Nitrogen	2·200	2·500	3·180	4·700	4·43
Change in %	+257	+170	+191	+157	—16

Animal received an injection of S^{35} methionine and after 24 hrs. the diaphragm or mitochondria were isolated. Half the preparation was frozen, the other half was incubated in a medium containing non-radioactive methionine in order to prevent the reutilization of breakdown products. After three hours' incubation the increment of free S^{35}-methionine was determined. The amount of this S^{35}-methionine liberated was then expressed in % of protein bound S^{35}-methionine (in % cpm liberated). This value then corresponds to the quantity of proteins broken down (Žák and Drahota 1960).

On the other hand investigation of intracellular protein degradation represents a far more physiological approach than the latter, purely enzymological, approach, since determinations are performed directly in living cells. Degradation of proteins is a process contrary to proteosynthesis, which it continuously accompanies, and these two processes together form the basis of protein turnover. Under conditions of the steady state the amount of amino acids newly incorporated into the protein molecule is balanced by liberation of amino acids already bound with the peptide bond. The total number of peptide bonds remains unchanged and the protein molecule undergoes continual renewal.

Intracellular degradation of muscle proteins *in vitro* is increased under conditions limiting the utilization of energy, such as, for example, incubation of the isolated diaphragm under anaerobic conditions or in a medium containing dinitrophenol or sodium azide (Tab. 18) — (Žák and Drahota 1960). Degradation in the muscle thus differs substantially from other tissues or cellular systems, such as liver and kidney slices, reticulocytes, macrophages or certain bacteria (see Žák 1959). In these systems degradation is, on the contrary, inhibited after application of inhibitors of oxidative processes. It is assumed that the splitting of peptide bonds is associated here with the existence of some complex acceptor rather than water which would be used for a simple hydrolytic reaction Formation of such a complex requires energy. (see Walter 1960) However, since protein degradation does not require energy, this indicates that degradation is the result of splitting of peptide bonds by tissue cathepsins.

In case tissue cathepsins are acting even in the living muscle cell, it will be important to determine their properties, which could also elucidate some physiological mechanisms. Two types of cathepsins have been found in the muscle: one group with a maximum in the acid reaction range (Snoke and Neurath 1950) and the other in the alkaline range (Koszalka and Miller 1958, 1960).

Activity of enzymes belonging to the first group is maximal at pH 4 with maximum stability at pH 6. As compared with analogous enzymes, for example, in the liver and spleen, their activity in the muscle is much smaller (Kamner et al. 1950). However, this can be explained by the fact that muscle cathepsins are far more specific than those of the liver and kidney. This means that they split endogenous substrate (i. e. tissue proteins themselves) more easily than casein which is used for measuring their activity (Goldstein and Milgram 1937). This question is not yet quite clear. According to Khritina (1942) muscle proteolytic activity is one of the highest when compared with other tissues, however, Haarman and Schroeder (1938) and Faulhaber et. al. (1961), on the other hand, reported very low proteolytic activity in muscle. As far as the intracellular localization of cathepsins of this type is concerned, activity with pepsin-like properties has been found in myosin, but none in myogen (Nardone 1952). Some proteolytic activity is also reported for action preparations (Drabikowski 1961).

The second group contains an enzyme with maximum activity at pH 9. At this pH the enzyme attacks exclusively endogenous substrate, while at pH 4 it is quite without effect on muscle proteins themselves (Koszalka and Miller 1958, 1960).

All physiological research has, so far, dealt with enzymes with maximum activity at pH 3—5. It is interesting that avitaminosis-C and B are associated with increased muscle cathepsin activity (Fomin and Romanyuk 1937). Similarly there is increased autolysis in avitaminosis-B and during hunger (Khritina 1942). This indicates that proteolytic enzymes are subjected in the muscle to physiological cell regulation. Interestingly enough, cross striation remains intact during autolysis of muscle tissue. Generally 70% muscle proteins are resistant to autolysis, which is considerably more than, for example, in the liver, where only about 15% of proteins are resistant (Bradley 1938).

2. Functional activity and muscle protein metabolism

Contraction represents a basic property of muscle. How does functional activity affect the metabolism of muscle proteins? Solution of this problem would at the same time give the answer to the question of the mechanism by which the nervous system affects muscle metabolism. In section A evidence has been presented demonstrating that not only denervation but also tenotomy and immobilization cause muscle atrophy. The common denominator of all these three experimental situations is muscle inactivity. Inactivity, i. e. elimination of nerve impulses, may therefore be the factor responsible for this radical change in metabolism of muscle proteins. On quantitative comparison it can be seen that denervation always leads to more extensive changes than immobilization. This difference is given by the exclusion of the trophic nervous influence. In our opinion this specific trophic effect of the nerve is especially conspicuous during the recovery phase following muscle activity. Changes of muscle proteins following contraction thus offer a possibility of investigating the trophic influence of the nervous system. Furthermore it is well known from clinical experience that electrotherapy retards the weight loss of denervated muscles. A deeper understanding of this therapeutic procedure also requires knowledge of muscle protein changes which are evoked by electrical muscle stimulation. Immediate effects of a single bout of stimuli are described in the following paragraphs, while changes resulting from systematic stimulation, i. e. electro-therapy, are given in Chapter X.

a) Changes of physico-chemical properties of muscle proteins

It has been known for a long time that the solubility of muscle proteins decreases in fatigued muscle. This effect of contraction on the solubility of muscle proteins is known as Deuticke's effect. This author reported in 1930 that direct stimulation of isolated frog muscle with electric current at a frequency of 30 c/min. lowers protein solubility. Later the same author found that this decrease already becomes manifested after thirty twitch contractions (Deuticke 1932). Similar changes were observed after insulin paroxysms, muscle rigidity (Deuticke 1930) and after intoxication with bromine or iodine-acetate (Embden and Metz 1930) or under phenylurethane anaesthesia (Mönninghoff 1949). A similar decrease of muscle protein solubility following stimulation was obtained by Meyer and Weber (1933), Mirski (1937) and Embden and Metz (1930). It is interesting that when phosphate, usually used for extraction, is exchanged for acetate, solubility is found to be increased in the same experiment (Hansey 1930). In two communications, however, negative results were reported (Blätter 1930, Smith 1937). If solubility changes are found, they are very quickly reversible during restitution following stimulation or after withdrawal of phenylmethane anaesthesia. Aerobiosis, however, is a necessary condition for restitution of solubility (Deuticke 1932). From the physiological point of view it is interesting that the decrease of protein solubility after fatigue is smaller in trained muscles of the rabbit than in untrained muscles (Seitz 1933).

The first more precise analysis of these solubility changes was given by Kamp in 1941 who confirmed the findings of Deuticke and furthermore demonstrated that decrease of solubility is indirectly proportional to the contraction ability of the muscle. The most important findings, however, concern the fact that these changes are only to be found in myosin and that they are induced neither by changes of pH or salt concentrations. Later Crepax and Hérion (1950, 1951b) pointed out that ATP concentration also does not play a role. A decrease of myosin solubility was also confirmed by Fischer and Ramsey (1946) after five-minute uninterrupted tetanic stimulation of the rabbit gastrocnemius. According to these authors the solubility of actomyosin is changed considerably less under the same conditions. The same result was obtained by Mrázek and Reed (1947) who stimulated the gastrocnemius muscle in the frog and rat to fatigue by induction shocks of 1 sec. duration. They found that changes of myosin solubility are dependent on the time required for attaining fatigue. As has been pointed out by Kamp (1941), myosin solubility changes are reversible under aerobic conditions. But no change in myosin solubility after tetanization was also reported (Seraydariyan and Williams 1960).

The electrophoretic method of Tisselius made further investigation of individual proteins possible. The main credit for this is due to Dubuisson and Crepax, who found in 1946—1951 that in muscles fatigued by electrical stimulation (but also after application of iodoacetate or rigor mortis.) myosin β_1 and β_2 and Y-protein are bound to muscle protein stroma more firmly. Both these proteins become less extractable or even non-extractable (Dubuisson 1950a, 1950b, 1950c). Since extraction solutions of greater dissociation strength, for example, KI (instead of the chloride) split this bond (Dubuisson 1950a, 1950b) it is evident that this is a case of a stronger bond between these proteins and the stroma, and does not mean that they have disappeared. Furthermore, it can be observed on the electrophoretogram extract of a fatigued muscle that the peaks of actomyosin (myosin α) and contractin (myosin γ) become intermingled (Dubuisson 1948).

Besides this, reduction of myosin and Y-protein, proteins with different properties from known muscle proteins, have been found in fatigued muscles. Contractin (myosin β) discovered by Dubuisson on extracting a muscle after contraction with solutions of high ionic strength (1948, 1950a) belongs to this category. In normal resting muscle contractin is present in insignificant amounts (5%), but it increases considerably after contraction. During the restitution phase it again disappears. It is a globulin in character. Another similar protein is metamyosin, discovered in 1955 by Raeb, Schapira and Dreyfus. There is about 0—5% of this protein in the normal muscle, 5—20% in the fetal muscle and it increases considerably after contraction, so that it may represent 50% of all myosin present.

It is important to note that these changes are not caused by a decrease of ATP in the muscle. This means that the same changes as after electrical stimulation will be found in muscles contracting after application of Na-monobromacetate when the levels of ATP do not change substantially (Dubuisson 1950c). On the other hand, contraction due to refrigeration leads to complete breakdown of ATP, without decreased solubility, however, being found (Crepax and Hérion 1950, 1951).

All the above changes are very quickly reversible at rest (Dubuisson 1950c). It is difficult answer the question of the character of these changes. Factors known to affect protein solubility such as pH, ions or ATP are excluded. Evidently changes occur in the protein molecule itself. It is not yet possible to decide to what extent denaturation plays a role in this phenomenon, as in proteins of a nerve after stimulation (v. page 320). The finding of Mirski (1937), who found that the number of S—S and SH-groups does not change after muscle stimulation, speaks against the denaturation of muscle proteins. On the other hand, Chagovets (1938) demonstrated that there is a decreased adsorption of dyes into muscle proteins following functional activity. However, both these methods are unsatisfactory by modern standards and the question of denaturation after functional activity still remains unsolved.

b) Protein degradation and muscle function

The relation between functional activity and the degradation of muscle proteins has been studied for more than a hundred years. These studies were at first connected with the investigation of the energy source for muscle activity (up to about 1930). In the second stage attention was concentrated on molecular changes of muscle proteins evoked by muscle function.

In principle two experimental approaches were used in both stages of investigation:

1. Functional activity of the whole organism was followed by analysis of the response in blood and urine composition.

2. Stimulation of one muscle or maximally a single muscle group was followed by determination of direct protein composition in the muscle itself.

1. In the first case the whole organism represents a most complicated system in which it is practically impossible to analyse all the factors which might affect the composition of blood or urine. It is also for this reason that in spite of great effort unanimous results have not been obtained. First reports were made in 1855 when Beigel in Germany and Smith in U. S. A. described increased excretion on nitrogen in the urine after physical effort. However, an exact approach to this problem was made in the classical papers of Parkes (1866—1871) and of Fick and Wislicenus (1866). According to Parkes increased nitrogen excretion in the urine after physical exercise is proportional to work performed. On the contrary, Fick and Wislicenus undertook an ascent of a mountain, performing about 13,000 kg.m. of work, and did not find any changes in nitrogen excreted. Further development of this problem is characterized by such contradictory findings. Negative results leading to the conclusion that work exertion does not affect nitrogen excretion in urine served as basis for the investigations of Voit and coworkers concerning the energetic role of lipids during muscle contraction. Chauveau and Seegen, on the contrary, tried to demonstrate the role of carbohydrates in energy metabolism. On the other hand the positive correlation between the amount of work done and nitrogen excretion served as evidence in favour of the hypothesis that proteins play a role in liberating energy for muscle function. For example, Liebig assumed that non-protein substances are the source of heat in the organism, proteins on the contrary being the source of energy for contraction. This hypothesis was later again revived in 1880 when Argutinski succeeded in demonstrating increased nitrogen excretion in the urine during ascent of a mountain. The principal advocate of the energetic role of proteins was Pflüger. This great controversy was finally solved in favour of carbohydrates in the classical experiment of Fletscher and Hopkins in 1907 (see p. 273).

It is necessary to evaluate experiments describing increased excretion of nitrogen after work from the present point of view. Why are these results so controversial? Great differences between the results of various authors are, to a certain extent, already bound with the form of functional activity to which the organism was subjected: nitrogen excretion was, for example, studied during marching of troops, in sportsmen during running and bicycle riding, work on the ergostat, after mountain climbing etc. In animals, dogs were usually made to pull a loaded cart.

The following factors could be responsible for the above contradictory results:

Diet: in some of the experiments increased nitrogen excretion was found only when the diet covering energy requirements necessary for the work performance was insufficient (Munk 1890, Cathcart 1925, Nocker 1951). A corresponding result has also been recently reported: non-protein nitrogen in the blood after exercise rises far more during starvation, while it is less prominent after administration of glucose (Usami 1957).

Time interval after functional activity of the organism: increased nitrogen excretion in urine was often not found until after a certain lag following work exertion (Rakestraw 1921, Caspari 1901). It is possible that in experiments where the whole time course was not investigated the interval chosen corresponds to the period prior to the increase in nitrogen excretion.

Respiration: insufficient respiration can serve as another factor affecting nitrogen excretion after work. For example, Oppenheim (1881) reported that nitrogen is excreted in urine in greater amounts only if dyspnoea occurs during the working period.

Temperature: it has been shown that nitrogen excretion in the urine is more dependent on temperature than on the actual amount of work performed (it is higher at higher temperature) (v. Caspari 1901).

Stress: it was found by Usami (1957) that non-protein nitrogen in the blood rose proportionally to stress evoked by heavy exercise. In closer analysis Usami showed that stress is caused by increased excretion of hydroxycorticoids. This then increased protein catabolism, which is followed by a rise in non-protein nitrogen in the blood. At a later training period adaptation to work appears and excretion of corticoids returns to normal. This finding is in agreement with older results of Caspari (1901), who reported that with progressing duration of work gradually more nitrogen is retained in the organism. On evaluation of various results it is necessary to bear in mind to what extent the respective methods evoke stress.

Reutilization of degradation products of proteins: Parkes (1866—1871) was the first to point out that reutilization of degradation products of proteins following functional muscle activity can affect nitrogen excretion in the urine This conclusion was made from observations of increased excretion only if the work performed was exhausting. Moderate work, or work interrupted with rest periods do not lead to this increase. Theoretically it can be postulated that under optimal conditions (such as, for example, sufficient storage of carbohydrates in the muscle) degradation products are utilized and the excretion of nitrogen in the urine remains unchanged. It is also possible, however, that during work nitrogen material excreted under normal conditions at rest is being utilized. Even in this case, nitrogen excretion on the whole would remain unchanged, since the production of nitrogen due to protein degradation is compensated by the utilization of nitrogen material mentioned above.

It is thus evident that functional activity of the whole organism results in very complicated for experimental work conditions. Nitrogen excretion in the urine which might serve as evidence of degradation of muscle proteins can be influenced in the course of different reactions by various factors which are difficult to standardize. Humoral factors, besides affecting the metabolism of muscle proteins, can also influence catabolism of other proteins. It is therefore very difficult to compare individual experiments. The effect of muscle protein degradation will probably be easier to investigate in the muscle itself.

2. Experiments in which the influence of muscle stimulation was studied of proteins and non-protein nitrogen in the muscle were originally a part of research investigating the source of energy for muscle contraction. Later, however, when the energy role of carbohydrates had been conclusively proved, this type of experiment served as an approach to studying molecular changes of proteins evoked by muscle function.

The oldest report dates back to 1866, when Nawrocki discovered that tetanisation of the frog gastrocnemius of 8·5 min. duration causes a loss of about 4% non-collagenous proteins. In agreement with this are the findings of Kuraiev (1886—1895) who worked on frog muscles using both direct and indirect stimulation. Changes of the protein content in these experiments are expressed in relation to wet muscle weight, as in the paper of Pugliese (1911), who stimulated the hind limb of dogs at tetanizing frequencies with a load of 1·5 kg., and of Blättler (1930). Both these authors also found a decrease of total muscle nitrogen. Similar results were also obtained by N. R. Chagovets (1957, 1958) and Rogozkin (1959), who demonstrated a 12% loss of total muscle nitrogen following 15 min. swimming in rats (expressed in mg%). His results were confirmed by Tsanev (personal communication). However, negative results have also been reported by Silakova (1957) who, after stimulation of rabbit muscles with a frequency 50—60 c./min. for 25—60 min., found no changes in the total protein content.

Already in 1897, however, Kashkadanov found that a lowered protein content, when present, is only relative and is given by increased water in the stimulated muscle. This was later again confirmed by Pugliese (1911) and R. Chagovets (1938). The relative decrease of proteins in some experiments, however, can hardly be explained by increased water content. For example, N. R. Chagovets (1957, 1958) described a 12% reduction of total nitrogen, while the water content increases only by 4%. It is also interesting that analogous stimulation of the muscle leads to the reduction of proteins in normal muscles only, but not in animals with lower levels of carbohydrates (by feeding non-carbohydrate diet containing phlorizin (Blättler 1930)). It is possible to assume a similar increase of water here, since the work performed is the same, but nitrogen is lost only in the first case. In spite of this, however, one cannot expect an answer to the problem of protein degradation from changes of their relative content in the muscle.

Far more convincing are the experiments in which it was possible to determine the absolute loss of proteins, i. e. net protein decrease per muscle. One muscle of an extremity can be stimulated, while the muscle on the contralateral limb serves as control: the difference between the right and left muscle have been found to be negligible. For example, the difference between total content of non-collagenous proteins in the anterior tibial muscle of the rat is 0·2 % and between total nitrogen it is 0·3 % (Žák et al. 1957a, 1957b). This approach was chosen by Kashkadamov (1897) and R. Chagovets (1938). Neither of these authors found an absolute decrease in proteins. These experiments, however, have a shortcoming in that blood was not removed from the muscle, which after stimu-

Table 19

The effect of functional activity on the content of muscle non-protein nitrogen.

Animal	Muscle	Type of activity	Frequency (cycle/min.)	Duration of activity (min.)	Change (%)	Author
Dog	Gastroc.	Running	120**	to fatique	+8·5	Blättler (1930)
Frog	***	Direct stimulation	60	15	+46	Popova (1951)
Rat	Tib. ant.	Direct stimulation	120	6	+20	Žák et al. (1957b)
Rat	Quadriceps	Swimming	120—150**	15	+ 6	Chagovets (1957)
Frog	Gastroc.	Faradic stimulation	***	***	0	Gefter and Kartashov (1937)
Rabbit	Quadriceps	Direct stimulation	50—60	15—60*	0	Silakova (1957)

* Irregularly.
** Approximation.
*** Values not given by the author.

lation contains large quantities of blood. Increased blood content in the stimulated muscle can even increase its dry weight and values of total nitrogen: for example, dry weight of anterior tibialis of the rat was 66·6 mg. before the experiment; immediately after six-minute stimulation at a frequency of 300 c./min. this increased to 67·7 mg. and total nitrogen rose from 9·36 to 9·62 mg. (Žák et al. 1957b). In removal of blood by bilateral perfusion of hind limbs of animals prior to excision of the muscles, an absolute decrease of total nitrogen and non-collagenous proteins will be found. This difference is even evident in dry weight (Žák et al. 1957a, 1957b). This decrease, however, is very small, about 1 %, and is at the limits of statistical significance.

Experiments with the contraction of isolated myofibrils could serve as a new approach. The problem here lies in evoking repeated contraction. Following a single contraction elicited by application of ATP, no changes of nitrogen contend have been found either in the myofibril or in the suspension medium (Ashley et al. 1956). A more detailed specification of individual nitrogen components would, however, be required.

If there is a decrease of proteins after stimulation, this must be shown by an increase in the content of degradation products in the muscle. However, it must be taken into account, that the blood flow through the stimulated muscle is increased and consequently the removal of degradation products will also be higher. It can thus be assumed that the increased content of non-protein nitrogen substances will be higher than that actually found.

It can be seen from the results in Table 19 that the increased content of non-protein nitrogen in the muscle after stimulation was found by various experimental approaches, but contrary findings have also been reported in which no changes were found. It should be pointed out, however, that changes in the non-protein nitrogen depend, among others, on the frequency of stimulation used. For example a loss of non-protein nitrogen was found in rat tibial muscle after stimulation with a frequency of 300 c./min., however, after a lower stimulation frequency (120 c./min.) it was, on the contrary, increased (Žák et al. 1957b). This fact, although unexplained itself, demonstrates that stimulation frequency must be taken into account when comparing various results. The degradation of proteins following muscular functional activity was also indicated in experimental results of Nechiporenko and Medovar (1957) who obtained increased non-proteins S^{35} in rabbit and frog muscles after stimulating to fatigue at a frequency 50 c./min. These authors, however, consider this to be due to increased content of blood which contains twice the specific activity of radioactive sulphur than the muscle.

Changes of non-protein nitrogen in the muscle after stimulation could be accompanied by a similar change in the blood. A rise of non-protein nitrogen in arterial blood speaks in favour of the production of non-protein nitrogen. An increase of 33% was found in the above mentioned experiments of Rogozkin (1959), Chagovets (1957) and Popova (1951), and furthermore in the blood of rabbits following running (Gefter and Kirian 1937) and finally in the blood of athletes following exhausting exercise (Usami 1957).

During muscle function deamination of ATP takes place and this liberates free ammonia and ITP (Embden et al. 1927—1930, Parnas et al. 1927—1929). This ammonia formation is a part of the non-protein nitrogen increase. However, Ferdman (1938) calculated that the production of ammonia in muscle exceeds the deamination of ATP. It is thus necessary to specify more closely the degradation products of proteins. This has been done in the following way: the content of free amino acids was investigated in muscles of salmon during the spawning migration, following long swimming periods, and it was found that their free amino acids levels are almost 100% higher (Ch. W. Greene 1919, C. H. Greene 1919). Similarly there is an increase of amino acids in muscles of rats running in a tread-mill (Williams et al. 1950). On the other hand, faradization of the frog gastrocnemius does not cause any changes (Gefter and Kartashov 1935). From among the other components of non-protein nitrogen, urea was found to be increased in muscles after work, but this finding was not confirmed (see Parnas et al. 1927). In the other hand, Usami (1957) found that urea is the main source of increased levels of non-protein nitrogen. It can thus be assumed on the basis of these findings that ammonia formed by ATP desamination is not the only nitrogen compound responsible for the increase of non-protein nitrogen in the muscle after stimulation, as is often erroneously stated.

Although the relationship between muscle function and degradation of its proteins is not yet satisfactorily explained, many results indicate a positive correlation between these processes. It is therefore important to know whether this is a general phenomenon or an exceptional case. Another type of tissue in which the relation between functional and protein metabolism can be studied are peripheral nerves and brain tissue.

Soula (1912—1913) was the first to investigate the relationship between nerve function and protein degradation and found that electrical stimulation lowers the content of proteins. However, this problem was widely and systematically studied more recently by Hydén in Sweden and Ungar in U. S. A. and their coworkers. As far as nerve cells and brain is concerned, their results clearly show degradation of proteins after stimulation. For example, Hydén (1955) demonstrated that acoustic stimulation of the nerve cell leads to a decrease of nucleoproteins. A similar effect was observed on nucleoproteins in anterior spinal horns under the influence of intensive muscular work (Hydén 1943). Similarly Vrba (1955, 1956) ascertained that swimming leads to a decrease of total brain proteins in the rat. It must be pointed out that in his experiments only protein nitrogen was determined and the specific reaction for determining the number of peptide bonds was not used. The loss of protein nitrogen can thus be caused in this case by a decrease of protein amide groups, and this was actually the result found by Vrba. It was further found in the brain by Ungar et al. (1957) that a 20-minute stimulation by electrical current of 40 V and at a frequency of 60 c./min. causes a loss of 16% of total proteins. Indirect evidence for protein degradation was already presented prior to this by Geiger et al. (1953). The content of non-protein nitrogen in the cerebral cortex after 7—20 sec. stimulation was increased approximately by 15%. However, this increase is not caused by amino acids, since amino-nitrogen is lowered. Neither ammonia nor volatile bases, even though increased, can explain the high increment of non-protein nitrogen found in their experiments.

The situation in the peripheral nerve is less clear when compared with these conspicuous changes. A decrease in total nitrogen in the sciatic nerve following electrical stimulation has not been demonstrated (Ungar et al. 1957). However, there was a significant 27% increase of non-protein nitrogen in the same experiment. Since individual components of non-protein nitrogen were not analysed, this cannot be considered as proof of protein degradation. This objection is very much to the point, as can be seen from the experiments of Geiger et al. (1953, 1956), who found no increase of free amino acids following brain stimulation even though non-protein nitrogen was increased.

If muscle and nerve function is associated with protein degradation, by what mechanism does this degradation occur? What role is played in this by proteolytic enzymes?

The influence of pH on autolysis of muscle tissue led Chen and Bradley (1924) to the conclusions that the production of acids during muscle contraction is sufficient to initiate autolysis. According to these authors, the restitution phase represents the time necessary for resynthesis of proteins degraded in the course of this autolysis. Direct studies, however, were not carried out until later: muscle functional activity results in an increase of autolysis both of the whole organism (Steppuhn et al. 1926) and of the muscle itself (Smorodintsev 1935). It has, however, already been pointed out on p. 308 that autolysis is only a very approximate indicator of protein degradation. Its increase can be given by various factors which do not act in the living cell (e. g. production of acids and activators during autolysis). The experiments of Yakovlev (1948) who investigated the proteolytic activity of minced muscle tissue before and after activation with hydrogen sulphide are more accurate for this reason. This author found that proteolytic activity in a trained muscle is decreased in the first case and moderately increased in the second (i. e. after activation with hydrogen sulphide). According to Yakovlev, training results in a decrease of activated cathepsin and in an increase of cathepsin available for activation. The differences, however, are extremely small. Changes can also be caused by structural transformation of muscle proteins, which can then be more easily attacked by tissue cathepsins. It would therefore be necessary to determine directly the cathepsin activity which has, of course, not yet been done after muscular activity. However, experiments of this kind have been performed on the brain (Ungar et al. 1957): 20-minute stimulation led to a 21% increase of cathepsin activity in the cat and dog. An similar change — a 15% increase — follows one-minute stimulation at a frequency of 60 c./min.

Functional activity thus leads to increased proteolysis in muscle and nerve. What is the triggering mechanism for this process? More experimental material in this direction has been obtained from investigations of nerve and brain. The alteration theory of nerve impulse conduction assumes that activity of nerve results in structural changes of its proteins (Nasonov, Meyer, Segal, Ungar). It rests chiefly on the work of Ungar et al. (1957) who have recently shown that these structural changes are analogous to denaturation changes. It is known that protein denaturation is characterized by unmasking of side-groups of amino acids in the polypeptide chain. Similarly, brain stimulation leads to an increase of side-group ionization ratio, which means that their number is increased.

Certain changes of protein properties occur even after muscle stimulation and these can be interpreted as denaturation changes, but this report is older (Chagowets 1935) and the method used is not nearly as conclusive as that used by Ungar et al.

Furthermore, it has been reported by Linderström-Lang (1953) that the first stage of enzymatic protein hydrolysis is represented by their denaturation. Hydrolysis does not occur until the enzyme can reach the peptide bond made accessible by denaturation. In this case stimulation compensates this first step of cathepsin activity and hydrolysis can progress at a faster rate. Ungar et al. (1957), however, pointed out still another possibility: it was found that protyrosinase can be transformed into the active form by denaturation. It is possible that stimulation similarly results in a certain enzyme denaturation and consequently in cathepsin activation. The enzyme thus activated then itself denaturates and splits nerve proteins, so that an increased side-group ionization ratio of amino acids can be found, as described above.

What experiments have been performed in this direction in the muscle tissue? It has been shown *in vitro* that it is possible to affect intracellular degradation of proteins without changing the proteolytic activity at the same time (Žák and Drahota 1960). As has been mentioned above on p. 310, the application of inhibitors of oxidative processes enhanced protein degradation in isolated rat diaphragm and muscle mitochondria. Nevertheless, the proteolytic activity measured at pH 3·6 and 9·0 is not altered in any way by this procedure. By what principle is this stimulating effect of inhibitors realized, when the activity of proteolytic enzymes remains unchanged?

Theoretically it is possible to assume that a certain role is played by denaturation of proteins: the maintenance of muscle proteins in a highly organized state requires a continuous supply of energy. If this is impaired, the structure breaks down and the protein can be partially denaturated and can thus become more accessible to the action of proteolytic enzymes. The other possibility, however, is more probable. Application of inhibitors of oxidative processes increases the degradation of muscle proteins. Besides, this proteosynthesis is blocked by these inhibitors. From this it is evident that protein degradation is stimulated under conditions which lower their synthesis. This is also demonstrated by the influence of antimetabolites of natural amino acids (see Table 19). In this case the blockade of proteosynthesis is associated with increased degradation of proteins. A similar relationship was also found under reversed experimental conditions. The addition of a complete mixture of amino acids into the medium is associated with marked activation of proteosynthesis. Protein degradation is inhibited at the same time. It thus seems that there is a reciprocal relationship between these two processes. A similar hypothesis has been stated by Dalgliesh (1957) who postulated that both synthesis and protenin degradation take place on the pattern of ribonucleic acids. Thus both processes compete for RNA. The above experiments concerning protein degradation following muscular activity fit into this hypothesis. Under normal conditions there is equilibrium between synthesis and degradation of proteins. However, muscle function utilizes most of the energy, so that energy requirement for continuous proteosynthesis is not met and the ribonucleic pattern is free for protein degradation.

Information concerning the balance between degradation and synthesis of proteins in the muscle during functional activity may also be obtained in incorporation studies. Unfortunately, results agained in this connection are contradictory. Nechiporenko and Medovar (1957) found no changes in incorporation of S^{35}-methionine into rabbit muscle or the neuromuscular preparation of the frog stimulated to fatigue. Results obtained on the brain during activity often disagree. For example, Gaitonde and Richter (1956) and Rosengart and Maslova (1956) showed that the incorporation of labelled amino acids into the brain after excitation is reduced. (It is interesting that the specific activity of nucleic acid phosphorus is lowered in stimulated nerve (Samuels 1957).) These results could thus indicate a predominancy of degradation over proteosynthesis. However, increased incorporation after brain stimulation was also described (see Folbergerová 1960). Generally speaking, it can be concluded that results obtained from incorporation studies are not sufficient to enable any definite conclusions to be made as yet.

A further necessary step towards research of quantitative changes in muscle proteins following contraction will be a closer identification of the protein fraction in which this change occurs. In the literature this aspect has only been dealt with by assumptions or circulation experiments. So, for example, Cathcart (1925) assumed that during muscle activity part of the contractile proteins is used up and excluded from further activity. Proteins "immobilized" in this way are then subjected to catabolism. During all these changes, however, the basic protein muscle structure is maintained. What evidence is there in favour of this hypothesis? Halperin and coworkers (1934) showed that the content of soluble proteins in frog's heart exhausted by jumping is increased by 200%. This result has, to a certain extent, some connection with an older finding of Pugliese (1911) who found that blood flowing from a stimulated hind extremity in the dog has a higher globulin content than that flowing into the muscle. On the other hand, however, data have been obtained recently which indicate that the solubility of muscle proteins is reduced after stimulation (p. 312). This, of course, concerns the increased content of structural proteins only. The increased content of soluble proteins in the fatigued muscle need not necessarily be the result of transformation of structural proteins into soluble proteins, but can be caused by increased amounts of plasma in the stimulated muscle. In experiments of Pugliese increased efflux of globulins in the blood from stimulated muscles nevertheless indicates that this artefact plays an insignificant role.

c) **Protein resynthesis following muscle function**

If a part of contractile proteins undergoes degradation during functional activity then it must be assumed that during the restitution phase their resynthesis again takes place. This means that in this phase proteosynthesis is temporarily accentuated, or, more precisely, that there is a shift in the equilibrium between degradation and protein synthesis in favour of synthesis. Since ribonucleic acids play an essential role in the production of the protein molecule, it would also be necessary to know changes of RNA connected with muscle activity.

It can be seen in Fig. 114 that muscle activity results in a decrease in RNA as was the case with proteins. The content of RNA is expressed in relation to non-collagenous proteins, so that changes in the water content are excluded. Six-minute stimulation of the rat anterior tibial muscle with electrical current at a frequency of 300 c./min. (i. e. 1800 contractions) leads to a 16% decrease of RNA. Changes are thus considerably larger than previously found in an analogous experiment for proteins (where an approximately 1% decrease was found) — (Žák 1958, Žák and Gutmann 1959, 1960).

A similar result was obtained in the nervous and brain tissue. A decreased RNA content has been known since the reports of Hydén. This author found, with the aid of Casperson's technique of ultraviolet absorption, that there is a considerable decrease of RNA after electrical as well as after acoustic stimulation of nerve cells (1955). The determination of RNA in nervous tissue is a very difficult problem and for this reason an especially valuable contribution is that of Causey and Stratmann (1956b) who, using the chemical method of Logan, found a decrease in RNA in ganglion cells of the rabbit. In their experiments the cervical sympathetic system was stimulated for three hours at a frequency of 300 c./sec. It is interesting that intensive motor activity also results in a decreased RNA in anterior spinal horns (Hydén 1943). In general it is necessary to stimulate a nerve cell for longer periods of time in order to obtain a RNA decrease than muscle.

In the peripheral nerve the situation is different. Stimulation, on the contrary, causes an increase of RNA (Causey and Stratmann 1956a). A decrease was only observed for DNA. This change masks the RNA increase, so that nucleic acids as a whole remain unchanged. This is contradictory, however, to the results of Abood (1956) who found a 15% decrease of total nucleic acids following one hour's tetanic stimulation of the frog sciatic nerve.

Stimulation of the brain leads to decreased content of RNA, in agreement with the above findings obtained on nerve cells. According to Mihailović et al. (1958) 150 V electroshocks of 0·9 sec. duration (evoking tonic spasms) lead to a decrease of RNA in all regions of the brain. A single shock has the same effect as four shocks.

The above results demonstrate that nerve and muscle function is not only associated with a loss of proteins, but also of RNA. According to Hydén, this decrease involves nucleoproteins in the nerve. From other experiments made on the nerve (Ungar et al. 1957, Causey and Stratmann 1956b) and especially on the muscle (Žák 1958, Žák and Gutmann 1959, 1960) it can be concluded that RNA decreases much more than proteins. This does not indicate that only a specific nucleoprotein is decreased, as has been postulated by Hydén. The question remains in what order of sequence and in what relationship are the changes of RNA and protein. In the literature only one model experiment demonstrates that RNA decrease follows after a protein decrease. This can be concluded from the effect of proteolytic enzymes on ribonucleoproteins of microsomata which not only leads to a decrease of proteins, but also of RNA. On the other hand, ribonuclease causes a decrease of RNA only, without alterations of proteins (Roth 1958a, 1958b). However, in this case more direct evidence is required. It has been shown by Vrba and Folbergerová (1958) that during incubation of brain slices endogenous pentose of ribonucleic acid is utilized. The above authors assume that utilization of the pentose is of significance for energy purposes.

Following the breakdown of proteins and ribonucleic acids their resynthesis occurs. It can be seen in Fig. 114 that the initial content of muscle proteins is again reached 1½ hours after stimulation. It is important that synthesis does not stop here, but progresses further, so that four hours after stimulation there is a 5% increase of non-collagenous proteins, of total nitrogen and dry muscle substance in the tibialis anterior of the rat as compared with the non-stimulated muscle. In the course of the next four hours the content of proteins again returns to values prior to the experiment (Žák et al. 1957a, 1957b). The time course of ribonucleic acid synthesis is similar in character, with the difference that maximum RNA content is attained two hours after stimulation and not four hours as in the case of proteins. It can thus be seen that RNA resynthesis precedes protein resynthesis (Žák 1958, Žák and Gutmann 1960). This relationship is often described in the literature. For example, Eliasson et al. 1951 found in the liver regenerating after hepatectomy a maximal turnover of RNA 30 hours after the operation, while the protein turnover did not reach a maximum until 30 hours later. Increased mitotic activity in the same case is also accompanied by RNA synthesis (Cater et al. 1957). A similar finding was observed in bacteria. Ribonuclease depresses considerably the level of RNA together with the incorporation of labelled aminoacids. Following the removal of ribonuclease, first RNA synthesis recovers and after this proteosynthesis (Groth 1956). It is interesting that this sequence in the restitution of RNA and proteins can be found only after their previous breakdown. This was the case in all the above examples. If, however, proteosynthesis is stimulated at normal RNA levels, for example, during tryptophane peroxidase induction (Dashman et al. 1958, Feigelton and Feigelton 1959), increased RNA turnover appears after the induction.

The character of RNA and protein resynthesis is also interesting from another aspect. It can be seen from Fig. 114 that the changes are a typical overshoot reaction (supercompensation). It is thus an analogous phenomenon to that found during the recovery phase of glycogen (see Chapter VIII, p. 238) and ions (Drahota, personal communication). The overshoot reaction was also postulated by Popova (1951) who presumed an increase of protein nitrogen two hours after electrical stimulation of a muscle in the cat on the basis of differences between total and non-protein nitrogen. In a direct experiment N. R. Chagovets (1957, 1958) found supercompensation in the content of protein nitrogen in rat muscles two hours after 15-minute swimming. Criticism of these results expressing the protein content per unit wet weight has already been mentioned above (p. 277). Furthermore, it must be pointed out that the difference between TN and NPN in Popova's experiments also includes the nitrogen of amide protein groups and nucleic acid nitrogen.

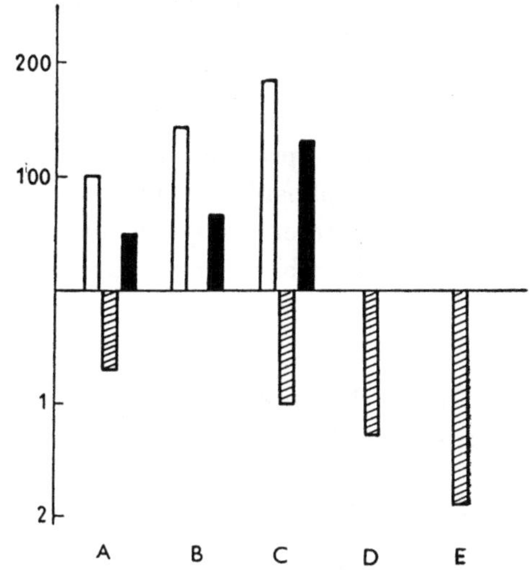

Fig. 112. Synthesis and degradation of proteins in muscles with different functional activity. Ordinate: above (100—200), on the right: incorporation, on the left: γRNA-P/10 mg. PN. Below (1—2): % of proteins broken down.
A — skeletal muscle; B — diaphragm; C — heart; D — flight muscle of pigeon; E — thoracal muscle of cockroach. White columns: incorporation of N^{15}-glycine into myosin (numerals indicate percentage increase of N^{15}-glycine above the values normally present in the protein). Black columns: RNA. Hatched columns: break-down of proteins in mitochondria incubated in vitro. (According to Schapira et al. 1953, Žák, unpublished, Žák and Drahota 1960.)

Increased protein synthesis during the restitution phase should be accompanied by increased incorporation of labelled amino acids. Similar studies have not yet been performed in skeletal muscle. According to Palladin et al. (1957), however, two and a half hours after brain excitation with phenamine, the incorporation of S-^{35}methionine is increased. Increase of labeled amino acids incorporation after stimulation was also reported in brain perfusion experiments (Geiger et al. 1960) and during peripheral nerve stimulation (Luxoro 1960).

If RNA and protein degradation and their subsequent synthesis are associated with muscle function, then this must also appear in the metabolic activity of muscles with different function. The results of various authors are given in Fig. 112. It can be seen that in agreement with increasing muscle functional activity the incorporation of N^{15}-glycine is also increased (Schapira et al. 1953). The content of RNA in whole muscle as well as in isolated myofibrils also increases (Žák — unpublished data). However, a positive correlation does not exist only between functional activity and protein synthesis in the muscle. In Fig. 112, furthermore, it can be seen that protein degradation in mitochondria increases in the same sequence of muscles. This is in agreement with an older finding of Bradley (1918) who reported that autolysis in heart muscle is considerably higher than in skeletal muscle.

The relationship between RNA and muscle function is evidently more marked than in the case of proteins. This can be concluded from the fact that the turnover of nucleic acid phosphorus in the muscle is high, even higher than in the liver (Hevesy and Ottesen 1943). The incorporation of labelled amino acids into muscle proteins, on the other hand, is far smaller than into the liver (Table 16).

3. Balance between protein degradation and synthesis in the denervated muscle

a) Protein degradation

A number of investigations has been carried out dealing with changes of proteolysis in the denervated muscle. The oldest of these is the report by Chen et al. (1924). These authors found that autolysis in the atrophying muscle was increased almost by 100%. This increase was the same one, two and nine weeks after section of the sciatic and femoral nerves in the hind extremity of the rabbit. This increased autolysis in their experiments was not, however, accompanied by a corresponding change in the content of amino acids in the atrophic tissue. Only tyrosine-reacting fragments were more abundant. As far as other amino acids are concerned, the levels of glycine, alanine, serine, asparagine and glutamine have recently been studied. Quantitatively they remain unchanged after denervation (see Schapira and Dreyfus 1950). On the contrary, the content of amino acids, with the exception of glycide and basic amino acids, is increased in muscles of rabbits with avitaminosis-E at early as well as late stages of dystrophy (Tallan 1955). Since increased autolysis was confirmed recently by Zubenko et al. (1950) and increased proteolytic activity was also found by Sheves (1955) and Sheves et al. (1956), it is necessary to explain why amino acids in atrophic muscles remain unchanged. It is possible that increased blood flow through the denervated muscle (see Chapter VII) is sufficient to maintain a constant level of amino acids.

While the increase of autolysis and proteolytic activity is conspicuous (e. g. in the experiments of Sheves the increase represented 50—250% five to twenty-four days after the operation), agreement has not been reached in studies of catheptic activity. Kruh (see Schapira and Dreyfus 1950) found no changes even when investigating the splitting up of endogenous myosin instead of casein. On the other hand, Zubenko et al. (1950) and Weinstock et al. (1958) described an increase. It will thus the difficult to comprehend increased autolysis, when the question whether the activity of proteolytic enzymes is actually increased, has not been answered clearly. It is probable that denervated muscle proteins are changed in such a way as to be more readily accessible to proteolytic enzymes. This would be borne out by the fact that denervated muscle proteins are more readily split by gastric juice (Zubenko et al. 1950). Similar changes in proteins are also assumed in an older report by Chen and Bradley (1924).

As far as other types of atrophy are concerned, increased proteolytic activity was found after tenotomy (Sheves 1955) and increased catheptic activity was described in muscles with hereditary, clinical and avitaminosis-E dystrophy (Weinstock et al. 1955, 1958). The results concerning cathepsins are the only changes from among the other enzymatic activities studied (aldolase, succinodehydrogenase and cytochromoxydase), which appear to be the same in all types of atrophy. The increased content of free amino acids in dystrophic muscle (Tallan 1955) is in agreement with increased catheptic activity. Contradictory findings have, however, been reported by Milman et al. (1953), who found that the production of free amino acids in the isolated diaphragm and skeletal muscle of the hamster is decreased in animals with avitaminosis. These authors explain their results by assuming that decreased protein breakdown is secondary to the decrease in protein synthesis. For this reason muscle weight is lost. A decrease of cell permeability in dystrophic muscles to amino acids has not, however, been excluded.

Intracellular protein degradation, as defined on p. 310, has not been studied in the denervated muscle. An analogous study, however, has been made by Simon et al. (1958) in muscles in mice with hereditary dystrophy. These authors found a larger decrease of radioactivity 20—30 days after injection. In their opinion, this is caused by protein degradation gaining predominancy over their synthesis. The possibility of increased turnover, however, has not been excluded.

b) Incorporation studies

The incorporation of radioactive amino acids into denervated muscle was first studied by Schapira and coworkers (1953). It was found that the incorporation of N^{15}-glycine into total proteins and myosin isolated from rabbit gastrocnemius is lowered ten days after denervation. Their results were later reinvestigated by Padieu (1959). According to this author, the decreased incorporation concerns not only myosin, but also soluble proteins, myosin being more affected. In Padieu's experiments apparent changes were already found 14 hours after the operation. As is evident from further work summarized in Table 20, changes in

Table 20

Incorporation of labelled amino acids into denervated muscle.

Animal	Muscle	Protein	Changes after denervation (%)		Amino acid	Note	Author
			I. stage*	II. stage*			
Rabbit	Gastroc.	Myosin	-18—38 (9—10)	—	N^{15}-glycine	—	Schapira et al. (1953)
Rat	Hind limb	Total protein	0 (to 14 days)	-21 (42) —48 (70)	C^{14}-glycine	—	Slack (1954)
Rabbit	Gastroc.	Total protein	-13—21 (5—10)	+35—82 (15—20)	S^{35}-methionine	Cystine not separated	Sheves (1955)
Rabbit	Soleus	Total protein	-32 (5—10)	0—87 (15—20)	S^{35}-methionine	Cystine not separated	Sheves (1955)
Rabbit	Calf muscles	Total protein	-38 (11—12)	+180 (30—40)	S^{35}-methionine	Cystine not separated	Ferdman (1957)
Rabbit	Calf muscles	Proteins sol. in water	-17 (11—12)	+300 (30—40)	S^{35}-methionine	Cystine not separated	Ferdman (1957)
Rabbit	Calf muscles	Myosin	—	+110 (30—40)	S^{35}-methionine	Cystine not separed	Ferdman (1957)
Rabbit	Calf muscles	Actin	—	+300 (30—40)	S^{35}-methionine	Cystine not separated	Ferdman (1957)
Rabbit	Gastroc.	Myosin	-45 (10—24 hrs.)	—	C^{14}-glycine	—	Padieu (1959)
Rabbit	Gastroc.	Proteins sol. in water	-24 (10—24 hrs.)	—	C^{14}-glycine	—	Padieu (1959)

* Days after operation in brackets (I. stage before, II. stage after 14 days of denervation).

incorporation after denervation are very complicated. Decreased incorporation during the early atrophic period (i. e. the first 14 days after denervation) was confirmed using S^{35}-methionine for myosin and soluble proteins by Ferdman (1957). On the other hand, Sheves et al. found this decrease only in the rabbit soleus, but there were no changes found in the gastrocnemius and plantaris muscles (Sheves 1955, Sheves et al. 1956). Similarly no differences in the specific activity of C^{14}-glycine isolated from total muscle proteins was found by Slack (1954) 14 days after denervation in the control and denervated extremity.

Results, however, are also not in agreement as far as late stages of denervation are concerned. While Slack (1954) reported decreased incorporation, Sheves et al. (1956) and Ferdman (1957) found a surprisingly high increase of incorporation into both the whole muscle and myosin, and especially into soluble proteins.

What is the reason for these differences? While Schapira et al., Padieu and Ferdman studied the incorporation into purified proteins, Slack and Sheves et al. studied incorporation into the whole muscle. Slack himself found increased incorporation into collagen. The percentage increase of collagen in the atrophic muscle can thus mask, to a certain extent, the decreased incorporation in proteins of the muscle phase. Furthermore, Slack, as well as Sheves et al. and Ferdman did not remove blood plasma. As is known, there is vasodilatation in denervated muscle. Plasma represents $^1/_{20}$ of soluble proteins and its radioactivity is higher by an order of magnitude than that of muscle proteins, especially so soon after administration of labelled amino acids (17—24 hrs.), as was done by Sheves et al. and Ferdman. It is thus possible that an increased amount of plasma in the denervated muscle masks the decreased incorporation into muscle proteins. In experiments in which decreased incorporation was found by Schapira et al. and Padieu, plasma was either removed by perfusion or myosin was carefully purified. This interference does not come into consideration here. It must further be pointed out that Sheves et al. and Ferdman determined S^{35}-methionine incorporation without having eliminated cystine and cysteine. It is known that radioactive methionine itself contains cystine and that methionine is even metabolized into cystine. This then becomes bound to proteins with the disulphide bond and not the peptide bond. Since free SH-groups of muscle proteins considerably increase after denervation (Popova 1958), this error is possible and can be held responsible for the increase of incorporation. It must, furthermore, be taken into account that the blood flow and consequently the supply of labelled amino acids is considerably increased at a late stage of atrophy.

However, not even decreased incorporation of amino acids into denervated muscle proteins can be considered as unequivocal evidence of decreased proteosynthesis. As was found by Padieu, the transfer of free C^{14}-glycine into the denervated muscle is lowered at the same time. Decreased specific activity of free amino acids inside the cell can thus be acounted for by decreased incorporation into the proteins.

It will probably be more suitable to follow the whole incorporation curve. This was the procedure adopted in dystrophic muscles by Simon et al. (1958), who found, as will be stated below, that incorporation is at first increased, however, later the labelled amino acid is released from the dystrophic muscle at a greater rate. Similar results were reported by Kruh et al. (1960). This experimental approach was not taken in the preceding reports. Up to now, incorporation has been studied shortly after administration of the radioactive amino acid. The specific activity of these labelled muscle proteins was not studied at a later stage.

What changes have been described in other types of atrophy? Sheves (1955) found that following section of the tendon there was at first a decrease in the incorporation of S^{35}-methionine (up to the tenth day), nevertheless, later an increase was observed (between 10 and 22 days after tenotomy). According to this author tenotomy causes a greater increase than denervation: a 40% greater increase in the gastrocnemius and as much as a 250% increase in the soleus. A 15% increase was found 30 days after tenotomy in the incorporation of S^{35}-methionine into soluble proteins, an 80% increase into myosin, 50% into actin and a 50% increase into total proteins. At the same time the S^{35}-methionine of the NPN fraction was slightly higher (Nechiporenko 1955). The incorporation of C^{14}-formate into dystrophic muscles of rabbits (Dinnig 1955) and monkeys (Dinning and Day 1958) with avitaminosis-E remains unchanged, while the incorporation of S^{35}-methionine into proteins is increased according to another author (Grigorieva 1955). This increment is given mainly by soluble proteins and actin. Collagen and myosin remain unchanged. An increased incorporation is also reported by Diehl (1960) for C^{14}-glycine, but not for C^{14}-leucine and C^{14}-lysine. Hereditary dystrophy is associated with increased incorporation of C^{14}-leucine, two to twenty-four hours after intravenous administration (Simon et al. 1958). There is a more rapid decrease of radioactivity in the dystrophic muscle later after the injection (20—30 days) and this is considered by the latter authors as evidence for the predominance of degradation processes. Increased turnover of myosin and water soluble proteins of mice with muscular dystrophy is also reported by Kruh et al. (1960) and by Coleman and Ashworth (1959).

c) Measurements of net protein synthesis

The method of direct determination of protein substance is much less sensitive than the incorporation of labelled amino acids. It can, however, give a better idea about proteosynthesis, the probability of artefacts is smaller and the results can be, generally speaking, more easily interpreted. What findings were obtained by this approach? It was shown by Schapira et al. (1950) that the growth of denervated muscles of young rabbits, although reduced, is maintained. These results were later confirmed by Vrbová (1952) and Zelená and Hník (1957). It is interesting to note that in these muscles all changes characteristic for atrophy are present, even though muscle weight increases (Schapira et al 1950): actomyosin of the growing denervated muscles of young rabbits decreases, collagen content increases. Hines et al. made the following experiment in adult rats (see Schapira and Dreyfus 1950): muscle weight was reduced by inanition and then unilateral denervation was performed. Only the normal muscle returned to initial weight during realimentation, while the denervated muscle did not regain weight. However, Gutmann (1959) demonstrated that weight of the denervated muscle can also increase during the first period of realimentation (i. e. three days) if the animals had previously been adapted to intermittent starvation for six weeks prior to denervation. After this period of first three days, however, sudden breakdown of synthetic ability occurs and the weight gained is again lost (Fig. 113).

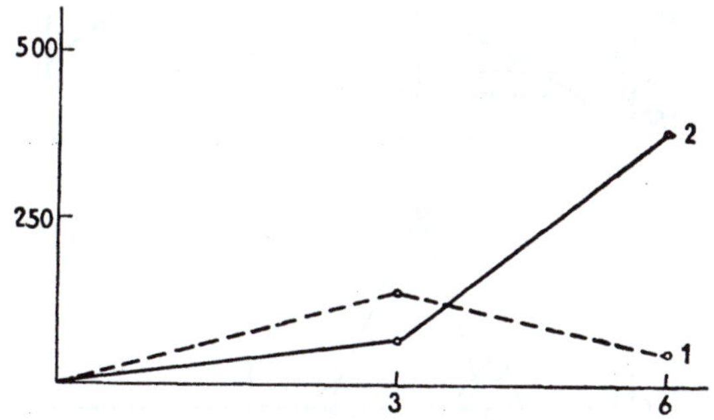

Fig. 113. Protein synthesis in the muscle during realimentation in animals adapted to intermittent starvation. Ordinate: increase of TN. Abscissa: time of realimentation (in days). 1 — denervated muscle (6 weeks), 2 — control muscle. The increase is calculated from the difference between absolute amount of TN in the extensor digitorum longus of the rat b fore and after realimentation (After Gutmann 1959).

The disturbance of proteosynthesis following preceding breakdown can also be demonstrated in another experiment. As has been shown on p. 323, electrical stimulation leads to the breakdown of proteins which is followed by resynthesis, the course of which is over-shoot in character. It can be seen in Fig. 114 that the breakdown of proteins in denervated muscle is the same as in normal muscle. During the recovery phase, however, the protein content does not return to initial values even eight hours after stimulation, while initial levels in the normal muscle are already attained 1½ hours after (Žák and Gutmann 1959, 1960). Indirect evidence for this change is offered by arteriovenous (A-V) differences of NPN. Following stimulation of the normal muscle the A-V difference rises during the recovery phase, indicating that nitrogen is being retained in the muscle. This is not the case in denervation: the A-V difference has a tendency rather to be reduced (Gutmann 1959).

d) The relationship between synthesis of proteins and nucleic acids

Proteosynthesis is gravely impaired in the denervated muscle, even though it is not wholly lost. What changes occur in the metabolism of nucleic acids?

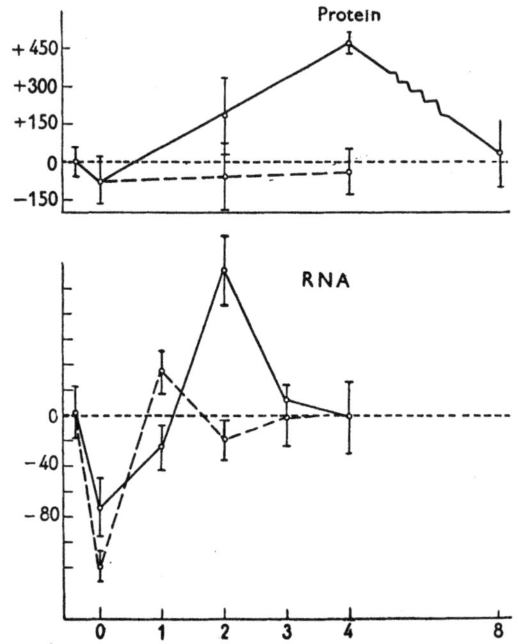

Fig. 114. The effect of electrical stimulation on the content of non-collagenous proteins and RNA in normal and denervated muscle. Ordinate: Proteins: γPN, RNA : γRNA-P/100 mg. PN. Abscissa: time after stimulation (in hours). Solid line: normal muscle. Interrupted line: muscle three days after denervation.
Anterior tibialis muscle of the rat stimulated directly at a frequency of 300 imp./min. for six min. Then perfusion was performed on both sides and the control and denervated muscles were excised and weighed. The results are expressed as the difference between control and stimulated muscle. Point 0 corresponds to values found immediately after stimulation (see Žák and Gutmann 1960).

Unfortunately the results are not in agreement even here. Mandel et al. (Mandel et al. 1949, Clavert et al. 1949) have demonstrated by histochemical and chemical methods that there is a decrease of RNA in denervated muscle. According to these authors, RNA is localized in the anisotropic discs and decreases by 30—60% by 10—30 days after denervation. DNA remains unchanged. The changes found in these experiments are in absolute units, i. e. content of nuclei acids per muscle. Similar results are reported by Hearn (1959).

In a second group of experiments it has been shown that there is a rise both of RNA and DNA. For example, Schmidt and Schlief (1956) found a 46% increase of RNA and a 115% increase of DNA 40 days after denervation of the rat muscle. The ratio between RNA and DNA is reduced from the normal value 2·5 to 1·6. These results are expressed as mg% per unit dry weight. A similar change can also be observed during the early period of atrophy. It was even found that the DNA content calculated either per wet weight or per non-collagenous protein is increased three days after cutting the nerve. This change becomes more and more evident until it represents a 70—80% increase 14 days after denervation (Gutmann and Žák 1961).

These results are in agreement with the finding of Azzone and Aloisi (1958), who reported an increase of RNA content in myosin and actomyosin of dystrophic muscles — almost by 100%. Hower, Dinning and his coworkers were not able to find any changes of RNA in the whole muscle. Only DNA was relatively higher (Young and Dinning 1951, Day et al. 1957). In muscles of mice with hereditary dystrophy an increase of both nucleic acids was found by Coleman and Ashworth (1959).

Degeneration in the distal part of a peripheral nerve is at first asso ciated with an increase of RNA and DNA by the second day after nerve section. Following this period, however, there is a reduction of RNA from the ninetieth day onwards, and after the sixteenth day for DNA (Logan et al. 1952, Causey and Stratmann 1956b). Changes are explained by increased numbers of Schwann cells and endoneural cells.

It can be seen from the above material that questions dealing with changes of RNA content in atrophic muscles have so far not been finally de ided. It is interesting that the ability of synthesis of ribonucleic acid is not impaired under these conditions, but even increased. This was demonstrated by direct determination of RNA, as well as by measuring the turnover of its separate components. Changes in the content of RNA at various intervals after electrical stimulation of a normal and a denervated muscle are given in Fig. 114. It can be seen that stimulation leads in both cases to a decrease of RNA, the decrease in the denervated muscle being 50% larger. Following stimulation resynthesis takes place and this is an over-shoot reaction in character. The maximal increase is attained two hours after stimulation of normal muscle, while in denervated muscle it is already reached after one hour. Although this maximum is lower in denervated muscle, the total amount of newly synthetized RNA in normal and denervated muscle is the same, since resynthesis in denervated muscle begins from a lower level. The rate of resynthesis, however, is faster in denervated muscle, since resynthesis is realized in half the time needed for normal muscle (Žák and Gutmann 1959, 1960).

The increased turnover of nucleic acids in dystrophic muscles of animals with avitaminosis-E is in accord with the above findings. It has, for example, been shown that incorporation of C^{14}-formate into RNA and DNA in muscles of the rabbit and monkey is increased (Dinning et al. 1955, Dinning and Day 1958), as well as that of C^{14}-guanylic acid and of P^{32} (Dinning et al. 1956a, 1956b). On the other hand the incorporation of C^{14}-glycine is lowered, but it has been shown that dystrophic muscles oxidize more glycine than normal muscles Dinning et al. 1955). Incorporation of glycine-1-C^{14} into nucleic acids of mice with hereditary muscular dystrophy is increased by 30—50% (Coleman and Ashworth 1959).

On comparing the ability of denervated muscle to synthetize proteins and nucleic acids, it can be seen that dissociation occurs between these two processes.

While proteosynthesis in adult animals is severely impaired or even completely lost, ribonucleic acids are synthetized at an even greater rate. This can be demonstrated during resynthesis following functional activity as has already been mentioned above (Fig. 114), as well as during realimentation. It is possible to lower the content of proteins and RNA by fasting. Both these values then return to initial levels after realimentation. The resynthesis of proteins is then markedly preceded by RNA resynthesis. However, if one muscle is denervated prior to realimentation, dissociation between these processes occurs. The level of RNA again returns to normal following three day's realimentation, while the content of proteins remains unchanged even after seven days' realimentation (Žák and Gutmann 1959, 1960).

Can this dissociation be a general result of the loss of physiological regulation? An analogous case has been described in the literature during embryonal development of hypophysectomized chick embryos. The content of RNA and proteins is smaller 16 days after the operation in comparison with the normal embryo. During further development, the levels of RNA remain unchanged, notwithstanding the fact that the content of proteins decreases still further. At first the level of RNA is not changed at all, however, later it begins to accumulate: on comparison with the control embryo its content is increased by 15—22% (Love and Konigsberg 1958). In order to be able to formulate a more general hypothesis about the biological significance of the dissociation between ribonucleic acids and proteins, more experimental material will have to be collected.

What general conclusions can be made from the above material?

The degradation of proteins in denervated muscle is apparently increased: this is borne out by autolysis and also by proteolytic activity. A change in catheptic activity has not been demonstrated unequivocally, as it has not been shown that the level of free amino acids is increased in denervated muscle.

Conclusions cannot yet be made concerning the disturbance of proteins on the basis of the incorporation of labelled amino acids where the conditions are extremely complex and the results are often contradictory. Net protein determination, nevertheless, indicates that a considerable disturbance of proteosynthesis is present. A certain ability of proteosynthesis, however, remains, this being expecially evident in denervated muscles of young animals.

The content of ribonucleic acids calculated in terms of the protein content of atrophic muscle is increased, as is the rate of their synthesis. At the same time, however, the synthesis of ribonucleic acids is not accompanied by proteosynthesis, as in the case of the normal muscle.

As far as other types of atrophy are concerned, it seems that tenotomy and avitaminosis E are associated with increased proteolysis. The content and turnover of RNA has up to now been studied only in avitaminosis E and here an

increase has been found similar to denervated muscle. The dissociation between the synthesis of RNA and that of proteins has, as yet, not been studied anywhere else but in denervated muscle.

References

Abood L. G.: Res. Commun. XX. Congress Int. Physiol. 9 : Brusel 1956.
Aloisi M., Ascenzi A., Bonetti E.: Experimentia 8 : 266, 1952.
— — — Biochim. Biophys. Acta 10 : 70, 1953.
Ashley C. A., Arasimavicius A., Háss G. M.: Exptl. Cell. Research 10 : 1, 1956.
Audova A.: Skand. Arch. Physiol. 44 : 1, 1923.
Avellone L., di Macco G.: Ann. di clin. Med. sper. 15 : 39, 1925.
Azzone G. F., Aloisi M.: Biochim. Biophys. Acta 18 : 451, 1955.
— — Biochem. J. 69 : 161, 1958.
Baechtel, W. R., Allen J. R., Dobson H. L.: Proc. Soc. Exptl. Biol. Med. 96 : 3, 1957.
Beigel viz Cathcart E. P.: Physiol. Rev. 5 : 225, 1925.
Bender A. D., Schottelius D. D., Schottelius B. A.: Proc. Soc. Exptl. Biol. Med. 102 : 362, 1959.
Beznák M.: 1st Congr. Biochem., Cambridge, Abst. Commun. 602, 1949.
Bidinost L. E.: J. Biol. Chem. 190 : 423, 1951.
Blättler E.: Biochem. Z. 221 : 359, 1930.
Bonetti E., Aloisi M., Merucci P.: Experimentia 8 : 69, 1952.
Bradley H. C.: J. Biol. Chem. 33 : XI, 1918.
— Physiol. Rev. 18 : 173, 1938.
Braun G., Dreyfus J. C., Kruh G.: Compt. rend. Soc. biol. 150 : 944, 1956.
Cahn T.: Ann. de Physiol. 3 : 4, 1927.
Caspari W.: P. Arch. ges. Physiol. 83 : 509, 1901.
Cater D. B., Holmes B. E., Mee L. K.: Biochem. J. 66 : 482, 1957.
Cathcart E. P.: Physiol. Rev. 5 : 225, 1925.
Causey G., Stratmann C. L.: Proc. Roy. Soc., B, 144 : 520, 1956a.
— — Biochem. J. 64 : 29, 1956b.
Chagovets E.: Ukr. Biokh. Zh. 12 : 241, 1938a.
Chagovets N. R.: Ukr. Biokh. Zh. 12 : 427, 1938b.
— Ukr. Biokh. Zh. 30 : 661, 1958.
— Ukr. Biokh. Zh. 29 : 450, 1957.
Chen K. K., Bradley H. C.: J. Biol. Chem. 59 : 151, 1924.
Chen K. K., Meek W., Bradley H. C.: J. Biol. Chem. 61 : 807, 1924.
Chepinoga O. P.: Ukr. Biokh. Zh. 10 : 1938.
Chor, H., Dolkart R. E., Davenport H. A.: Am. J. Physiol. 118 : 580, 1937.
Clavert J., Mandel L., Jacob M.: Compt. rend. Soc. Biol. 143 : 539, 1949.
Cole R. D., Coote J.: Work T. S.: Nature 179 : 199, 1957.
Coleman D. L., Ashworth M. E.: Am. J. Physiol. 197 : 839, 1959.
Crepax P.: Chem. Abstr. 45 : 243f, 1951, orig. 1949a.
— Experientia 5, 167, 1949b.
— Chem. Z. 127 : 3068, 1956, orig. 1951a.
— Biochim. Biophys. Acta 7 : 87, 1951b.
— Chem. Abstr. 46 : 9676f, 1952, orig. 1952a.
— Biochim. Biophys. Acta 9 : 385, 1952b.
— Boll. soc. ital. biol. sper. 29 : 1212, 1953a.

Crepax P.: Arch. sci. biol. 37 : 419, 1953b.
Crepax P., Hérion A.: Biochim. Biophys. Acta 6 : 54, 1950.
Csapo A., Herrman H.: Am. J. Physiol. 165 : 701, 1951.
Dalgliesh C. E.: Science 125 : 271, 1957.
Dashmann T., Feigelson P., Feigelson M., Fancher C.: IVth Int. Cong. Biochem., Ref. 6-38, Vienna 1958.
Day P. L., Young D. E., Dinning J. S.: Fed. Proc. 16 : 383, 1957.
Deuticke, J. H.: P. Arch. ges. Physiol. 224 : 1, 1930.
— Z. Physiol. Chem. 210 : 97, 1932.
Diehl J. F.: Arch. Biochem.: 87 : 339, 1960.
Dinning J. S.: J. Biol. Chem. 212 : 735, 1955.
Dinning J. S., Sime J. T., Day P. L.: J. Biol. Chem. 217 : 205, 1955.
— — — J. Biol. Chem. 222 : 215, 1956a.
— — — Biochim. Biophys. Acta 21 : 383, 1956b.
— — — J. Biol. Chem. 233 : 240, 1958.
Drabikowski W.: Acta Biochem. Polonica: 8 : 3, 1961.
Drahota Z.: Physiol. Bohemoslov. 8 : 501, 1959.
Dreyfus J. C., Joly M., Schapira G., Raeber L.: Compt. rend. 236 : 2351, 1953.
Dreyfus J. C., Schapira G., Kruh J.: Compt. rend. Soc. biol. 150 : 1145, 1956.
Dreyfus J. C., Schapira G.: S. de Physiol. 51 : 453, 1959.
Dreyfus J. C., Schapira G., Laurent R.: Compt. rend. Soc. biol. 153 : 928, 1959.
Dreyfus J. C., Kruh J., Schapira G.: Biochem. J. 75 : 574, 1960.
Dubuisson M.: Experimentia 2 : 258, 1946.
— Chem. Abstr. 46 : 9631g, 1952, orig. 1948.
— Nature 116 : 1116, 1950a.
— Biochim. Biophys. Acta 5 : 489, 1950b.
— Biol. Rev. 25 : 46, 1950c.
Dvornikova P. D., Grigoryeva V. A.: Ukr. Biokh. Zh. 32 : 192, 1960.
Eliasson N. A., Hammarsten E., Reichard P., Aqvist S., Theorell B., Ehrensvärd G.: Acta Chem. Scand. 5 : 431, 1951.
Embden G., Metz E.: Z. physiol. Chem. 192 : 233, 1930.
Embden G., Riebeling C., Selter G. E.: Z. physiol. chem. 179 : 147, 1928.
Embden G., Schmidt G.: Z. Physiol. Chem. 186 : 205, 1930.
Faulhaber I., Lehmann F. E., Hahn P.: Helv. Physiol. Pharmacol. Acta 19 : 214, 1961.
Feigelton P., Feigelton M.: Biochim. Biophys. Acta 32 : 430, 1959.
Ferdman, D. L.: Biokhimia 3 : 646, 1938.
— Voprosy biokhimii myshts. Kijev, 1954, p. 113.
— Voprosy Med. khim. 3 : 351, 1957.
Ferdman D. L., Mestechkina A. I.: Ukr. biokh. Zh. 23 : 103, 1951.
Feuer G., Frigyier A.: Acta Physiol. Hung. 3 : 1, 1952.
Fick A., Wislicenus J.: Zentralblatt med. Wiss. 4 : 309, 1866.
Fischer E.: Am. J. Physiol. 131 : 156, 1940.
— Arch. Physic. Med. 29 : 291, 1948.
— in Le Muscle. Etude de Biologie et de Physiologie. Royaumont, 1950, p. 1.
— Am. J. Physic. Med. 34 : 212, 1955.
Fischer E., Bower R. V., Skowlund H. V., Ryland K. W., Copenhaver N. J.: Arch. Phys. Med. 30 : 766, 1949.
Fischer E., Ramsey V. W.: Arch. Phys. Therapy 25 : 709, 1944.
— — Am. J. Physiol. 145 : 571, 1946.
Folberger J.: Čs. fysiol. 10 : 13, 1961.

Fomin S. V., Romanyuk N. M.: Chem. Abstr. 31 : 2660-4, 1937, orig. 1937.
Gaitonde M. K., Richter D.: Proc. Roy. Soc. B : 145 : 83, 1956.
Gefter I. M., Kartashov A. I.: Chem. Zentralblatt 1937 I : 3983, orig. 1935.
Gefter I. M., Kirian V. M.: Biokhimia 2 : 499, 1937.
Geiger A., Dobkin J., Magnes J.: Science 118 : 655, 1953.
Geiger A., Yamasaki S., Lyons R.: Am. J. Physiol. 184 : 239, 1956.
Geiger A., Horvath N., Kawakita Y.: J. Neurochem. 5 : 311, 1960.
Gligliotti G., Lamedica G.: Chem. Abstr. 52 : 15698g, 1958, orig. 1956.
Ginetskinskij A. G., Barbashova Z. I., Ginetsinskaya T. A., Shamarina N. M.: Trudy Fysiol. Inst. I. P. Pavlova I : 44, 1945.
Goldstein B., Milgram E. Yu. : Chem. Abstr. 31 : 74554, 1937, orig. 1937.
Greene Ch. W.: J. Biol. Chem. 39 : 435, 1919.
Greene H.: J. Biol. Chem. 39 : 457, 1919.
Grigorieva V. A.: Ukr. biokh. Zh. 27 : 477, 1955.
Groth D. P.: Biochim. Biophys. Acta 21 : 18, 1956.
Grund G.: Arch. exptl. Pathol. Pharmacol. 67 : 393, 1912.
Gutmann E.: Am. J. Physic. Med. 38 : 104, 1959.
Gutmann E.: Physiol. bohemoslov. 9 : 382, 1960.
Gutmann E, Zák R.: Physiol. Bohemoslov. 1961 in press.
Haan A. M. L.: Dissertation. Nijmegen 1952.
— Biochim. Biophys. Acta 11 : 258, 1953.
Haarmann W., Schroeder E.: Biochem. Z. 296 : 131, 1938.
Halperin L., Okun M., Simonson E., Suirkina G.: Chem. Abstr. 29 : 30107, 1935, orig. 1934.
Hearn G. R.: Am. J. Physiol. 196 : 465, 1959.
Heinberg M., Velick E. F.: J. Biol. Chem. 208 : 725, 1954.
Helander E.: Acta Physiol. Scand. 41, Supl. 141, 1957.
Hensay J.: P. Arch. ges. Physiol. 224 : 44, 1930.
Hermann, H., Nicholas J. S., Vorgian M. E.: Proc. Soc. exp. Biol. Med. 72 : 454, 1949.
Hetényi E.: Acta physiol. Hung. 10 : 421, 1956.
Hevesy G., Ottesen J.: Acta physiol. Scand. 5 : 237, 1943.
Hines H. M., Knowlton G. C.: Am. J. Physiol. 114 : 379, 1933.
Hoagland Ch. L.: Adv. Enz. 6 : 193, 1946.
Humphrey G. F.: see Fischer 1950, orig. 1949.
Hydén H.: Acta physiol. Scand. 6, Suppl. 17, 1943.
— in Neurochemistry, ed. by Elliot, K. A. C., Page, J. H., Quastel, J. H., Springfield, U. S. A., 1955., p. 204.
Ivanov I. I., Kasavina B. S., Chem. Abst. 42 : 69160, 1948, orig. 1948.
Kamner M. E., Peranio A., Bruger M.: Endocrinology 46 : 359, 1950.
Kamp F.: Biochem. Z. 307, 226, 1941.
Kashkadamov V. P.: Vrach 1 : 5, 1897.
Khritina K. M.: Chem. Abstr. 41 : 2474h,i, 1947, orig. 1942.
Kleinzeller, A.: Biochem. J. 36 : 729, 1942.
Kolínská J.: Čs. fysiol. 7 : 223, 1958.
Koszalka T. R., Miller L. L.: Fed. Proc. 17 : 257, 1958.
Koszalka T. R., Miller L. L.: J. Biol. Chem. 235 : 665, 1960.
Kruh J., Dreyfus J. C., Schapira G., Gey C. O.: J. Clin Invest. 39 : 1180, 1960.
Kuraiev D. L.: Jahrberichte Tier-Chem. 25 : 335, 1895.
— Jahrberichte. Tier-Chem. 26 : 487, 1896.
Langley J. N., Kato T.: J. Physiol. 49 : 432, 1915.
Linderstrøm-Lang K.: Bull. Soc. Chim. Biol. 35 : 100, 1953.

Lippmann R. K., Selig S.: viz Fischer E., Ramsey V. W.: 1946, orig. 1928.
Lipschütz A., Audova A.: J. Physiol. 55 : 300, 1921.
Logan J. E., Mannel W. A., Rossister R. J.: Biochem. J. 51 : 482, 1952.
Love D. S., Konigsberg I. R.: Endocrinology 62 : 378, 1958.
Luxoro M.: Nature 188 : 1119, 1960.

Mandel P., Jacob M., Mandel L.: Compt. rend. 228 : 957, 1949.
Martin A. W., Sola O. M.: Fed. Proc. 7 : 78, 1948.
McLean J. R., Cohn G. L., Brandt I. K., Simpson M. V.: J. Biol. Chem. 233 : 657, 1958.
Meyer H. H., Weber K.: Biochem. Z. 266 : 137, 1933.
Mihailović Lj., Janković B. D., Petrović M., Isaković K.: Experientia 14 : 144, 1958.
Mihalyi E., Laki K., Knoller M. I.: Arch. Biochem. 68 : 130, 1957.
Milman A. E., Siledes D. N., Milhorat A. T.: Proc. Soc. exptl. Biol. Med. 83 : 637, 1953.
Mirski A.: J. Gen. Physiol. 20 : 461, 1937.
Mönninghoff G.: Z. physiol. Chem. 284 : 79, 1949.
Mrázek G., Reed C. I.: Am. J. Physiol. 149 : 177, 1947.
Munk I.: Du Bois Arch. 1890 : 557.

Nardone R. M.: Proc. Soc. exp. Biol. Med. 80 : 756, 1952.
Nawrocki F.: Zentralblatt med. Wiss. 4 : 385, 1866.
Nechiporenko Z. I.: Ukr. biokh. Zh. 27 : 146, 1955.
Nechiporenko Z. I., Ferdman D. L.: Dokl. Akad. Nauk, USSR 92 : 803, 1953.
Nechiporenko Z. I., Medovar E. N.: Ukr. biokh. Zh. 29 : 65, 1957.
Neuberger A., Perrone J. C.: Biochem. J. 49 : 199, 1951.
Niklas A., Quincke E., Maurer W., Neyer H.: Biochem. Z. 330 : 1, 1958.
Nöcker J.: Deutsch. Ztschr. Verd. 11 : 69, 1951.

Ogata Y.: Nature 186 : 77, 1960.
Oppenheim N.: P. Arch. ges. Physiol. 23 : 446, 1881.
Orbeli L. A.: Trudy fysiol. Inst. I. P. Pavlova I : 3, 1945.

Padieu P.: Bull. Soc. Chim. biol. 41 : 57, 1959.
Palladin A. B., Belik I. B., Krachko L. I.: Biokhimia 22 : 358, 1957.
Parkes C.: Proc. Roy. Soc. B. 15 : 339, 1866; 16 : 44, 1868; 19 : 349, 1871.
Parnas J. K.: Biochem. Z. 206 : 19, 1929.
Parnas J. K., Mozolowski Wł.: Biochem. Z. 184 : 399, 1927.
Parnas J. K., Mozolowski Wł., Lewinski W.: Biochem. Z. 188 : 15, 1927.
Pennington R. S.: Biochem. J. 77 : 205, 1960.
Peters A. W., Guenther A. E.: J. Biol. Chem. 46 : XXII, 1921.
Popova N. K.: Fisiol. Zh. U. S. S. R. 37 : 103, 1951.
— see Koshtoyants Ch. S.: Sborník prací konference o nervovém řízení metabolismu. Praha 1958, str. 9.
Pugliese A.: Biochem. Z. 33 : 16, 1911.

Raeber L., Schapira G., Dreyfus J. C.: Compt. rend. 241 : 160, 1955.
Rakestraw N. W.: J. Biol. Chem. 47 : 565, 1921.
Renards S.: viz Fischer E. 1950.
Rogozkin B. A.: Ukr. Biokh. Zh. 31 : 489, 1959.
Rosengart, V. I., Maslova M. V.: Dokl. Akad. Nauk, USSR 109 : 1176, 1956.
Roth J. S.: Fed. Proc. 17 : 300, 1958.
— Arch. Biochem. 74 : 277, 1958.
Rychlík I., Šorm F.: Chem. listy 51 : 1738, 1957.

Samuels A.: Am. J. Physiol. 190 : 377, 1957.
Schapira G., Coursaget J., Dreyfus J. C., Schapira F.: Bull. soc. Chim. biol. 35 : 1309, 1953.

Shcapira G., Dreyfus J. C.: in Le Muscle Étude de Biollogie et de Pathologie, Ryaumont 1950, p. 299.
Schapira G., Dreyfus J. C.: Compt rend. Soc. biol. 152 : 1705, 1958.
Schapira G., Dreyfus J. C., Joly M.: Nature 170 : 494, 1952.
Schapira G., Dreyfus J. C., Kruh J., Labie D., Padieu P.: Bull. Soc. Chim. biol. 41 : 469, 1959.
Schapira G., Dra}fus J. C., Schapira F.: Compt. rend. Soc. biol. 144 : 829, 1950.
Schapira G., Kruh J., Dreyfus J. C., Schapira F., Tikhonitzky L., Alexandre Y.: J. Biol. Chem. 235 : 1738, 1960.
Schmidt C. G., Schlief H.: Z. ges. exptl. Med. 127 : 53, 1956.
Seitz W.: Z. physiol. Chem. 218 : 17, 1933.
Sheves G. S.: Biokhimia 20 : 152, 1955.
Sheves G. S., Epelbaum S. E., Ryumina V. I.: Biokhimia 21 : 71, 1956.
Seraydarian M. W., Williams E. G.: Fed. Proc. 19 : 255, 1960.
Siekevitz P.: Biol. Chem. 195 : 549, 1952.
Silakova G. I.: Ukr. biokh, Zh. 29 : 60, 1957.
Simon E. J., Lessell I., Gross Ch. S., Milhorat A. T.: Fed. Proc. 17 : 311, 1958.
Simpson M. V.: J. Biol. Chem. 216 : 179, 1955.
Simpson M. V., McLean J. R.: Biochim. Biophys. Acta 18 : 573, 1955.
Slack H. G. B.: Clin. Science 13 : 155, 1954.
Smith — see Cathcart 1925.
Smith E.: Proc. Roy. Soc. 124 B. : 136, 1937.
Smorodintsev A.: Usp. Khim. 4 : 632, 1935.
Snoke J. E., Neurath H.: J. Biol. Chem. 187 : 127, 1950.
Sola O. M., Martin A. W.: Am. J. Physiol. 172 : 324, 1953.
Soula C.: Compt. rend. Soc. Biol. 73 : 404, 1912; 74 : 330, 1913.
Spencer H. C., Morgulis S., Wilder V. M.: J. Biol. Chem. 120 : 257, 1937.
Steppuhn O., Rewsner J., Timofejewa A.: Biochem. Z. 175 : 471, 1926.
Stewart D. M.: Biochem. J. 59 : 553, 1955.
Stewart D. M., Martin A. W.: Am. J. Physiol. 186 : 497, 1956.
Steyrer A.: Beitr. chem. physiol. Pathol. 4 : 234, 1903.
Tallan H. H.: Proc. Soc. exptl. Biol. Med. 89 : 553, 1955.
Thomson J. D.: Am. J. Physiol. 180 : 202, 1955.
Thomson J. D., Morgan J. A., Brodish A.: Am. J. Physiol. 167 : 832, 1951.
Ungar G., Aschhein E., Psychoyos S., Romano D. V.: J. gen. Physiol. 40 : 635, 1957.
Usami S.: Chem. Abstr. 51 : 18613g, 1957.
Ussing H. H.: Acta physiol. Scand. 2 : 209, 1941.
Velick S. F.: Biochim. Biophys. Acta 20 : 228, 1956.
Verbolovich P. A., Aleksandrovich N.: Biul. eksp. biol. med. 29 : 196, 1950.
Vrba R.: Nature 176 : 1258, 1955.
— J. Neurochem. 1 : 12, 1956.
Vrba R., Folbergrová, J.: Nature 182 : 237, 1958.
Vrbová G.: Čs. fysiol. 1 : 38, 1952.
Walter H.: Nature 188 : 643, 1960.
Weinstock I. M., Epstein S., Milhorat A. T.: Proc. Soc. exptl. Biol. Med. 99 : 272, 1958.
Weinstock I. M., Goldrich A. D., Milhorat A. T.: Proc. Soc. exptl. Biol. Med. 88 : 257, 1955.
Westenbrink H. G. K., Krabbe H.: Chem. Abstr. 31 : 18617, 1937, orig. 1936.
Whipple G. H.: Am. J. Physiol. 76 : 708, 1926.
Williams J. W., Schur P. E., Elvehjem C. A.: J. Biol. Chem. 182 : 55, 1950.
Winnick R. E., Winnick T.: J. Biol. Chem. 235 : 2657, 1960.
Yakovlev N. N.: Fysiol. Zh. USSR 34 : 717, 1948.

Young J. M., Dinning J. S.: J. Biol. Chem. 193 : 743, 1951.
Žák R.: Čs. fysiol. 7 : 33, 1958.
— Čs. fysiol. 8 : 515, 1959.
Žák R., Babický A.: Čs. fysiol. 7 : 591, 1958.
Žák R., Babický A., Vereš K.: Biochem. Z. 331 : 117, 1959.
Žák R., Drahota Z.: Nature 186 : 973, 1960.
Žák R., Gutmann E.: Čs. fysiol. 8 : 264, 1959.
— — Nature 185 : 766, 1960.
Žák R., Gutmann E., Vrbová G.: Experientia 8 : 80, 1957a.
— — — Physiol. Bohemoslov. 6 : 309, 1957b.
Žák R., Klicpera M.: Biochem. Z. 329, 521, 1958.
Zelená J., Hník P.: Čs. fysiol. 6 : 159, 1957.
Zubenko P. M.: Voprosy biokhimii myshts, Kiev 1954, p. 149.
Zubenko P. M., Reva A. D., Plakhotishina E. P.: Biokhimia 15 : 79, 1950.

Chapter X

RATE OF DENERVATION MUSCLE ATROPHY

Hník P.

Research on factors affecting the rate of muscle atrophy represents a problem important from both the theoretical and practical points of view. From the theoretical aspect this study can represent an approach to elucidating mechanisms leading to muscle atrophy and to questions of the nervous regulation of normal muscle nutrition. This question is also of practical significance for the clinician, since precise knowledge of therapeutic measures will enable him to give optimal help to the muscle during the denervation period, so that regenerating nerve fibres find it in the best possible state during reinnervation. When experimental or therapeutic procedures aimed at retarding the rate of denervation atrophy are being evaluated, questions concerning objective evaluation of rate of muscle atrophy are of primary interest.

A. Biometric problems associated with the evaluation of muscle atrophy

Evaluation of the rate of muscle atrophy is related to the problem of speed in biological processes in general. A given pathological process is usually not restricted to local changes only, but is accompanied by changes of a more general character, which can again influence the original process. In the case of denervation atrophy the question arises how to express and evaluate changes in muscle weight. Is it better to evaluate and compare absolute muscle weights, or is it more advantageous to express the weight in relation to certain reference points? In the first case it is necessary to work with groups of animals of identical body weight and to perform a large number of experiments because of relatively large biological variations in muscle weights. Most authors have therefore chosen the second method, i. e. to use reference points for expressing their results. The most usual way of evaluating changes of muscle weight is to compare denervated muscles with muscles of the contralateral limb, which serves as control. It is only exceptionally that muscle weight is expressed in relation to other reference points, e. g. body weight etc. This problem is not specific for the evaluation of muscle weight changes. For example, various rates of nerve fibre regeneration in the peripheral nerve were found to be dependent on indicators used (Gutmann et al. 1942). However, it is necessary to bear in mind that even if the problem of the rate of atrophy is solved satisfactorily, wet muscle weight is only of limited significance. It does not give any information about the biochemical composition of the muscle and about the constituent, contractile or sarcoplasmatic, which is being affected by a given procedure. Dry muscle weight is perhaps more useful in obtaining data concerning muscle proteins (there is a fairly satisfactory correlation between dry weight and protein nitrogen — see Chapter IX). But, of course, information about functional muscle proteins can be obtained only by means of biochemical extraction of contractile proteins and by testing the functional capacity of the muscle (muscle strength and endurance), calculated per unit cross-section or unit muscle phase. For first orientation, however, wet and dry muscle weight determinations are sufficient. If positive results are obtained, however, this purely quantitative evaluation does not suffice and considerable error may be made if premature conclusions are made concerning the efficancy of a given method, without qualitative analysis of metabolic and functional properties of the muscle tissue.

If these questions of reference points and indicators were satisfactorily solved in the evaluation of the rate of muscle atrophy, then for clinical work there would be ideal quantitative criteria for evaluating therapeutic measures for affecting muscle atrophy and its prognosis: rate of regeneration of peripheral nerve fibres, retardation of atrophy, contractile strength of denervated muscle (to direct stimulation) or of reinnervated muscle (to indirect stimulation — or on attempting voluntary contraction).

1. Inaccuracies of currently used criteria for evaluating the rate of muscle atrophy

a) When evaluating muscle atrophy or factors influencing the rate of muscle atrophy, it is necessary to bear in mind that the most commonly used method according to which the contralateral muscles serve as control (D/C in %, or (C — D)/C in %) can introduce considerable error into the calculation. Usually one does not take into account that the weight of control muscles can change during the experiment either by further growth or, on the contrary, lose weight as a result of changes of body weight. This would mean, that the reference point in relation to which weight changes of denervated muscle were calculated is not stable and this factor can alter considerably any conclusions drawn.

This question has been studied by a number of authors (Langley and Kato 1915, Audova 1921, Hines et al. 1945). Knowlton and Hines (1936) came to the conclusion that contralateral muscles can be considered as control only if experiments are performed on adult animals in which it can be assumed that no changes either in body or muscle weight have taken place. It is difficult, however, to fulfill this condition, especially in the rat — most commonly used for these experiments — since growth in this animal is not arrested until a very late stage of life. This factor thus necessarily plays a role, at least in the rat, especially in long term experiments. This fact is rarely respected.

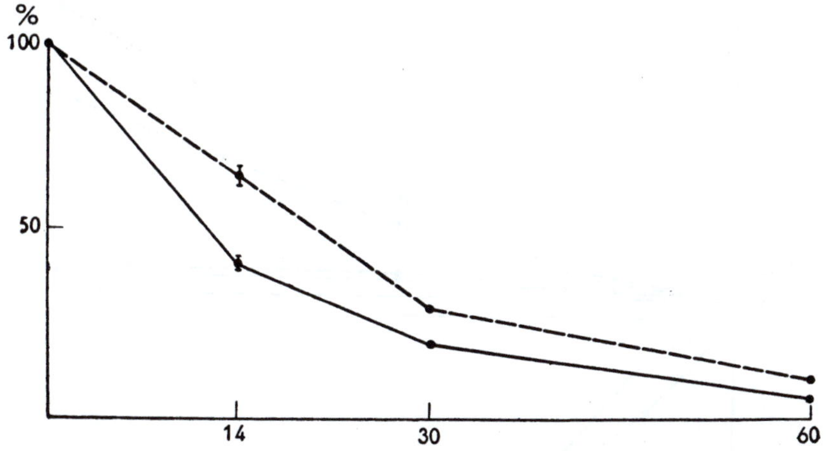

Fig. 115. Muscle atrophy of the soleus (solid line) and extensor digitorum longus (interrupted line) in young rats following section of the sciatic nerve immediately after birth. Abscissa: days after denervation. Ordinate: weight of the denervated muscle in percentage of the control side. Vertical lines give the standard error of the arithmetical mean (S. E.), as long as they are greater than the black points.

Muscle atrophy in young animals (Zelená and Hník 1957) can serve as illustration of the way in which evaluation in percentage of the control side can lead to distortion of the results. Following section of the sciatic nerve in newborn rats, atrophy of the soleus and extensor digitorum longus on evaluation D/C × 100 apparently progresses at dramatic speed as compared with muscle atrophy in adults (Fig. 115). However, on calculation of muscle weight at various denervation periods in terms of the initial muscle weight at the beginning of the experiment, it is obvious that changes occurring in young animals are quite different from those in adult animals (Fig. 116). While in adult animals denervated muscles lose weight, an absolute weight increment was actually registered in young animals. This increase, however, is far smaller than the growth increment on the other side, so that on comparison of the denervated and control sides the percentage difference after 60 days is larger than in adults, even though actually no decrease in weight has occurred. This finding corroborates earlier results of Schapira et al. (1950) who reported that an absolute increase of muscle proteins occurs following denervation in young rabbits.

b) Biometric equations elaborated by Schwartz et al. (1949) and Schapira and Dreyfus (1950) represented great progress. These authors investigated the re-

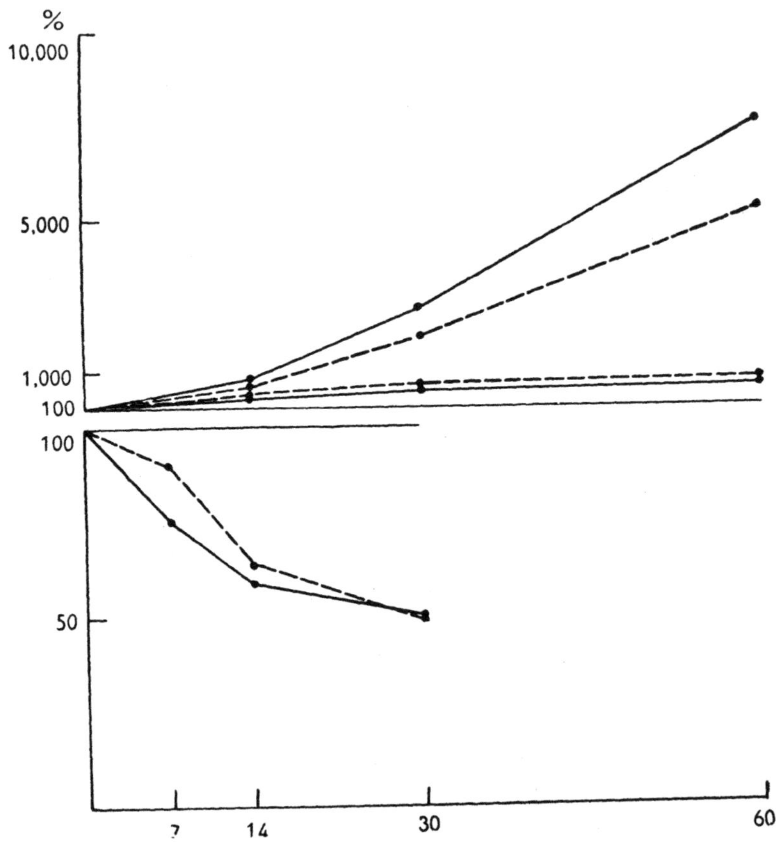

Fig. 116. Weight increase in normal and denervated muscles of young rats and the weight decrement in denervated muscles of adult rats. Section of the sciatic nerve was performed at birth. Abscissa: days after denervation. Ordinate: weight increment in muscles of young rats in percentage of initial weight at birth and the loss of muscle weight in adult rats expressed as percentage of the control side respectively. Solid line — soleus; interrupted line — extensor digitorum longus. Upper two pairs of curves — normal muscles in young rats; middle pair — denervated muscles in young rats; lower pair — denervated muscles in adult animals.

lationship between the weight of normal gastrocnemius and body weight of rats of different ages and found that, within the range of 20—225 g. body weight, this relation is given by a straight line on a logarithmic scale, the slope of which is 1·28.

The muscle constant $C_M = 1{\cdot}28 \log P - \log N = 0{\cdot}503$ (1), where P = body weight

N = weight of the normal muscle.

A similar relation was found for the denervated gastrocnemius ten days after section of the nerve in animals of different body weight — the gradient of the straight line relation was found in this case to be 1·35.

From these two equations the relationship between the normal and denervated gastrocnemius was computed for the ten-day period after denervation, independently of body weight — for rats within the range from 20—225 g. body weight. The constant of atrophy $C_A = 1{\cdot}07 \log N - \log A = 0{\cdot}354$ (2).

These equations have a number of advantages. Besides making the evaluation of atrophy possible without having to take into account differences in body weight (in the above range), equation (2) can, according to the above authors, help to determine the effect of therapeutic measures on the normal muscle (1). However, as is pointed out by these authors themselves, equation (2) can be used only if normal muscles remained unaffected by the therapeutic procedure, which would become evident by a significant difference from value C_M in equation (1).

It is almost impossible to fulfill this condition, especially where hormonal factors are concerned, since they usually influence the whole organism. If it is necessary to follow changes in the course of a longer period of time, then this evaluation is not too advantageous, since it means that these relations must be determined for each point of the gradient separately. Besides this it is known (see below) that when investigating factors influencing the atrophy of skeletal muscles, it is necessary to study more than one muscle, since differences between muscles make general conclusions about changes in weight obtained from one single muscle difficult and inaccurate. It would thus be necessary to calculate more equations and relationships for different muscles at different time intervals after denervation.

2. Investigation of atrophic gradients

The course of changes, i. e. the rate of atrophy, represents an important aspect of investigation of muscle atrophy. This is only possible if the whole gradient of changes is studied. Muscle weights can be evaluated in several ways:

a) As percentage of the contralateral muscle. This method can give objective information about the changes only if the weight of control muscles remains unchanged during the experiment.

b) Evaluation of muscle weight in absolute units (mg.). The assumption for this method is to have animals of the same body weight for individual points of the gradient. This condition is sometimes difficult to fulfill, especially in cases where the body weight changes under the influence of experimental conditions (either loss of weight or an increase in weight due to growth).

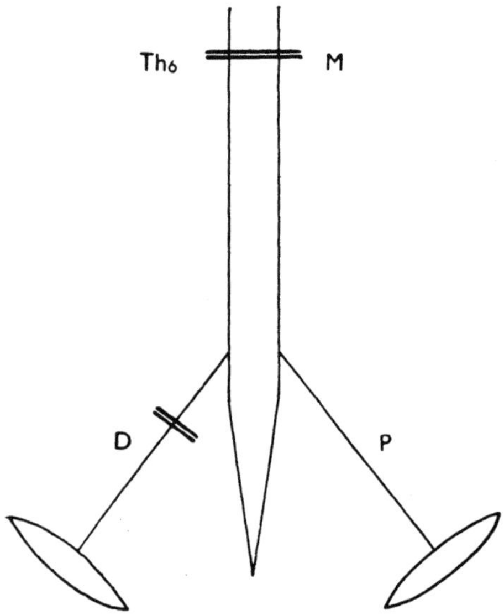

Fig. 117. Scheme of the experiment. T_6-M: transection of the spinal cord at T_6. D: section of the sciatic nerve. P: plegic side with intact peripheral innervation.

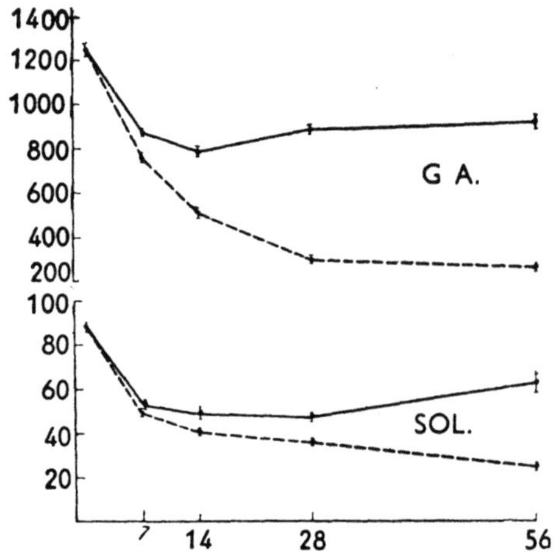

Fig. 118. Changes in weight of plegic (solid line) and denervated muscles (interrupted line) in mg. wet weight. GA. — gastrocnemius, Sol. — soleus. Abscissa: time after transection of the spinal cord and unilateral denervation in days.

c) This difficulty can be eliminated by comparing muscles of experimental animals with control animals of similar body weight — i. e. by expressing muscle weights as a percentage of initial values. This method is especially suitable when original body weight is the same, but has changed as the result of experimental conditions. The changes found, however, can be distorted to a certain extent, for example, as the result of a simultaneous reduction in muscle and body weight. It is also a disadvantage that such group experiments require relatively large numbers of animals.

d) A relatively good method of expressing results is their evaluation in relation to body weight, as an index:

$$\frac{\text{weight of muscle in mg.}}{\text{body weight in g.}}$$

This method ensures, to a great extent, the differentiation which of the changes found in the muscle can be considered as purely muscular in origin and which are probably associated with general changes in body weight.

e) It is also possible to express results as percentage of initial index values in control rats of the same original weight (i. e. again group experiments). This method of evaluation, however, appears to be superfluous.

These methods of evaluation require one basic condition: that the initial weight of all animals (control and experimental) should be the same as far as possible. When the weights (wet and dry) of atrophic and control muscles are obtained, then all the above methods of evaluation can be calculated from the same experimental material.

As will be demonstrated below, it is often necessary to proceed in this way when evaluating muscle weight changes.

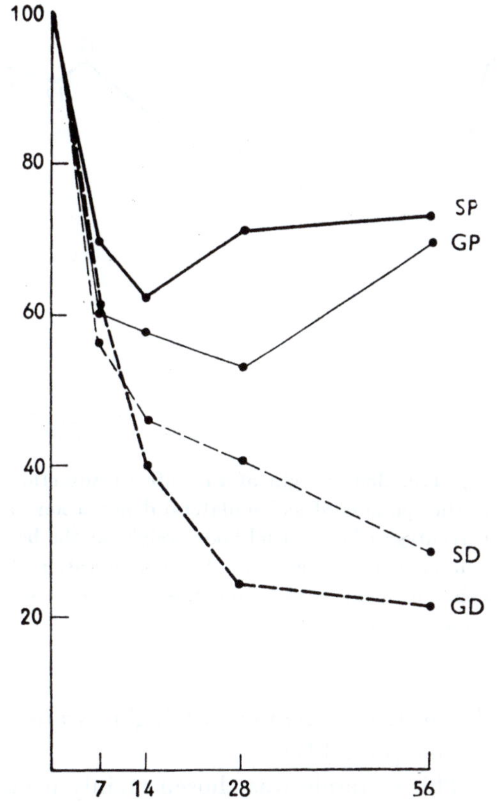

Fig. 119. Changes in weight of plegic (P) and denervated (D) muscles expressed as percentage of initial muscle weight (muscle weight in control animals). G — gastrocnemius, S — soleus. Abscissa: time after transection of the spinal cord and unilateral denervation in days.

It was shown by Fischer (1941) and Solandt and Magladery (1942,) that the atrophy of decentralized (plegic) gastrocnemius-soleus muscles after transection of the spinal cord takes place in two phases. During the first phase plegic muscles lose weight at the same rate as muscle on the contralateral extremity which is denervated in these experiments (see schematic representation of the experiment in Fig. 117). Later on, however, the atrophy on the plegic side is arrested and muscles gradually regain weight. This is the case when results are evaluated in absolute units (in mg.) — (Fig. 118), and on evaluation in % of initial muscle weight (group experiments — Fig. 119). It was found, however, that in the course of two months following transection of the spinal cord a transient loss of body weight takes place and this gradually returns to normal, so that at the end of two months experimental animals have the same body weight as at the beginning of the experiment (Fig. 120). When calculating muscle weight in relation to body weight, however, it was concluded that the increase in the weight of plegic muscles during the second phase is probably not a question of actual muscle changes, but that this increase is proportional to the increment of body weight and is thus given rather by a general state of nutrition than by local muscle changes (Hník 1960). Index values two months after the operation

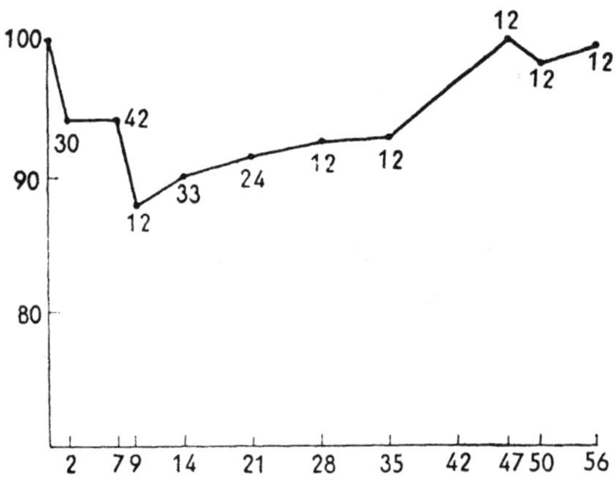

Fig. 120. Body weight of rats after transection of the spinal cord and unilateral denervation in percentage of the initial body weight at the beginning of the experiment. Abscissa: duration of experiment in days. Numerals on the curve: number of animals.

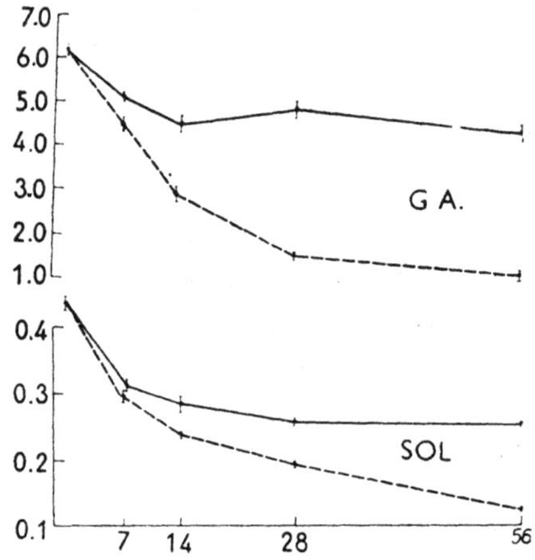

Fig. 121. Changes in weight of plegic (solid line) and denervated (interrupted line) muscles expressed in relation to body weight as an index (see text). GA — gastrocnemius, Sol. — soleus. Abscissa: time after transection of the spinal cord and unilateral denervation in days.

do not rise to significantly higher values as compared with values in the atrophic period (Fig. 121).

This example was chosen mainly because it illustrates the role which changes in body weight can play in the interpretation of some results. It is, of course, a case considerably more complex than the usual denervation studies with unilateral section of the nerve, since there are further accessory factors connected with the relatively grave operation on the spinal cord which necessarily lead to a reaction of the whole organism.

However, even in the case of unilateral denervation there is the question of the influence on the contralateral unoperated side. One factor, the gradual growth of control muscles, has already been mentioned. The weight of control muscles, however, can be affected far more rapidly by the exclusion of one extremity from function, which means that higher functional demands are laid on the normal, innervated limb, resulting in compensatory hypertrophy of its musculature. Both these factors would thus lead to an increase in weight of control muscles and this would mean that the atrophy evaluated in percentage D/C would appear to be faster than is actually the case. A number of authors have investigated the question, whether after unilateral denervation muscle hypertrophy does not occur and have answered negatively (Langley and Kato 1915, Audova 1921, Knowlton and Hines 1936, Sutfin et al. 1954 and others); on the other hand, Danielson (1955) described a loss in weight not only of denervated muscles, but also of muscles on the contralateral non-denervated extremity.

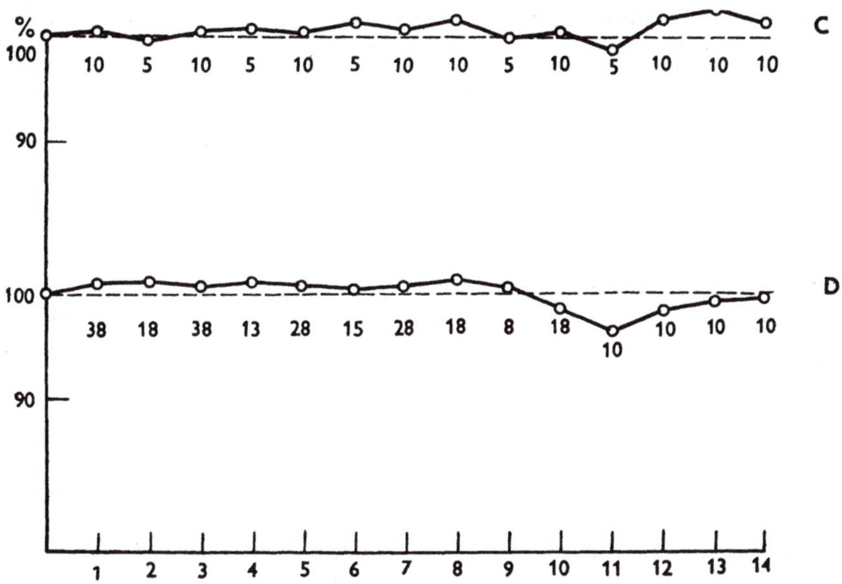

Fig. 122. Body weight of control rats (C) and rats with unilateral section of the sciatic nerve (D) in percentage of initial body weight at beginning of experiment. Abscissa: duration of experiment in days. Numerals on the curve: number of animals.

For this reason it was deemed necessary to control these findings experimentally and to evaluate changes in muscle weight from the aspect of the above criteria in denervated and control extremities.

Experiments were performed in female rats of 200·0 ± 4·85 g. average body weight. A group of control animals was taken into the experiment at the beginning and the weight of these muscles (soleus, extensor digitorum longus, tibialis anterior and gastrocnemius muscles) was taken as the initial value. On the same day unilateral section of the sciatic nerve was performed in the experimental animals and the above muscles were excised three, seven, ten and fourteen days after denervation. The rats were weighed daily; in addition one control group of the same initial weight was weighed in the course of 14 days in order to determine, whether denervated animals gained in weight in an analogous manner or differently from these unoperated animals. It was found that unilateral denervation did not lead to a decrease of body weight, so that both groups of animals — control and denervated — had the same weight at the beginning and at the end of the experiment (Fig. 122). It is also possible to say that the control muscles of the contralateral unoperated limb remain unchanged

Table 21

Wet muscle weight following unilateral section of the sciatic nerve in the rat. D = denervated muscles, C = control muscles. N = weight of muscles in normal non-operated rats. 3, 7, 10 and 14 D = days after denervation. Results are given in mg ± S. E. n = number of animals.

Muscle		N (n = 10)	3 D (n = 10)	7 D (n = 10)	10 D (n = 8)	14 D (n = 10)
M. soleus	D C	86·9 ± 1·68	73·6 ± 3·90 77·4 ± 3·69	61·2 ± 1·78 85·9 ± 3·03	58·6 ± 2·63 84·7 ± 3·77	51·4 ± 2·76 80·3 ± 2·45
M. ext. dig. l.	D C	107·5 ± 1·79	113·2 ± 3·27 110·2 ± 4·08	92·3 ± 3·13 111·8 ± 2·78	86·0 ± 2·63 110·3 ± 4·28	67·8 ± 2·47 108·8 ± 1·90
M. tib. ant.	D C	393 ± 3·89	351 ± 14·5 379 ± 10·8	294 ± 6·58 369 ± 9·45	280 ± 7·3 402 ± 15·95	208 ± 10·3 382 ± 13·2
M. gastrocnemius	D C	1226 ± 15·6	1106 ± 30·5 1205 ± 37·4	850 ± 23·5 1188 ± 32·9	794 ± 36·7 1265 ± 63·6	627 ± 29·9 1247 ± 38·8
Body weight		202 ± 1·88	203 ± 3·22	201 ± 3·78	203 ± 2·91	198 ± 2·32

during this period (Table 21 and 22), even though a certain tendency of control muscles to gain in weight is indicated and may become, perhaps, more evident at a later period. It is thus necessary to agree with the aforementioned authors that as long as the body weight and weight of control muscles does not change

Table 22

Dry muscle weight following unilateral section of the sciatic nerve. For explanation see Table 1. Muscle weight in mg.

Muscle		N (n = 10)	3 D (n = 10)	7 D (n = 10)	10 D (n = 8)	14 D (n = 10)
M. soleus	D C	21·0 ± 0·50	18·2 ± 0·97 19·8 ± 0·90	14·6 ± 0·49 21·4 ± 0·91	14·1 ± 0·71 20·6 ± 0·76	12·1 ± 0·69 19·1 ± 0·64
M. ext. dig. long.	D C	26·5 ± 0·49	27·2 ± 0·73 29·0 ± 0·58	21·6 ± 0·85 27·7 ± 0·86	20·0 ± 0·56 27·9 ± 1·20	16·0 ± 0·69 27·5 ± 0·74
M. tib. ant.	D C	98·3 ± 1·02	88·5 ± 2·30 95·9 ± 2·80	68·9 ± 1·62 92·7 ± 3·09	66·2 ± 1·52 101·6 ± 3·76	49·5 ± 2·46 95·8 ± 3·11
M. gastrocnemius	D C	307 ± 3·86	275·3 ± 7·30 303·4 ± 9·40	204·3 ± 5·60 295·8 ± 8·30	191·0 ± 7·68 318·0 ± 14·95	162·0 ± 11·85 313·0 ± 9·16

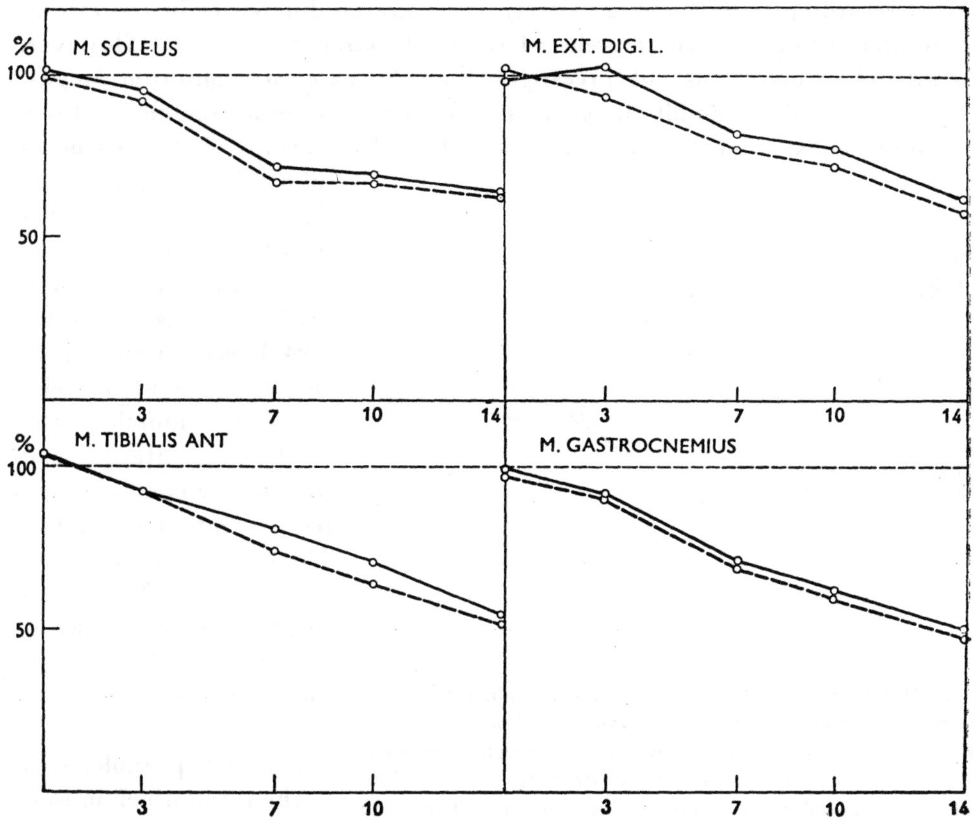

Fig. 123. Denervation atrophy in four muscles of the rat in percentage of the control side. Wet weight — solid line; dry weight — interrupted line. Abscissa: time after denervation in days.

during the experiment, it is possible to evaluate the percentage loss of the denervated side in terms of the control (Fig. 123) without introducing inaccuracy into the calculation.

At the same time it is evident that changes in wet and dry muscle weights do not differ conspicuously in these four muscles when evaluated in this way. It is characteristic that the percentage decrease of dry weight is always a little greater than that of wet weight, which is not surprising since there is a relative increase of extracellular water in the denervated muscle (Audova 1921, Hines and Knowlton 1934, Fenn 1937, Drahota 1959). Only in the case of the extensor digitorum longus is the difference between wet and dry weight more marked three days after denervation (Fig. 124). There is an evident difference between atrophy of the soleus and extensor digitorum longus which has already been pointed out in previous reports (Hník 1955, Beránek et al. 1959, Zelená and Hník

1957). According to Sutfin et al. (1954) there are no differences in the rate of denervation atrophy in the muscles of the hind extremity in the rat. However, the rate of atrophy was not investigated in the extensor digitorum longus, which was found to exhibit the greatest difference from other muscles in these and previous experiments (Beránek et al. 1957). The atrophy of the gastrocnemius also progresses at the fastest rate of all the muscles investigated. This observation was already made by Weiss and Edds (1946) according to whom this muscle contains relatively more muscle phase, so that percentage loss of this muscle can be expected to progress faster. These authors suggest a half-time value for expressing the rate of atrophy as a more suitable criterion for evaluating the rate of atrophy.

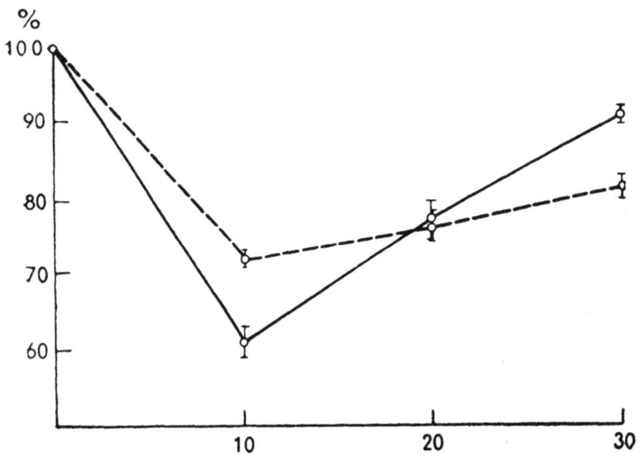

Fig. 124. The rate of atrophy in the denervation period and the recovery of weight during reinnervation following crushing of the sciatic nerve in the rat. Wet muscle weight expressed as percentage of the contralateral control. M. soleus — full line; m. extensor digitorum longus — interrupted line (Beránek et al. 1957).

It is thus possible, even in the rat in which muscle atrophy progresses at a relatively rapid rate, to observe certain differences between various muscles of the hind limb. Besides differences between different muscles of the same animal, which will be discussed in more detail below, the rate of muscle atrophy is to a large extent dependent on the animal species investigated.

B. Differences in the rate of denervation atrophy in different animal species

Knowlton and Hines (1936) were among the first to study systematically the question of the rate of atrophy in relation to animal species (see also Sutfin et al. 1954). The above authors investigated the rate of denervation atrophy of the gastrocnemius on relatively large numbers of mice, rats, cats, dogs and pigeons.

Their results were summarized in the following equation, which characterizes the rate of atrophy for a given species:

$$K = \frac{1}{t} \cdot \log 10 \cdot \frac{A}{A - x}$$

where K = constant for a given species, for example K = 0·028 in mice, 0·020 in rats, 0·0037 in dogs etc. and is probably dependent on the rate of growth and length of life (mouse, two years, rat, 3·5 years, dog 12—20 years),

t = time after denervation in days (from moment of loss of nerve viability),

A = percentage of the initial weight subject to atrophy (muscle phase in the gastrocnemius is 85·5%, 78·1% in the soleus, 85·9% in the anterior tibial, 83·6% in the rectus femoris, etc.).

X = percentage loss of weight in time t.

It was concluded by these authors from the above results that there is, perhaps, a direct correlation between the rate of muscle atrophy and the intensity of metabolism and/or the life span.

In man denervation atrophy progresses at a relatively slow rate (Bowden and Gutmann 1944). On the other hand, atrophy in rats is fast and the rat appears, for this reason, to be a particularly suitable experimental object for studying factors slowing down denervation atrophy. An interesting comparison of the rate of denervation atrophy in different animals species from the literature was made by Sunderland and Ray (1950), mainly, however, without data concerning the particular muscle investigated. As has been pointed out in the preceding section, however, denervation changes and atrophy progress differently even in the same animal in relation to the muscle, or functional group of muscles which is being studied.

C. Changes in the rate of denervation atrophy in various muscles

There are considerable differences between skeletal muscles as far as their functional, morphological, biochemical and other properties are concerned. Since this problem is relatively complicated it would be useful, before discussing differences between denervated muscles, first to mention differences already reported between normal skeletal muscles.

1. Differences between normal muscles

The classification of skeletal muscles according to their colour into red and pale (Needham 1926), on the basis of different content of myoglobin, stresses only one aspect of these differences.

a) Functional differences

From the functional point of view muscles can be classified into muscles with a predominantly static function (chiefly antigravity muscles, physiological extensor muscles) and muscles with a mainly dynamic or phasic function (chiefly physiological flexor muscles). The red muscles are associated with a static or tonic function and pale muscles with a dynamic or phasic function. This functional classification is considered, for example, by Yakovlev and Yakovleva (1953) to be more appropriate than muscle colour, since in mammals and man the latter is often independent of function.

Muscles with a predominantly static function are characterized by relatively slow contraction (long contraction time) and slow relaxation. The contraction and relaxation in dynamic muscles, on the other hand, takes place at a considerably faster rate. These differences also apply to their functional mobility (Sharipova and Zhukov 1954, Buller, Eccles and Eccles 1960).

b) Morphological differences

The above functional differences are accompanied by morphological differences characteristic for each group. Static muscles usually contain short muscle fibres mostly obliquely arranged with a relatively high content of connective tissue. These muscles usually have a broad insertion placed at some distance from the joint so that movements elicited by these muscles are rough, slow and not too precise. On histological examination it is evident that the muscle cells are rich in sarcoplasm and contain numerous lipid granules which make the sections of these fibres appear dark. This darkness is not caused by myoglobin but by these granules (Denny-Brown, in Bennett et al. 1958). As a result of the rich content of sarcoplasma and high level of lipids these fibres are of larger diameter.

Flexor muscles, on the other hand, have long muscle fibres longitudinally arranged which are relatively poor in sarcoplasma and lipid granules, but rich in fibrils. The insertion of these muscles is usually near the joint, so that they perform more precise movements without exhibiting much strength.

Red muscles also contain a large number of sarcosomata which are localized primarily in the anisotropic segments of fibrils, while in pale muscles there are fewer sarcosomata, they are smaller in size and are localized in the isotropic segments (see Slater 1957). An interesting observation was made by Denny-Brown (1958, see Bennett et al.) who reported that during development all muscle fibres are originally granulated and that pale muscles gradually lose these lipid granules concomitantly with functional differentiation. At the same time the contraction time shortens (Koshtoyants 1938, Denny-Brown — v. Bennett et al. 1958, Buller et al. 1960).

c) Biochemical differences

Muscles also differ from one another biochemically in relation to their functional differences.

Muscles performing rapid movements have a predominantly anaerobic type of metabolism. They thus contain more creatinephosphate, hexosephosphate, glycogen, carnosine and higher phosphorolytic activity. On the other hand, the slow-contracting muscles have a predominantly aerobic metabolism (high content of phospholipids, cholesterol, glutathione, ascorbic acid and oxidative enzymes e. g. cytochromes, etc.), so that the oxygen debt in these muscles, which are almost continuously in function (antigravity, tonic function for maintaining posture), is reduced to a minimum. The slow tonic muscles also have a higher consumption of oxygen than the fast muscles (Harman 1952). An extreme case of this appears to be the heart muscle which, according to Lawrie (1952), is highly dependent on oxygen supply for its function and is practically unable to work under conditions of oxygen debt.

These differences in biochemical characteristics of these two types of muscles are given, to a great extent, by their functional activity. According to Yakovlev and Yakovleva (1953) it is possible to obtain biochemical and functional modification of static muscles to the dynamic type by regular training of slow muscles by fast contractions, and vice versa. It would thus be possible to obtain adaptation of functional and biochemical properties of muscle through the influence of functional activity.

However, slow and fast muscles also differ as far as their content of ions, intracellular water and uptake of some labelled ions is concerned. It was found, for example, that extensor digitorum longus (fast muscle) in the rat contains more potassium than the soleus (slow muscle). On the contrary, the content of water and chloride ions is higher in the soleus than in extensor digitorum (Drahota et al. 1959). As far as the uptake of labelled ions is concerned, the uptake of Na^{24} and P^{32} is higher in the soleus than in the extensor digitorum (Hudlicka 1955).

d) Differences in the reactivity of muscles to humoral agents

The above biochemical differences between muscles most probably determine differences which have been demonstrated in their reactivity to humoral agents. It was found, for example, that various muscles behave differently under the influence of various hormones. An increment in weight after administration of

STH, for example, can be observed in most muscles. These changes, however, were found to be minimal in heart muscle and in soleus (Greenbaum and Young 1953). Differences in the reactivity of muscles to hormonal influences (hypophysectomy, castration and administration of testosterone) have also been reported by Karé (1953), namely between perineal muscles and rectus femoris. Korner (1955) and Korner and Young (1955) also described differences in the effect of methylandrostenediol on the weight of various skeletal muscles. Similar differences between various muscles were also found by Scow and Roe (1953) and Scow and Hagan (1955). The loss in muscle weight also differs from muscle to muscle after starvation. For example, the soleus and masseter in the rat lose practically no weight during starvation, while other muscles and tissues lose weight more or less in proportion to body weight (Greenbaum and Young 1953). Different rates of atrophy in various muscles of the guinea-pig during starvation was also described by Kochakian et al. (1956). It is interesting that soleus is again mentioned in connection with muscles which lose relatively little weight (together with adductor longus, posterior tibial and the diaphragm) as compared with other muscles investigated. The above authors do not explain these differences, but presume that proteins in some muscles are more labile than in others.

e) Vascular supply

There is a rich capillary network with numerous branchings and local capillary enlargements in static muscles. Fast muscles are usually supplied by one main artery from which smaller arteries branch off (Yakovlev and Yakovleva 1953).

f) Differences in innervation

According to Rexed and Thermann (1948) the gastrocnemius muscle receives only ⅓ as many afferent fibres as motor nerve fibres. In the anterior tibial muscle this ratio is 1 : 1, so that it appears that flexor muscles receive relatively more sensitive fibres than extensor muscles. No differences were found by these authors in the size of nerve fibres innervating flexor and extensor muscles. According to Häggquist (cit. Krüger 1952), however, there are relatively more large fibres in nerves innervating flexor muscles than the extensor muscles. According to Yakovlev and Yakovleva (1953) muscles with static function (tonic slow muscles) are relatively less densely innervated than dynamic, fast muscles. In flexor muscles one nerve fibre innervates 1·5 mg. of muscle substance, while in extensor muscles one fibre innervates 7·5 mg. of muscle substance.

Besides these morphological differences there have also been reported differences in the functional properties of neuromuscular synapses. For example, the soleus (red, slow muscle) is markedly resistent to the effects of decamethonium, while the anterior tibial (pale muscle) exhibits marks of neuromuscular block on application of doses which only have a slight stimulating effect in the soleus (Zaimis 1951). As far as the action of d-tubocurarine is concerned, these relations are reversed. The soleus is more sensitive to this substance than the tibialis anterior (Jewell and Zaimis 1953a). It is also very interesting that although the atrophic soleus after tenotomy retains the physiological properties of a red muscle, it behaves as a pale muscle towards neuromuscular blocking substances, i. e. it is possible to make a red muscle behave as a pale muscle towards decamethonium (Jewell and Zaimis 1953b, 1954). Since no changes were observed after tenotomy in the tibialis anterior in response to decamethonium, the above authors concluded that these changes occur at the muscle fibre membrane and that it is possible that the membrane or red muscle fibre is more specialized (Jewell and Zaimis 1954).

The whole problem of differences between muscles is all the more complicated if we take into account that one and the same muscle can have a different characteristic in various species, e. g. the gastrocnemius in mice is functionally a fast muscle, and a slow, static muscle in guinea-pigs (Yakovlev and Yakovleva 1953). By what factors are these differences between muscles governed? With the exception of Denny-Brown (Bennet et al. 1958) who describes, in the pigeon, innervation of a granulated as well as a pale muscle fibre from a single nerve fibre, most findings indicate that these differences are given by the relationship between the motoneuron and the muscle, and the connection of these differences with the type of functional activity, i. e. with slow, tonic activity or on the contrary with fast activity. These functional relations have been demonstrated by Yakovlev. These authors subjected fast muscles to the tonic type of function and slow muscles to fast, dynamic type of action. In these experiments it was found that it is possible to attain biochemical transformation of tonic muscles into fast, and vice versa. Buller et al. (1960) studied the question of differentiation of skeletal muscles in kittens, which are originally all slow, into fast and slow muscles. According to these authors, slow muscles (75 msec. to summit of twitches) are innervated by motoneurons which have a relatively prolonged afterhyperpolarization, and that fast muscles (25 msec. to summit of twitches) are innervated by neurons with a short afterhyperpolarization. On cross-innervation fast motoneurons then transformed slow muscles into fast and conversely slow motoneurons changed fast muscles into slow muscles. Connection with the type of function was then shown in experiments where on inactivation of motoneurons cross unions no longer had any effect on changes in the muscles; muscles retained the same rate of contraction as before the operation. If the frequency of activation of the muscle was the decisive factor for determining whether the muscle will be fast or slow, then all muscles innervated by inactivated motoneurons should gain characteristic properties of slow muscles. However, this is not the case and for this reason the above authors presume that motoneurons exert a specific influence on the muscles and consider it possible that "... fast and slow motoneurons secrete from their nerve terminals specific substances that act on muscle fibres causing and maintaining the differentiation into the fast or slow types and the pumping action of muscle contractions is required to propel these substances intracellularly along the whole length of the muscle fibres, where they exert their action".

2. Differences between denervated muscles

If the trophic influence of the nervous system determines the differences between muscles, then it can be expected that, following section of the nerve, the main differences between these two types of muscles will equalize out, so that during the early denervation period changes should be opposite in direction with a tendency to level out original differences. This presupposition is difficult to prove, since it is possible to talk about slow and fast muscle fibres, but individual muscles consist of both types of muscle fibres represented in different proportions, so that it is more correct to talk of muscles with predominantly tonic or dynamic function. This means that these changes can be masked in various muscles by the other component.

a) Differences in colour

Differences in colour and the quantity of lipid granules to a great extent disappear. Fast muscles containing only a few granules in the sarcoplasm under normal conditions, acquire the semblance of muscles ontogenetically older, i. e. granulated muscles (Denny-Brown, see Bennett et al. 1958).

b) Functional differences

Denervated muscles also return to ontogenetically lower forms as far as their functional properties are concerned, their rate of contraction becoming slower (Erb 1868); this fact has led Orbeli (1945) to formulate the hypothesis that the denervated muscle returns to a lower developmental stage. It is also in accord with the findings of Koshtoyants (1938) and Buller et al. (1960) concerning the original long contraction time in early development.

c) Biochemical changes

Perhaps the best documented equalization of differences between fast and slow muscle fibres are biochemical changes. Sheves and Ryumina (1956) were one of the first to demonstrate that the consumption of oxygen in flexor and extensor muscles equalizes out following denervation (antigravity muscles have a higher consumption of oxygen under normal conditions). It has been shown in a number of reports from our laboratory that this equalization of differences between fast (m. extensor digitorum longus) and slow (m. soleus) muscles after denervation also applies to other metabolic indicators.

Under normal conditions the extensor digitorum longus contains more glycogen than the slow soleus muscle. This is evidently due to the fact that its metabolism is predominantly anaerobic because of its dynamic function. These differences, however, equalize in the course of three days following denervation in the rat (Drahota et al. 1959).

As far as the ion composition of these two muscles is concerned, the soleus contains more potassium, while the extensor digitorum contains more sodium, chlorides and water. These differences again equalize out after denervation (Drahota et al. 1959). Similarly, the uptake of Na^{24}, K^{42} and P^{32} which is higher in the soleus equalize after denervation, so that the uptake in the soleus is reduced, while there is an increase in the extensor digitorum longus (Hudlická 1955).

d) Rate of atrophy

Question of the differences in rate of atrophy in various muscles is not yet clear. In literature contradictory findings have been reported. According to some authors there are differences in the rate of atrophy in various muscles. For example, according to Dreyfus and Schapira (1948) the rate of atrophy in pale muscles is slower than in red muscles, while Langley and Hashimoto (1919) reported that the tibialis anterior loses weight at a faster rate than the extensor digitorum longus. Eccles (1944) found the atrophy after tenotomy to progress faster in the soleus than in other crural muscles. On the other hand Sutfin et al. (1954) concluded that muscles of the hind limb in rats lose weight at the same rate.

When studying this question, it has been found that some muscles of the hind limb in rats do not atrophy at the same rate and that these differences are statistically significant, even though the rat is not the most suitable for this type of experiment, because of a relatively rapid rate of atrophy. On the basis of these results it is evident that the rate of atrophy equalizes after longer periods of denervation (see also Chapter II). The diaphragm seems to hold an exceptional position, since a weight increment has been found by several authors (Sola and Martin 1953, Stewart 1955, Stewart and Martin 1956) during the first period after denervation (first seven to fourteen days) which is not caused by an increased content of water, as was supposed by Thomson et al. (1951) and Thomson (1955), but by an increase of dry weight and proteins. The reason for this different course of changes from other skeletal muscles is not clear — the above authors themselves assume that perhaps a substance is liberated at nerve endings and is important for the maintenance of this muscle (Stewart and Martin 1956).

Gutmann's finding of a temporary increase of muscle protein nitrogen four hours after section of the nerve is interesting in this connection (Gutmann 1960). It is thus probable that there is a transient increase in weight in every muscle after denervation, however, the extent and duration of this increase differs in various muscles.

It can thus be concluded that although there exists a number of differences between innervated muscles, most of these differences equalize out after denervation. Differences between denervated muscles are present primarily during the early denervation period, i. e. the period when differences between muscles have not yet disappeared.

Some muscles seem to be more sensitive to nervous influences, others, on the contrary, to humoral agents. For example, soleus appears to belong to the former group in contrast to extensor digitorum longus. In the soleus changes after denervation set in faster and the recovery of weight during reinnervation is also faster than in the extensor digitorum (Beránek et al. 1957) — (Fig. 124). The rate of atrophy following tenotomy is faster at the beginning in the soleus. Following nociceptive stimulation (Gutmann and Vodička 1952, Vodička 1957) or deafferentation (Hník 1956), only atrophy of the soleus was observed, while the extensor digitorum longus exhibited no atrophy.

The soleus is relatively little sensitive to humoral agents. It has been shown, for example, that in comparison with a number of other muscles, the soleus loses the least weight after hypophysectomy and starvation, and practically gains no weight on administration of STH.

Exceptions have also been described in other muscles as far as reactibility to hormonal influences is concerned. For example, levator ani is especially sensitive to male sex hormones (testosterone and related substances) and its reaction to these hormones cannot serve as objective criterion of the reactibility of skeletal muscles to sex hormones in general (Nimmi and Geiger 1957).

Thus there exist differences between muscles which are, to a certain extent, myospecific. These differences are not apparently of a fundamental nature, but concern the rate at which denervation changes take place. For this reason it is necessary to evaluate factors influencing the rate of atrophy on more muscles than one. It is not usually possible to make conclusions about all muscles in general from results obtained on one muscle.

D. Methods affecting the rate of denervation atrophy

The rate of denervation atrophy is already given to a certain extent by the state of general nutrition of the organism. It was found, for example, that during starvation the rate of denervation atrophy is enhanced. Hines and Knowlton

(1934) concluded from their experiments that the atrophy of denervated muscles in starving rats is faster than would correspond to normally fed animals, but that these two catabolic processes (inanition and denervation atrophy) do not sum. From later experimental data, however, it seems that the ratio between denervated and non-denervated muscles of the contralateral side is the same, irrespective of the fact, whether starving or normally fed animals are investigated. From this it can be indirectly concluded that denervated as well as control muscles lost the same percentage of weight due to starvation and that the atrophy due to denervation and inanition do sum. However, since different muscles react differently to starvation, this conclusion would have to be investigated in more muscles. Nevertheless it is certain that the fundamental condition for obtaining on optimal result with methods affecting muscle atrophy, is a sufficient general state of nutrition (Throne et al. 1953).

All methods aimed at influencing denervation atrophy are intent on affecting the muscles undergoing atrophy and are based on the following assumptions:

1. Compensation of loss of specific function or efffective tension by external agents such as passive manipulations, massage, stretching and electrotherapy.

2. Influencing muscle metabolism by humoral agents in attempts to accentuate synthetic processes which are evidently depressed in the atrophic muscle.

1. Methods affecting muscle specific function

a) Passive movements and massage

Passive movements and massage represent an integral part of rehabilitation techniques and it is natural, therefore, that physiotherapists utilize this approach fully, even though no precise clinical data are available, whether passive manipulation with the muscle has any effect on trophic changes in the muscle, or not. The insignificant effect of massage was already reported by Langley and Kato in 1915. It was not until later, however, that negative results were conclusively proved. According to Suskind and Hines (1946), Suskind et al. (1946) and Hajek and Hines (1947), for example, passive stretching does not affect the rate of muscle atrophy even though it appears to retard loss of muscle strength. Similarly, Pollock et al. (1951 a, b) found no changes in the histological picture of denervated muscles nor in the rate of atrophy when muscles had been subjected to massage. According to these authors massage and passive movements favourably act in preventing the onset of muscle contractures which appear in denervated muscles (Sherman 1948) when they are immobilized especially in the short position.

Questions of immobilization of the denervated muscle are not yet clear. Roberts (1916, see also Tower 1939) described a greater rate of denervation atrophy in man when the denervated muscle was stretched. It is possible that this longlasting stretch resulted in a decreased blood flow through the denervated muscle and consequently in a reduction of blood supply. However, changes in muscle metabolism have been described on stretching the muscle without changes in circulation taking a more prominent role (Lilienthal et al. 1955, Stainsby et al. 1956). Of course, innervated muscles were investigated in these experiments. But changes in muscle metabolism have been recorded even *in vitro* after stretching the muscle, since Abbot and Aubert (1951) demonstrated that the heat of lengthening is negative. Immobilization of muscles in the long position has been shown to lead to muscle hypertrophy (Froboese 1922, Meyer 1922, Thompson 1934, Thomsen and Luco 1944, Thomsen et al. 1946, Summers and Hines 1951), of course, whith intact afferent innervation (Meyer 1922). This mechanism, however, can be excluded in denervation atrophy.

b) Electrotherapy

One of the first to recommend electrotherapy of paralysed muscles was Reid (1841). Positive results in rabbits, even though small, were also reported by Langley and Kato (1915). The main and fundamental material about the usefulness of electrotherapy in the care of denervated muscles were obtained during World War II. Muscle atrophy was found to be retarded by Fischer (1939), Gutmann and Guttmann (1942, 1944), Solandt (1942), Hines (1942), Grodins et al. (1942), Hines et al. (1943b), Solandt et al. (1943), Hines and Wehmacher (1944), Eccles (1944), Grodins et al. (1944a, 1944b), Hines et al. (1945), Wehrmacher et al. (1945), Kosman et al. (1945, 1947a,b, 1948a), Jackson (1945), Fischer and Ramsey (1945), Suskind and Hines (1946), Suskind et al. (1946), Hajek and Hines (1947a,b,) Kosman (1948b, 1948c), Osborne et al. (1949), Hines (1950), Osborne (1951), Obrosov and Livencev (1953), Zubenko (1954), Thom (1953, 1954, 1955), Shaffer et al. (1954), Wakim and Krusen (1955), Ginetzinsky (1956) and others in various experimental animals and in man. In spite of this, several older reports have adopted a negative attitude towards electrotherapy (Langley 1916, Hartmann and Blatz 1920, Hines and Knowlton 1939, Chor et al. 1939, Molander et al. 1941, Doupe et al. 1943, Pollock et al. 1951 and others).

What is the reason for these contradictory findings?

From the above reports, certain conclusions can be made, even though most authors have worked under widely differing conditions of application of electrical stimulation, and have investigated different muscles in various animal species etc. Electrical stimulation can retard the rate of muscle atrophy only under certain conditions.

Most authors seem to agree that to be effective, the stimulating current must be of sufficient intensity, in order that massive muscle contractions are obtained (e. g. Gutmann and Guttmann 1944, Wehrmacher et al. 1945, Ginetzinsky 1956 and others).

Thus the evoked contraction must lead to high effective tension. This means that the muscle should be stretched and fixed during stimulation, so that the contraction takes place under isometric conditions (see, for example, Eccles 1944, Fischer and Ramsey 1946 and others). Where this condition cannot be fulfilled, as, for example, in tenotomy, electrotherapy is ineffective (Eccles 1944).

The shape of stimulating shocks used in electrotherapy varies from author to author. According to Solandt et al. (1943) and Kosman et al. (1947) the most advantageous is sinusoidal current 25 c./sec. and next best the 60 c./sec. Galvanic and faradic stimulation is much less effective in the opinion of Solandt et al. In spite of this, however, most authors have worked with these currents and have obtained positive results. Faradic current, nevertheless, is not too suitable since the threshold of the denervated muscle to this kind of stimulation is considerably increased, so that a relatively high intensity is necessary for obtaining strong contractions. Interrupted galvanic current appears to be more suitable from this point of view, since during denervation the threshold (rheobase), on the contrary, is lowered. Pollock et al. (1951) object to the polarizing effect of galvanic current even when interrupted stimulation is used, and include exponentially rising currents in this group in case impulses of longer duration are applied. Of course, both the above types of stimulation have been found, under certain conditions, to have a positive effect, the exponential currents being more advantageous since they stimulate specifically the denervated elements (e. g. Kowarschik 1952, Thom 1953, 1954, 1955). Rectangular pulses, for example, have also been successfully used by other authors (Waking and Krusen 1955). On the whole it can be concluded in agreement with Throne et al. (1953) that the shape of the current used is not important as long as strong contractions are obtained.

In order to attain a positive result, it is absolutely necessary to begin electrotherapy as soon as possible. It is true that electrotherapy slows down the rate of denervation muscle atrophy, but weight already lost is not affected (Osborne 1951). An exception to this is the report of Ginetzinsky (1956) according to which it is possible to attain a weight increment in the denervated muscle during the denervation period. However, most authors agree with Osborne that this is not possible. This fact also explains why Chor et al. (1939) and Pollock et al. (1951) were not able to prove a positive effect of electrotherapy on muscle atrophy, since they began stimulating at a relatively late period after denervation (14 days later), i. e. at a time when considerable atrophy had already taken place. Although Pollock et al. found a considerable slowing down of atrophy in rats, this effect was ascribed to interspecies differences. It is probably not necessary to explain this by interspecies differences, however, because in rats stimulation was begun the next day, i. e. immediately after denervation, so that atrophy was not yet started and electrotherapy had the best chance of being effective.

It was found by a number of authors that electrical stimulation was already without effect during the reinnervation period, since it does not accelerate further the restitution of weight due to reinnervation. Although some effect was reported by Gutmann and Guttmann (1942, 1944) in the rabbit on the recovery of muscle weight even during the reinnervation period, other authors have not confirmed this in other experimental animals (mostly rats and cats) — (Hajek and Hines 1947, Hines 1950, Pollock et al. 1951). Similar observations have been made by Osborne (1951) in man. According to Pollock et al. (1951) electrotherapy in the reinnervation period is even detrimental, since it increases muscle contractures which it helped, together with massage and passive movements, to remove in the denervation period. Nevertheless, it seems that electrical stimulation has its effect even at late stages, even if only to serve as an auxilliary reeducation method (Osborne 1951 and others), as is also mentioned, for example, by Thom (1954).

Denervated muscles are liable to fatigue more than normal muscles and it is therefore necessary to choose a suitable regime of electrotherapy. According to Kosman et al. (1947b) at least 10 min. intervals should be made between individual muscle contractions. The comparison of results obtained by various authors is all the more difficult as various combinations and time relations between stimulation and rest periods have been tried.

Attempts at activating as many denervated muscle fibres as possible are important for obtaining optimal results of electrotherapy. This condition is usually difficult to fulfill on percutaneous stimulation. It is an advantage to use bipolar stimulation, since the electrical field covers a relatively large area. When the intensity of stimulating current is increased, often neighbouring innervated muscles are also stimulated and in addition nociceptive stimulation, which is detrimental to recovery of muscle weight as well as to restitution of function, comes into play.

This is probably one of the reasons why there are such widely differing opinions concerning electrotherapy, because some authors, in an attempt to avoid painful stimulation, did not use currents of sufficient intensity to activate a larger number of denervated fibres, while others avoided difficulties associated with supramaximal stimulation by using anaesthesia. It is all the more surprising then that the use of exponentially increasing currents which, when a suitable length of pulses is chosen, specifically stimulate denervated elements, is relatively little utilized (see Chapter XII). Accomodation of innervated muscle fibres and sensitive nerve fibres prevents these from being stimulated, so that use is made of the well-known pathophysiological finding of Skoglund (1942) about the loss of accommodation in the denervated muscle. It is thus possible to apply the required current intensity without giving rise to nociceptive stimulation.

It can thus be concluded that electrotherapy can be considered, under certain conditions, as a useful therapeutic method for slowing down the rate of denervation atrophy, thus contributing towards improvement of the state of denervated muscle at a time when reinnervation is to take place. From the pathophysiological aspect, however, it is not yet clear by what mechanism this improvement is realized.

A single bout of stimulation of a normal muscle results primarily in the breakdown of energy-rich substances with their consequent resynthesis which takes place as a typical overshoot reaction (see Chapter VIII and IX). The extent of changes during the catabolic and anabolic phase is dependent on the type of muscle and on the frequency and kind of stimulation. The glycogen overshoot reaction in tonic (slow) muscles of the rat, for example, is present after stimulation at 120 impulses/min., and not at lower frequencies (30 or 60/min.) — (Bass et al. 1955). Similarly Yampolskaya (1950) found differences in the course of the overshoot reaction, which she called "supercompensation", in relation to the frequency of stimulation. After stimulation a temporary increase in blood flow through the muscle was also found, followed by a late increase 60 to 90 min. after the original stimulation (Hudlická and Vrbová 1958).

Besides this, electrical stimulation of a muscle results in general changes in the organism, similarly as in voluntary movements such as changes of pulse rate, blood pressure, consumption of oxygen, etc. (Cogswell and Browner 1955).

Repeated stimulation of the muscle can then lead to muscle hypertrophy, as has been shown, for example, by Fischer (1940) who stimulated with faradic current (60 impulses/min.) twice daily for 20 min. by alternately stimulating and resting every other second under light anaesthesia. A weight increment and increase in total strength of the muscles stimulated was found by this author, but the ratio strength/muscle weight remained unchanged. This effect was observed, however, only when isotonic contractions were performed (Fischer and Ramsey 1946). This last condition is not clear, however, since some earlier authors found a positive result on indirect stimulation of muscles under isometric conditions (Siebert 1929 in the frog, Gukelberger and Keiser 1938 in the dog).

Biochemical changes reported in muscles under the influence of repeated electrical stimulation are considerably dependent on the type of muscle stimulated and on the type of stimulation (shortlasting, fast contractions or slow, endurance training) — (Yakovlev 1949, Zuyagina et al. 1951). According to these authors it is possible to change one type of metabolism (for example, a predominantly dynamic type — anaerobic metabolism) into the other (predominantly tonic, endurance — aerobic metabolism) and vice versa. The character of blood supply (capillary network) can also change in relation to the type of stimulation as has been shown by the above authors and also earlier by Vannotti and Pfister

(1933). Gukelberger and Keiser (1938) also reported an increase in tissue oxidase (labiler Geweboxydase) of muscles after electrical stimulation for several weeks.

It thus appears that muscle stimulation by electrical current, either direct or indirect, results in changes analogous to those described after normal voluntary movements in physical training. But electrical stimulation in the denervated muscle gives different results from the innervated muscle.

A single bout of stimulation of a denervated muscle by interrupted galvanic current gives rise to changes which can be characterized as the loss of the overshoot reaction which has been reported in glycogen metabolism (see Chapter VIII) as well as in the metabolism of nucleic acids and proteins (see Chapter IX). This disturbance of recovery (anabolic) metabolic processes following a single bout of stimulation evoking functional muscle activity could thus indicate the fundamental character of metabolic disturbances in the denervated muscle resulting in muscle atrophy. It is interesting to note that the restitution of glycogen, for example, above initial levels after administration of humoral agents (e. g. thyroxin, insulin, adrenaline) is also absent in the denervated muscle (Hines and Knowlton 1935).

If this was the final conclusion, a number of facts would remain to be explained in connection with repeated electrical stimulation of the denervated muscle. Repeated stimulation evidently slows down the rate of muscle atrophy. Nevertheless, with the exception of Ginetzinsky (1956), no author has described an absolute increment in the weight of muscles during the denervation period, even following very intense stimulation. From this it is evident that the denervated muscle reacts somewhat differently to functional activity evoked by electrical stimulation from the normal muscle which reacts by hypertrophy.

The answer to this question is not yet clear, but a number of findings dealing with the effect of repeated stimulation of denervated muscles are important in this connection. It was found, for example, by Bass et al. (1959) that the overshoot glycogen reaction is not present in the denervated muscle after the first bout of stimulation, it is, however, present following the second bout. Fischer et al. (1950) showed that there is an increased content of K^+ in denervated muscles following their repeated stimulation. Following electrical stimulation, Brodish et al. (1952) described an increase of glycogen in atrophic muscles (denervation, inanition) together with retardation of atrophy which are considered by these authors to be associated with the sparing protein effect of carbohydrate metabolism. The stimulation of denervated muscles also increases their working capacity and endurance in comparison with non-stimulated denervated muscles (Wakim and Krusen 1955). What mechanism then is responsible for the effect of electrotherapy on the rate of atrophy?

aa) It is possible that the effect of electrotherapy is realized through specific function of the muscle, i. e. via muscle contraction. If this was the case, it could be expected that electrotherapy should be more effective and that its effect should be more clearly dependent on the duration of stimulation. Although several authors have reported that the effect of electrotherapy is proportional to the length of daily stimulation (Gutmann and Guttmann 1944, Solandt et al. 1943, Wakim and Krusen 1955 and others), others maintain that several seconds' daily stimulation is enough to attain the same effect on retardation of muscle atrophy as after an hour's stimulation (Eccles 1944, Hines 1950).

bb) As has already been pointed above, electrical stimulation leads to increased blood flow in stimulated extremities. It could be presumed that the therapeutic effect acts by way of circulatory changes in the denervated muscle. This explanation would perhaps be plausible if the blood supply in denervated muscles was reduced, so that stimulation would substantially improve its blood supply. The denervated muscle, however, is relatively richly supplied with blood (Hudlická and Vrbová 1958), so that it is difficult to imagine that further increase in circulation could have any effect in slowing down denervation atrophy.

cc) It is also possible that contractile proteins gradually change when not activated at certain time intervals (thixotropic gel — Fischer see Hines 1950) and become irreversibly altered. This would be in agreement with the results of Eccles (1944) and Hines (1950) who have reported positive results even with very shortlasting electrotherapy (e. g. 10 sec. daily).

dd) The results of Bass et al. (1959) are of special interest when discussing the mechanism by which electrical stimulation affects the denervated muscle. These authors found that the recovery (anabolic) overshoot reaction of glycogen is present following a second bout of stimulation, even though it is absent after the first stimulation. The overshoot reaction has also been observed in denervated muscles following two hours' arrest of circulation (Hudlická and Gutmann 1959). Tentatively it can be postulated that certain substances, metabolic activators in character, are perhaps being liberated from the denervated or more probably from the normal muscles after stimulation, enter the circulation and affect the metabolic reaction of the denervated muscle in case of functional activity (electrical stimulation) or changes in the blood supply (arrest and restitution of circulation). However, this question will have to be studied further.

It can be stated in conclusion that the effect of electrotherapy — the method which has up to now been most successful in influencing denervation atrophy — is not yet clear. A possible approach, perhaps, to the whole question can be the finding that the denervated muscle reacts differently to repeated stimulation than to a single bout of stimulation. It is also possible that the finding of Gutmann concerning increased content of proteins after repeated stimulation without a corresponding increase of nucleic acids in the denervated muscle may be the fundamental question associated with the mechanism of electrotherapy (Žák and Gutmann 1960).

2. Methods influencing muscle metabolism

a) Methods increasing blood flow

Physical methods aimed at influencing blood flow through the denervated muscle are usually led by the intention of ensuring maximal blood supply for the denervated limbs. This appeared to be indicated from clinical observations that denervated extremities are paler, the skin is cold and clammy and exhibits symptoms of reduced circulation.

Most methods striving to increase blood flow attain this by increasing the temperature of deep tissues (see Licht 1958) by diathermy (Kemp et al. 1947, Wakim et al. 1948), ultrasonic radiation (Fischer et al. 1954, Paul and Imig 1955, Licht 1958), hot packs (Feucht et al. 1949. Randall et al. 1952) etc., or by electrically evoked muscle contractions (Wakim 1953, Randall et al. 1952) which mainly affect the innervated limb. The effect of electrotherapy does not, according to Randall et al. (1953), act by increasing the temperature.

However, questions of circulation through the denervated muscle are evidently not of fundamental importance in the pathogenesis of atrophy, since the blood flow in the muscle following section of the peripheral nerve is increased (Goltz 1874, Ricker 1901, Gilbert et al. 1950, Griffin et al. 1954, Hudlická and Vrbová 1958 and others) and concerns not only blood flow through arteriovenous anastomoses but also through the capillary bed (Hudlická et al. 1959).

Thus these methods are of no significance as far as the effect on denervation atrophy is concerned. Firstly, it was shown that neither decreased nor increased temperature affects the rate of muscle atrophy (Hines and Knowlton 1934, Hines 1942) and secondly, it is not clear whether, even if the intensity of metabolism was altered, blood flow would not be increased disproportionately to the increase of metabolic intensity so that a detrimental change in their relationship would take place (Randall et al. 1952, 1953).

b) Hormones accelerating muscle atrophy

When the possibility of influencing the rate of muscle atrophy was investigated, various substances were studied which might affect muscle metabolism and help to elucidate the metabolic character of the changes occuring in the denervated muscle.

Thyroxine is one of the foremost catabolic hormones increasing the intensity of metabolism and the rate of muscle atrophy (Hines and Knowlton 1933, Diaz et al. 1947a, Sheves et al. 1956, Selye and Bajusz 1958). Hyperthyrodism by itself causes atrophy in normally innervated muscles (Zierler 1951, Russell and Wilhelmi 1958). Since following thyroxine administration there is increased excretion of N in the urine (Rupp et al. 1949), it is presumed that thyroxine plays an important role in muscle protein metabolism. This similarity between denervation and hyperthyroidism led Huf and Fischer (1949) and Sheves et al. (1956) to the conclusion that changes, at least from the metabolic aspect, are analogous in these two states. This assumption is further strengthened by changes in the thyroid gland following denervation of an extremity (Schmiedt 1952). Hypothyroidism, on the other hand, slows down atrophy (Huf and Fischer 1949, Sheves et al. 1956), but also retards restitution processes in the course of reinnervation.

Cortisone is another potent catabolic agent which in large doses can cause even muscle necroses (Ellis 1956) and adrenal cortical extract has a conspicuous catabolic effect on proteins (Engel et al. 1949). It was also found by Selye and Bajusz (1958) and Bajusz (1958, 1959) that cortisone increases the rate of denervation and inactivation muscle atrophy. A similar, but less pronounced effect was also reported for estradiol (Koch et al. 1954, Selye 1955, Selye and Bajusz 1958).

c) Hormones retarding muscle atrophy

On the other hand a number of hormones exert an anabolic effect on the normal and denervated muscle. The administration of STH results in an increase of body and muscle weight (Greenbaum and Young 1953), even though the increase of muscle weight does not lead to a concomittant increase of strength (Hume 1952) and the myosin thus formed in consequence of STH application is functionally different from normal myosin. STH retards muscle atrophy (Gutmann and Vrbová 1955, Selye and Bajusz 1958), although there is not absolute agreement concerning this question (Schapira and Dreyfus 1950). It is also interesting that hypophysectomized animals do not exhibit cardiac hypertrophy (Beznák 1952, Russel and Wilhelmi 1958) so that it seems that STH is necessary for muscle hypertrophy.

The effect of testosterone is similar, to a certain extent, to that of STH, but the effect of both hormones summates (Eisenberg et al. 1949). When it is absent muscle atrophy appears and this can be prevented by administration of testosterone which can cause muscle hypertrophy (Weinmann and Shipounoff 1941, Eisenberg et al. 1949, Wright and Kochakian 1953, Kochakian and Tillotson 1957, Korner 1955, Leonard 1957, Russel and Wilhelmi 1958 and others). According to Leonard (1957) and Kochakian and Tollotson (1957), the administration of testosterone may retard the rate of denervation atrophy, this being in contradiction to the findings of Schapira and Dreyfus (1950) who have reported that testosterone accelerates muscle atrophy. When evaluating the effect of testosterone on skeletal muscles, special care must be taken, since, as has been shown by Nimmi and Geiger (1957), muscles directly connected with sexual functions are not quite comparable with other muscles as far as the effect of this hormone on muscles is concerned (e. g. levator ani).

Insulin, besides its well-known effect on carbohydrate metabolism, plays an important role in muscle protein metabolism. Relatively little is known, however, about this second effect. Incorporation of methionine into the muscle during diabetes is considerably decreased according to Folker et al. (1951) and evidently has an effect on protein SH groups (Ungar and Kadis 1959). It seems that even the nitrogen retaining effect of STH is dependent on insulin (Russell and Wilhelmi 1958).

When evaluating the effect of hormones on muscle metabolism, it is necessary to take into account the following questions:

1. What is the effect of a given hormone on the normal muscle and how has the sensitivity or reactivity of the denervated muscle changed towards this hormone?

2. Is the internal environment in which the denervated muscle exists the same as in normal organism, or is the function of at least some endocrine glands affected by the denervation? (e. g. thyroid gland). It is necessary to take results obtained by administration of hormones in high doses, which are not present in the organism under normal conditions, with reserve from the therapeutic point of view.

d) ATP and other substances

Further various substances have been tried which might in some way influence muscle atrophy. Thus experiments were performed with acetylcholine, parathormone (Hines et al. 1941), glycine (Hines and Knowlton 1934), vitamin B complex (Hines et al. 1944a), vitamin C (Hines et al. 1944b), vitamin E (Hines et al. 1943). All these substances were found to be without effect on the rate of atrophy. Positive results were also not found after biotin deficiency (Lazere et al. 1943) and thymectomy (Diaz et al. 1947).

The only substance, besides the above mentioned anabolic hormones, which has been reported to slow the rate of denervation atrophy, is ATP. Ferdman (1953) described a positive effect of subcutaneously injected ATP in animals with muscle dystrophy in avitaminosis E and after section of peripheral nerve. Similarly Bot et al. (1955) confirmed this result of ATP administered by intramuscular route. Analysis of the ATP mechanism in these experiments is very difficult. It is not possible to expect that ATP thus administered will act as adenosine-triphosphate, but rather as simpler breakdown products. Fischer (1955) and Dreyfus (1955) have not been able to reproduce these results. It is possible that this discrepancy can be explained by different purity of the ATP preparations used, or by some unidentified constituent and impurity.

When evaluating the results of methods designed to influence the rate of denervation atrophy, it should be pointed out that, taking into account the attention and energy given to questions of muscle atrophy, results are relatively poor. This is probably caused by the fact that the therapeutic influences on muscle atrophy attract practical interest — investigation of more or less accidently chosen substances and procedures — without solving problems and questions of basic research concerning the fundamental mechanism of trophic disturbance in the muscle following section of the nerve and about the possibilities of compensation of the trophic influence of motor nerve fibres. Until then it is not possible to talk about causal therapy of the denervated muscle. Physiology still owes clinical medicine the answer to this question.

References

Abbott B. C., Aubert X. M.: Proc. Roy. Soc. London Ser. B. 139 : 104, 1951.
Audova A.: Schweiz. Arch. Neurol. Psychiat. 9 : 270, 1921.
Bajusz E.: Canad. J. Biochem. Physiol. 36 : 825, 1958.
— Endocrinology 64 : 262, 1959.
Bass A., Gutmann E., Hudlická O., Vrbová G.: Physiol. bohemoslov. 8 : 93, 1959.
Bass A., Gutmann E., Vodička Z.: Physiol. bohemoslov. 4 : 267, 1955.
Bennett H. S., Szent-Györgyi A., Denny-Brown D., Adams R. D.: Neurology 8 : 65, 1958.
Beránek R., Hník P., Vrbová G.: Physiol. bohemoslov. 6 : 200, 1957.
Beznák M.: J. Physiol. 116 : 74, 1952.
Bot G., Köver A., Varga E.: Acta Physiologica Ac. Sc. Hung. 8 : 321, 1955.
Bowden R. M., Gutmann E.: Brain 67 : 273, 1944.
Brodish A., Thomson J. D., Hines H. M.: Am. J. Phys. Med. 31 : 3, 1952.
Buller A. J., Eccles J. C., Eccles R. M.: J. Physiol. 150 : 399, 1960.
Chor H., Cleveland D., Davenport H. A. Dolkart R. E., Beard G.: JAMA 113 : 1029, 1939.
Danielson A. K.: Fiziol. zh. SSSR 41 : 807, 1955.
Diaz-Guerrero R., Thomson J. D., Hines H. M.: Proc. Soc. exp. Biol. Med. 66 : 95, 1947a.
— — — Am. J. Physiol. 151 : No. 1, 1947b.
Doupe J., Barnes R., Kerr A. S.: J. Neurol. Psychiat. 6 : 136, 1943.
Drahota Z.: Physiol. bohemoslov. 8 : 501, 1959.
Drahota Z., Hudlická O., Hník P.: Konference o srovnávací fysiologii výměny látek. Biol. fakulta KU, Praha. April 1959.
Dreyfus J. C.: personal communication. 1955.
Dreyfus J. C., Schapira G.: Sem. Hôp. Paris 24 : 1999, 1948.
Eccles J. C.: J. Physiol. 103 : 253, 1944.
Eisenberg E., Gordan G. S., Elliott H. W.: Endocrinology 45 : 113, 1049.
Ellis J. T.: Am. J. Path. 23 : 993, 1956.
Engel F. L., Schiller S., Pentz E. I.: Endocrinology 44 : 458, 1949.
Erb W.: Dtsch. Arch. f. klin. Med. 4 : 535, 1868.
Fenn W. O.: Am. J. Physiol. 120 : 675, 1937.
Ferdman D. L.: Biokhimiya zabolevanij myshts, Kiev 1953.
Feucht B. L., Richardson A. W., Hines H. M.: Arch. Phys. Med. 30 : 687, 1949.
Fischer E.: Am. J. Physiol. 127 : 605, 1939.
— Am. J. Physiol. 131 : 156, 1940.
— Proc. Soc. exp. Biol. N. Y. 47 : 277, 1941.
— personal communication. 1955.

Fischer E., Moore J. W., Skowlund H. V., Ryland K. W., Copenhaver N. J.: Arch. Phys. Med. 31 : 429, 1950.
Fischer E., Ramsey V. W.: Am. J. Physiol. 145 : 57, 1945—46.
— — Am. J. Physiol. 145 : 583, 1946.
Fischer E., White E. A., Hendricks S. L., Chevalier B. A., Chevalier R. B.: Am. J. Phys. Med. 33 : 284, 1954.
Forker L. L., Chaikoff I. L., Entenman C., Tarver H.: J. Biol. Chem. 188 : 37, 1951.
Froboese C.: Mitt. Grenzgeb. Med. u. Chirurgie 35 : 683, 1922.
Gilbert D. L., Janney C. D., Hines H. M.: Am. J. Physiol. 163 : Nr. 3, 1950.
Ginetzinsky A. G.: Problems of the modern physiology of the nervous system and muscle systems. Ac. Sc. of the Georgian SSR, Tbilisi 1956, p. 409.
Goltz E.: Pflügers Arch. 9 : 174, 1874.
Greenbaum A. L., Young F. G.: J. Endocrinol. 9 : 127, 1953.
Griffin P. P., Green H. D., Youmans P. L., Johnson H. D.: J. Pharmacol. exper. Therapeutics 110 : 93, 1954.
Grodins F. S., Osborne S. L., Ivy A. C.: Arch. Physical Therapy 23 : 729, 1942.
Grodins F. C., Osborne S. L., Johnson F. R., Arana S., Ivy A. C.: Am. J. Physiol. 142 : 222, 1944.
Grodins F. C., Osborne S. L., Johnson F. R., Ivy A. C.: Am. J. Physiol. 142 : 216, 1944.
Gukelberger M., Keiser E.: Arbeitsphysiologie 10 : 94, 1938.
Gutmann E.: Physiol. bohemoslov. 9 : 382, 1960.
Gutmann E., Guttmann L.: Lancet 1 : 169, 1942.
— — J. Neurol. Neurosurg. Psychiat. 7 : 7, 1944.
Gutmann E., Guttmann L., Medawar P. B., Young J. Z.: J. exp. Biol. 19 : 14, 1942.
Gutmann E., Vodička Z.: Physiol. bohemoslov. 2 : 389, 1953.
Gutmann E., Vrbová G.: unpublished results. 1955.
Hajek N. M., Hines H. M.: J. Iowa State Med. Soc., April issue 1947.
— — Fed. Proc. 6 : 1, 1947.
Herman J. W.: Am. J. Phys. Med. 31 : 34, 1952.
Hartman F. A., Blatz W. E.: J. Physiol. 53 : 290, 1920.
Hines H. M.: JAMA 120 : 515, 1942.
— Extrait de "Le Muscle", Royaumont. France 1950.
Hines H. M., Knowlton G. C.: Am. J. Physiol. 111 : 243, 1935.
— — Proc. Soc. exp. Biol. Med. 31 : 1029, 1933—34.
— — Am. J. Physiol. 110 : 8, 1934.
— — Am. J. Physiol. 128 : 97, 1939.
Hines H. M., Lazere B., Thomson J. D.: Proc. Soc. exp. Biol. Med. 55 : 97, 1944.
Hines H. M., Lazere B., Thomson J. D., Cretzmeyer C. H.: Am. J. Physiol. 139 : 183, 1943.
— — — — J. Nutrition, 27 : 303, 1944.
Hines H. M., Melville E., Wehrmacher W. H.: Am. J. Physiol. 144 : 278, 1945.
Hines H. M., Thomson J. D., Lazere B.: J. Pharmacol. exp. Therap. 73 : 463, 1941.
— — — Am. J. Physiol. 140 : 115, 1943 a.
— — — Arch. Physical Therapy 24 : 69, 1943b.
Hines H. M., Wehrmacher W. H.: J. of the Iowa State Medical Soc. April Issue, 1944.
Hník P.: Candidate's Thesis, Institute of Physiology ČSAV, Prague 1955.
— Physiol. bohemoslov. 5 : 411, 1956.
— Physiol. bohemoslov. 9 : 390, 1960.
Hudlická O.: Physiol. bohemoslov. 4 : 276, 1955.
Hudlická O., Gutmann E.: Naunyn Schmiedebergs, Arch. exp. Path. u. Pharmakol. 234 : 501, 1958.

Hudlická O., Kovách A. G. B., Bodolay-Varga A.: Acta Physiol. Ac. Sc. Hung. 16 : 51, 1959.
Hudlická O., Vrbová G.: Physiol. bohemoslov. 7 : 38, 1958.
Huf E. G., Fischer E.: Am. J. Physiol. 159 : 6, 1949.
Hume B. J.: Biochem. J. Lond. 52 : 11, 1952.
Jackson S.: Brain 68 : 300, 1945.
Jewell P. A., Zaimis E. J.: J. Physiol., London 120 : 47P, 1953a.
— — J. Physiol., London 120 : 48P, 1953b.
— — J. Physiol., London 124 : 429, 1954.
Kare M. R.: Am. J. Physiol. 175 : 51, 1953.
Kemp C. R., Tuttle W. W., Hines H. M.: Am. J. Physiol. 150 : 705, 1947.
Knowlton G. C., Hines H. M.: Proc. Soc. exp. Biol. Med. 35 : 394, 1936.
Koch W., Heim E., Eschweiler J.: Acta endocr. Kbh. 16 : 369, 1954.
Kochakian C. D., Tillotson C.: Am. J. Physiol. 189 : 425, 1957.
Kochakian C. D., Tillotson C., Austin J.: Endocrinology 60 : 144, 1956.
Korner A.: J. Endocr., Lond. 13 : 90, 1955.
Korner A., Young F. G.: J. Endocr. Lond. 13 : 78, 1955.
Koshtoyants Kh. S.: Fiziol. zh, 24 : 221, 1938.
Kosman A. J., Osborne S. L., Ivy A. C.: Am. J. Physiol. 145 : 447, 1945—6.
— — — Arch. Phys. Med. 28 : 7, 1947a.
— — — Arch. Phys. Med. 28 : 12, 1947b.
— — — Arch. Phys. Med. 29 : 559, 1948.
Kosman A. J., Wood E. C., Osborne S. L.: Am. J. Physiol. 154 : Nr. 3, 1948a.
— — — Arch. Phys. Med. 29 : 489, 1948b.
Kowarschik J.: Brit. J. Phys. Med. 15 : 249, 1952.
Krüger P.: Tetanus und Tonus der quergestreiften Skeletmuskeln der Wirbeltiere und des Menschen. Leipzig, Akad. Verlagsges. 1952.
Langley I. N.: J. Physiol. 50 : 335, 1916.
Langley I. N., Hashimoto M.: J. Physiol. 52 : 15, 1919.
Langley I. N., Kato T.: J. Physiol. 49 : 432, 1915.
Lawrie E. A.: Nature, Lond. 170 : 122, 52.
Lazere B., Thomson J. D., Hines H. M.: Am. J. Physiol. 138 : 357, 1941.
Leonard S. L.: Proc. Soc. exp. Biol. N. Y. 94 : 458, 1957.
Licht S.: Therapeutic Heat. New Haven 1958.
Lilienthal J. L., Stainsby W. N., Fales J. T.: Fed. Proc. 14 : 92, 1955.
Meyer A. W.: Mitteil. Granzgebieten Med. Chirurgie 35 : 651, 1922.
Molander C. D., Steinitz F. S., Asher R.: Arch. Phys. Therapy 22 : 54, 1941.
Needham D. M.: Physiol. Rev. 6 : 1, 1926.
Nimni M. E., Geiger E.: Proc. Soc. exp. Biol. Med. 94 : 606, 1957.
Obrosov A. N., Livencev N. M.: Elektrodiagnostika i elektrostimulatsia myshts pri porazhenii perifericheskikh nervov. Medgiz, Moscow 1953.
Orbeli L. A.: Trudy Fiziol. inst. im. I. P. Pavlova AN SSSR 1 : 1, 1945.
Osborne S. L.: Arch. Phys. Med. 32 : 523, 1951.
Osborne S. L., Kosman A. J., Bouman H. D., McElvenny R. T., Ivy A. C.: Surg. Gynec. Obstet. 88 : 243, 1949.
Paul W. D., Imig C. J.: Am. J. Phys. Med. 34 : 370, 1955.
Pollock L. J., Arieff A. J., Sherman I. C., Schiller M., Tigay E., Hiller F., Liebert E., Yacorzynski G.: Arch. Phys. Med. 31 : 265, 1951a.
— — — — — — — — Arch. Phys. Med. 31 : 377, 1951b.
Randall B. P., Imig C. J., Hines H. M.: Arch. Phys. Med. 33 : 73, 1952.
— — — Am. J. Phys. Med. 32 : 1, 1953.

Reid J.: J. Med. Sci. 1 : 320, 1841.
Rexed B., Therman P.: J. Neurophysiol. 11 : 133, 1948.
Ricker G.: Arch. Path. Anat. Physiol. 165 : 263, 1901.
Roberts F.: Brain 39 : 297, 1916.
Rupp J., Paschkis K. E., Cantarow A.: Endocrinology 44 : 449, 1949.
Russell J. A., Wilhelmi A. E.: Ann. Rev. Physiol. 20 : 43, 1958.
Schapira G., Dreyfus J. C.: "Le muscle, étude de biologie et de pathologie", Royaumont, France, 1950.
Schapira G., Dreyfus J. Cl., Schapira F.: C. R. de la Soc. Biol. 144 : 829, 1950.
Schmidt C. G., Bogatzki M., Fischer L.: Arch. exper. Pathol. u. Pharmakol. 214 : 381, 1952.
Schwartz D., Dreyfus J. C., Schapira G.: C. R. Soc. Biol. 143 : 35, 1949.
Scow R. O., Hogan S. N.: Am. J. Physiol. 180 : 31, 1955.
Scow R. O., Roe J. H.: Am. J. Physiol. 173 : 22, 1953.
Selye H.: Acta endocr. Kbh. 20 : 1, 1955.
Selye H., Bajusz E.: Am. J. Physiol. 192 : 297, 1958.
Shaffer D. W., Branes C. K., Wakim K. G., Sayre G. P., Krusen F. H.: Arch. Phys. Med. 35 : 491, 1954.
Sharipova E. R., Zhukov E. K.: Fiziol. zh. SSSR 40 : 445, 1954.
Sherman I. C.: J. Bone and Joint Surgery 30 : 474, 1948.
Sheves G. S., Epelbaum S. J., Rjumina V. I.: Biokhimiya 21 : 71, 1956.
Sheves G. S., Rjumina V. I.: Biokhimiya 21 : 385, 1956.
Siebert W. W.: Ztschr. Klin. Med. 109 : 350, 1929.
Skoglund C. R.: Acta Physiol. Scand. 4 : Suppl. 12, 1942.
Slater E. C.: Symp. Soc. exp. Biol. 10 : 110, 1957.
Sola O. M., Martin A. W.: Am. J. Physiol. 172 : 324, 1953.
Solandt D. Y.: JAMA, 120 : 511, 1942.
Solandt D. Y., deLury D. B., Hunter J.: Arch. Neurol., Psychiat., Chicago 49 : 802, 1943.
Solandt D. Y., Magladery J. W.: J. Neurophysiol. 5 : 373, 1942.
Stainsby W. N., Fales J. T., Lilienthal J. L.: Bull. John Hopkins Hosp. 99 : 249, 1956.
Stewart D. M.: Biochem. J. 59 : 553, 1955.
Stewart D. M., Martin A. W.: Am. J. Physiol. 186 : 497, 1956.
Summers T. B., Hines H. M.: Arch. Phys. Med. 32 : 142, 1951.
Sunderland S., Ray L. J.: J. Neurol. Neurosurg. Psychiat. 13 : 159, 1950.
Suskind M. I., Hajek N. M., Hines H. M.: Arch. Phys. Med. 27 : 133, 1946.
Suskind M. I., Hines H. M.: Fed. Proc. 5 : Nr. 1, 1946.
Sutfin D. C., Thomson J. D., Hines H. M.: Am. J. Physiol. 178 : 535, 1954.
Thom H.: Ztschr. f. Orthopedie u. Grenzgebiete 84 : 1, 1953.
— III. Internat. Poliomyelitis Conference Rome, 1954, p. 542.
— Schweiz. Med. Wocheschr. 85 : 65, 1955.
Thomsen P., Luco J. V.: J. Neurophysiol. 7 : 245, 1944.
Thomsen P., Luco J. V., Frugone I. E.: Proc. Soc. exp. Biol. Med. 61 : 34, 1946.
Thompson T. C.: J. Bone Joint Surg. 16 : 564, 1934.
Thomson J. D.: Am. J. Physiol. 180 : 202, 1955.
Thomson J. D., Morgan J. A., Brodish A.: Am. J. Physiol. 167 : 832, 1951.
Throne E. M., Moore W. H., Psaki C. R.: U. S. Armed Forces M. J. 4 : 1043, 1953.
Tower S. S.: Physiol. Rev. 19 : 1, 1939.
Ungar G., Kadis S.: Nature Lond. 183 : 49, 1959.
Vannoti A., Pfister H.: Arbeitsphysiologie 7 : 127, 1933.
Vodička Z.: Physiol. bohemoslov. 6 : 62, 1957.
Wakim K. G.: Arch. Phys. Med. 34 : 291, 1953.

Wakim K. G., Gersten J. W., Herrick J. F., Elkins H. C., Krusen F. H.: Arch. Phys. Med. 29 : : 583, 1948.
Wakim K. G., Krusen F. H.: Arch. Phys. Med. 36 : 370, 1955.
Wainman P., Shipounoff G. C.: Endocrinology 29 : 975, 1941.
Wehrmacher W. H., Thomson J. D., Hines H. M.: Arch. Phys. Med. 26 : 261, 1945.
Weiss P., Edds Mac V.: Am. J. Physiol. 145 : 597, 1945—6.
Wright P. M., Kochakian C. D.: Am. J. Physiol. 173 : 217, 1953.
Yakovlev N. N.: Usp. Sovr. Biol. 27 : 257, 1949.
Yakovlev N. N., Yakovleva E. S.: Usp. Sovr. Biol. 35 : 134, 1953.
Yampolskaya L. I.: Fiziol. zh. USSR 36 : 749, 1950.
Zaimis E. J.: J. Physiol., 112 : 176, 1951.
Zelená J., Hník P.: Physiol. bohemoslov. 6 : 193, 1957.
Zierler K. L.: Bull. Johns Hopkins Hosp. 89 : 263, 1951.
Zubenko P. M.: Vopr. biokhimii myshts, Kiev 1954, p. 149.
Zuyagina F. E., Mnukhina E. S., Yakovlev N. N., Yampolskaya L. I.: Ukrainskii biokh. zhurnal 23 : 178, 1951.
Žák R., Gutmann E.. Nature 185 : 766, 1960.

Chapter XI.

METABOLIC REACTIBILITY OF THE DENERVATED MUSCLE

Gutmann E.

A. Denervation studies in relation to problems of nervous regulation of metabolism

1. Metabolic relations between nervous and non-nervous tissues

Studies concerning metabolic reactions of the denervated muscle are (or should be considered as) a part of research on nervous regulation of metabolism. Denervation demonstrates the trophic control of the nervous system on skeletal muscles by elimination (Tower 1939); it leads to serious metabolic disturbances and it is thus important to assume a very close relationship between the metabolism of the nervous and muscle tissue. Valuable material should thus be obtained from results of metabolic disturbances in denervated muscle, concerning the mechanisms by which the nervous system regulates the metabolism of innervated structures (see Gutmann, 1955, 1959).

It was stressed in the preceding chapter that these mechanisms are not realized only by impulse activity of the neuron. If metabolic interaction, which evidently regulates the "trophic state" of muscle, was realized by nerve impulse propagation only, then analogous changes to denervation could be expected after immobilization of the muscle. Of course, this is the main problem of denervation atrophy. It is true that part of the changes are similar in denervated and immobilized muscles, but characteristic differences have also been reported. In the first place, fibrillation activity is not present in the immobilized muscle (Tower et al. 1941) and no comparable increase in sensitivity of muscle fibres to acetylcholine can be found in muscles with isolated lower motor units (Johnsson and Thesleff 1961). Inactivity cannot therefore be responsible for the extension of the receptor surface in denervated muscle (Miledi 1960, Johnsson and Thesleff 1961). The deterioration of specific muscle proteins, found in the denervated muscle, is also not present in immobilized muscles (Fischer 1950). The rate of atrophy is practically the same in both these types of atrophy;

atrophy can at least partly be prevented by electrotherapy of denervated muscles, but the increase of less specific myoalbumin after denervation is little affected. Similarly it is possible to prevent, even though not completely, loss of muscle protein. However, changes in metabolism of nucleic acids, which represent an important indicator of disturbances in neurotrophic relations during denervation, remain unaffected (Gutmann 1960, Gutmann and Žák 1961). Anserine was found to decrease by Yudayev et al. (1953) together with an increase of alanine in the denervated, but not in immobilized muscle. All these results indicate that there are qualitative differences between denervation and immobilization atrophy, so that it is possible to assume that nervous regulation of muscle metabolism is realized also by other mechanisms than propagated impulse activity, and that denervation atrophy is not only "atrophia ex inactivitate".

Mutual trophic relations between neurons and the muscle develop gradually. These neurotrophic relations are doubtlessly determined by the metabolism of nerve and muscle cells, and their development is an expression of the growing dependency of muscle upon the nervous system. These relations are also stressed in the enzymo-chemical hypothesis of the nerve impulse (Koshtoyants 1957), according to which mediators of impulse transmission are specific products of nerve cell metabolism and these also exert their influence in metabolic processes which determine the functional activity of innervated organs. A clear expression of these relations are to be found during evolution of neuromuscular function, which gradually develops from slowly contracting tissue with a tendency to various forms of automatic activity to the fast activity of muscle fibres in mammals, with a quick contraction and relaxation phase and precisely limited characteristics of neuromuscular transmission in space and time (Voskresenskaya 1959).

Gradual muscle differentiation involves functional, electrophysiological, morphological and metabolic properties. In newborn animals all muscles are characterized by a slow contraction time, and they gradually differentiate into fast and slow muscles (Koshtoyants 1957, Eccles et al. 1958). According to Krüger (1952) fibres with different morphological structures correspond to these "tetanic" and "tonic" muscle fibres, i. e. fibres with fibrillar structure rich in sarcoplasmatic reticulum on the one hand, and fibres with a "field" structure, with insignificant sarcoplasmatic reticulum on the other. It is not clear whether this differentiation is complete. In any case, functional differentiation of muscle fibres is determined by various types of motoneurons (Eccles et al. 1958) and morphological differentiation is apparantly lost after denervation (Krüger and Günther 1956), so that all muscle fibres after denervation show the so-called "field" structure.

The metabolic subordination of muscle to the nervous system develops relatively late during embryonal life, and is, for example, characterized by the development of a typical protein composition, which is altered during denervation and is renewed in the course of reinnervation. It can be assumed that the absence of such metabolic specialization, normally induced to a great extent by the nervous system, is an important factor determining disturbances in differentiation of extra- and intrafusal muscle fibres following denervation at an early stage of development (Zelená 1959) and differentiation of regenerating muscle tissue (Studitskii and Striganova 1951, Studitskii 1959). Furthermore, a decrease in myoalbumin occurs in the course of embryonal development, while the amount of specific muscle proteins increases (Crepax 1952), the reverse process taking place in the regression changes following denervation (Fischer 1950, Crepax 1949). The ratio between carnosine and anserine increases during development (Severin and Yudayev 1951), decreases after denervation and again increases following reinnervation (Yudayev et al. 1953), the quantitative relations between these two peptides being, for a certain period of time, the same as found during early stages of development. Metabolic relations are evidently also realized by certain qualitative changes, an example of this being the decrease of anserine and appearance of alanine in connection with the decrease of carnosine in the denervated muscle, and the reverse process occurs during reinnervation (Yudayev et al. 1953).

Gradual specialization during development and regeneration, and its loss after denervation, underlines the role of the metabolic connection between nerve and muscle cells. Mediators evidently play an important role in this relationship. These metabolic relations are probably not bound only to the transmitter release notwithstanding the fact that subthreshold liberation of acetylcholine (Fatt and Katz 1952) represents an important factor in the metabolic connection mediating the trophic function of the nervous system.

2. A physiological approach to the problem of regulation of metabolism in denervated muscle

We may ask whether study of changes found in the denervated muscle can contribute to definite conclusions about the mechanism of neuromuscular metabolic relationships. Here we are dealing with metabolic reactions of denervated muscle, i. e. a physiological approach to the whole problem is chosen. Changes in the reactibility of the denervated muscle should involve data on neurotrophic relations, especially from the point of view of the mechanism of the trophic influence of the nervous system and the basis of metabolic disturbances in denervated muscle.

It does not seem possible, at first glance, that investigations of the reactibility of denervated muscle following various types of functional loads could substantially help in solving what seems primarily to be a biochemical problem. The physiological approach to problems of reactibility of an organ following various loads may, however, help in showing loci of least resistance or alternate metabolic pathways in denervated muscle. Furthermore, this approach assumes neurotrophic relations, and analysis of these regulatory mechanisms is difficult to approach from a purely biochemical aspect. Such investigations must analyse metabolic muscle changes taking place in the whole organism, i. e. they must not ignore the environment in which the denervated muscle, as a part of the organism, continues to react.

Studies on the regulation of metabolism in denervated muscle require data on changes in the ability of the denervated muscle to react to various functional loads within the organism as a whole. This is an important task for the physiologist. A number of such observations are already at our disposal. For example, the reports of Wertheimer concerning the influence of the nervous system on the mobilization of lipids (Wertheimer 1926), on the mobilization of glycogen in denervated muscle (Wertheimer 1927) and on the resynthesis of glycogen in denervated muscle during realimentation following starvation (Wertheimer 1927, Hines and Knowlton 1937). Furthermore, Levine et al. (1941) and others have investigated changes in energy metabolism of the denervated muscle following muscle stimulation. A further approach is represented by the important studies of the hypersensitivity of denervated structures (Cannon and Rosenblueth 1949, Axelsson and Thessleff 1959), but it seems that not enough attention has been paid to questions of the metabolic regulation in the denervated muscle.

It is thus our intention to demonstrate that the physiological approach to problems of denervation atrophy, i. e. the investigation of metabolic reactibility of denervated muscle in the whole organism, plays an important role in studies concerning the nervous regulation of muscle metabolism.

3. Changes in the internal environment of the organism in the course of muscle denervation

Studies of metabolic reactions in the denervated muscle have, so far, almost exclusively dealt with changes taking place in the muscle itself. However, denervation, by excluding a group of muscles, or even a whole extremity, from the normal reflex processes taking place in the organism, may seriously affect not only various functions of the organism, but also indirectly the metabolism of the denervated muscle itself. Changes in the motor analyser, i. e. disturbances in the formation of conditioned motor reflexes due to changes in afferent signalization from the denervated extremity (Vrbová and Gutmann 1956) can serve as an example of altered functions in the organism following denervation. Rehabilitation of motor functions must take these changes into account, since its main aim is to attain the best motor effect after lesions of the locomotor apparatus.

It is probable, however, that other functions, expecially respiratory and cardiovascular, are also altered, and these changes may develop as adaptative reactions of the organism to reduced motor activity.

Mention will be made here of those changes in the denervated muscle which may be due to changes in the internal environment, i. e. changes in homeostatic mechanisms. It is surprising that these changes have so far, attracted so little attention. Our knowledge of these changes is very limited, and a systematic study is lacking.

The findings of Hollán (1957) concerning the effect of peripheral nerve lesions upon the haematopoietic and reticuloendothelial systems are interesting in this connection. Following section of the sciatic and femoral nerves, a reduction in the number of erythrocytes and haemoglobin has been described; the elimination of the haematopoietic region by denervation is not the cause of this, since high amputation of the limb is not followed by anaemia. A trophic disturbance due to denervation of the extremity has to be assumed. A direct connection has been found between the degree of anaemia and the extent of the trophic (evidently cutaneous) lesion. Healing of trophic ulcers was followed by improvement of haematopoesis. It is thus concluded by Hollán (1957) that the main factor responsible is a local disturbance of tissue metabolism, or local breakdown of tissues (Hollán 1957b) which causes a disturbance in the regulation of haematopoiesis. These changes are extensive, erythropoiesis in the bone medulla being increased, with inhibition of maturation and pathological formation of erythrocytes accompanied by considerable reticulocytosis, indicating increased activity of the reticuloendothelial system (Hollán 1957c). It is assumed by the above author (Hollán 1957c) that the reason for this "postdenervation anaemia" is a disturbance in the metabolism of iron and nucleic acids. It is not probable that it is caused by a lack of iron. The amount of iron in the serum has been found to increase after denervation, this being due to myoglobin degradation in the denervated muscle which causes a rise in non-haemin iron in the serum (Dreyfus and Schapira 1958). It is also improbable that the serious changes in haematopoiesis are due to relative motor inactivity of the organism as a result of denervation. The main changes have been observed 15 days after denervation, and they later return to normal (Hollán 1957a). In our experiments a reduced content of haemoglobin was still found two months after denervation ($84 \pm 3{\cdot}2$ as compared with $95 \pm 1{\cdot}7$), while there were no differences found in the number of erythrocytes (4 900 000 as compared with 4 890 000). It is probable that this hypochromic anaemia following prolonged denervation is the result of adaptation to reduced motor activity due to denervation. It is improbable that these changes could substantially influence the metabolism of the denervated muscle, since metabolic disturbances already appear within the first few days after denervation, when changes in the blood have not yet appeared.

Little changes were found in blood glucose. The level of blood glucose in animals two months after bilateral denervation was 116 ± 5·1 mg% as compared with 134 ± 10 mg% in the controls (9 animals). The percentage increase of blood glucose two hours after oral administration of glucose was found to be the same in denervated and control animals, so that it is not possible to assume changes in glucose absorption, at least in animals a long time after denervation.

The liver is the main organ participating in the regulation of energy metabolism. For this reason the content of liver glycogen was determined in rats two months after bilateral denervation of the hind limbs. The content of liver glycogen was found to be higher in denervated animals (3061 ± 306 mg% as compared with 1392 ± 166 mg% in control animals) after 24 hours' fasting, and these differences were still apparent even after more prolonged starvation (451 ± 29·6 mg% as compared with 192 ± 14·4 mg% in control animals after 48 hours' starvation). The rate of glycogen breakdown is thus decreased in the liver of denervated animals. Glycogen synthesis from glucose, or gluconeogenesis, are not affected, however, as could be seen from a similar percentage increase of liver glycogen following realimentation with a protein diet or after glucose administration following 48 hours' starvation. (Glycogen levels 18 hours after protein realimentation were 3170 ± 244 mg% as compared with 2800 ± 169 mg% in control animals; the content of glycogen two hours after glucose administration of a 50% solution in amounts corresponding to 1·5% body weight was 986 ± 38 mg% as compared with 940 ± 30 mg% in control animals).

Different changes are, of course, to be found in muscles of denervated and control animals. In normal animals the content of muscle glycogen drops to 316 ± 27 mg% after 48 hours starvation and to 325 ± 27·8 mg% in muscles of animals with bilateral denervation. The rate of glycogen breakdown due to starvation is slowed down in muscles of denervated animals (see p. 398), while the net content of glycogen is relatively small when calculated in relation to muscle weight loss. Within 18 hours after realimentation with a protein diet, the content of glycogen rises to 669 ± 30 mg% in normal muscles, while the increase is insignificant in denervated muscles (to 363 ± 23 mg%).

It can thus be concluded that the amount of liver glycogen is relatively high in animals with bilateral denervation of both hind extremities, and that the rate of glycogen breakdown during starvation is relatively low, this being due to lower demands of peripheral tissues, especially of muscle. This is evidently the result of metabolic adaptation of the liver to reduced motor activity associated with denervation, the synthetic processes in the liver of denervated animals remaining unaffected. The above experiments indicate that changes in the internal environment or in organ functions are not so severe as to explain the main metabolic disturbances occuring in the denervated muscle. It is clear, however, that they do occur, and that no systematic investigations have been made, as yet, allowing definite conclusions to be drawn. Unfortunately, little attention up to now has been paid to these important aspects of research of denervation atrophy.

4. Alteration in the reactibility of denervated muscle to mediators and humoral agents

Changes in the internal environment thus cannot play a decisive role in the development of denervation atrophy. But alterations in the reactibility of denervated muscle to mediators and humoral agents might be of paramount importance. It is then a matter of further research to elucidate the question of whether these changes in reactibility of denervated muscle can explain the progress of muscle atrophy and the accentuation of catabolic processes, and furthermore, whether these changes are the result of inactivity only, or whether they are an expression of the loss of the "trophic" influence of the nervous system on muscle metabolism.

A certain time after denervation, hypersensitivity to acetylcholine develops in the muscle deprived of its motor innervation (Frank et al. 1922, Brown 1937, Rosenblueth and Luco 1937, and others). Excitability of the denervated muscle is increased about 100 times (Nicholls 1956) and about 1000 times in the mammalian muscle following an intraarterial injection of acetylcholine as compared with normal muscle (Brown 1937). It is not possible to explain this hypersensitivity by changes in the electrical properties of the muscle membrane (see Nicholls 1956), and it can be assumed that it is associated with increased sensitivity of the denervated muscle membrane to the depolarizing effect of acetylcholine (Axelsson and Thesleff 1959). However, this hyperesensitivity is not restricted only to acetylcholine, but also applies to adrenaline (Euler and Gaddum 1931) and other substances such as potassium (Frank et al. 1923), compounds with nicotine-like effects (Dale and Gasser 1926), ATP (Buchthal and Kahlson 1946) and others.

This hypersensitivity has also been reported in other denervated structures, so that it can be considered as a general phenomenon, which has been formulated in the law of denervation (Cannon and Rosenblueth 1949). It was pointed out by Cannon and Rosenblueth (1949) that the hypersensitivity of denervated organs has important consequences. Although these organs do not receive nerve impulses, they react under certain conditions, and participate in generalized reactions of the organism. It is therefore impossible to draw conclusions about the metabolic reactibility of the denervated muscle without knowledge of these humoral regulation. Increased mechanical activity was observed, for example, in denervated facial muscles of monkeys in the course of a normal reaction of the animals to, for example, fright and fear (Bender 1938). This activity evidently corresponds to contracture of the denervated muscle evoked by increased liberation of acetylcholine which reaches the denervated muscles via the blood stream (Bender and Kennard 1938).

Muscle atrophy due to inactivity is not associated with a marked hypersensitivity of the muscle to acetylcholine (Eccles 1941, Johnsson and Thesleff 1961).

It was demonstrated by Cannon and Rosenblueth (1949) that inactivity per se is of no significance in evoking the hypersensitivity in other denervated organs, e. g. the salivary glands, in which secretion is not lost following sympathetic denervation, but in spite of this the glands become hypersensitive to adrenaline.

The experiments of Luco and Eyzaguirre (1955), in which it was shown that the onset of hypersensitivity to acetylcholine is dependent on the length of the peripheral nerve stump, also indicate that inactivity is not the cause of the hypersensitivity. Increased sensitivity to acetylcholine was found to appear later if the peripheral stump of the sectioned nerve was longer. This observation is evidently related to the finding that the degeneration of motor endplates begins later in a muscle with a longer peripheral nerve stump (Gutmann et al. 1955). Thus, if hypersensitivity, fibrillation activity and degeneration of endplates are affected by the length of the degenerating axon, then the trophic influence of the nervous system must be exerted here, evidently by supplying peripheral tissues with some substances, maintaining the structure of the motor endplate. They can thus be considered to be exerting a "trophic" function. These mechanisms also come into play in other denervated structures. Synaptic transmission is also affected later in the denervated superior cervical ganglion, if the peripheral stump of the preganglionic fibres is longer (Davidovich and Luco 1956).

It is interesting to correlate the appearance of hypersensitivity, fibrillations and degeneration of nerve or of axon endings in the endplate. Spontaneous irregular fibrillation activity in denervated muscles begins about five days after denervation in the cat (Langley and Kato 1915, Denny-Brown and Pennybacker 1938) and three days after denervation in the rat (Hines and Knowlton 1933). According to these observations, fibrillation activity first appears at a time when the continuity of axons is lost, which has been considered to be a physiological criterion of degeneration, since it correlates precisely with the loss of excitability of nerve fibres (Gutmann and Holubář 1950). The onset of fibrillation activity and hypersensitivity correlate fairly accurately (Luco and Eyzaguirre 1955); both phenomena are elicited by loss of the same factor, apparently lack of transmitter release (Josefsson and Thesleff 1961, Thesleff 1960). However, terminal nerve branches in the endplate disintegrate earlier than in the peripheral stump of the sectioned nerve trunk (32—42 hours earlier in the rabbit — Gutmann and Holubář 1952). As disintegration of axons in the endplate correlates with the loss of indirect muscle excitability, fibrillations should actually appears sooner than is usually reported in literature. The results of Hník et al. (1958) confirm this assumption. The first signs of fibrillation activity in the rat already appeared 30 hours after nerve section. Indirect excitability in old animals is lost later, and it is thus possible that the later onset of fibrillation activity reported elsewhere is at least partly due to the age of the experimental animals. It is not clear how transmitter release can act in preventing the onset of fibrillation activity. Apparently we have to deal with trophic mechanisms not associated with the propagation of nerve impulses, since the development of these phenomena is not affected by the loss of transmission of propagated nerve impulses (these impulses are eliminated immediately after nerve section), but with a loss of mechanisms related to structural and metabolic connection between the nerve and muscle cell.

As far as the mechanism of the hypersensitivity in the denervated muscle to mediators is concerned, Cannon and Rosenblueth (1949) considered the increased permeability of muscle cells to be responsible in the first instance. However, (Ginetzinsky and Shamarina 1942, Axelsson and Thesleff 1959) following denervation, the muscle membrane attains the same sensitivity to acetylcholine as the endplate region prior to denervation. The acetylcholine sensitive region thus is extremely enlarged. While acetylcholine evokes membrane depolarization in the normal muscle only by application to the endplate region, such depolarization in den ervated muscle can be obtained from any part of the muscle membrane surface. This increase of the receptor field enables acetylcholine to depolarize the whole length of the muscle fibre and causes a contracture. This is an important finding, but it is still not clear by what mechanism this neurotrophic regulation of metabolism is realized, so that its elimination following section of the nerve leads to hypersensitivity and fibrillations in the denervated structure.

The reactibility of denervated muscle is often compared with the reactibility of muscles at early stages of ontogenetic development (e. g. Orbeli 1945, Voskresenskaya 1959) or with muscle reactibility at a lower evolutionary level. These conclusions are very suggestive when we consider, for example, the slow spontaneous and rhythmic contractions of the flight muscles of insects during development, which gradually disappear (Voskresenskaya 1959) on the one hand, and the slow, spontaneous and continuous activity of abdominal musculature in insects following denervation (Case 1956, Beránek and Novotný 1959), or the spontaneous activity of auricular muscles after section of the facial nerve which is similar to contractions of smooth muscles or of slow muscles in amphibia (Luco and Sanchez 1956) on the other hand. It is also interesting to point out in this connection that embryonal muscle tissue exhibits spontaneous activity which disappears when nerve fibres reach the muscle (Marinacci 1959). Similarly, spontaneous contractions of muscle fibres can be observed in tissue cultures (Lewis and Lewis 1917), indicating that spontaneous acticity is a general attribute of the muscle cell, and it seems that innervation depresses this activity by some, as yet unknown, mechanism. All this seems to indicate that the loss of the trophic regulatory mechanism leads to metabolic changes which correspond to a low stage of development.

Hypersensitivity of the denervated muscle to mediators is not restrictive to acetylcholine only. It also includes the effect of adrenaline. Since adrenaline can be liberated under various conditions, especially under conditions of "stress" — it can be presumed that its effect could be quite considerable. The course of muscle atrophy in animals which, following section of the peroneal nerve, were exposed daily to painful stimulation, can serve as an example of this. These animals (ten rats) were left in a cage for one hour per day and stimulated from the metal floor with painful shocks at a frequency of 30 c./min. Atrophy of the denervated tibialis anterior muscle progressed at a faster rate in these "neurotized" rats, so that the weight of the denervated muscle, expressed in percentage of the contralateral control, was 48 ± 0·079% in the stimulated animals and 62 ± 0·092% in non-stimulated controls. In these experiments it can be assumed that there is increased liberation of adrenaline and that this, together with hypersensitivity of the denervated muscles, evidently accelerated catabolic processes in the denervated muscle. It is also possible, however, that vasoconstriction is the primary factor here, and that it increases the catabolic effect of cortisone, since it has been shown by Bajusz (1958), that ligature of the iliac artery does not by itself cause muscle atrophy, but atrophy does occur when this "selectivity conditioning" procedure is combined with stress-evoking stimuli or with administration of cortisone.

Muscles are sensitive to a number of other humoral agents. In this connection only those hormones will be discussed which accentuate tissue catabolism, since it is a matter of interest whether hypersensitivity to these agents could not, at least in part, explain muscle atrophy, which is generally characterized by a predominance of catabolic processes. Thyroxin is the most likely catabolically active hormone which accelerates the progress of muscle atrophy (Hines and Knowlton 1933, Diaz-Guerrero at al. 1947, Huff and Fischer 1949, Sheves et al. 1956, Selye and Bajusz 1958). The increased production or administration of thyroid hormone increases the breakdown of glycogen (Russell and Wilhelmi 1958). Protein catabolism, however, is also increased, this being indicated by higher excretion of nitrogen in the urine (Rupp et al. 1949). If the sensitivity of the denervated

Table 23

Percentage decrease of weight of denervated muscle in control animals and in animals following administration of cortisone-ace.ate (C. A.) 1 month after unilateral denervation.

Muscle	Control animals	Animals after C. A.
M. ext. dig. long.	42 ± 1·0	47 ± 3·4
M. tib. ant.	29 ± 1·2	34 ± 2·7
M. soleus	27 ± 2·6	28 ± 1·4
M. gastrocnemius	21 ± 1·6	21 ± 1·5

muscle to thyroxin is increased, and this has not as yet been demonstrated, then this hormone could be of paramount importance in influencing the rate of denervation atrophy. Analogy of muscle changes after denervation and in hyperthyroidism have led to the opinion that some of the important disturbances in muscle metabolism are of common origin (Huff and Fischer 1949, Sheves et al. 1956). Because of these similarities, Huff (1948) characterized the metabolism of the denervated muscle as "local hyperthyroidism". However, these analogies cannot explain the basis of disturbances in denervated muscle.

Cortisone represents another important hormone accelerating catabolic processes in denervated muscle. Cortisone increases the rate of denervation atrophy (Selye and Bajusz 1958), and can be especially effective under conditions of stress (Bajusz 1958). In our experiments, however, we were unable to observe this effect of cortisone. Denervation atrophy was investigated in four different muscles one month after unilateral section of the sciatic nerve, and after daily intramuscular administration of cortisone-acetate (1 mg./100 g.) in rats of 60—80 g. body weight (in 20 rats). The results are given in Table 23.

Increased atrophy could not be demonstrated following cortisone-acetate. Comparison with the results of Selye and Bajusz (1958) is difficult, since these authors do not state duration of the denervation period. It is, of course, possible that the catabolic effect of cortisone also affected the normal muscle in our experiments, and that this mechanism equalizes eventual differences.

Studies on the effect of hormones upon the denervated muscle are difficult, and their interpretation is seldom absolutely clear. Various hormonal factors mutually enhance or inhibit each other's effects on metabolism; furthermore, the administration of the hormone evokes counterregulating mechanisms of other hormones (especially after administration of adrenaline and insulin). Finally, one hormone can have a different mode of action in various muscles of the same animal. (Russel and Wilhelmi 1960). Thus, for example, cortisone increases the content of glycogen in the rectus femoris muscle, but is without effect upon m. cremaster (Leonard 1953). A number of hormones exert an anabolic effect upon both normal and denervated muscles. When evaluating the role of hormones in muscle metabolism, it is necessary to take into account the effect of a given hormone upon normal muscle and on the denervated muscle with an altered reactibility. Furthermore, it is necessary to find out whether the environment of the denervated muscle is the same as in the normal organism, or whether denervation has led to any changes of at least some endocrine glands (for example, the thyroid). And finally, results obtained by administration of a hormone in high doses which do not occur in the organism under normal conditions, must be accepted with reservations. When considering the complex interrelations of hormonal regulations, it is not surprising that certain hormonal effects exert their influence only under certain conditions.

The effect of somatotrophic hormone (STH) is especially interesting from this point of view. Administration of STH leads to a rise in body and muscle weight (Greenbaum and Young 1953), even though the increase in muscle strength does not run in parallel with the increase in muscle mass (Hume 1952) and although the myosin formed under the influence of STH is functionally different from normal myosin.

STH administration slows muscle atrophy (Selye and Bajusz 1958), even though this finding has not been universally confirmed (Schapira and Dreyfus 1950). It is also interesting to note that heart muscle does not hypertrophy in hypophysectomized animals following ligature of the aorta (Beznák 1952, Russell and Wilhelmi 1958), so that it seems that STH is necessary for thiis type of muscle hypertrophy.

In our experiments the anabolic influence of STH was demonstrated in the sense that its administration slows down denervation atrophy, but only under certain conditions, i. e. during realimentation following starvation. In three groups of animals (18 animals in each group) the sciatic nerve was sectioned unilaterally and muscles were excised 14 days after denervation (m. tibialis ant.) — (Table 24).

Table 24

Type of experiment	Muscle weight (mg.)		% loss
	Normal	Denervated	
14 days' denervation	409 ± 11·3	213 ± 8·4	52 ± 1·4
14 days' denervation with 1 mg./100 g. STH daily	422 ± 11·9	219 ± 6·3	52 ± 0·6
14 days' denervation (7 days starvation, 7 days realimentation) with 1 mg./100 g. STH daily	326 ± 10·2	193 ± 5·9	59 ± 1·6

As can be seen from Table 24, after daily administration of 1 mg./100 g. body weight of somatotrophic hormone during the 14 days' denervation period, no slowing down of denervation atrophy occurred as compared with the control group. However, when realimentation was performed for seven days after seven days' starvation, under otherwise identical experimental conditions, then the STH administration resulted in a statistically significant lower rate of denervation atrophy. It thus appears that STH is being liberated in considerable amounts during realimentation, and under these conditions protein increment could be observed, even though temporary, in denervated muscle undergoing long term denervation (six weeks) see p. 331.

Denervated muscle is also hypersensitive to insulin. Three days following denervation, denervated muscle synthesizes more glycogen after insulin and glucose administration than after glucose administration alone, and this increase in glycogen synthesis occurs with very low glycaemic values (see Fig. 125).

Insulin, im addition to its effect on carbohydrate metabolism, also plays a role in muscle protein metabolism. Very little is known, however, about this latter effect. The incorporation of methionine into muscle, for example, is considerably reduced in diabetes (Forker et al. 1951) and insulin may enhance uptake of aminoacids (Russel and Wilhelmi 1960). It also seeems that the nitrogen retaining effect of STH is dependent on insulin (Russel and Wilhelmi 1958). Very little is known about these relations and their changes in the denervated muscle.

The effect of testosterone is similar, to a certain extent, to that of STH, but the influences of both hormons summate (Eisenberg and al. 1949). Castration is accompanied by muscle atrophy which can be prevented by administration of testosterone, and it has even been found possible to obtain muscle hypertrophy by administration of this hormone (Weinmann and Shipounoff 1941, Eisenberg et al. 1949, Wright and Kochakian 1953, Kochakian and Tillotson 1957, Korner 1955, Leonard 1957, Russell and Wilhelmi 1958, and others). According to Leonard (1957) and Kochakian and Tillotson (1957) denervation atrophy can be slowed down by testosterone in large doses; this finding, however, is not in agreement with the results of Schapira and Dreyfus (1950) according to whom muscle atrophy is, on the contrary, enhanced. When evaluating the effect of testosterone on skeletal muscles, it is necessary to bear in mind that muscles connected with sexual functions are not suitable for evaluating the effect of this hormone on skeletal muscles in general (Nimmi and Geiger 1957).

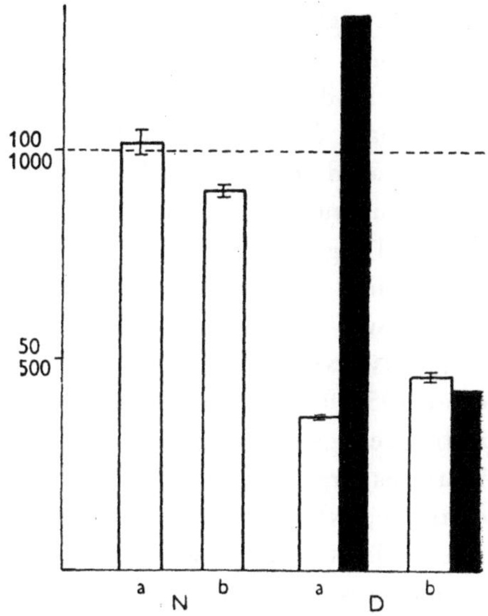

Fig. 125. Content of glycogen in mg% in M. soleus (white columns) 4 hrs. after glucose administration — 50% solution in amounts corresponding to 1·5% body weight (a) and after glucose administration with subcutaneous injection of insulin (b) in the normal soleus (N) and in the denervated muscle (D) three days after section of the nerve. Black columns: blood glucose in mg% four hours after glucose administration (a) and glucose administration with subcutaneous injection of insulin (b) respectively. Ordinate: Glucose and Glycogen levels in mg%.

Lack of testosterone can thus lead to atrophy, loss of muscle weight and muscle glycogen having been described after castration. This is an interesting case of muscle atrophy which can be observed mainly in perineal muscle which are especially sensitive to sexual hormones. The male sex hormone, for example, increases the inorganic phosphate fraction in these muscles (Kare 1953). However, it is without effect on the denervated muscle, not modifying its atrophy or changing the decrease of creatinephosphate (Kare et al. 1955). The characteristic action of this hormone on perineal muscles is lost after denervation. Both denervation and castration cause muscle atrophy. Atrophy seems to be evoked by some other mechanisms, since the increased turnover found in the denervated muscle is absent in muscles after castration (Kare et al. 1955).

Denervation atrophy cannot thus be explained by the action of hormones. Even though a number of hormones affect muscle metabolism under certain conditions, it is evident that in denervation atrophy they play only a minor role by modifying the course of atrophy.

Intricate interrelations between hormonal regulations are being gradually investigated in the metabolism of normal muscle, but very little is known as yet about humoral metabolic regulations in the denervated muscle. Nervous and humoral regulations are evidently closely coordinated in normal muscle, but this correlation is disturbed in denervated muscle. The fact that the denervated muscle was found to be hypersensitive to mediators and some hormonal agents, e. g. thyroxin (Hines and Knowlton 1935, Dreyfus and Schapira 1948, Polák et al. 1955) has led to the hypothesis that the nervous system exerts a protective influence against humoral agents (e. g. Charvát 1955). However, these interrelations are more complicated, as in the case of the male sex hormone, for example, which ceases to affect the metabolism of the denervated muscle. Conclusions cannot yet be made about the general mechanism of humoral regulations in denervated muscle because of insufficient evidence. It is indeed probable that humoral regulations are connected with the control of chemical sensitivity of muscle, mediated by the motor neuron apparantly related to subthreshold release of acetylcholine (Thesleff 1960). The important question remains, as to whether the hypersensitivity to mediators, especially to acetylcholine, together with the spontaneous fibrillation activity, can play a decisive role in the initiation of metabolic disturbances in denervated muscle. According to a number of authors (Langley and Kato 1915, Tower 1939) the energy requirements of fibrillation activity are so high that genesis of denervation atrophy could be explained by exhaustion of metabolic reserves in the muscle. The main factor in denervation atrophy would then be, according to these authors, the hypersensitivity to mediators and the fibrillation activity associated with this hypersensitivity. However, lack of transmitter release can be produced by botulinum poisoning without affecting nerve and muscle in other respect (Thesleff 1960). This finding suggests that also other mechanisms than subthreshold release of mediators are responsible for mediating the "trophic function of the nervous system".

5. *In vitro* experiments and their significance in metabolic studies of the denervated muscle

A large number of metabolic studies on muscle have been performed *in vitro*. These experiments have obtained fundamental facts about muscle metabolism and it is clear that it will be *in vitro* experiments which will make possible definite conclusions concerning the metabolic disturbance in denervated muscle, since by this approach, various factors affecting metabolism can be isolated and analysed separately. These analytical experiments cannot be successful unless correlated with metabolic reactions of the denervated muscle *in vivo*. The denervated muscle remains a part of the organism and does not cease to react to various stimuli from the internal environment. The reactibility to mediators is altered in a specific manner, and finally denervation itself may lead to changes in the organism which can in turn affect metabolic reactions of the denervated muscle itself. It is therefore not possible to ignore investigations of the physiology of denervated muscle, and results from this research must be correlated with biochemical in vitro studies.

Many controversial findings have been reported in *in vitro* experiments. This is due to the fact that isolated biochemical studies were in most cases concerned with late stages of denervation atrophy, i. e. stages when extensive morphological changes had already taken place. Early stages of denervation atrophy are especially important, and precise correlation must be made with changes in nerve and muscle excitability, as is documented by experiments on fibrillation activity.

These time relations cannot be studied so well in *in vitro* experiments. Differentiation of the tissue into muscle and interstitial phase has been emphasized by Fischer (1950). This has not been respected in many reports and consequently data concerning absolute changes in individual components of the "muscle phase" are not accurate.

In vitro studies of cellular metabolism are complicated and their primary aim is to demonstrate relations between substrate and enzymatic activity. A large number of interdependent heterogenous reactions participate in metabolic regulation, and the substrate *in vivo* is represented by an infinitely more intricate and changing complex of influences which is very difficult to recapitulate *in vitro*. It is necessary to bear in mind that determination of the enzyme content and activity are necessarily accompanied by changes in structural orientation, by elimination of some metabolic inhibitors and activators, and finally by dilution of enzyme and substrate, and in addition, by changes present in the organism.

When estimating the enzymatic activity under a given dynamic situation considerable difficulties arise, since it is necessary to arrest the reaction before the next stage in the cycle. Difficulties are also encountered when estimating protein fractions in the denervated muscle, the extraction of the muscle proteins being a function of the ionic concentration in the extraction medium which can vary with the state of the muscle (see Schapira and Dreyfus 1950).

In vitro experiments do not usually take into account considerable metabolic differences between various muscles. Large differences, of an order of magnitude, were, for example, demonstrated in red and pale insect muscles in cytochromoxidase activity, glycogen content, resting oxygen consumption and the type of glycolysis (Kubišta 1959). Considerable differences have also been reported between pale and red muscles in mammals (Yakovlev 1955) which are due to differences in oxidative metabolism. These differences, evidently caused by functional adaptation of the nervous system to various functional requirements of the innervated structures, are convincingly documented by changes in the functional type of red and pale muscles following cross-unions of nerves (Buller et al. 1959). Differences have also been reported in the content of cholinesterase in various animal species and different muscles. The content of cholinesterase is reduced in the denervated gastrocnemius of the guinea-pig (Couteaux and Nachmansohn 1940), but not in the serratus muscle (Brooks and Myers 1952). Furthermore, various muscles have also been shown to react differently to hormonal agents (Leonard 1955). All these differences, which have a functional genesis, are often ignored in *in vitro* metabolic studies. Investigation of metabolic reaction of denervated muscle in the organism as a whole can correct some discrepancies of *in vitro* studies, and can in this way help to lead and orientate biochemical analysis. It is thus necessary to stress the importance of metabolic studies in denervated muscle *in vivo*, and to point out the necessity of a critical approach to some of the results concerning *in vitro* metabolism of denervated muscle.

6. Adaptation mechanisms in studies of the metabolic reactibility of denervated muscle

Comparative studies have demonstrated that muscle metabolism is effectively adapted to various functional requirements. Expecially instructive results have been obtained from insect muscles, which are adapted in various species to various functional conditions. This is apparent, for example, in the special type of metabolism in insect flight muscles. The extremely high contraction frequency of insect flight muscles requires other neuromuscular mechanism than those of vertebrates (see Pringle 1954). The high efficiency of insect muscles is given by the fact that oxygen requirements must be met very quickly, this necessarily being a characteristic property of insect muscle. Very low values of oxygen debt (Chadwick 1947) and a special type of glycolysis (Kubišta 1959) are in agreement with this.

Functional requirements of flight muscles, characterized by highly effective aerobic metabolism, are reflected in a high content of sarcosomes, the size of mitochondria, high activity of cytochromoxidase, high content of glycogen, and a relatively small role of anaerobic glycolysis, especially of lactic acid. Further, it was shown that muscles with a high cytochromoxidase activity also have a higher resting oxygen consumption. Thus there exists a direct correlation between functional requirements and intensity of oxygen metabolism, which is the outcome of adaptation processes and which are reflected in muscle metabolism. Similarly, increased activity of all enzymatic systems connected with oxygen metabolism has been found in mammalian muscles undergoing repeated functional "training", which requires a high efficiency of oxidative processes (Yakovlev 1955).

What adaptation mechanisms can take place in denervated muscle? It is interesting that following a relatively rapid loss of weight and of protein nitrogen during the first two months after denervation, these changes later progress at a very slow rate (see Chapter II). The difference in structural changes between muscles, denervated 4—6 months and 12 months, is relatively small. The denervated muscle may become adapted to metabolic changes evoked by the loss of innervation.

The loss of the trophic influence of the nervous system results in metabolic changes, some of which indicate an increased energy consumption in the denervated muscle. On the one hand, continuous irregular fibrillation activity may be the cause of increased oxygen consumption. On the other hand, there may be an early absolute as well as relative increase in the number of nuclei in the denervated muscle (Gutmann 1960, Gutmann and Žák 1961). The rise in the content of desoxyribonucleic acids (DNA), calculated per non-collagenous protein nitrogen, is especially marked, and this means that individual muscle cells may contain an exceptionally large number of nuclei. The energy requirements in the denervated muscle must thus be considerably increased and these requirements are probably further increased by parallel degenerative changes and disintegration of some of the nuclei, so that it is probable that the total metabolic turnover in the nuclei is increased.

It seems that these processes are reflected in the increased consumption of oxygen in denervated muscle (see Chapter VIII). All this, however, is taking place under conditions of reduced utilization of oxygen, as can be deduced, for example, from a decrease in the concentration of cytochrome C and in the activity of cytochromoxidase, succinic dehydrogenase and succinoxidase (Schmidt and Schlieff 1956), and furthermore, from an increased content of lactic acid and probably by a decrease of the P/O quotient (Grigorieva 1958, Slozhenikina 1957). Some of these changes can be considered as the result of muscle adaptation to inactivity. This probably applies to cytochrome changes in the muscle, the content of which, in contrast to the changes in the denervated muscle, is increased in trained animals (e. g. Guckelberger and Keiser 1939).

Denervated muscle must apparently adapt itself to increased consumption of oxygen which is probably associated with increased energy requirements of fibrillation activity (see Langley 1917, Tower 1939) and with the increase of DNA in the denervated muscle (Schmidt and Schlieff 1955, Gutmann 1960

Gutmann and Žák 1961). Some of the metabolic changes in the denervated muscle indicate that the adaptation mechanism involves increased oxygen economy. Glycogenolysis and glycogen synthesis in denervated muscle are reduced, and blood glucose is utilized more as an energy substrate and for lipid synthesis (Bass and Hudlická 1960). Considerably less oxygen is needed for this lipid synthesis, which is a beta oxidation process, so that from the point of view of oxygen consumed, lipid synthesis from glucose is more economical (see Bass and Hudlická 1960). Finally, it seems that in the denervated muscle lactic acid is also being utilized for synthesis of glycogen (Sato and Kasugai 1935, Hudlická and Gutmann 1958).

The shift in the relative content of sarcoplasmatic and specific structural proteins represents an important change in the metabolism of denervated muscle and possibly reflects an important adaptation reaction of the muscle to inacticity. It appears that a relative increase in the content of sarcoplaasmatic proteins takes place after denervation (Stewart 1955), the specific structural proteins decreasing due to lack of muscle contraction. It is, of course, not yet possible to decide whether this is the result of an adaptation to inactivity, or whether it is due to the absence of substances transported along the nerve, which are important for normal proteosynthesis.

It is thus difficult to apply adaptation studies in explaining the mechanism of denervation atrophy. Two adaptation reactions, to a certain extent antagonistic, probably play a role. Muscle inactivity must necessarily lead to a reduction in enzymatic systems associated with oxidative processes and also, perhaps, to a relative increase of sarcoplasmatic proteins. On the other hand, the absence of the trophic influence of the nervous system, which is connected with fibrillation activity and with proliferation of nuclei, leads to conditions favouring increased oxygen consumption to which the denervated muscle adapts itself by means of mechanisms saving oxygen. It is not yet possible to say to what extent these mechanisms participate in, and are responsible for, the specific disturbance of denervated muscle metabolism. However, evidence about the important role of adaptation mechanisms for normal muscle metabolism indicates that study of adaptation is also useful in expaining the disturbance of metabolism in the denervated muscle.

B. Metabolic disturbances in denervated muscle

Experimental evidence

Loss of cellular muscle constituents represents a characteristic sign of denervation, and it would be natural to suppose that the primary mechanism leading to muscle atrophy is a disturbance in synthetic processes in the muscle. This general conclusion must, of course, be documented by concrete evidence. This disturbance can affect either all anabolic processes in general, or it can be a "primary" disturbance of a precisely defined link in metabolic reactions. Such a disturbance of anabolic processes can be due to a) vasomotor changes, so that the supply of important substances is reduced, b) changes in permeability of the cell, c) a reduction in the content and activity of enzymes, d) a disturbance in the utilization of metabolites.

It is not probable that the cause lies in vasomotor disturbances which would restrict the supply of metabolic substrates. Blood flow through the denervated extremity is increased (Doupe et al. 1943, Keller, Loeser and Rein 1930, Hudlická and Vrbová 1958 and others). The supply of oxygen is sufficient and oxygen consumption has often been reported to be increased in the denervated muscle (Langley and Igataki 1917, Sato and Kasugai 1935, Knowlton and Hines 1934, Sheves et al. 1956, Bass and Hudlická 1960). Vasomotor changes cannot thus be the main factor causing a disturbance of anabolic processes in muscle metabolism.

It seems more probable that changes in permeability are a more likely link in the development of metabolic disturbances in the muscle. The membrane resistance is increased (Nicholls 1956) and a corresponding drop in the conductivity to potassium ions has been reported in denervated muscle (Harris and Nicholls 1956). Considerable changes have also been found in the content of water (Hines and Knowlton 1937) and in the distribution of electrolytes (see, for example, Fenn 1937, Fischer 1949, 1952, Schapira et al. 1947, Drahota 1959, 1960) and these could be expected to alter membrane permeability. However, the situation is more complicated when changes in permeability of the muscle cell membrane are to be investigated *in vivo*. Acetylcholine, for example, increases the permeability to potassium ions in the motor end-plate region (Castillo and Katz 1955), as well as in the denervated muscle membrane in the absence of sodium ions in the medium (Axelsson and Thesleff 1959). The receptive surface in the denervated muscle is considerably enlarged, so that the whole muscle membrane becomes sensitive to acetylcholine. These conditions should increase permeability to potassium ions *in vivo*. In spite of this, the muscle cell maintains a very constant intracellular concentration of potassium after denervation and in the course of muscle reinnervation. This doubtlessly requires special energy and a disturbance of potassium becomes apparent when electrolyte metabolism is functionally loaded (Drahota et al. 1957).

It is thus possible that grave alterations in permeability, especially to potassium ions, are not manifested until electrolyte metabolism is functionally loaded, and such changes could affect synthetic processes in the cell. So far, no precise correlation is known between changes in electrolyte composition and the capacity of cells to synthetize metabolites. It was thought, for example, that synthesis of glycogen depends on transfer of potassium into the cell (Hastings et al. 1952). However, glycogen synthesis is renewed very soon in the reinnervated muscle, at a time when the potassium content is still low (Drahota 1959), and glycogen synthesis takes place *in vitro* even in a potassium-free medium (Drahota et al. 1958). Furthermore, the transport of glucose into the denervated muscle cell cannot be presumed to be reduced (Bass and Hudlická 1960), even though glycogen synthesis is considerably decreased in denervated muscle.

Permeability changes can thus alter energy requirements of the cell, which are important for the maintenance of normal osmotic gradients, but they are unable to explain grave metabolic disturbances in the denervated muscle.

The main metabolic link is thus to be found in a disturbance of substrate utilization (Bass and Hudlická 1960). The disturbance must be defined more precisely. It has been shown that decrease of glucose utilization for glycogen synthesis appears very early after denervation (see Chapter VIII) and is associated with structural changes of glycogen (Koshtoyants and Janson 1950, Bass and Vodička 1955). The disturbance of metabolism could be explained by changes in the content and activity of enzymes. The activity of phosphorylase (Levine et al. 1941) and amylase (Zubenko 1954) is actually decreased in the denervated muscle. The role of amylase in glycogen metabolism has been stressed by Petrova (1954). Nevertheless, hexokinase and phosphorylase activity increase in denervated muscle at a time when glycogen synthesis is already considerably reduced (Mirski and Wertheimer 1942, Epelbaum and Kantor 1954, Kubišta and Gutmann 1957) and the reduction of phosphorylase and glycolytic activity does not appear until late stages of denervation atrophy. It has been found that ATPase activity is decreased very soon (12—16 hrs.) after denervation. Mitochondrial ATPase would be very important, since it participates in oxidative phosphorylations. ATPase activity of myosin is also reduced (Fischer 1955). The reduction of ATPase activity in denervated muscle should lead to a decrease in the effectiveness of oxidative phosphorylations. It is, of course, not known to what extent ATPase can be the limiting factor in oxidative phosphorylations. At a later period of denervation, cytochromoxidase (Schmidt 1952, Humoller et al. 1952) and succindehydrogenase activity (Knowlton and Hines 1941, Humoller et al. 1951) were found to decrease, this being the result of gradual adaptation of the denervated muscle to a lack of functional activity. This change cannot, of course, explain the early disturbances of oxidative processes in the denervated muscle.

As has already been shown, oxidative metabolism undergoes a characteristic change in denervated muscle, since increased consumption is apparently accompanied by decreased utilization of oxygen. It is necessary to bear in mind that muscle respiration can be increased by as much as one hundred times and yet its oxidation capacity is not used to the full. It is thus not probable that oxygen concentration could be a limiting factor in muscle respiration (see Krebs 1959). However, diphosphopyridinenucleotide (DNP) and adenosinediphosphate (ADP) which play an important role as catalysts, are the limiting factor for normal "utilization" of oxygen in the first place. They are both present in small quantities and require constant regeneration (Krebs 1959). Regeneration requirements are very high, as has already been stressed because of fibrillation activity and a large number of nuclei, so that it is possible that the main factor in the disturbance of denervation metabolism is insufficient regeneration of these substances.

Changes in nucleic acids and protein nitrogen can serve as further examples of the disturbance in utilization of substances in denervated muscle. The relative content of nucleic acids in the denervated muscle increases, in spite of the fact that protein content is progressively reduced. This may mean that nucleic acids cannot be "utilized" normally for protein synthesis in denervated muscle, and it is probable that "catalysts" are missing or are incompletely regenerated.

Disturbance of proteosynthesis in denervated muscle becomes manifest in marked changes in both structural (Fischer 1948, 1955, Dreyfus et al. 1953) and sarcoplasmatic proteins. These changes can also be due to alterations in enzyme systems. For example, Fischer (1950) found that the ratio between myosin-bound ATPase to total muscle ATPase is reduced during the denervation period. It is thus a reverse process to that observed during development (Hermann and Nicholas 1948), when the ratio between myosin-linked ATPase and total ATPase gradually increases. But these enzyme changes have been described at a relatively late period of denervation, while protein synthesis is altered earlier. More detailed knowledge concerning proteosynthesis in denervated muscle is needed, especially from the point of view of a transitory increase of sarcoplasmatic and contractile proteins in the denervated diaphragm (Stewart 1955). It is not unlikely that the primary change lies in a decrease of enzyme activities. Incorporation studies with labelled nitrogen should obtain more accurate information. Incorporation of nitrogen into myosin was actually found to be decreased in the denervated muscle (Schapira et al. 1953). According to Ferdman (1955, 1957), on the other hand, incorporation of methionine into denervated muscle is increased, and it is therefore concluded that protein turnover is higher in denervated muscle in the sense of increased compensatory metabolic activity.

The rate of protein degradation or proteolysis in denervated muscle has also not been elucidated as yet. Proteolytic enzyme activity can be increased in denervated muscle. In our experiments, autolysis *in situ* has been found to be accelerated after denervation. Normal and denervated muscles were left for two hours *in situ* and then quantitative determinations were made of amino acids liberated during 20 min. action of proteolytic enzymes, according to the method of Lowry-Hunter (1945). The results obtained in the normal and denervated muscle after two hours were expressed as percentage of the initial value in the contralateral muscle. The percentage increase was equal to $28 \pm 4{\cdot}4\,\%$ in normal muscles and $49 \pm 6{\cdot}4\,\%$ in denervated muscles (three days after denervation). Increased autolysis in vitro was also reported by Bradley (1938), but proteolytic activity was not found to be in creased by Schapira and Dreyfus (1950).

Definite data concerning the rate of metabolism, i. e. the rate of breakdown and synthesis, are missing, and it will only be possible in this way to determine the limiting factors which can explain the specific metabolic disturbances which occur in muscle after denervation. Although our knowledge of intimate metabolic processes in muscle has progresssed considerably, it has not so far been possible to elucidate the primary or main link in the metabolic disturbances of denervated muscle by biochemical analysis only.

Let us then return to investigation of the metabolic reactions of the denervated muscle, with the hope that this physiological approach will, perhaps, offer some clues for solving this problem. When discussing metabolic regulatory mechanisms, it is necessary to differentiate autoregulatory, hormonal and nervous mechanisms. These control mechanisms affect the rate of metabolism and adapt cellular metabolic activity to existing requirements. How do these mechanisms participate in the metabolic disturbance which prevails after denervation?

1. Autoregulatory mechanisms

If nervous regulation of metabolism is to be defined, or if it is to be demonstrated that after denervation changes are due to the absence of nervous mechanisms, then the first question is to differentiate autoregulatory mechanisms, i.e. those "primitive" control mechanisms which can also be demonstrated in unicellular systems and on to which hormonal and nervous influences are superimposed (see Krebs 1957). The enzyme-substrate relationship is the basic element which influences the rate of metabolism. Assuming that there is sufficient enzyme activity, autoregulatory mechanisms, for example, in glycogen synthesis, are determined by enzyme activity, the content of the metabolite (glycogen) and concentration of the substrate (glucose). A direct relationship is then expected between glucose concentration and glycogen synthesis, as has been shown, for example, in the case of rat diaphragm *in vitro* (Hechter et al. 1941). However, the content of a given metabolite is already the result of a dynamic equilibrium between processes of breakdown and synthesis, which is altered following denervation. The equilibrium between breakdown and synthesis of glycogen is established at different levels in the rabbit and the rat, explaining species differences in glycogen changes during the first days after denervation (Martínek 1956).

The content of glycogen is reduced in both the rabbit and the rat in later stages of denervation, but three days after section of the nerve glycogen levels are still higher in the rabbit, although they are already lower in the rat. Glycogen synthesis following administration of glucose is decreased in both the rabbit and rat after denervation. However, less glycogen is broken down in the course of three days' starvation in the denervated muscle of the rabbit (Fig. 126)

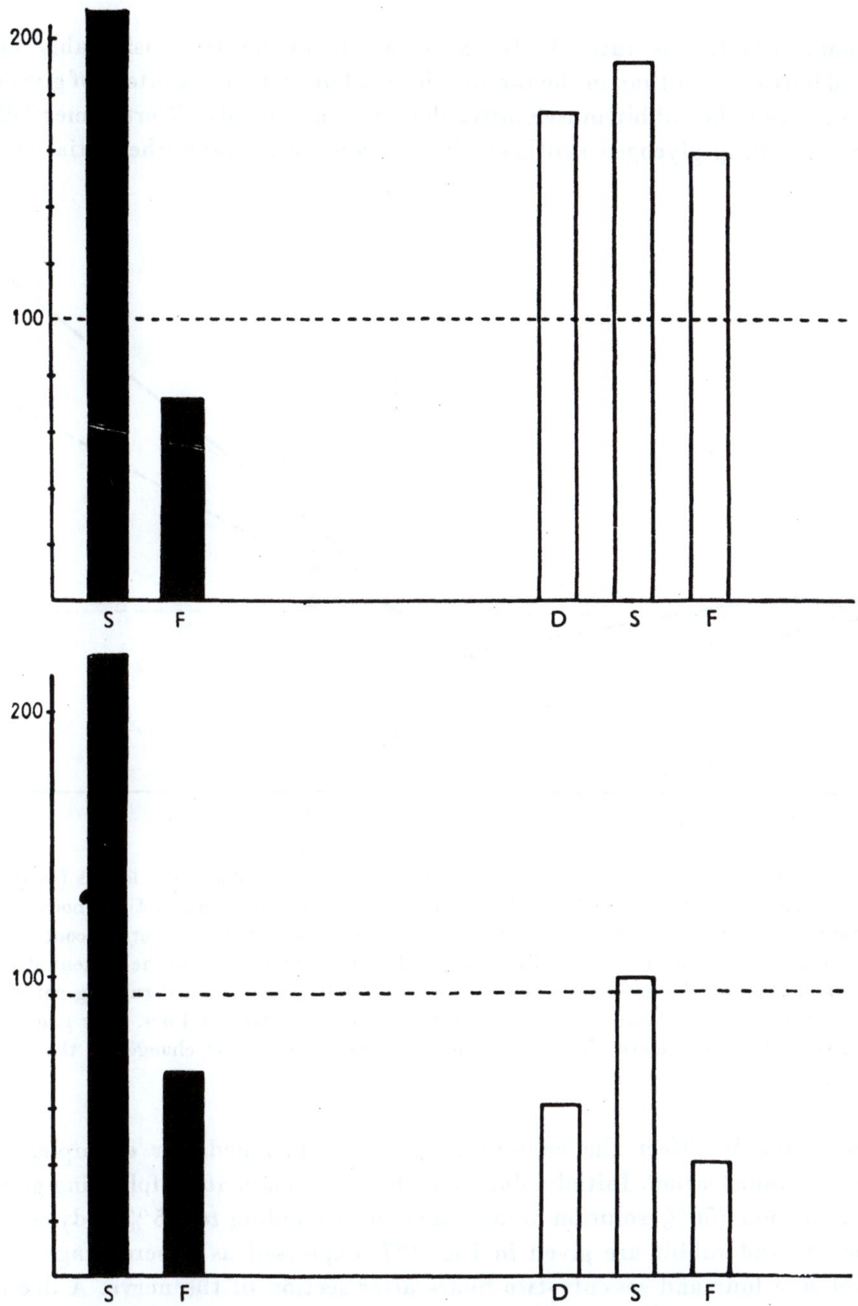

Fig. 126. Percentage changes in glycogen content in the extensor digitorum longus of the rabbit (A) and rat (B). Black columns: glycogen content in normal muscle 4 hrs. after administration of 50% solution of glucose corresponding to 1·5% body weight (S) and following 3 days fasting (F). White columns: glycogen content in denervated muscles 3 days after section of the nerve (D), 4 hrs. after glucose administration (S) and following 3 days' fasting (F).

(Gutmann 1959). The rate of glycogenolysis is evidently considerably more reduced in the rabbit than in the rat and this explains why the content of glycogen is increased in the rabbit muscle after denervation (see also Wertheimer 1927). When evaluating glycogen synthesis, it is necessary to know the initial values

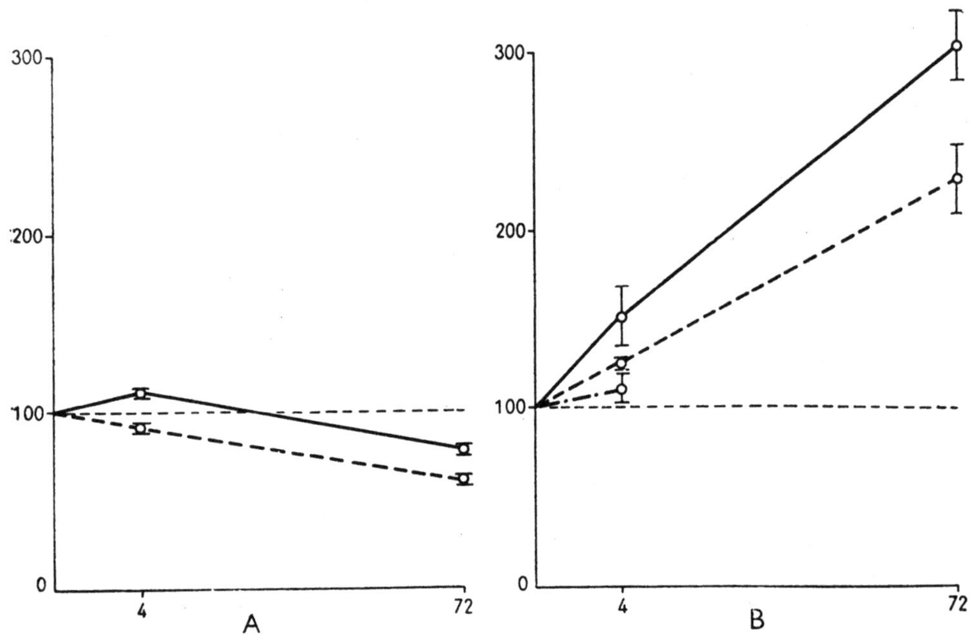

Fig. 127. Percentage changes in glycogen content in the extensor digitorum longus (solid line) and following administration of 50% glucose in amounts corresponding to 1·5% body weight (interrupted line) four and 72 hrs. after denervation as compared with the normal contralateral muscle in the rat (A) and the rabbit (B). Dashed and dotted line: changes in the content of glycogen in rabbits in which the sciatic nerve was sectioned following a 5 hours' running period in a rotating drum on one side and extensor digitorum longus was excised 4 hrs. after glucose administration. Abscissa: hours after denervation. Ordinate: percentage changes in the content of glycogen.

which markedly affect the content of glycogen attained, for example, after glucose administration. Initial values and the glycogen content following glucose administration (50% solution in amounts corresponding to 1·5% body weight) in the rat and rabbit are given in Fig. 127, expressed as a percentage of the control side four and seventy-two hours after section of the nerve. A decrease of glycogen synthesis can already be demonstrated four hours after nerve section in the rat (Fig. 127A), while this decrease is masked in the rabbit by a higher initial level of glycogen in denervated muscle at this time (Fig. 127B).

Glycogen synthesis is reduced in the rabbit four hours after nerve section if the initial values are lowered by functional activity. The initial levels of glycogen were

150 ± 19·7 mg% (6 animals) following running in a rotating drum. The sciatic nerve was sectioned unilaterally immediately after this activity, and muscles were excised four hours after glucose administration (50% solution in amounts corresponding to 1·5% body weight). The percentage change of glycogen in the denervated muscle was 110 ± 17·3% as compared with 124 ± 7·1% in animals without functional activity prior to denervation. The increased content of glycogen which has been found following section of the nerve is thus associated with a relatively decreased glycogen synthesis, and is evidently the result of decreased glycogen breakdown in the denervated muscle of the rabbit.

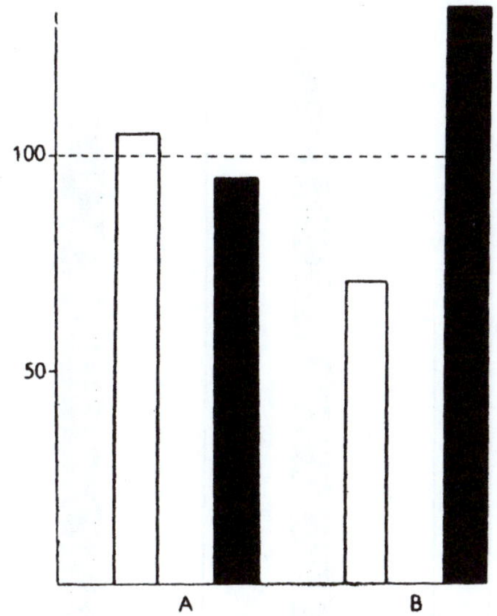

Fig. 128. Glycogen changes in the denervated extensor digitorum longus of the rat following subcutaneous injection of 0·02 mg. adrenaline per 100 g. body weight expressed as percentage of the glycogen content in normal control muscle. White columns: glycogen content before; black columns: immediately after adrenaline injection; immediately (A) and three days after denervation (B).

It is also possible to explain different changes in glycogen content following an injection of adrenaline in various animal species by different glycogen levels in denervated muscles. According to Hines and Knowlton (1935) administration of adrenaline in rats leads to the same reduction of glycogen content in normal and denervated muscles. On the other hand, Yakovlev (1951) reported a smaller reduction of glycogen in the denervated muscle of the cat following an injection of adrenaline. It is concluded from these experiments that denervated muscles are less sensitive to the effects of adrenaline. But the initial levels of glycogen in the denervated muscle, which differ in mucles of different animal species, are again an important factor in this connection. The decrease of glycogen in the denervated muscle of the rat following administration of adrenaline has been found to be the same as reported earlier by Hines and Knowlton (1935). But this applies to short intervals after denervation only, i. e. at a time when glycogen levels are still high (Fig. 128). Three day after denervation, i. e. when the glycogen content has already considerably diminished, adrenaline caused only a small decrease of glycogen. On the other hand, in guinea-pigs in which there is an increase of glycogen during the first few days after denervation, injection of adrenaline results in a considerable reduction in glycogen. This is evident from the relation between glycogen content of the denervated and normal muscle of the rat as compared with this ratio in guinea-pigs (Fig. 129). The breakdown of glycogen following muscle stimulation is also dependent on the initial content of glycogen (see chapter VIII). It can thus be concluded that the autoregulatory mechanism related to the initial level and the degree of glycogen breakdown comes into play after muscle stimulation, as well as following adrenaline administration. It is not until this autoregulatory mechanism is excluded that it is possible to draw conclusions concerning the different effect of adrenaline on normal and denervated muscle. If the initial values in normal and denervated muscles are the same (due to starvation or glucose administration), then the relative decrease of glycogen content following injection of adrenaline is smaller in the denervated than in the normal muscle.

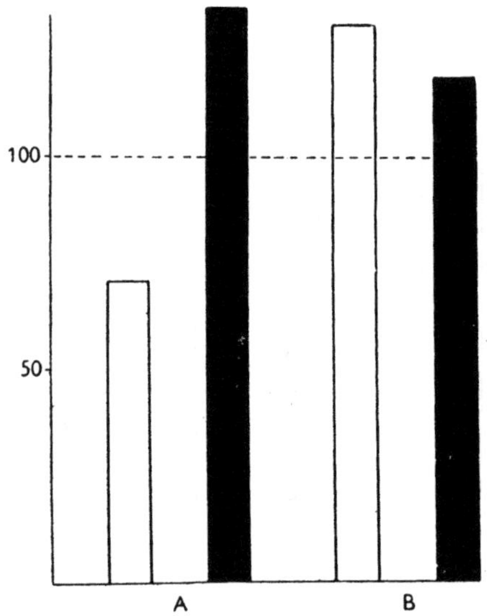

Fig. 129. Glycogen changes in the extensor digitorum longus in the rat (A) and guinea-pig (B) three days after denervation and 4 hrs. after subcutaneous injection of adrenaline expressed as percentage of the glycogen content in normal control muscle. White columns: percentage changes in glycogen 3 days after denervation. Black columns: after injection of adrenaline.

Autoregulatory mechanisms are also involved in glycogen synthesis following functional activity (Bass et al. 1955). Stimulation of muscle with a higher frequency, which results in more extensive glycogen breakdown, leads, under otherwise identical experimental conditions, to faster and more intensive resynthesis in the postfunctional period (see Chapter VIII). Different initial levels of metabolites following denervation in various animal species also explain, as has already been pointed out, the different course of glycogen synthesis after glucose administration. The lower the initial level of glycogen, the greater the percentage increase of glycogen following glucose administration. Glycogen synthesis is impaired very soon in both the rat and the rabbit, but it is much more evident in the rabbit or the guinea-pig, since very high initial glycogen levels prevent further glycogen synthesis to take place in the cell (Fig. 130).

This can be documented in experiments in which glycogen synthesis in the denervated muscle of guinea-pigs was determined following glucose administration, where the content of glycogen remains at high levels for a considerable time after denervation. It is evident that following glucose administration, there is practically no glycogen synthesis in the denervated muscle (Fig. 130).

The influence of this autoregulative mechanism, i. e. of the effect of initial levels on the intensity of glycogen synthesis, is evident when the initial glycogen content is reduced by starvation. In a muscle three days after denervation, glycogen synthesis following glucose administration after 24 hours' starvation is decreased to $60 \pm 3{\cdot}5\%$ (Vodička et al. 1957). In further experiments, the

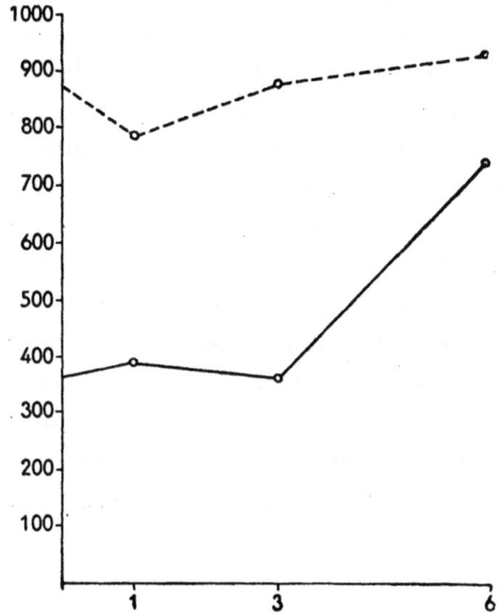

Fig. 130. Glycogen changes in mg% in the extensor digitorum longus of the guinea-pig in normal (solid line) and denervated muscle three days after section of the nerve (interrupted line) following administration of 50% glucose solution corresponding to 1·5% body weight. Abscissa: glycogen content in mg%. Ordinate: hours after glucose administration.

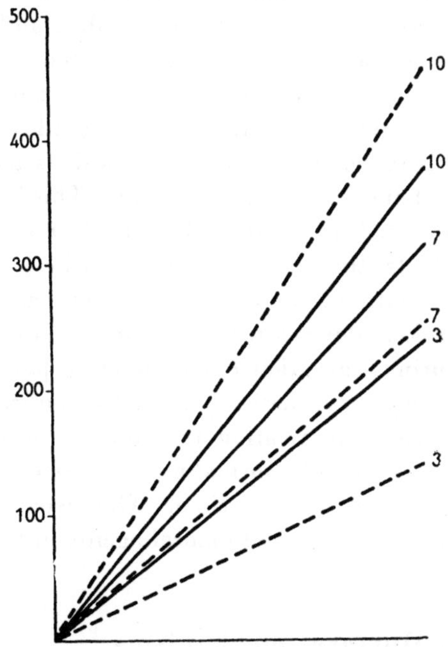

Fig. 131. Percentage glycogen increase in the normal (solid line) and denervated (interrupted line) muscle in animals after realimentation following 3, 7 or 10 days' starvation. The initial value corresponds to glycogen content following various periods of starvation. The second value represents the percentage increase of glycogen 24 hrs. after realimentation.

Table 25

Content of glycogen 24 hours after realimentation (number of animals in brackets).

Type of experiment	Normal muscle		Denervated muscle	
	Before re-alimentation	After re-alimentation	Before re-alimentation	After re-alimentation
3 days' starvation (9)	319 ± 14	671 ± 37 821 ± 62	319 ± 14	436 ± 13
7 days' starvation (7)	337 ± 40	1062 ± 152	337 ± 40	865 ± 96
10 days' starvation (8)	280 ± 36	1068 ± 46	267 ± 38	1205 ± 50
Adaptation to intermittent starvation (9)	296 ± 38	1980 ± 197 2189 ± 159	386 ± 22	1804 ± 122

relations between initial glycogen values and glycogen synthesis were studied in normal and denervated muscles three days after section of the nerve, by prolonging the duration of starvation to three, seven and ten days respectively. After this, animals had free access to food. Prior to realimentation the sciatic nerve was sectioned unilaterally, and glycogen was determined in the normal muscle and the muscle denervated for three days, 24 hrs. after realimentation.

As is demonstrated in Table 25 and Fig. 131, after three days' starvation and 24 hours' realimentation glycogen levels also rise in the denervated muscle, and this increased glycogen synthesis is more marked when the starvation period is prolonged to seven or ten days. This is caused by the fact that prolonged starvation leads to a further decrease in muscle glycogen and this low glycogen content affects the intensity of resynthesis. However, this decrease during further starvation is very small, and most glycogen available for mobilization is already broken down after theree days' starvation. Other humoral factors activated during starvation must therefore participate, and these affect the intensity of glycogen synthesis. This is also confirmed by the fact that the rise in content of glycogen in relation to duration of starvation is more marked in denervated muscle than in normal muscle, since the ability to synthesize glycogen is already decreased at this time. This would indicate that the denervated muscle is more sensitive to humoral agents in the course of starvation and realimentation.

2. Humoral mechanisms

As an example of autoregulative mechanisms in denervated muscle, the effect of the initial levels of metabolites was especially demonstrated in the extent of synthesis and breakdown of metabolites following various experimental conditions. Of course, biochemical study of these primitive control mechanisms is much wider and more extensive, and should help to determine the factors limiting the rate of metabolism, e. g. cellular respiration (see Krebs 1959). It is evidently a question of complex "feed-back" mechanisms taking place in "multi-reaction" systems, and their application in physiological experiments which are aimed at analysing or differentiating nervous regulation of metabolism is, as yet, too complicated. We have therefore attempted to demonstrate in the first place some primitive autoregulatory mechanisms on which humoral and nervous mechanisms are superimposed.

Hormonal regulation of metabolism in denervated muscle can best be studied under conditions of adaptation, which are necessarily realized in denervated muscle by means of humoral mechanisms. Such an approach was chosen in rats which were subjected to repeated intermittent starvation (three days' starvation alternating with one day of realimentation for six weeks). Under these conditions, the liver (Fábry 1955, Fábry and Hrůza 1956) and normal muscle are capable of synthetising extremely high glycogen levels during realimentation (Gutmann and Vrbová 1957, Vrbová and Gutmann 1958). This increased synthesis, which is apparently the result of "training of anabolic processes", is also present in acutely denervated muscle, with the difference that values in normal muscle return to normal in the course of three days of final realimentation, while this return to normal is evidently slowed down in denervated muscle (Fig. 132). It can be seen that the denervated muscle in animals adapted to intermittent starvation is capable of glycogen synthesis, which is practically the same as that in the normal muscle. This cannot be due to lower initial levels of glycogen. A change in activity of enzyme systems in denervated muscle has undoubtedly also taken place, and this evidently lasts for a certain time after denervation, and probably participates in increasing glucose utilization in tissues during adaptation (Gutmann 1959). It has been demonstrated, for example, in the liver that hexokinase activity is increased under these conditions of adaptation (Petrásek 1959). The factors limiting the rate and intensity of synthetic processes can thus be modified by a process of adaptation.

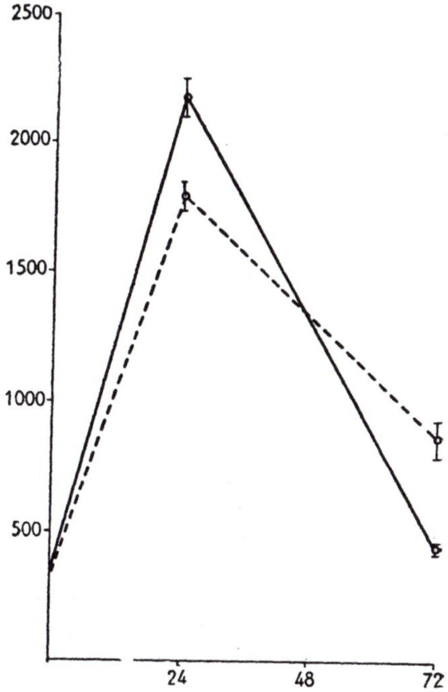

Fig. 132. Glycogen changes in mg% in the soleus of rats adapted for six weeks to three-day periods of intermittent starvation, always followed by one day of realimentation. Abscissa: time after final realimentation. Ordinate: glycogen content in mg% in animals with normal, innervated muscle (solid line) and in animals with section of the sciatic nerve, performed immediately prior to realimentation.

However, glycogen synthesis is also accentuated during realimentation following a single starvation period. It is evident that during single and intermittent starvation, certain hormonal agents are activated which make the increased glycogen synthesis in the denervated muscle possible. Since adapted animals maintain higher levels of blood glucose during starvation (Gutmann 1959), it is probable, that glucocorticoids are being activated, thus making increased gluconeogenesis possible. It is also probable that somatotrophic hormones also participate, since its secretion has been found to be stimulated during adaptation to intermittent starvation (Fábry 1958). It has already been demonstrated that the rate of denervation muscle atrophy is slowed down after growth hormone administration, but only during realimentation following starvation. Insulin is another hormone which could participate in these adaptation reactions during starvation and realimentation. This hormone has been found to increase glycogen synthesis in denervated muscle, blood glucose levels being very low (Fig. 125). The amount of substrate, so important in autoregulatory mechanisms, seems to be of secondary importance in this case, and the rate of synthesis is apparantly modulated by hormonal mechanisms. Insulin increases the permeability of the muscle cell to glucose and this process may be accentuated in denervated muscle.

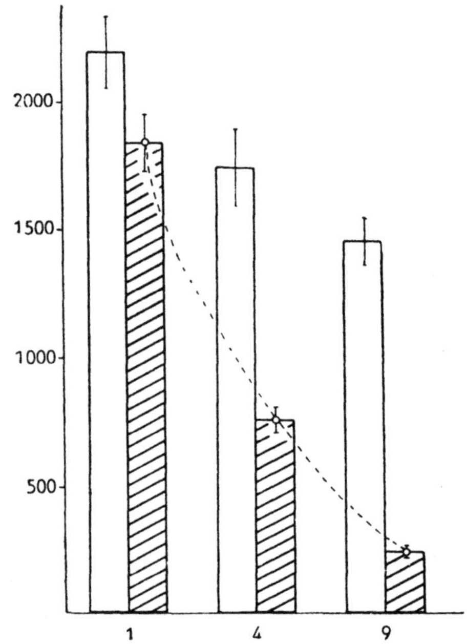

Fig. 133. Content of glycogen in mg % in the soleus of animals adapted to intermittent starvation for six weeks 24 hrs. after final realimentation. The sciatic nerve was sectioned 1, 4 or 9 days prior to final realimentation. Ordinate: glycogen content in normal (white columns) and denervated muscle (hatched column).

Adaptation processes can thus explain, to a certain extent, the role of hormonal mechanisms in anabolic processes taking place in denervated muscle, and may help to differentiate these mechanisms from autoregulatory ones.

After interruption of the adaptation cycle, denervated muscle also loses the ability to increase glycogen synthesis. However, increased synthesis of glycogen is lost gradually (Fig. 133). This has been shown in experiments in which the nerve was sectioned unilaterally towards the end of the adaptation cycle, one, four and nine days before final realimentation. Increased glycogen synthesis 24 hours after final realimentation lasts for a certain period of time, and is lost between four and nine days after nerve section. Persistence of the adaptation reaction cannot be caused by increased glucose supply, which is naturally the same in each realimentation phase. It is evidently caused by metabolic changes in peripheral tissues which are clearly associated with altered reactibility to humoral agents.

Hormonal factors thus participate in the process of persistence of the adaptation reaction in denervated muscles, and they may be able to produce this adaptation reaction in chronically denervated muscle. The higher levels of glycogen found in denervated muscles of adapted animals during final realimentation were not observed in control animals in which glycogen content was determined during realimentation following a single starvation period. However, this increase of glycogen levels need not be the result of the adaptation process. It was assumed to be due to a gradual replenishment of energy stores mobilized during repeated starvation, because this glycogen synthesis (which is absent in

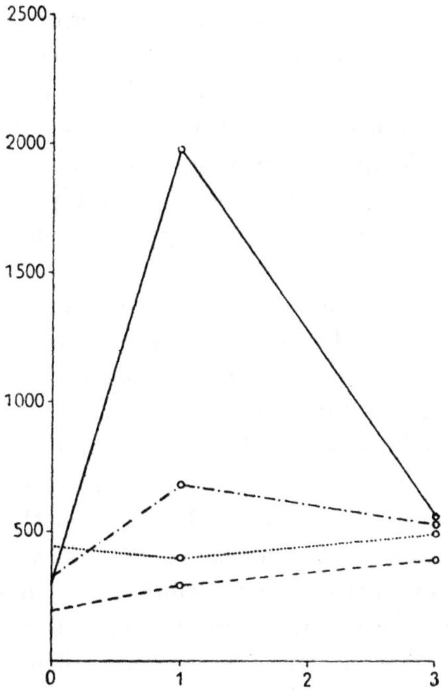

Fig. 134. Glycogen changes (in mg %) in the soleus of rats adapted to intermittent starvation for six weeks 24 and 72 hrs. after final realimentation in muscles of normal adapted animals (solid line), in normal muscles of control animals after a single 3-day's starvation period (dashed and dotted line), in denervated muscles of adapted animals (interrupted line) and in denervated muscles of control animals (dotted line) after a single starvation period of three days' duration. Ordinate: glycogen content in mg%. Abscissa: days after final realimentation.

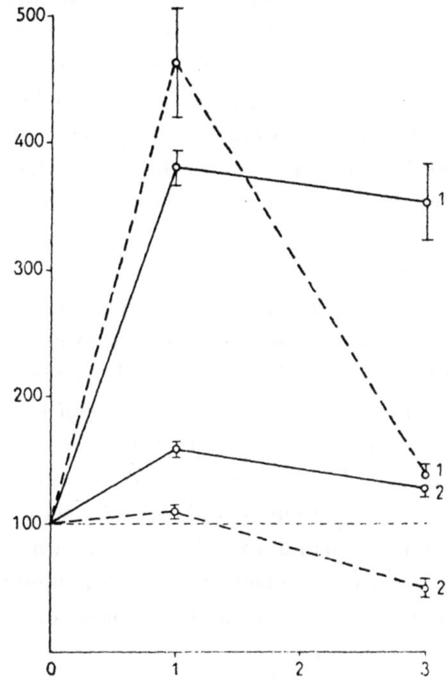

Fig. 135: Glycogen changes in the extensor digitorum longus (1) and changes in dry weight in the tibialis anterior (2) in animals adapted to intermittent starvation (interrupted line) and in animals after 10 days' starvation (solid line) and three days after final realimentation following starvation. The sciatic nerve was cut before realimentation. Ordinate: changes in glycogen (100%) and dry weight(10%) expressed as percentage of initial values obtained before realimentation. Abscissa: days after realimentation.

control denervated muscles) does not surpass initial values of control muscles. Figure 134 shows that the gradual increase of glycogen levels to initial values may be evoked by autoregulatory mechanisms, gradually replenishing energy reserves in denervated muscle (Gutmann and Vrbová 1958) to the initial value. At any rate, the experiments show that glycogen synthesis is not completely lost even in a muscle denervated for six weeks, if autoregulatory mechanisms are subjected to conditions of extreme metabolic stress, in our case to intermittent fasting.

A different situation was during acute denervation encountered in protein metabolism. The metabolic disturbance due to the exclusion of the trophic influence of the nervous system is much more severe and autoregulatory or hormonal mechanisms under conditions of starvation cannot sufficiently increase the synthetic abilities of the muscle cell during the period of acute denervation. Following denervation, a gradual decrease of muscle mass, wet and dry weight, and protein nitrogen take place. Percentage changes in dry weight and glycogen levels in the anterior tibial muscle of animals starved for ten days are given in Fig. 135. After this period the tibialis anterior and extensor digitorum longus were excised on one side, and the animals were realimented. Twenty-four and seventy-two hours respectively after denervation and the beginning of realimentation, the tibialis anterior and extensor digitorum longus of the other side were excised and their dry weight (tibialis ant.) and glycogen content (extensor dig. long.) determined. It can be seen that the realimentation after fasting even after adaptation cannot substantially affect the course of weight loss in denervated muscle. On the other hand, an exceptionally marked increase in glycogen levels occurs in denervated muscle, which are higher 24 hrs. after denervation than in the normal muscle, this again demonstrating the marked reactibility of denervated muscle to humoral agents. Differences in the reactibility of denervated muscle from the point of view of these autoregulatory or humoral mechanisms are also apparent when absolute amounts of protein nitrogen are determined (see Table 26) in an experiment in which one muscle was excised after nine days' starvation, and the other three days after realimentation, which was started immediately after denervation.

Protein synthesis was not observed in a chronically denervated muscle (six weeks) during realimentation after three days' starvation. It is only under

Table 26

Type of experiment	M. tib. ant. (protein N) in mg.		Percentage change in protein N
	1 initial	2 after 3 days' realimentation	
3 days' denervation	8·0	7·3	91.2
9 days' starvation (1) and 3 days' realimentation (2)	8·2	10·0	122
9 days' starvation (1), 3 days' denervation and realimentation (2)	8·0	6·8	85

exceptional conditions, i. e. during adaptation to intermittent starvation, that it is possible to induce protein synthesis even in such a chronically denervated muscle following starvation and realimentation. Considerable differences have

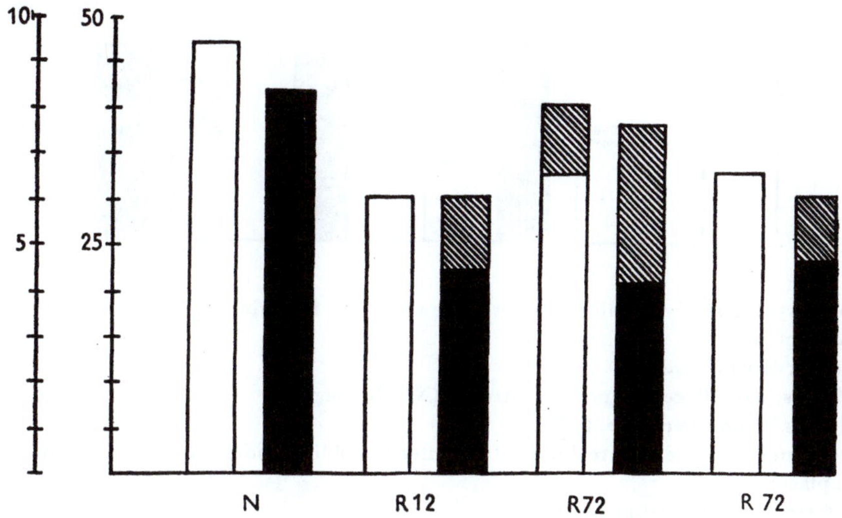

Fig. 136a. Resynthesis of proteins and RNA following realimentation after fasting.
White columns: non collagenous protein nitrogen (mg) N: normal muscle
Black columns: ribonucleic acids. (γP)
Hatched columns: increase after realimentation.
R 12: 10 days fasting, 12 hours realimentation.
R 72: 10 days fasting, 72 hours realimentation.
DR 72: 10 days fasting followed by denervation and 72 hours of realimentation.

been found, however, between changes in glycogen levels and total nitrogen in the course of six days of final realimentation in animals subjected to adaptation to intermittent starvation. The content of glycogen in normal muscle returns to normal values, but this process is slowed in denervated muscle. Dry weight and total nitrogen gradually rise in normal muscle during the six days of final realimentation, but in denervated muscle there is a decrease to values before realimentation after a transient increase in proteins during the first three days of realimentation (Gutmann 1959). It is assumed that energy reserves stored in glycogen, and rapidly renewed during the first realimentation phase in normal muscle, are made available for proteosynthesis, but that this process, which needs inclusion of muscle metabolism into reflex regulation of metabolism of the organism as a whole, is evidently impaired in denervated muscle. Further studies will have to be made of these relationships between energy and protein metabolism in normal and denervated muscle. The important conclusion is that adaptation is able to induce proteosynthesis in chronically denervated muscles and apparently to abolish the dissociation between synthesis of ribonucleic acids and proteins, which appears to be characteristic for denervation atrophy (see chapter IX). During realimentation following starvation resynthesis of RNA, preceeding proteosynthesis takes place. If the muscle is denervated before realimentation

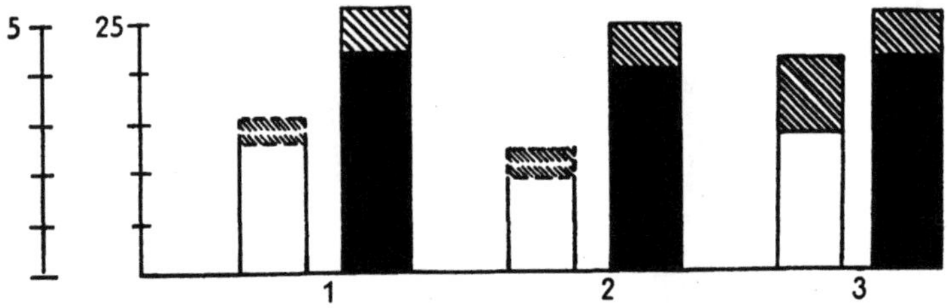

Fig. 136b. Resynthesis of noncollagenous proteins and RNA following realimentation in denervated muscle of rats.
Time od denervation: 6 weeks.
White columns: noncollagenous protein nitrogen (NCPN—mgN).
Black columns: ribonucleic acids. (γP)
Hatched columns: increase after realimentation (interrupted lines indicate that decrease of NCPN has taken place).
1: 3 days fasting, 3 days realimentation.
2: 8 days fasting, 3 days realimentation.
3: adaptation to intermittent starvation (3 days fasting followed by 1 day realimentation — repeated for a period of 6 weeks), 3 days fasting, 3 days realimentation.

resynthesis of RNA but not of proteins takes place (Žák and Gutmann 1960). However, following adaptation to intermittent starvation proteosynthesis is induced, resynthesis of RNA being followed by proteosynthesis in denervated muscle (Fig. 136 a and b) (Gutmann and Žák 1961).

Changes in proteosynthesis present in denervated muscle can thus be influenced by hormonal mechanisms only under special conditions. According to Hines (1950) nutritional factors do not play an important role in the course of muscle atrophy, and denervated muscles in starving animals lose weight more rapidly than in animals only starving or in animals with only denervated muscles (Hines and Knowlton 1935). Muscle atrophy can be more successfully influenced by functional activity, as has been demonstrated by favourable results with electrotherapy of denervated muscles (see Hines 1950).

It can thus be concluded that humoral mechanisms, which participate in adaptation processes, affect energy more than protein metabolism in denervated muscle. Hormones primarily activate or inhibit enzymes which are affected at a relatively late denervation stage. On the other hand, it seems that protein metabolism is more dependent upon the nervous system, changes in the metabolism of nucleic acids probably being of special significance in this connection (Žák and Gutmann 1960).

3. Nervous mechanisms

Autoregulatory mechanisms of energy metabolism are apparently modulated in the first place by hormonal agents. The main aim of denervation experiments, however, is to solve the question of the nervous regulation of metabolism. The role of nervous regulation will be further documented by an example concerning glycogen metabolism. It is, of course, well-known that catabolic processes are reflexely regulated by the nervous system, i. e. by the number and frequency of nerve impulses, as has been demonstrated by Yampolskaya and Yakovlev (1951) for muscle glycogen levels. What is the role of the nervous system in the regulation of postfunctional resynthesis? If it was only the autoregulatory mechanisms, then the extent of glycogen breakdown would directly determine the degree of resynthesis, as is the case *in vitro* (Hansen et al. 1951). However, the rate of resynthesis is also affected by nervous mechanisms, and these nervous regulations can best be investigated by determining the time gradient of glycogen resynthesis following stimulation of the normal and denervated muscle.

A higher frequency of stimulation (120 imp./min. as compared with 30 imp./min.), of course, results in a higher glycogen breakdown after stimulation with the same number of impulses, but this relationship is the same in both normal and denervated muscles (Bass et al. 1955). Resynthesis is accelerated sixty minutes after stimulation of normal muscle, so that there is a temporary increase in the level of glycogen above initial values 60—120 min. after muscle stimulation. Following stimulation with the lower frequency (30/min.), glycogen levels return to normal within 240 min. after stimulation, while the over-shoot reaction due to stimulation at a higher frequency (120/min.) is still present 360 min. after stimulation. This over-shoot reaction in glycogen levels was also described by Yampolskaya (1950). Accentuation of resynthesis is not present in denervated muscle; the over-shoot reaction under the same experimental conditions is absent after low and high frequency of stimulation, and curves of glycogen resynthesis are practically the same following stimulation by either frequency (Bass et al. 1955). The threshold frequency for eliciting this accentuation of resynthesis in normal innervated muscle is different for different muscles, and is evidently given by various functional types of muscle. A higher threshold frequency has, for example, been found in tonic muscles. Postfunctional resynthesis associated with an over-shoot reaction was present in the diaphragm after tetanic stimulation only. However, these functional relations, which are apparently the product of the adaptation function of the nervous system, are lost in denervated muscle. Similarly, differences in the incorporation of Na^{24} and P^{32}, found under normal conditions, equalize out in denervated muscle (Hudlická 1955). It is probable that this equalizing of differences in muscle reactibility, or the loss of specific reactibility after denervation, can be correlated with the striking change in the time constant of accomodation, as measured with progressive currents of various slopes (see Desmedt 1959). Denervated muscle accomodates less than the normal muscle, i. e. the stimulating current can be established more and more slowly without losing its stimulating efficiency (Desmedt 1950).

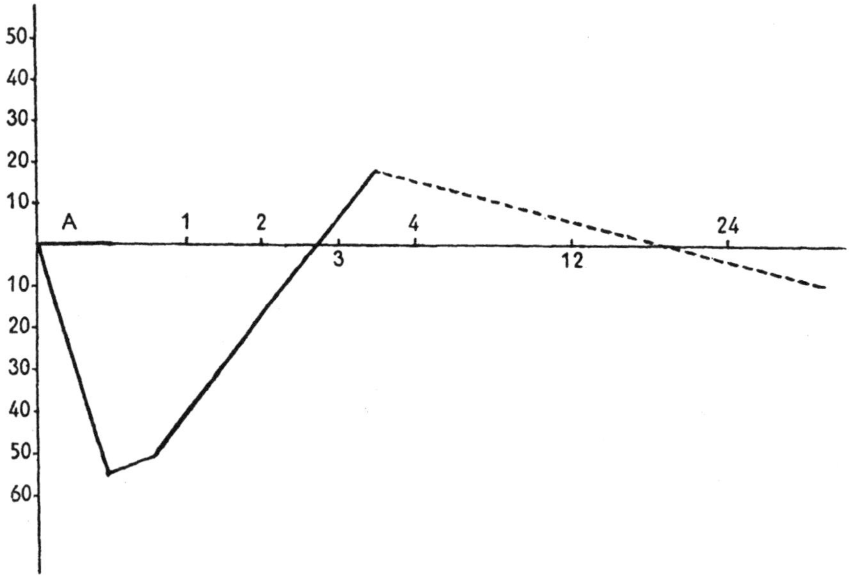

Fig. 137. Changes in glycogen content in femoral muscles of Periplaneta americana during anoxia lasting for one hour (A) and during the restitution period. Abscissa: time after exposure to nitrogen atmosphere in hours. Ordinate: glycogen changes in percentage of control side.

The overshoot type of reaction following muscle stimulation is not lost immediately after denervation (Gutmann et al. 1954). The gradual loss of accentuation of metabolic recovery processes indicates that this phenomenon may be due to substances or axonal transport mechanisms, which are dependent upon an intact neuromusclular junction.

Overshoot reactions are evidently of general biological significance (Gutmann 1955). A transient increase of dry weight, total and protein nitrogen, by 4%, 4% and 5·2% of the initial values respectively, was observed following muscle stimulation (Žák et al. 1957). Similar changes were, furthermore, found in increased glycogen levels during reinnervation (Beránek et al. 1954) and in insect muscles following anoxia (Fig. 137) (Kubišta and Gutmann 1957) and also in dry muscle weight following temporary hydration of the muscle (Fig. 138) (Gutmann and Vrbová 1952). They are evidently the result of nervous regulation, and are absent in denervated muscle under identical experimental conditions. There is also a transient increase in the content of ribonucleic acids (RNA) which is related to a decrease in level immediately after muscle stimulation (Žák and Gutmann 1960), this increase in RNA resynthesis in the denervated muscle not being followed by increased proteosynthesis, as in the case in normal muscle. Similarly RNA synthesis is increased during realimentation following prolonged starvation, and this is followed by higher proteosynthesis in the normal muscle.

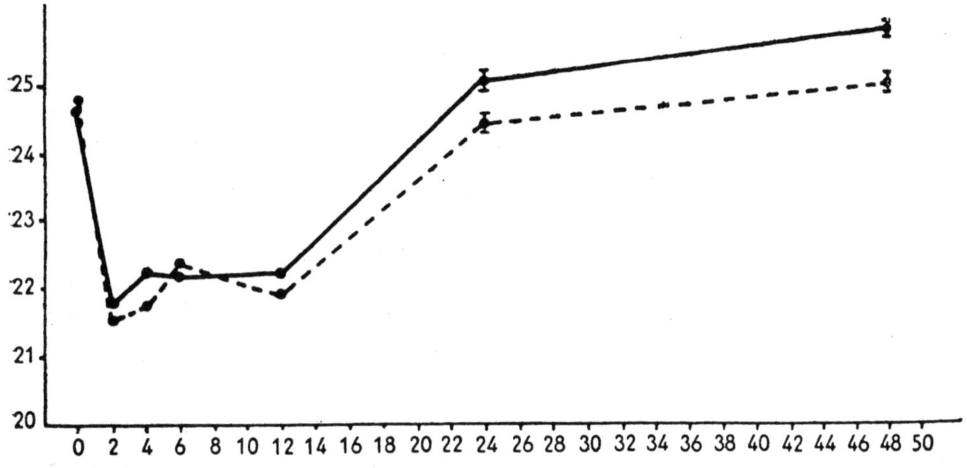

Fig. 138. Changes in dry weight of normal (solid line) and denervated muscle (interrupted line) after intraperitoneal injection of distilled water in amounts corresponding to 15% body weight. Abscissa: hours after injection of distilled water. Ordinate: muscle dry weight as percentage of muscle weight.

In denervated muscle, however, although RNA synthesis is increased after three days' realimentation and denervation, proteosynthesis is not accentuated (Žák and Gutmann 1959).

A disturbance of proteosynthesis already manifests itself in the course of the first few hours after denervation, i. e. long before degeneration of motor endplates takes place. On the other hand, glycogen synthesis is impaired (in the sense that accentuation of metabolic recovery processes does not take place and the over-shoot reaction is lost) relatively much later. The disturbance in protein metabolism during postfunctional resynthesis and during realimentation following starvation appears sooner than in energy metabolism.

A further example of the disturbance of synthetic processes following functional activity can be demonstrated by the fact that utilization of non-protein nitrogen after stimulation of denervated muscle is not increased (Bass et al. 1955), as is the case in normal muscle (Popova 1951). This utilization of nonprotein nitrogen was studied by determining arteriovenous differences following muscle stimulation. Positive A—V differences were obtained after stimulating normal muscle, however, these values were negative in denervated muscle, evidently because the latter is unable to utilize non-protein nitrogen accumulating in the muscle after stimulation. The disturbance of proteosynthesis can thus be documented by the different course of changes in arteriovenous difference of non-protein nitrogen in the normal and denervated extremity (Fig. 139).

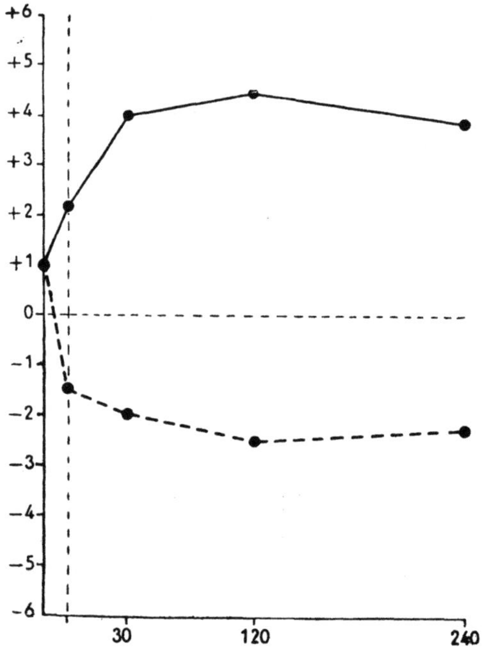

Fig. 139. Changes in arterio-venous differences of non-protein nitrogen following direct muscle stimulation (tibialis ant. stimulated for 15 min. at frequency of 60 imp./min.). Ordinate: a—v differences of non-protein nitrogen in %mg before, immediately after (vertical dotted line) 120 min. and 240 min. after muscle stimulation.

Denervated muscle loses more potassium after dialysis by intraperitoneal injection of potassium-free Tyrode solution than normal muscle, and recovery of normal potassium values was found to be impaired after subsequent injection of potassium chloride (Drahota et al. 1957). Metabolic recovery processes after functional activity are also impaired as far as oxygen consumption is concerned. The consumption of oxygen increases five times after stimulation of normal muscle, this rise being very small in the denervated extremity (Hudlická and Vrbová 1958), although resting oxygen consumption in the denervated muscle is higher.

Thus it seems that the nervous system influences muscle metabolism primarily by activating processes of resynthesis following functional activity. When this "regulation" of trophic processes is missing, trophic disturbances occur so that it can be concluded that the metabolic disturbance in denervated muscle is characterized especially by a disturbance of metabolic recovery processes. Of course, this general conclusion will have to be further documented.

C. Loss of the trophic influence of the nervous system and its significance for development of metabolic disturbances in denervated muscle

Studies concerning metabolic reactibility of denervated muscle have had to cope repeatedly with the problem, whether the metabolic disturbance is due to a lack of specific muscle function, or to loss of the trophic function of the nervous system. There can be no doubt that prolonged inactivity of an organ results in certain metabolic disturbances which are also evident in the denervated muscle. It should be possible to define the "trophic influence of the nervous system", which is eliminated in denervated muscle, when effects of inactivity can be differentiated from the loss of the "trophic effect" of the nervous system. How can this differentiation be made? It has already been pointed out that various mechanisms come into play in the course of denervation, the most important from the point of view of trophic processes being those which can be correlated with degeneration of the nerve. For this reason the early denervation period has been stressed, and that is why metabolic disturbances in the denervated muscle are studied from the aspect of their early development.

1. Stages of development

The disturbance of metabolism in denervated muscle develops in a certain sequence, this being due to the fact that some changes are directly associated with degeneration of terminal nerve endings and the loss of neuromuscular transmission. This, seems to apply for example, to the development of fibrillation activity. However, other metabolic changes have been described in denervated muscle which appear even earlier than morphological changes in axons, and which could possibly be ascribed to acute and complete immobilization of the muscle. Finally, some of the metabolic changes, especially those associated with changes in the content and activity of enzymes, gradually appear at later stages of denervation atrophy which could be explained by adaptation of muscle metabolism to inactivity. Thus, metabolic disturbances in denervated muscle undergo various stages of development which may be due to different mechanisms. It is, of course extremely difficult to delimit exactly the individual stages of denervation metabolic disturbances, since there must be considerable overlapping of different mechanisms, but an attempt to differentiate these should aid in defining the "trophic influence of the nervous system". It is possible to divide the denervation changes in muscle metabolism into the following three stages:

a) Acute muscle immobilization,
b) Loss of the trophic influence of the nervous system,
c) Metabolic adaptation to muscle inactivity.

a) Acute muscle immobilization

Following section of the nerve, increased glycogen levels have been found in denervated muscles (Vodička et al. 1957) and these return to initial values in 4—24 hrs. after denervation. A similar transient increase has been reported in the content of potassium (Drahota 1960). Four hours after denervation, the wet and dry weight and protein nitrogen in the denervated muscle were also shown to be increased. Protein nitrogen was still increased 24 hrs. after denervation when muscle atrophy is studied by the method of "successive" examination of muscles, i. e. the muscle is excised on one side and at the same time the contralateral nerve

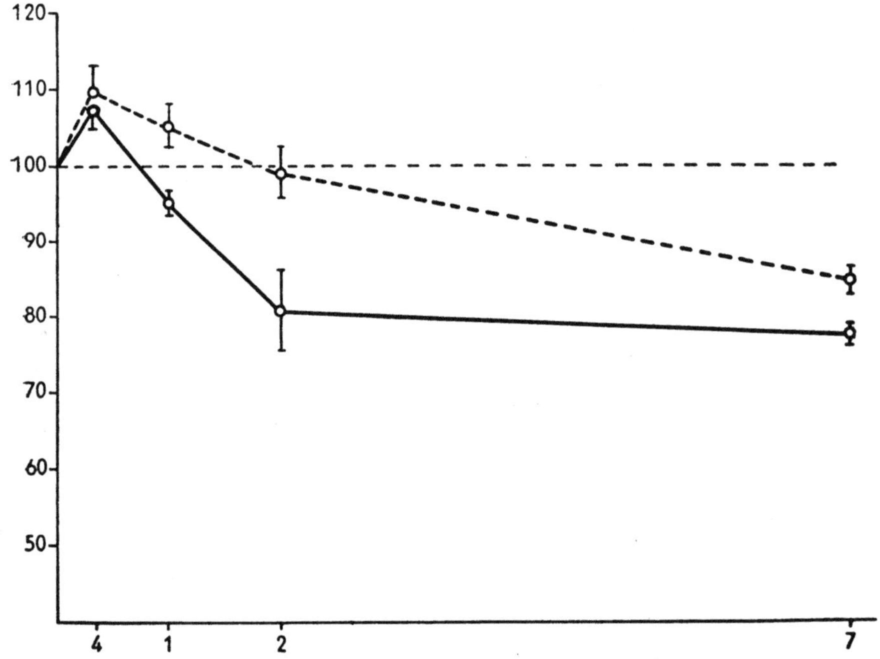

Fig. 140. Changes in non-collagenous protein nitrogen in the tibialis ant. 4 hrs. and 1, 2, and 7 days after denervation. The sciatic nerve was either sectioned on one side and muscles of both the denervated and control side were excised on the same days (solid line) or the nerve was sectioned on one side after excision of the contralateral "control" muscle and the denervated muscle was excised at various intervals after nerve section (int. line). Ordinate: changes in non-collagenous protein nitrogen expressed as percentage of control muscle. Abscissa: time after nerve section.

is sectioned. The contralateral muscle is then taken for estimation at various time intervals after denervation. When using this method of successive examination, the loss of protein nitrogen in denervated muscle is slower than after excision of both muscles (denervated and control) "simultaneously" (see Fig. 140). This slowing down of muscle atrophy is evidently evoked by the absence of compensatory muscle hypertrophy during "successive" excision of the muscle, which occurs after "simultaneous" excision of the normal and denervated muscles. This conclusion has been confirmed in control experiments in which excision of the muscle on one side was combined with section of the ipsilateral nerve. Compensatory hypertrophy of the contralateral muscle was found to occur by the seventh day, hypertrophy being evidently evoked by increased activity of the muscles on this side (Gutmann 1960).

There is thus a transient increase in protein nitrogen during the first hours following nerve section, which is subsequently followed by a gradual decrease in protein mass. "Denervation hypertrophy" described in the diaphragm after section of the phrenic nerve (Stewart 1955), often considered to be the result of temporary edema, also takes place in the skeletal muscle. It is possible that this temporary increase in proteins is caused by a decrease of breakdown processes in denervated muscle and that it is a consequence of acute muscle immobilization after nerve section prior to the onset of fibrillation activity, which appears after indirect muscle excitability has been lost.

The transient increase in protein nitrogen and glycogen levels four hours after denervation was found in both the rabbit and the rat (Table 27). It appears that denervation hypertrophy and acute immobilization can be attained only when all propagated nerve impulses are eliminated. It has been found after novocaine block of the nerve, but not in immobilization in which the continuity between nerve and muscle cells is preserved, as for example after tenotomy (see Table 27).

Table 27

Percentage changes in the absolute content of protein nitrogen and glycogen in the extensor digitorum longus 4 hrs. after denervation, novocaine block and tenotomy respectively in rabbits and rats.

Type of experiment	Percentage change in protein nitrogen	Percentage change in glycogen levels
Denervation (rabbit)	105 ± 1·2	151 ± 15·8
Novocaine block (rabbit)	108 ± 10	180 ± 24·7
Tenotomy (rabbit)	99 ± 3·7	112 ± 5·7
Denervation (rat)	110 ± 3·0	111 ± 4·3
Tenotomy (rat)	96 ± 4·7	83 ± 4·9

It thus seems that denervation hypertrophy is a characteristic phenomenon of the first stage of denervation atrophy and can be interpreted as the result of acute complete immobilization of the muscle occurring before the start of fibrillation activity (Gutmann 1960).

b) Loss of the trophic influence of the nervous system

Following disintegration of terminal endings in the motor end-plates, and the simultaneous loss of indirect muscle excitability, metabolic changes occur in denervated muscle which are to be found exclusively after nerve section, and which are evidently due to the loss of the trophic influence of the nervous system.

This includes the occurrence of fibrillation activity and the rise in nucleic acids which are a very important part of the metabolic disturbance taking place in denervated muscle.

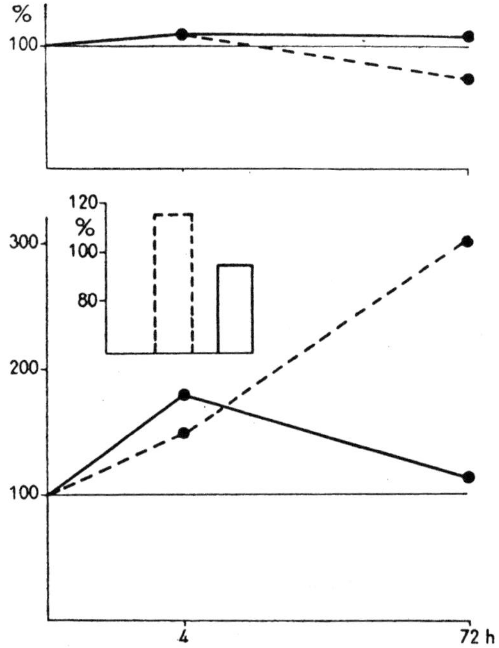

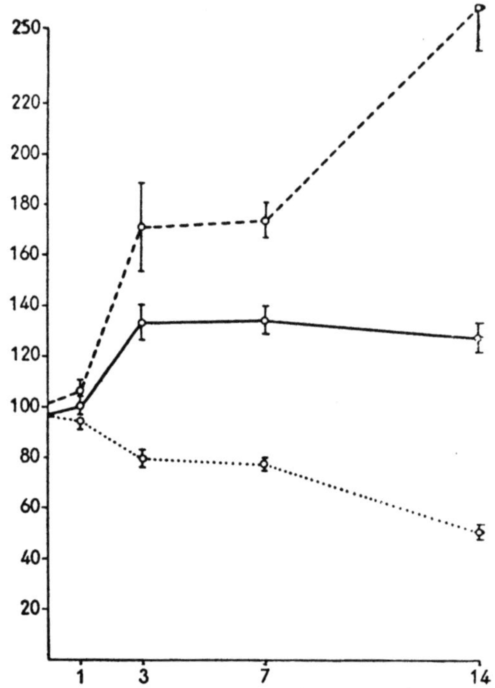

Fig. 141. % Changes in non-collagenous protein nitrogen (upper diagram), amount of desoxyribonucleic acids (middle diagram) and amount of glycogen (lower diagram) in the M. tibialis ant. of the rabbit 3 days after section of the nerve (interrupted line) and after blocking the nerve by novocaine for 3 days (full line). Abscissa: Hours after denervation and beginning of the nerve block. Ordinate: Changes in non-collagenous protein nitrogen, DNA and glycogen, expressed in percentage of the control side.

Fig. 142. Changes in absolute (solid line) and relative amounts of desoxyribonucleic acids (γP-DNA/mg. N) (interrupted line) and in the content of non-collagenous protein nitrogen (dotted line) after denervation. Abscissa: days after section of the sciatic nerve. Ordinate: percentage changes in absolute and relative amounts of DNA and non-collagenous nitrogen as compared with the contralateral control muscle.

Experiments in which the propagation of nerve impulses is blocked without interrupting the continuity of the nerve trunk, provide a good opportunity to differentiate changes due to the loss of the trophic influence of the nerve fibres from those due to inactivity only. A reversible block of transmission of nerve impulses lasting 3 days and leading to a temporary paresis of the muscle can be induced by a continuous novocaine application using a drip infusion (Gutmann and Žák 1961). Fig. 141 shows that 3 days after denervation there is a considerable loss of the proteins, increase of the amount of desoxyribonucleic acids and (in the rabbit) a considerable increase of the amount of glycogen in the muscle. No such changes and no degeneration can be found in the muscle with complete blokade of transmission of nerve impulses (see Plate III).

Histological studies, qualitatively evaluated, indicated that there are pathological nuclear reactions present, and have tentatively set forth the possibility that the number of nuclei in the denervated muscle is increased. However, quantitative histological studies have partly disproved (Sunderland and Ray 1950, Willard and Graw 1924) and partly confirmed this (Altschul 1947). Similar controversies have been reported concerning the content of desoxyribonucleic acids (DNA) determined chemically; when calculated in mg% in relation to dry weight their content was found to be increased (Schmidt and Schlieff 1956), while Mandel et al. (1949) found no changes in the DNA content in the muscle. In our experiments (Gutmann 1960, Gutmann and Žák 1961) changes in DNA were expressed in both absolute units and in relation to protein nitrogen (γP/mg N). One to three days after denervation there is a gradual increase in the desoxyribonucleic acids (DNA) in absolute units and as relative values (Fig. 142). With progressing atrophy, the relative increase of DNA content, progressively rises. Absolute amounts of DNA decrease hand in hand with a loss of proteins in denervated muscle, while the relative content of DNA, i. e. percentage composition of DNA in muscle fibres, progressively increases. With progressive loss of proteins, there is thus a considerable increase in the number of nuclei in denervated muscle. It is however to decide whether the changes are due only to increase of number or also to volume of the nuclei (see chapter II.). It is important to note that changes in the number of nuclei appear approximately at the time when motor endplates degenerate, i. e. when indirect muscle excitability is lost. This already indicates that these changes in the metabolism of nucleic acids are the result of a disturbance of neurotrophic relations between the nerve and muscle cell, caused by denervation.

c) Metabolic adaptation to muscle inactivity

It has already been pointed out that there are probably two types of adaptation mechanisms at play in denervated muscle. On the one hand, adaptation mechanisms to inactivity which are associated with, for example, a decrease of the cytochrome content which has also been reported in muscle after tenotomy (Humoller et al. 1952), or by reduction in glycolytic enzymes, also reported in both denervated muscle and muscle after tenotomy (Mirski and Wertheimer 1942, Fischer 1955). On the other hand, it is assumed that the denervated muscle adapts itself by economizing in oxygen consumption, the increase of which is probably due to the energy requirements of fibrillation activity, and by a rise in the content of DNA in the denervated muscle. The utilization of blood glucose for lipid synthesis can serve as an example of this. Of course this is an exceptional situation caused by the loss of function. For example, increased uptake of P^{32} into phospholipids was found in denervated muscle, but a still greater increase was shown in muscle after tenotomy (Samuels 1957). It would be very interesting to compare lipid and oxygen metabolism in denervated and immobilized muscle. So far it is not possible to reconstruct an accurate picture about the onset and the time course of adaptation processes taking place in denervated muscle, but it is quite clear, of course, that an intermingling of various processes must be assumed at different stages of denervation atrophy. A precise definition of the adaptation stage of muscle metabolism will only be possible when mechanisms of these adaptation processes will be investigated in more detail.

2. Specific characteristics of denervated muscle metabolism

It can be assumed that part of the metabolic disturbances in denervated muscle is due to inactivity, or to loss of impulse activity of the motoneuron. However, part of these disturbances cannot be explained by the absence of nerve impulses causing contraction of muscle fibres, but are probably due to the loss of the "trophic influence of the nervous system". Analysis of these disturbances should demonstrate specific characteristics of denervated muscle metabolism, which cannot be found in immobilized muscle. Comparison of denervation and immobilization atrophy is thus very important. However, it is first necessary to choose the correct model for immobilization. From the experimental aspect, the most suitable would be a muscle with intact motor innervation, completely excluded from reflex influences which could lead to muscle contraction. It would seem that an ideal model would be the isolated spinal cord with bilateral section of the dorsal roots (Tower 1937, 1939).

In this method the distal or lumbar stump of the cord is isolated from all afferent and all descending nerve impulses. The muscles supplied by those isolated cord segments are paralysed and completely flaccid. Fibrillation activity characteristic for denervated muscle was not found under these conditions.

Loss of muscle weight is considerable (40% in 3 weeks) after inactivation of the muscles by this method (Eccles 1941) and could be prevented for the most part by electrical stimulation of the nerve supplying the muscles. It would be interesting to know more about metabolic reactions of muscles thus immobilized.

Inactivation has further been obtained by skeletal fixation, or by application of a cast to restrict activity. All these methods represent imperfect models, since the degree of inactivity cannot be controlled in animals, and expecially in rats reflex defensive movements of the "immobilized" extremity may lead to strong muscle contractions. Comparing the changes in muscles immobilized by casts with changes in denervated muscle, Chor and Dolcart (1936) found no differences iu relative amounts of water, nitrogen or protein. Hoagland (1946) notes that lactic acid, glutathione content and collagen are increased both in denervated and immobilized muscle. Also protein, creatine, glycogen, and phospholipid content and succinic dehydrogenase activity, are decreased in both groups. We may assume, that these changes are due in both cases to immobilization, i. e. to adaptive metabolic reactions to inactivity.

It seems that it is important to study the changes taking place during the first days which are related to loss of neuromuscular connections, the later changes being complicated by secondary processes. Few studies of this kind are available. No fibrillation and no increase in DNA content of muscle can be found in "immobilized" muscles during this time.

A third method, often used to produce immobilization, has been tenotomy of muscles. But even here, some scepticism is necessary, especially as increased reflex activity has been found in spinal segments innervating muscles tenotomized for a long time (Beránek and Hník 1959).

Decrease in muscle proteins as well as adenosine triphosphate activity corresponded to denervation atrophy (Fischer and Ramsey 1946, Fischer 1948). Fischer (1950) points out that relatively quick deterioration of myosin and myogen can be detected in tenotomized muscles, and assumes that contracture of tenotomized muscle is responsible.

We may conclude that many changes taking place in denervated muscle are of the same kind as in immobilized muscle. Some changes such as deterioration of proteins (Fischer 1950) or "acute vacuolar and granular degenerative changes" (Tower 1939) are thought to be typical for denervation atrophy only, but more data should be available before ascribing significance to these findings.

However, there are fundamental differences between denervated muscle metabolism and that of immobilized muscle and it is thus necessary to seek further mechanisms which may explain the metabolic disturbance of denervated muscle.

Comparative experimental studies have not so far given satisfactory results. However, absence of the "trophic influence" of the nervous system leads to several characteristic disturbances in the metabolism of the denervated muscle:

a) Increased sensitivity of denervated muscle to mediators, especially to acetylcholine, which is not to be found in other metabolic disturbances of the muscle; by Cannon and Rosenblueth (1949) defined as "Law of denervation".

b) Fibrillation activity of denervated muscle, which is not present in immobilized muscle, nor in muscles innervated by isolated spinal segments (Tower 39).

Increase of the acetylcholin-sensitive region in size (Axelsson and Thesleff 1959) and fibrillation activity in the denervated muscle are apparently due to lack of subthreshold release of acetylcholine and cannot be reproduced by lack of transmission of nerve impulses causing contraction of muscle fibres. The nervous control of chemical sensitivity of the muscle related especially to subthreshold release of acetylcholin is apparantly a very important "trophic" function of the nervous system. Study of this important function of the motor nerve (see Thesleff 1960) will have considerable consequences for the whole concept of the trophic function of the nervous system. However, control of the size of the chemosensitive area can only be one of the trophic functions, not connected with impulse transmission, e. g. lack of transmitter release and not nerve degeneration is responsible for the increase in size of the chemo-sensitive area of the muscle membrane. However, nerve degeneration is related to different metabolic changes in the muscle fibre, whose occurrence cannot be connected with lack of transmitter release.

c) Increased number of nuclei in denervated muscle.

Even though a moderate increase in the number of nuclei has been reported after tenotomy, the changes are not as extensive as in denervated muscle, and it is important to note that these changes can be correlated with the loss of neuromuscular transmission. It is characteristic that the increase in the number of nuclei progresses further even in the late stages of atrophy. The content of DNA increases in a muscle nine months after denervation by 400%, while there is an 80% loss of protein nitrogen at this time. It can be assumed that this change, similarly as fibrillation activity, must lead to serious consequences in denervated muscle metabolism. Further studies will be necessary to define more exactly the changes characteristic for denervation atrophy alone.

d) Disturbance of metabolic recovery processes

Soon after section of the nerve, anabolic processes in the denervated muscle are impaired, e. g. glycogen synthesis after glucose administration. This disturbance, however, was also found in immobilized muscle (Gutmann and Jakoubek 1960), so that it cannot be considered to be specific for the loss of the trophic influence of the nervous system. Postfunctional resynthesis, however, is very markedly decreased in denervated muscle, It is possible to attain accentuation of metabolism, characterized by an over-shoot reaction, under exceptional conditions only. The disturbance in "metabolic recovery processes" in protein metabolism is especially marked, together with characteristic "dissociation" between metabolism of nucleic acids and proteins, which can be observed after stimulation of denervated muscle as well as during realimentation following starvation. Increased RNA content in denervated muscle with a concomittant gradual decrease in protein nitrogen and increased RNA in the course of resynthesis following muscle stimulation or starvation, which is not associated with increased proteosynthesis, indicates that the inability to utilize RNA for proteosynthesis in denervated muscle represents an important role in the disturbance of "metabolic recovery processes". Similarly the utilization of non-protein nitrogen following stimulation is impaired in denervated muscle. What then is the fundamental mechanism responsible for the loss of "metabolic activation" during postfunctional resynthesis? Changes in muscle enzyme activity do not seem to be of decisive importance, since disturbances in the recovery processes of protein metabolism appear earlier than changes in enzymatic activity in muscle.

Let us return again to the experiments indicating that the proximo-distal transport of axoplasmatic substances may represent the basic mechanism of the "trophic" influence of the nervous system. Lack of these substances in denervated muscle could partly explain specific metabolic disturbances in denervated muscle. That such transport exists has been demonstrated by a number of authors (Young 1942, Weiss and Hiscoe 1948, Samuels et al. 1951, Lubinska 1952, Ochs and Burger 1958). These substances could induce metabolic changes in the muscle, and experiments in which it was shown that the rate at which glycogen synthesis is decreased after denervation is dependent upon the length of the sectioned peripheral nerve stump (Fig. 4) indicate that substances transported along the axon are exhausted sooner when the peripheral nerve stump is short (Gutmann et al. 1955). It is also interesting to note in this connection that the accentuation of proteosynthesis, which is normally present after muscle stimulation, is considerably smaller when the nerve is sectioned even immediately after muscle stimulation.

The metabolic connection between the nerve and the muscle is lost gradually, and substances in the peripheral nerve stump may prolong for a certain time the existence of this connection. However, it appears that this metabolic connection between the nerve and muscle cell can be lost very quickly by a reflex route. For example, glycogen synthesis and total nitrogen are decreased immediately after nociceptive stimulation, and these changes correspond to the metabolic disturbance found after three days' denervation (Fig. 143) (Gutmann and Vodička 1953, Vodička 1957), and consequently a direct "inhibition of the metabolic connection" was assumed (Vodička 1957). Of course, a more concretely defined mechanism of this "inhibition of metabolic connections" is, as yet, lacking. Glycogen synthesis in muscle was also found to be reduced rapidly following administration of tubocurarine.

Two hours after glucose administration of a 50% solution in amounts corresponding to 1·5% body weight by stomach tube glycogen levels increase to 150 $\pm$ 20% as compared with the control muscle. If tubocurarine is injected prior to glucose administration, glycogen synthesis is seriously affected (the increase being 105 $\pm$ 4·6%). These experiments indicate that the rapid drop in glycogen synthesis following nociceptive stimulation can be caused by a block of neuromuscular transmission, but these experiments do not allow to differentiate clearly the trophic influence of the nervous system.

From this aspect experiments investigating the nature of these substances transported intracellularly, as well as the disturbances in the utilization of substrates (RNA in the first place) for muscle proteosynthesis, are important. According to Vodička (1956, 1959), the content of total nitrogen calculated per unit dry weight increases in the distal portion of the nerve following stimulation

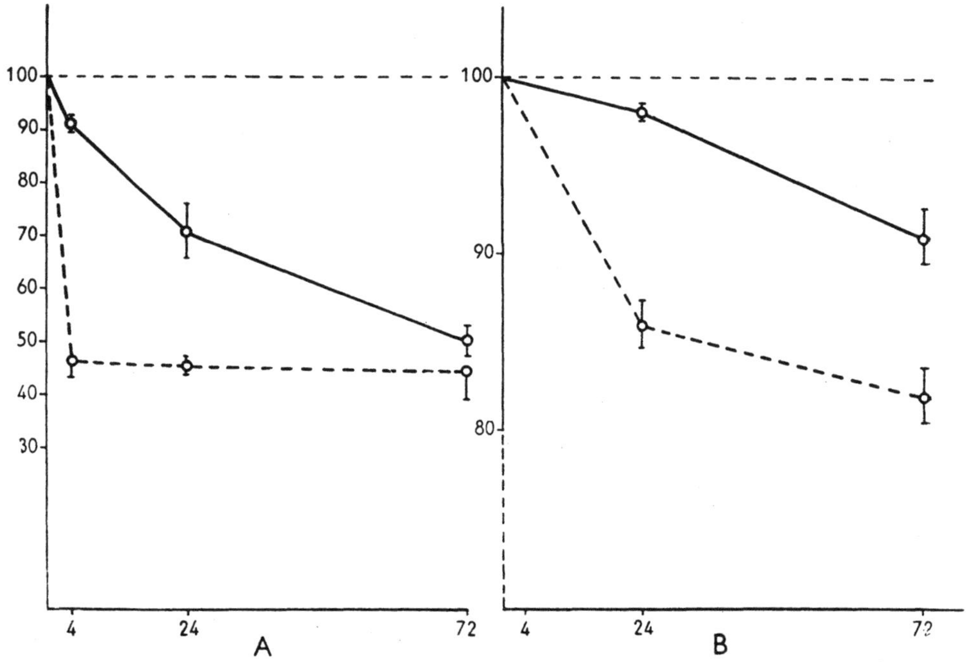

Fig. 143. Changes in glycogen synthesis in the soleus muscle following denervation (solid line) and nociceptive stimulation evoked by injection of turpentine oil into planta pedis (interrupted line) four hours after administration of 50 % solution of glucose in amounts corresponding to 1·5 % of body weight. (A) Changes in total nitrogen in the denervated soleus (solid line) and after nociceptive stimulation (interrupted line) B. Individual values are expressed in percentage of contralateral control muscles.

or functional activity. It seems that the proximo-distal transport of substances is accelerated during activity of the nerve, and it is probable that these substances activate muscle metabolism and that their absence could be responsible for disturbances in postfunctional resynthesis in denervated muscle. Substances participating in the activation of metabolic recovery processes, and in maintaining the normal trophic state of the muscle, need not necessarily be restricted to the group of mediators, even though it is possible that continuous subtreshold liberation of quantal amounts of acetylcholine may play an important role (Fatt and Katz 1952). The transport of subcellular particles, evidently associated with metabolism of nucleic acids (Young 1956), indicates that metabolic products of the nerve cell nucleus can play an important part in the "mediation of activation" of metabolic recovery processes.

In this connection it is necessary to return again to problems of neurosecretory activity of neurons. The characteristic properties of certain nerve cells which have been found to posses some structural properties of gland cells and which produce and secrete hormones, has led to the concept that these cells unite two main integrating systems — nervous and endocrine (see Scharrer and Scharrer 1954). According to this assumption, the nervous system can exert control over the endocrine system by a special type of cell which has the characteristic properties of nerve and gland cells, and which is therefore able to transform nerve impulses into hormonal activity. The activity of the supraoptic nuclei, associated with the production of antidiuretic hormone, is a typical example of such neurosecretory activity. But the finding of neurosecretory cells in invertebrates and vertebrates already indicates that this is an important regulatory function which does not only apply to mediation between the nervous and endocrine system only.

Especially interesting from this point of view is the growth and moulting hormone in insects, which are activated by neurosecretory cells of the brain and which provoke in the course of a few hours rapid synthesis of nucleic acids and proteins. Injection of the moulting hormone results in the development of cross-striated muscle fibres in the ventral abdominal muscle (Wiggelsworth 1957). It is interesting to recall in this connection that "gland like" cells were first observed in spinal cord nerve cells of the shark (Dahlgren 1914). In regeneration processes, proximo-distal flow of substances is also evidently the leading mechanism by which the "trophic state" of muscle fibres is renewed. When these facts are taken into account, it is then not unreasonable to presume that neurosecretory activity of nerve cells is, perhaps, a universal nerve mechanism which mediates the "trophic activity" of the neuron in the sense of continuous "growth" and renewal of cellular components, especially of proteins, after every functional activity.

When investigating the role of the growth and moulting hormone in insects for proteosynthesis, it is controversial whether it is a substance essential for the activity of enzymes associated with synthesis of proteins, or whether it affects cellular permeability so that enzymes obtain an easier approach to their respective substrates (Wiggelsworth 1957). It is again the question of the primary link which mediates the trophic influence of the nervous system, the elimination of which causes the loss of metabolic connection between the nerve and the terminal organs and leads to grave metabolic disturbances in the tissue. An analogous problem is encountered when attempts are made to determine the primary link in disturbances of metabolism in denervated muscle.

It is not yet possible to point out the primary metabolic disturbance in denervated muscle even though the utilization of nucleic acids seems to be the important link in protein metabolism. It is possible that the lack of a product of the metabolism of nucleic acids in the nerve cells, which is normally transported towards the muscle and which activates its proteosynthesis, is of fundamental importance. A number of further metabolic disturbances have been discovered such as increased oxygen and glucose consumption, increased synthesis of lipids from glucose etc., but even though they are characteristic for the metabolism of denervated muscle, they cannot be considered as specific. It is natural that a study of the metabolic reactibility of denervated muscle cannot indicate precisely the primary link of the disturbance in denervated muscle metabolism. It can, however, give data which point out new problems for biochemical research. The fact that specific metabolic disturbances appear at the period when neuromuscular transmission is lost represents one of these clues. This period will thus have to be investigated in greater detail.

3. The reversibility of changes in the metabolism of denervated muscle

Complete restitution of metabolism in denervated muscle to normal is only possible when reinnervation takes place, i. e. at a time when the metabolic connection between the muscle and the nerve is renewed. Fibrillation activity disappears, hypersensitivity to mediators is lost, and the ratio of nucleic acids returns to normal (Drahota and Gutmann 1961). All this indicates that neurotrophic relations are reestablished. Some qualitative changes probably also occur in muscle proteins (Yudayev et al. 1953) and the structure of glycogen (Koshtoyants and Janson 1950). In the course of reinnervation, glycogen reserves are replenished first, and this occurs at a period when metabolism of water in the muscle is still considerably impaired (Drahota 1959) and when muscle atrophy is still considerable. It is not until considerably later that the protein content and electrolyte composition return to normal, but so far data are missing concerning complete restitution of these processes to normal.

Electrotherapy, which at least partly substitutes for the specific function of the nervous system, evokes contraction of muscle fibres and should prevent at least some of the metabolic disturbances due to inactivity, but it should remain without effect on specific disturbances which are due, in our opinion, to the disturbance of neurotrophic relations, or to the absence of the "trophic influence of the nervous system". Intensive electrotherapy can, for example, maintain the normal diameter of a major part of muscle fibres for as long as two months after denervation (Gutmann and Guttmann 1944). Fibrillation activity of denervated muscle, however, is unaffected by intensive electrotherapy.

An attempt was made to elucidate this question by experiments in which changes were investigated in denervated muscle which was stimulated daily

(Gutmann and Žák 1961). Considerable loss of protein nitrogen (—50%) can be observed 14 days after denervation, while there is an absolute, as well as a relative, increase in the DNA content. When bilateral denervation is performed, and the denervated muscle of one side is intensively stimulated (3 times daily for 4 min.), the content of protein nitrogen in the stimulated muscle is increased by 14% as compared with the non-stimulated denervated control. There is no difference in the content of DNA between the stimulated and non-stimulated denervated muscles (Fig. 144). It is thus not possible to influence the changes of nuclei by means of electrotherapy of denervated muscle, they are evidently due to lack of trophic influences of the nervous system.

Disturbances characteristic of the loss of the "trophic" influence of the nervous system are thus irreversible if reinnnervation does not take place, i. e. if neuromuscular metabolic relations are not renewed.

Investigation of the metabolic reactibility of the denervated muscle should demonstrate and analyse the trophic influence of the nervous system, but the the analysis of the concrete mechanism is domain of further biochemical studies. The significance of neurosecretory activity of nerve cells, which would most probably correspond to the "trophic influence" of the nervous system, has up to now remained a generalization of the problem on the basis of physiological data. In spite of this, it is possible that they could represent an important approach to investigation of the mechanism by which the "trophic influence" of the nervous system is realized.

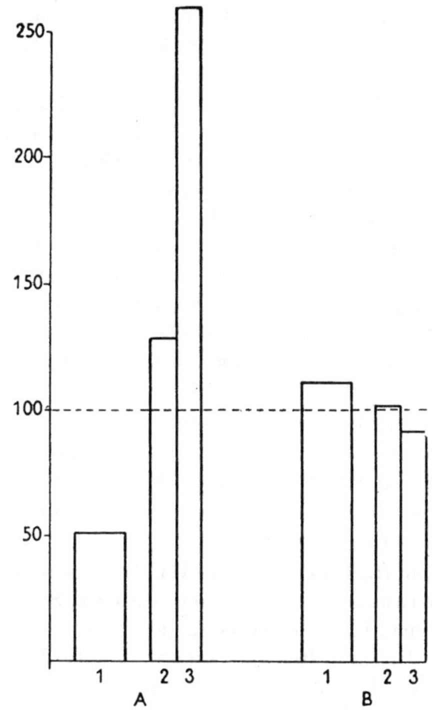

Fig. 144. Percentage changes in non-collagenous protein nitrogen (1), absolute (2) and relative (3) amounts of DNA in the tibialis ant. 14 days after section of the sciatic nerve as compared with the normal muscle of contralateral extremity (A). Changes in noncollagenous protein nitrogen (1) and absolute amounts of DNA (3) in the denervated tibialis ant. subjected to electrotherapy compared with the denervated contralateral control (B).

References

Altschul R.: Rev. Canad. Biol. 6 : 485, 1947.
Axelson J., Thesleff S.: J. Physiol,. 147 : 178, 1959.

Bajusz E.: Rev. Canad. de Biol. 17 : 483, 1958.
Bass A., Gutmann E., Vodička Z.: Physiol. bohemoslov. 4 : 267, 1955.
— — — cit. Gutmann 1959.
Bass A., Hudlická O.: Physiol. bohemoslov. 9 : 408, 1960.
Bass A., Vodička Z.: Physiol. bohemoslov. 4 : 1955.
Bender M. B.: Amer. J. Physiol. 121 : 609, 1938.
Bender M. B., Kennard M. A.: 1 : 431, 1938.
Beránek R., Hník P.: Science 130 : 981, 1959.
Beránek R., Gutmann E., Vrbová G.: Physiol. bohemoslov. 3 : 47, 1954.
Beránek R., Novotný I.: Physiol. bohemoslov. 8 : 87, 1959.
Beznák M.: J. Physiol. 116 : 74, 1952.
Bradley H. C.: Phys. Rew. 18 : 173, 1938.
Brown G. L.: J. Physiol. 89 : 438, 1937.
Brooks V. B., Myers D. K.: J. Physiol. 116 : 158, 1952.
Buchthal F., Kahlson G.: Acta Physiol. Scand. 11 : 284, 1946.
Buller A. J., Eccles R. M., Eccles J. C.: Abstracts of Communications XXI. Int. Congress of Physiological Sciences, 1959.
Cannon W. B., Rosenblueth A.: The Supersensitivity of Denervated Structures. Macmillan, New York, 1949.
Case F. J.: Science 124 : 1079, 1956.
Castillo J. del, Katz B.: J. Physiol. 128 : 396, 1955.
Chadwick L. E.: Biol. Bull. 93 : 229, 1947.
Charvát J.: In Symposium on: Nervous Regulation of Metabolism and Active Transport of Ions. Prague 1955.
Chor H., Dolkart R. E.: Am. J. Physiol. 117 : 626, 1936.
Couteaux R., Nachmansohn D.: Proc. Soc. Exper. Biol. Med. 43 : 177, 1940.
Crepax P.: Experientia 5 : 167, 1949.
— Biochim. Biophys. Acta 9 : 385, 1952.
Desmedt J. E.: Arch. int. Physiol. 58 : 23, 1950 — 58 : 125, 1950.
Dahlgren V.: Science 40 : 862, 1914.
Dale H. H., Gasser H. S.: J. Pharmacol. Exp. Therap. 29 : 53, 1926.
Davidovich A., Luco J. V.: Acta Physiol. Latinoamer. 6 : 49, 1956.
Denny-Brown D., Brenner C.: Arch. Neurol. Psychiat. 51 : 1 ,1944
Denny-Brown D.: Pennybacker J. B.: Brain 61 : 311, 1938.
Desmedt J. E.: Arch. int. Physiol. 58 : 23, 1950 — 58 : 125, 1950.
— Am. J. Phys. Med. 38 : 90, 1959.
Díaz-Guerrero R., Thomson J. D., Hines H. M.: Amer. J. Physiol. 151 : 91, 1947.
Doupe J., Barnes R., Kerr A. S.: J. Neurol. Neurosurg. Psychiat. 6 : 136, 1943.
Drahota Z.: Proc. Soc. exp. Biol. Med. 103 : 849, 1960.
— Physiol. bohemoslov. 8 : 501, 1959.
Drahota Z., Gutmann E.: Gerontologia 5 : 88; 1961.
Drahota Z., Gutmann E., Vrbová G.: Physiol. bohemoslov. 6 : 77, 1957.
Drahota Z., Klicpera J., Žák R.: Biochim. Biophys. Acta 28 : 31, 1958.
Dreyfus J. C., Joly M., Schapira G., Roeler L: Compt. rend. Acad. sc. 256 : 2351, 1953.
Dreyfus J. C., Schapira G.: Le fer, Paris 1958.
— — Sem. Hop. Paris. 24 : 1999, 1948.
Eccles J. C.: Med. J. Australia 2 : 160, 1941.
Eccles J. C., Eccles R. M., Lundberg A.: J. Physiol. 142 : 275, 1958.
Eisenberg E., Gordan G. S., Elliot H. W.: Endocrinology 45 : 113, 1949.
Epelbaum S. E., Kantor L. E.: Biokhmiya 19 : 660, 1954.

Euler U. S., Gaddum J. H.: J. Physiol. 73 : 54, 1931.
Fábry P.: Physiol. bohemoslov. 4 : 33, 1955.
Fábry P., Hrůza Z.: Physiol. bohemoslov. 5 : 142, 1956.
Fatt P., Katz B.: J. Physiol. 117 : 109, 1952.
Fenn W. O.: Am. J. Physiol. 120 : 575, 1937.
Ferdmann D. L.: Primeneniye radioaktivnykh izotopov dlya issledovanii v oblasti biokhimii myshts. Moscow, 1955.
— Vopr. mediz. khimii 3 : 351, 1957.
Fischer E.: Arch. Phys. Med. 29 : 291, 1948.
— Arch. Phys. Med. 30 : 375, 1949.
— Changes in structural proteins and enzyme during muscular atrophy of nervous origin, in "Le Muscle, étude de Biologie et Pathologie" — Colloque tenue à Royaumont, 1950, L'Expansion Scientifique Francaise, 1952.
— Arch. ges. Physiol. 255 : 1, 1952.
Fischer E., Ramsay W. W.: Amer. J. Physiol. 145 : 571, 1946.
— Am. J. Phys. Med. 34 : 212, 1955.
Forker L. L., Chaikoff I. L., Entenmann G., Tarver H.: J. Biol. Chem. 188 : 37, 1951.
Frank E., Nothmann M., Gutmann E.: Pflügers Arch. ges. Physiol. 199 : 567, 1923.
Frank E., Nothmann M., Hirsch-Kaufmann H.: Pflügers Arch. ges. Physiol. 197 : 270, 1922.
Ginetzinsky A. G., Shamarina N. M.: Usp. sovr. biol. 15 : 283, 1942.
Greenbaum A. L., Young F. G.: J. Endocrin. G. 127 : 135, 1953.
Grigorieva V. A.: Ukr. Bikhim. Zh. 30 : 356, 1958.
Guckelberger M., Kaiser E.: Arbeitsphysiol. 10 : 93, 1939.
Gutmann E.: Čas. lék. čes. 94 : 452, 1955.
— Physiol. bohemoslov. 8 : 405, 1959.
— Am. J. Phys. Med. 38 : 104, 1959.
— Čs. fysiol. 9 : 231, 1960.
Gutmann E., Guttmann L.: J. Neurol. Psychiat. 7 : 7, 1944.
Gutmann E., Holubář J.: Physiol. bohemoslov. 1 : 220, 1952.
— — J. Neurol., Neurosurg., Psychiat. 13 : 89, 1950.
— Physiol. bohemoslov. 9 : 382, 1960.
Gutmann E., Vodička Z.: Physiol. bohemoslov. 2 : 389, 1953.
Gutmann E., Vodička Z., Vrbová G.: Physiol. bohemoslov. 3 : 182, 1954.
Gutmann E., Vodička Z., Zelená J.: Physiol. bohemoslov. 4 : 200, 1955.
Gutmann E., Vrbová G.: Physiol. bohemoslov. 1 : 157, 1952.
— — Physiol. bohemoslov. 1 : 205, 1952.
— — J. Physiol. 140 : 50P, 1957.
— — Physiol. bohemoslov. 7 : 424, 1958.
Gutmann E., Jakoubek B.: 1960, unpublished.
Gutman E., Žák R.: Physiol. bohemoslov. 10 : 20, 1961.
Hansen R. G., Rutter W. R., Samuels L. T.: Jour. Biol. Chem. 192 : 243, 1951.
Harris E. J., Nicholls J. G.: J. Physiol. 131 : 1, 1956.
Hastings A. B., Ching, Tseng Teng, Nesbett F. B., Sines F. M.: Biol. Chem. 194 : 69, 1952.
Hechter O., Levine R., Soskin S.: Proc. Soc. Exper. Biol. Med. 46 : 390, 1941.
Herman H., Nicholas J. S.: J. exp. Zool. 107 : 165, 1948.
Hines H. M.: Physiological basis of the treatment of muscular atrophy of nervous origin. The Muscle. Symposium Royamont-France, 1950.
Hines H. M., Knowlton G. C.: Am. J. Physiol. 111 : 243, 1935.
— — Am. J. Physiol. 120 : 719, 1937.
— — Am. J. Physiol. 104 : 379, 1933.

Hník P., Škorpil V., Janda V.: Čs. fysiol. 7 : 468, 1958.
Hoagland C. L.: Advances in Enzymol. 6 : 193, 1946.
Hollán S. R.: Acta physiol. hung. 12 : 215, 1957.
— Acta physiol. hung. 12 : 257, 1957.
— Acta physiol. hung. 12 : 230, 1957.
Hudlická O.: Čs. fysiol. 4 : 433, 1955.
Hudlická O., Gutmann E.: Arch. exp. Path. u. Pharmak. 234 : 501, 1958.
Hudlická O., Vrbová G.: Physiol. bohemoslov. 7 : 38, 1958.
Huff E. G., Fischer E.: Am. J. Physiol. 159 : 6, 1949.
Hume B. J.: Biochem. J. Lond. 52 : p. XI, 1952.
Humoller F. L., Groswold B., McIntyre A. R.: Am. J. Physiol. 167 : 656, 1951.
Humoller F. L., Hatch D., McIntyre A. R.: Am. J. Physiol. 170 : 371, 1952.
Johnsson T. R., Thesleff S.: Acta physiol. scand. 51 : 136, 1961.
Josefsson J. O., Thesleff S.: Acta physiol. scand. 51 : 163, 1961.
Kare M. R.: Am. J. Physiol. 51 : 175, 1953.
Kare M. R., Medway W., Brandt C. S.: Am. J. Physiol. 183 : 416, 1955.
Keller M. J., Loeser A., Rein H.: Ztschr. Biol. 90 : 260, 1930.
Knowlton G. C., Hines H. M.: Am. J. Physiol. 109 : 200, 1934.
— — Am. J. Physiol. 132 : 336, 1941.
Kochakian C. D., Tillotson C.: Am. J. Physiol. 189 : 425, 1957.
Korner A.: J. Endocr., Lond. 13 : 90, 1955.
Koshtoyants Ch. S.: Osnovy sravnitelnoi fiziologii. Tom II. Moscow, 1957.
Koshtoyants Ch. S. Janson E. A.: DAN SSSR 75 : 81, 1950.
Krebs H.: Rate limiting Factors on Cell Respiration. Ciba Found. Regulation of Cell Metabolism. 1959.
— Control of metabolic processes. Endeavour 16 : 125, 1957.
Krüger P.: Tetanus und Tonus der quergestreiften Skelettmuskeln der Wirbeltiere und des Menschen. Leipzig 1952.
Krüger P., Günther P. G.: Zeitschr. f. Biol. 109 : 41, 1956.
Kubišta V.: Metabolismus hmyzího svalu. Disertation 1959.
Kubišta V., Gutmann E.: Physiol. bohemoslov. 6 : 188, 1957.
Langley J. N., Igataki M. J.: Jour. Physiol. 51 : 202, 1917.
Langley J. N., Kato T.: J. Physiol. 49 : 410, 1915.
Leonard L. S.: Endocrinol. 43 : 226, 1953.
— Endocrinol. 57 : 393, 1955.
— Proc. Soc. Exp. Biol. IV. Y. 94 : 458, 1957.
Levine R., Hechter O., Soskin S.: Am. J. Physiol. 132 : 336, 1941.
Lewis W. H., Lewis M. R.: Am. J. Anat. 22 : 169, 1917.
Lowry E., Hunter K.: J. Biol. Chem. 159 : 564, 1945.
Lubinska L.: Acta Biol. Exper. 16 : 73, 1952.
Luco J. V., Eyzaguire C.: Jour. Neurophysiol. 18 : 65, 1955.
Luco J. V., Sanchez P.: Acta Physiol. Latinoamer. 6 : 171, 1956.
Mandel P., Jacob M., Mandel L.: Compt. Rend. 228 : 457, 1949.
Marinacci A. B.: Arch. Neurol. 1 : 243, 1959.
Martínek J.: Physiol. bohemoslov. 5 : 78, 1956.
Miledi R.: J. Physiol. 151 : 1, 1960.
Mirski A., Wertheimer E.: Biochem. J. 36 : 221, 1942.
Nicholls J. G.: Jour. Physiol. 131 : 473, 1956.
Nimmi M. E., Geiger E.: Proc. Soc. Exp. Biol. Med. 94 : 606, 1957.
Ochs S., Burger E.: Am. J. Physiol. 194 : 499, 1958.

Orbeli L. A.: Trudy Fiziol. Inst. I. P. Pavlova, AN SSSR 1 : 1, 1945.
Petrásek R.: Nature 183 : 329, 1959.
Petrova A. N.: Voprosy biokhimii myshts, Kiev, p. 168, 1954.
Polák H., Küchel O., Engelberth O.: Čas. lék. čes. 94 : 1150, 1955.
Popova N. K.: Fiziol. Zh. 37 : 103, 1951.
Pringle J. W. S.: J. Physiol. 124 : 269, 1954.
Rosenblueth A., Luco J. V.: Amer. J. Physiol. 120 : 781, 1937.
Rupp J., Paschkis K. E., Cantarow A.: Endocrin. 44 : 449, 1949.
Russel J. A., Wilhelmi A. E.: Ann. Rev. Physiol. 20 : 43, 1958.
— — in Structure and Function of Muscle. Vol. II., 1960.
Samuels A.: Am. J. Physical Med. 36 : 78, 1957.
Samuels A. J., Boyarsky R., Gerard W., Libet B., Brust M.: Am. J. Physiol. 164 : 1, 1951.
Sato I., Kasugai F.: Tohoku J. Exp. Med. 26 : 336, 1935.
Schapira G., Coursaget J., Dreyfus J. C., Schapira F.: Bull. Soc. Chim. Biol. 35 : 1309, 1953.
Schapira G., Dreyfus J. C.: Metabolismes de muscle atrophié.
— — In Le Muscle, étude de Biologie et Pathologie, Colloque tenue à Royaumont, 1950, L'Expansion Scientifique Francaise, 1952.
Schapira G., Dreyfus J. C., Schapira F.: Bull. Acad. Méd. 20 : 361, 1947.
Scharrer E., Scharrer A.: Progress in Hormone Research vol. 10 : 1954.
Schmidt C. G.: Biochem. Z. 323 : 266, 1952.
Schmidt C. G., Schlief H.: Zeitschr. f. d. ges. experiment. Med. 127 : 53, 1956.
Selye H., Bajusz E.: Amer. J. Physiol. 192 : 297, 1958.
Severin S. E., Judajev N. A.: Biokhmiya 16 : 286, 1951.
Sheves G. S., Epelbaum S. E., Rjumina V. J.: Biokhimiya 21 : 71, 1956.
Slozhenikina L. V., Vopr. Med. Khim. 3 : 137, 1957.
Studitskii A. N., Striganova A. P.: Vostanovitelnye protsesy v skeletnoi muskulature. Moskva 1951.
Stewart D. M.: Biochem. J. 59 : 553, 1955.
Studitskii A. N.: Eksperimentalnaya kkirurgija myshts. Moskva 1959.
Sunderland S., Ray J. L.: J. Neurol. Neurosurg. Psychiat. 13 : 159, 1950.
Thesleff S.: J. Physiol. 151 : 598, 1960.
Tower S. S.: J. Comp. Neurol. 67 : 241, 1937.
Tower S.: Physiol. Rev. 19 : 1, 1939.
Tower S., Howe H., Bodian D.: Neurophysiol. 4 : 398, 1941.
Vodička Z.: Physiol. bohemoslov. 5 : suppl. 55, 1956.
— Physiol. bohemoslov. 8 : 285, 1959.
— Physiol. bohemoslov. 6 : 62, 1957.
Vodička Z., Gutmann E., Bass A.: Physiol. bohemoslov. 6 : 343, 1957.
Voskresenskaya A. K.: Funktsionalnye svoistva nervno-myshechnogo pribora nasekomych. 1959.
Vrbová G., Gutmann E.: Physiol. bohemoslov. 5 : 1, 1956.
— — Physiol. bohemoslov. 7 : 221, 1958.
Weinmann P., Shipounoff J. G.: Endocrin. 29 : 975, 1941.
Weiss P., Hiscoe H. B.: J. Exp. Zool. 107 : 315, 1948.
Wertheimer E.: Pflügers Archiv f. Physiol. 215 : 796, 1927.
— Pflügers Archiv f. Physiol. 213 : 287, 1926.
Wertheimer E., Hoffmann A.: Pflügers Arch. f. d. Ges. Physiologie 216 : 337, 1927.
Wiggelsworth V. B.: Symposium of the Soc. Exp. Biol. vol. 11, 1957.
Willard W. A., Graw E. C.: Anat. Rec. 27 : 192, 1924.
Wright P. M., Kochakian C. D.: Amer. J. Physiol. 173 : 217, 1953.
Yakovlev N. N.: DAN 81 : 709, 1951.

— Biochimija sporta. Moskva 1955.
Yampolskaya L. I.: Fiziol. Zh. SSSR 36 : 749, 1950.
Yampolskaya L. I., Yakovlev N. N.: Fiziol. Zh. SSSR, 38, 110, 1951.
Yudayev N. A., Smirnov M. I., Razina R. G., Dobbert N. N.: Biokhimiya 18 : 732, 1953.
Young J. Z.: Physiol. Rev. 22 : 318, 1942.
— Endeavour 15 : 5, 1956.
Zelená J.: Vliv inervace na vývoj svalu. Praha 1959.
Zubenko P. M.: Voprosy biokhimii myshts. Kiev, p. 149, 1954.
Žák R., Gutmann E.: 1959, unpublished.
— — Čs. fysiol. 8 : 264, 1959.
— — Nature, 185 : 766, 1960.
Žák R., Gutmann E., Vrbová G.: Experientia 13 : 80, 1957.

Chapter XII.

DIAGNOSIS AND THERAPY OF DENERVATION MUSCLE ATROPHY

P. Hník, V. Škorpil, L. Vyklický

The clinical approach to problems of denervated muscle differs somewhat from the experimental approach. The primary aim of the experimental investigator is to elucidate the factors determining the progress of reinnervation and the causes initiating denervation muscle atrophy. In experimental work the whole problem can be subdivided into a number of partial questions which can then be progressively analysed and on the basis of which complex functional and trophic relations between the nerve cell and the muscle can be deduced.

On the other hand, the clinician must necessarily regard every case of muscle atrophy from a wide diagnostic and therapeutic aspect. His decisions are dictated by contemporary experimental and clinical knowledge, and his main aim is to attain the best possible therapeutic result.

In the chapter dealing with clinical problems of denervated muscle, it was thought pertinent to discuss questions concerning topical and aetiological diagnosis. Attention is mainly directed to electrodiagnosis, especially to electromyography, since it represents the most important auxilliary method for determining the diagnosis and course of muscle atrophy. In the second part, therapeutic problems of denervation muscle atrophy are reviewed and open problems of muscle atrophy from the clinical aspect are discussed.

A. Diagnosis and evaluation of the course of denervation muscle atrophy

On encountering a case of muscle atrophy the clinician is forced to decide its aetiology. Case history and objective neurological symptoms, including muscle tests, serve as the basis for his diagnosis. The clinical examination gives information about the development and extent of the lesion and thus obtains fundamental data for differential diagnostic purposes and for determining which auxilliary examination methods are necessary.

These auxilliary examination methods can help to decide whether the muscle atrophy is of denervation origin and whether it is a case of complete or partial denervation. If it is deduced that the muscle atrophy is the result of a peripheral motor neuron lesion, it is necessary to determine the localization of the lesion. Denervation atrophy is caused by lesions of any part of the peripheral motor neuron, of its cell body, the axon or the neuromuscular junction. In lesions of the axon, it must be decided at which place the lesion is localized. Determination of the aetiology of denervation atrophy is a very important problem, since it affects the therapy as well as prognosis of a given case.

Fundamental diagnostic questions are closely associated with problems of objective evaluation of the development of denervation muscle atrophy and therapeutic procedures affecting it. For this reason the clinician is interested in auxilliary examination methods, since they enable him to evaluate the functional state of the peripheral motor neuron from the aspect of development or regression of the pathological process, to judge the trophic state of the muscle in the course of denervation and to follow the therapeutic effects aimed at retarding the course of muscle atrophy. In cases of prolonged denervation it is important to be able to decide whether the muscle is still capable of functional restitution in case reinnervation takes place. During reinnervation it is important to follow not only the trophic state of the muscle, but also to evaluate motor functions, especially coordination of movement and endurance and to observe all compensatory mechanisms.

At the present time the most useful auxilliary examination methods belong to the group of electrodiagnostic methods, which together with other methods of investigation, especially X-ray and cerebrospinal fluid examinations are helpful in determining the aetiology of muscle atrophy.

Currently used clinical electrodiagnostic methods either investigate the electrical excitability of neuromuscular structures or register the electrical manifestation of the muscle at rest or during contraction.

Combination of both these methods, i. e. registration of action potentials in motor units evoked by the stimulation of mixed peripheral nerves, is very valuable. The registration of nerve action potentials according to Dawson and Scott (1949) has not come into wide use because of considerable methodological difficulties, even though the clinical significance of this method has been stressed by Gilliat and Sears (1958) and others.

1. Indirect muscle excitability

Even though this is one of the oldest electrodiagnostic methods, the investigation of indirect muscle excitability remains a useful method for quick orientation examination (Anokhin and Mayorchik 1952, Ritchie 1954. Woodhall et al. 1956).

The examination can be carried out either with galvanic current or with electrical pulses of various shapes. The nerve can be stimulated with cutaneous electrodes or needle electrodes introduced intradermally into the vicinity of the nerve or during surgical operation by applying the electrodes directly to the nerve trunk. In cases where a large resistance of tissues above the nerve can be expected, e. g. a thick layer of skin or extensive connective tissue reaction around the nerve, stimuli of high intensity must be used. For this reason Woodhall et al. (1956) recommend the use of intradermal needle electrodes introduced into the neighbourhood of the nerve trunk whenever it is not possible to stimulate the nerve percutaneously. According to Anokhin and Mayorchik (1952), it is advantageous to observe the amplitude of the motor response of the muscle to nerve stimulation with stimuli of increasing intensity during surgical revisions. The motor reaction to such stimulation increases in the normal muscle and muscle contractions become stronger. In partially denervated muscles and especially where only a few viable nerve fibres are preserved, increasing stimulus intensity does not lead to stronger muscle contractions. When early reinnervation sets in, it is more suitable to stimulate the nerve above the lesion. Threshold of nerve fibres below the lesion is higher, evidently because the regenerating nerve fibres are markedly smaller in diameter (Gasser and Grundfest 1939, Gutmann and Sanders 1943, Gutmann and Young 1944, Hodes et al. 1948). According to Woodhall et al. (1956) the finding of indirect muscle excitability is as valuable a symptom of early reinnervation as the appearance of voluntary contraction.

2. Direct muscle excitability investigated with faradic and galvanic current

The normal muscle responds to faradic stimulation with a tetanic contraction, while the denervated muscle loses this excitability at comparable intensities. The excitability to faradic current of high intensity is preserved, but such stimulation is very painful for the patient and cannot thus be used in practice. On the other hand galvanic current stimulates the denervated muscle, usually at lower intensities than under normal conditions. The contraction and decontraction of the denervated muscle is, however, prolonged and maximal excitability is not localized at the motor point but at the border-line between muscle and the tendon. The difference in the response to galvanic and faradic stimulation is related to differences in the duration of the respective stimuli. The denervated muscle requires for its excitation a stimulus of considerably longer duration (at the same current intensity) and is therefore inexcitable by faradic current, the impulses of which are of very short duration.

Threshold values of faradic and direct current cannot be considered as absolute measures of electrical excitability of the muscle because of the great variability in tissue resistance above the muscle, especially of the skin.

If the muscle is found to be inexcitable to galvanic stimulation of the usual stimulus intensity, it need not necessarily be a fibrotic muscle which has lost its ability to contract. This is often the case when the tissue is very oedematous, scarified with connective tissue or when the skin is exceptionally thick. It is then necessary to use considerably stronger currents, this being in most cases very painful for the patient.

According to Ritchie, stimulation with faradic current is not suitable for determining early reinnervation of the muscle since voluntary movement appears considerably sooner than normal faradic excitability (in 60% of cases examined by him). The examination based on differences in the reaction to switching galvanic current on and off is no longer used in clinical practice because of its unreliability (Parry 1958, Lesný 1953, Ritchie 1954).

Galvanic and faradic currents are of little significance in clinical examination at the present time and other more reliable examination methods have taken their place.

3. Chronaxie

Chronaximetry, clinically instituted especially by Lapicque (1921) and Bourguignon (1923) has become one of the most widely used electrodiagnostic methods for examination of muscle atrophies. One of the main reasons for this is the fact that it is a very quick examination, and not unpleasant for the patient. Its advantages and disadvantages for clinical work have been reviewed in numerous reports (Anokhin and Mayorchik 1952, Lesný 1953, Abrikosov and Darkshewich 1954, Walthard and Tchicaloff 1954, Grundfest et al. 1956, Woodhall et al. 1956 and others).

Chronaxie, which does not, under physiological conditions, reach values above 1 msec. and is usually below 0·5 msec., increases considerably in the denervated muscle and attains values even above 10 msec. Lesný (1953, 1954) recommends measurement of chronaxie at the anode as well as the cathode and maintains that during denervation, the chronaxie values at the cathode rise sooner than those at the anode, and that anode chronaxie returns to normal sooner during reinnervation. Woodhall et al. (1956) consider chronaxie to be a suitable method for determining denervation, however, it was not advantageous for investigating the early reinnervation stage, since return to normal chronaxie values does not precede clinical restitution of voluntary movements. Our observations confirm this finding (Škorpil and Vyklický 1957). According to Grundfest et al. (1956) and Anokhin and Mayorchik (1952), one of the main failings of chronaximetry lies in the fact that normal chronaxie, corresponding to a small number of innervated muscle fibres, may be found in a partially denervated muscle, even though the majority of fibres are denervated. A further disadvantage is that from the whole range of the strength-duration curve, chronaxie expresses only one value of the relation of muscle excitability to strength of stimulus and its duration.

4. Neuromuscular excitability in relation to time and stimulus intensity (I/t curve)

This examination came into general use in clinical work at the time of World War II. It was clinically evaluated by Marble et al. (1942), Ritchie (1954), Pollock et al. (1944). Its clinical usefulness was stressed by Ritchie (1954) and Parry (1958).

For this examination either condensor or rectangular electric pulses are used: the strength of the stimulus can either be expressed as the current intensity (in mA) or its voltage (in V.). Time is measured in msec. The curve expressing the relationship of current intensity or voltage to the duration of the stimulus can be obtained from a number of individual measurements, nine to ten being considered as optimum (Ritchie 1954), each of which comprises determinations of the threshold response to electrical stimuli the voltage or intensity of which is gradually decreased and its duration progressively increased.

In the normal muscle the curve is an asymptotic hyperbola (Fig. 145), while in the denervated muscle the curve is moved to the right and upwards. In partial denervation and during reinnervation the curve is indented, the break being especially significant from the clinical aspect (Fig. 145). The break in the curve during reinnervation appears earlier than the clinical onset of voluntary movements (a difference of six weeks according to Parry — 1958). This type of curve represents the excitability of two different substrates. First signs of reinnervation in the I/t curve appeared 15—18 days earlier then restitution of voluntary movements in 80% of the 483 muscles of Ritchie's material (Ritchie 1954). According to this author, voluntary movement must appear within about

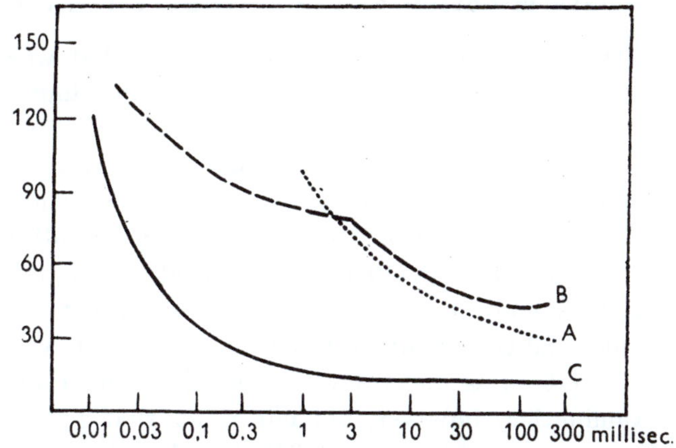

Fig. 145. Time-duration curves. A — denervated muscle, B — during reinnervation, C — normal muscle. (According to Ritchie A. E.: Electrical diagnosis of peripheral nerve injury. London, H. M. Stationery Office 1954.)

one month of the appearance of the break in the I/t curve. If it does not appear, it can be anticipated that further restitution will be unlikely especially if the muscle is localized distally. Newman and Livingston (1947) found the break on the curve in 54 %, the average interval between electrical and clinical restitution being 97 days. It is also possible to discover very soon whether the reinnervation process is progressing, or not (Parry 1958). Rectangular pulses are more suitable for obtaining the compound curve, since I/t curves plotted from condensor discharges exhibit a less evident break in the curve. After reinnervation of the muscle there is also a transfer of the curve downwards and the curve is flattened. With progressing reinnervation the break in the curve usually disappears before the restitution of voluntary movement.

When tissue resistance above the muscle examined alters, the curve is shifted but its course and general shape as a whole do not change (Ritchie 1954). The curve has also a high threshold for long and short pulses in late and incompletely reinnervated muscles.

The disadvantage of this examination lies in the fact that it is fairly time consuming.

However, it is possible to say that this method represents one of the most reliable electrodiagnostic examinations based on excitability measurements.

5. Determination of the galvanic-tetanic ratio

This examination is based on the fact that when a muscle is stimulated by galvanic current of increasing intensity, the single twitch response changes into a tetanic contraction. The threshold for eliciting the tetanic response in the normal muscle is 3—5 times higher than normal rheobase. This ratio decreases below 2 and is about 1·0 in the denervated muscle.

Pollock et al. (1945, 1948) and recently Grundfest et al. (1956), Ritchie (1954) and Woodhall et al. (1956) have applied this method to clinical material. According to Grundfest, the ratio rapidly rose during reinnervation, surpassed normal values and then returned to normal. Values above 2 are considered by Woodhall et al. (1956) as uncertain evidence for reinnervation. A negative result of the examination, however, did not mean that reinnervation is not taking place. Newmann and Livingston (1947) found an increase of this ratio in 47 % of their cases, 109 days before clinical recovery on the average.

This test, however, is not agreeable for the patient since it is painful and if sufficient care is not taken, skin burns may occur.

6. Progressive current testing

This method was mainly elaborated by Solandt (1935), Kugelberg (1944), Pollock et al. (1945) and has also been used by Obrosov and Livencev (1953), Henge (1953), Kowarschik (1952), Jantsch (1952), Janča (1957) and others.

For stimulation of muscle, exponentially or linearly rising currents can be used. When stimulating with these pulses of different duration, it is possible to plot an I/t curve in the same way as on stimulation with rectangular pulses. The curve differs from that of the normal muscle obtained with rectangular pulses by the fact that when pulses of longer duration are used, higher intensity currents must be applied in order to stimulate the muscle, since the normal muscle accomodates considerably to these currents. The denervated muscle, on the other hand, ceases to accomodate and the I/t curve to exponentially increasing stimuli is similar to the I/t curve obtained by stimulating with rectangular pulses. For practical purposes in clinical work the so-called accomodation index is calculated, and this expresses the ratio between the threshold intensity of the progressive current stimulus to the threshold intensity of the rectangular stimulus of 1000 msec. duration. This index reaches values around 3—6 in normal muscle. In complete denervation, when the muscle is unable to accomodate, it is about 1·0 and about 3 in partial denervation. Pollock et al. (1945) made a diagnosis of denervation in 50 cases of nerve trauma by this method and this finding was later always confirmed. Reinnervation was always discovered by this test several days before the return of voluntary movements (150 days prior to restitution at the most).

At the present time factory-made stimulators are available for stimulation with progressive currents and these can also be used for electrotherapy. The method to-day is considered to be very valuable by clinicians (Thom 1953, Gilbert 1957, Janča 1958 and others), since exponentially rising currents stimulate selectively denervated muscle fibres only, without stimulating afferent nerve fibres. In this way it is possible to apply currents of the required intensity without also stimulating innervated muscle fibres and without painful sensations for the patient.

7. Alternating current intensity — frequency curves

When a nerve or a muscle is stimulated with sinusoidal current, there is a certain optimum frequency of stimulation at a given stimulus intensity (Vvedenskii 1901). This question was later studied by Hill et al. (1936), Osborne et al. (1944), Grodins et al. (1944).

The optimal frequency at minimal current intensity values is about 60 c/sec. in the normal muscle. Following denervation the optimal frequency is markedly reduced (between 0·5—2 c./sec.).

It is a disadvantage of this method that a generator producing low frequency current (0·5—5 c./sec.) of sufficient intensity for stimulating is expensive and complicated in construction.

8. Electromyography

The first EMG action potentials from human muscle were recorded by Piper (1907). Clinical foundations of electromyography were laid by a number of authors (Proebster 1928, Wachholder 1928, Adrian and Bronk 1929, Denny-Brown and Pennybacker 1938, Hoefer and Putman 1939), but it was not until the time of World War II that this method became widely used in clinical practice especially in investigations of denervated muscles (Buchthal and Clemmesen 1940, 1941, 1943, Buchthal 1949, Kugelberg 1946, 1947, 1950, Kugelberg and Petersen 1949a, 1949b, Denny-Brown 1949, Bowden 1954, Woodhall 1956, Lesný 1953, Grundfest et al. 1956 and a number of others).

Usually needle electrodes are used in diagnosis of the denervated muscle. It was not until lately that it has become possible to record, with suitably adapted apparatus, intracellular action potentials from a single muscle fibre with the aid of microelectrodes also in man (Beránek 1959, 1961).

a) Electromyography during voluntary contraction

aa) EMG findings in the normal muscle

No action potentials can be found at rest in a completely relaxed muscle. Action potentials do not appear until volitional contraction takes place. As far as the action potentials are concerned, the following characteristics are usually described: 1. amplitude, 2. duration, 3. shape, 4. frequency of discharges and their grouping. The evaluation of these represents the basis for differentiating the normal and pathological electromyogram. Individual characteristic features of action potentials change considerably even under physiological conditions and are dependent on several physiological conditions. It is necessary to know these before evaluating pathological changes.

Amplitude of action potentials in all muscles is very variable. Maximum amplitude of the action potential — 4 mV — was found by Buchthal (1957) in the biceps brachii muscle. The amplitude is dependent primarily on the distance of the tip of the electrode from muscle fibres the potentials of which are being recorded, and on the position of the electrode to the surface of the muscle fibre. It was found by Häkansson (1956) in experimental studies on an isolated frog muscle that amplitude of the action potential is reduced to 10% of its original maximum value at a distance 60 μ from the fibre. At larger distances the reduction in amplitude is considerably less conspicuous. Similarly in the human muscle *in vivo*, it was found by Buchthal et al. (1957) that amplitude of action potentials is also very much reduced in relation to the distance of the electrode from the muscle fibre. At 0·15—0·3 mm the amplitude is lowered by a half and 0·6 mm. away from the muscle fibre it is reduced to $^1/_{10}$—$^1/_{20}$ of the original value. At greater distances the drop in amplitude is not so rapid. The largest drop in amplitude of the action potential is to be found when the electrode is placed in the extracellular space in close vicinity to a small group of active fibres.

Peaks of the action potentials are very sharp in the region of the motor unit, while those of more distant motor units have neither a steep ascending slope nor a sharp peak. It was further shown on isolated muscle fibres in a conducting medium that the amplitude of individual action potentials is directly proportional to the diameter of muscle fibres and to the position of electrode tips in relation to the surface of the muscle fibre. *In situ* this factor probably plays an insignificant role only, the resistance conditions of the surrounding medium being the most important factor in determining the amplitude of the spike.

The duration of action potentials under physiological conditions in various muscles is relatively constant, so that a number of authors have elaborated parameters for different skeletal muscles and there are but slight differences between the results of different authors. According to Buchthal et al. (1957) the duration of action potentials is dependent, among others, on the time dispersion in summation of action potentials of individual muscle fibres belonging to a given motor unit which is affected by the following factors:

a) differences in the arrival of nerve impulses at the motor end-plates,

b) length of the innervation zone, i. e. the region containing motor end-plates of a single motor unit,

c) conduction rate of different muscle fibres,

d) distance traversed by action potentials in muscle fibres.

Differences in the time of arrival of nerve impulses at the motor end-plates can be ignored, since the rate of conduction in motor nerve fibres is so great under physiological conditions that it cannot practically influence the duration of the action potential. The length of the innervation region is a much more important factor. The innervation zone in the biceps brachii is about 4 cm. long according to EMG and histological reports, i. e. it represents about 15 % of the whole length of the muscle. The innervation zone in the biceps muscle in man, for example, is localized in the proximal third of the caput breve and in the distal third of the caput longum.

Action potentials recorded in the innervation zone have a steep negative phase. This steep slope of the action potential indicates the simultaneous onset of contraction in single muscle fibres in a given motor unit. When recording action potentials outside this innervation region, differences in conduction rate of different muscle fibres already play a role, the beginning of the contraction is not synchronous and the ascending phase of the action potential is no longer steep. The rate of conduction in striated muscles has been studied relatively extensively. First attempts at determining the rate of conduction in human muscle fibres were made by Denslow and Hasset (1943). Their results 1—12 m./sec., however, are complicated by a number of methodical inaccuracies, so that values obtained cannot be considered as reliable. In methodically up-to-date experiments by Buchthal et al. (1955, 1957) the rate of conduction in muscle fibres of man *in situ* was found to be $4{\cdot}02 \pm 0{\cdot}13$ m./sec. in caput breve of the biceps brachii muscle. These authors did not find any significant differences in conduction rate of muscle fibres in persons of different ages and thus there does not seem to be a closer relation between the rate of conduction in situ and muscle fibre diameter. This is an unexpected result considering the fact that in adults the muscle fibre diameter is very variable, ranging from 7—250 μ, with a mean value of 153 μ (the average in a year-old infant being 30 ± 5 μ). This can be explained by different relations between internal and external resistance during conduction of muscle fibre potentials under various experimental conditions.

The conduction rate in denervated muscle is markedly smaller than in normal muscle (Jarcho et al. 1954, 1958, Buchthal and Rosenfalck 1958. Determinations of muscle fibre conduction rate has not come into wide clinical use because of the considerable technical difficulties of this method.

Duration of action potentials is independent of the distance between the electrode and contracting muscle fibres and strength of contraction. Potentials are wider in muscles with large motor units than in muscles with small motor units. The relation between action potential duration and the size of motor units was also confirmed histologically (Feinstein et al. 1956). The duration of action potentials represents one of the most important aspects when evaluating the EMG record for diagnosis of muscle atrophies of neural and muscular origin. Several authors have elaborated parameters of the average duration of action potentials (Peterson and Kugelberg 1949 et al.) in different muscles. Action potentials are especially brief in extraocular and facial musculature (Björk and Kugelberg 1953).

Shape of action potentials. Action potentials are usually bi- or triphasic under physiological conditions. These two types represent 80—90% of all potentials. The remaining potentials are mono- or polyphasic. In the first dorsal interrosseal muscle, for example, there are only about 2—4% polyphasic potentials. However, polyphasic potentials are common in facial muscles. An increased number of polyphasic potentials appears in extreme fatigue (Pinelli and Buchthal 1954). Polyphasic potentials appearing during longlasting maximal contractions are probably caused by an increasing tendency to synchronous discharges of motor units or longer time dispersion of contractions in individual muscle fibres. Polyphasic potentials with very narrow spikes are characteristic for early reinnervation. Very wide polyphasic potentials are an expression of the formation of large motor units during late reinnervation.

Frequency of action potentials. It is very difficult to determine the frequency of action potentials under physiological conditions. Only in exceptional cases do isolated action potentials of one motor unit appear during slight voluntary contraction. When increasing this volitional effort more action potentials appear from different motor units with varying frequency. Thus mutual interference occurs and it is very difficult, or perhaps impossible, to analyse the shape of single potentials and their frequency. Most data on the frequency of action potentials of a single motor unit have been obtained in muscle atrophy of nervous origin, so that it is possible to register action potentials even during maximum voluntary contractions. Under such conditions the maximum frequency of action potentials reached 60—90 c./sec. According to Björk and Kugelberg (1953) the frequency in extraocular muscles under physiological conditions is 150 c./sec. Attempts were also made to determine the frequency of discharge using special metal microelectrodes, for example, by Norris and Gasteiger (1955), which made possible the isolated registration of action potentials from a single motor unit even during maximum voluntary contraction. In the rectus femoris and gastrocnemius a frequency of 10—20 c./sec. was found during weak voluntary contraction, 10—50 c./sec. during medium strength contraction and 60—140 c./sec. during maximal voluntary contraction. The frequency of discharge during maximum voluntary contraction, however, was very inconstant and maximum frequency appeared in volleys only. Neither apparatus nor electrodes used for clinical electromyography have, as yet, been standardized, so that EMG findings can also be influenced by technical differences in recording action potentials. The shape of electrodes affects the duration of potentials. Potentials of longest duration are to be found when recording with monopolar electrodes, they are shorter when concentric needles are used and shortest in bipolar recording. Guld (1951) demonstrated that among physical factors which can affect the duration of action potentials, an important role is played by the division of potential difference between electrode resistance and input resistance of the amplifier. Therefore, when using electrodes with small surface area a high input impedance is necessary.

bb) EMG in the denervated muscle

The denervated muscle is characterized by fibrillation potentials which appear in man 13—21 days after denervation (Denny-Brown and Pennybacker 1938, Feinstein et al. 1945). At first they can be observed only shortly after insertion of the needle, later they last longer as spontaneous activity. They are present in the denervated muscle even several years after denervation. We have found these fibrillations frequently in poliomyelitic patients 5—13 years after the acute stage (Janda et al. 1959).

Besides these typical fibrillations, atypical denervation potentials have also been described in human muscles. These are of longer duration (8—100 msec.)

and can attain an amplitude of several millivolts (Lerique 1944, Buchthal 1957, Thiebaut et al. 1956, Janda et al. 1959).

The presence of fibrillation potentials and atypical denervation potentials indicate that the muscle under investigation is denervated. Only exceptionally fibrillations have also been described in primary muscle amyotrophies (Bowden and Gutmann 1944, Buchthal 1957). Even in this type of disease, however, it is assumed that the fibrillation potentials only appear in muscle fibres which have lost connection with their motor nerve fibre. Fibrillations have been found in no other type of atrophy and for this reason are considered very valuable from the diagnostic point of view. They disappear gradually from the denervated muscle when reinnervation takes place, or after complete loss of the contractile muscle phase in a muscle denervated for a long time. Fibrillation activity is reduced with progressing atrophy.

cc) *EMG in the partially denervated muscle*

During maximum volitional effort at movement there is only a small reduction in the number of action potentials in early stages of partial denervation, i. e. EMG recording of maximum voluntary contraction changing from interference of action potentials into the so-called intermediary state. This early denervation stage is difficult to determine with the EMG method, since a similar finding can be found in the normal muscle when volitional effort is reduced. In late muscle denervation isolated action potentials of motor units are to be found during maximum volitional effort, and these can differ, as far as the shape of action potentials is concerned, from normal potentials. They can be polyphasic, of longer duration (above 10 msec.) and of higher frequency (about 40—50 msec.). Fibrillations are not a constant phenomenon in partially denervated muscles.

In chronic progressive muscle atrophies, especially of the spinal type, it is possible to register action potentials of the fasciculation type at rest. These potentials are usually polyphasic, similar to action potentials of motor units. Their electromyographic registration is of no diagnostic significance, since they cannot be found sooner in the EMG than by clinical examination.

In chronic spinal neurogenic muscle atrophies the so-called synchronization has been described in the EMG record (Buchthal and Clemmensen 1941, Buchthal 1949, Buchthal and Madsen 1950). This is characterized by a synchronous discharge of several motor units, this being demonstrated electromyographically by multielectrode recording. The explanation of this synchronization of action potentials during voluntary contraction in spinal neurogenic muscle atrophies has been criticized by Denny-Brown (1949), Kugelberger and Taverner (1950), Lesný (1953) and others. The significance of synchronization for differential diagnosis of neurogenic muscle atrophies of peripheral and spinal origin is small, so that this method is not commonly used.

dd) EMG in early reinnervation

According to Feinstein et al. (1945) fibrillation potentials disappear at the site of reinnervation 1—2 weeks prior to the appearance of first reinnervation

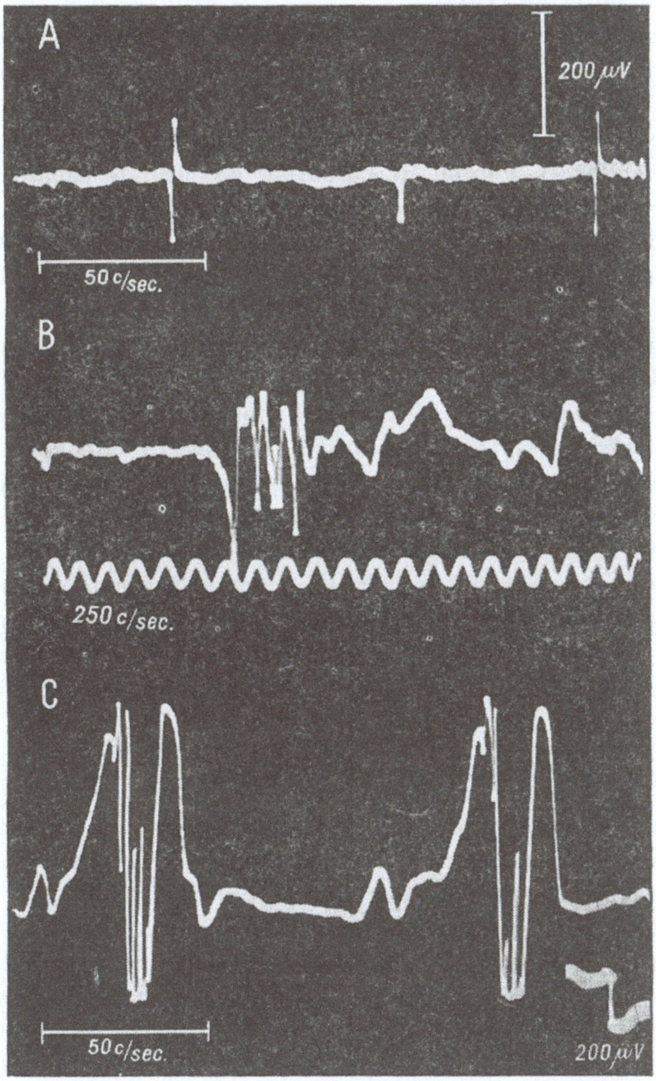

Fig. 146. A — First reinnervation potentials from the tibialis ant. during maximal voluntary effort in a 20-year-old patient following axonotmesis of the common peroneal nerve. B — Reinnervation polyphasic potentials from the tibialis ant. during maximal voluntary effort in a 24-year-old patient with axonotmesis of the common peroneal nerve. C — Polyphasic action potentials from the deltoid muscle during incomplete and late reinnervation following traumatic paresis of the upper portion of the brachial plexus.

potentials. Such proof is technically very difficult. We have ourselves often seen fibrillation potentials together with first reinnervation potentials. Early reinnervation potentials are characterized, among others, by a very small amplitude and short duration (Fig. 146A). They have been called "nascent motor units" by Weddell et al. (1944).

In the beginning they are present in the record for only short periods of time during volitional effort. At this time they are similar to fibrillation potentials because of their short duration and low amplitude, later they become more numerous, however, they last longer, their amplitude rises and they are often polyphasic (Fig. 146B). Polyphasic potentials are usually caused by different conduction rates in muscle fibres and different rate of maturation of terminal nerve fibres of one motor unit during early reinnervation. Giant polyphasic potentials found at late reinnervation stages are caused by nerve fibres growing into denervated muscle fibres which originally belonged to a neighbouring motor unit, the original unit thus increasing in size (Fig. 146C). (Van Harreveld 1945, Wohlfahrt 1949, 1958, Yahr et al. 1950, Wohlfahrt et al. 1954). In early reinnervation action potentials are first to be found in the vicinity of the motor point, later they spread to the whole muscle (Marinacci 1955). The clinical value of these potentials is considerable, since they appear several days or even weeks before first symptoms of restitution of motor function (Bowden 1954, Woodhall 1956, Grundfest 1956, Feinstein et al. 1945).

If, however, the number of reinnervation potentials in the muscle does not increase for some time, and if voluntary movements do not reappear within several weeks of their first occurrence, they cannot be evaluated as prognostically favourable from the clinical aspect (Buchthal and Hoenke 1954).

b) Stimulation electromyography

Stimulation electromyography has the advantages of both graphic registration and the examination of electrical excitability. Even though this method has been known to physiology for a long time (Piper 1907, Hoffmann 1922), it has come into wider clinical use since the work of Hodes et al. (1948).

Stimulation electromyography obtains information not only about the state of muscle fibres in the respective motor units, but also about the functional state of corresponding motor nerve fibres, since it is possible to measure their rate of conduction by stimulating the peripheral nerve. For this reason stimulation electromyography has proved itself especially useful in the differential diagnosis of muscle atrophies of spinal origin and chronic affections of peripheral nervous origin (Vyklický 1959).

Every analysis of electrical manifestations of peripheral nerve stimulation assumes that nerve fibres of larger diameter conduct nerve impulses at a faster rate than small nerve fibres (Erlanger and Gasser 1937). This has been demon-

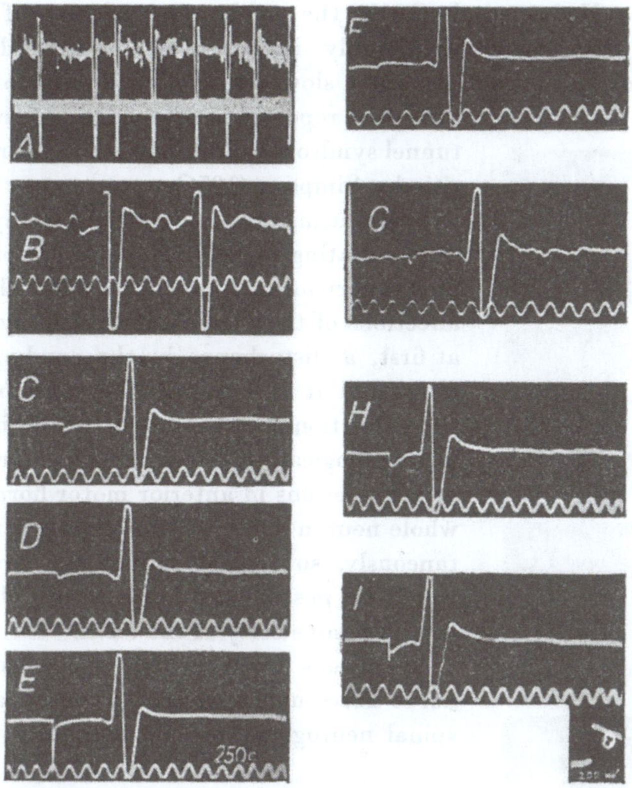

Fig. 147. EMG. records from the first dorsal interosseal muscle in a 15-year-old patient with marked muscle atrophy of small muscles of the hand following myeloradiculitis six years previously. A and B: record of maximal voluntary contraction. C and I: electromyographic response to stimulation of the ulnar nerve with increasing intensity of electrical stimulus. Stimulation on the arm C: 100 V, D: 150 V, E: 200 V; at sulcus olecrani F: 125 V, G: 200 V; above the wrist H: 180 V, I: 200 V. Amplitude and shape of the evoked action potential does not change with increasing intensities of the electrical stimulus even with changes in the place of stimulation.

strated and confirmed by a number of authors. The nerve fibre diameter during regeneration of peripheral nerves remains markedly smaller than under physiological conditions (Gutmann and Sanders 1943). Berry et al. (1944) have shown that the rate of conduction in regenerating nerves of the cat at the earliest phase of restitution is less than 5% of the original conduction rate. Ninety days after suture the conduction rate of nerve fibres attained about 33% of the original value. Berry and Hinsey (1946) repeated the same observations 1363 days after suture and found that the nerve fibres conducted impulses at a rate 80% of the initial velocity.

According to Hodes et al. (1948) the conduction velocity of regenerating nerve fibres in man is also considerably slowed down. Considerable asynchrony of evoked motor unit potentials, and a slowed rate of conduction of motor nerve fibres was reported in the median nerve, in the carpal tunnel syndrome and ulnar nerve in traumatic neuritis by Simpson (1956).

When using stimulation electromyography for differentiating muscle atrophies of spinal and peripheral nervous origin we assumed that in chronic affections of the peripheral nervous system there is, at first, a disturbance in the conduction of nerve fibres and it is not until considerably later that their function is lost completely. On the other hand, in pathological processes of spinal origin there are primary lesions in anterior motor horn cells and the whole neuron loses its function practically simultaneously, so that it cannot be expected that it would be possible to discover any changes in the conduction velocity of the axon.

We always found the rate of conduction in motor nerve fibres unchanged in patients afflicted with spinal neurogenic muscle atrophies of the amyotrophic lateral sclerosis type, in syringomyelia or in residua after myeloradiculitis. Evoked potentials of motor units appeared at threshold intensities according to the all-or-none reaction, and in case larger number of action potentials in motor units appeared during maximum voluntary contraction, their contractions on submaximal or maximal stimulation were synchronous (Fig. 147). Supramaximal values of the electrical stimulus often evoked repeated discharges of motor units (Fig. 148 and 149).

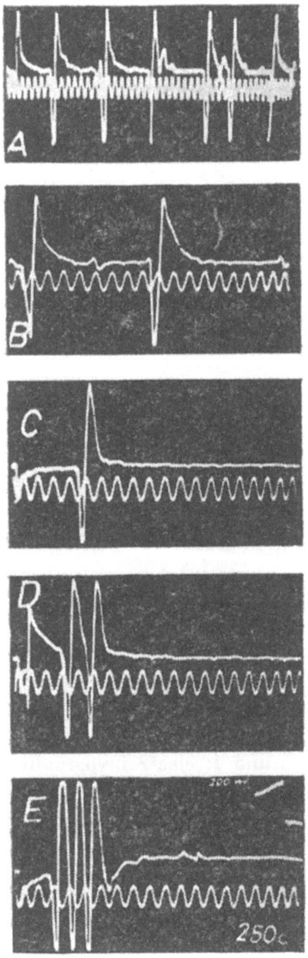

Fig. 148. EMG. record in a 56-year-old patient with amyotrophic lateral sclerosis. All records are taken from the same place in the first dorsal interosseal muscle. A and B: EMG record during maximal voluntary contraction. On stimulation of the ulnar nerve at sulcus olecrani the evoked potential appears according to the all-or-none law. C: 140 V. Supramaximal values of the electrical stimulus elicit a double (D: 150V) or a triple (E: 200 V) evoked potential of the same motor unit.

Peripheral neurogenic muscle atrophies result from pressure paresis of the ulnar nerve, for example, at the wrist or in late posttraumatic paresis of the

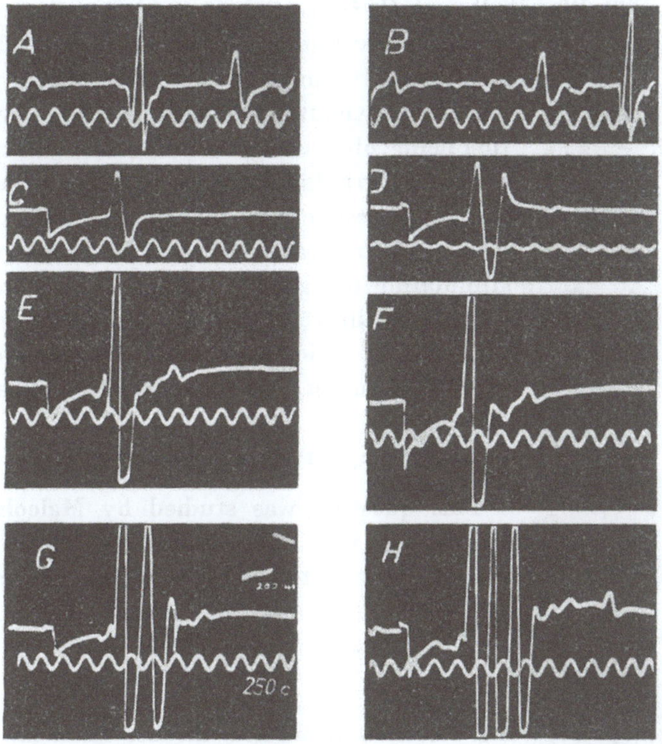

Fig. 149. EMG. record from the first dorsal interosseal muscle in a 56-year-old patient with extensive muscle atrophy and residual peripheral paresis of the right upper extremity following a paretic form of Czechoslovak encephalitis. A and B: during maximal voluntary contraction individual action potential, appear from three motor units. On stimulation of the ulnar nerve at the elbow with stimuli of increasing intensities (C: 115 V, D: 145 V, E: 150 V, F: 200 V) an action potential of increasing amplitude appears which corresponds to the synchronous contraction of all stimulated motor units. When duration of the electrical stimulus is increased, a double or a triple action potential appears (G: 145 V, H: 200 V).

ulnar nerve at the elbow and are characterized by unequal conduction rates of the motor nerve fibres, so that stimulation of the peripheral nerve gives rise to a completely asynchronous contraction of motor units in the periphery (Fig. 150). The localization of the nerve lesion can be deduced from the stimulation EMG records. If the nerve lesion is localized in the proximal part of the nerve, the difference in conduction velocity of motor nerve fibres is more evident when the proximal parts of the nerve are stimulated (Fig. 151). On the other hand, if the lesion lies below both places of stimulation, the evoked complex potential is of the same duration in both cases (proximal or distal site of stimulation), since differences in conduction rate affect the potential along the same length of nerve (Fig. 152). Woodhall et al. (1956) consider stimulation electromyography to be valuable for investigating early reinnervation symptoms. These authors did not find a single case where voluntary movement appeared without any motor response to nerve stimulation being observed. According to our experience the electromyographic response to nerve stimulation and the EMG response to volitional effort at contraction both appear practically at the same time. When stimulating a regenerated nerve, however, it must be stimulated above the site of the lesion, since regenerated fibres, evidently because of their considerably smaller diameter, have a markedly higher threshold.

c) EMG registration of tendon reflexes

This question was studied by Malcolm (1951) in connection with partial muscle denervation. This author evoked the patellar reflex and reflex of the

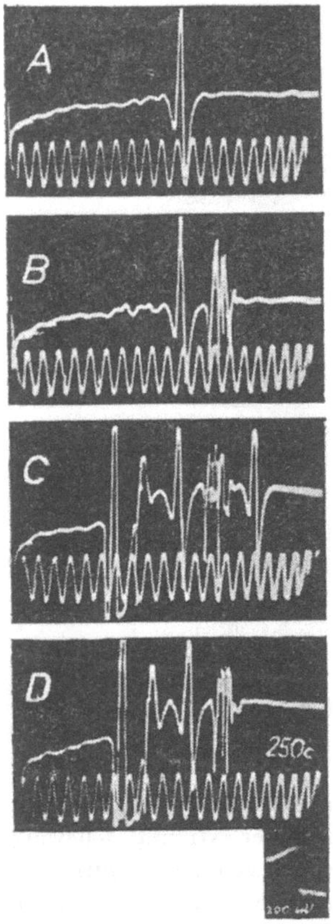

Fig. 150 EMG. record from the first dorsal interosseal muscle on stimulation of the ulnar nerve in a 59-year-old patient with late posttraumatic paresis of the ulnar nerve at the elbow. Nerve stimulation at sulcus olecrani: A: 170 V, B: 195 V, C and D: 200 V. Stimulation of the ulnar nerve with submaximal and maximal intensities evokes a completely asynchronous contraction of motor units.

Achilles tendon in patients with a painful lumbosacral syndrome and at the same time recorded the reflex contraction of the appropriate muscles electromyographically. On comparison with the control side, it was found that there is a prolongation of the latent period of the reflex response registered electromyographically and this prolongation was considered by Malcolm to be the result of spinal root compression.

d) EMG registration of incoordinations and disturbances of reciprocal innervation

Following peripheral lesions, partial or complete, incoordinations and disturbances of reciprocal innervation appear during the restitution phase, and these can be demonstrated electromyographically without difficulty. These changes can suitably be registered by surface electrodes on a multichannel apparatus. The

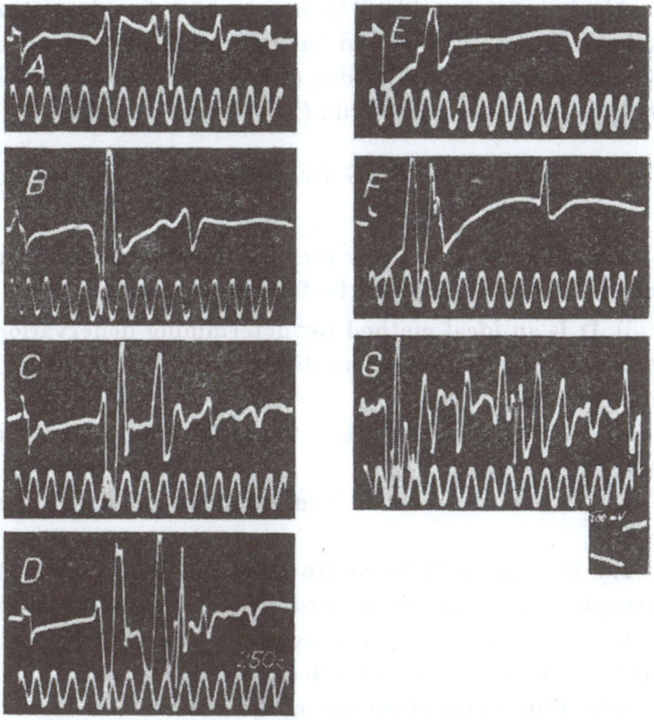

Fig. 151. EMG. record during stimulation of the ulnar nerve in a 58-year-old patient with grave trophic affection of the elbow joint. Stimulation on the arm: A: 140 V, B: 143 V, C: 150 V, D: 180 V. Stimulation of the ulnar nerve above the wrist: E: 170 V, F: 200 V, G: record of maximal voluntary contraction. Duration of an evoked series of action potentials is substantially longer on stimulation on the arm than when stimulating above the wrist.

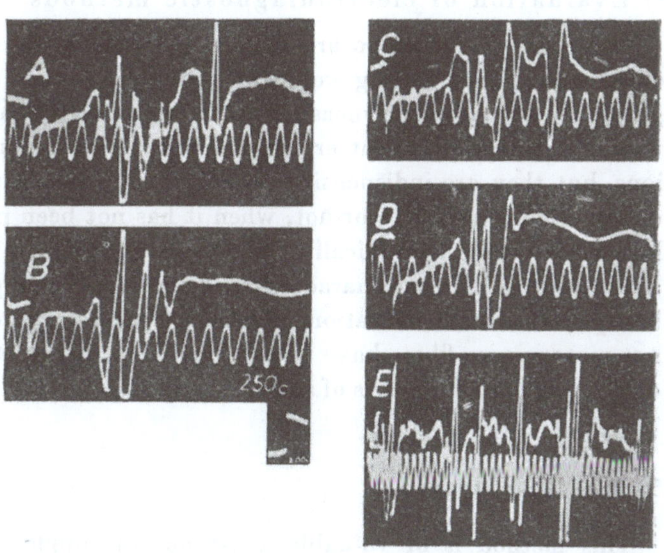

Fig. 152. EMG. record from the first dorsal interosseal muscle in a 35-year-old patient three years after suture of the ulnar nerve above the wrist. Records A: 180 V and B: 190 V on stimulation of the ulnar nerve above the wrist. Records C: 170 V and D: 180 V on stimulation on the arm. E: maximal voluntary contraction. Duration of the evoked series of potentials is practically the same after nerve stimulation above the wrist and on the arm. All EMG. records taken with a coaxial needle electrode.

method is of considerable importance for determining the therapeutic and rehabilitation regimen and has been applied for investigating peripheral traumatic nerve lesions by Bowden (1954) and poliomyelitis by Schwartz and Bouman (1942), Drechsler and Obrda (1955) and others.

e) Evaluation of EMG in clinical problems concerning denervated muscles

The main advantages of the electromyographic examination, from this aspect, can be summarized into the following points:

i) It is an ideal method for determining denervation when there are fibrillations present in the denervated muscle. It can thus be of decisive diagnostic importance.

ii) It is a valuable method for discovering early reinnervation and for evaluating its progress.

iii) Partial denervation can safely be determined at late stages of development.

iv) It helps to differentiate between denervation atrophy and other types of atrophy, especially from atrophy accompanying primary muscle dystrophies.

v) Stimulation electromyography enables the conduction of motor nerve fibres and the extent of the lesion to be evaluated.

vi) Stimulation electromyography assists in localizing the site of the nerve lesion and represents a very valuable examination method for differential diagnosis of peripheral and spinal muscle atrophies of nervous origin.

f) Evaluation of electrodiagnostic methods

According to contemporary clinical experience the most useful methods are electromyography during voluntary contraction and stimulation electromyography, with chronaxie measurements and strength-duration curves as supplementary methods. The latter two methods serve primarily as control examinations, but they are indispensible in cases when decision is to be made whether denervation is complete or not, when it has not been possible to find any fibrillations electromyographically. Electrodiagnostic examination methods have a disadvantage in that characteristic denervation changes appear at a relatively late period after denervation. When it is necessary to state early whether or not motor nerve fibres have been interrupted, nerve stimulation above the site of the anticipated lesion is of special significance.

9. Muscle biopsy

This method is of valuable assistance in supplementing auxilliary clinical methods when other methods cannot be expected to elucidate the cause of muscle atrophy. A number of reports have demonstrated its considerable usefulness in clinical practice (Bowden and Gutmann 1944, 1949, Adams et al. 1954 and others). With the aid of muscle biopsy, it is possible to obtain valuable information about the progress of regenerating nerve fibres and about changes taking place in the denervated muscle during reinnervation. The clinical significance of muscle biopsy has diminished since the wider use of electrodiagnostic methods, especially of electromyographic examinations, which are less exacting and troublesome for the patient and obtain information which cannot be gained by muscle biopsy.

Muscle biopsy, however, is still of paramount importance in differentiating early denervation atrophy from myopathies and myositis, where electrodiagnostic examinations are not usually characteristic.

B. Therapy of the denervated muscle

In traumatic lesions the therapeutic efforts are directed towards obtaining optimum conditions for regeneration of nerve fibres in the peripheral nerve. None the less important, however, are therapeutic attempts at maintaining a relatively good trophic state of the denervated muscle through the period preceding reinnervation. When reinnervation is taking place, therapy should ensure most efficient movement rehabilitation making use of all compensatory mechanisms which can be of assistance, especially in partially reinnervated muscle.

The question whether in open wounds primary or secondary sutures should be made is at present solved, due to antibiotics, in the sense that primary suture is performed in every case where anatomical connection between the two nerve stumps is possible.

Indication for surgical revision is given by observing early signs of reinnervation in relation to time elapsed since the suture or injury. For every suture about 30 days' delay should be allowed and the rate of 1 mm. per day calculated for progress of the regenerating nerve fibres. It is true that Bowden and Gutmann (1944) have demonstrated that nerve fibres in man grow at a velocity of 3 mm. per day, but during clinical examination we are left to consider functional regeneration which evidently lags behind anatomical regeneration of axons. If first signs of regeneration do not appear at the period necessary for functional nerve regeneration, surgical revision of the wound is indicated. In some cases surgical revision may be indicated even earlier. Such is the case in nerves superficially placed under the skin where a forming neuroma can be palpated. If the neuroma grows and its palpation or rapping evokes paraesthesia corresponding to the appropriate innervation region of the nerve, surgical revision may be indicated in case that this paraesthesia cannot be evoked from progressively more distal parts. Testing of the progress of regenerating nerve fibres by this method is known in the literature as Hoffmann's or Tinell's sign. This sign can easily be observed in nerve trunks which are placed relatively superficially under the skin. Otherwise the clinician has no method at his disposal for examining the progress of regenerating nerve fibres before actual reinnervation of the terminal organ has taken place.

Sometimes it is very difficult to decide during surgical revision, whether it is necessary to perform resection of the nerve and a new suture, or whether neurolysis or nerve transposition will be sufficient. Unfortunatelly there is, as yet, no method giving reliable information about the progress of regenerating nerve fibres through the site of the lesion, until restitution of muscle motor function has taken place. X-ray contrast photography of endoneural pathways with radioopaque substances, as has been attempted, for example, by Funaoka (1930), Brunner (1944) and others, has not come into wide clinical use, especially because of toxic action of the contrasting substances, which cause local inflammatory reactions and may lead to proliferation of connective tissue. Electrophysiological registration of nerve action potentials below the lesion to stimulation above the lesion has, as yet, not been clinically elaborated. The neurosurgeon is thus forced to decide whether to perform a new resection and suture from evaluation of anatomical conditions at the site of the lesion only. Where there is a neuroma formed above the original lesion or where there are fibrous changes in the vicinity of the suture, there is no need for hesitation and a sharp resection of pathologically altered nerve stumps and a new suture is indicated. When a neuroma is not found on surgical revision and the continuity of the nerve is not interrupted, it can be presumed that only a delay of nerve fibre regeneration has occurred and neurolysis or nerve transposition will be considered as sufficient. Uncertainty is felt when anatomical conditions found during surgical revision are not clear. For this reason, whenever the possibility cannot be excluded that nerve fibres have grown into the peripheral stump, it is correct not to undertake nerve resection, neurolysis being sufficient, and to continue in conservative therapy. Repeated revisions of the peripheral nerve will not damage the patient nearly as much as incorrectly indicated nerve resection.

Care of the denervated muscle is an important part of therapy and its aim is to ensure minimum atrophy during the denervation period. The smaller the atrophy, the faster and easier will be the rehabilitation after reinnervation. Where regeneration of the nerve takes place under adverse conditions, so that late reinnervation of the muscle can be expected, care of the trophic state of the muscle becomes of primary importance since it determines, whether muscle tissue will disappear altogether by the time regenerating nerve fibres begin to reinnervate the muscle. Under unfavourable conditions when fibrous transformation of the muscle takes place, functional restitution cannot occur, even though regenerating motor fibres reinnervated the muscle (Bowden and Gutmann 1949).

Thus the main task during the denervation period is to make use of all therapeutic possibilities to maintain the muscle tissue in an optimum trophic state for expected reinnervation.

1. Electrotherapy

To date, electrotherapy represents the most effective method of influencing the rate of denervation atrophy. It has been used in clinical practice since the first half of the last century (see, for example, Reid 1841).

Evaluation of its significance as a therapeutic measure in denervation atrophy passed alternately several times from enthusiastic propagation to categoric rejection (Egen 1958 and the historical review on electrotherapy by Licht 1959). Differences of opinion about the clinical results of electrotherapy in influencing denervation muscle atrophy are due, to a certain extent, to the fact that almost every author used a different electrotherapeutic procedure. Electrical stimuli used for stimulation of denervated muscles were not only of different intensity, voltage and frequency, but different forms of application were given, especially as far as intervals between individual contractions, or series of contractions and different length of stimulation during a single therapeutic session are concerned. The method of clinical evaluation of the progress of muscle atrophy is important for judging the effect of electrotherapy. Objective evaluation of the trophic state of a single muscle is practically impossible in man. The clinician can evaluate volume changes only in several muscle groups at the same time. The most commonly used method is to measure the circumference of an extremity at a given point. Small volume changes are better registered from the content of liquid displaced from a calibrated cylinder, into which the affected extremity is immersed (Jackson 1945, Osborne 1951). However, even this method is not satisfactory, so that the clinician is often in doubt as to whether he is evaluating the results of therapy quite objectively.

Experimental work dedicated to questions of the electrotherapeutic effects on denervation atrophy has convicingly proved that electrotherapy is indicated and that so far better results can be expected by this method than by any other. Contraversies from the clinical literature can also be found in physiological literature. To-day, however, the main factors which affect the results of electrotherapy in denervation atrophy are known. One of the most important conditions for obtaining positive results with electrotherapy is supramaximal stimulus intensity and stimulation of all denervated muscle fibres. Experiments on animals demonstrated that it is actually only the supramaximum electrical stimuli which can successfully slow the development of muscle atrophy. Muscles of experimental animals are usually considerably smaller than in man, so that it is not necessary to use high intensities for stimulating deep muscle layers. On the contrary in man, muscle groups are bulky, so that supramaximal stimuli which stimulate muscle fibres even of deep denervated groups must reach considerable intensity and may thus be quite painful and lead to unfavourable side-effects on regeneration processes in the peripheral nerve (Beránek and Gutmann 1953), as well as to adverse effects on metabolism and trophic state of the muscles (Gutmann and Vodička 1953, Vodička 1955).

Most authors agree that the strong contraction evoked by the electrical stimulus should take place under isometric condition. It is thus important that the muscle be fixed during stimulation in such a way so as to prevent isotonic contraction which is not therapeutically effective (Hines 1950, Eccles 1954). Even though the therapeutic effect of electrotherapy is unequivocal, it is quite clear that even effectively performed electrotherapy of the denervated muscle can only retard the development of muscle atrophy, but never cause an absolute weight increment. For this reason it is necessary to start with electrotherapy as soon as possible. The fact that isometric contractions are most effective may be clinically advantageous, since they can be performed even under conditions when movement during isotonic contractions could adversely affect healing of the wound or suture of the nerve.

A number of authors have investigated the question of the most suitable form of electrical current for use in electrotherapy (Radier 1957, Licht 1959). According to contemporary experimental results it seems that the shape of electrical current is of no importance for the effectiveness of electrotherapy. The important point is to attain strong contractions of as many denervated muscle fibres as possible. Commonly used currents, however, have the disadvantage of being very painful at the effective intensities and often neighbouring innervated muscles are also stimulated. The use of exponentially rising currents has meant considerable progress in electrotherapy. This method of stimulation is based on the observation that the denervated muscle loses its accomodation ability and is selectively stimulated by electrical stimuli of progressively increasing intensity. Innervated muscle fibres and afferent nerve fibres, however, are not stimulated by these pulses. Thus a further advantage of exponentially rising currents is that they are not painful. Already favourable clinical reports have been published about the results of muscle stimulation with these currents (Kowarschik 1952, Hensage 1952, Thom 1955, Gilbert 1958, Janča 1957, 1959 and others).

2. Drug therapy

The possibility of influencing denervation atrophy with drugs is, as yet, very limited and the effect in clinical therapy is usually uncertain. Although experiments on animals seem to indicate that some substances, e. g. somatotropic hormone (Gutmann and Vrbová 1955, Selye and Bajusz 1958), thiouracil (Huf and Fischer 1949) or ATP (Ferdman 1953, Bot et al. 1955) retard the rate of denervation muscle atrophy, yet prolonged clinical administration of effective substances is considerably limited by their undesirable side-effects. A favourable effect on the course of muscle atrophy in clinical cases was described by Müller and Hagemann (1954) after administration of small doses of insulin. It is, however, not quite clear whether positive results are due to the direct effect of insulin on protein synthesis in the denervated muscle, or whether it is only a question of improved general state of nutrition from hyperphagia resulting from long-lasting insulin hypoglycaemia.

3. Positioning

In early stages of denervation muscle atrophy considerable effect may be achieved by suitable positioning begun in time. This can prevent muscle and tendon retractions and maintain a position of the limb appropriate for further rehabilitation. Its significance is not limited to denervated muscles only, but also for muscles of the antagonistic group. Positioning is especially important in traumatic peripheral nerve lesions and in neuroinfections, e. g. poliomyelitis.

Splinting is usually used for positioning during the acute and subacute phase of the disease. Later braces are more suitable, these enabling the most necessary movements to be made in cases where reinnervation had not taken place, or where reinnervation is insufficient.

4. Massage and passive movements

Passive movements can affect the denervated muscle in two ways. If the muscle is also deafferented, i. e. after section of the mixed peripheral nerve, passive movements play a role in preventing muscle and tendon contractures and ankyloses of joints. However, if the denervated muscle has intact afferent innervation, then passive movements can also signalize and enhance active movements probably by a facilitatory effect of proprioceptive discharges on motor nerve centres. Such is the case, for example, in poliomyelitis.

Massage is a very old therapeutic procedure used for various lesions of the locomotor apparatus. It seems that its significance for denervation atrophy has been somewhat overestimated. Massage combined with passive movements reduces the extent of contractures (Pollock et al. 1951a) and according to some authors acts favourably on the strength of the denervated muscle (Suskind et al. 1946, Hájek and Hines 1947). Massage is without effect on denervation atrophy or blood circulation through the denervated muscle (Ebel et al. 1952, Randall et al. 1952).

5. Therapeutic heat

Therapeutic heat in the first place improves blood flow through the skin and subcutaneous tissues. In order to increase the blood flow through deeper tissues more effective methods are necessary, e. g. diathermy, hot paraffin packs, ultrasonic radiation etc. The effect of therapeutic heat on denervation atrophy is uncertain. Increased blood flow by means of heat is only brief and it is known as well that the blood flow through the denervated muscle is increased in comparison with the normal muscle, so that it is problematical whether any further increase is useful. It must be borne in mind that heat application besides increasing blood flow also increases the intensity of metabolism, so that it is not clear whether the resultant effect is not rather detrimental to the trophic state of the muscle than otherwise (Randall et al. 1952, 1953).

Heat therapy, however, can favourably affect other symptoms accompanying peripheral nerve lesions (see Licht 1958). Heat loosens and softens retracted connective tissue structures and has a marked analgetic effect. This effect can be explained by improved circulation where pain is of ischaemic origin. One of the most widely used methods of influencing painful muscle spasms is hot packing according to Kenny. They represent the most effective procedure for relieving spasms in acute poliomyelitis. However, heat can also exert an unfavourable influence since it elevates the threshold of sensitive nerve endings to painful stimuli and leads to central interference of sensory stimuli (Fischer 1958).

Therapeutic heat is commonly used for supplementing the effect of massage prior to passive and active movement exercises. Its application is effective especially where chronic trophic changes of the skin are present and in cases where there are contractures of muscles, tendons and fascia.

C. Rehabilitation of the reinnervated muscle

When reinnervation of the denervated muscle begins, electrotherapy ceases to be effective. Several authors have agreed on this point (Hájek and Hines, 1947, Hines 1950, Pollock et al. 1951b, Osborne 1951). It seems that the restitution of the trophic state in the reinnervated muscle is already progressing at a maximum velocity and it cannot be accelerated any further by electrical stimulation. It must be pointed out, however, that reinnervation is a gradual process, i. e. that for some time after the first signs of functional restitution have appeared, the majority of muscle fibres may still be denervated. Premature termination of electrotherapy will, especially in cases where reinnervation progresses slowly, leave denervated muscle fibres without therapy. On the other hand, excessive electrotherapy of the reinnervated muscle can lead to the production of contractures, so that electrotherapy in early reinnervation is suitable only in muscles where reinnervation is incomplete and progresses at a slow rate. In cases where reinnervation has progressed to such a stage that there is evident movement performed, active movement exercises are to be preferred.

Muscle reinnervation does not only mean the restitution of functional connections between the nerve cell and the muscle, but also the reintegration of motor units into reflex patterns, in which they could not participate during the denervation period. Vrbová and Gutmann (1956), who studied the conditioned spreading reflex in rats, demonstrated that inhibitory processes are to be expected in the motor analyser during the denervation period. It was shown by these authors that the conditioned spreading reflex is very firm in the rat and that it is possible to obtain 100% positive responses to the conditioned stimulus even when the experiments were interrupted for three months. In rats in which the sciatic nerve had been crushed, the unconditioned reflex response appeared several days prior to the reappearance of the conditioned motor response. It thus seems that in early reinnervation a certain period of time is necessary for the reintegration of the conditioned motor response. Proprioceptive signalization plays an important role in the elaboration of conditioned motor reflexes.

Active exercises with reinnervated muscle represent one of the most important forms of rehabilitation. During the early reinnervation period the newly formed neuromuscular connections are easily fatigued, it is not clear, however, whether volitional contractions should be performed till complete exhaustion of the muscle or whether the number of contractions should always be limited.

When prescribing the extent of active movements, aetiology of the affection must be taken into account. In poliomyelitis, where denervation is caused by lesions of the motoneurons themselves, it is possible to assume that it will affect other functions of the whole motoneuron as well. Its exertion in early stages of restitution could have adverse effects on its recovery. For this reason a limited number of reeducation movements during the early phase of muscle function restitution seems to be a logical principle (Pohl and Kenny 1949). On the other hand, in traumatic lesions of peripheral nerves the region of the cell body is not affected by the pathological process, so that its functional activity during early reinnervation is fully indicated. Strenuous active exercises favourably influence the trophic state of reinnervated muscle fibres and aid reintegration of motoneurons into reflex patterns.

It is commonly known from experimental and neurophysiological reports that if a voluntary movement is to take place, a certain minimum of excitability of spinal motoneurons is necessary. For this reason rehabilitation practice has devised a number of facilitation methods with the aim of obtaining an optimum functional state of motoneurons during active reeducation. Even though in some cases the mechanism of its action is not fully understood, they are usually very effective rehabilitation methods. They are based on physiological mechanisms which affect the excitability of spinal motoneurons.

At the present time a number of facilitatory manoeuvres are used such as, for example, resistance exercises (De Lorme 1945, Kabat 1952), exercises with controlateral extremities (Scripture 1894, Hellebrandt et al. 1947, Kabat 1952, Gregg et al. 1957 and others) and facilitation mechanisms associated with the degree of muscle stretch (Gellhorn 1953) and various rehabilitation methods stimulating proprioceptors (Pohl and Kenny 1949). A favourable effect has also been found by using the so-called active recuperation first described by Sechenov (1907). It was observed in a large number of cases of acute and chronic poliomyelitis that the number of active contractions of which the affected muscle is capable is increased when work is performed by healthy muscle groups in the interval between the first and second bout of functional activity carried out to exhaustion (Beránek et al. 1953, 1957).

A commonly used method of evaluating muscle strength by muscle testing does not afford a complete picture of the functional capacity of the muscle. This is borne out by the fact that, for example, two muscles with muscle test 2 (able to perform active movement with elimination of gravity) may differ very substantially from each other as far as their ability of repeated contractions is concerned. One may, for example, be capable of performing only a very limited number of contractions, while the other may perform as many as several hundreds without appearance of fatigue. Active recuperation is effective only in the first case, where marked fatigue is present. In the second case active recuperation is without effect. It can thus be assumed that in the former case the function of motoneurons can be favourably affected by facilitation maneouvres, while the latter rather requires systematic functional exercise and further improvement can be expected from muscle fibre hypertrophy.

Besides simple muscle hypertrophy which is primarily connected with systematic muscular work, there are still other compensatory mechanisms which can come into play during reinnervation as well as during restitution and reintegration of motor patterns. There is always a certain loss of nerve fibres during reinnervation. This loss is mainly dependent on the type of lesion of the peripheral neuron. Axonotmesis, which corresponds to experimental nerve crushing, has optimum conditions for regeneration, since endoneural cylinders are not interrupted and it can be presumed that maximum regeneration of nerve fibres into the appropriate "canal" will take place (see Gutmann 1955). Otherwise allowance must be made for certain losses which are caused by incorrect growth of nerve fibres. In young animals a large number of motoneurons has also been reported to disintegrate following section of the peripheral nerve (Romanes

1951). As a result of this, the corresponding region is reinnervated by a smaller number of motor nerve fibres than is the case under physiological conditions.

During motor rehabilitation it is always necessary to bear in mind that painful stimuli in the denervation and reinnervation period are detrimental to the final result. It was found that painful, nociceptive stimulation affects unfavourably processes which take place during denervation and reinnervation. Nociceptive stimulation results in very rapid muscle atrophy which affects, in the first place, antigravity muscles (Charcot 1889, Gutmann and Vodička 1953, Vodička 1955, Jirásek 1957) and retards restitution of function during reinnervation by about 50 % (Beránek and Gutmann 1953). However, painful stimulation can also lead to a break down of compensatory mechanisms built up previously by the central nervous system (Beránek 1953).

It is not yet clear, whether the patient should attempt voluntary movement prior to reinnervation, or whether exercises should be begun after first signs of restitution of motor function had appeared. Experimental results on animals seem to indicate that volitional movements should not be attempted before first signs of reinnervation (Vrbová and Gutmann 1956). Direct analogy between man and these experimental results cannot be made, however, since in man very complicated compensatory mechanisms come into play.

It is furthermore, not clear whether isotonic or isometric contractions are the most advantageous for restitution of motor functions. Darcus (1956) in an illuminating review on purposeful rehabilitation pointed out that muscle function cannot be considered from the aspect of muscle strength only, but also from the point of view of endurançe, speed and movement coordination. According to this author, rehabilitation exercises should be directed by the result which is to be attained. For example, exhausting resistance exercises result mainly in muscle hypertrophy, rapidly repeated contractions affect the speed movements, light muscle work improves motor coordination. Nevertheless, functional, biochemical and other differences between individual muscles and muscle groups must not be left out of consideration. Rehabilitation training must take the function of a given muscle into account, i. e. whether it is a muscle predominantly with a tonic function, or on the contrary with a dynamic, fast function. Therapeutic physical training and therapeutic work must be utilized as much as possible from this point of view.

The psychological state of the patient is very important in all rehabilitation procedures, since his confidence and optimism concerning the final result are an indispensible part of therapy.

Conclusions

When evaluating present day achievements in clinical therapy and diagnosis of denervated muscles, it is evident that they are far from satisfactory. There are still a number of questions the solution of which would greatly assist clinical practice.

One of the basic diagnostic questions is the problem of objective evaluation of the progress of peripheral nerve regeneration during the period before they reach the effector organ, i. e. the myoneural junction in the muscle. Since there is no examination method which could help in this problem and in early discovery of the arrest of nerve growth, the denervation period is thus often unnecessarily prolonged.

Problems connected with the therapy of denervated muscle do not concern peripheral nerve injury cases only, since analogous disturbances in muscle metabolism exist in a number of clinical syndromes, such as neuroinfections, painful afflictions, immobilization etc. Even problems of muscular dystrophy and its therapy will not be properly understood unless the pathogenetic mechanisms of denervation atrophy are clear.

The main aim of therapy is to keep the muscle in a relatively optimal state. For this purpose electrotherapy has been widely advocated and it is actually the most successful method of retarding muscle atrophy so far known. However, this is an empirical therapeutic approach, founded on the assumption that muscle atrophy in denervation is only the result of inactivity. As has been pointed out by Thesleff (1960), motor innervation influences the skeletal muscle in more ways than one. Besides controlling its mechanical activity, it exerts a regulating influence upon the chemical composition, muscle metabolic reactions, electrical excitability and chemical sensitivity. It is thus evident that it is necessary to study the trophic function of the nervous system, i. e. those influence which are not directly associated with the triggering of muscle contraction, but the phenomena named above.

A number of experimental findings presented in this monograph speak in favour of the fact that the motoneuron is able to influence the muscle by other mechanisms besides impulse transmission or liberation of subthreshold quanta ACh. One of these is most probably neurosecretory in character. Of course, this mechanism cannot be expected to be solved by elimination experiments performed on the denervated muscle only, and consequently direct studies of the motoneuron will be required to solve this question definitely.

References

Abrikosov I. A., Darkshevich N. V. Fiziol. zh. USSR 40 : 504, 1954.

Adams R. D., Denny-Brown D., Pearson G. M.: Diseases of Muscle, New York, Paul B. Hoeber, 1953.

Adrian E. D., Bronk D. W.: J. Physiol. 67 : 119, 1929.

Anokhin P. K., Mayorchik V. E.: Opyt sovetskoi meditsiny v velkoi otechestvennoi vojne 1941 to 1945. Medgiz, Moscow 1952.

Bajusz E.: Canad. J. Biochem. Physiol. 36 : 825, 1958.
Bass A., Vodička Z.: Čs. fysiol. 4 : 427, 1955.
Beránek R.: The part played by the central nervous system in recovery processes in disturbances of the motor apparatus. Candidate's Thesis, Institute of Physiology ČSAV, Prague 1953.
— Čs. fysiol. 8 : 172, 1959.
— Physiol. bohemoslov. 10 : 94, 1961.
Beránek R., Graf V., Pokorný F.: Čas. lék. čes. 96 : 385, 1957.
Beránek R., Gutmann E.: Čs. fysiol. 1 : 38, 1953.
— — Čs. fysiol. 2 : 140, 1953.
Beránek R., Hník P., Vrbová G.: Čs. fysiol. 2 : 51, 1953.
Berry C. M., Grundfest H., Hinsey J. C.: J. Neurophysiol. 7 : 103, 1944.
Berry C. M., Hinsey J. C.: Ann. New York acad. Sc. 47 : 559, 1946.
Björk A., Kugelberg E.: EEG Clin. Neurophysiol. 5 : 271, 1953.
Bot Gy., Köver A., Varga E.: Acta physiol. hung. 8 : 321, 1955.
Bouman H. D., Schwartz R. P.: N. Y. State J. of Med. 44 : 147, 1944.
Bourgouignon G.: La chronaxie chez l'homme, Paris, Dion, 1923.
Bowden R. E. M.: Peripheral Nerve Injuries. Ch. II, London, Her Majesty's stationary office, 1954.
Bowden R. E., Gutmann E.: J. Bone and Joint Surg. 31B : 356, 1949.
— — Brain 67 : 20, 1944.
Brunner 1944: Cit. Gutmann E.: Rozhledy v chirurgii 29 : 1, 1950.
Buchthal F.: IV. Congress Neurologiques International. Vol. I. Raports 35—46, 1949.
— Introduction to electromyography. København 1957.
Buchtal F., Clemmesen Sv.: Acta Psychiat. Neurol. Scand. 16 : 143, 1941.
Buchthal F., Guld Ch., Rosenfalck P.: Acta Physiol. Scand. 34 : 75, 1955.
— — — Acta physiol. Scand. 39 : 83, 1957.
Buchthal F., Hoencke P.: Acta med. Scand. 116 : 148, 1944.
Buchthal F., Madsen A.: EEG Clin. Neurophysiol. 2 : 425, 1950.
Buchthal F., Rosenfalck P.: Acta Psychiat. Neurol. Scand. 30 : 125, 1955.
— — EEG Clin. Neurophysiol. 10 : 521, 1958.
Charcot J.: Lectures on diseases of the nervous system III. 1889.
Darcus H. D., Brewerton D. A.: Proc. R. Soc. Med. London 49 : 999, 1956.
Dawson G. D., Scott J. W.: J. Neurol. Neurosurg. Psychiat. 12 : 259, 1949.
De Lorme T. L.: J. Bone and Joint Surg. 27 : 645, 1945.
Denny-Brown D.: Arch. Neurol. Psychiat., Chicago 66 : 99, 1949.
Denny-Brown D., Pennybacker J. B.: Brain 61 : 311, 1938.
Denslow J. S., Hasset C. C.: Am. J. Physiol. 139 : 652, 1943.
Drechsler B., Obrda K.: Čas. lék. čes. 94 : 319, 1955.
Ebel A., Wisham L. H.: Arch. phys. med. 33 : 399, 1952.
Eccles J. C.: J. Physiol. 103 : 253, 1944.
Egen K.: Hippokrates. Stuttg. 29 : 44, 1958.
Erlanger J., Gasser H. S.: Electrical signs of nervous activity. Philadelphia, Univ. of Pennsylvania Press, 1937.
Feinstein B., Lindegärd B., Nyman E., Wohlfahrt G.: Acta anat. 23 : 127, 1955.
Feinstein B., Pattle R. E., Weddell G.: J. Neurol., Neurosurg., Psychiat. 8 : 1, 1945.
Ferdman D. L.: Biokhimiya zabolevaniji myshts. Izd. AN Ukrajinskoi SSR, Kiev 1953.
Fischer E.: Changes in structural proteins and enzymes during muscular atrophy of various origin. Colloque tenu à Royaumont, France, 1950.
— Physical therapy Reviews 38 : 468, 1958.
Funaoka 1930: Cit. Gutmann E.: Rozhledy v chirurgii 29 : 1, 1950.

Gasser H. S., Grundfest H.: Am. J. Physiol. 127 : 393, 1939.
Gellhorn E.: Physiological foundations of neurology and psychiatry. The University of Minnesota Press, Minneapolis 1953.
Gilbert O.: Aus Theorie und Praxis der Krankengymnastik. R. Pflaum Verlag. München 1958.
Gilliatt R. W., Sears T. A.: J. Neurol. Neurosurg. Psychiat. 21 : 109, 1958.
Gregg R. A., Mastellone A. F., Gersten J. W.: Am. J. Physical Med. 36 : 269, 1957.
Grodins F. S., Osborne S. L., Johnson F. R., Arana S., Ivy A. C.: Am. J. Physiol. 142 : 222, 1944.
Grundfest H., Oester Y. T., Becke G. W.: Peripheral nerve regeneration. Ch. IV., Washington, V. Medical Monograph 1956.
Guld Ch.: Acta Physiol. Scand., 25 : 30, 1951, Suppl. 89.
Gutmann E.: Funkční regenerace periferních nervů, Praha NČSAV 1955.
Gutmann E., Guttmann L.: Lancet 7 : 169, 1942.
Gutmann E., Sanders F. K.: J. Physiol. 101 : 489, 1943.
Gutmann E., Vodička Z.: Čs. fysiol. 2 : 400, 1953.
Gutmann E., Vrbová G.: Unpublished results. 1955.
Gutmann E., Young J. Z.: J. Anat. 78 : 15, 1944.
Hajek N. M., Hines H. M.: Fed. Proc. 6 : 1, 1947.
Häkansson E. H.: Acta physiol. Scand. 37 : 14, 1956.
Hellebrandt F. A., Parrish A. M., Houtz S. J.: Arch. Phys. Med. 28 : 76, 1947.
Hensage E.: Selektive niederfrequente Reitztromtherapie und Elektromyographie. G. Thieme, Leipzig 1952.
Hill A. V., Katz B., Solandt D. Y.: Proc. Roy. Soc., Sec. B., 121 : 74, 1936.
Hines H. M.: Physiological basis of the treatment of muscular atrophy of nervous origin. Extrait de "Le Muscle étude de biologie et de patologie", compte rendu du Collegue tenu à Royaumont, France, du 31. août au 6. septembre 1950.
Hodes R., Larrabee M. G., German W.: Arch. Neurol. Psychiat., Chicago, 60 : 340, 1948.
Hoefer P. F. A., Putmann T. J.: Trans. Amer. Neurol. Ass. 18 : 1939.
— — Arch. Neurol. Psychiat., Chicago, 42 : 201, 1939.
Hoffmann P.: Untersuchungen über die Eigenreflexe (Sehnenreflexe) menschlicher Muskeln. Berlin. J. Springer, 1922.
Huf E. G., Fischer E.: Am. J. Physiol. 159 : 6, 1949.
Jackson S.: Brain 68 : 300, 1945.
Janča V.: Sborník věd. prací VLA JEvP 6 : 63, 1958.
— Fysiatr. věstník 35 : 214, 1957.
Janda V., Škorpil V., Véle F.: Čas. lék. čes. In press.
Jantsch H.: Brit. J. Phys. Med. 15 : 252, 1952.
Jarcho L. W., Berman B., Dowben R. M., Lilienthal J. L.: Am. J. Physiol. 178,: 129, 1954.
Jarcho L. W., Vera C. L., McCarthy C. G., Williams P. M.: EEG Clin. Neurophysiol. 10 : 527, 1958.
Jirásek A.: The Traumatology of the Knee Joint (in Czech), Prague, SZN, 1957.
Kabat H.: Arch. Phys. Med. 33 : 521, 1952.
Kovarschik J.: Brit. J. Phys. Med. 15 : 249, 1952.
Kugelberg E.: Acta physiol. Scand. 8 : 1944. Suppl. 24.
— J. Neurol. Neurosurg. Psychiat. 10 : 122, 1947.
Kugelberg E., Petersen I.: J. Neurol. Neurosurg. Psychiat. 12 : 268, 1949.
Kugelberg E., Skoglund C. R.: J. Neurophysiol. 9 : 391, 1946.
— — J. Neurophysiol. 9 : 399, 1946.
Kugelberg E., Taverner D.: EEG Clin. Neurophysiol. 2 : 125, 1950.
Lapicque L.: L'excitabilité en fonction du temps. Paris, Dion 1921.
Lérique J.: L'électro-myographie élementaire chez l'homme. Thése, Paris 1944.

Lesný I.: Klinická chronaximetrie, Praha, SZN, 1954.
— Základy neurologické elektrodiagnostiky, Praha, SZN, 1953.
Licht S.: History of Electrotherapy. Therapeutic Electricity and Ultraviolet Radiation, 1959.
Malcolm D. S.: J. Neurol. Neurosurg. Psychiat. 14 : 15, 1951.
Marble H. C., Hamlin E., Watkins A. L.: Am. J. Surg. 55 : 274, 1942.
Marinacci A. A.: Clinical Electromyography, Los Angeles, San Lucas Press, 1955.
Müller-Hegemann D.: Dtsch. Gesundheitswesen 9 : 589, 1954.
Newmann H. W., Livingston W. K.: J. Neurol., Neurosurg. Psychiat. 10 : 118, 1947.
Norris F. H., Gasteiger E. L.: EEG Clin. Neurophysiol. 7 : 115, 1955.
Obrosov A. N., Livencev N. M.: Elektrodiagnostika i elektrostimulaciya myshts pri porasheniyi perifericheskikh nervov. Medgiz, Moskva 1953.
Osborne S. L.: Arch. Phys. Med. 32 : 523, 1951.
Osborne S. L., Grodins F. S., Mittleman E. E., Milne W. S., Ivy A. C.: Arch. Phys. Ther. 25 : 338, 1944.
Parry C. B. W.: Canad. M. Ass. J. 77 : 842, 1958.
Petersen I., Kugelberg E.: J. Neurol. Neurosurg. Psychiat. 12 : 124, 1949.
Pinelli P., Buchthal F.: EEG Clin. Neurophysiol. 5 : 589, 1953.
Piper H.: Arch. f. ges. Physiol. 119 : 301, 1907.
Pohl J. F., Kenny E.: The Kenny Concept of Infantile Paralysis and its treatment, Minneapolis, 1949.
Pollock L. J., Arieff A. J., Sherman J. C., Schiller M., Tigay E., Hiller F., Liebert E., Jacorzynski G.: Arch. Phys. Med. 31 : 265, 1951a.
— — — — — — — — Arch. Phys. Med. 31 : 377, 1951b.
Pollock L. J., Golseth J. G., Arieff A. J., Mayfield F.: J. Surg. Gynec. Obstet 81 : 192, 1945.
Pollock L. J., Golseth J. G., Arieff A. J., Sherman J. C., Schiller M. A., Tigay E. L.: Arch. Neurol. Psychiat. Chicago 51 : 147, 1944.
Pollock L. J., Golseth G. J., Mayfiled F., Arieff A. J., Oester Y. T.: Arch. Neurol. Psychiat., Chicago 60 : 1, 1948.
Proebster R.: Klin. Wschr. 7 : 297, 1928.
Radier G.: Montpelliér Méd. 52 : 184, 1957.
Randall B. F., Imig C. J., Hines H. M.: Arch. Phys. Med. 33 : 73, 1952.
— — — Am. J. Phys. Med. 32 : 1, 1953.
Reid J.: J. Med. Sci. 1 : 320, 1841. Cit. Gutmann E., Gutmann L., Lancet 7 : 169, 1942.
Ritchie A. E.: Peripheral nerve injuries. Ch. VI., London, Her Majesty's Stationary Office, 1954.
Romanes G. J.: J. Anat. 80 : 117, 1946.
Schwartz R. P., Bouman H. D.: J. Amer. med. Ass. 119 : 923, 1942.
Scripture E. W., Smith T. L., Brown E. M.: Studies Yale Psychological Laboratory 2 : 114, 1894.
Sechenov I. M.: Mozkové reflexy (1863), Praha 1952.
— 1907: Fiziologiya nervnoi systemy III., p. 155, Izd. med. lit., Moscow 1952.
Selye H., Bajusz E.: Am. J. Physiol. 192 : 297, 1958.
Simpson J. A.: J. Neurol. Neurosurg. Psychiat. 19 : 275, 1956.
Solandt D. J.: J. Physiol. 85 : 5P, 1935.
Suskind M. J., Hajek N. M., Hines H. M.: Arch. Phys. Med. 27 : 133, 1946.
Suskind M. J., Hines H. M.: Fed. Proc. 5 : 1, 1946.
Škorpil V., Vyklický L.: Acta Univer. Carol., Suppl. IV: 274, 1957.
Thesleff S.: Physiol. Rev. 40 : 734, 1960.
Thiebaut F., Isch F., Isch-Treussard C.: L'encéphale, 5 : 1300, 1956.
Thom H.: Ztschr. f. Orthopedie u. ihre Grenzgebiete 84 : 1, 1953.
Van Harreveld A.: Am. J. Physiol. 144 : 477, 1945.
Vodička Z.: Metabolic effects of nociceptive stimulation in the skeletal muscle (In Czech).

Candidates Thesis, Institute of Physiology ČSAV, Prague 1955.
— Physiol. bohemoslov. 1 : 62, 1957.
Vvedenskij N. E.: Vozbuzhdenie, tormozhenie i narkoz (1901). Fiziologiya nervnoi sistemy II. p. 314, Izd. med. liter. Moscow, 1952.
Vrbová G., Gutmann E.: Physiol. bohemoslov. 5 : 1, 1956.
Vyklický L.: EMG man.festations of stimulation of the ulnar neıve in man. Candidate's Thesis, Medical faculty of the Charles University, Plzeň 1959.
Wachholder K.: Ergeb. d. Physiol. 26 : 568, 1928.
Walthard K. M., Tchicaloff M.: La Chronaximetrie. Proc. of the II. Internat. Congress of Physical Medicine, København 1956.
Weddell G., Feinstein B., Pattle R.: Brain 67 : 178, 1944.
Wohlfahrt G.: Neurology 7 : 124, 1957.
Wohlfahrt G., Feinstein B., Fex J.: Arch. f. Psych. u. Neurol. 191 : 478, 1954.
Woodhall B., Beebe G. W.: in: Peripheral nerve regeneration. A follow-up study of 3, 656 World War Injuries Washington 1956.
Woodhall B., Nulsen F. G., White J. C., Davis Lee: in Peripheral nerve regeneration. A follow-up study of 3,656 World war II. Injuries. Washington 1956.
Yahr M. D., Hertz E., Moldaver J., Grundfest H.: Arch. Neurol. Psychiat. Chicago 63 : 728, 1950.

AUTHOR INDEX

G

H

I

J

K

L

M

SUBJECT INDEX

N

O

P

U

V

W

Y

MIX
Papier aus verantwortungsvollen Quellen
Paper from responsible sources
FSC® C105338

If you have any concerns about our products,
you can contact us on
ProductSafety@springernature.com

In case Publisher is established outside the EU,
the EU authorized representative is:
Springer Nature Customer Service Center GmbH
Europaplatz 3, 69115 Heidelberg, Germany

Printed by Libri Plureos GmbH
in Hamburg, Germany